PRODUCTION AND NEUTRALIZATION OF NEGATIVE IONS AND BEAMS

CONFERENCE PROCEEDINGS NO. **210**

PARTICLES AND FIELDS SERIES 40

PRODUCTION AND NEUTRALIZATION OF NEGATIVE IONS AND BEAMS
FIFTH INTERNATIONAL SYMPOSIUM
BROOKHAVEN, NY 1990

EDITOR:
ADY HERSHCOVITCH
BROOKHAVEN NATIONAL LABORATORY

American Institute of Physics   New York

Authorization to photocopy items for internal or personal use, beyond the free copying permitted under the 1978 US Copyright Law (see statement below), is granted by the American Institute of Physics for users registered with the Copyright Clearance Center (CCC) Transactional Reporting Service, provided that the base fee of $2.00 per copy is paid directly to CCC, 27 Congress St., Salem, MA 01970. For those organizations that have been granted a photocopy license by CCC, a separate system of payment has been arranged. The fee code for users of the Transactional Reporting Service is: 0094-243X/87 $2.00.

© 1990 American Institute of Physics.

Individual readers of this volume and non-profit libraries, acting for them, are permitted to make fair use of the material in it, such as copying an article for use in teaching or research. Permission is granted to quote from this volume in scientific work with the customary acknowledgment of the source. To reprint a figure, table or other excerpt requires the consent of one of the original authors and notification to AIP. Republication or systematic or multiple reproduction of any material in this volume is permitted only under license from AIP. Address inquiries to Series Editor, AIP Conference Proceedings, AIP, 335 E. 45th St., New York, NY 10017.

L.C. Catalog Card No. 90-53316
ISBN 0-88318-775-2
DOE CONF 891055

Printed in the United States of America.

Contents

Preface xi

Organizing Committee xvii

FUNDAMENTAL PROCESSES

Surface Processes in the Production of Negative Hydrogen Ions 3
    A. W. Kleyn

H^- Production from Non-Cesiated Converter-Type Negative Ion Sources 17
    C. F. A. van Os, K. N. Leung, A. F. Lietzke, J. W. Stearns, and W. B. Kunkel

Surface Production of Negative Hydrogen Ions by Reflection of
Hydrogen Atoms from Cesium Oxide Surfaces 30
    M. Seidl, S. T. Melnychuk, S. W. Lee, and W. E. Carr

Ion–Ion Collisions Involving H^- Ions 40
    W. Debus, F. Melchert, M. Liehr, and E. Salzborn

Recombination of Atomic Hydrogen on Metal Surfaces 49
    R. I. Hall, M. Landau, F. Pichou, C. Schermann, and I. Cadez

Vibrational Relaxation of Highly Excited H_2 Molecules in Gas-Phase
and Gas–Surface Interactions 62
    M. Cacciatore, M. Capitelli, and G. D. Billing

Coupled Solution of Boltzmann Equation and Non Equilibrium
Vibrational Kinetics in H_2 Volume Discharges:
An Analysis of the Input Data 74
    M. Cacciatore, M. Capitelli, R. Celiberto, P. Cives, and C. Gorse

Generation of H^-, H_2 (v″), and H Atoms by H_2^+ and H_3^+ Ions Incident
Upon Barium Surfaces 88
    J. R. Hiskes and A. M. Karo

Electron Energy Distributions and Vibrational Population Distributions 95
    J. R. Hiskes

Dissociative Electron Attachment to Light Molecules: A Comparative
Study of H_2, LiH, and Li_2 114
    H. H. Michels and J. M. Wadehra

Isotope Effect in Vibrational Excitation of H_2 by Low Energy
Electron Impact 121
    D. E. Atems and J. M. Wadehra

A Comparison of Experimental and Theoretical Electron Energy
Distribution Functions in the Driver of a Multicusp
Ion Source 129
    J. Bretagne, W. G. Graham, and M. B. Hopkins

H^- Formation from Collision Induced Dissociation of H_3^+ in Noble
Gases at keV Energies 135
    I. Alvarez, H. Martinez, C. Cisneros, A. Morales, and J. de Urquijo

Gas Phase Alkali–Hydrogen Interactions in Negative Ion Sources 142
    H. H. Michels and J. M. Wadehra
The Dissociative Recombination of H_3^+ and D_3^+ 152
    F. B. Yousif, D. Van der Donk, and J. B. A. Mitchell
Vibrational Population of H_2 Produced by a Discharge 159
    C. Schermann, R. I. Hall, M. Landau, F. Pichou, and I. Cadez

H^-/D^- ION SOURCES

Cesium Mixing in the Multi-Ampere Volume H^- Ion Source 169
    Y. Okumura, M. Hanada, T. Inoue, H. Kojima, Y. Matsuda, Y. Ohara,
    M. Seki, and K. Watanabe
Plasma-Volume Stationary H^- Ion Source with Hollow Cesium Cathode 184
    S. P. Antipov, L. I. Elizarov, M. I. Martynov, and V. M. Chesnokov
Long-Pulsed Surface-Plasma Sources with Geometric Focusing 198
    Yu. I. Belchenko and A. S. Kupriyanov
Negative Ion Production in an Ion Source Operating in H_2 and D_2 214
    W. G. Graham and A. A. Mullan
Optimization of the Sheet Plasma Negative Ion Source 223
    A. Ando, T. Kuroda, Y. Oka, O. Kaneko, Y. Takeiri, T. Kawamato,
    and A. Karita
Operation of a Large Negative Ion Source in Deuterium 233
    L. M. Lea, A. J. T. Holmes, and M. F. Thornton
The Effects of Electron Suppression Fields on D^- Production 244
    M. F. Thornton, A. J. T. Holmes, L. M. Lea, and G. O. R. Naylor
Physics Tests of an Electron Suppressor with Variable Electric
and Magnetic Fields... 255
    R. McAdams, R. F. King, and A. F. Newman
Enhancement of Negative Ion Extraction and Electron Suppression
by a Magnetic Field 266
    J. Bruneteau, R. Leroy, M. Bacal, and J. H. Whealton
A Low Frequency rf Discharge: A Possible Source of D^-? 278
    C. A. Anderson and W. G. Graham
Negative Hydrogen Volume Source with rf Multipole Ion Containment 285
    G. Brautti, A. Boggia, A. Rainò, V. Stagno, V. Variale, and V. Valentino
Selection of Conditions for Production of Maximum H^- Beam Current
Density from a Multicusp Source... 290
    A. I. Krylov, V. V. Kuznetsov, D. V. Penkin, and N. N. Semashko
Magnetic Field of a Toroidal Volume H^- Source 298
    C. R. Meitzler
The BNL Volume H^- Ion Source 304
    K. Prelec
A Compact DC Cusp Source... 323
    D. H. Yuan, M. McDonald, P. W. Schmor, and K. Jayamanna

Continuously Operated Negative Ion Surface Plasma Source 329
    A. A. Bashkeev and V. G. Dudnikov
Steady-State Production of H^- Ions by Reflex-Type Cs Free Ion Source 340
    P. M. Golovinsky, V. P. Goretsky, A. N. Mosijuk, I. A. Soloshenko,
    A. F. Tarasenko and A. I. Tschedrin
Negative Ions from Magnetically Insulated Diodes 354
    R. Prohaska, H. Lindenbaum, A. Fisher, G. Sheperd, and N. Rostoker

POLARIZED AND HEAVY NEGATIVE ION SOURCES

Investigation of Spin-Exchange Processes in the Optically
Polarized Ion Source 373
    A. N. Zelenskii, S. A. Kokhanovskii, V. G. Polushkin, and
    K. N. Vishnevskii
Recent Developments in the BNL Intense Polarized H^- Source Program 385
    A. Kponou, A. Hershcovitch, J. G. Alessi, B. DeVito, and C. R. Meitzler
Intense Negative Heavy Ion Sources 392
    Y. Mori, A. Takagi, K. Ikegami, A. Ueno, and S. Fukumoto
Design Features of an Axial-Geometry, Plasma-Sputter, Heavy
Negative Ion Source 412
    G. D. Alton

DIAGNOSTICS

Spectroscopic Study of Hydrogen–Cesium Discharge Plasma of
Surface-Plasma Ion Sources 427
    V. V. Antsiferov, V. V. Beskorovaynyy, A. M. Maximov, P. G. Sova,
    L. P. Skripal', Yu. I. Belchenko, and G. E. Derevyankin
VUV Laser Diagnostics of H^- Ion Sources 450
    A. T. Young, G. C. Stutzin, K. N. Leung, and W. B. Kunkel
$H^°$ Temperature and Density Measurements in a Penning Surface-Plasma
H^- Ion Source. II 462
    H. V. Smith, Jr., P. Allison, E. J. Pitcher, R. R. Stevens, Jr.,
    G. T. Worth, G. C. Stutzin, A. T. Young, A. S. Schlachter,
    K. N. Leung, and W. B. Kunkel
Dynamics of Negative Hydrogen Ions in a Volume Source 474
    R. A. Stern and M. Bacal
Measurement of the H^- Thermal Energy by Two Laser Pulse Photo
Detachment 489
    M. Bacal, P. Berlemont, J. Bruneteau, P. Devynck, C. Konieczny,
    R. Leroy, and R. A. Stern
Atomic Temperature and Density in Multicusp H^- Volume Sources 504
    A. M. Bruneteau, G. Hollos, R. Leroy, P. Berlemont, M. Bacal,
    and J. Bretagne

Mass Spectrometry in a Multicusp Ion Source 516
    A. A. Mullan and W. G. Graham
Emittance Measurements on a Volume H$^-$ Ion Source 526
    J. G. Alessi
Neutral Beam Detectors 534
    U. von Wimmersperg

EXTRACTION LOSSES IN AND MODELING OF VOLUME SOURCES

Extraction Induced Emittance Growth for Negative Ion Sources 539
    J. H. Whealton, P. S. Meszaros, R. J. Raridon, and K. E. Rothe
Modeling of Volume Hydrogen Negative Ion Sources 557
    D. A. Skinner, P. Berlemont, and M. Bacal
Modeling JAERI 1 and 5 A Tandem H$^-$ Volume Sources 572
    J. H. Fink
Generalized Multibody Computer Simulations of Plasma-Wall Desorption
and Energy-Transfer Processes 585
    A. M. Karo, J. R. Hiskes, T. M. DeBoni, and J. R. Hardy
Transport Processes Through a Magnetic Filter in a Negative Ion Source 596
    M. Ogasawara, T. Yamakawa, F. Sato, and Y. Okumura
Computerized Analysis of Hydrogen Plasma in a Compact H$^-$
Cusp Source 603
    D. H. Yuan, K. Jayamanna, and P. W. Schmor
Simulation of Charged Particle Dynamics in Gas Medium of a
Fusion Reactor Injector 614
    V. P. Sidorov, S. Yu. Udovichenko, A. M. Astapkovich, P. N. Afanasyev,
    Yu. V. Zuev, and Yu. A. Svistunov
Panel Session: Extraction Loss and Modeling of Volume Sources 626
    J. R. Hiskes, J. Bretagne, M. Capitelli, L. Elizarov, L. M. Lea,
    and M. Ogasawara

ACCELERATION AND NEUTRALIZATION

The Matching of SPS to an Electrostatic Negative Ion Accelerator 651
    G. I. Dimov
Fusion Applications of rf Accelerators 660
    R. Thomae, H. Deitinghoff, H. Hopman, H. Klein, and A. Schempp
Injector for RFQ Using Electrostatically Focused Transport and Matching 676
    O. A. Anderson, L. Soroka, J. W. Kwan, and R. P. Wells
Test of a Compact 750 keV H$^-$ Preinjector 690
    C. R. Meitzler, P. Datte, F. R. Huson, R. Kazimi, C. Kronke, S. Machida,
    W. MacKay, S. Ohnuma, D. Raparia, D. Sun, P. Tompkins, and J. Ziegler
HESQ, A Low Energy Beam Transport 699
    D. Raparia

H⁻ Source and Low Energy Transport for the BNL RFQ Preinjector 711
    J. G. Alessi, J. M. Brennan, and A. Kponou
Practical Considerations for a Plasma Neutralizer 717
    K. G. Moses, J. R. Trow, and J. C. Dooling
Plasma Neutralizers 729
    P. M. Vallinga, D. C. Schram, and H. J. Hopman
Comments on Velocity Space Relaxation in High Charge-State
Plasma Neutralizers 741
    A. Hershcovitch
Space Charge Neutralization 746
    U. von Wimmersperg

APPLICATIONS AND PROGRAMS

Tritium Inventory Considerations for Beam-Fuelled Tokamaks 753
    L. R. Grisham
Negative Ion-Based Neutral Injection on D III D 761
    L. D. Stewart, D. K. Bhadra, A. P. Colleraine, and J. Kim
Assessment of Possible Implications of a Neutral Beam
Configuration for CIT 771
    L. R. Grisham, W. S. Cooper, P. Purgalis, and T. Brown
Negative Ion Beam Programs at JAERI 776
    Y. Ohara
Concluding Remarks 786
    R. Hemsworth

APPENDICES

Appendix I: List of Participants 797
Appendix II: Symposium Program 803
Author Index 813

Preface

It was a most rewarding experience for me to be Chairman and Proceedings Editor for the Fifth International Symposium on the Production and Neutralization of Negative Ions and Beams. Krsto Prelec started this Symposium series in 1977. Since then, this series has tracked the progress in the field, as documented in the published Proceedings. During the past decade, changes in the major funding sources dictated the course of this field. From the Proceedings of the 1977 and 1980 Symposia, it is obvious that the main thrust of the field was to develop multi-Ampere, steady-state negative ion-based neutral beams for fusion. Although the one Ampere level was reached in surface plasma sources (as reported in the 1983 Symposium), by the end of FY 1984 the U.S. program lost all its fusion funding. Consequently, the emphasis shifted to volume sources due to the need for low emittance rather than for high current. Now, the main thrust of the field is again fusion oriented. Very impressive results were presented at this meeting by researchers from Japan and the USSR. Once again, surface effects are believed to play a major role in the production of negative ions. Strong commitment to develop negative ion-based neutral beam systems for fusion was shown in programs from Japan, the USSR, and the European community at this meeting. Therefore, we can look forward to continued progress during this decade. One issue that was emphasized by Ron Hemsworth in the concluding remarks is that a switch from H^- to D^- is of paramount importance.

Many thanks to the sponsors of this Symposium—the Office of Fusion Energy at the U.S. Department of Energy, the Air Force Office of Scientific Research, and the U.S. Army Strategic Defense Command. Ray McKenzie-Wilson and Pierre Grand from BNL's Department of Nuclear Energy facilitated funding availability from the latter sponsor. The continuing support from the AGS Department at BNL is gratefully acknowledged. Members of the Program Committee provided many helpful suggestions. Special thanks to Marthe Bacal for her advice on how to facilitate attendance of so many participants from outside the U.S. John Hiskes was instrumental in making the panel session a buoyant forum for ideas on a very important topic. I would like to thank Jim Alessi, the Co-chairman, and the local Organizing Committee for their help. The experience of Ron Clipperton, the Symposium Coordinator, and Marion Heimerle, the Symposium Secretary, was invaluable. Ron and Marion are veterans of a number of symposia; they were dedicated and performed their tasks with perfection. Barbara Cox, the Assistant Coordinator, extended herself well beyond the call of duty by taking on the additional task of Proceedings Secretary. Barbara very skillfully handled all the difficult tasks, which she performed with great enthusiasm.

When editing the transcribed discussion tapes, I felt compelled to maximize the amount of information that is retained in the text. I tried to ensure that the content and intent remain unaltered even if it resulted in retaining spontaneous oral language (in a very small number of cases), rather than a well written text. I apologize for the delay in the publication of these Proceedings. Finally, I would like to thank the participants for making it a truly productive and enjoyable Symposium.

Ady Hershcovitch
Brookhaven National Laboratory

LOCAL ORGANIZING COMMITTEE

Ady Hershcovitch, Chairman
Jim Alessi, Co-chairman
Ron Clipperton, Symposium Coordinator
Barbara Cox, Assistant Coordinator
Ahovi Kponou
Krsto Prelec

INTERNATIONAL PROGRAM COMMITTEE

A. Hershcovitch	Brookhaven National Laboratory Upton, NY, USA Chairman
P. W. Allison	Los Alamos National Laboratory Los Alamos, NM, USA
M. Bacal	Ecole Polytechnique Palaiseau, France
Yu. Belchenko	Institute of Nuclear Physics Novosibirsk, USSR
L. R. Grisham	Princeton Plasma Physics Laboratory Princeton, NJ, USA
J. R. Hiskes	Lawrence Livermore Laboratory Livermore, CA, USA
H. J. Hopman	NET Team, G.F.R. and FOM Institute Amsterdam, The Netherlands
A. J. Holmes	Culham Laboratory Abingdon, England
K. N. Leung	Lawrence Berkeley Laboratory Berkeley, CA, USA
K. Prelec	Brookhaven National Laboratory Upton, NY, USA

SYMPOSIUM SECRETARY

Marion V. Heimerle

PROCEEDINGS SECRETARY

Barbara Cox

FUNDAMENTAL PROCESSES

Surface processes in the production of negative hydrogen ions

Aart W. Kleyn

FOM-Institute for Atomic and Molecular Physics, Kruislaan 407,
1098 SJ Amsterdam, The Netherlands

Abstract

Surface processes are very important in the production of intense beams D$^-$-ions. This is obvious for the surface conversion sources. By using Ba as converter material current densities for H$^-$-beams exceeding 40 mA/cm^2 have been achieved. But also in volume sources surface effects are important, since the walls affect the vibrationally excited molecules and hydrogen atoms, which are very important components of the source plasma. In this report surface processes in the production of negative hydrogen ions will be discussed in the context of existing knowledge about these phenomena at the surface science level.

Introduction

Intense beams of negative deuterium atoms with energies of typically 1 MeV are needed to drive the current of next generation tokamaks like NET, ITER or FER. The beams are produced by acceleration of negative ions which are subsequently neutralized and injected into the torus. D$^-$ ions are used because the neutralization cross section is sizable at 1 MeV, in contrast to that for D$^+$. Because it is easy to detach the extra electron of D$^-$, which is bound by only 0.7 eV, it is rather hard to produce these ions from an ion source. At present two principles are used to produce the ions, volume production and surface production.

In the former case D$^-$ formation is thought to occur in a two step process. At first energetic (primary) electrons (> 12 eV) from the source plasma excite the D$_2$ molecules to an electronically excited state. After radiative decay ground state D$_2$ molecules are formed which are vibrationally excited. D$^-$ formation then proceeds by dissociative attachment of vibrationally excited D$_2$ by slow (thermal, plasma) electrons via: D$_2$(v*) + e$^-$ → D$^-$ + D, see e.g. 1,2.

Surface conversion implies the direct double electron capture into D$^+$ at a surface to yield D$^-$ ions. This process can occur either in direct reflection of D$^+$ or after sputtering of D^0 present at the surface. For this process to occur surfaces with a very low work function are required, see e.g. 3,4.

D$^-$ ion sources can be based upon either principle. In pure volume sources only the dissociative attachment process is assumed to occur whereas in surface conversion sources the D$^-$ ions are produced at low work function convertor surfaces. In addition, both principles can operate simultaneously in a single source. The aim of this review is to show that surface processes are important in all types of D$^-$ ion sources. Therefore, formation and quenching of vibrationally excited molecules, formation and decay of atoms, neutralization of ions, sputtering, implantation and direct negative ion formation, all occuring at surfaces will be discussed. In view of the available space, this report will by no means be exhaustive. The emphasis will be on the comparison between observations made in ion sources and in surface science experiments. It will be attempted to explain the first in terms of the second to get a detailed understanding of the wall processes occuring in negative hydrogenic ion sources. Because much of the work has been performed with H$_2$ rather than D$_2$ molecules, most of the discussion will concern studies with H$_2$.

Vibrationally excitation of H_2

Vibrational excitation of H_2 molecules in volume sources in operation has been studied at the Ecole Polytechnique, FOM and LBL.[5-9] In all cases vibrational temperatures higher than that of the ambient gas have been detected. At FOM an onset of plateau formation indicating nonthermal behaviour was found.[7,8] Vibrational excitation has also been found in D⁻ ion sources with heated filaments, but without running the discharge. Also in this case spectacular vibrational excitation has been observed by Hall et al., which is shown in figure 1.[10] The FOM group showed that the production of this vibrationally excited hydrogen is directly connected to production of H atoms at the filaments.[11] A linear dependence of the $H_2(v^*)$ density (for v=4,5) with H-atom density was found. Recombination at the surface of H-atoms is most likely the cause for the vibrational excitation found by Hall et al. and at FOM.

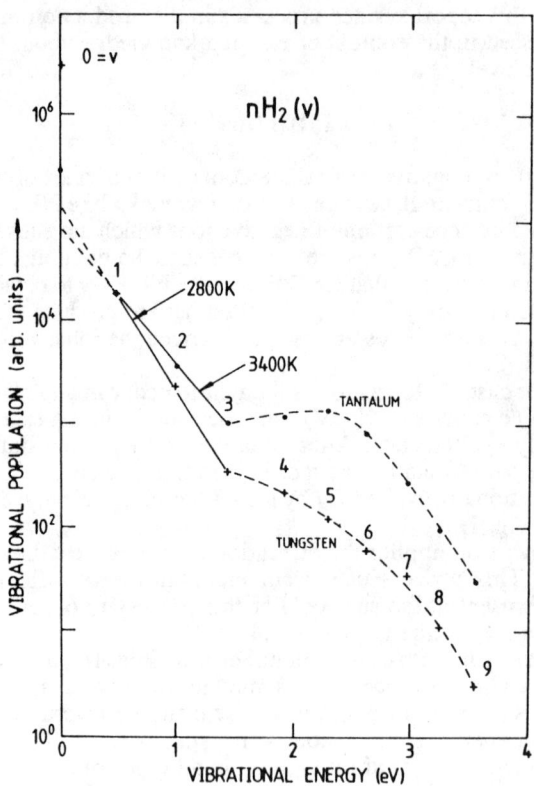

FIG. 1. Vibrational populations against vibrational energy on a semilog scale for tungsten and tantalum filaments. From ref. 10.

It would seem not unreasonable to assume that the observed vibrational excitation in the running discharge would be due to recombination of H-atoms at the wall of the volume source. However, Eenshuistra et al. could exclude this explanation in a rather direct way.[7,8] When the discharge was switched on a large enhancement of the $H_2(v^*)$ density was observed. An almost linear dependence of the $H_2(v^*)$ density on the discharge current was observed. After the discharge was switched off, the $H_2(v^*)$

density essentially dropped to zero. After a few minutes the $H_2(v^*)$ signal has recovered and reaches the value without discharge on. During all experiments the filaments are kept hot. This observation clearly shows that surface production of $H_2(v^*)$ is unlikely if the discharge is running. Similar observations on the disappearance of $H_2(v^*)$ with $v^*>4$ have been made in the group of Hall.[12] These authors didnot observe the recovery of the wall produced $H_2(v^*)$. In their analysis of the scaling of the $H_2(v^*)$ density as a function of discharge parameters Eenshuistra et al. concluded that the wall destroys $H_2(v^*)$ with v=4,5 in effectively a single collision with the wall.[8] A rather rapid quenching has also been observed in theoretical studies by Karo, Hiskes and coworkers and to a lesser extend at the Ecole Polytechnique.[13-15] Therefore, the wall not only does not act as a source for $H_2(v^*)$, it acts as a sink. By contrast, Hall et al. have concluded that when the discharge is off $H_2(v^*)$ survives many wall collisions.[10] Clearly the wall composition changes by turning on the discharge. In the case of Eenshuistra et al. this change is reversible, but in the case of Hall et al. it is irreversible and reconditioning of the wall is required.[10-12]

H-atoms

Another important constituent of the source plasma are H-atoms. Also their concentration is strongly affected by the walls and the filaments. In their analysis Eenshuistra et al. show that atomization of H_2 on the filaments is a source of atoms, secondary processes like dissociative neutralization of positive molecular ions being an additional source.[8] Bonnie et al. showed that wall collisions are the major sink for H-atoms.[6] Eenshuistra et al. have described the observed scaling of the H-atom density as a function of discharge current as shown in figure 2. It is seen that the trend measured experimentally is reproduced in the model. Although the absolute determination of the density is subject to a rather large error, quantitative agreement is observed at low current. An essential parameter to obtain good agreement is the recombination coefficient, $\gamma=0.12$. Eenshuistra et al. remark that if their estimate of the H-atom density is too high, γ should be even lower. Thus most of the H-atoms leave the surface without recombination. Similar values of γ have been obtained for other 'bulk' systems.[16]

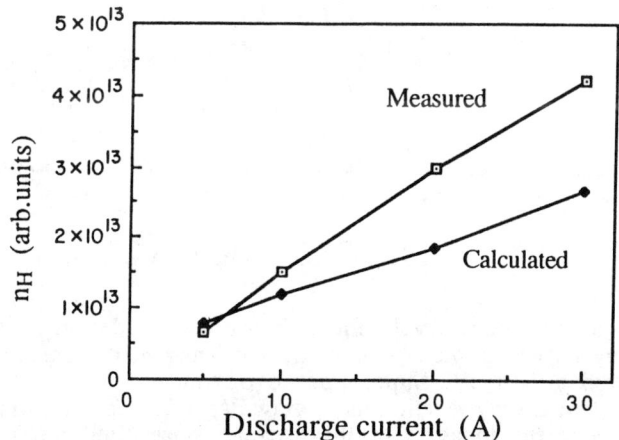

FIG. 2. Experimental and calculated atomic density as a function of discharge current. The recombination coefficient γ is 0.12. The discharge voltage and pressure are 100 V and 1 Pa. From ref. 8.

Summarizing, several important observations have been made concerning the role of the walls in volume sources. These are:
1) Vibrational excitation of H_2 molecules at the walls has been observed when the filaments are hot and the discharge is off. It is due to recombination of atomic H at the walls.
2) No appreciable vibrational excitation of H_2 molecules at the walls has been observed when the filaments are hot and the discharge is running. Volume production is seen instead.
3) The walls quench the vibrationally excited molecules under discharge conditions, but not when the filaments are on and the discharge is off.
4) The recombination of atomic H at the walls under discharge conditions is not very efficient.

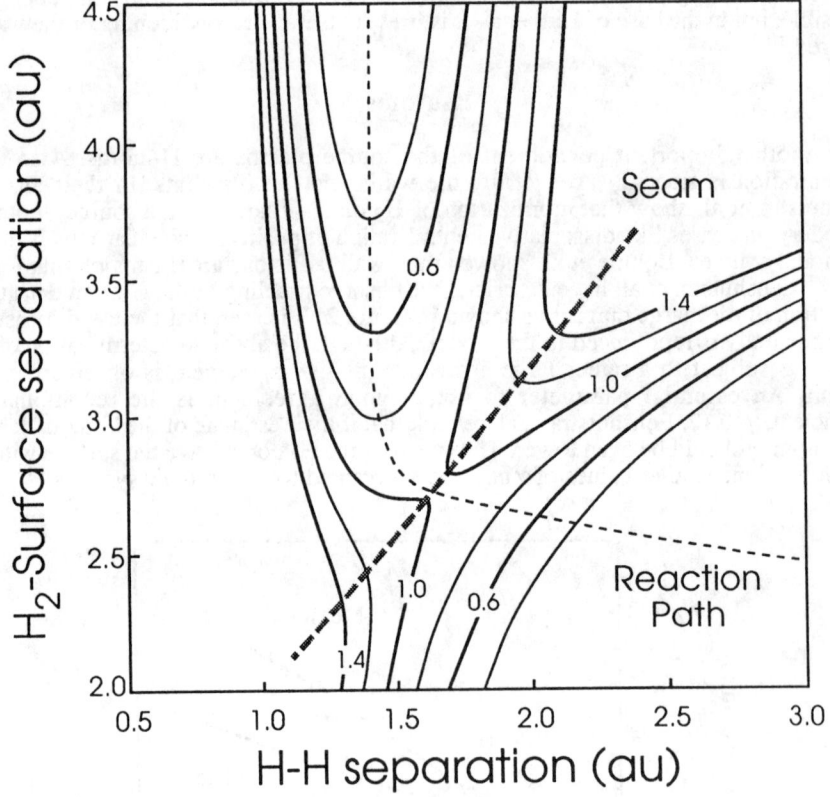

FIG. 3. An energy level contour plot of the model potential energy surface for the H_2-Cu interaction. The potential energies marked are in eV. Superimposed on the contours is the minimum energy reaction path, drawn as a dashed line. Also shown is the location of the "seam", separating the two electronic configurations. From ref. 24.

Surface science interpretation

To better understand the role of the walls in D⁻-sources an explanation of the phenomena listed above is highly desirable. The details of the physics involved are much better explored in experiments with well defined surface conditions, i.e., surface science experiments performed under ultra high vacuum (UHV) conditions. However, directly related experiments are not available and one should try to try to understand the observations listed using the physical pictures developped for the interpretation of the UHV experiments.

Vibrational excitation of H_2 molecules has also been studied by several groups under ultra high vacuum (UHV) conditions.[17,18] Also in this case a (drastic) overpopulation of v=1 with respect to a thermal distribution has been found. Population of higher lying states has not been observed in contrast to the measurements by Eenshuistra et al. and Hall et al.[10,11] The underlying physics of the UHV studies can perhaps most easily be seen from an investigation of the potential energy surface that governs the interaction. A contour representation of one computed for the H_2-Cu interaction is depicted in figure 3. This type of potential energy surfaces is for instance discussed in refs. 19-21. Karo et al. have used potentials that are obtained by summing pair-potentials.[13] It is not clear how the potential surfaces obtained compare to the one shown in figure 3. Most presumably there will be a strong orientational dependence of the barrier towards dissociation. In the top left of figure 3 the situation for H_2 far from Cu is seen. The H_2 potential is essentially undisturbed. At the lower right the binding of two widely separated H-atoms at the surface is seen. A very deep well binds the atoms to the surface. Between these regions a mixing region is seen, the strongest interaction being at the socalled seam. The lowest energy path going from free H_2 to two bound H-atoms is given by the dotted line, labeled reaction path. From this diagram several points are directly clear. There is a barrier towards dissociative chemisorption and conversely to recombination, the barrier height in this particular case being 1 eV. The H-atoms are very strongly bound by on the order of several eV per atom. Vibrational excitation may be effective to overcome the barrier and recombining H-atoms may lead to vibrationally excited molecules. Recombination takes place between two adsorbed and equilibrated atoms in a so-called Langmuir-Hinschelwood mechanism, see e.g. refs. 22. All of the points mentioned have been verified theoretically both using classical and quantum mechanics by Holloway, Harris and coworkers.[23,24] Obviously, the discussion given is very much simplified but the points made also appear in studies by DePristo or Cacciatore and coworkers.[21,25] The points mentioned have also been verified experimentally, for instance in the measurement of final translational energy of desorbing molecules or vibrational excitation.[17,18,26] Very recently, the predicted enhancement of the dissociative sticking probability with vibrational excitation has been verified by Hayden and Lamont.[27] From the observed onset of the dissociative sticking probability for molecules with v=1 one can infer that the barrier depicted in figure 3 is too high and that quenching for highly excited molecules will be probable.

Can the results of vibrational excitation for volume sources be interpreted in this way, assuming that the walls of the devices consist of pure metal, being the material of the filaments that is evaporated on the walls (W or Ta)? Since the barrier between molecular and atomic hydrogen in figure 3 is about one eV (which already appears to be high), one cannot understand the observed very high vibrational excitation, v=5 corresponding to more than 2 eV internal energy. The observed de-excitation when the discharge is running is very natural, because for molecules with v>2 the barrier almost vanishes and immediate dissociation can occur. In this process most of the internal energy will be dissipated to the surface and the emerging molecule after recombination will be vibrationally cold. Only molecules with v≤2 can survive wall collisions. This has been concluded for volume sources by Bonnie et al. and has been seen for NO molecules colliding with e.g. Ag(111) or graphite.[6,28,29]

The vibrational excitation observed by Eenshuistra et al. and Hall et al. cannot result from fully bound H-atoms, because in this case only the barrier height would be available for excitation. Therefore, the H-atoms should not be chemisorbed. One way of avoiding chemisorption is to assume an Eley-Rideal reaction, in which an unbound H-atom from the gasphase directly impinges on an adsorbed H-atom, see e.g. refs. 22. This mechanism may explain why the formation of vibrationally excited H_2 depends linearly on the H-atom density.[11] The linear density dependence implies that the walls are saturated with H-atoms. Supporting evidence for such an Elea-Rideal mechanism comes from UHV studies by Kay and coworkers on the D-H recombination reaction on single crystal surfaces.[30] At this meeting Karo and Hiskes gave beautiful examples of the occurence of Elea-Rideal reactions in their classical trajectory studies.

A second explanation is that the recombination involves more loosely bound or even physisorbed H-atoms.[31] There is evidence for the presence of such loosely bound H-atoms. Kasemo and Törnqvist have observed in an UHV experiment that H-adsorption leading to larger coverages than the saturation coverage for low pressure (< 10^{-9} mBar) are possible at higher pressures.[32] In addition, Kasemo et al. found that the H-D exchange reaction proceeds readily at higher pressures.[33] A loosely bound delocalized H-adatom has also been proposed by Sinniah et al. to explain the first order recombinative desorption of H_2 from Si(100).[34] Assuming that the walls are saturated with a chemisorbed layer, on top of which loosely bound H-atoms are present, nearly the entire recombination energy is available for vibrational excitation. Therefore, the observed vibrational excitation, in particular for $v^*>5$, may reasonably explained by recombination of H-atoms loosely bound on chemisorbed layers in a Langmuir-Hinschelwood mechanism. The distinction between the Langmuir-Hinschelwood and the Elea-Rideal mechanism is not easy to make. Both might be active and under certain conditions they virtually coincide.[35]

Under discharge conditions the wall production of vibrationally excited molecules seems to disappear. This indicates that the walls change when the discharge is running. A reasonable explanation could be that in this case the loosely bound films are removed. This seems somewhat unreasonable, because the discharge does not significantly heat the well-cooled walls. Therefore, sputter cleaning must be invoked. Recent experiments under UHV conditions by Ceyer et al. indicate that sputter or recoil cleaning of surfaces by particles with energies of an eV is quite possible.[36] The ions reaching the wall are accelerated by the plasma potential to an energy of around 2 eV, but their abundance in the discharge is at least two orders of magnitude lower than that of molecules and H-atoms.[37] However, the wall is intensively bombarded by H-atoms with a translational temperature of about 3000K, that would sputter loosely bound H-atoms efficiently.[38-40] In case the discharge is off, the H-atom temperature is lower (2000 K), the density is about an order of magnitude lower and thus cleaning will not or much less occur. This cleaning then would lead under discharge conditions to either a pure metal wall, which quenches vibrationally excited molecules and does not produce them, as discussed before, or prevents the building up of a layer of physisorbed H-atoms on top of a possibly chemisorbed layer. If the sputter cleaning can remove the strongly bound chemisorbed H-atoms remains to be seen. Thus the assumption of a physisorbed H layer, which is removed under discharge conditions seems to be capable to explain most experimental observations; the assumption cannot rule out the occurence of an Elea-Rideal reaction for $v^*\leq 5$. The long time needed to create the H-layer after cleaning may have to do with the fact that part of the H-atoms will diffuse into the bulk of the wall, either directly or along grain boundaries, see e.g.ref. 41.

Low values of γ have been seen before under discharge conditions.[16] However, in an extensive review of the UHV litterature, Christmann states that atomic hydrogen is always strongly bound to most metal surfaces and that therefore atomic desorption has never been observed for any metal surface at low temperature.[41] Also in the very recent work by Sinniah et al. invoking loosely bound H-atoms, no H-atom desorption is claimed.[34] Therefore, the low values of γ observed under discharge conditions are very surprising from an UHV surface science point of view. Reflection of atomic H has only been observed in UHV from weakly interacting surfaces such as graphite and LiF.[43,44] In the very shallow well no acceleration of the H-atom towards the surface occurs, preventing sufficient energy transfer to the solid and subsequent trapping of the atom. This is in contrast to what one would expect for all metal surfaces because of their deep wells. Therefore, only physisorption or loosely bound systems can have a low γ. Above it has been argued that the presence of a physisorbed layer is unlikely when the discharge is on. However, if the chemisorbed H-atoms are still present the binding energy for additional H-atoms will be small, leading to the observed low γ. In addition, if atoms have a sizable energy the amount of energy transferred in the collision with the surface may be insufficient to lead to trapping even in case of a deep well and the atoms will scatter back. This is discussed e.g. in ref. 45. This may particularly apply to H-atoms because the mass ratio with respect to almost any wall material is unfavorable for energy transfer. Recently, the disappearence of trapping of H-atoms with increasing translational energy has been seen in trajectory calculations by Seidl and coworkers.[46] The vibrational excitation discussed previously should be proportional to γ. In view of the rather low absolute yield of vibrationally excited molecules, the low γ is in agreement with the absence of significant vibrational excitation when the discharge is running.

Ion neutralization

In spite of the confinement of the plasma that is applied in some cases, the wall will always be bombarded by ions. It is generally assumed that these ions are neutralized with an efficiency of 100%. Recently, however, Murata and collaborators have observed that for some ions like Ar^+, N^+, and N_2^+ the neutralization decreases drastically at energies below 100 eV.[47] If this would be true as well for H_x^+ our ideas about the effects of neutralization in wall collisions have to be modified drastically. Large ion survival probabilities have not been observed for H_2^+ neutralizing at low perpendicular velocities, but having large parallel components of the velocity and thus total energies above 1 keV.[48] Clearly further study is needed here. If one accepts that neutralization occurs with 100% for the moment it is very interesting what the outcome of the collision would be. In medium density volume sources, like the one studied by the FOM group, H_3^+ is assumed to be the dominant species. In several experiments on H- formation in collisions with incident H_3^+ complete dissociation of the molecular ion prior to negative ion formation has been observed.[4,49-51] By contrast Imke et al. have shown that neutralization into groundstate molecules occurs for H_2^+ scattered from Al(110).[52] In addition, Hiskes and coworkers have shown in theoretical work that neutralization of H_x^+ can result in vibrationally excited H_2.[53] Therefore, neutralization of positive ions at surfaces can be the cause of unexpected behaviour.

Negative ion formation at Ba surfaces

It is well known that negative hydrogen ions can be formed in collisions of energetic positive ions with surfaces having a low work function. Los, van Wunnik, Geerlings and coworkers have performed extensive experiments on formation of H- when a proton beam of energies from 50 to 2000 eV is grazingly incident on a Cs covered W(110) crystal.[4,54] Van Bommel et al. demonstrated that also for Cs covered polycrystalline W a sizable H- yield can be obtained.[55] These authors have explained their results in terms of classical and semi-classical models.[3,4] Basically, the affinity level of the H- ion is stabilized as the ion approaches the surface and at separations of 4-6 a.u. the level becomes degenerate with the Fermi level of the metal, thus facilitating resonant charge transfer to form H-. At very low velocities the electronic system will adiabatically readjust itself when the H- ion recedes from the surface. However, the experiments and model calculations show that at energies of eV's the reneutralization efficiency differs from unity and conversion efficiencies from H^+ to H^- of 60% can be observed.[56] The negative ionization probability is plotted in figure 4 as a function of the energy normal to the surface.[57] Very high yields are predicted even at energies of 1 eV. Indeed Seidl and coworkers have shown that even for thermally reflecting H-atoms negative surface ionization can be observed at cesiated surfaces.[46,58].

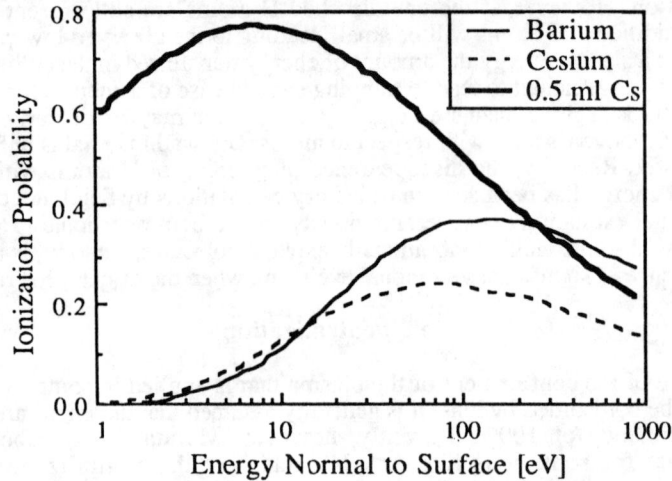

FIG. 4. The calculated negative-ion formation probability for H-atoms leaving the surface along the normal versus energy. The lines correspond to three coverages of the W(110) crystal as labeled. From ref. 57.

Although several D^- sources presented at this meeting employ Cs as an agent to enhance the yield, the use of Cs should be avoided since it has a high vapor pressure and an electron affinity, which could lead to contamination of the accelerator and the tokamak. Therefore, other materials to enhance the negative ion yield are to be preferred. Los, van Os and collaborators have convincingly demonstrated that Ba is a very good choice.[57,59-61] Ba has a considerably higher work function (2.5 eV) than Cs (2.15 eV) or Cs/W (1.45 eV), but it has a larger density of states at the Fermi level and this results in H^- yields at Ba comparable to those at Cs, as is clear from figure 4. Van Os et al. demonstrated this effect experimentally in an UHV experiment, in which it was verified that the models developped earlier are still valid in case of Ba surfaces, even though one of the basic assumptions of these models, that the affinity level crosses the Fermi level at large distances, breaks down.[57] Therefore, Ba convertors seem an excellent candidate for a non poluting negative ion source and this has been investigated in detail by van Os et al.

In their experiments van Os et al. brought a Ba surface in contact with a dense plasma generated by a hollow cathode arc.[57,59,61] The convertor surface was biased negatively with respect to the plasma. Therefore, the negative ions formed are "self-extracted" from the plasma and can be easily separated from plasma electrons. The energy distributions of the resulting ions have shown that the H^- ions have an energy slightly above the converter potential, indicating that the H^- ions are formed by sputtering or direct recoiling and not by reflection. Presence of H at the surface and in the bulk of the Ba is therefore a prerequisite for H⁻ formation. In the experiments van Os et al. were able to show that the conversion efficiency actually depends on the discharge current and thus on the current density on the convertor surface.[57,61] This result is shown in figure 5. The data is obtained in two arrangements. The closed symbols labeled ALICE have been obtained in a low pressure hollow cathode arc and the H⁻ yield is directly measured at the detector. The open symbols, labeled MEGALICE, have been obtained in a high pressure hollow cathode arc and the H⁻ yield has to be corrected for severe stripping losses. Nevertheless it is seen that for high current densities conversion efficiencies of about 8% are obtained. Van Os et al. attribute the decrease in the conversion effiency at current densities > 0.5 A cm^{-2} to overheating of the converter.[61] Figure 5 clearly shows that surface conversion has a great potential.

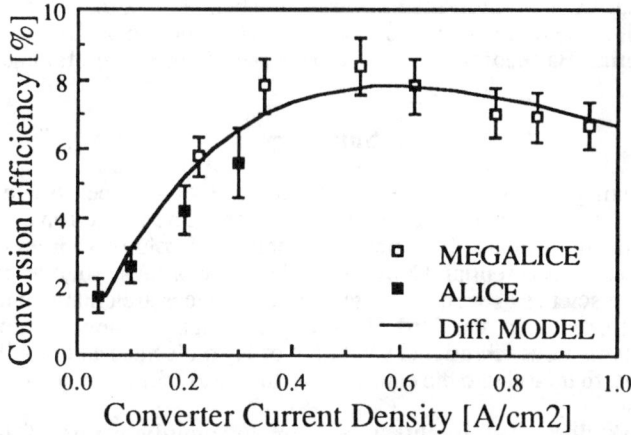

FIG. 5. The conversion efficiencies obtained for two different experimental arrangements as a function of the incident positive ion current density. The converter potential is in all cases -300 V. The efficiencies have been corrected for stripping losses if necessary. The solid line is the result of a model calculation. From ref. 61.

If the H⁻ yield would depend linearly on current density figure 5 should have displayed a constant conversion efficiency. That is it not the case at lower values of j is due to the fact that the yield depends non-lineairly on the current density. It can easily be shown that to first approximation the conversion efficiency for sputtered H⁻ is the product of the charge transfer probability for an individual atom leaving the surface times the sputter yield times the H-density at the surface. The first two terms are assumed to be independent of current density. Only the last term can depend on current density if one assumes that implantation is the major source of hydrogen at the surface and that sputtering and desorption are the drains for H-atoms. Van Os et al. have constructed a set of coupled equations to determine the relative H-atom concentration as a function of distance in the Ba from the surface and time.[62] It appears that indeed the equilibrium concentration depends on current density. Computing the H-atom concentration in the steady state as a function of current density, using the charge transfer probabilities for individual reflection as given in figure 4 and taking the rise of surface temperature with current density into account, the full line in figure 5 has been obtained. The agreement between theory and experiment is quite satisfactory. To verify that indeed implantation at a high current density leads to an enhanced concentration of

H-atoms at the surface and in the bulk preliminary experiments have been carried out for hydrogen implantation into Ti, which was a fashionable material in the second quarter of 1989. Experiments using Elastic Recoil Detection of D atoms recoiled by 2 MeV He$^+$ ions indeed show that a D:Ti ratio of about 0.5 can be obtained by plasma implantation.[63] Currently the sputter process is investigated by means of measurements of optical emission.[64] Also the production of D$^-$ ions at Ba surfaces at high current densities and low pressures, unlike for the experiments shown in figure 5, are being performed at the FOM-Institute. Earlier van Os et al. have attempted to obtain very high current densities at a Ba convertor placed in a large volume source developped at Culham.[60] In these experiments the current densities needed could not be obtained in spite of the very large arc currents used. This type of experiment will be repeated at the FOM-Institute using hopefully less polluting LaB$_6$ filament developped at LBL.[65] At this meeting van Os et al. presented new results obtained at LBL indicating that at the moment Ba still is the best material for a surface converter.[66] In addition to using Ba as a converter material, Ba seeding of has also been seen to be successful and increases the yield.[67]

Summary

Summarizing, I hope to have demonstrated that surfaces processes are very relevant in the development of intense D$^-$ ion sources. Direct electron attachment at Ba surfaces is promising and also for a proper modelling of volume sources the knowledge of surface processes is essential. Conversely, the study of these sources can be relevant for basic surface science as well. In these sources surfaces are created under conditions that are hard to make in standard UHV experiments and new phenomena can be observed like anomalous vibrational excitation of H$_2$ molecules and possibly even Elea-Rideal reactions, that fascinate the catalysis world already for many years.

The work described in this report is the mainly result of the work and collaboration of and stimulating discussions with Peter Eenshuistra, Hans Geerlings, Ron Heeren, Henk Hopman, Cor Leguit, Joop Los, Ron van Os, Helene van Pinxteren, Paul Reijnen, Udo van Slooten, Henk Timmer, John van Wunnik. Stimulating discussions with many participants of the Brookhaven Symposium are greatfully acknowledged. Bruce Kay and Bengt Kasemo are also thanked for illuminating discussions. Dragoslav Ciric, Henk Hopman and Ron van Os are thanked for their careful reading of the manuscript. This work is part of the research program of the Euratom/FOM association and is financially supported by NWO and Euratom.

REFERENCES
1   J.R. Hiskes, Comm. At. Mol. Phys. **19**, 59 (1987).
2   M. Bacal and D.A. Skinner, Comm. At. Mol. Phys. in press.
3   R. Brako and D.M. Newns, Vacuum **32**, 39 (1982);
4   J.N.M. van Wunnik and J. Los, Physica Scr. **T6**, 27 (1983);
5   M. Pealat, J-P.E. Taran, M. Bacal and F. Hillion, J. Chem. Phys. **82**, 4943 (1985).
6   J.H.M. Bonnie, P.J. Eenshuistra, and H.J. Hopman, Phys. Rev. A **37**, 1121 (1988).
7   P.J. Eenshuistra, A.W. Kleyn and H.J. Hopman, Europhys. Lett. **8**, 423 (1989).
8   P.J. Eenshuistra, R.M.A. Heeren, A.W. Kleyn and H.J. Hopman,
    Phys. Rev. A. **40**, 3613 (1989).
9   G.C. Stutzin, A.T. Young, A.S. Schlachter, K.N. Leung and W.B. Kunkel,
    Chem. Phys. Lett. **155**, 475 (1989).
10 R.I. Hall, I. Cadez, M. Landau, F. Pichou and C. Scherman,
    Phys. Rev. Letters **60**, 337 (1988).
11 P.J. Eenshuistra, J.H.M. Bonnie, J. Los and H.J. Hopman, Phys. Rev. Letters **60**, 341 (1988).
12 R.I. Hall et al., these Proceedings.
13 A.M. Karo, J.R. Hiskes and R.J. Hardy, J.Vac.Sci.Tech. A **3**, 1222 (1985).

14 J.R. Hiskes, these Proceedings.
15 D.A. Skinner and M. Bacal, these Proceedings.
16 H. Wise and B.J. Wood, Adv. At. Mol. Phys. **3**, 291 (1967).
17 H. Zacharias, Appl. Phys. A **47**, 37 (1988); L. Schröfer, H. Zacharias and R. David, Phys. Rev. Lett. **62**, 571 (1989).
18 G.D. Kubiak, G.O. Sitz and R.N. Zare, J. Chem. Phys. **83**, 2538 (1985).
19 J. Harris, S. Andersson, C. Holmberg and P. Nordlander, Phys. Scr. T **13**, 160 (1986).
20 P. Madhaven and J.L. Whitten, J. Chem. Phys. **77**, 2673 (1977).
21 C-Y. Lee and A.E. DePristo, J. Chem. Phys. **84**, 485 (1986); J. Chem. Phys. **85**, 4161 (1986); J.Vac.Sci.Tech. A **5**, 485 (1987).
22 B. Kasemo and B.I. Lundqvist, Comm. At. Mol. Phys. **14**, 229 (1984); R. Gasser, Introduction to chemisorption and catalysis, Oxford Science Publ., Clarendon Press, 1985.
23 J. Harris, S. Holloway, T.S. Rahman and K. Yang, J. Chem. Phys **89**, 4427 (1988)
24 S. Holloway, J. Vac. Sci. Tech. A **5**, 476 (1987);
    M. Hand and S. Holloway, J. Chem. Phys. in press.
25 M. Cacciatore, M. Capitelli and G.D. Billing, Surf. Sci. 217, **L391** (1989); and these Proceedings.
26 G. Comsa and R. David, Surf. Sci. Rep. **5**, 145 (1985).
27 B.E. Hayden and C.L.A. Lamont, Phys. Rev. Lett. **63**, 1823 (1989).
28 J. Misewich, P.A. Roland and M.M.T. Loy, Surf. Sci. **171**, 483 (1986).
29 H. Vach, J. Häger and H. Walther, J. Chem. Phys. **90**, 6701 (1989).
30 B.D. Kay, K.R. Lykke and S.J. Ward, to be published.
31 N. Richardson, unpublished idea.
32 B. Kasemo and E. Törnqvist, Phys. Rev. Lett. **44**, 1555 (1980).
33 B. Kasemo, K-E. Keck and T. Högberg, J. Catalysis **66**, 441 (1980).
34 K. Sinniah, M.G. Sherman, L.B. Lewis, W.H. Weinberg, J.T. Yates JR and K.C. Janda, Phys. Rev. Lett. **62**, 567 (1989).
35 J. Harris and B. Kasemo, Surf. Sci. **105**, L281 (1980).
36 J.D. Beckerle, A.D. Johnson and S.T. Ceyer, Phys. Rev. Lett. **62**, 685 (1989).
37 P.J. Eenshuistra, M. Gochitashvili, R.Becker, A.W. Kleyn and H.J. Hopman, J. Appl. Phys. in press.
38 M.P.S. Nightingale, M.J. Forrest and R. McAdams, Proc. 2nd Europ. workshop on production and application of light negative ions, Palaiseau, France, 1986, Eds. M. Bacal and C. Mouttet, p. 123.
39 F. Launay, M. Bacal, A.M. Bruneteau and F. Hillion, Proc. 2nd Europ. workshop on production and application of light negative ions, Palaiseau, France, 1986, Eds. M. Bacal and C. Mouttet, p. 129.
40 G.C. Stutzin, A.T. Young, A.S. Schlachter, J.W. Stearns, K.N. Leung, W.B. Kunkel, G.T. Worth and R.R. Stevens, Rev. Sci. Instr. **59**, 1363 (1989); ibid. **59**, 1479 (1989).
41 B.D. Kay, C.H.F. Peden and D.W. Goodman, Phys. Rev. B **34**, 817 (1986).
42 K. Christmann, Surf. Sci. Rep. **9**, 1 (1988).
43 H. Frankl, H. Hoinkes and H. Wilsch, Surf. Sci. **64**, 362 (1977).
44 T.H. Ellis, S. Iannotta, G. Scoles and U. Valbusa, Phys. Rev. B **14**, 2307 (1981);
    T.H. Ellis, G. Scoles and U. Valbusa, Surf. Sci. **118**, L251 (1982).
45 J.A. Barker and D.J. Auerbach, Surf. Sci. Rep. **4**, 1 (1984).
46 M. Seidl et al., these Proceedings.
47 H. Akazawa and Y. Murata, Phys. Rev. Lett. **61**, 1218 (1988).
48 B. Willerding, H. Steininger, K.J. Snowdon and W. Heiland, Nucl. Instr. Meth. B **2**, 453 (1984);
    K.J. Snowdon, B. Willerding and W. Heiland, Nucl. Inst. Meth. B **14**, 467 (1986);
    K.J. Snowdon, Nucl. Inst. Meth. B **33**, 365 (1988).
49 E.H.A. Granneman, J.J.C. Geerlings, J.N.M. van Wunnik, P.J. van Bommel, H.J. Hopman and J. Los, Proc. Third Int. Conf. on Production and Neutralization of Negative ions and Beams, Ed. K. Prelec, Brookhaven (1983).
50 M. Shi, J.W. Rabalais and V.A. Esaulov, Rad. Eff. Def. Sol. **109**, 81 (1989).
51 W. Eckstein, H. Verbeek and R.S. Bhattacharya, Surf. Sci. **99**, 356 (1980).
52 U. Imke, S. Schubert, K.J. Snowdon and W. Heiland, Surf. Sci. **189/190**, 960 (1987).
53 J.R. Hiskes and A.M. Karo, in: Dissociative Recombination, Eds: J.B.A. Mitchell and S. Guberman, World Scientific, 1989, p. 204; and these Proceedings.
54 J.J.C. Geerlings, R. Rodink, J. Los and J.P. Gauyacq, Surf. Sci. **186**, 15 (1987).

55 P.J.M. van Bommel, J.J.C. Geerlings, J.N.M. van Wunnik, P. Massmann, E.H.A. Granneman and J. Los, J. Appl. Phys. **54**, 5676 (1983).
56 J.J.C. Geerlings, P.W. van Amersfoort, L.F.Tz. Kwakman, E.H.A. Granneman, J. Los and J.P. Gauyacq, Surf. Sci. **157**, 151 (1985).
57 C.F.A. van Os, P.W. van Amersfoort and J. Los, J. Appl. Phys.**64**, 3863 (1988).
58 A. Pargellis and M. Seidl, Phys. Rev. B **25**, 4356 (1982).
59 C.F.A. van Os, R.M.A. Heeren and P.W. van Amersfoort, Appl. Phys. Letters **51** (1987) 1495.
60 C.F.A. van Os, A.W. Kleyn, L.M. Lea, A.J.T. Holmes and P.W. van Amersfoort, Rev. Sci. Instr. **60** (1989) 539.
61 C.F.A. van Os, C. Leguit, R.M.A. Heeren, J. Los and A.W. Kleyn, SPIE **1061** (1989) 568.
62 C.F.A. van Os, C. Leguijt and J. Los, J. Appl. Phys.in press. J.H.M. Bonnie, E.H.A. Granneman and H.J. Hopman, Rev. Sci. Instr. **58** (1987) 1354.
63 R.M.A. Heeren, J.P. van Maaren, C.F.A. van Os and A.W. Kleyn, to be published.
64 R.M.A. Heeren, D. Ciric, S. Yagura, C.F.A. van Os, H.J. Hopman and A.W. Kleyn, to be published.
65 K.N. Leung, D. Moussa and S.B. Wilde, Rev. Sci. Instr. **55**, 1064 (1984).
66 C.F.A. van Os, K.N. Leung, A.T. Lietzke, J.W. Stearns and W.B. Kunkel, these Proceedings.
67 S.R. Walther, K.N. Leung and W.B. Kunkel, Appl. Phys. Lett. **54**, 210 (1989); K.N. Leung, S.R. Walther and W.B. Kunkel, Phys. Rev. Lett. **62**, 764 (1989).

DISCUSSION

Hopman: From your discussion, do I understand that at the moment you favor the explanation given by Hall, et al for the formation of vibrationally excited molecules by physisorptions above the explanation of the FOM Institute which is an Eley-Rideal reaction.

Kleyn: Intuitively I have two reasons to prefer the Hall explanation. First: it is more pleasing, it looks more natural because an Eley-Rideal reaction is something that the catalysis community has been looking for decades and they have never found it. Actually this was the first sort of claimed observation under some sort of defined conditions. The issue of the sputtering is very confusing again, I think it would be hard to say what is happening under discharge conditions. Another thing is that the linear pressure dependance which favored the Eley-Rideal has been observed now for various other systems, for instance, hydrogen and silicone. Also, an Eley-Rideal type exchange mechanism has been seen recently at Sandia, so basically it is still an open question, I think the physisorbed layer has to be there. I don't believe that the metal is clean. I simply don't believe it. In some cases these two mechanisms almost coincide and that may be the case that we're facing.

Hiskes: You had a reference to Harris H_2 on copper, is that a theoretical paper or is that an experimental paper?

Kleyn: That is a theoretical paper.

Hiskes: Do you know what the maximum magnitude of those physisorption and chemisorptions are?

Kleyn: Do you mean the barrier height.

Hiskes: Yes

Kleyn: The barrier height in this particular case is about a volt (which may very well be an artifact). I guess it is basically a calculation of Hartree-Fock levels. So the barrier height may be wrong, you can easily fit the data by lowering it to half a volt.

Hiskes: I think there is some experimental data by DePristo that shows that minimum and maximum by more than 20 to 30 volts?

Kleyn: Well, DePristo is not an experimentalist, so it can

not be experimental data. These things depend on the surface. I guess DePristo's is on nickel, to my knowledge it isn't done for tungsten.

Hiskes: You showed wall recovery in one of your earlier slides, after you turned the discharge off there was a wall recovery, is that recovery specific to certain vibrational levels.

Kleyn: The problem is that discovery has not been documented very well. He has noted it, so he wrote a note about that in his notebook, how funny it disappears and suddenly reappears. Now, in view of what I said here he should have actually made some graphs.

York: You showed that the production conversion efficiency on barium was dependent on the hydrogen and the loading of the barium surface as a function of the discharge current. We also see on cesiated surface that there is a maximum arc current at which we are not going to get more H^- off the cesiated surface. The question is can you calculate the loading of the barium surface that would predict a maximum loading of the surface of which you could not put more hydrogen on the surface or at which the efficiency of putting hydrogen on the surface would go down.

Kleyn: I don't think that such a maximum has been observed theoretically.

York: Well it is easy to reach a maximum arc current, at which the efficiency goes down. We don't know whether we cannot keep optimum cesium coating on the surface or if it has something to do with the hydrogen. But right now it is an unknown.

Kleyn: Well in the case of Ron van Os, he studied the same set of equations for cesium on tungsten and here the hydrogen does diffuse very rapidly into barium but doesn't in tungsten. In the tungsten case you basically have to assume you have a monolayer, or very slight hydrogen containing layer, and the hydrogen sinks into the cracks. Essentially it is grain boundary diffusion, nevertheless, I think the observations there were different.

H⁻ PRODUCTION FROM NON-CESIATED CONVERTER-TYPE NEGATIVE ION SOURCES*

C.F.A. van Os, K.N. Leung, A.F. Lietzke, J.W. Stearns and W.B. Kunkel
Lawrence Berkeley Laboratory, University of California,
Berkeley, CA 94720, USA

ABSTRACT

Recent results of surface produced negative ions are presented. Two low work function metal surfaces have been studied, barium and magnesium, in combination with several plasma generators; RF- and DC-filament discharges. The negative ion yield for barium is about 5 to 6 times larger than magnesium. This ratio is confirmed by model calculations on resonant charge exchange.

INTRODUCTION

The two fundamentally different techniques to produce or generate negative ion beams are generally referred to as *volume production* [1,2] and *surface conversion*.[3,4] Each technique has its specific advantages and drawbacks and at the moment it is not clear which technique will prove the most useful. Therefore, a large research effort is still dedicated to the development and understanding of both production methods. This paper discusses the ongoing research in the field of surface conversion at Lawrence Berkeley Laboratory with two separate research objectives: (i) An experimental and theoretical survey of low work-function metals applicable in a surface conversion source and (ii) a study of the integration of plasma generation with surface production.

One of the first surface conversion experiments was done by Belchenko et al., who extracted a beam of negative hydrogen ions from a magnetron source.[5] Some of these ions were produced on the cathode surface. It was generally believed that the conversion process taking place at the cathode could be better controlled if it was separated from the plasma production process. Therefore, in most subsequent experiments the negative ions are produced on an isolated electrode, the so-called converter.[6,7,8,9] In a typical surface-conversion negative-ion source, the converter is biased at a negative potential of a few hundred volts, so that it draws a flux of positive hydrogen ions from the surrounding plasma. A fraction of these ions is back-scattered and some cause adsorbed hydrogen atoms to be sputtered from the converter surface. The back-scattered and sputtered hydrogen atoms can form negative ions via electron capture from the metal surface. Negative ions thus formed are accelerated across the

---

* This work was supported by the Director, Office of Energy Research, Office of Fusion Energy, Development and Technology Division, of the U.S. Department of Energy under Contract No. DE-AC03-76SF00098.

plasma sheath and are "self-extracted" from the source.[4] The formation process is operative over a distance of a few times the Bohr radius, a_o, which is small with respect to the sheath thickness. The latter is typically a few tens of a μm in an intense discharge. Therefore, the charge exchange process between metal surface and hydrogen atom is not influenced by the sheath potential.

The converter should have a low work function to obtain a high negative ion yield. This can be achieved via adsorption of a sub-monolayer of cesium on a metal substrate.[10] The cesium coverage can be obtained by admitting cesium vapor into the discharge chamber. However, cesium in the source migrates to the accelerator structure and causes electrical breakdown. This encouraged a search for other, less contaminating, methods for producing a low work function surface.

Recent experiments at the FOM institute in Amsterdam showed that, using a pure barium-metal converter, negative ion conversion efficiencies could be obtained that were of the same order as for cesiated surfaces.[11,12,13] The conversion efficiency, defined as the ratio between the produced negative ion current denisty and the positive ion current density striking the converter surface, provides us with a means to compare different types of converter materials in combination with different plasma generators.

The materials we have investigated experimentally are barium, magnesium, copper and molybdenum. The latter two metals are routinely used to verify the operation of the diagnostics, and to provide us with the necessary base line data. Basically three different plasma generators are used; a magnetically confined plasma column (sheet plasma), a filament discharge (employing both tungsten and barium oxide cathodes) and a radio-frequency discharge (operating at around 1.7 MHz). This paper reports on the results obtained so far and compares the conversion efficiencies with calculated ionization probabilities for the various metals used in the experiments.

THEORY ON SURFACE CONVERSION

Let $\eta_H(v)$ be the velocity dependent probability for a hydrogenic atom leaving a surface to escape as a negative ion. Then the negative ion flux is,

$$J_{H^-} = \int_V dv\ \eta_H(v)\ J_H(v) \qquad [1]$$

where $J_H(v)$ is the flux of hydrogenic particles leaving the surface and the integral is taken over all velocity space with a component that exits the surface.[14,15] In a practical surface conversion source, only negative ions leaving more or less normal to the surface are collected. In view of the difference in angular distribution of sputtered and reflected particles, $J_H(v)$ can be simplified with the assumption that only sputtered or recoiled hydrogen atoms contribute.[16,12] Both processes result in hydrogen atoms leaving the surface with a relatively low energy, of the order of a few tenth's of an eV. This reduces the integration in Eq. [1] to a small energy range which can be further simplified by assuming that the particles leave with an average energy <E>;

$$J_{H^-} \approx \eta_H(<E>)\ J_H^* \qquad [2]$$

Here <E> represents the average energy of the sputtered and other low energy particles and J_H^* the sputtered flux. The latter is equal to the incident flux of positive particles on the converter multiplied by a coefficient, Γ_{H-H} which determines the stimulated desorption or sputtered yield;

$$J_H^* = \Gamma_{H-H} J_{H+} \qquad [3]$$

For a solid body metal converter (e.g. Barium), van Os, van Amersfoort and Los have argued that the coefficient Γ_{H-H} is equal to the sputter coefficient multiplied by a probability of hitting a hydrogen atom sitting in the surface layers.[13] The latter probability is assumed to be equal to the relative hydrogen concentration in the surface layers. A comparison of different converter surfaces in terms of conversion efficiency gives an indication of whether relative hydrogen concentrations differ for different converter materials, provided that the ionization probability in Eq. [2] is known for these materials. In the following, we will briefly discuss a model with which this ionization probability can be calculated.

The process of forming a negative ion via interaction with a low work function surface is generally referred to as a "resonant" charge exchange.[17,18,19,20] The electron affinity of a negative hydrogen ion increases when it comes into the vicinity of a metal surface because of the attractive interaction with its image in the metal plane. This lowering of the potential well can be expressed as a function of the atom metal surface separation, z, by a $0.25 (z + k_s)^{-1}$ dependence, where k_s denotes the screening length of the metal electrons expressed in atomic units.[21] Due to this lowering, at short distances, the potential drops below that of the metal surface. At this point electrons from the metal may tunnel through the potential barrier and be shared between the metal surface and the ion. The tunneling frequency depends exponentially on the height and the width of the barrier. In other words, electrons tunnel between the hydrogen atom and the metal conduction band with a certain transition frequency $\omega(z)$ which is an exponentially decaying function of the distance, z, between atom and metal.

Via this process of resonant population, the atom has a certain charge probability, $N^-(z)$, in the vicinity of the surface. When this atom moves away from the surface, this probability decays, depending on the normal velocity of the atom and the transition frequency. If the atom moves very slowly, almost adiabatically, it will lose all of its charge. On the other hand if the atom moves too rapidly with respect to the transition frequency no charge will be transferred and no negative ion will be formed. In the intermediate range we expect some probability of forming a free negative ion. This formation probability can be calculated using the stationary phase approximation which yields[22]

$$\eta_{H^-}(v_\perp) = \frac{1}{v_\perp} \int_{z_0}^{\infty} N^-(z)\omega(z)\exp\left(\frac{-1}{v_\perp}\int_z^{\infty}\omega(z')dz'\right) dz. \qquad [4]$$

Where $N^-(z)$ is an occupation probability depending mainly on the position of the affinity level with respect to the Fermi level, and $v_\perp$ represents the normal component of velocity of the hydrogen atom. Note, this expression has constraints on the region over which it is valid. However, it has been shown that meaningful results can be obtained with this model for the metal surfaces considered here.[23,24] With this expression, the free negative-ion production probability can be calculated as a function of the energy of the hydrogen atom leaving normal to the surface.

Using the experimental data from scattering experiments at grazing angles of incidence, van Os et al. were able to determine the screening length and use Eq. [4] to calculate the production probability; as has been done previously for cesiated surfaces.[25,26] For barium a value of 2.4 a_0 for the screening length has been reported. This value is about 1 a_0 longer than a value based on the assumption that the affinity level of the hydrogen atom should join the bottom of the conduction band at zero atom-metal separation. Since no experimental data is available for the other materials we are interested in, we added this difference of 1 a_0 to the previous values to calculate the production probability at the surface of these metals. Since most of the metals are in the alkali earth metal group, it is anticipated that this is a reasonable assumption. In Figure 1 we have collated the results of these calculations for barium, magnesium, strontium, copper and lithium for a 'normal' energy ranging from 1 to 20 eV. The calculated probabilities for barium and strontium more or less coincide, which was expected in view of the minor differences in their electronic properties.[27] The H⁻ production

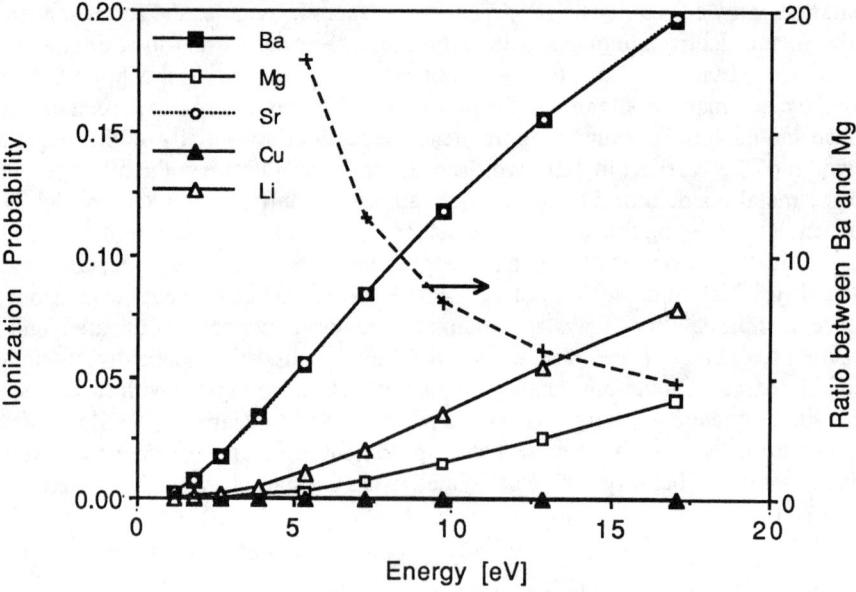

Fig. 1. The calculated production probability for the formation of a negative hydrogen ion as a function of the energy with which the atom leaves normal to the surface for five different metals. The dotted line represents the ratio in ionization probability for barium and magnesium.

probability for copper in this energy range is less than 10^{-4}. The calculated results for molybdenum are not plotted but are essentially the same as for copper. So far, experimental data has been obtained for barium, magnesium and copper surfaces. Therefore, the ratio of the probabilities for barium and magnesium is also shown in Figure 1 for comparison with experimental data.

EXPERIMENTS

This section is subdivided in three parts which describe the set-up, operation and results of individual experiments related to surface conversion using a barium and a magnesium metal converter. The source pressures are all of the order of 1 mTorr and, before measurments, the converters are routinely cleaned by argon sputtering.

I. .RF-Plasma generation

A pulsed radio-frequency discharge, with a pulse length of 1 ms and a repetition rate of about 100 Hz, is created in a multicusp bucket with a diameter of 15 cm. The set-up is schematically shown in Figure 2. The antenna is made of anodized aluminum and connected to the RF generator using a matching network. The RF frequency is around 1.7 MHz. The surface produced negative ion current is analyzed with a small 180° magnetic analyzer located just behind a 1 mm extraction aperture. No extraction voltage is applied, so the analyzer collects only self-extracted negative ions.

Since these ions are formed as a result of incident positive ions that desorb or sputter hydrogen atoms located at the surface layers, it is worthwhile to measure the

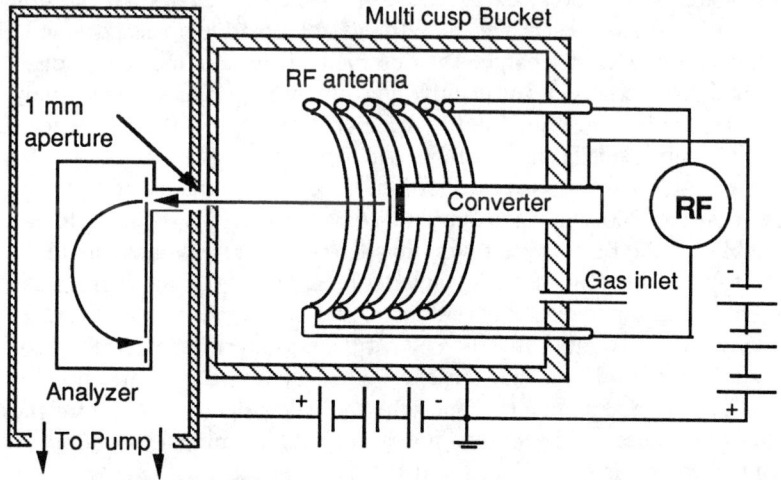

Fig. 2. Schematic diagram of the RF plasma generator. Also shown is the position of the biasable converter and the magnetic analyzer.

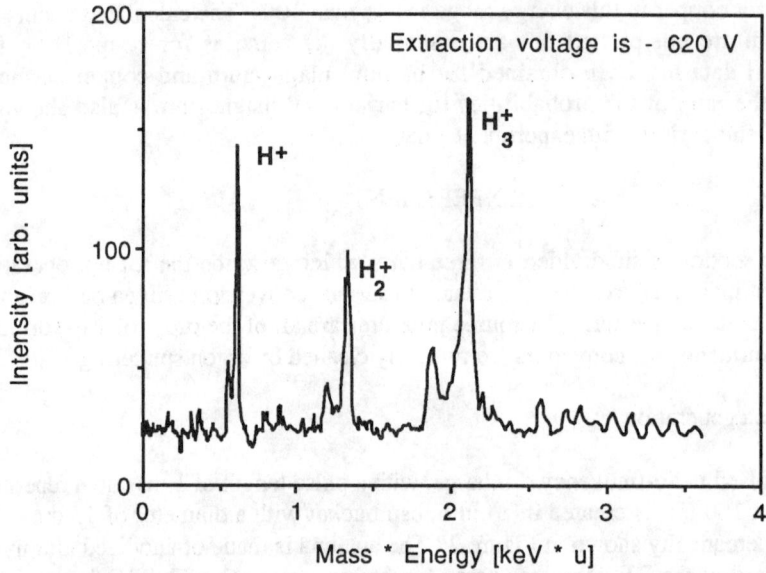

Fig. 3. The energy distribution of positive ions directly extracted from the RF plasma bucket. The energy resolution of the analyzer is of the order of 0.5 %.

energy distribution of the ions hitting the surface. This has been done by applying a positive extraction voltage to the source and using the magnetic analyzer to measure the square of the momentum of the extracted ions as depicted in Figure 3. The structure of the positive ion peaks resembles that found for positive ions extracted from a capacitively-coupled RF-discharge. The observed structure in the energy distribution, which should also be present for positive ions impurging upon the negatively biased converter, results in a less well defined incident energy on the converter surface. However, it is only a small fraction of the total incident energy.

Figure 4 shows the energy distributions of self extracted ions for a barium converter biased at -200 V and a magnesium converter biased at - 400 V. In the present set-up the bias on the Ba converter was limited to -200V to minimize the formation of converter spots. With a magnesium converter biased at -200 V the average area under the sputtered peak for several measurements was compared with an average for the barium converter. The ratio in peak area for the barium surface and the magnesium surface was 6.6 ± 3. Note, that the energy distribution for a barium converter is dominated by sputtered particles whereas the equivalent distribution for magnesium is not. This is because of the higher probability for forming low energy negative hydrogen ions in the vicincity of a barium surface.

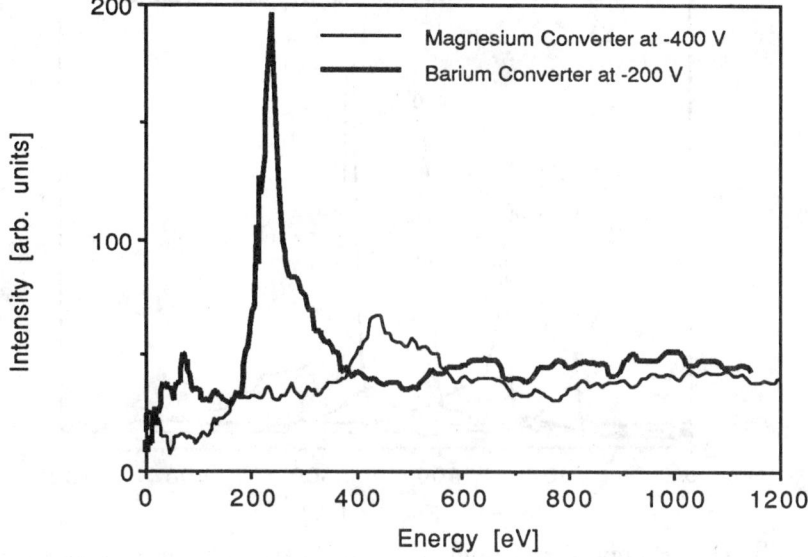

Fig.4. The energy distribution for the "self-extracted" negative ions for a barium (thick line) and a magnesium (thin line) converter. The source operating conditions are the same in both cases.

II. Discharges with thermionic emitting cathodes

An arrangement similar to the one shown Figure 2 is employed, without the RF antenna and with a multicusp bucket with a diameter of 15 cm. The same analyzer is used for measuring the energy distribution of the surface produced negative ions. The bucket can be equipped either with a tungsten filament or a barium oxide cathode. With the tungsten filament installed we measured the energy distribution for a barium converter as a function of its bias, as depicted in Figure 5. These spectra confirm that a large fraction of the negative ions are produced by ion impact stimulated desorption (sputtering and recoiling).[28] Furthermore, the increase in converter bias tends, preferentially, to increase the desorbed fraction of the negative ion yield.

The same bucket has been used to measure the energy distribution for a magnesium converter. The result of this measurement is depicted in Figure 6 together with the data for barium. The ratio in peak area, obtained using the method described in the previous section, is 4.6 ± 1.5 at a converter bias of -400 V.

In doing the experiments it appeared that the results are very sensitive to the preparation method of the converter surfaces and how well one is able to clean the surface in-situ. This led to the suspicion that the material coming from the tungsten filament was contaminating the surface. Therefore, the bucket was operated with a barium oxide cathode to check the extent of this contamination. This cathode is an ohmically heated, coaxially driven, porous tungsten cylinder impregnated with barium

24 H⁻ Production from Negative Ion Sources

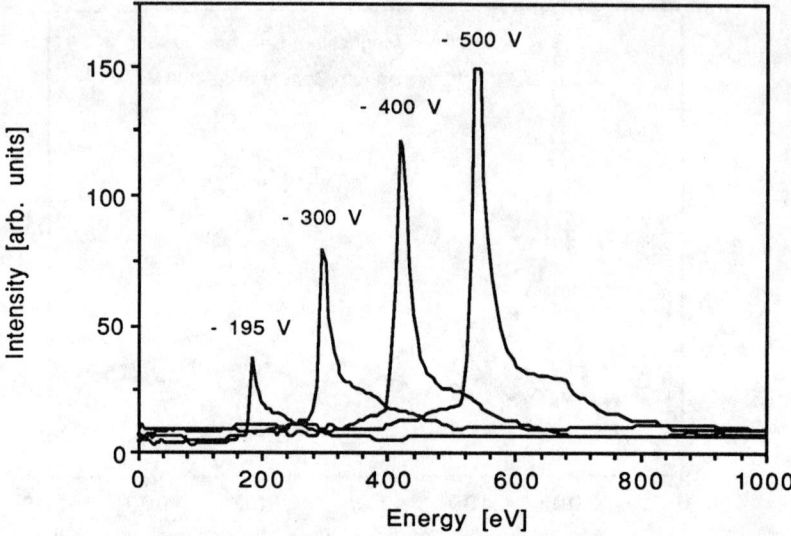

Fig. 5. The energy distributions of the self extracted negative ions for a converter bias ranging from -195 V to -500 V. The source pressure is of the order of 1 mTorr and a moderate arc current of 10 A

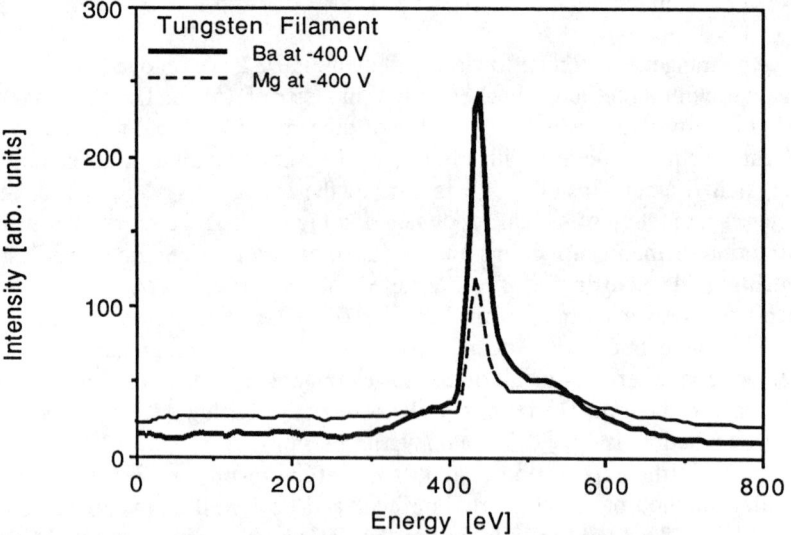

Fig. 6. The energy distribution of the self extracted negative ions for a converter bias of -400 V for a barium and a magnesium surface. Operating conditions are the same as Figure 5.

oxide, that is of similar construction as cylindrical LaB$_6$ cathodes previously reported.[29,30] The energy distribution, obtained with this cathode, for a converter bias of -220 V is depicted in Figure 7 along with the previously determined distribution for a barium converter biased at -195 V employing a tungsten filament. There is a distinct difference in shape but the yield is of the same order which indicates that the tungsten contamination problem is not serious at the low power levels used for thesemeasurments.

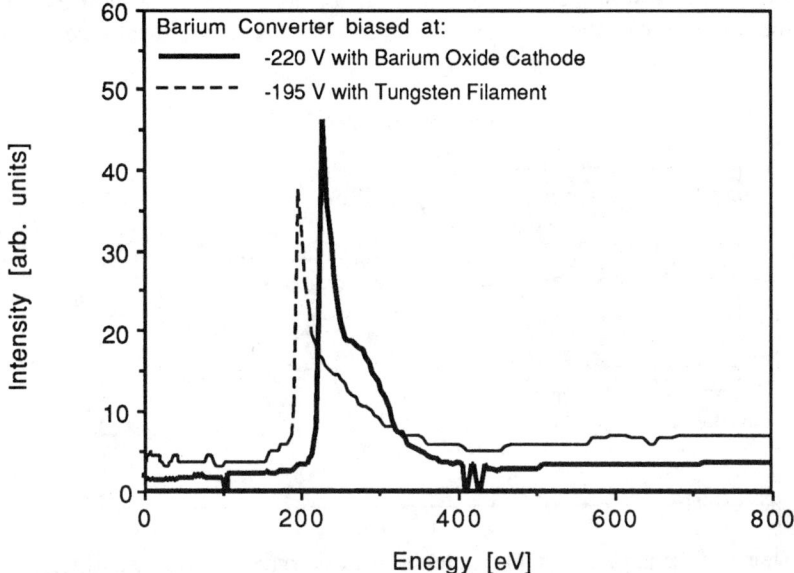

Fig. 7. The energy distribution of the self extracted negative ions for a converter bias of the order of -200 V for a barium converter operated in a bucket equipped with a tungsten filament and a barium oxide cathode. Operating conditions are the same as Figure 5.

III. Sheet Plasma discharge

A different approach to reduce the surface pollution due to material evaporated and sputtered from the emitting cathode is to position it far from the converter and create a magnetically confined plasma column. The arrangement used is essentially a variation of an Uramoto sheet plasma source[31] modified to permit a converter in the vicinity of the column. A schematic arrangement of the converter module and negative ion collector are shown in Figure 8. The negative ion accelerator has an aperture of 1 cm^2, and the converter we used was 2.5 cm in diameter. The obtained negative ion current density as a function of the arc current are depicted in Figure 9. The positive ion current density ranges from 50 mA/cm^2 for an arc current of 10 A, to 400 mA/cm^2 for 60 A.

The associated conversion efficiencies are 2.5 % for barium and 0.45 % for magnesium for a positive ion current density of 50 mA/cm^2. For larger positive ion current densities the conversion efficiency for barium and magnesium drop to a value of 0.7 % and 0.13 %, respectively. This drop in conversion efficiency with increasing positive ion current density is in sharp contrast with the results obtained at the FOM Institute in Amsterdam,[13] who reported an increase of the conversion efficiency with increasing positive ion density. We attribute this scaling difference to the higher electron temperature and higher primary electron density observed in the LBL sheet plasma. Both aspects will increase the barium ion density in the discharge and the

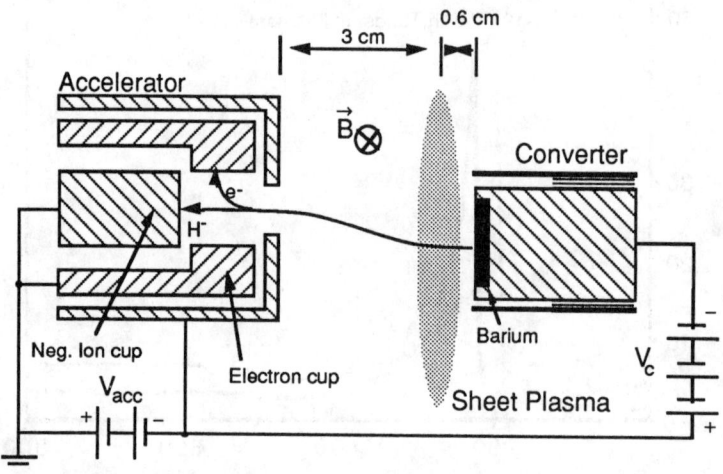

Fig. 8 Detail of the experimental arrangement of the converter and diagnostics.

the barium ion bombardment (and sputtering) of the converter surface. This is suspected of having two deleterious effects upon the delivered negative ion current. i) The presence of barium ions in the discharge is expected to noticeably increase the ion and electron density immediately in front of the converter, thereby aggravating the negative ion destruction processes[32], and (ii) the increased barium ion bombardment of the surface is postulated to reduce the (target) density of accumulated hydrogen in the surface layers. The hydrogen atoms desorbed in this process are believed to be to slow to have a significant probability for forming a negative ion.[13] Both effects are increasing functions of plasma density and converter voltage. Further work is needed to unravel the relative influences of the two mechanisms.

The ratio of the negative ion current density for barium and magnesium, at a converter bias of -200 V, amounted to 5.2 ± 0.6. This ratio agrees with the value obtained from the RF source which was taken at the same converter bias.

CONCLUSIONS

The contamination problem of the active low work function surface (converter) can be avoided in different ways. The use of a RF plasma is clearly favorable if one is able to create sufficiently high plasma densities to feed the converter with enough positive ions. The RF created plasma shields the converter sufficiently to prevent direct coupling of the RF field to the converter and support structure. A second approach is to place the filaments out of sight of the converter surface to inhibit material from being depositied. However, ionized cathode material may sill reach the converter and

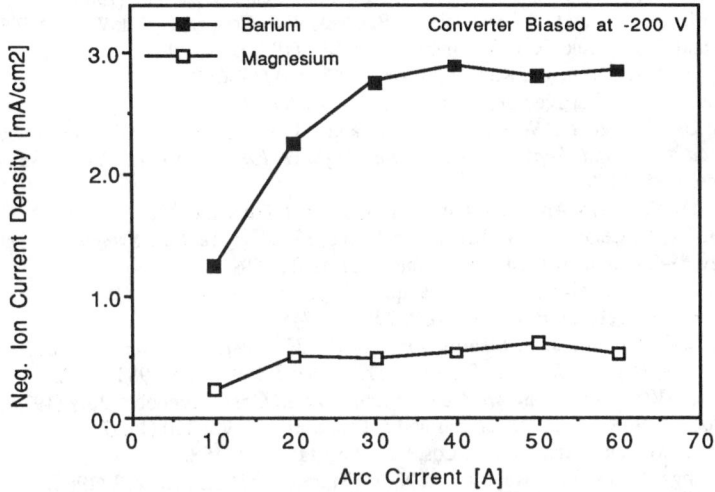

Fig. 9. The self extracted negative ion current density vs arc current. The source pressure is 2 mTorr, the arc voltage -80 V, the converter voltage -200 V, the acceleration voltage 3.3 kV, the distance between converter surface and sheet center is 0.6 cm and for the accelerator this distance is 3 cm.

contaminate the surface. This resulted in a third alternative; the use of a cathode made of the same material as the converter itself. However, too many heavy ions in the discharge can reduce the negative ion yield either through stripping collisions or depletion of the surface-accumulated hydrogen.

Qualitatively, the empirically determined ratios of the negative ion yield for barium and magnesium, follow the trend in the calculated ratios in attachment probability, depicted in Fig. 1, with energy. However, qualitatively the empirical ratios are lower when compared to the theoretically determined ratios. The latter observation might we attribution to a different behavior in terms of the hydrogen desorption yield upon ion impact.

ACKNOWLEDGEMENTS

The authors would like to thank Cheryl Hauck and Don Williams for supplying results from their experiments prior to publication. We would also like to express our gratitude to Steve Wilde and members of his shop for their valuable support.

REFERENCES

1. K.W. Ehlers, B.F. Gavin and E.L. Hubbard, Nucl. Instrum. Methods **22**, 87 (1963).
2. Yu.I. Belchenko, G.I. Dimov and V.G. Dudnikov, Novosibirsk Report No. 66-72 (1972).
3. A.I. Herscovitch and K. Prelec, Rev. Sci. Instrum. **52**, 1459 (1981).
4. K.N. Leung and K.W. Ehlers, Rev. Sci. Instrum. **53**, 803 (1982).
5. Yu.I. Belchenko, G.I. Dimov and V.G. Dudnikov, Nucl. Fusion **14**, 113 (1974).
6. K.W. Ehlers and K.N. Leung, Rev. Sci. Instrum. **51**, 721 (1980).
7. A.I. Hershcovitch, V.J. Kovarik and K. Prelec, Rev. Sci. Instrum. **57**, 827 (1986).
8. J. Kwan, G.D. Ackerman, O.A. Anderson, C.F. Chan, W.S. Cooper, G.J. deVries, A.F. Lietzke, L. Soroka and W.F. Steele, Rev. Sci. Instrum. **57**, 831 (1986).
9. B. Piosczyk and G. Dammertz, Rev. Sci. Instrum. **57**, 840 (1986).
10. J.L. Desplat and C.A. Papageorgopoulos, Surf. Sci. **92**, 97 (1980).
11. C.F.A. van Os, C. Leguyt, P.W. van Amersfoort and J. Los, *Proc. of IIIrd European workschop on the Production and Application of Light Negative Ions*, Februari 17-19, Amersfoort, Netherlands, 1988, p149.
12. C.F.A. van Os, P.W. van Amersfoort and J. Los, J. Appl. Phys. **64**, 3863 (1988).
13. C.F.A. van Os, C. Leguyt, A.W. Kleyn and J. Los, *Proc. of the 15th Symposium on Fusion Technology*, 19-23 september, Utrecht, Netherlands, 1988, p589.
14. M.S. Huq, L.D. Doverspike and R.L. Champion, Phys. Rev. A **27**, 2831 (1983).
15. J.S. Risley and R. Geballe, Phys. Rev. A **9**, 2485 (1974).
16. M. Wada, R.V. Pyle and J.W. Stearns, *Proc. of the 3rd Inter. Conf. on the Production and Neutralization of Negative Ions and Beams*, Ed. K. Prelec, Brookhaven, 1983, p247.
17. J. R. Hiskes, *XIVth Intern. Conf. on Phenomena in Ionized Gases*, Grenoble, July (1979).
18. E.G. Overbosch, B. Rasser, A.D. Tenner and J. Los, Surf. Sci. **92**, 310 (1980).
19. B. Rasser, J.N.M. van Wunnik and J. Los, Surf. Sci. **118**, 697 (1982).
20. J.J.C. Geerlings, J. Los, J.P. Gauyacq and N.M. Temme, Surf.Sci. **172**, 257 (1986).
21. R. Gomer and L.W. Swanson, J. Chem. Phys. **38**, 1613 (1963).
22. J.J.C. Geerlings, P.W. van Amersfoort, L.F.Tz. Kwakman, E.H.A. Granneman, J. Los and J.P. Gauyacq, Surf. Sci. **157**, 151 (1985).
23. C.F.A. van Os, H.M. van Pinxteren A. W. Kleyn and J. Los, submitted to Appl. Surf. Sc.
24. C.F.A. van Os, H.M. van Pinxteren and J. Los, *Proc. of IIIrd European workschop on the Production and Application of Light Negative Ions*, Februari 17-19, Amersfoort, Netherlands, 1988, p166.
25. J.N.M. van Wunnik, J.J.C. Geerlings, E.H.A. Granneman and J. Los, Surf. Sci. **131**, 17 (1983).
26. P.W. van Amersfoort, J.J.C. Geerlings, R. Rodink, E.H.A. Granneman and J. Los, J. Appl. Phys. **59**, 241 (1986).
27. V.L. Moruzzi, J.F. Jonak and A.R. Williams, *Calculated Properties of Metals*, Pergamon Press, 1978.
28. P.J. Sneider, K.H. Berkner, W.G. graham, R.V. Pyle and J.W. Stearns, Phys. Rev. B**23**, 941 (1981).
29. S. Tamaka, K.N. Leung, P. Purgalis and M.D. Williams, Rev. Sci. Instrum. **59**, 120 (1988).
30. K.N. Leung, P.A. Pincosy and K.W. Ehlers, Rev. Sci. Instrum. **55**, 1064 (1984).
31. A.F. Leitzke and G. Guethlein, *Proc. of the 12th Symp. on Fusion Engineering*, Montery, CA, October 12-16, 1987, p..
32. S.R. Walter, K.N. Leung and W.B. Kunkel, Appl. Phys. Lett. **54**, 210 (1989).

DISCUSSION

Kleyn: Do I understand from your results and the remarks you made that the convertors get dirty very quickly from tungsten filaments and that it does not make a whole lot of sense to pursue experiments in which you try to work with real clean cathodes like LaB_6.

van Os: The big advantage in running a surface convertor source with LaB_6 filaments would be that the amount of material evaporated from these filaments is lower, but I think that running a bucket source with what ever kind of filament that sputtering material from filaments is larger than the evaporation. So the temperature of the thing is not really important. So I think, in my view, that it is important to have a filament which is quite pure so you at least know what kind of contamination you can expect at your surface and that is why barium oxide cathodes come in handy. If you are expecting a pollution of the surface why not take a cathode which produces the same metal as the surface. But then again, we observe from this contaminating the surface at high powered densities. When we run at moderate power levels, at 10A or lower we were able to get reproducible results and we think our surface was fairly clean.

SURFACE PRODUCTION OF NEGATIVE HYDROGEN IONS BY REFLECTION OF HYDROGEN ATOMS FROM CESIUM OXIDE SURFACES

M. Seidl, S.T. Melnychuk, S.W. Lee, and W.E. Carr
Department of Physics and Engineering Physics
Stevens Institute of Technology, Hoboken, NJ 07030

ABSTRACT

Negative hydrogen ions are produced by backscattering a thermal distribution of hydrogen atoms from a converter surface coated with a mixture of cesium oxides. The thick film of cesium oxide is produced by thermal decomposition of cesium carbonate and subsequent thermal activation aided by atomic hydrogen. About 60% of atoms with energies greater than 1.5 eV are reflected as negative hydrogen ions for a temperature of 0.22 eV in the thermal distribution. The H$^-$ ions have a Maxwellian parallel energy distribution with a temperature equal to the atomic temperature. Replacing the thermal source of hydrogen atoms with a discharge source results in H$^-$ ions of 1 eV temperature and 100 times lower intensity.

INTRODUCTION

At the last Brookhaven meeting we reported on experiments involving surface production of negative hydrogen ions by simultaneous bombardment of metal targets with cesium and hydrogen ions [1]. In pursuing this work, we noticed that a fraction of the negative hydrogen ion population had a low energy spread of about 0.2 eV. Further experiments have shown that the slow H$^-$ ions were due to backscattering of hydrogen atoms produced by thermal dissociation of hydrogen gas on hot tungsten filaments. In the past three years we have studied production of H$^-$ ions by backscattering the Maxwellian tail of thermally produced hydrogen atoms from several low work function converter surfaces [2-4]. The following conclusions have come out from this work:

a) H$^-$ ions can be produced with high efficiency (> 10%) by backscattering low energy (> 1eV) hydrogen atoms from low work function (< 1.5 eV) surfaces. This has been demonstrated experimentally by scattering hydrogen atoms. However, theory indicates that the same result holds for proton scattering.

b) Low incident energy of hydrogen atoms guarantees low energy spread of the backscattered negative hydrogen ions.

c) An additional bonus of low incident energy is the elimination of physical damage to the converter surface. This has opened up new avenues for the optimization of the converter surface.

Possible applications of low energy surface production in H$^-$ ion sources are schematically shown in Fig.(1). In principle, the converter surface can be placed in two different positions with respect to the hydrogen plasma. The "internal converter", shown in Fig.(1a), is in direct contact with the plasma, the voltage across the plasma sheath being close to the floating potential (a few volts). The converter surface is exposed to a flux of H$^+$, H$_2^+$ and H$_3^+$ ions in addition to hydrogen atoms and excited molecules. All these species contribute to surface production of H$^-$ ions which enter the plasma with low kinetic energy and must be extracted from the plasma.

The "external converter", shown in Fig. (1b), is separated from the plasma (e.g. by means of a magnetic field). Only hydrogen atoms and excited molecules contribute to surface production of H⁻ ions which are extracted in a plasma- free region with a high extraction voltage.

Apparently, surface production of H⁻ ions by means of an internal converter has been recently observed in volume sources in which the work function of the converter was reduced by adding cesium [5,6] or barium [7] to the discharge. Low electron temperature in the extraction region of the volume source made the extraction of the H- ions possible.

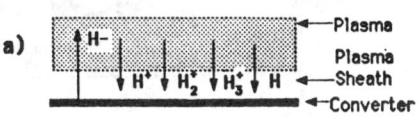

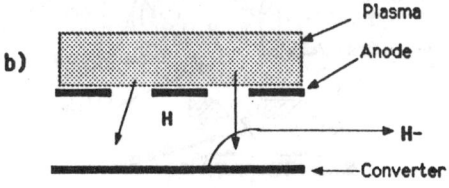

Fig. 1 (a) Internal Converter
(b) External Converter

An H⁻ ion source using an external converter has yet to be built. It critically depends on the availability of a source of hydrogen atoms with a kinetic energy in the 1 to 2 eV range. According to our estimates [4], these atoms will be backscattered as H⁻ ions with a probability better than 50% assuming a converter work function between 1.2 and 1.5 eV. Dissociative electron collisions with hydrogen molecules produce hydrogen atoms with a minimum energy of 2.2 eV. More studies are needed to determine the fraction of atoms reaching the converter without substantial energy loss. Temperature measurements of hydrogen atoms in discharges show that a large fraction of the atoms have an energy higher than 1 eV [8].

Most important for low energy surface production of H- ions is the converter surface. A good converter must have a work function of 1.5 eV or less, it must be chemically stable under exposure to atomic hydrogen and must be easy to use. The first two conditions are met with some single crystal surfaces ,like Si(100) [9], W(110) [10], covered with about half monolayer of cesium. The high vapor pressure of cesium makes this type of converter somewhat inconvenient to use.

The production of surfaces with work functions near or below 1.0 eV have been reported by several groups [10-16]. These surfaces can be divided into two classes. The first represented by Si/Cs/O [12,13] and W/Cs/O [10,11] require atomically clean and structurally perfect surfaces with precise dosing of Cs and O_2 in the submonolayer regime. These surfaces are difficult to produce and maintain and would not be suitable for a negative ion source environment. Recently we have experimented with converters consisting of thick Cs/O and Ba/O layers. The advantage of oxide layers is the fact that their vapor pressures as well as their work functions are lower than the values corresponding to their metals. The Cs/O surface can have a work function as low as 1.1 eV [14-16]. The work function of the Ba/Sr/O cathode is 1.4 eV at 1000° K [17]. In this paper we describe our experiments with cesium oxide converters.

EXPERIMENTAL APPARATUS

The experimental apparatus is shown in Fig.(2). It consists of a planar diode H⁻ surface conversion source, a movable Faraday cup, a rotating magnetic sector mass spectrometer, two interchangeable atomic hydrogen sources, and a tuneable light source.

The planar diode H⁻ source consists of a cesium oxide converter surface, a tungsten mesh anode, a cover plate with a 1.25 mm exit aperture, and a pair of vertical and horizontal deflector plates. The atomic hydrogen beam impinges on the target through a hole in the front cover plate. The converter is a 0.002" thick Mo ribbon which is coated with a layer of cesium oxide. The Mo ribbon can be ohmically heated and its temperature can be monitored with a thermocouple spotwelded to the ribbon.

Total negative ion and electron currents leaving the exit aperture are detected with the Faraday cup. The electron current is separated from the ion current by imposing a magnetic field perpendicular to the beam by external Helmholtz coils. The magnetic sector is used for checking for impurity ions and for measuring the angular profile of the H⁻ beam.

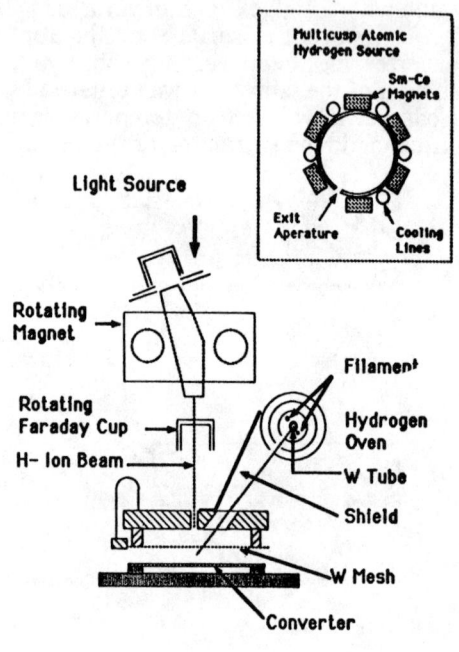

Fig. 2. Experimental Apparatus

PRODUCTION OF ATOMIC HYDROGEN

Two methods for H atom production are used. The first method consists of thermal dissociation of hydrogen gas in a heated tungsten tube similar to that used by Pargellis and Seidl [18], while the second method relies on dissociation of hydrogen gas in a multicusp plasma source similar to that of Leung et. al. [19].

The thermal dissociation source consists of a W tube 3.6 mm O.D., 2.38 mm I.D. and 64 mm long with a 0.7mm hole in the side of the tube. The tube is closed on one end while the other end is Cu brazed into a water cooled SS 304 base through which gas is admitted into the tube. A 20 mm long section is heated by electron bombardment. The tube and filament are enclosed by a water cooled heat shield.

Atomic and molecular hydrogen are assumed to be in thermal equilibrium in the tube. For effusive molecular flow [20] we can write:

$$1/4 * CA * [1/2 * n_a v_a + n_m v_m] = Q/kT_0 \qquad (1)$$

where n is the number density of atoms or molecules in the tube, v is the mean velocity in the tube, Q is the H_2 gas throughput, T_0 is the H_2 gas temperature in the chamber, A is the area of the exit aperture, and C is the Clausing factor [21]. The molecular and atomic hydrogen pressures in the tube are related by the equilibrium constant for the H_2 dissociation reaction

$$K = P_H/(P_{H_2})^{\frac{1}{2}} \qquad (2)$$

where K is given in the JANAF tables [22]
Combining equations (1) and (2) and using the ideal gas law p = nkT we can solve for the atomic density

$$n_a = K^2/(2\,2kT) * \{-1 + (1 + 8Q/CA * 1/K^2 * (4\pi mkT)^{\frac{1}{2}}/kT_0)^{\frac{1}{2}}\} \qquad (3)$$

where T is the tungsten tube temperature. The H effusion flux density impinging on the target is given by

$$\Phi = 1/4 * Cn_a v * \cos(\alpha) * (r/R)^2 \qquad (4)$$

where $\alpha = 35.5$ is the angle between the target normal and the exit aperture of the tube, r = 0.35 mm is the radius of the exit aperture in the tube, R = 75.8 mm is the distance between the converter and the H source, and the Clausing factor C = 0.61. Under typical operating conditions 1 < Q < 50 sccm, and the tube temperature is in the range 1900° K < T < 2700° K. At a flow rate of 11 sccm, and T = 2500° K the atomic hydrogen pressure in the tube is 1.4 torr, the chamber pressure is 2.4 x 10^{-4} torr, and the atomic hydrogen flux density impinging on the target is 8.4 x 10^{15} $cm^{-2}sec^{-1}$.

For a thermal dissociation source where the atoms have a Maxwellian distribution only 1% of the beam atoms have energies above 1.5 eV at a temperature of 2600° K. This illustrates the limitations of this type of source for producing large fluxes of energetic atoms.

The second source of H atoms used in this experiment was a bucket type source using a 0.5 mm tungsten filament, and Sm-Co permanent magnets, see inset in Fig.(2). The source was operated in a constant current mode, and the discharge voltage was changed by varying the pressure and filament temperature. Typical operating parameters were I(discharge) = 2.0 to 3.5 Amps, V(discharge) = 40 to 100 volts and $P(H_2)$ = 1 to 50 mtorr. The H energy distribution and density in our source are presently unknown. In the current experiment this source was used for comparison with the thermal dissociation source.

CESIUM OXIDE CONVERTER

Previously three methods have been reported for producing low work function thick Cs/O layers [14]. The first method involves the deposition of Cs on a substrate followed by alternate exposure to Cs and O_2 until a minimum work function is reached. The second method consists of simultaneous exposure to Cs and O_2 with careful control of the pressures. The third method used in this experiment involves the production of Cs/O by thermal decomposition of Cs_2CO_3.

The converter surface is prepared in a way analogous to the conventional way of making BaO thermionic cathodes. A suspension of organic binder and finely ground Cs_2CO_3 powder is brush painted onto the Mo ribbon producing a 0.10 mm thick coating. Since the finely ground powder is unstable in air and absorbs large amounts of water, the entire grinding and coating procedure is carried out in a nitrogen filled enclosure.

The coated converter is initially heated to 720° K for approximately 1 hour to evaporate the binder. During the initial heating the temperature is ramped slowly to keep the chamber pressure below 10^{-6} torr.

After degassing, the substrate is heated to 883° K for 50 - 60 sec to convert the Cs_2CO_3 to some mixture of cesium oxides and suboxides presumed to be Cs_2O and Cs_2O_2 [14,15]. Following this procedure the substrate is allowed to cool down to 300° K and H is admitted into the chamber at a flux of approximately 10^{16} $cm^{-2}sec^{-1}$ and an oven temperature of 2500° K. The H⁻ ion current from the converter as monitored by the mass spectrometer increases by several orders of magnitude after initial exposure to atomic hydrogen. The H⁻ current saturates in 2500 to 3000 sec. After this initial growth a further increase in the H⁻ current can be realized by adjusting the substrate temperature to an optimum value of 475° K. An additional increase in the H⁻ signal can be achieved by repeating the overheating procedure in the presence of atomic hydrogen. The converter is said to be "activated" when the H⁻ current reaches its maximum steady state value. The activation history of the converter is shown schematically in Fig.(3). The data show that exposure to atomic hydrogen activates the surface.

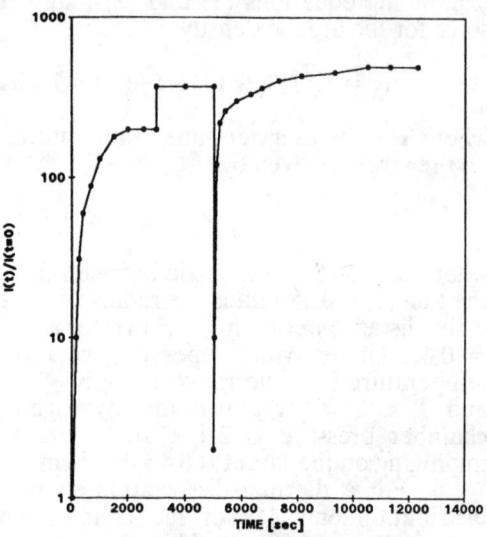

Fig. 3. Converter Activation. 60 to 3000 sec. initial activation by atomic hydrogen, T = 300 K. 3000 to 5000 sec. T = 475 K. 5000 to 13,000 sec., additional activation by heating in atomic hydrogen.

SURFACE PRODUCTION OF H⁻ IONS

The H⁻ ion current was measured as a function of the H oven temperature, the extraction voltage, the chamber pressure, and the converter temperature. Yields were calculated by taking the ratio of the negative ion flux at the exit aperture to the incident atomic flux from the H oven. Fig.(4) shows the H⁻ yields at several chamber pressures and an extraction voltage of 600 volts as a function of 1/T(oven). These data were taken at an optimum substrate temperature of 475° K. The effect of changing the substrate temperature will be discussed later. We attribute the variation in yields as a function of pressure to variations in the daily activation of the converter. The data at pressures of 6.24 x 10^{-4}, 2.4 x 10^{-4}, and 2.9 x 10^{-5} torr all show a maximum in the yield curves. The calculated yields depend on the incident atomic flux and on the target properties. Assuming that the ionization probability continues to increase with atomic beam temperature then the apparent drop in the yield would indicate a drop in the H flux on the order of 1.3 times the calculated value. Variations in the H flux may be due to a departure from equilibrium conditions in the tungsten tube. These effects are currently being investigated. No maximum was observed for the data taken at 1.1x10^{-3} torr. All of the yield curves exhibit a exp(-C/kT) dependence at temperatures below the turnover point. The highest measured yield was 6.5x10^{-3} at an oven temperature of 2593° K and a chamber pressure of 6.2x10^{-4} torr. The solid lines in the figure are the calculated fractions of atoms leaving the W tube with energies greater than 1.0 eV and 1.5 eV. It can be seen that about 62% of the atoms with energies higher than 1.5 eV are reflected as H- ions at a temperature of 2600° K.

Fig.(5) shows the H⁻ yield at several extraction voltages and a chamber pressure of 6.24×10^{-4} torr as a function of 1/T(oven). The yields show an increase with extraction voltage due to a Schottky dependence on the extraction field. The usefulness of an atomic reflection type negative ion source is demonstrated here since large extraction fields can be applied which are not possible in conventional H- sputter sources due to excessive sputtering of the extraction electrodes.

The dependence of the H⁻ yield and the ratio of electron to ion current on the converter temperature is shown in Fig.(6). An optimum in the converter temperature is observed where the negative ion yield reaches a maximum and the corresponding ratio of electrons to ions is a minimum. This optimum occurs in the neighborhood of 475° K and is observed at all atomic oven temperatures and pressures. This optimum was stable and reproducible in day to day operation and for several different cesium oxide targets. We attribute the optimum in the substrate temperature to maintenance of a minimum work function surface due to an optimum coverage of Cs and Cs_2O/Cs_2O_2 on the surface.

The dependence of the ratio of electron to ion current on 1/T(oven) for several pressures is shown in Fig.(7). An exponential increase in the ratio is observed for decreasing oven temperature. At high oven temperatures in the range of 2500° K to 2700° K electron to ion ratios on the order of 1 can be achieved.

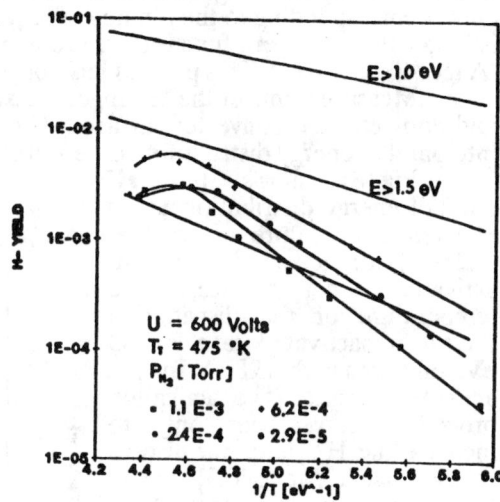

Fig. 4. Negative Hydrogen Ion Yield vs. 1/T(oven) at several chamber pressures. Extraction voltage = 600 volts.

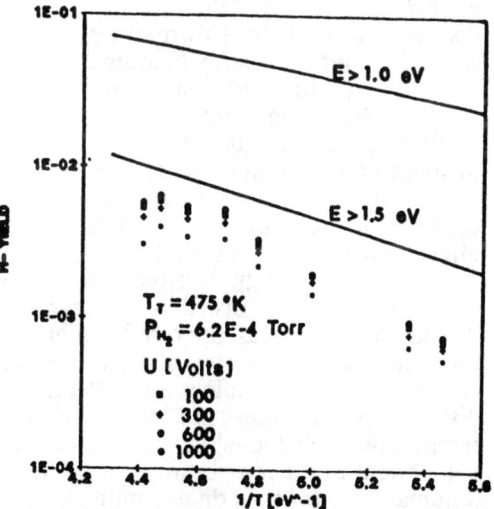

Fig. 5. Negative Hydrogen Ion Yield vs. 1/T(oven) at several extraction voltages. Chamber pressure = 6.2E-4 torr.

The increase in the ratio with decreasing oven temperature may be due to trapping and recombination of slow atoms on the surface. The slow atoms are chemisorbed by the image force and recombine on the surface releasing 4.4 eV of energy corresponding to the binding energy of the H_2 molecule. Since this energy is larger than the work function electrons may be emitted from the surface by an "Auger-like" process. This process has not yet been investigated.

Measurements of the H^- angular distributions were done for the activated and non-activated converter surfaces. The angular distributions were converted into parallel energy distribution as described in ref [3].

Fig.(8) shows the H^- parallel energy distribution at an H oven temperature of 2523° K (0.22eV) for activated and non-activated target surfaces. The temperature of the distribution for the nonactivated case is 0.43 eV compared with 0.28 eV for the activated surface. The activation procedure serves not only to increase the H^- yield, but also to remove patch effects [23] due to Cs coverage on the surface. After activation the energy spread was independent of converter temperature variations from 325° K to 550° K. It has previously been shown by Melnychuk et.al.[3] that the parallel energy distribution of H^- ions formed by reflecting thermal energy H atoms from cesiated Mo and n or p type Si (100) surfaces have temperatures equal to the incident atomic temperature. The Cs/O surface has a similar dependence except that patch effects are more pronounced.

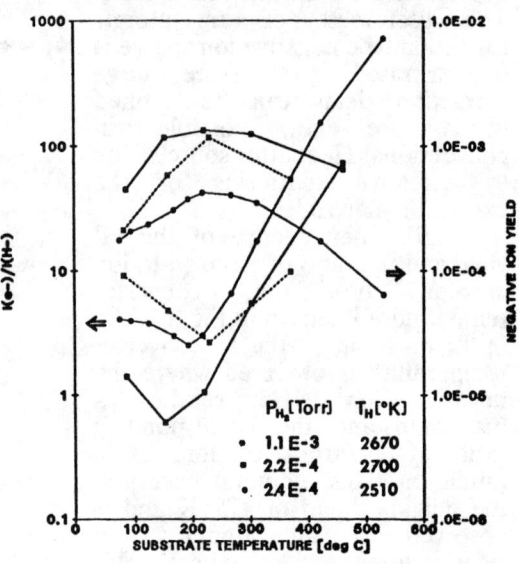

Fig. 6. Ratio of electrons to ions and H^- Yield vs. Converter Temperature.

The H^- energy distributions obtained with the tungsten tube source were compared with the distributions from the plasma H source. See Fig.(8). The discharge source was operated under conditions where there were no positive ions extracted from the plasma, therefore all the H^- ions produced on the converter are due solely to reflected hydrogen atoms. The parallel energy distribution indicates the existence of a slow and a fast component. The temperature of the slow component is approximately 0.5 eV and the fast component has a temperature of 1.1 eV. This is a temperature well above that obtainable by thermal dissociation.

The H^- current was optimum at a source pressure of 30 mtorr and increased with the discharge voltage. The maximum H^- currents obtained were approximately 100 times smaller than those obtained with the W tube source at the same chamber pressure. This indicates a low production rate of fast hydrogen atoms in the discharge.

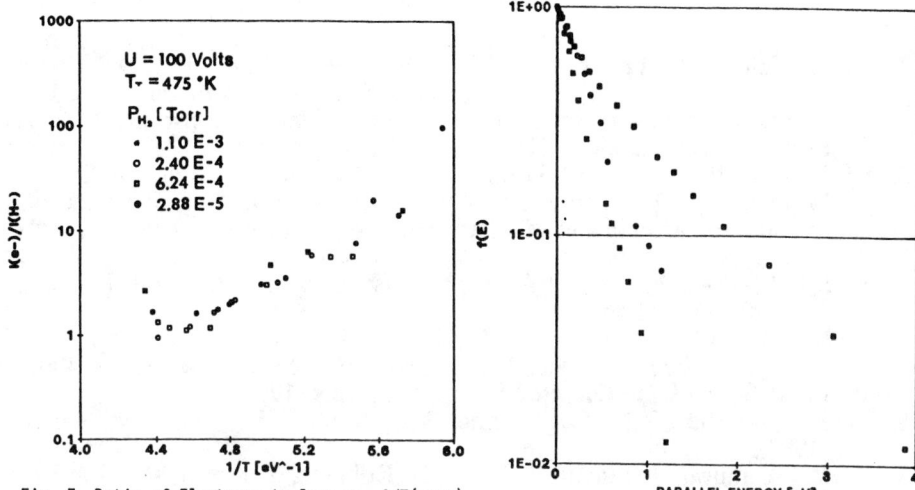

Fig. 7. Ratio of Electrons to Ions vs. 1/T(oven).

Fig. 8. H⁻ Parallel Energy Distribution. Thermal H atom source: ○ non-activated surface T = 0.43 eV, ■ activated surface T = 0.28 eV. Plasma H atom source: □ T_{slow} = 0.5 eV, T_{fast} = 1.1 eV.

CONCLUSIONS

Thick films of cesium oxides produced by thermal decomposition of cesium carbonate have been successfully used as converter surfaces for production of negative hydrogen ions by backscattering hydrogen atoms. These surfaces are easy to use, and after activation are stable in a flux of atomic hydrogen of 10^{16} atoms/cm²/sec and of 0.22 eV temperature.

About 60 % of the atoms with an energy higher than 1.5 eV are reflected as H⁻ ions for a temperature of 0.22 eV in the thermal distribution. Under optimum conditions the electron to ion ratio is smaller than one. Large extraction fields can be used since the converter surface is not exposed to ion bombardment.

For a thermal distribution of the incident hydrogen atoms the H⁻ ions have a Maxwellian parallel energy distribution with a temperature nearly equal to the incident atoms (0.22 eV). For a discharge source of atomic hydrogen a fraction of the H⁻ ions have a parallel temperature of 1 eV but an intensity 100 times lower.

ACKNOWLEDGEMENT

This work has been supported by the Air Force Office of Scientific Research. We would like to thank George Wohlrab for machinig the apparatus which made this work possible.

REFERENCES

1. M. Seidl, J.L. Lopes, S.T. Melnychuk, W.E. Carr, and G.S. Tompa, Production and Neutralization of Negative Ions and Beams, edited by J. Alessi (Brookhaven National Laboratory), American Institute of Physics Conference Proceedings No. 158, p.432 (New York,NY, 1987).
2. M.Seidl, W.E. Carr, J.L.Lopes, and S.T. Melnychuk, Proc.3rd European Workshop on Production and Application of Light Negative Ions, FOM Institute for Atomic and Molecular Physics, Amsterdam, February 1988, p 157.
3. S.T. Melnychuk, M.Seidl, W. Carr, J. Isenberg, and J. Lopes, J. Vac. Sci. Tech. **A7**, 2127 (1989).
4. M.Seidl, W.E.Carr, S.T. Melnychuk, A.E. Souzis, J.Isenberg, and H. Huang, Conference Proceedings, SPIE Vol. 1061, Microwave and Particle Beam Sources and Directed Energy Concepts (1989) p. 547.
5. S.R. Walther, K.N. Leung, and W.B. Kunkel, J. Appl. Phys. **64**, 3424 (1988).
6. Y. Okumura, M.Hanada, T. Inoue, H. Kojiman, Y. Matsuda, Y. Ohara, M. Seki, and K. Watanabe, This Symposium.
7. K.N. Leung, S.R. Walther, and W.B. Kunkel, Phys. Rev. Lett. **62**, 764 (1989).
8. A.S. Schlachter, A.T. Young, G.C. Stutzin, J.W.Stearns, H.F. Dobele, K.N. Leung, W.B. Kunkel, in ref. 4, p. 610.
9. J.D. Levine, Surf. Sci. **34**, 90 (1973).
10. J.L. Desplat and C.A. Papageorgopoulos, Surf.Sci. **92**, 97 (1980).
11. J.L. Desplat, Surf.Sci. **34**, 588 (1973).
12. R.U. Martinelli, J. Appl. Phys. **45**, 1183 (1974).
13. A.E. Souzis, M. Seidl, W.E. Carr, and H. Huang, J. Vac. Sci. Tech. **A7**, 720 (1989).
14. T.R. Briere and A.H. Sommer, J. Appl. Phys. **48**, 3547 (1977).
15. A.H. Sommer, J. Appl. Phys. **51**, 1254 (1980).
16. B. Woratschek, W. Sesselmann, J. Kuppers, G. Ertl, and H. Haberland, J. Chem. Phys. **86**, 2411 (1987).
17. G. Herrman and S. Wagener, The Oxide Coated Cathode vol. 2 (Chapman & Hall, London, 1951) p. 223.
18. A. Pargellis and M. Seidl, Phys. Rev. **B25**, 4356 (1982).
19. K.N. Leung, K.W. Ehlers, C.A. Hauck, W.B. Kunkel, and A.F. Lietzke, Rev. Sci. Inst. **59**, 453 (1988).
20. E.H. Kennard, Kinetic Theory of Gases (McGraw-Hill, New York 1938) p. 60-64.
21. S. Dushman, Scientific Foundations of Vacuum Technique, Second Edition (John Wiley & Sons, New York, 1962) p. 91.
22. JANAF Thermochemical Tables (Dow Chemical, Midland, MI, 1965) p.91.
23. C. Herring, M.M. Nichols, Rev. Mod. Phys. **21**, 185 (1949).

DISCUSSION

Jaquot: Do you observe any production of O^- in these configurations.

Seidl: Yes, if it is dirty.   I would say that in the initial phase if the system is not clean there may be some, but in this hydrogen oven there are not many, it doesn't happen very much.

Kleyn: What is the binding energy that you put in for the atomic hydrogen that you use in your calculations.

Seidl: In the atomic scattering calculation?   One eV

Kleyn: What is the scattering model you use, do you have multiple collisions.

Seidl: No, we don't have the multiple collisions, we just have but one single binary collision, and, if the atom comes out it, it is counted, if it is reflected inside, it is not counted.

ION-ION COLLISIONS INVOLVING H^- IONS

W. Debus, F. Melchert, M. Liehr and E. Salzborn
Institut für Kernphysik, Justus-Liebig-Universität Giessen
D-6300 Giessen, F.R.G.

ABSTRACT

A report is given on recent crossed-beams experiments studying single- and double-electron removal from H^- in energetic collisions with Ar^{q+} ($q \leq 8$) ions. The data is discussed with respect to conversion efficiencies of H^- to neutral H^0 in plasma neutralizers proposed for efficient neutral beam heating of future fusion devices.

INTRODUCTION

In magnetic fusion, there is presently much interest in plasma neutralizers[1-3] which offer considerably higher neutralization efficiencies than gas targets ($\approx$ 85 % vs 60 %) for conversion of energetic, multi-megawatt H^- beams into neutral H^0 beams needed for auxiliary heating of next generation fusion plasmas. Further, the use of a plasma neutralizer based on multiply-charged ions is expected[4] to offer the advantage of a substantially reduced optimum line density due to the larger electron removal cross sections. However, design and modelling studies for plasma neutralizers have suffered from the lack of experimental cross sections for the single-electron removal

$$H^- + X^{q+} \rightarrow H^0 + \ldots \quad (\sigma_{-0}) \qquad (1)$$

and the double-electron removal reactions

$$H^- + X^{q+} \rightarrow H^+ + \ldots \quad (\sigma_{-+}). \qquad (2)$$

Reaction (2) is especially important since it directly degrades the efficiency of a plasma neutralizer.

EXPERIMENTAL TECHNIQUE

In atomic collision experiments usually neutral atoms - either free in gases, or bound in solids - are used as targets for the incident charged projectiles. The principal difficulties of experiments wherein free ions are used as targets arise from the tenuous thickness provided by the ionic target.

We have employed the Giessen crossed-beams apparatus to measure absolute cross sections for reactions (1) and (2) in H^- + Ar^{q+} ($q \leq 8$) collisions. The experimental arrangement has been described previously in detail[5,6]. Briefly, two well-collimated (< 2 mm diam.) and charge-analyzed ion beams of adjustable energies (H^- beam: (8-300) keV, (4-70)nA; Ar^{q+} beam: (20-80)keV, (2-200)nA) are arranged to intersect at an angle $\theta=45°$ in an ultrahigh vacuum region. The collision products (H⁰, H⁺) formed in the H⁻ beam are separated immediately after the interaction region by electrostatic deflection and counted individually by a channeltron-based single-particle detector.

Some crucial apparatus improvements, however, have been necessary to achieve the present measurements. A newly designed[7] small 5 GHz ECR ion source has been utilized for the production of intense and stable beams of multiply-charged argon ions. The most serious experimental problem has been provided by background events due to H⁰ and H⁺ particles, respectively, which result from the interaction of H⁻ ions with the residual gas. In order to reduce this background the ultrahigh vacuum in the interaction region has been lowered to about $2 \cdot 10^{-11}$ mbar with both beams "on". Furthermore, the H⁻ beam has been cleaned, shortly before intersection, by electrostatic deflection. Even under these conditions, however, a beam modulation technique[8] had to be employed for signal recovery since the signal of 4 to 200 counts/s was masked by a background which was a factor 2000 to 50 higher, respectively. Typical measurement times were up to 4 hours for one cross section.

RESULTS

In Fig. 1 the experimental cross sections for single- and double-electron removal in H⁻ + Ar^{q+} (q ≤ 8) collisions at a center-of-mass energy E_{cm} = 50 keV are displayed. Also shown are CTMC calculations[9] for the respective processes along with a Bethe-Born calculation[10] for reaction (1).

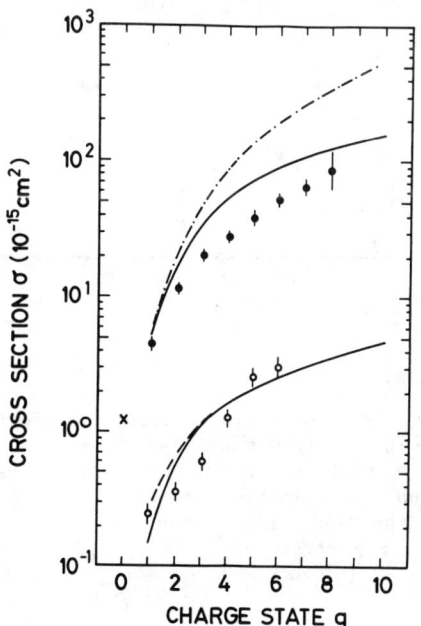

Fig. 1. Experimental single (●) and double (o) electron removal cross sections for H⁻+Ar^{q+} (q ≤ 8) collisions at E_{cm}=50keV. An experimental data point (x) for single-electron removal in H⁻ + Ar⁰ collisions (Ref. 11) is shown for comparison. The solid lines are CTMC calculations[9] for the respective processes. The dot-dashed line is the Bethe-Born calculation of Ref. 10. The dashes line is the electron removal cross section for neutral H⁰ (Ref. 12).

It can be seen from Fig. 1 that the q^2 scaling predicted by the first-order Born approximation for the single-electron removal cross section becomes increasingly invalid as the charge state of the ion increases. In fact, the experimental cross sections scale, within error bars, with a $q^{1.3}$ dependence on the ionic charge. The measurements are also overestimated to some extend by the CTMC calculations[9]. These calculations, however, lead to the interesting prediction that the single-electron removal cross sections will be independent of the ion species of the multiply-charged ion and simply be a function of its charge state. This is realized by examining calculated transition probabilities for removing the weakly bound 1s' electron on H^- (binding energy E_b=-0.754 eV). It is found that the transition probabilities for ionization saturate close to unity for impact parameters as large as several tens of a_o (Bohr radius) for even low-charge state ions. Thus, all ions of charge q tend to "look much the same" for a distant H^- ion.

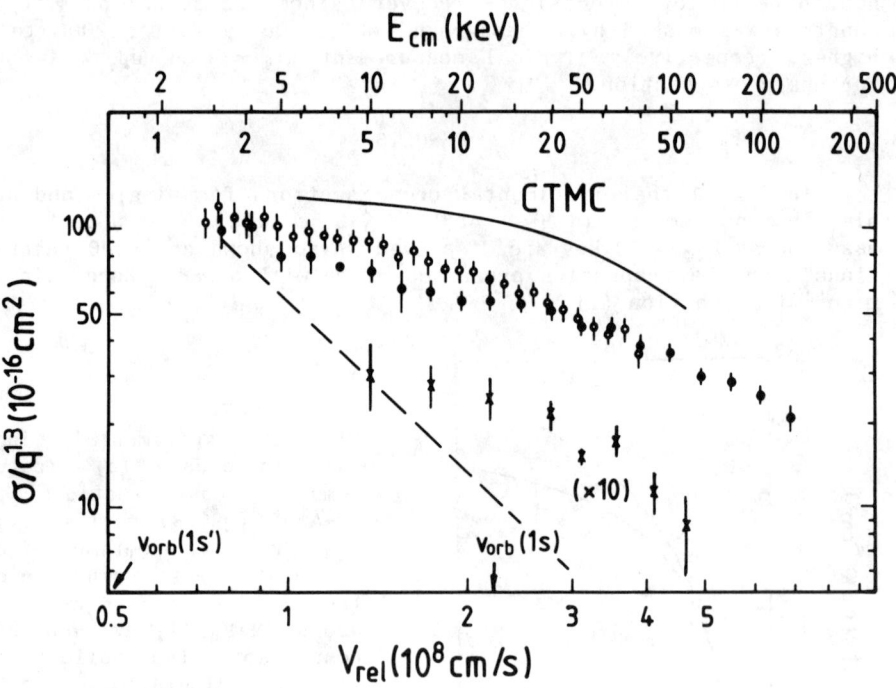

Fig. 2. Scaled experimental single-electron removal cross sections for $H^- + Ar^{4+}$ (●) and $H^- + H^+$ (o, Ref.13). Scaled double-electron removal cross sections (multiplied by a factor 10) for $H^- + Ar^{3+}$ are shown by crosses. The dashed line indicates a $1/v^2$ dependence for $H^- + Ar^{q+}$ collisions. The solid line is the CTMC result[9] for the $H^- + Ar^{4+}$ system. The upper x-axis scales show the center-of-mass energies for the systems $H^- + Ar^{q+}$ (above) and $H^- + H^+$ (below), respectively.

For the double-electron removal cross sections shown in Fig. 1 there is qualitative agreement between experiment and the CTMC calculations[9]. The latter show that the transition probabilities for removing both the loosely bound outer and the tightly bound inner electron concentrate at small impact parameters where the outer 1s' electron is ionized with an $\approx$ 100 % probability. This leads to the expectation that the double-electron removal cross section from H⁻ is approximately equal to the single-electron removal from neutral H⁰, i.e. $\sigma_{-+}(H^-) \approx \sigma_{0+}(H^0)$. This is supported by the data shown in Fig. 1.

In Fig. 2 are shown the velocity dependence of the H⁻ + Ar⁴⁺ single-electron removal cross section along with that for H⁻ + H⁺. When scaled by $q^{1.3}$, the results are nearly identical, thus displaying the lack of dependence on ion species. The behavior of the cross sections in velocity is far removed from the $\sim 1/v^2$ scaling predicted by asymptotic theories. The CTMC results[9] for H⁻ + Ar⁴⁺ are also shown in Fig. 2. The calculated energy dependence is similar to that observed, however, the absolute magnitude is $\approx$ 50 % too large.

Taking into account the present data, both the maximum neutralization efficiency η_{max} and the corresponding optimum target thickness π for a plasma neutralizer can be calculated. At high collision energies, where electron attachment reactions can be neglected, η_{max} and π are given by[1]

$$\eta_{max} = \frac{<\sigma_{-0}>}{<\sigma_{-0}>+<\sigma_{-+}>} \left[\frac{<\sigma_{0+}>}{<\sigma_{-0}>+<\sigma_{-+}>} \right]^{\frac{<\sigma_{0+}>}{<\sigma_{-0}>+<\sigma_{-+}> - <\sigma_{0+}>}} \quad (3)$$

$$\pi = \frac{1}{<\sigma_{-0}>+<\sigma_{-+}> - <\sigma_{0+}>} \ln \left[\frac{<\sigma_{-0}>+<\sigma_{-+}>}{<\sigma_{0+}>} \right] \quad (4)$$

where

$$<\sigma_{ij}> = \frac{\sigma_{ij}^e \, n_e}{n_e + n_i + n_a} + \frac{\sigma_{ij}^i \, n_i}{n_e + n_i + n_a} + \frac{\sigma_{ij}^a \, n_a}{n_e + n_i + n_a} \quad (5)$$

The subscripts i and j refer to initial and final charge states, respectively, and the superscripts e, i, a refer to electrons, ions and atoms, respectively. For a fully ionized plasma containing ions in a single charge state q, we have for the particle densities

$$n_a = 0 \quad \text{and} \quad n_i = \frac{n_e}{q} \quad (6)$$

The electron impact ionization cross sections in eq. (5) are taken from Ref. 14-16. All target particles are assumed at rest in the laboratory frame.

Calculations based on the present data show that at 50 keV/u maximum H⁻ neutralization efficiencies between 82% (q=5,6) and 89% (q=2,3) are realizable (Fig. 3). For q=1 we obtain η_{max} = 86%. It is obvious that a large ratio of cross sections σ_{-0}/σ_{-+} for ions in a given charge state results in a high value for η_{max}.

In order to calculate neutralization efficiencies at higher energies we have extrapolated the energy dependence of the single-electron removal cross sections for H⁻ + Ar^{4+} shown in Fig. 2. Furthermore, it was assumed that all cross sections σ_{-0} and σ_{-+} show the same energy dependence, regardless of q. It can be seen from Fig. 3 that the neutralization efficiencies calculated decrease by only about 3% in the energy range from 50 keV/u up to 1 MeV/u.

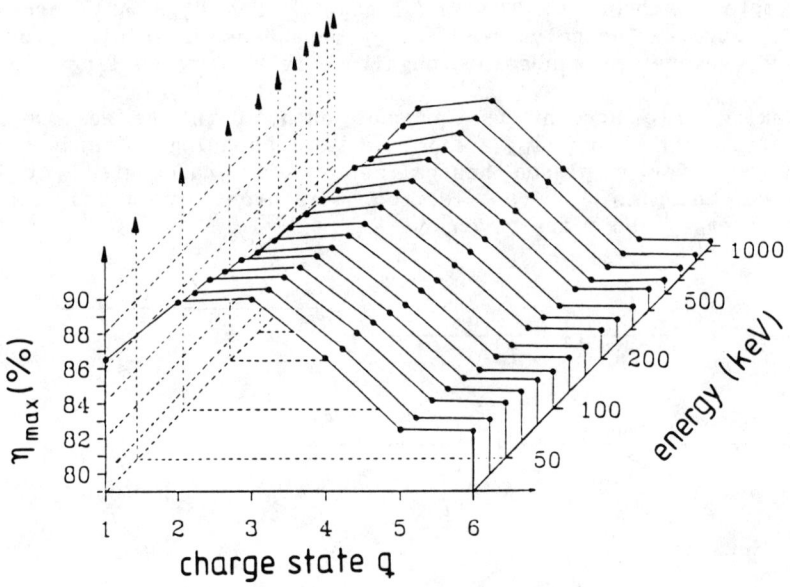

Fig. 3. Calculated maximum neutralization efficiencies η_{max} vs charge state q and H⁻ energy. The solid lines are drawn to guide the eye.

The above results are consistent with previous predictions[4]. The significant superiority of a plasma target for the neutralization of energetic D⁻ beams can be readily seen from Fig. 4 where present η_{max}-values for D⁻ ions in an Ar^{4+} plasma target are compared to corresponding efficiencies[17] for D⁻ and D⁺ beams, respectively, in a D₂ gas target. Note that future fusion devices call for injection energies at hundreds of keV.

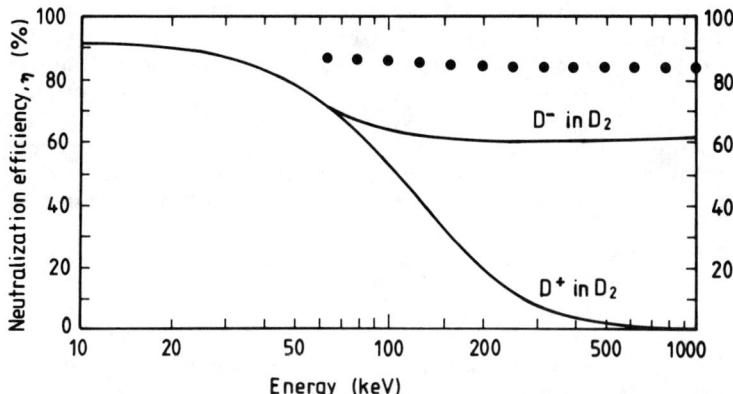

Fig. 4. Maximum neutralization efficiencies vs beam energy.
● D⁻ in an Ar^{4+} plasma target (present results). The solid lines refer to D⁻ and D⁺ beams, respectively, in a D_2 gas target[17]

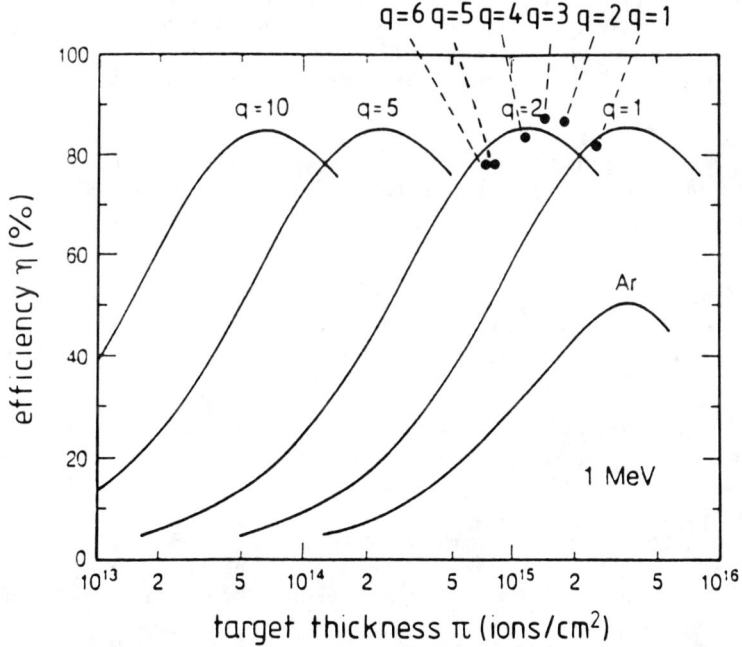

Fig. 5. Calculated neutralization efficiencies as a function of target ion (atom) line density for 1 MeV H⁻ in an Ar gas neutralizer and in a plasma neutralizer with ion charge states q = 1,...,10 (from Ref. 4). The solid circles show the present results for q = 1,...,6 calculated from eqs. (3) and (4).

However, our measurements do not bear out the prediction[4] that plasma neutralizers can be made significantly more compact by employment of multiply-charged ions compared to singly-charged ions which are much easier to produce. This is a result of our observed $q^{1.3}$ scaling of the single-electron removal cross section which is in contrast to the expected q^2 scaling. For example, the optimum target ion line density for 1 MeV H⁻ ions calculated from eq. (4) yields a reduction of 3 compared to 15 for using a q = 5 ion rather than a q = 1 ion (Fig. 5).

CONCLUSIONS

Absolute cross sections for single- and double-electron removal from H⁻ in collisions with multiply-charged argon ions have been measured, using the crossed-beams technique. The cross sections for charge states from q = 1 to q = 8 and energies from 3 to 250 keV/u do not follow predicted scaling rules. H⁻ neutralization efficiencies of plasma targets calculated from the present data show that η_{max}-values of up to 89 %, only slightly decreasing with energy, are realizable. However, a major implication of our measurements is that plasma neutralizers based on multiply-charged ions gain little in expected reduced optimum target thickness. Since the measurements and CTMC calculations[9] indicate that the species of ions does not relevantly change the efficiency of H⁻ neutralization, a plasma neutralizer based on hydrogen or low Z alkali ions appears to be most advantageous.

This work has been funded by the German Federal Minister for Research and Technology (BMFT) under the contract no. 06 GI 307.

REFERENCES

1. K. H. Berkner, R. V. Pyle, S. E. Savas and R. K. Stalder, Proceedings of the Second International Symposium on the Production and Neutralization of Negative Hydrogen Ions and Beams, ed. Th. Sluyters, Brookhaven 1980, BNL-Report 51304, p. 291
2. A. I. Hershcovitch, B. M. Johnson, V. J. Kovarik, M. Meron, K. W. Jones and K. Prelec, Rev. Sci. Instrum. 55, 1744 (1984)
3. J. H. Fink, in Production and Neutralization of Negative Ions and Beams, ed. J. G. Alessi, Brookhaven 1986, AIP Conference Proceedings No. 158, p. 618
4. A. S. Schlachter, K. N. Leung, J. W. Stearns and R. E. Olson, in Production and Neutralization of Negative Ions and Beams, ed. J. G. Alessi, Brookhaven 1986, AIP Conference Proceedings No. 158, p. 631.
5. K. Rinn, F. Melchert and E. Salzborn, J. Phys. B 18 3783 (1985)
6. E. Salzborn, Invited Review Paper, Proceedings of the 16 th International Conference on the Physics of Electronic and Atomic Collisions, New York (July 26 - Aug 1, 1989), in print
7. M. Liehr, G. Mank and E. Salzborn, Proceedings of the International Conference on ECR Ion Sources, ed. J. Parker, NSCL Report No. MSUCP-47 (1987), p. 292

8. K. Rinn, F. Melchert, K. Rink and E. Salzborn, J. Phys. B 19, 3717 (1986)
9. F. Melchert, W. Debus, M. Liehr, R. E. Olson and E. Salzborn, Europhys. Lett. 9, 433 (1989)
10. Y.-K. Kim and M. Inokuti, Phys. Rev. A 3, 665 (1971)
11. C. J. Anderson, R. J. Girnius, A. M. Howald and L. W. Anderson, Phys. Rev A 22, 822 (1980)
12. R. E. Olson, K. H. Berkner, W. G. Graham, R. V. Pyle, A. S. Schlachter and J. W. Stearns, Phys. Rev. Lett. 41, 163 (1978)
13. S. Krüdener, F. Melchert, W. Schön and E. Salzborn, to be published
14. B. Peart, D. S. Walton and K. T. Dolder, J. Phys. B: At. Mol. Phys. 3, 1346 (1970)
15. P. Defrance, W. Claeys and F. Brouillard, J. Phys. B: At. Mol. Phys. 15, 3509 (1982)
16. M. B. Shah, D. S. Elliot and H. B. Gilbody, J. Phys. B: At. Mol. Phys. 20, 3501 (1987)
17. K. H Berkner, R. V. Pyle and J. W. Stearns, Nucl. Fusion 15, 249 (1975)

DISCUSSION

Jacquot: I have seen very quick the ratio of the σ_{-0} over σ_{0+}. What is the dependence of the charges of the ions. Is it more suitable to use Ar^{+q} with very high q in order to increase that ratio, because for fusion it is important to decrease the fraction of H^+ particles after neutralization.

E. Salzborn: At 50 KeV, this ratio, the ratio of double removal to single removal, directly effects, or inversely effects the η. So if you have a very low ratio here you get a rather high neutralization efficiency. This is just the consequence of the little structure you see in the cross section. Well, what we finally calculated is that all q's are more or less giving the same efficiencies. So I think I would rather go to singly charged ions.

Von Wimmersperg: I would like to ask you how do you compare your results for intersecting beams with the situation that you have in a plasma where you have high density and high temperature electrons. Did you take that into account when you use your results from the intersecting beam to calculate neutralization.

E. Salzborn: In calculating here the optimum target thickness and neutralization efficiencies, we had to use the electron impact ionization data, certainly. But what we assume here is that our plasma was standing, so the relative velocity came from the incident beam. We used cold electrons, and the relative velocity just came from the incoming ions. All we can do is provide cross sections which are a basis for model calculations.

Alton: If you are using a plasma stripper to strip H^- ions to convert them to neutral particles, is that a practical means for converting heavy negative ions to multiple charged ions? I visualize that as the possible extrapolation between conventional gas strippers and foil strippers, notoriously bad for scattering high energy beams.

Salzborn: I thought about this problem and certainly you can neutralize them. Whether you can, lets say strip off additional electrons, that is another question. I am not sure if you get the densities you need to release more than one electron.

RECOMBINATION OF ATOMIC HYDROGEN ON METAL SURFACES

R.I. Hall, M. Landau, F. Pichou, C. Schermann
Laboratoire de Dynamique Moléculaire et Atomique
Université P. et M. Curie. 75252 Paris Cedex 05. France.
I. Cadez
Institute of Physics
P.O.B. 57, 11000 Belgrade. Yugoslavia.

ABSTRACT

Recombination of atomic hydrogen in a gas filled cell where the walls are covered with metal evaporated from either tungsten or tantalum filaments has been studied. The high internal energy (vibration) of the product H_2 molecule indicates that recombination occurs via a new mechanism. This mechanism involves the initial trapping of the H atoms in very weakly bound or physisorbed states on the top of the usual layer of strongly bound or chemisorbed layer of atoms which cover most metals when in presence of hydrogen.

INTRODUCTION

The process of recombinative desorption of atomic hydrogen on a metal surface to produce H_2 in the gas phase can be written :

$$H + H + \text{surface} \longrightarrow H_2^* + \text{surface}$$

where the recombination energy of 4.5 eV can be shared between the surface and translation, vibration and rotation of the product H_2.

That this process liberated large amounts of energy was appreciated in the early nineteen hundreds soon after Langmuir in his work on the light bulb had discovered the reverse process of atomisation on an incandescent filament[1]. At that time the General Electric Co. marketed welding equipment based on flowing atomic hydrogen over the metal to be welded.

The parameters which are of interest in this reaction are (i) : the recombination coefficient (γ) which is defined as the probability that an H atom colliding with a surface will yield an

H_2 molecule in the gas phase. (ii) : the accomodation coefficient (β) which is the fraction of recombination energy deposited in the surface during the reaction. The values of these coefficients used today in modelling were obtained in the sixties using gas flow techniques. Atomic hydrogen was produced in a discharge prior to flowing it over the metal under study in the form of a filament. The γ and β were obtained from the amount of heat absorbed by the filament from recombination and a study of the reaction kinetics. Low γ glass tubes were employed with gas pressures on the order of 100 to 1mT. The motivation for this work came from a desire to understand gas phase catalytic processes for which recombination of atomic hydrogen is the prototype.

Figure 1 shows the values of γ thus determined for a variety of metals[2,3]. In general γ is low, about 0.1, although Wood and Wise determined considerably higher values for Ni, Al and Ti. At lower pressures (10^{-2} to 10^{-7} mT) where the surfaces were considered to be free of any adsorbed layer of gas, γ was found to be ≈1 for Pt, W and Mo over a wide temperature range (400K to 1100 K)[4].

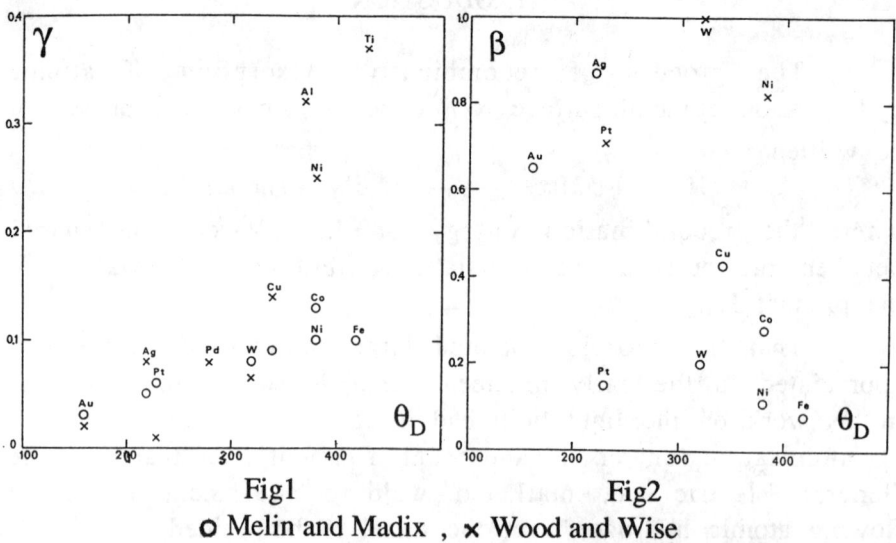

Fig 1 Fig 2

○ Melin and Madix , × Wood and Wise

The β coefficients of many metals[3,5] are plotted in figure 2 against the Debye temperature (θ_D) of the metal. This was done at that time in an attempt to understand the recombination mechanism. γ was seen to increase with θ_D whereas there would appear to be a rough tendency for β to decrease with θ_D. A high value of θ_D represents a rigid structure with a high force constant for the lattice and it was argued that metals with high values of θ_D could only weakly absorb energy from a nascent H_2 molecule. Hence it was considered that high values of γ were correlated with low values of β and high values θ_D.

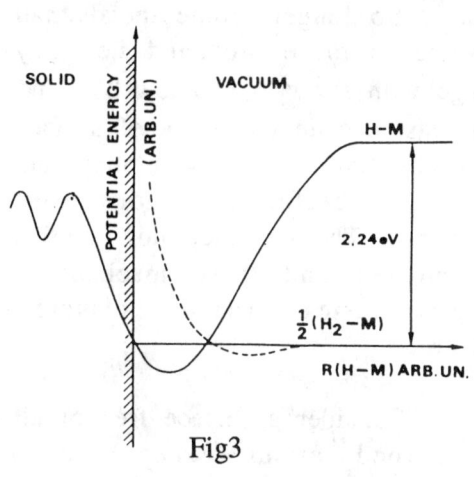

Fig3

The interaction of H_2 and H with a metal surface can be rationalised using the potential curves of figure 3 first proposed by Lennard-Jones[6] in 1932. Impiging H_2 molecules (dashed line) can first be trapped in the shallow well (physisorbed) before dissociating and leaking through into the chemisorbed state of atomic hydrogen. The reverse reaction takes place when the chemisorbed atoms take energy from the heat bath and are "boiled off" into the gas phase. The coverage of atomic H a surface may have, depends on the bulk temperature, the depth and shape of the chemisorption well and the gas pressure. Most of the metals in figures 1 and 2 in the presence of hydrogen would be covered with a monolayer of atomic hydrogen at room temperature down to pressures below 1mT. The process invoked for recombination in the gas flow experiments, where the coverage is high, is that known as Eley-Rideal (ER). The incident H atom which has about 2eV of potential energy available "picks off" a chemisorbed atom to yield a gas phase H_2 molecule. This process would be associated with low values of β, i.e. a process where the energy in the different degrees of freedom of the H_2 molecule does not correspond to the surface temperature. Hot molecules would be the signature of the ER process and γ would be low. The contrasting process, known as the Langmuir-Hinshelwood (LH) process, would occur for low coverages. The incident H atom would first thermalize and be chemisorbed on the clean surface prior to being boiled off as mentioned above when it met another atom. The LH process would be characterized by "cold" molecules at the surface temperature i.e. high values of β. The accomodation coefficient γ was also expected to be high.

As further experiments were performed, particularly with atomic nitrogen[7], the simple picture based on these two types of process was found to hold no longer. Some metals had behaviours which were the opposite to those predicted i.e. they revealed high β's for high coverage with low γ's, and low β's and high γ's for low coverages. In the last decade the study of surface dynamics has evolved dramatically but is still in its infancy and clean experiments are now being performed under well controlled and characterized conditions. The evidence now is that the ER-LH picture must be tempered and these mechanisms considered as two extreme cases. Instead a more plausible picture can be extracted from figure 4.

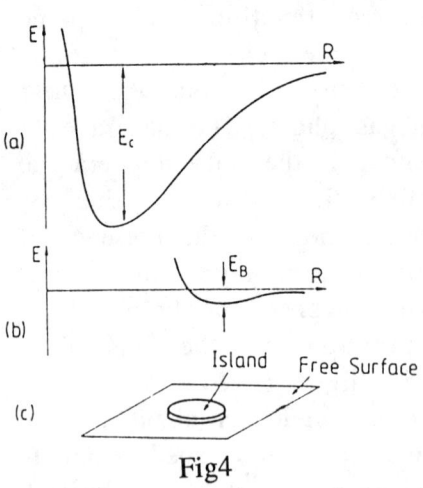

Fig4

Consider a surface free of all adsorbed atoms except for a region where the adatoms have formed into an island (adsorbing atoms do in fact seem to congregate in patches). The potential energy curve (a) represents the interaction of H with the free surface and (b) the interaction with the island of atoms. Three situations where considered[8]. The atom meets the surface far from the island, is thermalized and eventually migrates to the island where it can be boiled off (this would be the LH case). If the point of impact is near the island and as thermalization is not instantaneous then the atom can migrate to the island and before loosing all its energy to the bulk, can recombine without needing energy from the heat bath. Similarly an atom impacting the island can be trapped in a physisorbed state, would have a large amount of potential energy and can recombine with the chemisorbed atoms below it or on the edge of the island. In fact there is a fourth situation which has been brought to light in the observations described below, viz recombinations of atoms in the physisorbed state. Here the available energy is near the maximum value of 4.5 eV.

RESULTS AND DISCUSSION

Recombination of atomic hydrogen has been studied using the cell shown in figure 5. This cell is manufactured from stainless steel and the walls can be water cooled. Atomic hydrogen is produced when H_2 molecules are atomized on the incandescent tungsten filament. These atoms subsequently recombine on the cold walls and the H_2 molecules thus produced diffuse out of the cell through a 6mm diameter orifice.

The internal energy of the H_2 molecules is probed by a low energy electron beam using the dissociative attachment process i.e.

$$e + H_2^* \longrightarrow H + H^-$$

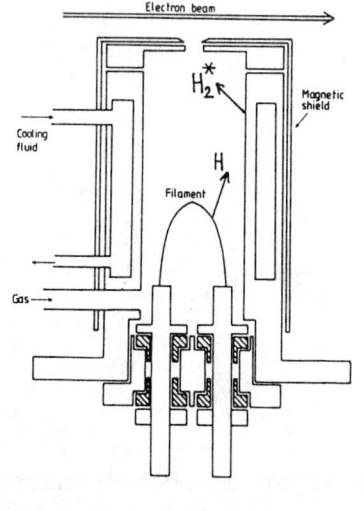

Fig5

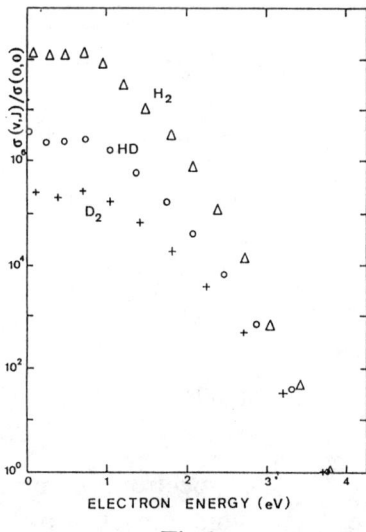

Fig6

A complete set of cross sections for this process has been established for all the v,J levels of H_2 and its isotopes[10,11] following the initial experimental observations of Allan and Wong[12]. These cross sections have the special quality that they increase very rapidly with increasing vibrational level going from 10^{-21} cm^2 for v=0 to near 10^{-16} cm^2 for v$\geqslant$5. Furthermore they peak at threshold which implies that cross section maxima occur at lower impact energies as v increases (see figure 6).

A schematic diagram of the apparatus is displayed in figure 7. A high resolution (60 meV) electron beam of about 50 nA is produced from a tungsten cathode by a cylindrical electrostatic filter. An electrostatic lens focuses the electron beam on the escaping molecules some 10 mm above the orifice. The ion optics generate a shallow potential well in the collision volume which selectively traps the H$^-$ ions with near zero kinetic energy before focusing them into a quadrupole mass spectrometer prior to detection by an off-axis channel electron multiplier. Typical count rates for the v=0 level are around 50 sec^{-1}.

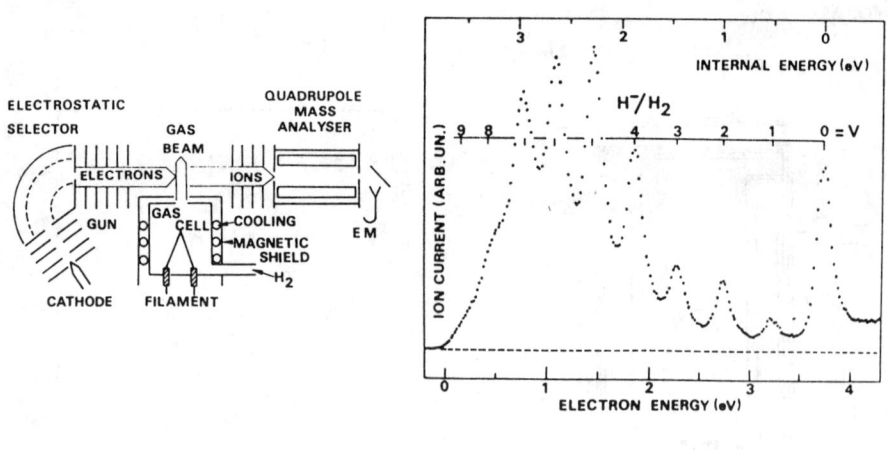

Fig7   Fig8

An H$^-$ spectrum is presented in figure 8. The association of the threshold peaks in the dissociative attachment cross sections with the potential well trap produces peaks in the spectrum corresponding to the individual vibrational levels. In this way levels up to v = 9 are clearly seen to be populated. Following a study in a mixture of H_2 and D_2, the mechanism for populating the vibrational levels was deduced to be the recombinative desorption process mentioned above. The important fact to be noticed in this spectrum is that H_2 is formed with internal energies up to 3.5 eV. The maximum energy available in the ER mechanism is 2.2 eV, insufficient to account for the observations. A plausible explanation for the origin of this large amount of energy is that the atoms recombine from weakly bound states i.e. the atoms will be trapped in physisorbed states on top of the chemisorbed layer of atomic hydrogen. Atoms in this states behave like a two-dimensional gas having a high mobility along the surface then making recombination an efficient process. In fact this process reduces to that of LH with atoms weakly bound on a metal-hydride type surface.

Another experimental observation is that the vibrational population of H_2 ceases when the wall temperature of the cell is allowed to rise above ≈ 200°C. At these temperatures the chemisorbed layer is being boiled off which would then remove the basis for the weakly bound states. Instead the atoms would be strongly chemisorbed before recombining via the LH mechanism to yield H_2 molecules with virtually no internal energy other than that corresponding to the wall temperature.

Further evidence for the existence of these physisorbed states emanated from the experiments performed by Wilsch and co-workers[13]. Under ultra-high vacuum conditions they first covered a metal surface with a mono-layer of chemisorbed H atoms. This surface was then bombarded by a beam of D atoms and the resulting molecules were monitored by a mass spectrometer. The only molecular species detected was that of D_2, the HD molecule being absent. This would indicate that the D atoms recombine on the metal-H atom surface without perturbing the strongly bound H atoms.

The vibrational populations derived from an H^- spectrum

using the cross sections of figure 6 are shown in figure 9. Also shown are the populations obtained using a tantalum filament and by consequence tantalum covered walls. It is clear from this that the nature of the metal covering the walls is an important ingredient in determining the internal H_2 energy. The population distributions are strongly non-Boltzmannian with the high levels (v>3) being super-populated. It must be kept in mind that these distributions are partially relaxed from whatever they were for the nascent molecules by the collisions they undergo with the walls before escaping from the cell. It should also be noticed that the rotational excitation is weak, in fact it has been estimated to be about 500K and would be the consequence of either fast collision relaxation for rotation or a low rotational temperature when the molecules are generated.

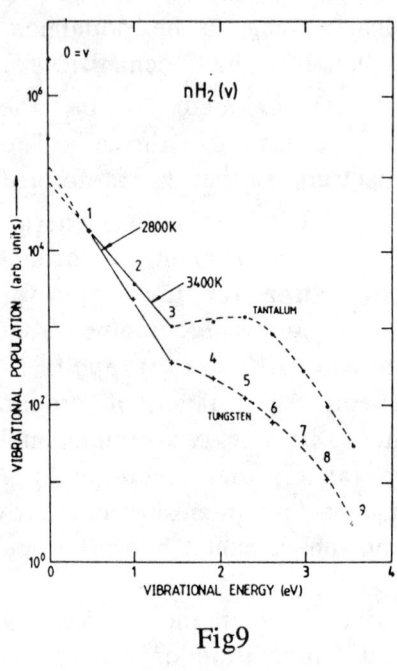

Fig9

With the 6mm diameter orifice, the molecules, on average, make 75 collisions with the walls yet molecules in levels such as v=6 or more are still present in the gas flux leaving the cell. In fact when the orifice is closed down to 1mm, the number of collisions then rises to 2700 yet the vibrational distribution is still not fully relaxed.

The use of tantalum is more efficient in producing high populations of high v levels but involves initially some conditioning of the cell. The conditioning amounts to an overnight bake-out at 200°C in the absence of cooling followed by a short overheating of the filament to evaporate metal onto the surfaces once the cooling is restored. The vibrational populations then increase gradually over a period of several hours. From these observations it is not clear whether the role of tantalum is to populate the high levels more efficiently during recombinative desorption or be less efficient in relaxing the hot molecules when they collide with the surface. However, as is related in the accompagnying paper[14], these qualities of the tantalum film can be destroyed if a plasma is generated in hydrogen gas.

CONCLUSION

Experimental evidence would indicate that clean metal surfaces (at least Ni, Pt and W) under high vacuum conditions have recombination coefficients γ approaching unity. However once these metals are put in the presence of hydrogen gas at pressures above $\sim$1mT, γ drops to values around 0.1 presumably due to the presence on the metal surface of a layer of strongly bound hydrogen atoms.

The situation regarding the accomodation coefficient β is even less clear. Measured values would range from near 1 (Ag, W, Ni) to around 0.1 (Pt, Fe), however different determinations give both 1 and 0.1 for the same metals (W, Ni)! In our study we have observed that recombination of H on tungsten and tantalum surfaces can yield highly vibrationally excited molecules. Simple conservation of energy arguments demonstrate that here recombination does not fit into the usual Eley-Rideal or Langmuir-Hinshelwood picture and has lead to the proposal that the process occurs via a new mechanism involving weakly bound surface states. These states would correspond to physisorption of the H atoms on top of the chemisorbed or strongly bound layer of H atoms which cover the metal. The consequence of this mechanism is that the nascent molecule would interacts only weakly with the surface and β is low. We have also observed, and this may be related to the low β, that the collisional relaxation of vibrationally excited H_2 is also low on these surfaces.

REFERENCES

[1] I. Langmuir J. Am. Chem. Soc. 37, 1139 (1915) .
[2] B.J. Wood and H. Wise J. Phys. Chem. 65,1976 (1961).
[3] G.A. Melin and R.J. Madix Trans. Faraday Soc. 67, 2711 (1971).
[4] G.A. Beitel and S. Fossum in D.E. Rosner Annual Review of Materials Science (1972), 2, 573. G.A. Beitel Ph. D. University of Wisconsin, Madison, (1969).
[5] B.J. Wood, J.S. Mills and H. Wise J. Phys. Chem. 67, 1462 (1963) ; Errata J. Phys. Chem. 68, 3911 (1964).
[6] J.E. Lennard-Jones Trans. Faraday Soc. 28, 333 (1932).
[7] B. Halpern and D.E. Rosner J.C.S. Faraday I 74, 1883 (1978).
[8] J. Harris and B. Kasemo Surf. Sci. 105, L281 (1981).
[9] R.I. Hall, I. Cadez, M. Landau, F. Pichou and C. Schermann Phys. Rev. Letters 60, 337 (1988).
[10] J.N.Bardsley and J.M. Wadhera Phys. Rev. A 20,1398 (1979).
[11] J.P. Gauyacq J. Phys. B 18, 1859 (1985).
[12] M. Allan and S.F. Wong Phys. Rev. Letters 41, 1791 (1975).
[13] H. Kaarmann, H. Hoinhes and H. Wilsch unpublished (1981). H. Kaarmann Doctoral thesis. University of Erlangen-Nüremberg (1981).
[14] C. Schermann, R.I. Hall, M. Landau, F. Pichou and I. Cadez in this volume

DISCUSSION

Hemsworth: I seem to be somewhat confused about how you turn on the discharge and the highly vibrational excited state disappear. So how do we make negative ions if there are no highly vibrational excited H_2

Hall: I don't know if there is any real relationship between the sources we have in our laboratory and the source you are using. Ours is a very small 40 millimeter source, no magnetic confinement and when I put this to Bill Graham some months ago, he said that is not a plasma, you are not even tickling the gas. So this is just what we see in our very simple system.

Hiskes: Would you say again which levels disappeared after you turned on the discharge.

Hall: Let's say above 4.

Graham: With the tungsten there was no plasma affect, you saw no difference.

Hall: It was much weaker. It was particularly spectacular for tantalum. The problem is that we have a hard time finding out what is going on here. Because in our experiment as soon as you put the plasma in the source then we get so much noise looking for our negative ion, we can't measure anything. So we put baffles in on top of the plasma and then got a proper gas further down the path.

Bacal: We have also been interested in the effect of tantalum vs. tungsten, which Richard Hall and collaborators reported. In an experiment we have done in JAERI, Japan with Ohara and Okumura we have tantalum filament and compared to results of tungsten filament. We have seen a difference, we have seen the tantalum filament was better and gave us more negative ions. Then I had, for some reason, a doubt and a colleague, Fukumasa repeated this experiment and he really found this difference. He said the ratio was different from that which we had, but the effect was there, so indeed your observation lead to something and I don't think the effect disappeared, anybody can repeat it and find it. I think the study does not finish here. I mean you don't have the end of the story, the end of the story maybe, I feel, is a big difference between the different materials we use on the wall, even on the plasma conditions.

Holmes: Your current is very small, so I would expect the plasma density is not significant compared to the larger devices in terms of the density. You have already implied that there is very little evidence for significant quantities of highly vibrationals levels particularly if you turn the discharge off. So where do the negative ions come from?

Hall: You tell me.

Kleyn: First of all the comment that you made, how clean is the surface. We did experiments with the single crystal silicon at pressure of 1×10^{-10} Torr and we could not get all the hydrogen off, it is extremely hard to get hydrogen off certain materials. You apparently grow a film on the walls and then you turn on the discharge, and apparently you modify the films such that it doesn't recover, it doesn't start to work again and produce these vibrational excited molecules. Could it be just due to the fact that you are using diffusion pumps and you get contamination. If you have oil diffusion pumps you get carbon in the system and you carbonate your films, so I guess the question you are really asking is what do these films really look like. So the question may not be of material as such because they may be very different but the structures of the films that you get. If there is carbon around, which you would sort of say there has to be, until everything is very carefully cryopumped, then it may just be that you sort of crud the wall and you really have to regenerate. So then it may really depend upon system to system, from history to history that determine what you will get.

Hall: I said that. We do, in fact, use oil diffusion pumps. There probably is all sorts of things wrong we use for surface experiments.

Kleyn: Oil diffusion pumps can be extremely good in principle if you can trap.

Hall: I hope that in the future they will be doing clean air experiments, using turbo pumps. And the point is that can you still see a difference between tungsten and tantalum.

Bacal: I would like to add something, actually the end of the discharge is a dramatic moment which you all know, because during the discharge you have a lot of ion pumping and all your surfaces get full of gas and when you stop the discharge all the gas tends to go out, it is enough to look at the gauge to see, if you have a gauge on the discharge. So that is a very strange moment and I don't think we have to conclude anything about, well anything very decisive about what happens at the end of the discharge. I mean, you destroy gases and a lot of things you had on the surfaces. I believe this is something that has to be considered.

VIBRATIONAL RELAXATION OF HIGHLY EXCITED H_2 MOLECULES IN GAS-PHASE AND GAS-SURFACE INTERACTIONS.

M.Cacciatore, M.Capitelli and G.D.Billing(*)

CNR Centro di Studio per la Chimica dei Plasmi,
Dip. di Chimica, Università di Bari, v.Amendola 173, Bari (Italy)
(*)Chemistry Lab.III, H.C.Orsted Institute, University of Copenhagen
2100 Copenhagen (Denmark)

ABSTRACT

The relaxation of highly vibrationally excited H_2 molecules in gas-phase and interacting with copper surfaces has been studied by using time-dependent semiclassical models for the collisional dynamics. The best available potential for H_2-H_2 and H_2-Cu has been used in the calculations, and the most important features of the collisional interactions have been taken into account.

INTRODUCTION

It is known that a variety of collisional processes of interest for the production of negative hydrogen atoms such as dissociative electron attachment[1], electron excitation to dissociative electronic states[2], as well as dissociation of neutral and charged molecules at the surface[3], can be activated when the vibrationally excited states are sufficiently populated. Therefore it is important to correctly describe the dynamics of processes in the gas-phase and in the wall region responsible for the redistribution of vibrational quanta in H_2.

As far as the collisional processes in the gas-phase are concerned, the processes through which the vibrational energy is redistributed in the vibrational ladder are primarily due to vibration-to-translation (1) and vibration-to-vibration (2) collisional exchanges:

$$H_2(v) + H_2(v_1) \longrightarrow H_2(v) + H_2(v_1') \qquad (1)$$

$$H_2(v) + H_2(v_1) \longrightarrow H_2(v') + H_2(v_1') \qquad (2)$$

The rate constants for these processes critically depend on the vibrational quantum number v and v_1 of the colliding molecules. Moreover we have recently shown[4] that the V-T rates for process (1) depend themselves on the vibrational energy level v' of the scattered molecule.

Collisions with solid surfaces can affect the vibrational kinetics in the gas-phase as well. Although heterogeneous interactions involving H_2 molecules have been studied in the past, quantitative dynamical informations on the energy relaxation processes at surfaces are almost completely absent[5]. On the other hand, the importance of inelastic (3) and reactive (4)-(5) processes in the wall region

$$H_2(v,j) + wall \longrightarrow H_2(v',j') + wall \quad (3)$$

$$H + H + wall \longrightarrow H_2(v',j') + wall \quad (4)$$

$$H_2(v) + wall \longrightarrow H + H \quad (5)$$

has emerged in several works on the kinetics of H^- production under plasma conditions[3,6]. Therefore a 'collisional' problem must be solved in the context of non-equilibrium kinetics, that is the search for 'accurate' state-to-state rate constants for gas-phase and gas-surface processes involving molecules in vibrationally excited states. The persisting difficulties in explaining the H_2 vibrational distributions detected in various sources[6] are also due to the approximate knowledge of the rate constants for most of the relevant processes introduced in the kinetic equations.

It is purpose of this paper to present some of the results obtained in our recent work[4,7,8,10] on the dynamics of processes (1),(2),(3),(5). Refined semiclassical models have been developed in order to describe the dynamics of the collisional processes under considerations. This, toghether with the use of accurate potential energy surfaces (PES), allowed us to obtain detailed state selected information on the vibrational relaxation in H_2.

V-V AND V-T RATE CONSTANTS.

According to the semiclassical coupled state method[9] used in this study the translational and rotational motions of the colliding molecules are treated classically, while a quantum mechanical description is assumed for the molecular vibrations. The quantum amplitudes for the vibrational transitions are obtained by solving the time dependent Schroedinger equations for perturbed anharmonic (Morse) oscillators. An analytical expression for the H_2-H_2 interaction potential has been derived[10] which fits the ab initio SCF-CI points and the potential recently determined in molecular beams experiment[11]

In Fig.1 it is reported the V-T deactivation rate out of the first excited levels as a function of temperature.

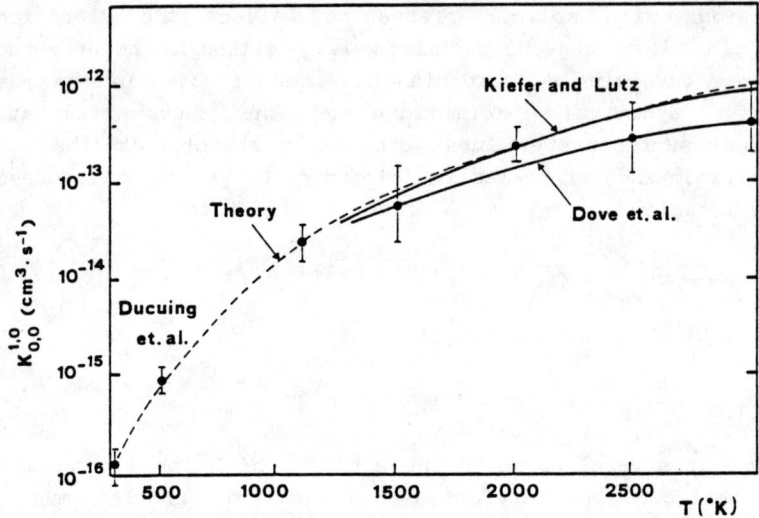

Figure 1: V-T rate constant as a function of temperature. Dashed curve: this work. Experimental values taken from Ref.12.

We can see a generally good agreement with the existing experimental results. Of special interest is the agreement with the experimental results of Ducuing et al at 300K. The region of low temperatures is crytical because the rotational motion can enhance the energy transfer from translation into vibration.

In Table 1 we have reported the deactivation rate constants for the process:

$$H_2(v=1) + H_2(v) \longrightarrow H_2(v=0) + H_2(v) \tag{6}$$

Table 1: V-T rate constants ($cm^3 s^{-1}$) for process (6) as a function of v and the translational temperature.

T(K)/v	1	2	3	5	7	9	
300.	1.25	1.51	1.89	2.50	6.60	19.1	x10(-16)*
500.	5.81	7.50	8.60	10.8	26.1	51.5	"
700.	2.63	2.87	3.28	3.62	8.40	11.2	x10(-15)
1000.	1.13	1.38	1.40	1.59	2.74	3.01	x10(-14)

* $10(-16) = 10^{-16}$

The energy mismatch for the processes (6) does not depend on the v-state, but nevertheless the rate constants reported in Table 1 show such a dependence. This behaviour has an explanation and it has never been considered in the kinetic modelling of H_2. This result could open new perspectives in that context.

In Figure 2 we have reported the V-V $K(1,0|v,v-1)$ rate constant for the deactivation of the first excited level as a function of v, toghether with the V-T $K(0,0|v,v-1)$ rate for the process

$$H_2(v=0) + H_2(v) \longrightarrow H_2(v=0) + H_2(v-1) \tag{7}$$

$$H_2(v=1) + H2(v) \longrightarrow H_2(v=0) + H_2(v-1) \tag{8}$$

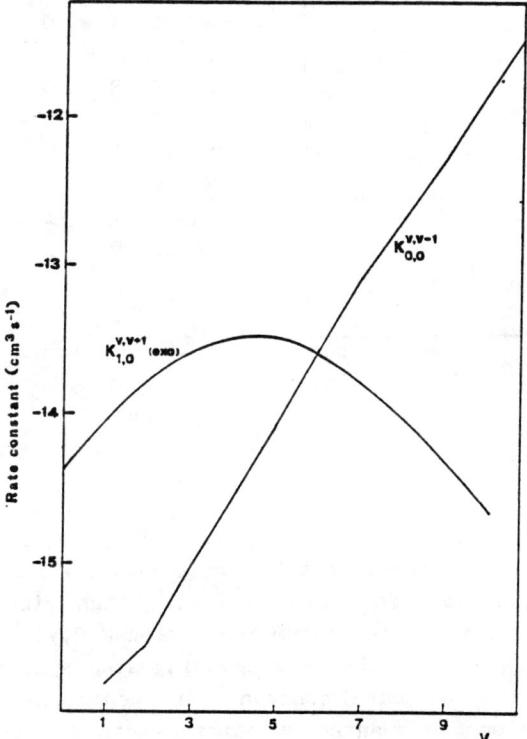

Figure 2: V-V (exothermic) $K(1,0|v,v-1)$ and V-T $K(0,0|v,v-1)$ rate constants as a function of the vibrational quantum number v. T=300K.

Figure 2 shows that the V-V energy exchanges dominate in the region of medium lying levels, while for the excited vibrational levels the V-T relaxation processes are predominant.

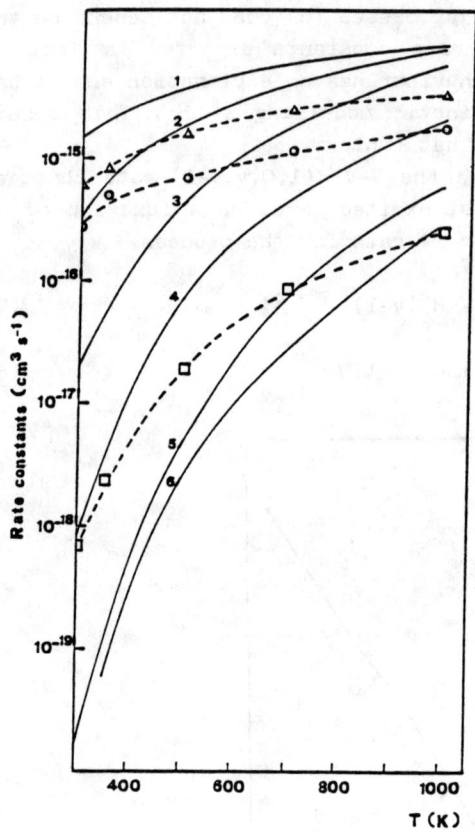

Figure 3: Single versus two-quantum transitions for the processes:

$$H_2(v) + H_2(v_1) \longrightarrow H_2(v') + H_2(v_1')$$

	v	v'	v_1	v_1'
1	3	2	0	1
2	5	4	0	1
3	7	6	0	1
4	9	8	0	1
5	10	9	0	1
6	12	11	0	1
○	9	7	0	1
□	5	7	1	0
▲	11	9	1	0

In Figure 3 we have reported the V-V rate constants for single quantum and double quantum transitions involving high vibrationally excited levels. At 300K the two-quantum transition K(9,7|0,1) is one order of magnitude higher then the corresponding single quantum rate K(9,8|0,1). As far as the rate constants are concerned, in the region of high v the double quantum transitions can be of the same importance as one quantum transitions.

$H_2(v,j)$ RELAXATION ON A Cu SURFACE.

Although the interactions of Cu surfaces with H_2 have been investigated both theoretically[13-15] and experimentally[16,17], a number of fundamental aspects are still open to question.
When the dynamical probabilities for the collisional processes (3) and (5) are to be computed, then the first step is the search for an accurate PES. An analytical PES[8] has been fitted by us to the ab initio points recently reported in a cluster SCF-CI calculation[18]. It turns out that the potential surface is very corrugated and dependent on the H-H bond distance. The energy barriers for H_2 in the vibrational ground state is high in most configurations, but may in some configurations be as low as 1.eV. The barriers are in accord with the ab initio 1.5eV[18], 0.9eV of Ref.15 and the 1.0eV jellium adjusted[14]. It is appreciably higher than the 0.2-0.5eV claimed in earlier investigations[13,16]. The barrier heights in the PES show a remarkable dependence on the H-H bond distance, so that H_2 interacts differently depending whether the molecule is in the ground or in vibrationally excited states. In Ref.7 we have reported the dissociation probabilities due to quantum tunnelling through the potential barriers for H_2 in different vibrational states.

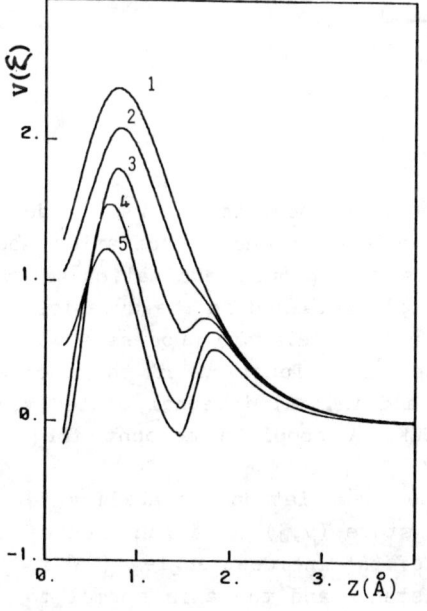

Figure 4: H_2-Cu(111) interaction potential $V(\mathcal{E})$ with H_2 interating perpendicular in the bridge site as a function of H_2-Cu distance and R_{HH} bond distance.
$R_{HH}(\text{Å})=0.8, 0.9, 1.0, 1.2, 1.4$ for curve 1 to 5.
1. =100Kcal/mol.

We have also run a 3D semiclassical trajectory calculations for H_2 propagating in the gas phase or in the surface region. While the translational motion of H_2 is treated classically, via Hamilton equations in an 'effective' potential, the phonon and electron-hole eccitation mechanism is treated quantum mechanically[8,19]. Figure 5 shows the energy transfered to the phonon modes ΔEph and to the electrons ΔEe-h as a function of the interaction time.

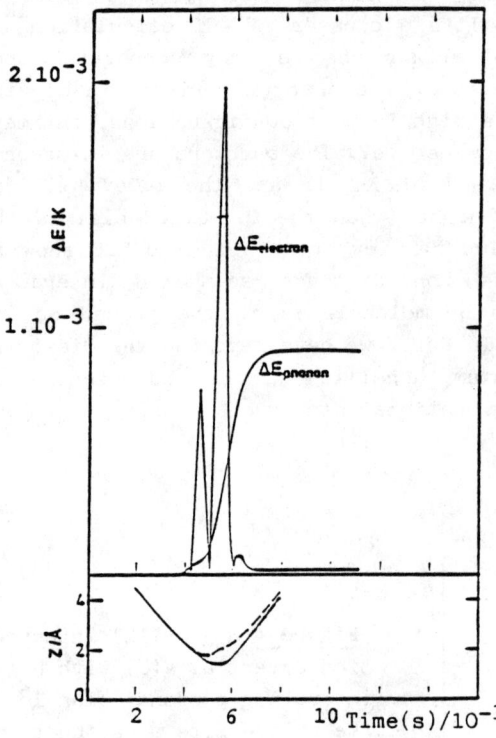

Figure 5: Lower panel: the trajectory z(t) for two hydrogen atoms as a function of time. H_2 is in the initial state (v=2, j=1). Ekin=1.79x10^4 K.
Upper panel: Energy transferred to the phonons ΔEphonons and to the electrons ΔEelec. as a function of time for the trajectory shown in the lower panel.

In the long time scale the energy accomodation is mainly due to phonon excitation. Nevertheless the electron-hole excitation above the Fermi level indirectly affects the phonon excitation[8] so that both mechanisms should be in principle retained in the dynamics.

The dynamical probabilities for the inelastic processes (3) and dissociation (5) have been computed as a function of the incident angles ϑ, φ, impact kinetic energy and initial internal state (v,j). The surface temperature is Ts=300K. A complete account for such calculations can be found in Ref.8.

In Table 2 we have reported the dissociation probability for H_2 in a given initial ro-vibrational state (v,j) as a function of the impact energy Ekin for two values of the incident angle (ϑ is the angle between H_2 centre of mass distance and the axis normal to the surface).

Table 2.: Dissociation probability P_{diss} for $H_2(v,j)$ as a function of the kinetic energy E_{kin}. ϑ is the approaching angle(see text). Eint is the average energy transfered to the phonon modes, v' and j' are, respectively, the average vibrational and rotational actions (in unit of $\hbar$) for the reflected trajectories.

$\vartheta = 0°$

v	j	E_{kin}/eV	P_{diss}	v'	j'	E_{int}/eV
0	0	1.55	0.00	0.	1.1	0.124
2	0	1.55	0.30	1.3	7.4	6.78(-2)
5	0	1.00	0.44	4.3	4.0	3.99(-2)
5	5	1.00	0.90	2.7	10.0	3.91(-2)
6	0	0.20	0.00	5.9	0.8	6.46(-3)
6	5	0.20	0.00	5.7	5.5	6.96(-3)

$\vartheta = 45°$

v	j					
2	0	1.55	0.00	2.0	1.2	2.54(-2)
5	0	1.00	0.00	4.9	0.9	1.60(-2)
5	5	1.00	0.17	4.4	5.4	1.84(-2)
6	0	0.20	0.00	5.8	0.6	2.40(-3)
6	5	0.20	0.00	5.7	5.5	1.78(-2)
6	0	0.50	0.00	6.0	0.7	7.71(-3)

For $\vartheta = 0°$, the dissociation probabilities are higher than for $\vartheta = 45°$, and this is in agreement with the experimental findings. For H_2 in the ground state the molecule does not dissociate in the energy range explored here (up to 1.5eV). The data reported in Table 2 show that there are small changes in the vibrational energy for the reflected molecules. The trajectory analysis show that a complex excitation-deexcitation mechanism is acting when the molecule comes closer to the surface. Figs.6,7 show a trajectory where the hydrogen molecule is vibrationally quenched from the surface in the v=5, and rotationally excited to j=4.In figure 6 (x,y)

70 Highly Excited H$_2$ Molecules

is the crystal plane and z is the axis normal to the surface.

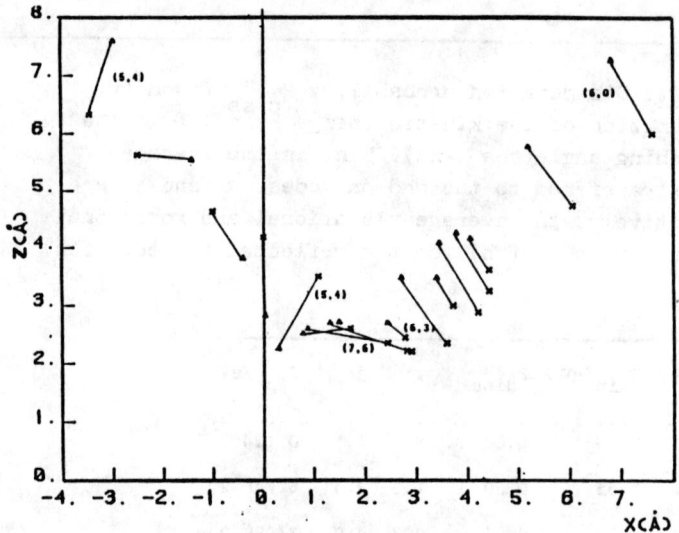

Figure 6: A quenched trajectory for H$_2$ in the initial v=6,j=0 states. The incident kinetic energy is Ekin=0.2eV, ϑ=45°, Ts=300K. The numbers in parentesis are the instantaneous v and j values.

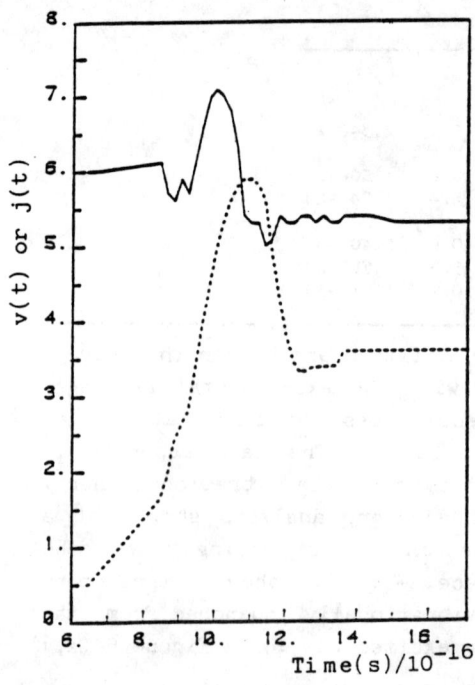

Figure 7: The rotational j and the vibrational v action in unit of h are plotted as a function of time (in unit 10^{-16} s) for the trajectory plotted in Fig.6.

On the basis of the trajectory analysis the general conclusion we can draw is that the Cu surface is quite inert toward dissociation and vibrational quenching of H_2. This is in agreement with recent experimental findings[20]. H_2 has a low dissociation probability unless it is produced in the gas phase in hot vibrational and translational states. This is due to the high barriers to dissociation exhibited by the PES, from 1.eV up to 3.0eV depending on the absorption site and the H-H bond distance.

Morover , the dissociation mechanism is direct: the molecule dissociates in the gas phase ,before a precursur absorption state would be required.

REFERENCES.

1. J.N.Bardsley and J.M.Wadehra Phys.Rev.$\underline{A20}$,1298(1979)
    J.M.Wadehra Phys.Rev.$\underline{A29}$,1061(1984)
2. M.Cacciatore and M.Capitelli Chem.Phys.$\underline{55}$,67(1981)
    R.Celiberto,M.Cacciatore and M.Capitelli Chem.Phys.$\underline{133}$,369,1989)
    J.R.Hiskes Comments At.Mol.Phys.$\underline{19}$,59(1987)
3. J.R.Hiskes and A.M.Karo J.Appl.Phys.$\underline{56}$,1927(1984)
    J.R.Hiskes : this conference
4. M.Cacciatore,M.Capitelli,G.D.Billing Chem.Phys.Lett.$\underline{157}$,305(1989)
5. K.Christmann Surface Sci.Report $\underline{vol.9}$(1988)
6. M.Bacal: this conference
7. M.Cacciatore,M.Capitelli,G.D.Billing Surf.Sci.$\underline{217}$,L391(1989)
8. M.Cacciatore and G.D.Billing 'Dynamical Relaxation of $H_2(v,j)$ on a Copper Surface': submitted for pubblication
9. G.D.Billing Chem.Phys.Lett. $\underline{97}$,188(1983)
    G.D.Billing and M.Cacciatore Chem.Phys.Lett. $\underline{86}$,20(1982)
10. M.Cacciatore and G.D.Billing:'Semiclassical Theoretical Study on the Vibrational Energy Transfer by $H_2(v)-H_2(v')$ Collisions': submitted for the pubblication.
11. M.J.Norman,R.O.Watts and U.Buck J.Chem.Phys. $\underline{81}$,3500(1984)
12. M.M.Audibert,R.Vilaseca,J.Lakasik and D.J.Ducuing Chem.Phys.Lett. $\underline{31}$,232(1975)
    J.E.Dove,D.G.Jones,H.Teitelbaum IV Symposium on Combustion,University Park,Pennsylvania 1972(Pittsburg 1973)
    J.H.Kiefer J.Chem.Phys. $\underline{48}$,2332(1968)
13. A.Gelb and M.Cardillo Surf.Sci.$\underline{59}$,128(1976)
14. A.DePristo,C.Y.Lee and J.M.Hutson Surf.Sci.$\underline{169}$,451(1986)
15. J.Harris and S.Andersson Phys.Rev.Lett.$\underline{55}$,1583(1985)
16. M.Balooch,M.J.Cardillo,D.R.Miller,R.E.Stickney Surf.Sci.$\underline{46}$,358(19

17. G.Anger,A.Winkler,K.D.Rendulic Surf.Sci. 220,1(1989)
18. P.Madhavan and J.L.Whitten J.Chem.Phys. 77,2673(1982)
19. G.D.Billing Chem.Phys. 74,143(1983);70,223(1982)
    G.D.Billing and M.Cacciatore Chem.Phys. 103,137(1986)
    G.D.Billing and M.Cacciatore Chem.Phys.Lett. 113,23(1984)
20. R.I.Hall,I.Cadez,M.Landau,F.Pichou and I.Schermann Phys.Rev.Lett. 60,337(1988)
    P.J.Eenshuista,J.H.M.Bonnie,J.Los and H.J.Hopman Phys.Rev.Lett. 40,341(1988)

DISCUSSION

Roberts: I don't think I understood what you said about the vibrational translational relaxation as a function of the high quantum high rotation, the vibration quantum numbers. It looks to me like these results would be strongly dependent on whether the molecule is anharmonic or not and it would be strongly dependent on the translational temperature, the temperature of the bath that the vibration has to exchange energy with and on the difference of vibrational temperature and the bath temperature. Your result of this doesn't apply to anything. What in the world could it possibly mean.

Cacciatore: Are you talking about the energy transferring in the gas phase.

Roberts: Gas phase.

Cacciatore: So the question is.

Roberts: This result that nobody has seen before. The change with two quantum numbers. What condition does this happen. Surely it doesn't happen under all conditions. If you had a very large vibrational temperature compared to the bath the bath won't take too much energy at one time. You even get a ladder walking up.

Cacciatore: So you don't think this is surprising that the v dependance of the rate constants.

Roberts: I not only think it is surprising, I think it is not right.

Cacciatore: Well that is what we actually found, and I think it is right. We have an explanation for such a behavior, it depends on the coupling matrix elements for the two colliders. This matrix element depends on the v states of the two molecules. There is quite a bit of different changes in this matrix if you look the first order probability of vibrational excitation. The excitational probability is given as an integral and under integral there are two terms, one term which depends upon the energy mismatch in the process that you are looking at and it depends on the coupling matrix element. Of course in the second term, the energetic term it doesn't change with v and v' it is constant does not depend on the v states.

Roberts: I want to thank you because I understand it as well as I am going to.

COUPLED SOLUTION OF BOLTZMANN EQUATION AND NON EQUILIBRIUM VIBRATIONAL KINETICS IN H_2 VOLUME DISCHARGES : AN ANALYSIS OF THE INPUT DATA

M. Cacciatore, M. Capitelli, R. Celiberto, P. Cives and C.Gorse

Centro di Studio per la Chimica dei Plasmi del CNR and Dipartimento di Chimica dell'Università di Bari (Italy).

ASTRACT

New sets of direct electron impact excitation cross sections involving vibrational excited states of H_2 (including dissociation, ionization and E-V processes) have been calculated by using Gryzinski formulation. These rates are then inserted in a numerical code based on the simultaneous solution of Boltzmann equation, of vibrational kinetics and H, H⁻ production rates. Comparison of theoretical and experimental results shows satisfactory agreement for the electron temperature and electron density, while some discrepancy does exist on the negative ion concentration.

INTRODUCTION

An international collaboration between our group and two french groups was able to build up in the last few years a numerical code describing the complex physical phenomena occurring in multicusp magnetic H_2 discharges[1-3].

In particular the code, which is based on the simultaneous solution of Boltzmann equation, vibrational kinetics, dissociation and negative ion production, predicts the electron energy distribution function (EDF), the non equilibrium vibrational distribution of H_2 (N_v), the dissociation and ionization degrees as well as the concentration of negative ions (N_{H^-}).

The numerical results were validated by comparing them with existing experimental results obtaining in general satisfactory agreement[1,2].

Application of the code to new experimental situations[4] once again seems to give satisfactory agreement.

The satisfactory agreement between theoretical and experimental results could appear surprising in view of the accuracy of the input data entering the code. Keeping in mind in fact that the code treats each vibrational level of as an indipendent species with its own cross sections, we can realize that a large part of the relevant cross sections come from calculations. In particular the electronic excitation cross sections including dissociation and ionization are based on the semiclassical Gryzinski approximation as developed by our group some years ago[5].

Recently, however, these sets of electronic cross sections have been recalculated[6] by modifying some parameters contained in the Gryzinski formulae. The new cross sections better reproduce the v=0 experimental results. Moreover the excitation of b $^3\Sigma_u^+$ state involving v≠0 vibrational levels closely follows the behaviour of most accurate quantum mechanical results[7,8]. Insertion of these new sets of cross sections in the general code gives us the possibility to better understand the reliability of part of our input data, in view of the extension of our approach to D_2 discharges.

ELECTRONIC CROSS SECTIONS

Table I reports the microscopic processes inserted in our kinetic model.

Table I

e-V	$e + H_2(v) \rightarrow e + H_2(w)$
E-V	$e + H_2(v=0) \rightarrow e + H_2(B^1\Sigma_u^+, C^1\Pi_u) \rightarrow e + H_2(v') + h\nu$
V-V	$H_2(v) + H_2(w) \rightarrow H_2(v-1) + H_2(w+1)$
V-T	$H_2(v) + H_2 \rightarrow H_2(v-1) + H_2$
V-T	$H_2(v) + H \rightarrow H_2(w) + H$
e-D	$e + H_2(v) \rightarrow e + H_2^* \rightarrow e + 2H$
e-D	$e + H_2(v) \rightarrow e + H_2^* \rightarrow e + H + H^*$
e-I	$e + H_2(v) \rightarrow e + H_2^+ + e$
e-da	$e + H_2(v) \rightarrow H + H^-$
e-E	$e + H_2(v) \rightarrow e + H_2^*$
wall	$H_2(v) \rightarrow H_2(w)$
wall	$H + H \rightarrow H_2(v)$
wall	$H_2^+ + (e)_{surf} \rightarrow H_2(v)$

We focus our attention on some electronic transitions. Let us start with the dissociation process

$$e + H_2(X^1\Sigma_g^+, v) \rightarrow e + H_2^*(b^3\Sigma_u^+) \rightarrow e + 2H \quad (1)$$

involving each vibrational level of H_2. We have recalculated the entire set of cross sections according to ref. 5. Contrary to this reference, however, we consider the number of equivalent electrons in H_2 $N_e=2$, while the next state is considered being the potential curve of H_2^+. Another modification comes from a better description of the vibrational wave functions of $H_2(v)$ which have been obtained by solving the Schrodinger equation for the nuclear motion.

Figure 1 reports a comparison of the calculated cross sections for v=0 with theoretical and experimental results[9,10]. We can note that the agreement between the present results and experimental ones is indeed very good, showing that a judicious choice of parameters in the Gryzinski formulae yields cross sections in satisfactory agreement with most accurate data. Note that the present data are a factor two larger than the old data utilized in refs. 1, 2, 4. Figure 2 compares dissociation cross sections with

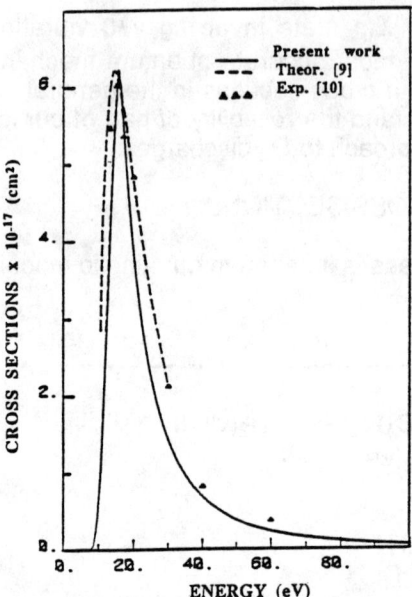

Fig. 1. Cross sections for process (1) and v=0.

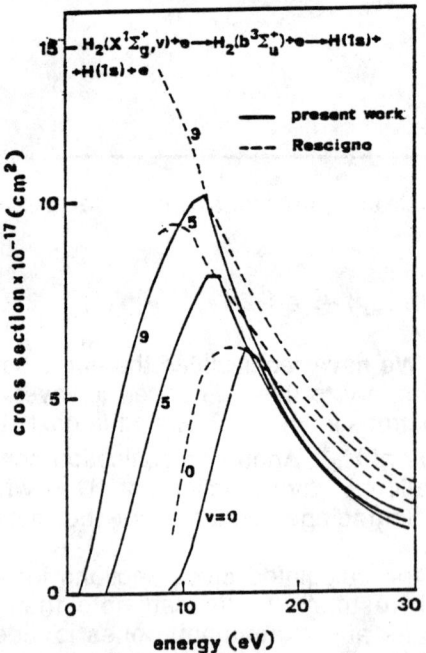

Fig. 2. A comparison between theoretical cross sections for process (1) at different vibrational quantum numbers.

the corresponding Rescigno[7] results for different vibrational levels. We can note that the agreement between our results and Rescigno ones remains good for excited vibrational levels, even though some disagreement is present in v=9 probably due to some difficulties met by the accurate quantum mechanical method to calculate cross sections near the threshold energy of the process. This point can be appreciated by looking at fig. 3 where we compare Gryzinski, Rescigno and Huo cross sections for v=4. These last cross sections utilize an approach similar to that one used by Rescigno, trying however to better describe quantum mechanically the target. Inspection of figure 3 shows that the agreement between Gryzinski and Huo cross sections near the threshold process is indeed satisfactory.

Let us consider now an other important process affecting the vibrational distribution of H_2, the so called E-V process, i.e.

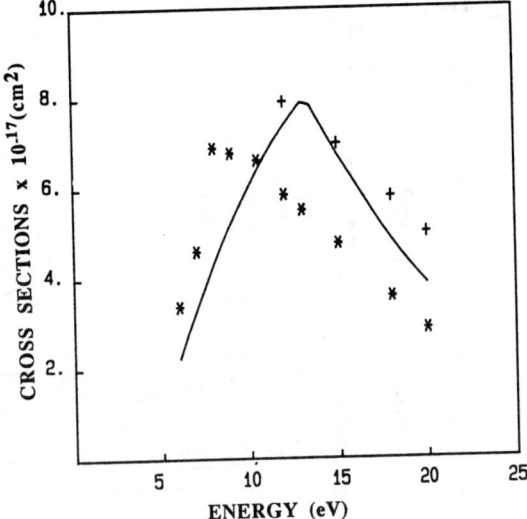

Fig. 3. Comparison of cross sections for process (1) and v=4. (— Gryzinski, +++ Rescigno, *** Huo)

$$e + H_2(X^1\Sigma_g^+, v_i) \longrightarrow e + H_2^* \longrightarrow e + H_2(X^1\Sigma_g^+, v_f) + h\nu \qquad (2)$$

Cross sections for this process have been calculated by Hiskes[11] some years ago and utilized in all codes. The calculation of these cross sections requires a two step procedure. First we need the cross sections for the excitation of singlet states, this cross section is then modulated by Franck-Condon factors and Einstein probabilities to obtain the cascade cross section over the vibrational levels of ground state.

Figure 4a compares our calculated cross sections (obtained with Ne=1) for the excitation of $B^1\Sigma_u^+$ state with recent experimental data[12,13] as well as with a theoretical curve[14]. We note that our results agree within the error bar with the experimental results, being a factor two less than the other theoretical curve. Figure 4b gives the excitation cross sections of $B^1\Sigma_u^+$ for different vibrational levels.

Figure 5 reports the E-V cross sections (summed over $B^1\Sigma_u^+$ and $C^1\Pi_u$ states) for populating the different vibrational levels of H_2, while figure 6 reports some of these cross sections for different initial vibrational levels of H_2. A comparison of the results of fig. 5 with the corresponding ones calculated by Hiskes shows that the present results are a factor two less than the Hiskes data, this difference being essentially due to the different choice of excitation cross sections for $B^1\Sigma_u^+$ and $C^1\Pi_u$ states. Note however that present results as well as Hiskes ones neglect contributions coming from other singlet states of H_2 radiatively coupled to the ground state.

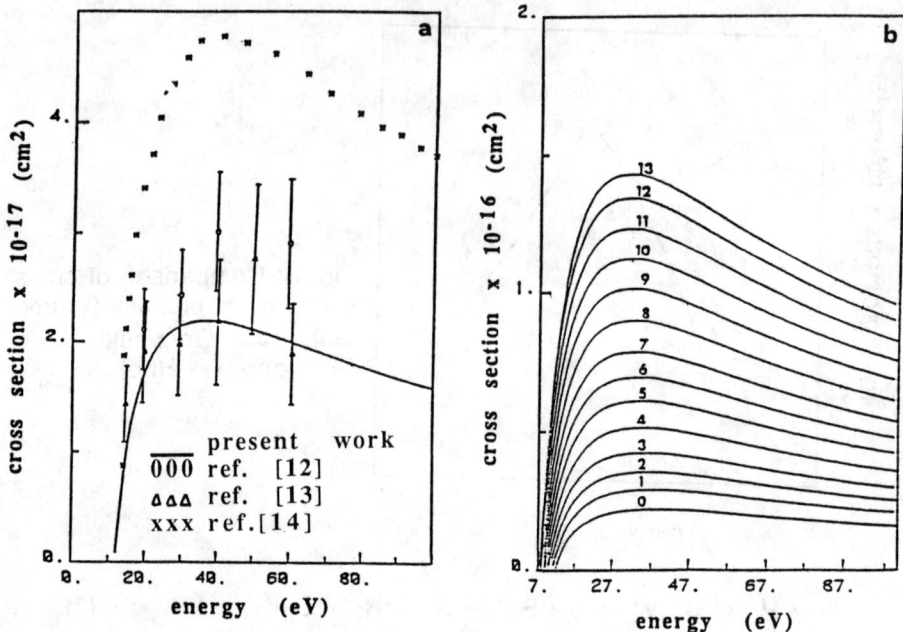

Fig. 4a-b. Cross sections for the excitation of $B^1\Sigma_u^+$ state : a,v=0 ; b,all v. Comparison between experimental and theoretical results.

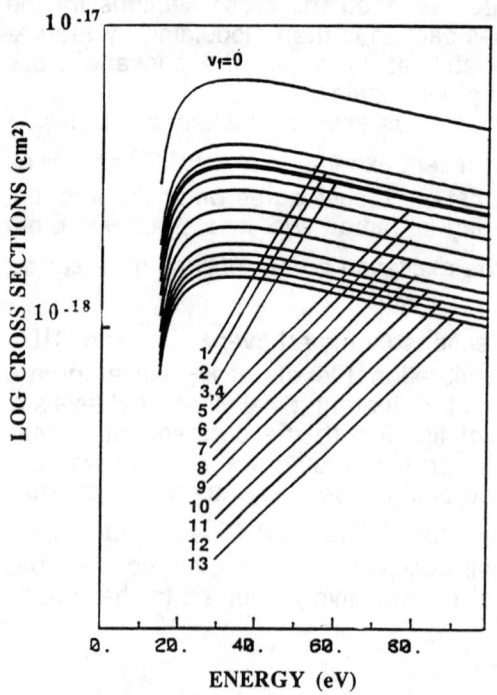

Fig. 5. Cross sections for process (2) and $v_i=0$.

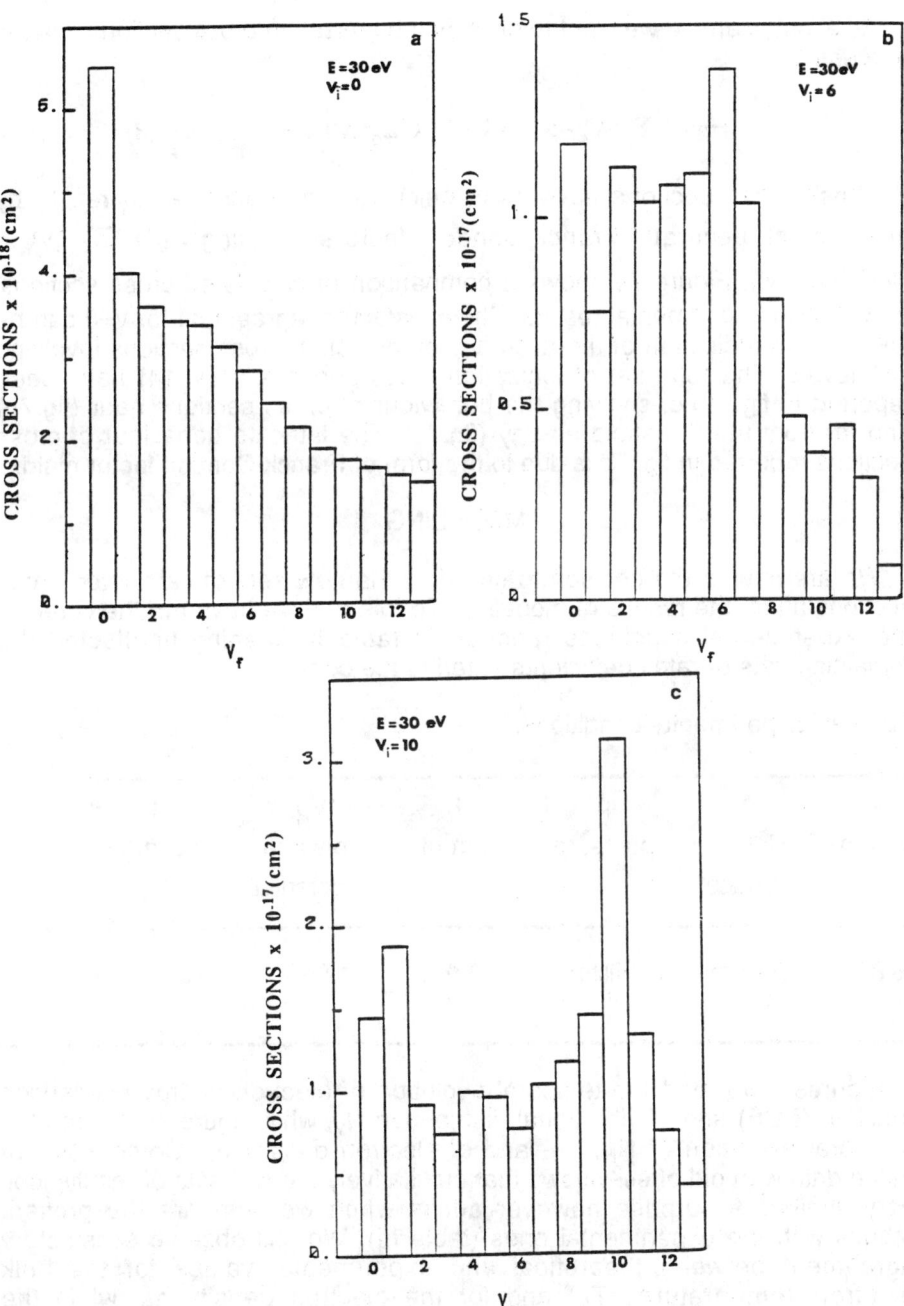

Fig. 6a-b-c. Cross sections for process (2) and E_i=30eV and for v_i=0 (a), v_i=6 (b) and v_i=10 (c).

As a last example we want to discuss the ionization cross sections, i.e. the process

$$e + H_2 (X^1\Sigma_g^+, v) \longrightarrow e + H_2^+ (X^2\Sigma_g^+, v') + e \qquad (3)$$

These cross sections have been calculated as described in ref. 5, by using most accurate Franck-Condon factors[15] linking $H_2(X^1\Sigma_g^+,v)$ to $H_2^+(X^1\Sigma_g^+,v')$. Figure 7a shows a comparison of calculated cross sections for v=0 with experimental results. The satisfactory agreement for v=0 can be taken as an indication of the accuracy of ionization cross sections involving v≠0 levels. The total set of ionization cross sections (0⩽v⩽14) have been reported in fig. 7b-c, showing the behaviour of cross sections near (fig.7b) and far from the threshold energy (fig.7c). The intricate behaviour of cross sections reported in fig. 7c is due to the form of Franck-Condon factor matrix.

MODELING

We are now in the position to see how the new sets of calculated cross sections affect the results of modeling. To this end we have run the code for the experimental conditions reported in table II, keeping unaffected the remaining sets of rate coefficients entering the code.

Table II Experimental conditions.

V Volume	A loss surface	p pressure	I current	Vp plasma potential	Vt discharge voltage
8.8 l	831 cm²	2 millitorr	5 A	1.96 V	50 V

Figures 8-9 report the temporal evolution of electron energy distribution function (EDF) and of vibrational distribution N_v, while figure 10 reports the temporal evolution of N_H, N_{H^-} and of electron density n_e. Comparison of these data with old ones[2] shows that qualitatively the two sets of results look very similar. A surprise however arises when we compare the present results with the experimental ones (table III). We still observe satisfactory agreement between theoretical and experimental values for the bulk electron temperature (T_e) and for the electron density n_e, while the theoretical H⁻ concentration is a factor 10 less than the experimental value. The agreement between calculated and experimental H⁻ concentration was better[2] when using a worse set of cross sections.

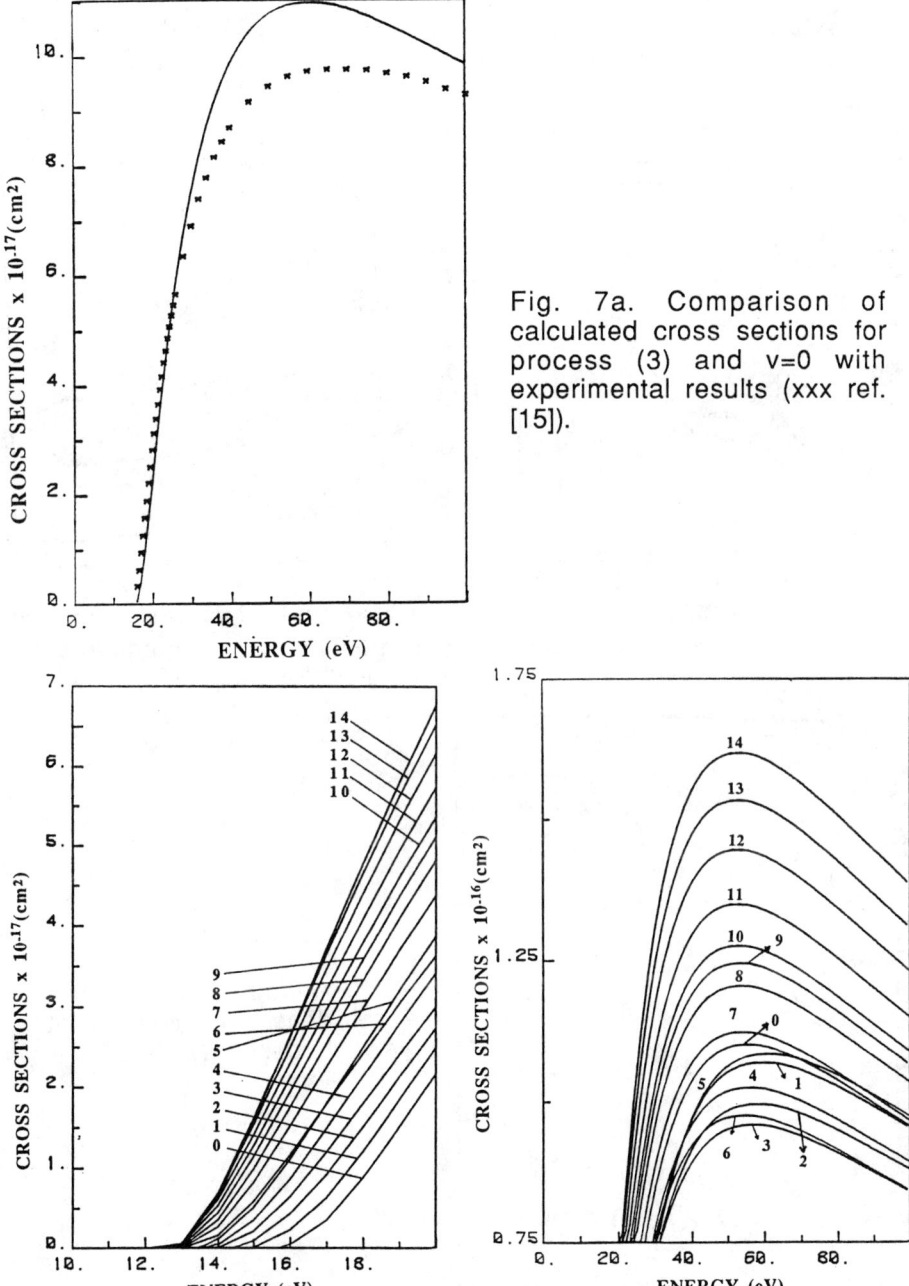

Fig. 7a. Comparison of calculated cross sections for process (3) and v=0 with experimental results (xxx ref. [15]).

Fig. 7b-c. Cross sections for process (3) near (b) and far (c) from the threshold energy.

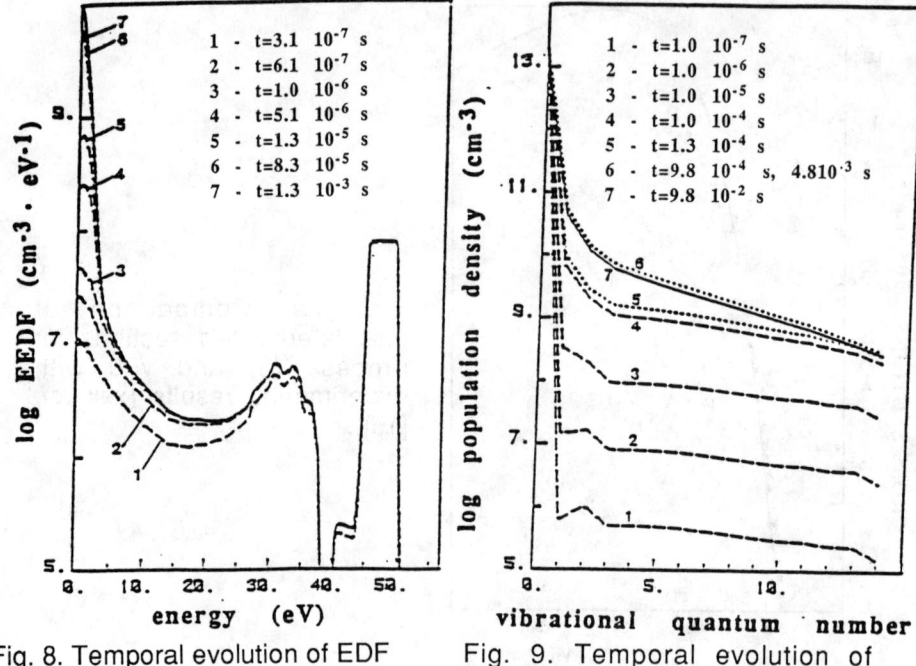

Fig. 8. Temporal evolution of EDF

Fig. 9. Temporal evolution of vibrational distribution N_v

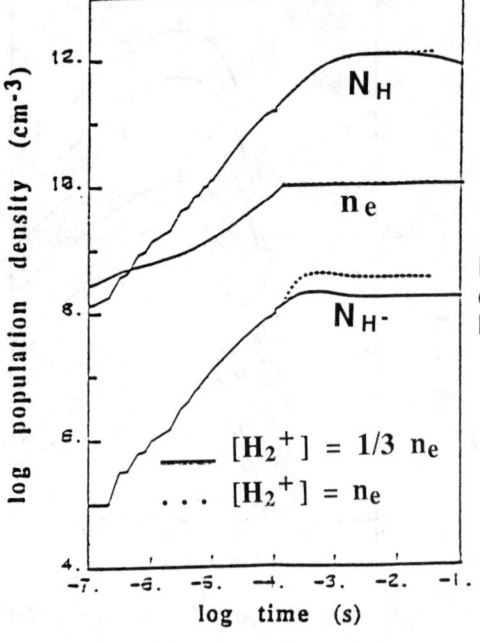

Fig. 10. Temporal evolution of electron density n_e, atom density N_H and ion density N_{H^-}.

Table III   Comparison between experimental and numerical results.

	n_e (cm^{-3})	T_e (eV)	N_{H^-} (cm^{-3})
exp.	2.25 10^{10}	0.6	3.2 10^9
theor. ($[H_2^+] = \frac{1}{3} n_e$)	9.8 10^9	0.47	2.0 10^8
theor. ($[H_2^+] = n_e$)	9.8 10^9	0.47	4.0 10^8
theor. ($[H_2^+] = n_e$) (e$^-$losses)/3	1.7 10^{10}	0.73	1.13 10^9

This apparent contradiction is due to the fact that the new calculated cross sections as compared with old ones go in the direction to decrease the vibrational plateau responsible for the formation of H$^-$ through dissociative attachment. In fact the decrease by a factor two of E-V rates decreases by the same factor the population of plateau, while the increase of dissociation cross sections increases the concentration of atomic hydrogen which is a very effective quencher of vibrational energy.

COMMENTS

The results presented in the previous pages show that some caution must be excercised in the comparison of experimental and theoretical results. Improvement of cross sections has in fact led to considerable disagreement of N_{H^-} which strongly depends on N_v.

To improve the agreement we can follow two directions. First we can change some parameters which affect N_v and therefore N_{H^-}. As an example we can increase the concentration of H_2^+ which increases N_v through heterogeneus ricombination obtaining a better agreement between calculated and experimental H$^-$ values (see table III and fig. 10). On the other hand we can decrease the loss of fast electrons in the plasma, obtaining again better agreement with the experimental results (see table III).

The other direction we can follow is to continue the improvement of input data of table I. In particular we have now a better set of V-T rates involving $H_2(v)-H_2$[16].

Comparison between new and old rates shows that the new ones are larger than the old ones over the high lying vibrational levels up to a factor 10 (T=300 k). The new rates therefore should still decrease the formation of H⁻ by increasing deactivation of high lying vibrational levels.

At the same time we have performed dynamical calculations on the deactivation of $H_2(v)$ on copper walls[17]. These calculations have shown that at low kinetic energies the deactivation of $H_2(v)$ on copper is very small. In the previous modeling on the contrary we have inserted the deactivation rates calculated by Hiskes and Karo[18], who on the contrary have found strong deactivating rates of $H_2(v)$ colliding with iron surfaces. Further investigations on the role of walls in the formation of $H_2(v)$ seems to be of paramount importance for understanding the formation of H⁻ in volume sources, specially at the light of new results of Hiskes and Karo[19] on the recombination of H_3^+ on surfaces yielding vibrationally excited H_2 molecules.

In conclusion we can say that future extention of the present code to D_2 discharges should solve the numerous problems still existing in the selection of rates of elementary processes listed in table I.

ACKNOWLEDGMENTS

The authors thank M. Bacal and J. Bretagne for useful discussions.

This work has been partially supported by CNR through the "Progetto Finalizzato Chimica Fine II".

REFERENCES

1. C.Gorse, M.Capitelli, J.Bretagne, M.Bacal, Chem. Phys. 93, 1 (1985).

2. C.Gorse, M.Capitelli, M.Bacal, J.Bretagne and A.Laganà, Chem. Phys.117, 177 (1987).

3. J.Bretagne, G.Delouya, C.Gorse, M.Capitelli and M.Bacal, J.Phys. D 18, 811 (1985); 19, 1197 (1986).

4. M. Bacal, D.A. Skinner and P. Berlemont, Report PMI 2158 (1989)

5. M.Cacciatore, M.Capitelli and M.Dilonardo, Chem. Phys 34, 193 (1978); M.Cacciatore, M.Capitelli, C.Gorse, J. Phys. D: Appl. Phys.13, 575 (1980); M.Cacciatore and M.Capitelli, Chem. Phys 55, 67 (1981).

6. P. Cives, Tesi Università di Bari (1989)

7. T.N. Rescigno and B.I. Schneider, UCRL-99210 (to be published on J. Phys. B.).

8. W.M. Huo, 42nd Annual Gaseous Electronics Conference, Palo Alto(CA) E-40, 80 (1989) and private communication.

9. M. A. P. Lima, T. L. Gibson, W. M. Hero and V. Mckoy, J. Phys. B 18, 2865 (1985).
10. Nishimura, J. Phys. Soc. Japan 55, 3031 (1986).
11. J. R. Hiskes, J. Appl. Phys. 51, 4592 (1980).
12. M. A. Khakoo and Trajmar, Phys. Rev. A 34, 146 (1986).
13. S. K. Svrivastava and S. Jensen, J. Phys. B 10, 3341 (1977).
14. Hazi, private communication (1982).
15. M. R. Flannery, H. Tai and D. L. Albritton, Atomic and Nuclear Data Tables 20, 565 (1978).
16. M.Cacciatore, M.Capitelli and G.D. Billing, Chem. Phys. Lett. 157, 305 (1989).
17. M.Cacciatore, M.Capitelli and G.D. Billing, Surface Science 217, 391 (1989)
18. J.R.Hiskes and A.M.Karo, J. Appl. Phys. 56, 1927 (1984).
19. J.R.Hiskes and A.M.Karo, "Non Equilibrium Processes in Partially Ionized Gases", Eds. M. Capitelli and J.N. Bardsley Plenum (1990)

DISCUSSION

Hiskes: You pointed out the fact of that there are two discrepancies in the big eV excitation rates, there is a certain degree of arbitrariness there, whether you fit the Braginskii calculations to the experimental data or whether you fit it to the Abnecio calculations and I choose the Abnecio reference point and you choose the experimental point. I think that is the basis for our discrepancies.

Capitelli: The experimental result it seems it is smaller than the Abnecio equation. Anyway you know that there is also other excited states which can cascade, another big problem.

Hiskes: The difference is not between you and me, but between the experimental and the Abnecio calculations.

Wadehra: I have one comment and one question. The comment is that very recently we came up with numerical algorithm for solving the Boltzmann equation exactly without making any two-time expansion like you have done. And we have tested the algorithm on a microcomputer for electrons in the real gasses, we have only been through a very few processes like elastic, inleatic and ionization. If we have a lot of processes like in your case for hydrogen, I am sure our algorithm would have generated exactly the distribution function. Now the question, at the initial time you start with some Maxwellian distribution I believe, how does that change with time? Do you still maintain the distribution or is the distribution very different from experiment.

Capitelli: We did not start this kind of calculation. But for the comment, it is good to have an analytical solution for the Boltzmann equation. For the starting point of our Boltzmann calculation, we don't start in this case with a Maxwellian distribution function because we have a source of electron, for example at 50 or 100 eV. In any case, if you start with a Maxwellian distribution what you can change is only the early part of the temporal evolution. But after a few milliseconds your distribution function doesn't depend on the initial distribution function.

Bretagne: A very short comment about the problem of the comparison between the results of a modeling experiments. I think there is some problem in fact about what is included in the code. Another question is when we compared the experiments, the experiments have real values in the plasma volume and a real velocity. When we model a discharge we make assumption of these values, for us it remains a problem.

Capitelli: I agree with you because I am assured that the loss is very important in the shaping the electron energies that we have found and it depends on electron densities.

Graham: Is it possible to measure the effect of volume and the effect of source area in the ion source? If you put that into the EDF's then you could normally get better agreements probably, and I think your using a factor of three less for the electron loss to be justified in terms of the volume surface ratio, the actual volume surface ratio in the ion source. I think it is important that people start to fold in the physical characteristic of the ion source into the atomic physics calculation.

Capitelli: That is exactly the decrease of our loss. I think it would be better linking between plasma physics and the atomic physics field. I think you are right.

Generation of H⁻, H_2 (v"), and H Atoms by H_2^+ and H_3^+ Ions Incident Upon Barium Surfaces

J.R. Hiskes and A.M. Karo

Lawrence Livermore National Laboratory

I. Introduction

The generation of vibrationally excited molecules by electron excitation collisions and the subsequent generation of negative ions by dissociative attachment to these molecules has become a standard model for volume source operation.[1] These processes have been supplemented recently by the demonstration of atom-surface recombination to form vibrationally excited molecules[2,3], and enhanced negative ion formation by protons incident upon barium electrodes.[4] In this paper we consider the additional processes of molecular vibrational excitation generated by recombination of molecular ions on the electrode surfaces, and negative ion formation by vibrationally excited molecules rebounding from low work-function electrodes.

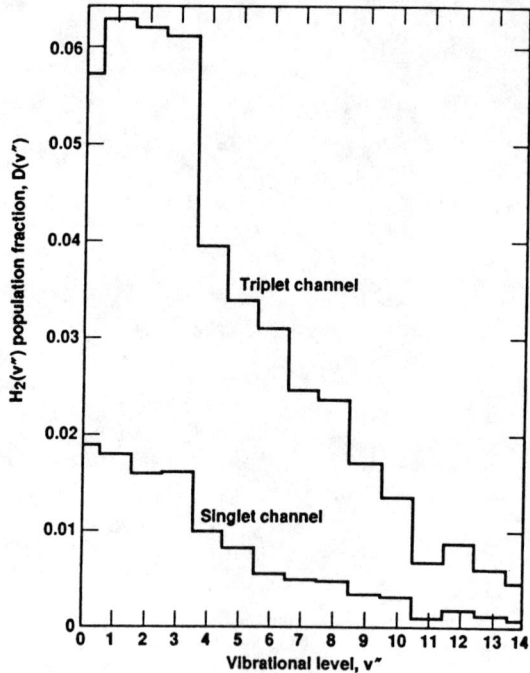

Figure 1: Population distribution per incident H_2^+ ion

88 © 1990 American Institute of Physics

II. Recombination and Dissociation of H_2^+ and H_3^+ on Surfaces

A four-step model has been developed for the formation of excited molecules, $H_2(v")$, generated either by H_2^+ ions or H_3^+ ions incident upon metal surfaces.[5,6,7] The final vibrational distribution per incident H_2^+ ion is shown in Fig. 1 for the case of neutralized ions impinging on the surface with 4eV of translational energy, and an internal excitation distribution appropriate to H_2^+ neutralization. The distinguishing feature of this figure is the broad vibrational distributions extending across the entire spectrum of bound vibrational levels. The singlet channel refers to H_2^+ neutralization directly into the singlet ground state of the H_2 molecule, the triplet channel corresponds to neutralization proceeding through the lowest excited triplet state prior to ground state formation. The total production is the sum of these two distributions.

In Fig. 2 is shown the population distribution per incident ion for incident H_3^+ neutralized to form 6eV H_3, which in turn spontaneously dissociates into a 4eV H_2 molecule and a two eV H atom immediately prior to the "hard" collision with the surface. Here again the final population distribution spans the bound spectrum. A comparison of the distributions of Figs. 1,2 shows that the final population for $v" \geq 5$ is larger in the case of H_3^+ recombination than for H_2^+ recombination. For this reason the H_3^+ ion appears to be a more useful ion leading to H^- formation than does the H_2^+ ion.

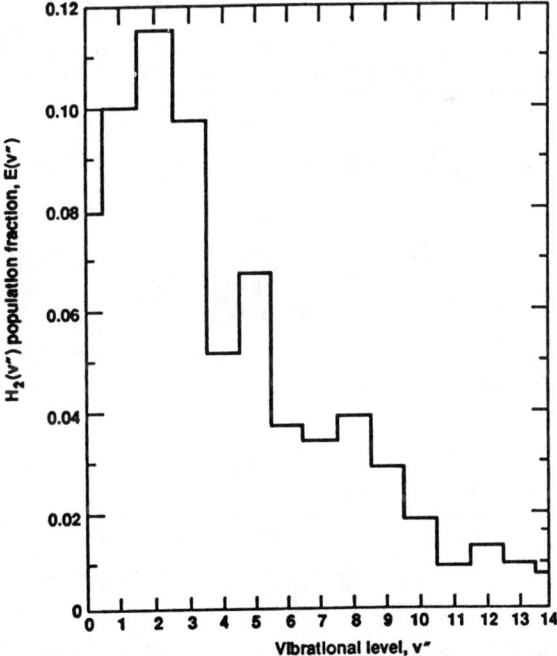

Figure 2: Population Distribution per Incident H_3^+ Ion

III. H_3^+ Ions Incident Upon Barium Surfaces

The material of this section is discussed in more detail in Refs. 6,7. We present here only a summary.

Molecular ion collisions with low work-function surfaces will allow for resonant captures into higher electronic states. These captures in turn will be Auger relaxed to provide an additional contribution to the ground state vibrational distribution. Low work-function collisions also allow for two additional mechanism for negative ion generation: Direct H^- production by electron capture to recombination product H atoms backscattering from the surface, and H^- production by dissociation of H_2^- ions formed from rebounding $H_2(v'')$ moving outward through the surface selvage.

The generation of H^- ions by H^+, H incident upon barium surfaces has been characterized experimentally by van Os et al.[4] In Fig. 3 their data is re-plotted for comparison with the product function

$$N(-) = a (H) \exp - v_\perp^0/v_\perp \qquad (1)$$

plotted against the outgoing perpendicular velocity component, $v_\perp$.

From this figure the asymptotic dependence on $v_\perp$ indicates a formation probability, a (H), equal to 0.3; the survival probability, given by the exponential factor, remains as large as 0.33 for perpendicular energy components as low as 2.0 eV.

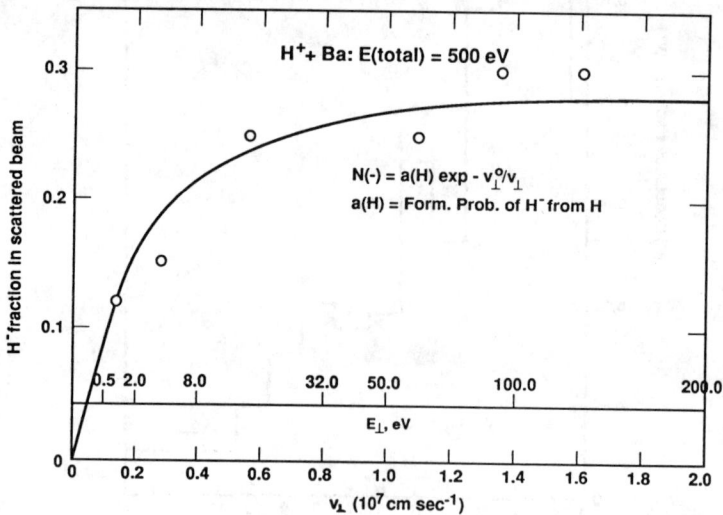

Figure 3: H^- generation vs outgoing normal velocity or energy, Eq.1.

In general one expects the energy level of a neutral atomic or molecular system to be raised as it experiences an increasing image field. Bruch and Ruijgrok[8] have shown however, that there is

In general one expects the energy level of a neutral atomic or molecular system to be raised as it experiences an increasing image field. Bruch and Ruijgrok[8] have shown however, that there is an insignificant level shift for image-plane separations $z - z_0 > 3a_0$, the range where resonant electron capture would be expected to occur. For the downward shift of the negative ion level we use the expansion,

$$\Delta E = -\frac{1}{4}\frac{1}{z - z_0} - \frac{1}{2!}\alpha\frac{1}{16(z - z_0)^4} - \frac{1}{4!}\beta\frac{1}{256(z - z_0)^8} + \cdots \quad (2)$$

The dipole and quadrupole polarizabilities, α, β for H^- are 215.5 and 7765.0, respectively.[9,10] No information is available for the polarizability of the H_2^- ion, but in a first approximation the H_2^- configuration is an $H^- + H$ configuration from the level crossing outward, and for this discussion we shall approximate the H_2^- level shift by Eq. (2) and using the H^- polarizabilities.

The energetics of the barium H_2^-, $H_2(v")$ system is illustrated isometrically in Fig. 4 where is shown the energy variations as a function of image-plane separation. The relative positions of the H_2^-, $H_2(v")$ potentials are shown in the right hand plane for infinite surface-molecule separation. At this separation the H_2^- level lies too high to allow resonant capture from the barium ($\emptyset = 2.7$ eV) to $H_2(v")$ to form H_2^-.

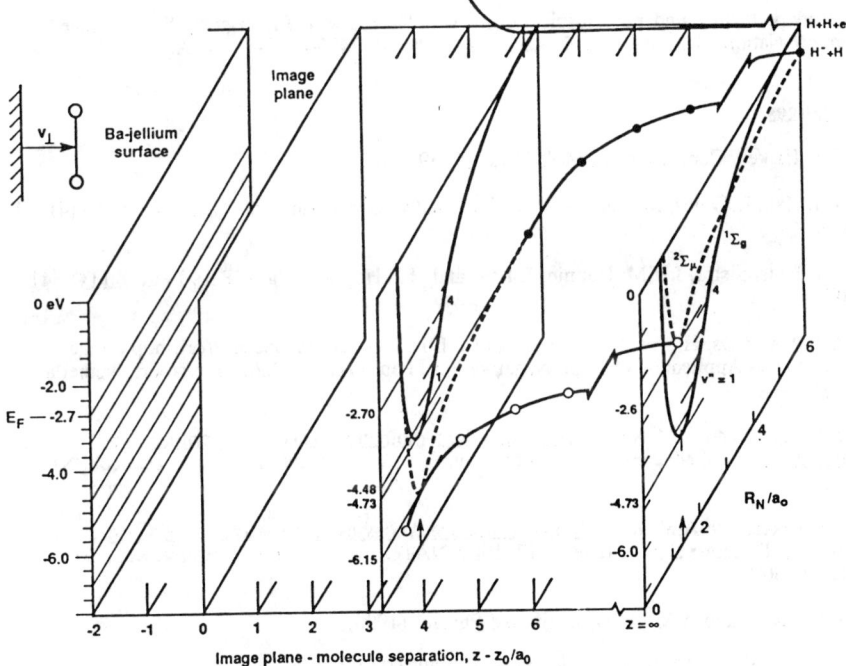

Figure 4: Energetics of the barium H_2^-, $H_2(v")$ system.

At an image-plane molecule separation of 3.2 a_0 the H_2^- configuration has shifted downward with respect to $H_2(v'')$ such that the asymptotic H_2^- potential lies as low as the $H_2(v'' = 0)$ level. At this separation resonant electron capture can occur from the barium to levels $v'' = 0, 1, 2$, and 3 of H_2, resulting in H_2^- dissociation into $H^- + H$.

We can now summarize the resonant capture possibilities for a 6 eV incident H_3^+ ion rebounding as 4 eV $H_2(v'')$ and 2 eV H fragments. In the initial recombination process 20% of the H_3^+ leads to 3H dissociation and 80% to $H + H_2$ to give a yield $1.4H/H_3^+$. From the van Os data of Fig. 3 we conclude that 10% of these atoms will appear as H^- ions to give $0.14 H^-/H_3^+$.

From Ref. 7 we have that 32% of the rebounding $H_2(v'')$ are in levels $v'' = 0, 1, 2$, and 3. These molecules are subject to H_2^- formation and dissociation in their flight from $z - z_0 = 2$ to 3.2 a_0. From Fig. 3 we find that 0.33 of these H^- survive to give an additional $0.10\ H^-/H_3^+$. The remaining flux is then $0.54\ H_3^+$. For those $H_2(v'' > 3)$, representing a fraction $0.48\ H_2/H_3^+$, no resonant capture to form H_2^- can occur.

One can now take inventory of the subsequent capture and dissociation processes for 6.0 eV incident H_3^+ on barium. The relative ratios $H : H_2 : H^-$ per incident H_3^+ ion are found to be 1.80:0.48:0.24. The new contribution offered by the low work-function barium surface is the $0.24\ H^-/H_3^+$. The H_2^- intermediate state has effectively converted the low-lying $H_2(v'')$ population into an additional source of H^- and H.

Work performed under the auspices of the U.S. Department of Energy by the Lawrence Livermore National Laboratory under contract number W-7405-ENG-48 and AFOSR-ISSA-89-0039.

References

1. J.R. Hiskes, Comments At. Mol. Phys. 19, 59 (1987).

2. R.I. Hall, I. Cadez, M. Landau, F. Pichou, and C. Schermann, Phys. Rev. Lett. 60 (4) 337 (1988).

3. P.J. Eenshuistra, J.H.M. Bonnie, J. Los, and H.J. Hopman, Phys. Rev. Lett. 60 (4) 341 (1988).

4. C.F.A. van Os, H.M. van Pinxteren, and J. Los, Proc. III European Workshop on the Production and Application of Light Negative Ions, Feb. 17-19, p. 266, Amersfoort, Netherlands (1988).

5. J.R. Hiskes and A.M. Karo, Dissociative Recombination: Theory, Experiment, and Applications, J.B.A. Mitchell and S. Guberman, Eds., World Scientific, p.204, Teaneck, NJ (1989).

6. J.R. Hiskes and A.M. Karo, Non-Equilibrium Processes in Partially Ionized Gases, Aquafredda di Maratea, Italy, June 14-17, 1989, NATO ASI M. Capitelli, Editor, Rept. No. UCRL-101309.

7. J.R. Hiskes and A.M. Karo, J. Applied Physics (1990).

8. L.W. Bruch and Th.W. Ruijgrok, Surf. Sci. 79, 509 (1979).

9. S.A. Adelman, Phys. Rev. A**5**, 508 (1972).
10. K.T. Chung and R.P. Hurst, Phys. Rev. **152**, 35 (1966).

DISCUSSION

Kleyn: On this did you put the autho-detachment of the H_2^- as it recedes from the surface.

Hiskes: The auto detachment of the H_2^-.

Kleyn: Because the H_2^- itself is unstable (I guess its normal lifetime is 10^{-15} seconds). So you have to stretch it considerably before it is stable and will eventually fall apart.

Hiskes: As you form the H_2^- it undergoes about a vibration or two, in the first vibration it moves out across the H - H_2 barrier, so there is a possibility of loosing the electron back to the metal. I think that is what you mean by the auto-detachment.

Kleyn: What I mean is if it recedes from the metal, first of all the electron can tunnel back then the next thing is that you have a H_2^- which didn't quite make it to H⁻ + H then that can shake up the electron and goes anywhere.

Hiskes: In free space it can drop off the electron leaving with a residual vibrationally excited molecule. But as you move in towards the surface and that is because the H_2^- level lies above the level of the H_2. As you move in towards the surface it now lies below the level of H_2 so the possibility of auto-detachment is no longer there.

Kleyn: But when the level comes back up?

Hiskes: By that time you have made the transition into H + H⁻.

Kleyn: So in that time the molecule has completely disassociated.

Hiskes: Right

ELECTRON ENERGY DISTRIBUTIONS AND VIBRATIONAL POPULATION DISTRIBUTIONS

J.R. Hiskes
Lawrence Livermore National Laboratory
University of California
Livermore, CA 94550

I. INTRODUCTION

This last year two experimental groups have reported on the vibrational population distributions of hydrogen molecules, $H_2(v'')$, generated in a hydrogen discharge.[1,2,3] These distributions extend up to the $v'' = 5$ vibrational level corresponding to the threshold excitation for significant H^- production. A comparison of theoretical modelling distributions[4] with the measured distributions of Stutzin et al.[1] has shown a qualitative agreement taking into account known excitation processes, and fair quantitative agreement when account was taken of the calculated wall relaxation parameters. In this paper these comparisons are extended to include the distributions of Eenshuistra et al., and provide further insights into the formation processes. A novel feature of the Eenshuistra distributions is a five-fold enhancement of the high v'' levels with increased discharge current. These data are shown in Fig. 1 for the "5 Amp" and "30 Amp" cases summarized in Table II, page 104, of Ref. 3. The sources of this enhancement are considered in this paper.

II. DISCHARGE CONFIGURATIONS

A schematic of the Eenshuistra multicusp hydrogen discharge[2,3,5] is shown in Fig. 2. Electrons are emitted from two filaments at -105 and -115 Volts respectively, pass through the discharge and are collected at the far end on a front plate that is maintained at -100 Volts. If an electron undergoes an excitation or ionization event in its passage through the discharge, it will lose sufficient energy to be reflected from the front plate and will be "injected" into the discharge. From the published system parameter one concludes that 30% of the electrons emitted from the 105 Volt filament are degraded in energy and "injected", and 19% from the second filament are injected.

Electrons injected into the discharge fall into one of two categories: they follow either adiabatic or non-adiabatic trajectories. The distinction between these two zones is determined by the adiabaticity parameter,

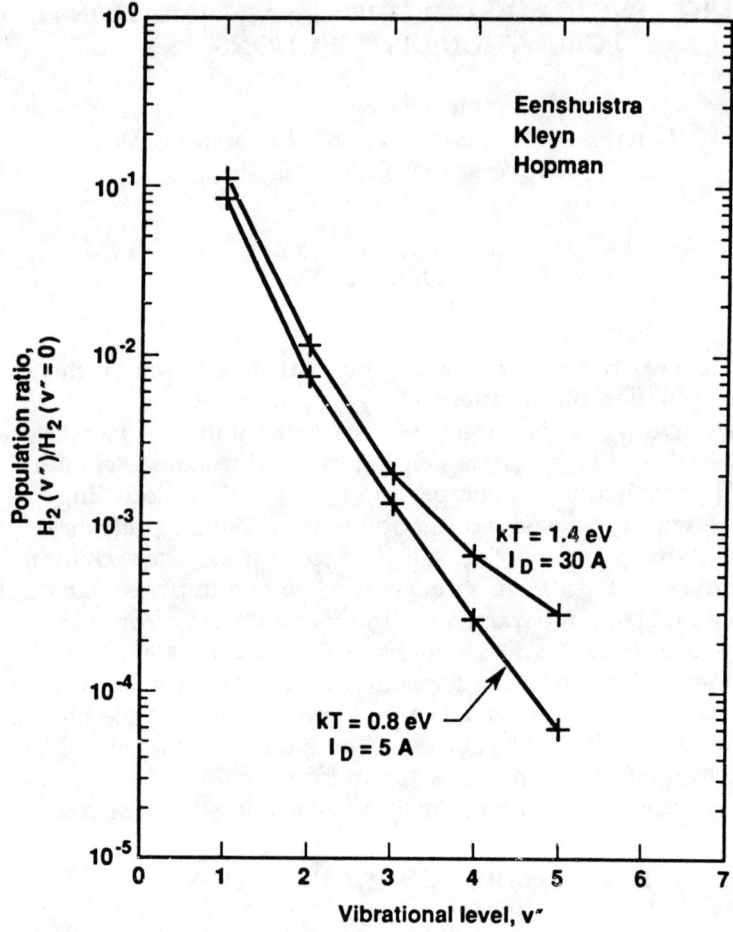

Fig. 1. Measured population distributions for I_D = 5A, 30A.

$$\frac{\rho}{B}\frac{dB}{ds} \quad <0.22, \text{ adiabatic zone}$$
$$>0.22, \text{ non-adiabatic zone} \qquad (1)$$

Here ρ is the local electron radius of curvature and dB/ds the local magnetic field gradient. The boundaries of these zones are quite distinct and have been extensively explored both computationally and analytically.[6,7,8,9]

In the adiabatic zone electrons are reflected from magnetic mirroring points and confined. Electron trajectories in the non-adiabatic zone however are not described by a simple algorithm, and these

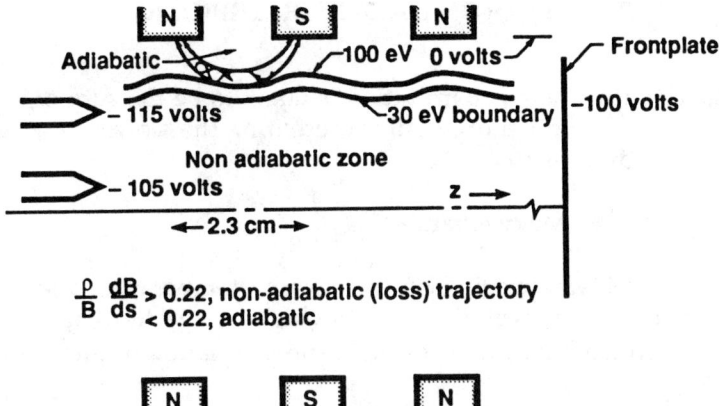

Fig. 2. Schematic of the Eenshuistra et al. discharge.

electrons, after one or two transits between magnets, will cross the boundary and can be lost to the magnet pole faces. The boundary of these two zones is located using the lowest-order solution for the magnetic multicusp configuration:

$$B_r = B_0 \frac{I_1\left(\frac{2\pi r}{l}\right)}{I_0\left(\frac{2\pi a}{l}\right)} \cos \frac{2\pi z}{l},$$

$$B_z = -B_0 \frac{I_0\left(\frac{2\pi r}{l}\right)}{I_0\left(\frac{2\pi a}{l}\right)} \sin \frac{2\pi z}{l}, \qquad (2)$$

The coefficient B_0 is the field intensity at the boundary (700 gauss) $r = a$, and l is the length of the multipole periodicity. The 30 eV and 100 eV electron boundaries obtained from (1) and (2) are shown on the figure.

When an energetic electron crosses the adiabatic boundary it sees a mirror ratio determined by the ratio of B_0 and the field value at the crossing point. Because this mirror ratio is quite large only a small fraction ~5%, of an isotropic or randomized angular distribution will fall in the loss cone and reach the pole tips; the remaining electrons will be reflected backwards across the boundary. Since the position of the boundary and therefore the apparent mirror ratio are functions of the electron energy, the loss cone will be energy dependent, with the solid angle of the loss cone increasing approximately as $E^{1/2}$.

III. ELECTRON ENERGY DISTRIBUTIONS

The electron energy distribution is taken to be the sum of a Maxwellian plus a high energy tail representing the slowing injected electrons, according to

$$f(E) = f(\text{Maxwellian}) + f_s. \qquad (3)$$

With the Maxwellian specified by the electron temperature and thermal density, we require an expression for the high-energy term, f_s. This term can be obtained from a first-order differential equation of the form[10,11]

$$\frac{\partial}{\partial E}\left(f_s \frac{dE}{dt}\right) = -\frac{J(E)}{eV} + \frac{A(E)}{V}. \qquad (4)$$

The energy loss rate, dE/dt, appearing in the left hand bracket is due to slowing of the fast electrons by collisions with the thermal electrons, atoms, and molecules, and H_2^+ and H_3^+ ions, respectively. This rate is given by

$$-\frac{dE}{dt} = B(E)\, n(e) + G(E)\, N_1 + H(E) N_2 + K(E) N_{2+} + M(E) N_{3+}. \qquad (5)$$

The electron loss coefficient, $B(E)$, is written

$$B(E) = 4.9 \times 10^{17} \frac{4\pi e^4 \ln\Lambda}{\sqrt{2mE}}\ \text{eV sec}^{-1}, \qquad (6)$$

where $\ln\Lambda$ is the coulomb logarithm.[12] The expression given in Ref. 11 for $B(E)$ is too small by the factor $\sqrt{2}$. The correct expression is given in the original derivation developed in Ref. 10. The remaining loss coefficients are shown in Fig. 3.

The first term on the right hand side Eq. (4) is the injection current density, $J(E)$ amperes eV^{-1}, injected into plasma volume, V. The second term is the non-adiabatic loss to the walls of the slowing electrons. This term can be shown to be equal to

$$A(E) = I_D \left(\frac{dE}{dt}\right)^{-1} \frac{v(\text{inj})}{22\, L} \left(\frac{E}{E(\text{inj})}\right)^{p+1/2} \qquad (7)$$

The quantities I_D, $v(\text{inj})$, $E(\text{inj})$ are the discharge current, injection velocity and energy, respectively, and L is a non-adiabatic trapping length between boundary crossings. The factor $(22)^{-1}$ is an isotropic loss-cone

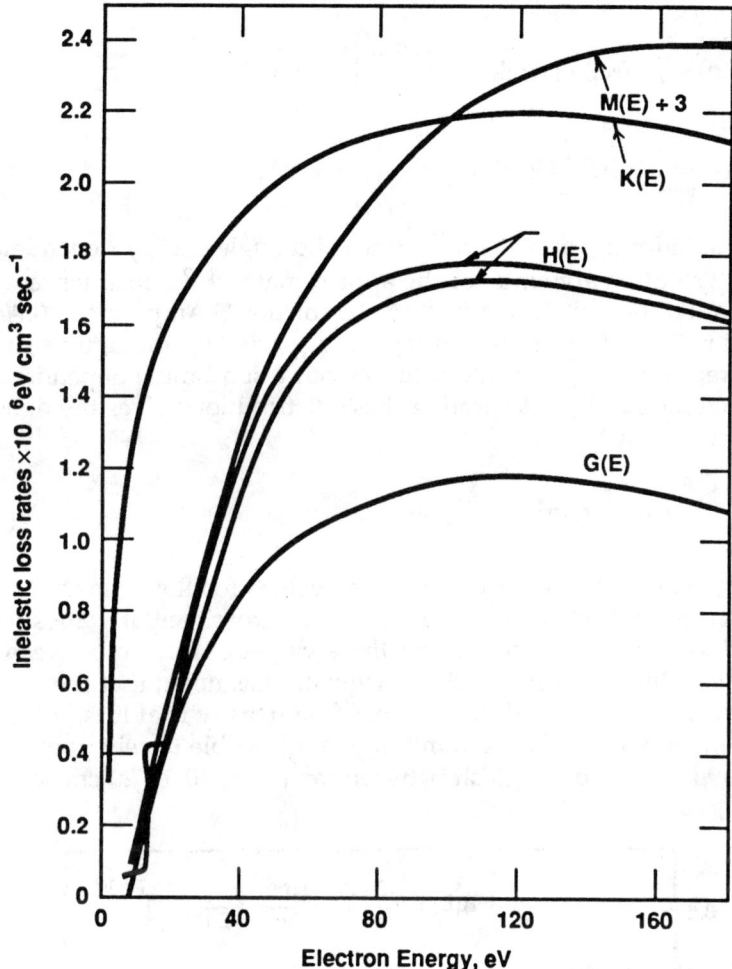

Fig. 3. Electron energy loss rate coefficients.

factor, and p is the exponent of the energy dependence for this factor. Integrating Eq. (4),

$$\frac{V}{I_D} f_S = -\left(\frac{dE}{dt}\right)^{-1} + \frac{v(inj)}{22\,L} \left(\frac{dE}{dt}\right)^{-2} g(E,p)$$

with

$$g(E,p) = \frac{2}{3} E(\text{inj}) \left[(E/E(\text{inj}))^{3/2} - 1 \right], \quad p = 0$$

$$= \frac{1}{2} E(\text{inj}) \left[(E/E(\text{inj}))^2 - 1 \right], \quad p = 1/2. \tag{8}$$

The p = 0 solution ignores the loss-cone solid-angle energy dependence, the p = 1/2 solution accounts for the approximate $E^{1/2}$ dependence. The solutions of Eq. (8) are shown in Figs. 4,5 for the "5 Amp" and "30 Amp" data listed in Table II, page 104, of Ref. 3. Inspection of the figures shows little difference in the p = 0, 1/2 solution, but a substantial dependence on the trapping length, L. Intergrating these distributions gives the density of fast electrons, n(f).

$$n(f) = \int f_s \, dE. \tag{9}$$

Comparing these integral values with the values for the primary electron densities given in Table II of Eenshuistra, the experimental values for n(f) are reproduced by the integral (9) for those values of L lying between the L = 5, 10 cm solutions of Figs. 4,5. Eenshuistra has noted that the uncertainty in the measured primary electron density is at least a factor of two. With this degree of uncertainty it is not possible to rely upon the experimental data to distinguish between the L = 5, 10, or 20 cm solutions.

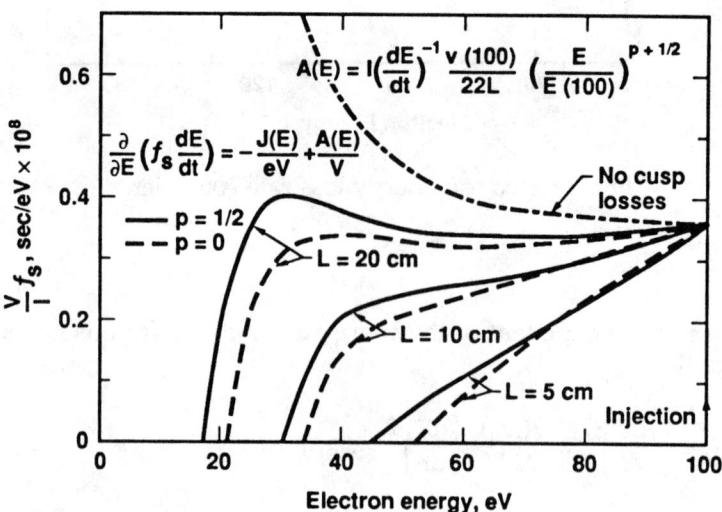

Fig. 4. High energy electron distribution functions for I_D = 5A.

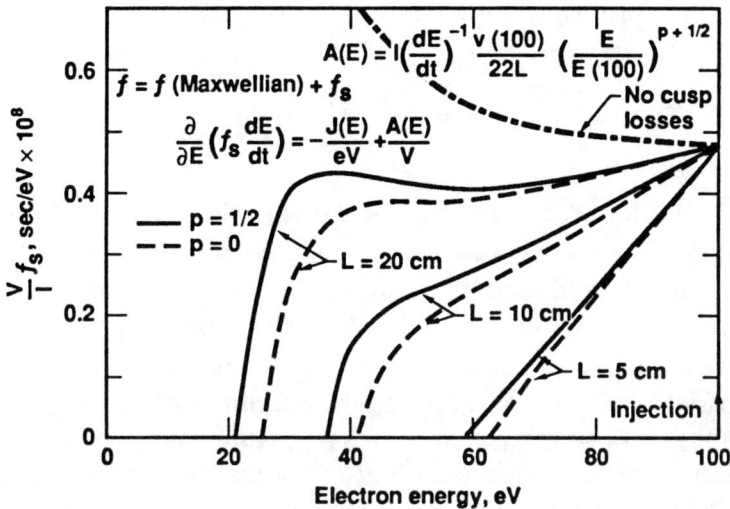

Fig. 5. High energy electron distribution functions for $I_D = 30$ A.

Also shown in the figures are the solutions for $A(E) = 0$, corresponding to no cusp losses, and appropriate to electrons confined to the adiabatic zone. These electrons are lost from the high energy distribution only through their being degraded in energy via inelastic collisions.

IV. VIBRATIONAL EXCITATION AND LOSS PROCESSES

The interpretation of the distributions shown in Fig. 1 is based upon the following four excitation processes:

1. The e-V process, proceeding through the H_2^- ($^2\Sigma_u$) resonance,[13-17]

$$e(\text{thermal}) + H_2(v'') \rightarrow H_2^- \rightarrow H_2(\underline{v}'') + e' \, . \tag{A}$$

2. The E-V process, proceeding through the electronically excited singlet states,[18]

$$e(\text{fast}) + H_2(v''=0) \rightarrow H_2^* + e' \rightarrow H_2(\underline{v}'') + e' + h\nu \, . \tag{B}$$

3. The S-V process, generated by H_3^+ wall collisions[19-21]

$$H_3^+ + \text{Surface} \rightarrow H_2(\underline{v}'') + H \, . \tag{C}$$

4. The A-V process, generated by atom recombination on the walls[22,23]

$$H + H/Surface \rightarrow H_2(v'') . \qquad (D)$$

The rates for process (A) are based on the Ehrhardt et al.[24] $o \rightarrow v''$ cross sections at low v'', and the theoretical cross sections of Mundel et al.[14] and Domcke and Mundel[15] at higher v'', but adjusted to fit smoothly onto the Ehrhardt cross sections at low v''. The Allen[25] cross sections, $0 \rightarrow v''$, are used as a guide for extrapolation to the high v'' levels. The Mundel-based rates have been divided by a factor of 2.3 for $v'' \geq 3$ to join with the Ehrhardt-based rates. In a first approximation the $0 \rightarrow v''$ cross sections can be parameterized as a function of the level energy difference and the $\underline{v}'' > 0 \rightarrow v''$ chains obtained from parameterization. The anharmonic form of the H_2 potential, however, has the effect of enhancing the cross sections of the higher chains. Bardsley and Wadehra[16] and Wadehra[17] have evaluated cross sections at specific energies for several members of these higher chains. These cross sections have been utilized to extend the first approximation results to provide a second-approximation matrix of cross sections. Using this matrix, the rates $\underline{v}'' \rightarrow v''$, have been evaluated for $kT = 0.8$ and 1.4 eV, appropriate to the "5 Amp" and "30 Amp" cases, and listed in double entry in Table I; the 0.8 eV value is listed above the 1.4 eV value. The rates in the table are given in units of 10^{-11} cm^{3}sec^{-1}, and are displayed through $v'' = 10$. It is evident from the table that principal excitations will occur near the diagonal.

The rates for process (B) are evaluated using the distribution functions shown in Figs. 4,5.

The rates for process (C) are taken equal to the flux of H_3^+ moving with the sound velocity in the direction of the anode surface, and averaged over the volume. Fractional contributions to the different levels, v'', are given by the $E(v'')$ factors discussed in the previous paper.[20] With R as the discharge radius, the formation rate is then equal to

$$+ N(3+) \sqrt{\frac{kT}{\pi N(3+)}} \frac{3}{R} E(v'') . \qquad (10)$$

Table I. The e-V rates used for this analysis.

v''/v'	1	2	3	4	5	6	7	8	9	10
0	130	6	.60	.10	.017	.0052	.0017	.0006	.0002	.00007
	240	19	1.9	.29	.065	.018	.006	.0021	.0007	.00025
1		360	33	2.7	.73	.20				
		570	61	5.2	1.4	.39				
2			706	87	9.3	3.1				
			940	140	12	4.9				
3				1100	210	30				
				1200	290	40				
4					1100	300				
					1400	400				
5						1000	300			
						1400	400			

There is not sufficient data available for process (D) to make a modelling comparison with the experimental vibrational distributions. The most complete information is given by Hall et al.[22] who show an $H_2(v'')$ distribution that is an equilibrium distribution caused by $H_2(v'')$ formation in equilibrium with wall relaxation. For this analysis the formation distribution itself is required.

The principal loss process for the $H_2(v'')$ is wall relaxation. This rate is taken to be

$$-N_2(v'') \alpha \frac{V(v'')}{R} \frac{1}{b(v'')}, \qquad (11)$$

with V(v") the mean velocity of a molecule with excitation v". The b(v") values have already been reported.[4] If the H_2(v") mean free path is short compared to the system dimension α has the value 0.75. If the mean free path is long α is approximately 1.8. For the discharge parameters appropriate to Eenshuistra's experiment the value for α is intermediate between these extremes but closer to the long mean-free-path limit. In this analysis two values for α are used, 1.8 and 1.3.

In the limit where only e-V processes are active and fast compared with wall relaxation rates the velocity V(v") is the mean gas velocity appropriate to the gas temperature. The gas temperature here is taken to be 540°K. Alternatively, when E-V and S-V rates are operative the excited molecules are heated in the wall relaxation process and the V(v") will be larger than indicated by the mean gas temperature. The wall relaxation process for the high v" is assumed to lead to an equipartition among the rotational, translational, and vibrational temperatures, and the sum of contributions from all higher levels leads to a new mean value for V(v"). The new mean velocity can be written V(v") = β(v") V_0(v"), where V_0 in the 540°K value and the β(v") values are listed in Table II below.

Table II.

v''	1	2	3	4	5	6	7	8	9	10	11	12	13	14
β(v')	1.1	1.3	1.8	2.0	2.0	1.8	1.6	1.5	1.4	1.3	1.2	1.2	1.1	1.0

With these input values together with other secondary processes mentioned already in Refs. 26,27, the set of coupled rate equations for the fourteen bound vibrational levels is solved using the kT = 0.8 and kT = 1.4 eV input data given in Table II, p. 104, of Eenshuistra.[3]

V. COMPARISON OF THE MODELLING AND EXPERIMENTAL VIBRATIONAL DISTRIBUTIONS

In Fig. 6 the kT = 1.4 experimental data is compared with the modelling results taking into account only the e-V excitation process. The model results are shown in double entry corresponding to the two values for α. The e-V process alone gives a distribution that is approximately a factor of two to three lower than the experimental data, although the trend is similar. For this comparison β is taken equal to unity for all levels. These calculated vibrational distribution with e-V only have been compared for the kT = 0.8 eV data (not shown) and the

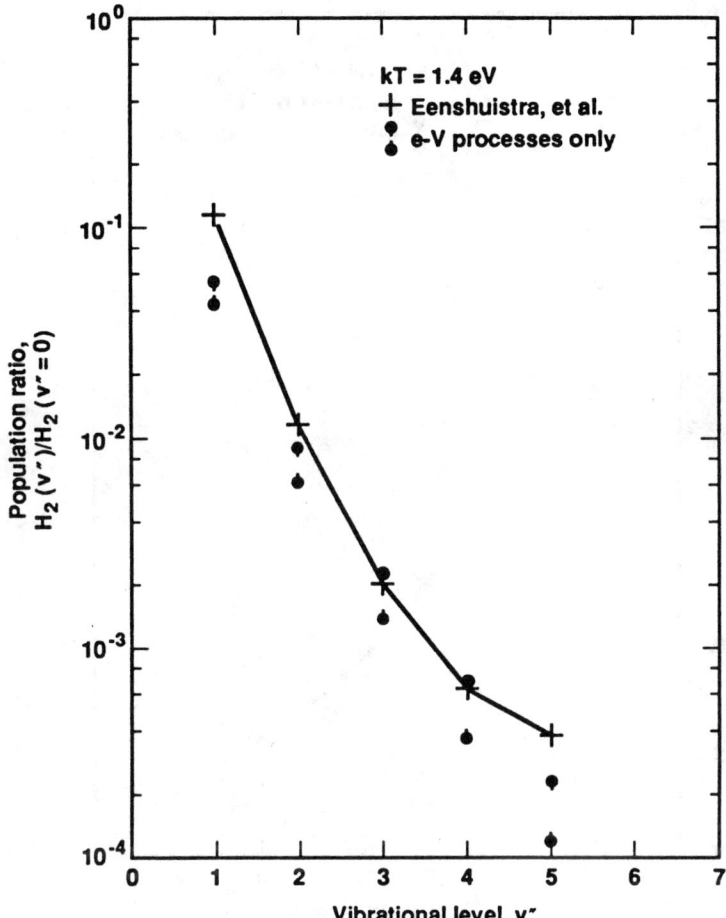

Fig. 6. Exp. and calc. distributions, kT = 1.4 eV, e-V processes only.

1.4 eV data. At kT = 0.8 eV the ratio for v" = 5 is quite small and is enhanced a factor of twenty-three for the 1.4 eV data. This large enhancement when compared with enhancement shown in Fig. 1 is further evidence that the e-V process alone is not the principal source of excitation near v" = 5.

In Fig. 7 both e-V and E-V processes are included, and the β values of Table II are used. The new β values tend to lower the calculated ratios but the effect of the E-V excitations is to bring the calculated ratios almost into coincidence with the experimental data for the higher levels. The v" = 1 ratio remains a factor of two or three too low.

In Fig. 8 both the e-V and the H_3^+ S-V process are included. The H_3^+ density is taken to be 60% of the thermal electron density as assumed

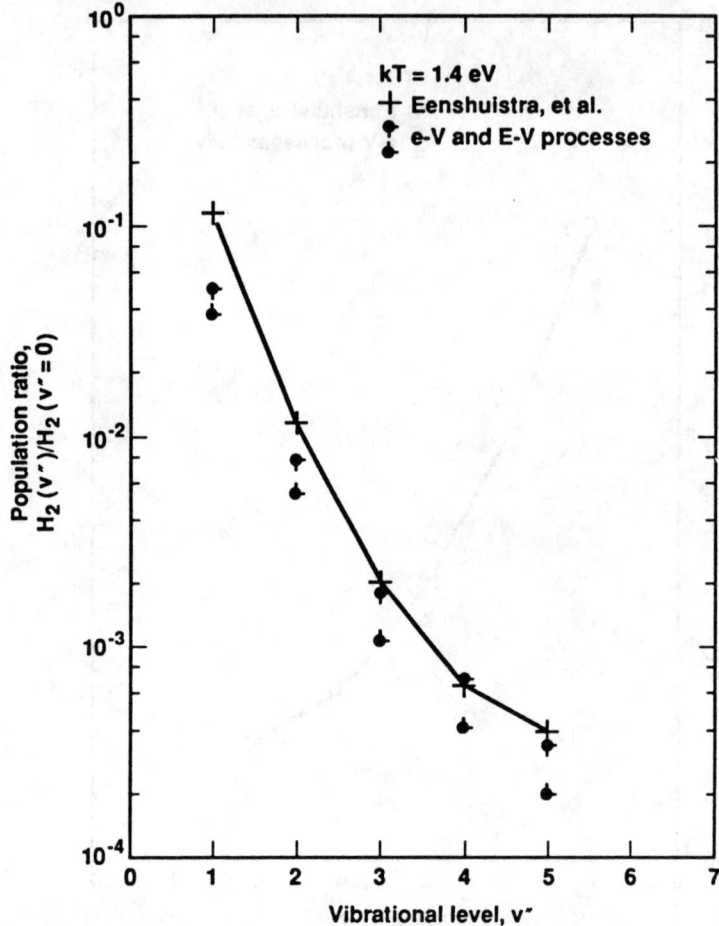

Fig. 7. Exp. and calc. distributions, kT = 1.4 eV, e-V and E-V processes.

in Ref. 3. Inspection of the figure and comparing with Fig. 7 shows that the S-V process has a more pronounced effect at raising the higher v'' portion of the spectrum than does the E-V process. The e-V still dominates at the lower levels and the $v'' = 1$ ratio remains a factor of two or three too low.

The e-V, E-V, and S-V process are included in the solutions shown in Fig. 9. The calculated ratios for the uppermost levels now lie well above the experimental distribution and the slope of the calculated distribution is somewhat more shallow. Eenshuistra has specified on uncertainty in the quoted fast electron density of more than a factor of two. To account for this uncertainty there is shown on the figure solutions for either double or half the primary (fast) electron density,

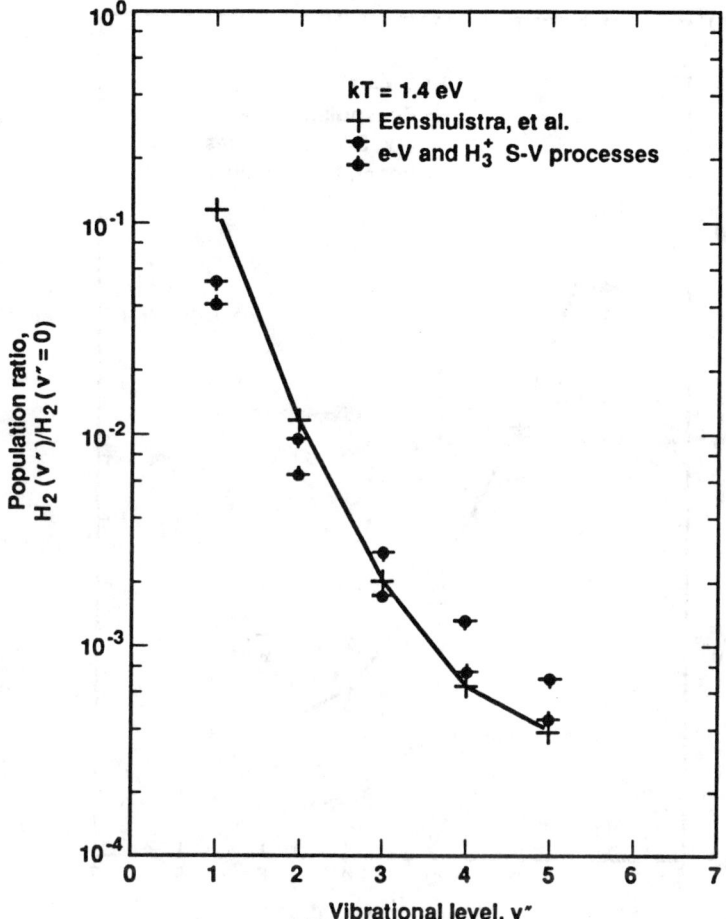

Fig. 8. Exp. and calc. distributions, kT = 1.4 eV, e-V and S-V processes.

these solutions indicated by the horizontal bars. No uncertainties are quoted for the thermal electron density deduced from the probe measurements, but if one were to assign a plausible uncertainty value of thirty percent the lower limits for $v'' = 4,5$ would come into coincidence with the experimental values.

The $v'' = 1$ population is scarcely affected by the E-V or S-V precesses. The question remains as to the role of the atom-wall recombination, process (D), in affecting the population of the $v'' = 1$ level. As aforementioned, only equilibrium data is available, there being no explicit data for the formation distribution. Hall et al.[22] deduce a very slow wall relaxation rate for their cold gas, cold wall system, and there is some implication in their data that the wall relaxation changes slowly with v'' across the spectrum. In this case the shape of the equilibrium distribution will approximate the shape of the formation distribution.

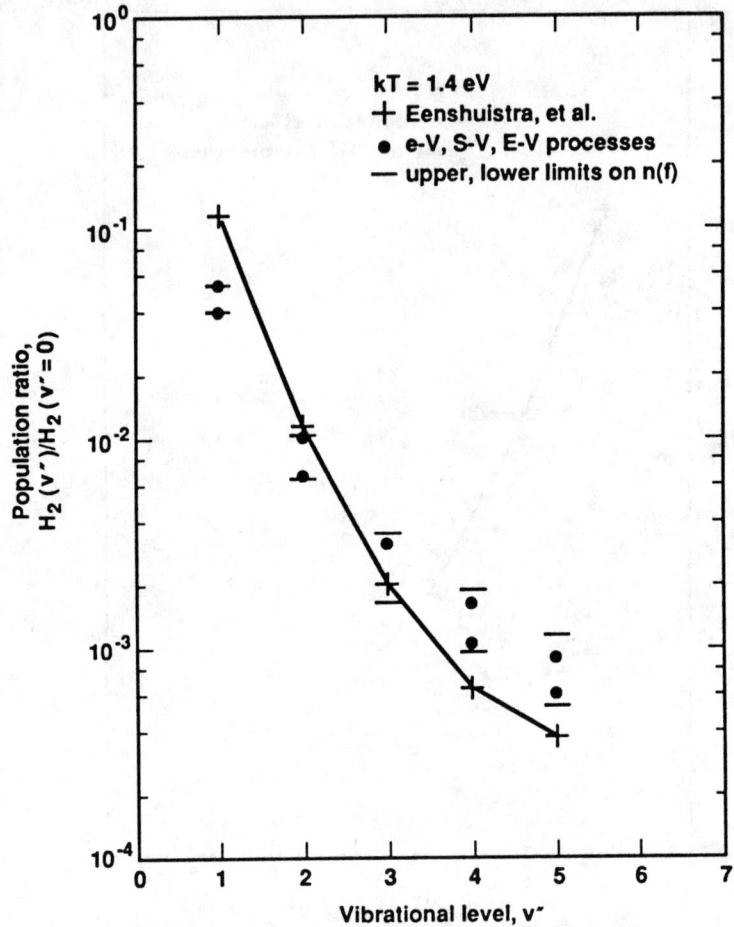

Fig. 9. Exp. and calc. distributions, kT = 1.4 eV, e-V, E-V, and S-V processes.

The 2800° tungsten distribution shown in Ref. 22 has been employed in the modelling code and the distribution shifted to obtain a fit to the $v'' = 1$ ratio. A fit is obtained if process (D) is normalized such that 1-2% of the atom flux given in Table II of Ref. 3 is converted to $H_2(v'' = 1)$ molecules. This new distribution is not displayed here since it represents a fitting procedure rather than a comparison of theoretical and experimental distributions.

A similar sequence of calculations has been performed using the "5 Amp" data with kT = 0.8 eV. The distribution including e-V, S-V, and E-V processes is shown in Fig. 10. The discrepancy at $v'' = 1$ is larger than for the kT = 1.4 eV case. Again fitting the $v'' = 1$ ratio with process (D), with the distribution normalized such that 5-10% of the atom flux of

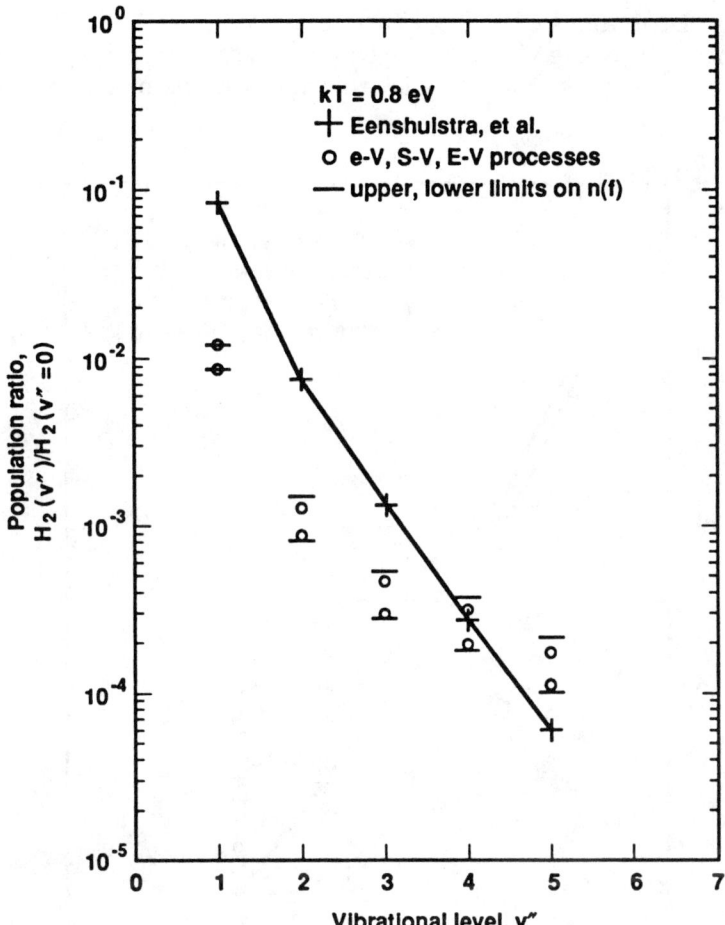

Fig. 10. Exp. and calc. distributions, kT = 0.8 eV, e-V, S-V, and E-V processes.

Table II, Ref. 3 is converted to $H_2(v'' = 1)$, one finds that this is sufficient to provide an approximate fit for the $v'' = 2,3$ levels as well.

The atom flux given in Table II of Ref. 3 differ for the 0.8 and 1.4 eV cases. When this difference is taken into account the fitting to $v'' = 1$ is accomplished for the same total atom flux in either case. Also, only a small fraction of the atoms present are required to fit the low v'' portion of the distribution. There is a hint here, but as yet not a demonstration, that some sub-group of the total atomic population may be populating the lower spectrum. This is all the more plausible if this sub-group is generated by the hot filaments and is little-changed for the two discharge currents.

The distributions of Figs. 9,10 are compared in Fig. 11. The five-fold enhancement of the $v'' = 5$ ratios with increased discharge current as displayed in the experimental data is reproduced in the modelling results.

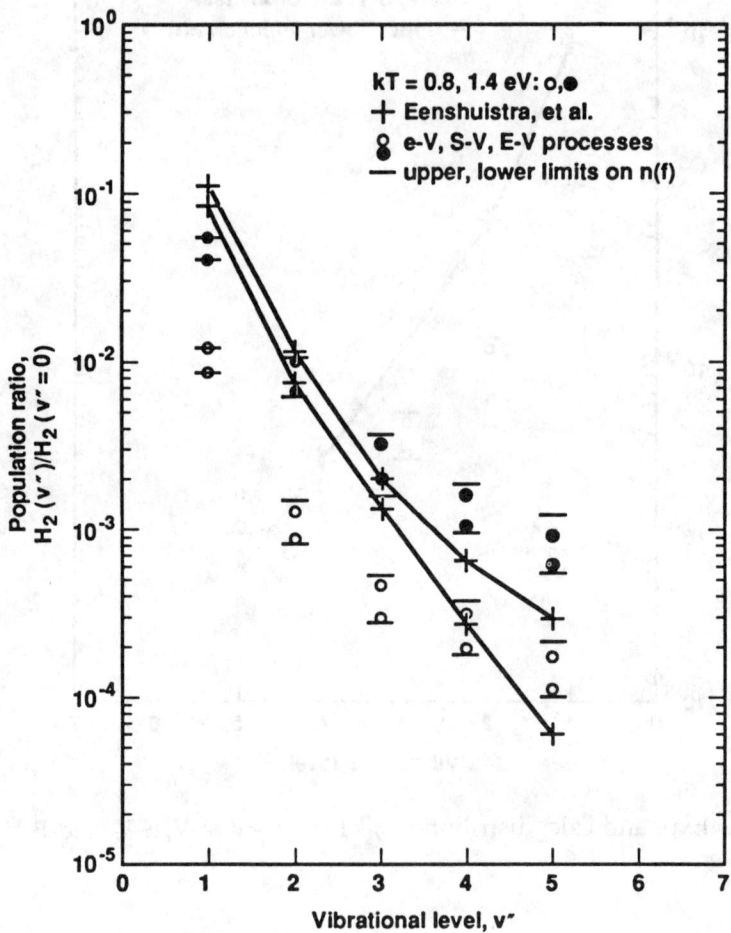

Fig. 11. Comparison of 0.8 eV and 1.4 eV distributions.

VI. CONCLUSIONS

The comparison of the theoretical modelling distributions with the experimental population distributions of Eenshuistra et al. exhibits qualitative agreement. Taking into account plausible uncertainties in the experimental data, these comparisons are in quantitative agreement over most of the spectrum. The observed five-fold enhancement with discharge current is generated by a combination of E-V and S-V processes. For this particular discharge the H_3^+ S-V process makes a greater contribution to the upper portion of the vibrational spectrum than does the fast-electron E-V process. Atom-wall recombination processes may make an important contribution to the populations of the lower lying levels.

ACKNOWLEDGEMENT

Work performed under the auspices of the U.S. Department of Energy by the Lawrence Livermore National Laboratory under contract number W-7405-ENG-48 and AFOSR-ISSA-89-0039.

REFERENCES

1. G.C. Stutzin, A.T. Young, A.S. Schlachter, and W.B. Kunkel, Chem. Phys. Letters 155, 475 (1989).
2. P.J. Eenshiustra, A.W. Kleyn, and H.J. Hopman, Europhys. Lett. 8(5), 423 (1989).
3. P.J. Eenshiustra, "Vibrational Excitation In A Hydrogen Volume Source," Thesis, FOM Institute, Amsterdam (1989).
4. J.R. Hiskes and A.M. Karo, J. Appl. Phys. Lett. 54(6), 508 (1989).
5. A.W. Kleyn, Private Communication.
6. C.S. Leffel, Jr. and E.P. Gray, Phys. Fluids 12, 1008 (1969).
7. J.E. Howard, Phys. Fluids 14, 2378 (1971).
8. R.H. Cohen, G. Rowlands, and J.H. Foote, Phys. Fluids 21, 627 (1978).
9. R.H. Cohen, Intrinsic Stochasticity In Plasmas, p. 181, G. Laval and D. Gresillon Eds. (Editions de Physique, Orsay, 1979).
10. J.R. Hiskes and A.M. Karo, "Electron Energy Distributions, Vibrational Population Distributions, and Negative Ion Concentrations in Hydrogen Discharges," presented at the NATO Advanced Study Institute on Atomic and Molecular Processes in Thermo. Research, Palermo, Italy, July 19-30, 1982, Rept. No. UCRL-87779, June 1982.
11. J.R. Hiskes and A.M. Karo, J. Appl. Phys. 56, 1927 (1984).
12. L. Spitzer, Jr., Physics of Fully Ionized Gases (Interscience, New York, 1956).
13. J.M. Wadehra, Phys. Rev. A29, 106 (1984).
14. C. Mundel, M. Berman, and W. Domcke, Phys. Rev. A32, 181 (1985).

15. W. Domcke and C. Mundell, Invited Papers of the XIV Int. Conf. on Physics of El. and Atomic Coll. Palo Alto, CA, 24-30 July, 1985, North Holland Amsterdam (1986) p. 195.
16. J.N. Bardsley and J.M. Wadehra, Phys. Rev. A 20, 1398 (1979).
17. J.M. Wadehra, Private Communication.
18. J.R. Hiskes, J. Appl. Phys. 51, 4592 (1980).
19. J.R. Hiskes and A.M. Karo, Dissociative Recombination: Theory, Experiment, and Applications, World Scientific, Teaneck, NJ (1989) p. 204.
20. J.R. Hiskes and A.M. Karo, Previous Paper, this Symposium.
21. J.R. Hiskes and A.M. Karo, J. Appl. Phys. (1990).
22. R.I. Hall, I. Cadez, M. Landau, F. Pichou, and S. Schermann, Phys. Rev. Lettr. 60(4) 337 (1988).
23. P.J. Eenshuistra, J.H.M. Bonnie, J. Los, and H.J. Hopman, Phys. Rev. Lettr. 60(4) 341 (1988).
24. H. Ehrhardt, D.L. Langhans, F. Linder, and H.S. Taylor, Phys. Rev. 173, 222 (1968).
25. M. Allen, J. Phys. B18, L451 (1985).
26. J.R. Hiskes and A.M. Karo, J. Appl. Phys. 56(7), 1927 (1984).
27. J.R. Hiskes, A.M. Karo, P.A. Willmann, J. Appl. Phys. 58, 1759 (1985).

DISCUSSION

Bacal: What about the negative ions?

Hiskes: We haven't had time to model negative ions in this particular case.

Bacal: Thank you.

Jacquot: I don't understand exactly what is the reason for which we increase the discharge current and you increase the production of very high v. That means you produce more H_3^+.

Hiskes: If you raise the discharge current you raise the thermal electron density, you raise the thermal electron temperature, you raise the number of electrons in the upper end of the electron energy distribution.

Jacquot: Why the change of the ground charging ions? What are the losses for the first level.

Hiskes: You mean the first level, you don't have as many fast electrons contributing to the upper end of the spectrum.

Jacquot: That means the increase is due to the fast electrons.

Hiskes: No, there is a secondary effect, as you raise the discharge currents you also raise the density of the thermal distribution, the thermal distribution contributes to the upper portion of the spectrum with little eV process, so that all the mechanisms here are enhanced in the higher power discharge, and one just has to look at each contribution in detail to see what happens.

DISSOCIATIVE ELECTRON ATTACHMENT TO LIGHT MOLECULES: A COMPARATIVE STUDY OF H_2, LiH AND Li_2

H. H. Michels
United Technologies Research Center
East Hartford, Connecticut 06108

and

J. M. Wadehra
Department of Physics and Astronomy
Wayne State University, Detroit, Michigan 48202

ABSTRACT

In this paper we have compared the energetics of three light molecules (H_2, LiH and Li_2) to form negative ions by the process of dissociative electron attachment. For two light homonuclear molecules (H_2 and Li_2) we have done explicit calculations, using the local formalism of the resonance model, to obtain the electron attachment cross sections and the rates of negative ion formation.

INTRODUCTION

Much has been said in the past[1] about the formation of H^- via the process of dissociative attachment of low energy electrons to molecular hydrogen. It has also been well established that the cross sections for dissociative electron attachment to H_2 are quite significantly enhanced if the molecule H_2 is initially rovibrationally excited. The purpose of the present paper is to provide a *comparative study* of the energetics of a few light molecules (including H_2) for forming negative ions by the process of dissociative electron attachment. Such a comparative study will explicitly show the similarities among several molecules and will, thus, provide a general insight about the degree of enhancement of negative ion formation when the molecule under investigation is rovibrationally excited. Furthermore, these investigations will have practical utility since other light negative ions (such as Li^- ions) could possibly play, in the future, the same roles as have been played by H^- ions for neutral beam formation. Finally, the observation that the production rate of H^- is enhanced in discharge type negative ion sources by seeding an alkali (Cs) into the system suggests that now is the appropriate time to investigate dissociative attachment to alkali hydrides.

ENERGETICS OF H_2, LiH AND Li_2

The three molecules H_2, LiH and Li_2 are isovalent and, therefore, exhibit similarities in their electronic configurations. For example, the two homonuclear molecules H_2 and Li_2 in the ground state have configurations $(1\sigma_g)^2$ and $(1\sigma_g)^2 (1\sigma_u)^2 (2\sigma_g)^2$, respectively. Each of these two states has symmetry $^1\Sigma_g^+$ and these two states dissociate into $H(^2S) + H(^2S)$ and $Li(^2S) + Li(^2S)$, respectively. The heteronuclear molecule LiH, on the other hand, has the lowest energy state that dissociates into

Li(^{2}S) + H(^{2}S) and has configuration which in the united-atom limit can be expressed as $(1s\sigma)^2 (2s\sigma)^2$ with symmetry $^1\Sigma^+$.

The lowest electronic states of the negative molecular ions with configuration $(1\sigma_g)^2 (1\sigma_u)^2 (2\sigma_g)^2 (2\sigma_u)$ for Li_2^- and $(1\sigma_g)^2 (1\sigma_u)$ for H_2^- also have the same symmetry, namely, $^2\Sigma_u^+$. However, compared to the hydrogen molecule, the lithium molecule possesses a large polarizability and a weak bond strength which makes^2 the ground state of Li_2^- a true bound state for all values of the internuclear separation R. In the case of H_2^-, on the other hand, the $^2\Sigma_u^+$ state is a true bound state only for internuclear separations larger than 2.9 a.u. and is an autodetaching state for smaller values of R. The first excited state of the negative molecular ions Li_2^- and H_2^-, each with symmetry $^2\Sigma_g^+$ and configurations $(1\sigma_g)^2 (1\sigma_u)^2 (2\sigma_g) (2\sigma_u)^2$ for Li_2^- and $(1\sigma_g) (1\sigma_u)^2$ for H_2^-, is a partly Feshbach and a partly shape resonance in nature. For both H_2 and Li_2, the potential curve of the $^2\Sigma_g^+$ resonance crosses the potential curve of the ground X $^1\Sigma_g^+$ state of the corresponding neutral molecule such that the $^2\Sigma_g^+$ state becomes a true bound state for internuclear separations larger than 5.81 a.u. for H_2^- and 6.50 a.u. for Li_2^-. This $^2\Sigma_g^+$ state is the essential channel for dissociative attachment of low energy electrons to lithium molecules. Similarly, the ground state of LiH^-, that dissociates into Li(^{2}S) + H$^-$(^{1}S), has configuration $(1s\sigma)^2 (2s\sigma)^2 (2p\sigma)$ and symmetry $^2\Sigma^+$. This state of LiH^-, analogous to the ground state of Li_2^-, is a true bound state^3 for all values of the internuclear separation R. The first excited state of LiH^- dissociates into Li^-(^{1}S) + H(^{2}S) and is the lowest energetic channel for dissociative electron attachment to LiH. This state of LiH^- is a true bound state only for internuclear separations larger than 5.58 a.u. and is an autodetaching resonance for smaller values of R. Incidentally, the electron affinities (EA) of atomic hydrogen and atomic lithium are 0.75 eV and 0.62 eV, respectively.

Figure 1 shows the potential curves of the ground states of the molecules H_2, LiH and Li_2 as well as the potential curves of the states of corresponding molecular anions that are primarily responsible for dissociative electron attachment to these molecules. As one progresses from H_2 to LiH to Li_2 the equilibrium internuclear separation increases from 1.40 a.u. to 3.08 a.u. to 5.06 a.u. and the dissociation energy (D) of the molecule decreases from 4.73 eV to 2.33 eV to 1.06 eV. Furthermore, the energy spacing between the same two low-lying adjacent vibrational levels *decreases* as one proceeds from H_2 to LiH to Li_2, in such a manner that the total number of bound vibrational levels *increases* on increasing the molecular mass. These energetics have different quantitative effects on the threshold for dissociative electron attachment to these molecules as well as on the factor by which the attachment cross section is enhanced on vibrationally exciting these molecules. The threshold energy for dissociative attachment to a molecule that is rovibrationally excited to a particular level (v, J) is $E_{th}^{DA} = D - EA - E_{vJ}$ if $E_{vJ} < D - EA$, and $E_{th}^{DA} = 0$ otherwise.

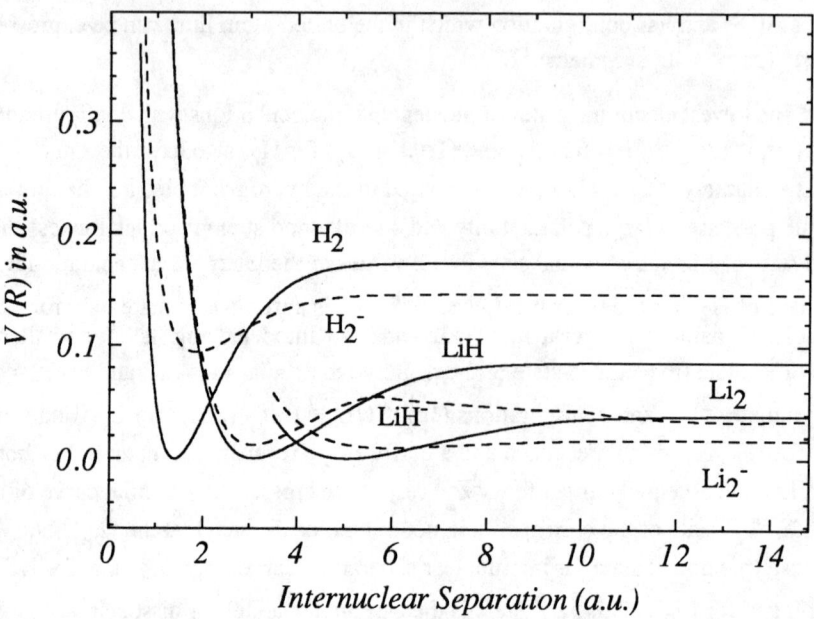

Figure 1. The potential curves (solid lines) of the ground states of the molecules H_2, LiH and Li_2 as well as the potential curves (dashed lines) of the states of corresponding molecular anions that are primarily responsible for dissociative electron attachment to these molecules.

DISSOCIATIVE ATTACHMENT TO H_2 AND Li_2

The process of dissociative electron attachment to a molecule is understood[4] to proceed via an intermediate formation of the electron-molecule resonance that is capable of autodetaching the temporarily bound electron with a finite lifetime (related to the width, Γ, of the resonance). The potential curves of the neutral molecule (H_2 or Li_2) and of the resonant state (H_2^- or Li_2^-) cross at an internuclear separation $R = R_s$ such that, for $R \geq R_s$, the autodetachment of the electron is energetically not possible and the resonance turns into a stable bound state. Essentially, then, the cross section for dissociative electron attachment is determined by the asymptotic value of the radial nuclear wave function $\xi(R)$ of the resonant state (assuming energy normalized continuum functions):

$$\sigma_{DA} = \frac{\pi}{k_i^2} \left\{ \frac{2\pi \hbar^2 K}{M} \lim_{R \to \infty} |\xi(R)|^2 \right\}$$

Here $\hbar K/M$ is the relative velocity of the ion-atom pair after the attachment process. Note that the 'geometrical' cross section π/k_i^2 provides an upper bound[5] to the attachment cross section since the quantity within the curly parenthesis can be interpreted as a product of two probabilities, one for the capture of an electron to form the resonant state and the other (called the survival probability) for the nuclei in the resonant state to separate out to the stabilization radius R_s. Using a semiclassical analysis the cross section for dissociative electron attachment, σ_{DA}, can also be written as[6] a product of two factors:

$$\sigma_{DA} = \sigma_{cap} S.$$

The first factor, σ_{cap}, is interpreted as the cross section for the formation of a resonant anion state by the capture of the incident electron and the second factor S (the survival probability) is interpreted as the probability that the separation of the nuclei in the resonant anion state increases up to the stabilization radius R_s without electron autodetachment having occurred. This separation ensures the process of dissociative electron attachment to occur. The survival probability S, in the semiclassical analysis, is approximated by

$$\exp(-\bar{\Gamma}\tau/\hbar)$$

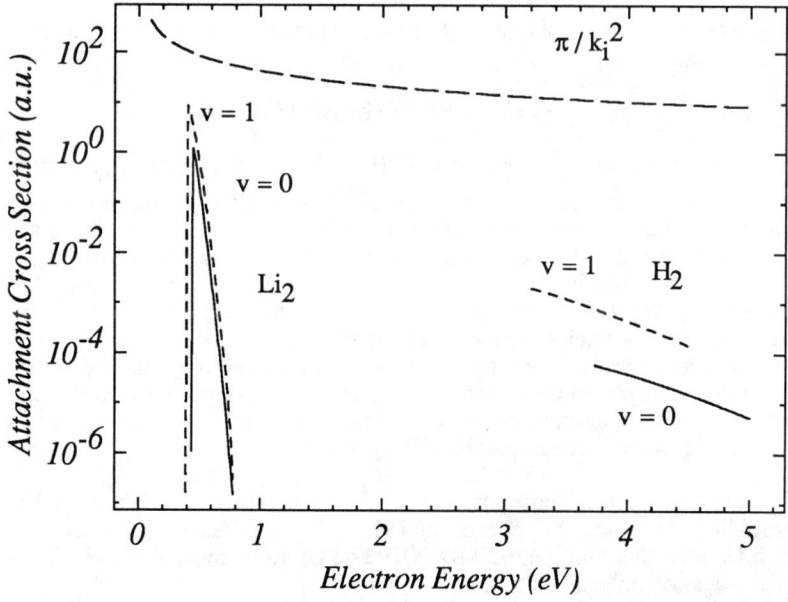

Figure 2. Cross sections for dissociative attachment to H_2 as well as to Li_2 when the molecules are initially in vibrational levels $v = 0$ and $v = 1$. π/k_i^2 represents the upper bound to the attachment cross sections.

where τ, the time taken by the nuclei to separate after resonance formation to the stabilization radius R_s, is directly proportional, by simple kinematical considerations, to $M^{1/2}$. $\bar{\Gamma}$ is the 'average' width of the resonance. As a measure of $\bar{\Gamma}$ we note that the values of the actual width of various resonant states, at the equilibrium internuclear separation of the neutral molecule, are[2,7]:

$$\Gamma(H_2^- ; {}^2\Sigma_u^+ ; R = 1.40 \text{ a.u.}) = 0.306 \text{ a.u.}$$
$$\Gamma(H_2^- ; {}^2\Sigma_g^+ ; R = 1.40 \text{ a.u.}) = 0.0612 \text{ a.u.}$$
$$\Gamma(Li_2^- ; {}^2\Sigma_g^+ ; R = 5.06 \text{ a.u.}) = 0.00223 \text{ a.u.}$$

Now, since $\bar{\Gamma}$ is smaller for Li_2^- than for H_2^-, the survival probability as well as the cross section for dissociative electron attachment would be expected to be larger for Li_2 than for H_2. Moreover, the lower energy threshold for dissociative electron attachment to Li_2 (essentially because of the lower dissociation energy for Li_2) would further enhance the cross section for attachment to Li_2 over that for H_2.

Figure 2 explicitly *compares* the cross sections for dissociative attachment to H_2 as well as to Li_2 and the factors by which the attachment cross sections are enhanced when the molecules H_2 and Li_2 are initially vibrationally excited from $v = 0$ to $v = 1$. As expected, the cross sections for attachment to Li_2 are larger than the cross sections for attachment to H_2. The factor by which the peak attachment cross section is enhanced on vibrationally exciting the molecule from $v = 0$ to $v = 1$ is 33.8 for H_2 and 6.8 for Li_2. The upper bound to the attachment cross section (namely, π/k_i^2) is reached when the process of dissociative electron attachment becomes exoergic for either molecule.

FUTURE POSSIBILITIES

The radial nuclear wave function $\xi(R)$ of the resonant state, in the most general theoretical description, satisfies[4] an integrodifferential equation with nonlocal potentials. If either the electron energy is large or the vibrational spacing is small, then the integrodifferential equation can be approximated by an ordinary differential equation with local potentials. The cross sections for dissociative electron attachment to H_2 and Li_2 presented above were obtained using the local formalism. In the immediate future we intend to investigate the effect of nonlocal potentials on the cross sections for electron attachment to molecular hydrogen and lithium. We are also planning to investigate in the future, using the insights gained from the present comparative study, the formation of negative ions from other molecules containing atomic hydrogen such as LiH, NaH, CH, SiH etc.

It is a pleasure to thank Mr. Dale E. Atems and Professor Walter E. Kauppila for providing assistance. Support of the US Air Force Office of Scientific Research through Contract Number F49620-89-0019 and Grant Number AFOSR-87-0342 is gratefully acknowledged.

REFERENCES

1. J.M. Wadehra, Phys. Rev. **A29**, 106 (1984).
2. H.H. Michels, R.H. Hobbs and L.A. Wright, Chem. Phys. Lett. **118**, 67 (1985).
3. B. Liu, K. O-Ohata and K. Kirby-Docken, J. Chem. Phys. **67**, 1850 (1977).

4. J. M. Wadehra, in *Nonequilibrium Vibrational Kinetics*, edited by M. Capitelli (Springer - Verlag, Heidelberg, 1986), p. 191.
5. J. P. Gauyacq, *Dynamics of Negative Ions* (World Scientific, Singapore, 1987).
6. J. N. Bardsley, A. Herzenberg and F. Mandl, in *Proc. 3rd Int. Conf. Atomic Collisions* (North-Holland, Amsterdam, 1964), p. 415.
7. J. N. Bardsley and J. M. Wadehra, Phys. Rev. **A20**, 1398 (1979).

DISCUSSION

Hiskes: Can you comment on the effects of the choice of the local potential versus the non-local potential and what influence it will have on the little eV rates.

Wadehra: eV rate is going to be mentioned in my second talk. And in that connection I am going to talk about the local vs. non-local equation, which I didn't quite show you in this talk.

ISOTOPE EFFECT IN VIBRATIONAL EXCITATION OF H_2 BY LOW ENERGY ELECTRON IMPACT

D. E. Atems and J. M. Wadehra
Department of Physics and Astronomy
Wayne State University, Detroit, Michigan 48202

ABSTRACT

Cross sections for the vibrational excitation of molecular hydrogen and its five heavier isotopes by the impact of low energy electrons are calculated using both the local and the nonlocal versions of the resonance model. It is demonstrated that, for a given incident electron energy, the cross sections for the vibrational excitation of heavier isotopes can be obtained from those of molecular hydrogen by a simple scaling procedure. It is seen that the scaling law holds for cross sections with values ranging over eight orders of magnitude.

The vibrational excitation of a molecule (such as H_2) occurs via the formation of an intermediate resonant state (H_2^-) which can either autodetach the temporarily bound electron (and, then, leaves behind a vibrationally excited molecule) or can dissociate into $H + H^-$ (that is, leads to dissociative electron attachment). Schematically, these two complementary processes are:

$$e^- + H_2(v_i, J_i) \rightarrow H_2^-(^2\Sigma_u^+ \text{ or } ^2\Sigma_g^+) \rightarrow \begin{cases} e^- + H_2(v_f, J_f) \\ H + H^- \end{cases}$$

The nuclear wave function ξ of the resonant state, in the most general theoretical description, satisfies[1] a integrodifferential equation with nonlocal potentials:

$$[T_N(\vec{R}) + V^-(R) - E]\xi(\vec{R}) = -V(E - E_{v_i}, R)\chi_{v_i}(\vec{R})$$

$$- \int d\vec{R}' \sum_v \chi_v^*(\vec{R}')\chi_v(\vec{R}) \left\{ P\int d\varepsilon \frac{V^*(\varepsilon, R') V(\varepsilon, R)}{E - E_v - \varepsilon} - i\pi V^*(E - E_v, R')V(E - E_v, R) \right\} \xi(\vec{R}').$$

(1)

Here χ_v are the vibrational bound state wave functions of the neutral molecule, T_N is the nuclear kinetic energy operator and $V^-(R)$ is the potential curve of the resonant anion state. The sum on the right hand side includes all energetically open channels. The matrix elements $V(\varepsilon, R)$ couple the discrete resonant state with the background continuum; this coupling leads to an energy-shift of the resonance (the principal part integral on the right hand side) and provides a width to the resonance (the imaginary part on the right hand side) which determines the lifetime of the resonance. The total cross section for vibrational excitation, in atomic units, is[1] (assuming momentum normalized continuum functions),

$$\sigma(v \rightarrow v') = \frac{16\pi^4 k'}{k} \int d\hat{k}' |T(v \rightarrow v')|^2.$$

Here k and k' are the wave numbers describing the incident and the outgoing electron motion. The transition matrix element $T(v \rightarrow v')$ is obtained as:

$$T(v \rightarrow v') = \int d\vec{R}\, \chi_{v'}^*(\vec{R})\, V^*(\varepsilon',R)\, \xi(\vec{R}).$$

There are two points to be noted:
(i) One could either numerically solve the integrodifferential equation (Eq. (1)) with nonlocal potentials (and the results for cross sections would be referred to as the nonlocal results) or one could approximate the nonlocal potentials by some appropriate local potentials which will reduce the integrodifferential equation into an ordinary differential equation (and the results for cross sections would, then, be referred to as the local results).
(ii) The actual potentials to be used in the integrodifferential equation (for the nonlocal case) or in the ordinary differential equation (for the local case) could be obtained either by some separate *ab initio* calculations or by some judicious choice of parameters in semiempirical potentials.

Our previous calculations[2] of vibrational excitation cross sections of H_2 were done using a local formalism of the theory with an *ab initio* potential curve for H_2 (from the work of Kolos and Wolniewicz[3]) and a semiempirical potential curve for H_2^-; the complex potential energy curve of H_2^- was adjusted to ensure that, first, the calculated cross section for dissociative electron attachment to H_2 in the (v = 0, J = 0) level, as well as the isotope effect (that is, the ratio $\sigma_{DA}(H_2)/\sigma_{DA}(D_2)$) in attachment cross sections agreed with the experimental values and, second, the real part of the potential curve of H_2^- agreed with the *ab initio* curve calculated by Bardsley and Cohen[4].

The nonlocal equation (1) can be reduced to a local equation by noting that the level shift and the level width functions depend on $E - E_v$, the energy of the scattered electron when the target molecule undergoes the transition $0 \rightarrow v$. Now, if either the electron energy is large or the vibrational spacing is small, then during the vibrational transition the energy of the electron is not significantly changed. In such a situation one can replace $E - E_v$ by either the incident electron energy or the local classical electron energy, $V^-(R) - V_0(R)$. Here $V_0(R)$ represents the potential energy curve of the neutral molecule. On using the closure property for vibrational wave functions the sum over the vibrational levels on the right-hand-side of Eq. (1) then reduces to a delta function $[\delta(R - R')]$ and the integrodifferential equation (1) reduces to an ordinary differential equation.

As mentioned earlier, the process of vibrational excitation is closely related to the process of dissociative attachment in that it, too, proceeds via an intermediate resonant anion state. Using the nonlocal equation described above (Eq.(1)), we have obtained cross sections for vibrational excitation of molecular hydrogen via the $^2\Sigma_u^+$ resonance for electron energies up to 5 eV. Our objective in studying vibrational excitation has been twofold: to compare the cross sections obtained on using the nonlocal theoretical formalism with those obtained using the local approximation and also to examine the dependence of the cross sections on the nuclear mass (that is, the isotope effect).

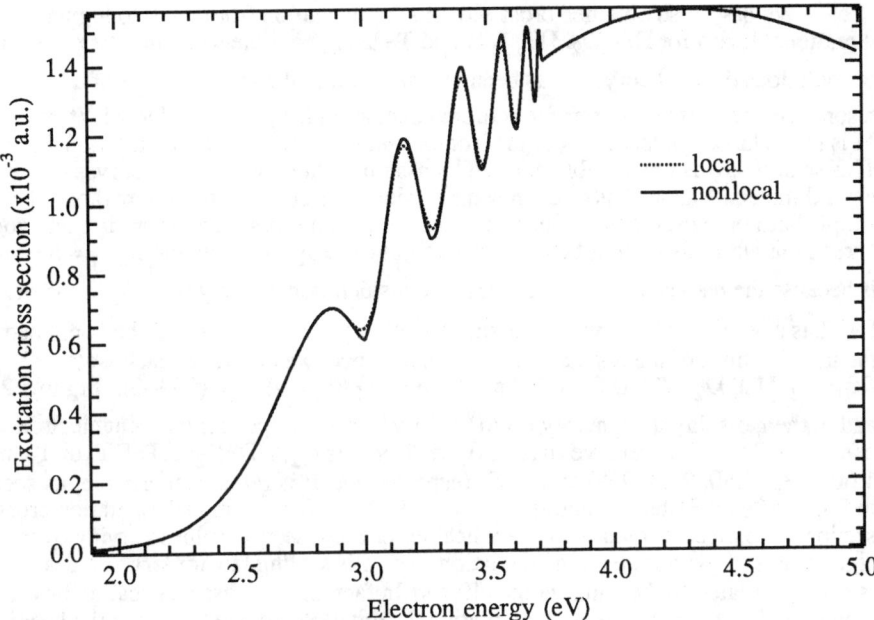

Figure 1. Cross sections for vibrational excitation ($v = 0 \rightarrow 4$) of H_2 using both nonlocal and local formalisms of the theory.

Figure 1 shows, as a typical example, cross sections for excitation from the ground level ($v_i = 0$) to the $v_f = 4$ level of H_2 using both the local and the nonlocal formulations of the theory. The nonlocal effects are seen to be small here, although differences of up to a factor of two are seen for excitation of the highest vibrational levels. The prominent feature here, however, is the well-defined structure in the cross sections which, we emphasize, is by no means a nonlocal effect, being present in the local cross sections as well. This structure is seen only for electron energies below the threshold for dissociative attachment and is entirely absent, for excitation from the ground level, for $v_f < 3$; the structure becomes more pronounced with increasing v_f. This behavior is in qualitative agreement with experimental observations[5], as well as in quantitative agreement (in terms of the location and relative values of peaks) with theoretical results reported by other authors[6]. We have numerically confirmed that as the width is made smaller, the peaks in the cross section become sharper and occur at values of the electron energy which approach those at which an anion bound state could be excited. The same structure is also seen in excitation from higher vibrational levels, and it occurs there for lower values of v_f, appearing in the superelastic cross sections by $v_i = 2$. The exact reason for the behavior of the structure as a function of v_f is, at present, under investigation.

We have also obtained cross sections for excitation from v = 0 to higher vibrational levels for HD, D_2, HT, DT, and T_2 using both the local and the nonlocal formulations of the theory. It has been shown[5] that, in the impulse limit of a resonance, the cross section for vibrational excitation is proportional to $M^{-v/2}$, where M is the reduced nuclear mass of the isotope and v refers to the final vibrational level. This scaling law is presumably derived by assuming that the potential curves of the neutral molecule as well as of the resonant anion state can be replaced by those of simple harmonic oscillators. Our calculated excitation cross sections are in quite good agreement with this scaling behavior for values of v_f up to and including $v_f = 5$. This is because the resonant state under present consideration, namely the $^2\Sigma_u^+$ state of H_2^-, has a relatively large width (or small lifetime) and, therefore, can be construed as the impulse limit of the resonance. For convenience, we define for each isotope X (X = H_2, HD, D_2, HT, DT, or T_2) a ratio $\rho = \sigma_X(0 \to v) / \sigma_X(0 \to 0)$. Figures 2 and 3, then, display the quantity $\rho \cdot M^{v/2}$ for various isotopes for incident electron energy of 5 eV. The relative masses of the isotopes H_2, HD, D_2, HT, DT, or T_2 are 1.00, 1.33, 1.50, 2.00, 2.40 and 3.00, respectively. It is rather remarkable, as seen in Figures 2 and 3, that the above isotope scaling law for vibrational excitation cross sections of molecular hydrogen is applicable for cross section values varying over eight orders of magnitude! Any deviation from this scaling law for large values of v_f is perhaps related to the anharmonic effects. In fact, one can use this scaling law to obtain, within a factor of two, the vibrational excitation cross sections for the heavier isotopes of molecular hydrogen from the corresponding cross sections for H_2 for any v_f.

More recently, Mundel et al[6] calculated the potential curves of H_2 and H_2^- using an *ab initio* technique and used those potentials, in a nonlocal theoretical formalism, to obtain the cross sections for dissociative attachment and vibrational excitation. The difference between their local calculations and our previous local calculations is the use of different potential curves of H_2 and H_2^-. In the present course of investigations we have completed our own fully nonlocal calculations for both dissociative attachment and vibrational excitation of H_2 and, indeed, we do not see the kind of difference between local and nonlocal results for cross sections that Mundel et al have obtained. More interestingly, since the apparent difference between Mundel et al's calculation and our calculation is the use of different potential curves of H_2^-, we made a few computer runs using their potential curves in our nonlocal computer code and we did not get the same results for attachment cross sections that Mundel et al have reported in their paper. This has been very intriguing as well as puzzling. Incidentally, our nonlocal as well as local results for the cross sections for vibrational excitation to higher levels of H_2 indeed show the structure, for electron energies below the threshold for dissociative attachment, that was recently seen[5] by Allan in his experiments.

In conclusion, we have obtained cross sections for vibrational excitation of molecular hydrogen and its five heavier isotopes by electron impact using both local and nonlocal formulations of the resonance model. Using both formulations one gets a well defined structure, in the excitation cross sections as a function of incident electron energy, for electron energies below the dissociative electron attachment

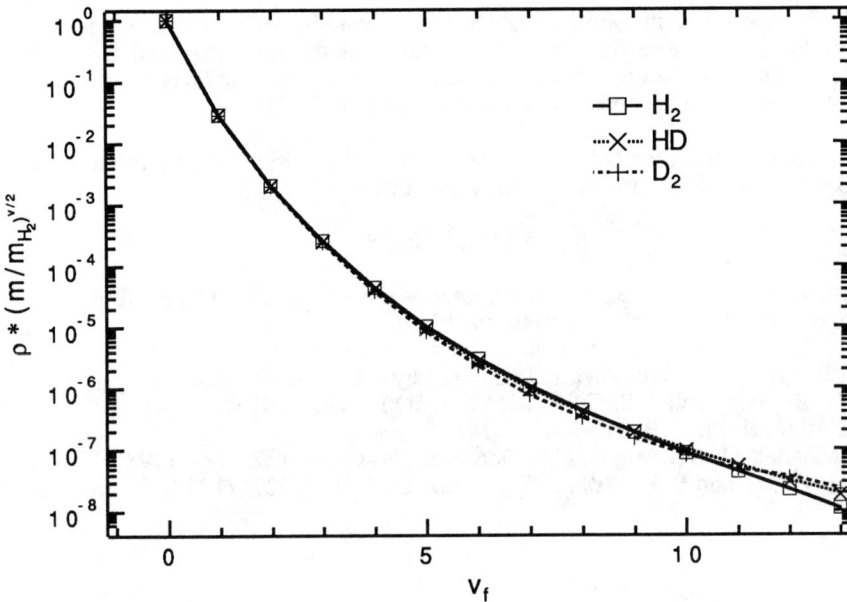

Figure 2. Isotope effect for vibrational excitation of molecular hydrogen.

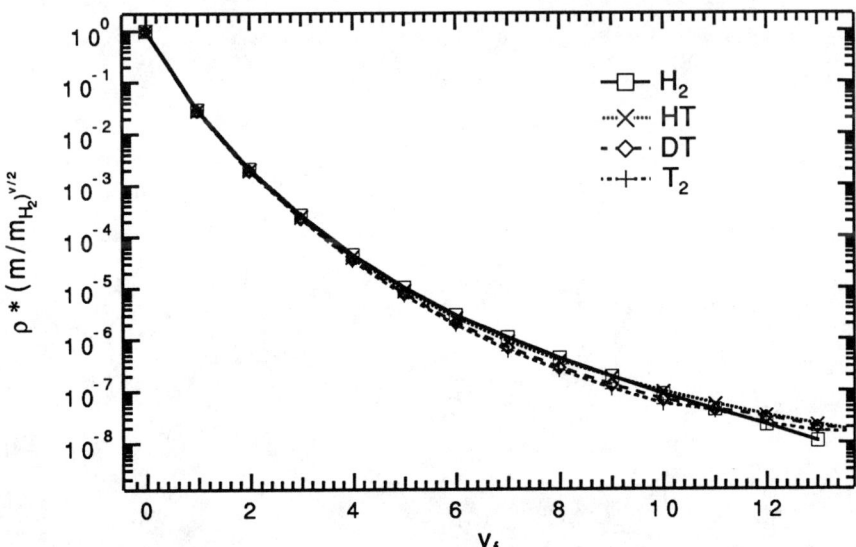

Figure 3. Isotope effect for vibrational excitation of heavier isotopes of molecular hydrogen.

threshold. In any case the nonlocal effects are found to be small. All the excitation cross sections satisfy a useful isotope scaling law such that the vibrational excitation cross sections for the heavier isotopes of molecular hydrogen can be obtained, within a factor of two, from the corresponding cross sections for H_2.

The support of the U.S. Air Force Office of Scientific Research through Grant Number AFOSR - 87 - 0342 is gratefully acknowledged.

REFERENCES

1. J. M. Wadehra, in *Nonequilibrium Vibrational Kinetics*, edited by M. Capitelli (Springer - Verlag, Heidelberg, 1986), p. 191.
2. J. N. Bardsley and J. M. Wadehra, Phys. Rev. **A20**, 1398 (1979).
3. W. Kolos and L. Wolniewicz, J. Chem. Phys. **43**, 2429 (1965).
4. J. N. Bardsley and J. S. Cohen, J. Phys. **B11**, 3645 (1978).
5. M. Allan, J. Phys. B. **18**, L451 (1985).
6. C. Mundel, M. Berman, and W. Domcke, Phys Rev **A32**, 181 (1985).
7. E. S. Chang and S. F. Wong, Phys. Rev. Lett. **38**, 1327 (1977).

DISCUSSION

Hopman: I just have a comment. The main application is for deuterium sources, that is why I am very glad that you are embarking on this kind of work. Cross-sections for all hydrogen processes are reasonably well known but to describe what happens in deuterium sources, we need lots of cross-sections and I am very glad, thank you for what you have started.

Skinner: In our simulations we have noticed that when the little eV process seems to be dominant, or be very important, in determining the vibrational distribution function we get a Boltzman, a straight line. Is there some fundamental reason why the little eV process is the only process that determines the vibrational distribution. Is there any reason why you get a Boltzman distribution out of that.

Wadehra: For that I'll have to digress a little bit. We have not shown in our results any solution of Boltzmann equation, it is a separate project we have right now with my other graduate student. We are trying to solve exactly the Boltzmann equation. We have tested that particular solution, as responding to Professor Capitelli a while back or so, that particular algorithm works for electrons and real gases, we have tested that very usefully and the paper appeared in Physical Rev. of August 1989, a couple of months ago. We would be interested in trying that algorithm for obtaining the distribution function of electrons in hydrogen sources eventually. The problem with hydrogen, as we know, is that in a computer code we have to put a mixture of two gases, we have atomic hydrogen and molecular hydrogen and then you have to put so many different cross sections. One set of cross sections for atomic species one for molecular species and that means the whole program is rather expensive for us. But we have such intention in the future and maybe next time when we meet I will be able to tell you whether the distribution is indeed mixing well or not.

Capitelli: The question asked by Skinner was a little different. He was asking you when the eV rates dominate the distribution he found is a Boltzman distribution for the vibrational levels, This is clear because the eV rates backward and forward are related by the electron temperature so it must find the Boltzmann distribution along the level when the eV rates dominates all the other rates. This is a different problem: Boltzmann equation, Boltzmann distribution.

Wadehra: Yes, if eV rates is the dominant.

Bacal: Is there a scaling law for the attachment cross-sections.

Wadehra: Yes, in fact you were in Los Angeles earlier this year as you remember in a January meeting, and I did present a scaling law for the associative attachment which also works. There are three different scaling laws for the dissociative attachment. One which corresponds to the replacing of molecular hydrogen by its isotope and in any case the molecule is in $v = 0$ level and seeing how much the cross-section changes as you go from H_2 to HD to D_2. There is another scaling law for associative detachment which tells us the factor by which the cross section is enhanced as you go from $v = 0$ to $v = 1$ in the case of H_2. In the case of HD or D_2, it is a separate scaling law.

A COMPARISON OF EXPERIMENTAL AND THEORETICAL ELECTRON
ENERGY DISTRIBUTION FUNCTIONS IN THE DRIVER OF A
MULTICUSP ION SOURCE

J. Bretagne
Universite de Paris–Sud, 91405 Orsay, Cedex, France

W.G. Graham
Physics Department, Queen's University, Belfast, Northern Ireland

M.B. Hopkins
Physics Department, Dublin City University, Dublin, Ireland

ABSTRACT

Experimental and theoretical electron energy distribution functions (EEDF) measured and calculated for the driver of a multicusp ion source operating in hydrogen are compared. The results indicate that atomic physics based theoretical models can accurately predict the EEDF in such discharges if some appropriate experimentally determined quantities are used as input parameters. The magnitude and shape of the EEDF is found to be particularly sensitive to the effective surface to volume ratio for electrons.

INTRODUCTION

There has been considerable interest in determining the electron energy distribution functions (EEDF) in negative ion sources since they play an important role in the determining the reaction rates in the discharge and hence in determining the H^- ion density. Recently a theoretical calculation[1,2] of EEDFs, generally based on a solution of the Boltzmann equation, and some reliable experimental EEDF measurements[3] have been reported. To make the theoretical calculations applicable to a particular ion source accurate experimental measurements of some plasma properties are required as input conditions. As yet there have been no direct comparisons between these theoretical calculations and experimental measurements under the same operating conditions.

In this paper theoretical calculations and experiment measurements of the EEDF in the "driver" region of a magnetic multipole hydrogen discharge are compared.

APPARATUS

The ion source is a tandem multipole typical of those used to produce beams of H^- ions for neutral beam injection. This multipole source has been described in detail elsewhere[3,4]. It is basically a rectangular stainless steel chamber with dimensions 19 x 19 x 24 cm. Permanent magnets with surface pole strengths of 3.6 kG are arranged to create a line cusp symmetry, broken on opposing walls to create a virtual filter, a one–dimensional field across the source which divides it into two regions. The present measurements will refer to the driver region, ie, that region which contains two hot tantalum filaments, biased negatively by 60V with respect to the vessel wall which forms the anode. The present measurements were made in hydrogen gas at pressures of between 2 and 4.7 mTorr with discharge currents of 6 and 10A and a discharge voltage of 65V.

CALCULATION OF EEDF

A detailed description of the atomic physics based theoretical calculations has been reported elsewhere[1,2]. Briefly the EEDF is assumed to be isotropic and the plasma to be homogeneous. The time–dependent Boltzmann equation, written in the form

$$\frac{\partial n(\epsilon,t)}{\partial t} = -\left[\frac{\partial J_{e\ell}}{\partial \epsilon}\right]_{e-m} - \left[\frac{\partial J_{e\ell}}{\partial \epsilon}\right]_{e-e} \quad (1)$$
$$+ \text{In} + \text{Ion} + \text{Sup } S - L$$

is then solved to find $N(\epsilon,t)d\epsilon$ which represents the number density of electrons with energy between ϵ and $\epsilon + d\epsilon$. The terms on the right hand side of equation (1) represent the flux of electrons along the energy axis due to elastic electron–molecule collisions, electron–electron collisions, inelastic collisions, ionizing collisions and superelastic collisions respectively. Explicit expressions for each of these terms can be found in reference 2.

In this discussion the last two terms are of particular interest. The source term (S) represents the number of electrons emitted from the filament per second per unit energy interval

$$S = \frac{I_d}{e} \times \frac{1}{\Delta\epsilon_p} \times \frac{1}{V} \quad (2)$$

Experimentally determined values of the discharge current, I_d, the energy spread of the primary electrons $\Delta\epsilon_p$ and the plasma volume V are available as initial parameters for the calculation. The loss term (L) includes the loss of electrons by diffusion towards the wall and by gas phase recombination and is written as

$$L = \tfrac{1}{4} n(\epsilon) v(\epsilon) \frac{A}{V} (1 - \frac{eV_p}{\epsilon}) + n(\epsilon) n + v(\epsilon) \sigma_V(\epsilon) \quad (3)$$

Here experimental measurements of the effective surface area to volume ratio for electrons, $\frac{A}{V}$, and the plasma potential, V_p, are available as input parameters. The values used for these parameters in the calculations are presented in Table 1. The ion density, n+, is assumed to consist mainly of H_3^+ [5].

EXPERIMENTAL MEASUREMENTS OF EEDF

Details of the technique used to measure the EEDF have already been published[3,4]. Briefly cylindrical Langmuir probes are used to determine the EEDF. The current–voltage (I–V) characteristic is recorded in the region $-90 < V < V_p$ using a high–speed large dynamic–range data acquisition system, where V is the probe voltage and V_p is the plasma potential relative to the anode. The second derivative of the electron current to the probe $I_e''(V)$ is related to the EEDF by the Druyvestyn equation[6]

$$n(\epsilon) = n_e f(\epsilon) = \frac{2I_e''(V)}{eS} \left(\frac{2m\epsilon}{e}\right)^{1/2} \quad (4)$$

where the energy $\epsilon = (V - V_p)$, I_e is electron current to the probe, S is the probe surface area and m and e are the mass and charge of the electron respectively. $n(\epsilon)$ is the number of electrons per unit energy interval per unit volume.

The Langmuir probe can also be used to measure a wide range of plasma parameters[3] including the electron density n_e, thermal electron temperature kT_e and plasma potential V_p. Of particular significance to the present work is the observation in previous measurements that the discharge was non–homogeneous[7]. The present EEDF measurement was therefore made in the centre of the driver region of the discharge but at least 2cm from the filaments.

As discussed above particular experimental measurements are necessary as input parameters for the calculation. The plasma potential was measured in the same position as the EEDF. The energy spread of the primary electrons emitted from the filament $\Delta\epsilon_p$ was determined to be about 15eV from EEDF measurements in a low density, low gas pressure (5×10^{-5}T) discharge[3] where there is little degradation of the primaries and so the energy distribution measured at the probe reflects directly that of the electrons leaving the filaments. This energy spread is due to the voltage drop across the filaments. The effective volume of the source can be estimated by measuring either the spatial variation of the positive ion density or the density of energetic electrons. These are found to follow the structure of the confining magnetic fields[3,7]. It was found that in a 10A, 65V, 2mT discharge the plasma production was confined to about 50% of the "driver" volume (17% of the total source volume), ie, 1.5ℓ.

The effective wall loss area for electrons is complicated by the presence of the magnetic fields. A calculation based on the Larmor radius of full energy electrons predicts an effective wall loss area of approximately 100 cm². The surface area to volume ratio is related to the electron confinement time, τ_e, by the expression[8]

$$\tau_e = \frac{4V}{A v_e} \quad (5)$$

where v_e is the electron velocity. In the present source τ_e has a value of 0.56×10^{-6} s, determined experimentally from the dependence of the fast electron density on gas pressure[8].

In Table 1 experimentally derived values are compared with those used in the theoretical calculations. The experimentally determined values for the plasma potential are close to those used in the theoretical calculations. The volume used in the theoretical calculations is the geometric size of the driver rather than the smaller, experimentally determined, effective plasma volume. The surface area to volume ratio is a significant input parameter and in the calculations for a 6A discharge the A/V ratios used is close to the experimental value. For 10A discharges different A/V values have been used.

In all cases, except where mentioned the width of the energy distribution of electrons from the filament was taken to be 6eV.

RESULTS

In Figure 1 experimental and theoretical EEDFs under similar plasma conditions, as detailed in Table 1, are compared. The agreement below electron energies of 50 eV is excellent. This is reflected in the agreement between the values of n_e and kTe given in Table 1. Close to and above eVd there is considerable disagreement when an initial energy distribution of 6 eV is used in the calculation. The peak in the theoretical EEDF around the discharge voltage is not observed in the experimental measurements, which is smooth and indicates the presence of electrons with energies significantly higher than the discharge voltage. Increasing the energy spread in the primary electrons from the filament to 20 eV improves the agreement above 50 eV considerably (Figure 1).

In Figure 2 and Table 1 using theoretical calculations with different surface area to volume ratios are compared with experimental EEDFs and plasma parameters. Since the electron confinement is overestimated when using A/V values lower than the measured value the calculated electron densities and electron temperatures are found to be larger than the experimental values. Likewise using an A/V value higher than the measured value in the calculation, underestimates the electron density and electron temperature, indicating that the secondary electrons are not thermalised before being lost to the wall.

DISCUSSION

The experimental EEDFs are found to be non–Maxwellian in the driver region of the ion source and can be conveniently represented by a bi–Maxwellian with the bulk of the electrons having a low temperature and a second group of fast electrons represented by an apparent higher temperature[3]. When some appropriate experimentally determined parameters are used as the initial condition for the calculation of the EEDF, the agreement between the theoretical and experimental EEDFs is excellent.

The magnitude and shape of the theoretically determined EEDF is particularly sensitive to the surface area to volume ratio used as an input parameter. This has important implications in modelling multicusp ion sources since we find that the effective plasma volume and surface loss area cannot be readily determined from simple geometric considerations.

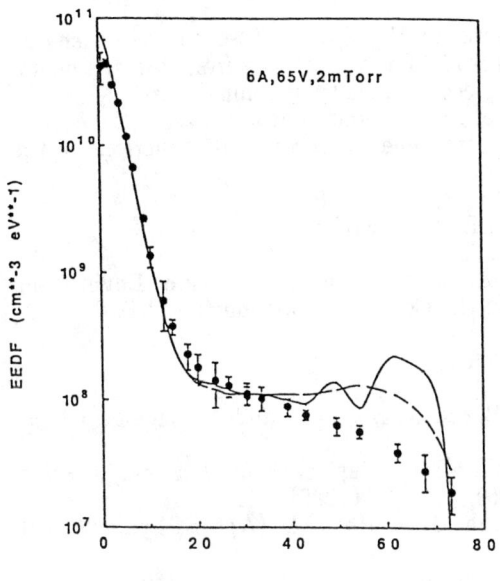

Figure 1

Electron energy distribution function: ●, measured, full curve: calculated with electron energy spread of 6 eV, dashed curve: calculated with electron energy spread of 20 eV.

Figure 2

Electron energy distribution function. ●, measured, full curve: calculated with surface area to volume ratio = 0.033, dashed curve: calculated with surface area to volume ratio = 0.004.

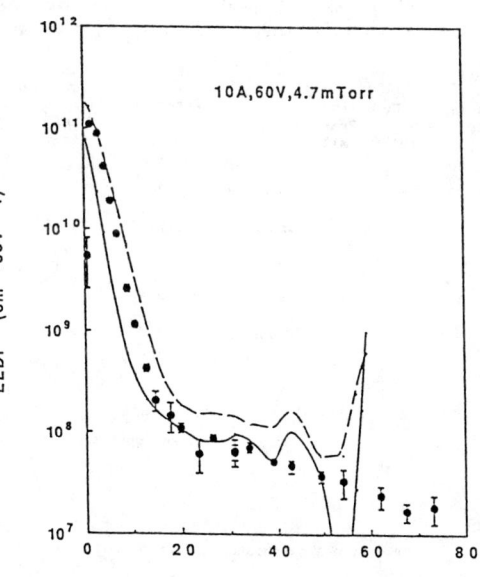

We find that, while the shape of the EEDF close to the discharge voltage is sensitive to the energy spread of the electrons from the filaments, there is little or no effect on the bulk electron density or temperature.

The present results indicate that atomic physics based theoretical models can accurately predict the electron energy distribution functions in the driver regions of negative hydrogen ion sources.

ACKNOWLEDGEMENTS

This work was supported in part by the Science and Engineering Research Council of Great Britain and the Conseil Scientifique du CCVR.

REFERENCES

1. C. Gorse, M. Capitelli, M. Bacal, J. Bretagne and A. Lagana, Chem. Phys. 117, 177 (1987).
2. J. Bretagne, G. Delouya, C. Gorse, M. Capitelli and M. Bacal, J. Phys. D (Appl. Phys.) 18, 811 (1985); 19, 1197 (1986).
3. M.B. Hopkins and W.G. Graham, J. Phys. D (Appl. Phys.) 20, 838 (1987).
4. M.B. Hopkins and W.G. Graham, Rev. Sci. Instrum. 57, 1697 (1986).
5. A.A. Mullan and W.G. Graham, (these proceedings).
6. M.J. Druyvesteyn, Z. Phys. 64, 781 (1930).
7. M.B. Hopkins and W.G. Graham, Vacuum 36, 874 (1986).
8. M.B. Hopkins, PhD thesis (University of Ulster) 1987.

Plasma Id (a)	Condition Vd (V)	Condition Press (mT)	Parameters	n_e (10^{11}cm^{-3})	kT_e (eV)	V_p (V)	$\Delta\epsilon_p$ (eV)	Vol(v) (L)	Area(A) (cm^2)	A/V
6	65	2	Experiment	2.1	1.82	3.1		1.5	22	0.015
			Theory	2.7[a]	1.65[a]	3.6	6	3.0	37	0.012
			Theory	2.5[a]	1.65[a]	3.6	12	3.0	37	0.012
10.8	60	4.7	Experiment	4.6	1.43	2.7		1.5	22	0.015
			Theory	6.5[a]	2.05[a]	2.0	6	3.0	12	0.004
				1.84[a]	1.88[a]	2.0	6	3.0	100	0.033

Table I. Plasma operating conditions and parameters

[a] Theoretical calculated values, other parameters are used as inputs to calculation

H^- FORMATION FROM COLLISION INDUCED DISSOCIATION OF H_3^+ IN NOBLE GASES AT keV ENERGIES.

I. Alvarez, H. Martinez, C. Cisneros, A. Morales and J. de Urquijo
Instituto de Fisica, UNAM, P.O. Box 139-B, Cuernavaca, Mor., 62191
Mexico

ABSTRACT

We have measured the energy distributions of H^- ions produced by the collision-induced dissociation process of H_3^+ molecular ions in noble gas targets. The most probable H^- c.m. energy was measured and over the energy range studied it was found to be dependent of the target species. Inelastic energy loss of 22 ± 6 eV was measured and found to be the same for all cases.

INTRODUCTION

Detailed knowledge of the characteristics of reaction products leads to a deeper understanding of the mechanisms and states responsible for their creation. An interesting molecular dissociation channel that leads to a negative ion product is a process called polar dissociation. The mechanism occurs via a ionic state of the molecule composed, for example in two body dissociation, of a positive and negative ions, i.e., $XY \rightarrow X^+ + Y^-$. Considerable experimental and theoretical work has been done on the two body polar dissociation of neutral molecules, where the excitation energy is supplied by electron or photon impact. However, no detailed studies of polar dissociation of molecular ions exist, due to the considerable amount of additional experimental difficulties associated with the production and characterization of ionic targets. Montgomery and Jaecks[1] have suggested an alternative method for obtaining information on polar dissociation of molecular ions, and have presented initial measurements of H_3^+ on He, and very recently[2] the energy distribution of H^- for the same colliding system. Alvarez et al[3] have reported the absolute total cross sections for this process by numerical integration of the laboratory differential H^- production cross sections, and also some preliminary results of the energy distribution of the negative products[4]. In this method the excitation energy is furnished to the molecular ion via an energetic ion-atom collision, thereby simplifying the experimental difficulties significantly. Other characteristics are apparent: a) Only one dissociation channel predominates; in other words, when measuring the H^- fragment, the other two fragments are restricted to be H^+, since charge transfer process are very unlikely with He as a target, and b) on bombarding the He target with H_3^+ at keV energies, the molecule can be considered as "frozen" and consequently, the resulting fragments carry the molecular state at the time of collision. In Ref. 4 it was assumed that a two step model together with the assumption of a binary dissociation can be applied when account was taken on the symmetry of this three-body Coulomb break up.

The purpose of this paper is to present the new results obtained in this laboratory with a modified detection system and using the same method of analysis independent of the break up model used in Ref. 2.

We also present the H^- energy spectra obtained with other noble gases.

EXPERIMENTAL METHOD

The experimental apparatus has been described elsewhere[5]. Molecular ions were formed in an arc discharge source, electrostatically accelerated to energies 2.75, 3.8 and 4.8 keV and selected by a Wien filter. The selected ionic species was bent 10° and allowed to pass through a series of collimators before entering the interaction cell, containing He at a pressure of about 0.1 mTorr. The cell was a cylinder of 2.5 cm in length and diameter, with a 1 mm diameter collimator at its entrance, and a rectangular exit aperture of 2 x 6 mm. This cell was fixed at the center of a computer controlled, rotatable vacuum chamber. Upon rotation, the chamber moves the detection system, located 47 cm away from the interaction cell. Vacuum base pressures in the system were $\sim 10^{-8}$ Torr without gas in the cell and $\sim 8 \times 10^{-7}$ Torr with gas. The detection system consisted of a retarding field, parabolic flight electrostatic analyzer with a channel electron multiplier attached to its exit end. For the H^- energy distribution measurements a rectangular slit of 0.01 mm x 10 mm was placed in front of the CEM. Under these conditions the calculated energy resolution of the analyzer is $\Delta E/E = 4.5 \times 10^{-3}$. The energy spectra were recorded at zero angle only. The energy analyzer was calibrated as follows: prior to any energy distribution measurement, the H_3^+ beam was allowed to pass through the analyzer with no gas in the target cell and the voltage (V) was adjusted to correspond with the center of the energy distribution; the energy spread of this beam was about 10 eV for a 3.8 keV acceleration energy. Then, helium was introduced in the target cell and upon application of one third of a reverse voltage across the analyzer plates (-V/3), the H^- fragments could be detected.

Counting of H^- ions produced in the interaction cell containing a noble gas as a function of the energy gained or lost was performed by sweeping the analyzer voltage with a computer-controlled operational amplifier power supply, and simultaneously counting the H^- signals from the CEM for each voltage step. Figures 1, 2, and 3 show typical energy distributions which are the average of 10 independent runs of about 400 scans each.

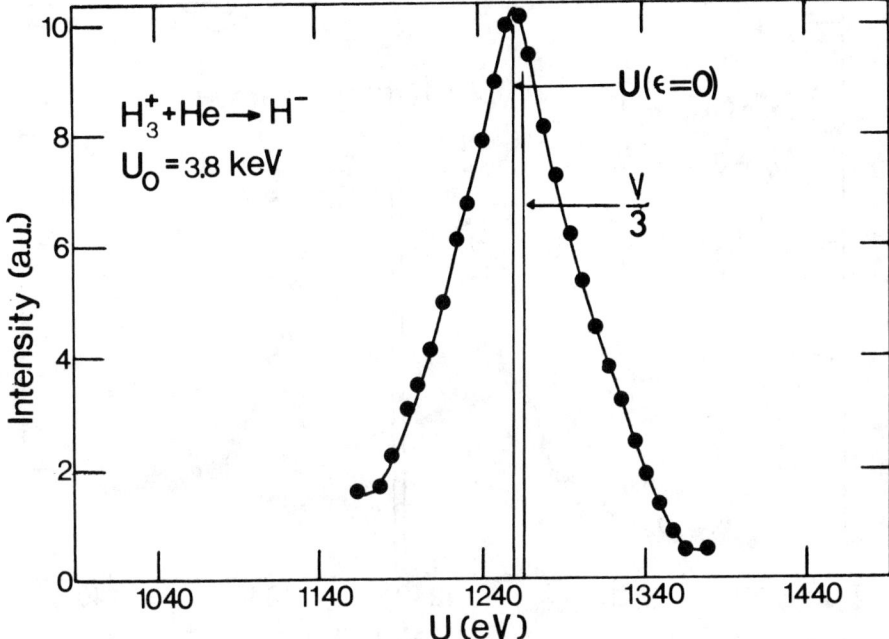

Fig. 1. Laboratory energy distributions of H⁻ at 3.8 keV of the primary beam energy interacting with He. No significant changes were found at other energies.

DISCUSSION

The data were analyzed by using the relation:

$$(M+m) U = M (V-E) + (M+m) \varepsilon \pm 2 [M (M+m)(V-E) \varepsilon]^{\frac{1}{2}} \qquad (1)$$

derived from energy conservation considerations. U is the energy of the detected fragment of mass M at zero degrees in the LF system; V is the primary beam energy of the molecular ion of mass (M+m); E is the internal energy increment of the system, and ε is the kinetic energy at which the dissociated H⁻ are ejected. Since the dissociation is isotropic the kinetic energy can be added or substracted to give the slow (-) or fast (+) products.

Figure 1 shows the energy spectra of H⁻ obtained at 3.8 keV energy of the primary beam interacting with He. This broad distribution is due to the isotropic characteristic of the dissociation process and also to the different excitation states of the molecular ion at the time of the collision. Consequently

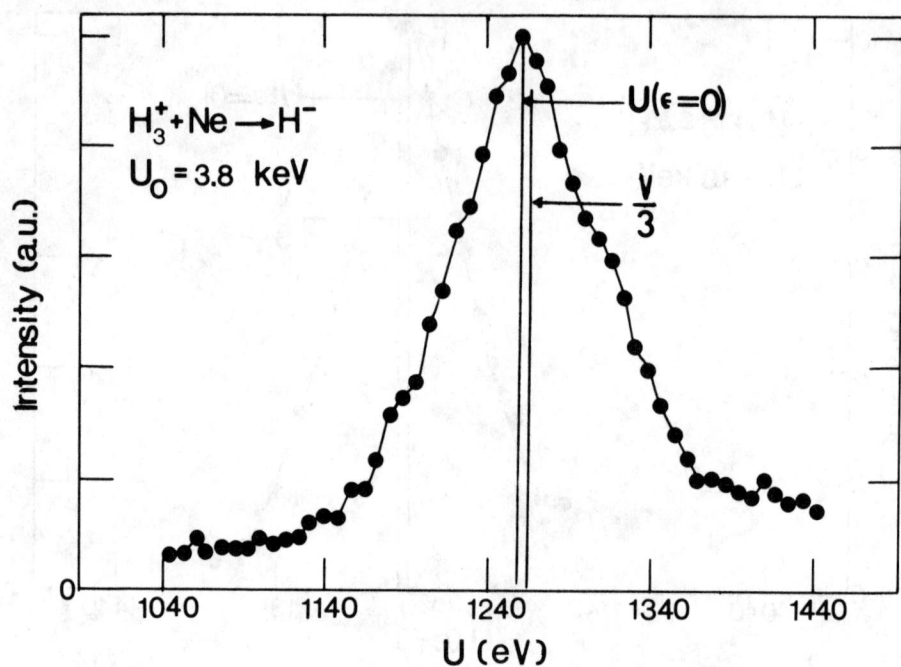

Fig. 2. Laboratory energy distributions of H⁻ at 3.8 keV of the primary beam energy interacting with Ne. No significant chages were found at other energies.

it was impossible to determine exactly the information about the inelastic energy loss necessary to produce the excited H_3^+ state producing H⁻, and also the kinetic energy of this fragment. Nevertheless useful information can still be extracted from the spectrum. A careful inspection of the energy distribution indicates that their highest peaks are slightly shifted towards the left of the position that would correspond to zero energy loss, namely, one third of the incident energy. This shift is a measure of the increment in internal energy of the molecule, E, and corresponds also at ϵ equal zero. E can be calculated from eq. (1) by setting $\epsilon = 0$, taking the U value at the maximum of the distribution and solving for E. In the energy range of the present experiment we obtained E = 22 ± 6 eV. The estimated error is one standard deviation. It is important to note that the same value was reported by us in Ref. 4 using a different method of analysis and different experimental procedure.

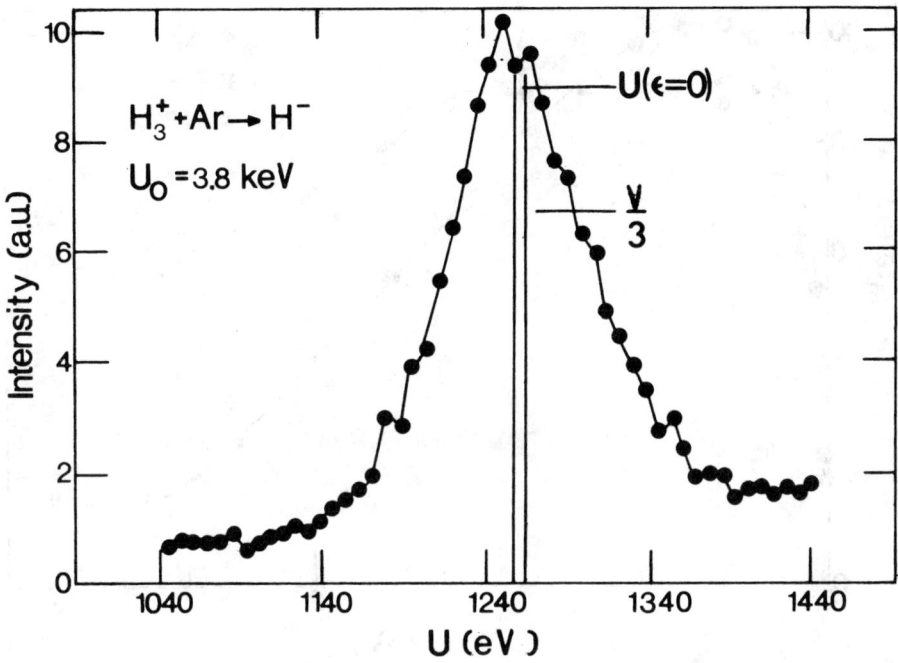

Fig. 3. Laboratory energy distributions of H^- at 3.8 keV of the primary beam energy interacting with Ar. No significant changes were found at other energies.

Figures 2 and 3 show the energy distributions of H^- using Ne and Ar as targets and at the same projectile energy respectively. From the shift of the center of the distributions the same E value was obtained, but the distributions are thicker and some structures are evident as a consequence of other channels acting in the dissociation process, such as charge transfer. This fact is more evident in Figure 4 where the deconvolution of the three spectra are shown for comparison. From this c.m. energy distribution it is also possible to assign the most probable kinetic energy of H^- in the dissociation process. For the various targets, these values are: $\varepsilon = 0.5$ eV (helium), $\varepsilon = 0.6$ eV (argon) and $\varepsilon = 0.85$ eV (neon) at 3.8 keV of projectile energy.

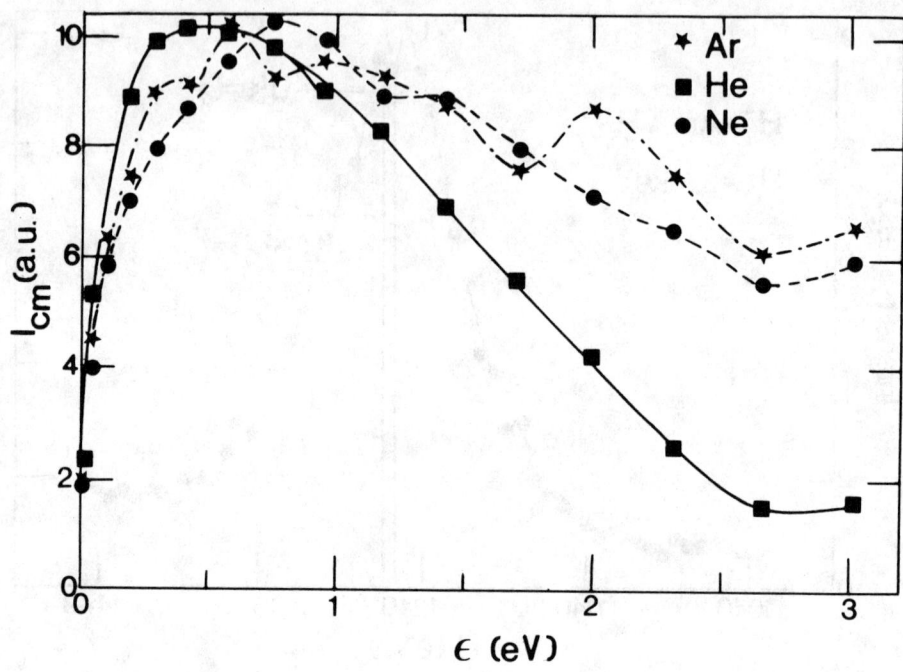

Fig. 4. Deconvolution of the spectra of Figs. 1, 2, and 3. The presence of a structure originated by more dissociation channels open in the production of H^- using Ar or Ne as targets are evident.

In conclusion, we have measured the energy distribution of H^- product produced by the collision induced dissociation of H_3^+ in several noble gases. An analysis of the spectra based on energy conservation rules allowed the determination of the inelastic energy loss or energy taken by the system to be dissociated which was found to be $E = 22 \pm 6$ eV and is the same as that reported in Ref. 4. In this reference a binary model of dissociation was assumed. This energy is found to be the same when He was either replaced by Ar or Ne as targets but with these gases a structure in the energy distributions is present because more channels are open in the potential energy curves of the system. The characterization of these channels is presently under study. The most probable kinetic energy of the H^- was determined, the value of 0.75 eV reported by Yenen et al[2] is the same we found at 4.8 keV but their E value of 60 ± 12 eV

differs strongly from the present one. We can not provide an
explanation for the source of this discrepancy at this time but it
has motivated us to repeat these measurements with a different
particle analyzer and detector and also a different method of
analysis. Moreover the fact that the present results and those
reported by us in Ref. 4 are the same give support to the strong
hypotesis that, on bases of the symmetry of this three-body
coulomb break up, the application of a binary dissociation model
is a good approximation.

REFERENCES

1. D.L. Montgomery and D.H. Jaecks, Phys. Rev. Lett. $\underline{51}$, 1862 (1983).
2. O. Yenen, D.H. Jaecks and L.M. Wiese, Phys. Rev. $\underline{A39}$, 1767 (1989).
3. I. Alvarez, C. Cisneros, J. de Urquijo and T.J. Morgan, Phys. Rev. Lett. $\underline{53}$, 740 (1984).
4. I. Alvarez, H. Martinez, C. Cisneros, A. Morales and J. de Urquijo, Nucl. Instr. and Methods Phys. Res. $\underline{B40/41}$, 245 (1989).
5. H. Martinez, I. Alvarez, J. de Urquijo, C. Cisneros and A. Amaya-Tapia, Phys. Rev. $\underline{A36}$, 5425 (1987).

GAS PHASE ALKALI-HYDROGEN INTERACTIONS IN NEGATIVE ION SOURCES

H. H. Michels
United Technologies Research Center, East Hartford, CT. 06108

J. M. Wadehra
Wayne State University, Detroit, MI 48202

ABSTRACT

The role of Li_xH_y (Cs_xH_y) molecules, which can be formed from seeding an alkali into a hydrogen plasma, is examined through *ab initio* calculations of the structure and stability of such species. The simplest alkali hydride, LiH, supports a bound anion for all internuclear separations. Larger structures, such as Li_2H_2 and Li_3H, do not support stable anions but exhibit thermodynamic stability as gas phase molecules. Their possible roles in dissociative attachment to form H^- and/or Li^- is examined.

INTRODUCTION

The negative ions of light atoms are currently being studied for their possible application in gaseous discharges, fusion plasmas and gas lasers.[1] One source for the volume production of atomic anions is the process of dissociative electron attachment to molecules.[2] This process has been studied experimentally for low energy electron-hydrogen molecule collisions by Allen and Wong.[3] A parallel theoretical study has been reported by Wadehra and Bardsley.[4] More recently, McGeoch and Schlier[5] have examined dissociative attachment (DA) in electron-lithium molecule collisions and have found large DA rates for attachment to highly vibrationally excited Li_2 molecules. The effect of vibrational excitation on the DA rates has been studied by Hiskes[6] for $e + H_2$ collisions and by Wadehra and Michels[7] for the $e + Li_2$ system. A theoretical study of the absolute cross sections for DA in the $e + Li_2$ system, including enhancement by rotational excitation, has also recently been reported by Wadehra.[8,9]

In contrast to the homopolar case, the role of Li_xH_y (Cs_xH_y) molecules, which can be formed from seeding an alkali into a hydrogen plasma, is presently not well understood. The addition of an alkali such as cesium has been shown[10] to enhance the volume production rate of H^-. This observation is interesting in light of the study by Gauyacq, et al,[11] which indicates that charge transfer and collisional detachment processes should reduce H^- production in sodium seeded plasmas. Jordan[12] has suggested that if the dimer of LiH were a linear configuration, the resultant large dipole field would result in enhanced DA of e + $(LiH)_2$ to yield $Li_2H + H^-$ or possibly, LiH^- + LiH. In the present study, we assess the stability of several Li_xH_y clusters and examine their possible role in dissociative attachment to form negative ion products.

THEORETICAL CONSIDERATIONS

In order to examine the stability of Li_xH_y clusters, *ab initio* calculations were carried out for several species at the MP2 level of theory. The basis set chosen was the Gaussian 6-311G triple split-valence set,[13] augmented by d-polarization functions [α=0.20] for Li and p-polarization [α=0.75] for H. In addition, a diffuse s-function [α=0.036] and a diffuse sp-shell [α=0.0074] were added to hydrogen and lithium respectively, to better describe the negative ion charge distributions. All calculations were performed using CADPAC,[14] an electronic structure code that can perform geometry optimizations and calculate force constant matrices analytically. Optimized geometries were calculated at the MP2 level of theory. Harmonic vibrational frequencies were subsequently calculated at the MP2 stationary points.

In addition, the LiH and LiH^- systems were studied using the DIATOM molecular structures code with a [13σ, 6π, 2δ] STO basis. These STO basis calculations were required to examine the relative behavior of the LiH and LiH^- species at very short internuclear separations.

CALCULATED RESULTS AND DISCUSSION

The results of our calculations for the LiH and LiH^- systems, using the DIATOM code, are shown in Fig. 1 for states of $^2\Sigma^+$ symmetry of LiH^-. In Fig. 2, we illustrate the short range behavior of the lowest and first excited $^2\Sigma^+$ states of LiH^-. It is evident that the ground $X^2\Sigma^+$ state of LiH^- is bound, relative to $X^1\Sigma^+$ of LiH, for all internuclear separations. Our calculated adiabatic electron affinity is 0.33 eV, in good agreement with previous theoretical studies.[15] The first excited $^2\Sigma^+$ state of LiH^-, which asymmetrically correlates to $Li^- + H$, exhibits repulsive behavior in the region $3.0 \leq R \leq 6.0$ Å. Thus, DA of e + LiH mainly forms Li^- for $E_{coll} \geq 3.0$ eV. The formation of $Li + H^-$ products may occur by non-adiabatic coupling of the $X^2\Sigma^+$ state of LiH^- to the continuum of e + LiH [$X^1\Sigma^+$] for $E_{coll} \geq 2.1$ eV. However, the cross section for such an indirect electron capture mechanism is predicted to be very small.

The possible role of Li_xH_y clusters was examined through *ab initio* calculations of Li_2H, Li_3H, Li_2H_2 and their respective negative ion states. All cluster calculations were carried out using the CADPAC code. In Fig. 3, we illustrate the stationary geometry of Li_2H and the corresponding negative ion. We find that the C_{2v} structure for Li_2H has vibrational stability and is thermodynamically stable relative to $Li_2 + H_2$. In addition, the corresponding C_{2v} anion is slightly bound (0.06 eV) relative to the neutral species. These results are in qualitative agreement with previous theoretical studies of Li_2H by Cardelino, et al,[16] that were carried out at the SCF level of theory.

The Li_3H species was examined both as a C_{3v} (trigonal pyramid) and as a C_{2v} (kite-like) structure. Our previous studies[17] of Li_3H indicated that, although both structures exhibited stable vibrational frequencies, only the C_{2v} structure was thermodynamically stable. In Fig. 4, we illustrate the two stable geometries of neutral Li_3H. Of these, the C_{2v} structure represents the global minimum for the system. The Li_3H^- anion is not stable as a C_{3v} structure and collapses to $Li_3 + H^-$. The C_{2v} structure of Li_3H^- has the stationary structure shown in Fig. 4, but a vibrational analysis indicates that this is probably a saddle region on the potential energy surface. In addition, this region is thermodynamically endothermic by 0.24 eV, relative to the neutral Li_3H.

Finally, calculations were carried out for the $(LiH)_2$ dimer, as a linear ($C_{\infty v}$), rhomboid (D_{2h}) and y-structure (C_{2v}). The results are shown in Fig. 5 where we show that the linear conformation has one imaginary frequency, and thus represents a saddle point on the surface. The C_{2v} structure is totally unstable and collapses to separate $Li_2 + H_2$ molecules. The only vibrationally stable geometry in the $^1A_1$ state as a D_{2h} structure. The corresponding negative ion Li_2H_2- is vibrationally unstable and unbound by over 2.0 eV, relative to Li_2H_2 (D_{2h}).

A summary of the thermodynamics of these Li_xH_y clusters is given in Table 1. All energies are given relative to Li_2 and H_2 as gas phase species. As previously known, the formation of LiH is endothermic by 0.740 eV. However, higher clusters are all thermodynamically stable: Li_2H (–0.280 eV), Li_3H (–1.034 eV) and Li_2H_2 (–1.263 eV).

CONCLUSIONS

The trend of our studies to date indicates that large Li_xH_y (and by analogy, Cs_xH_y) clusters are thermodynamically stable. In particular, the Li_2H_2 species, as a C_{2v} structure, may be an important component of alkali-hydrogen mixtures. This species can dissociatively attach an electron to form $Li_2H + H^-$ for $E_{coll} \geq 2.0$ eV. The Li_3H (C_{2v}) species should also exhibit DA to form $Li_3 + H^-$, but the concentration of this molecule will be lower than that of the more stable Li_2H_2 cluster. The mechanism for DA to these Li_xH_y clusters can be understood by following their intrinsic reaction coordinate pathways leading to dissociation. In addition, the role of higher order clusters in volume DA processes is uncertain. Additional studies of these systems are currently in progress.

We thank Drs. J. A. Montgomery, Jr. and J. R. Peterson for helpful discussions and J. B. Addison for assistance in carrying out the calculations. This work was supported in part by AFOSR under Contract F49620-89-C-0019 and Grant AFOSR-87-0342. Use of the computational facilities at AFSCC-Kirtland is also acknowledged.

REFERENCES

1. K. Prelec, ed., Proceedings of the Third International Symposium on the Production and Neutralization of Negative Ions and Beams, AIP Conf. Proc. 111, (AIP, New York, 1984).
2. M. Bacal, Physica Scripta, T2/2, 467 (1982).
3. M. Allan and S. F. Wong, Phys. Rev. Letters. 41, 1791 (1978).
4. J. M. Wadehra and J. N. Bardsley, Phys. Rev. Letters, 41, 1795 (1978); Phys. Rev. A20, 1398 (1979).
5. M. W. McGeoch and R. E. Schlier, in: Proceedings of the Third International Symposium on the Production and Neutralization of Negative Ions and Beams, AIP Conf. Proc. 111, ed. K. Prelec (AIP, New York, 1984), p.291; Phys. Rev. A33, 1708 (1986).
6. J. R. Hiskes, J. Appl. Phys. 51, 4592 (1980).
7. J. M. Wadehra and H. H. Michels, Chem. Phys. Letters, 114, 380 (1985).
8. J. M. Wadehra in Proceedings of the Fourth International Symposium on the Production and Neutralization of Negative Ions and Beams", J. G. Alessi, ed., AIP Conf. Proc. 158 (AIP, New York, 1986).
9. J. M. Wadehra, "Rotationally Enhanced Dissociative Electron Attachment to Molecular Lithium", J. Chem. Phys., in press.
10. K. N. Leung, S. R. Walther and W. B. Kunkel in "Microwave and Particle Beam Sources and Directed Energy concepts", H. E. Brandt, ed., SPIE Proc. 1061 (SPIE, WA, 1989).
11. J. P. Gauyacq, et al., Phys. Rev. A38, 2284 (1984).
12. K. D. Jordan, Chem. Phys. Letters, 40, 441 (1976).
13. R. Krishnan, J. S. Binkley, R. Seeger and J. A. Pople, J. Chem. Phys. 72, 650 (1980).
14. R. D. Amos and J. E. Rice, CADPAC: The Cambridge Analytic Derivatives Package -Issue 4.0, Cambridge, 1987.
15. K. D. Jordan, et. al., J. Chem. Phys. 64, 4730 (1976).
16. B. H. Cardelino, W. H. Eberhardt and R. F. Borkman, J. Chem. Phys. 84, 3230 (1986).
17. J. A. Montgomery, Jr., H. H. Michels, O. F. Güner and K. Lammertsma, Chem. Phys. Letters, 1989, in press.

Table I Thermodynamics of Li_xH_y clusters

Reaction	ΔE (eV)	
Li_2 (g) + H_2 (g) → $2LiH$ (g)	+0.740	(1)
Li_2 (g) + $1/2\,H_2$ (g) → Li_2H (C_{2v})	−0.280	(2)
$3/2\,Li_2$ (g) + $1/2\,H_2$ (g) → Li_3H (C_{2v})	−1.034	(3)
Li_2 (g) + H_2 (g) → Li_2H_2 (D_{2h})	−1.263	(4)

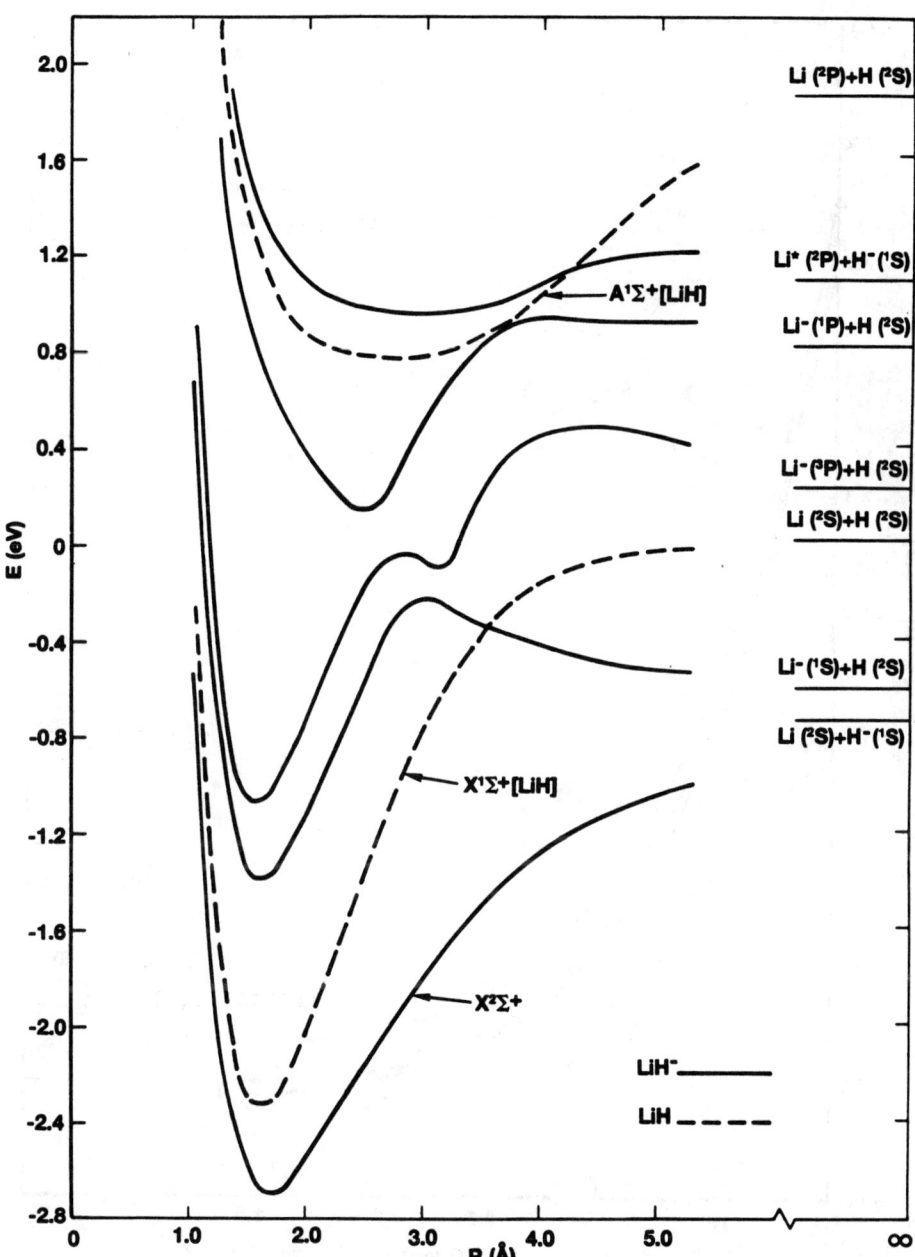

Fig. 1 LiH/LiH⁻ potential energy curves.

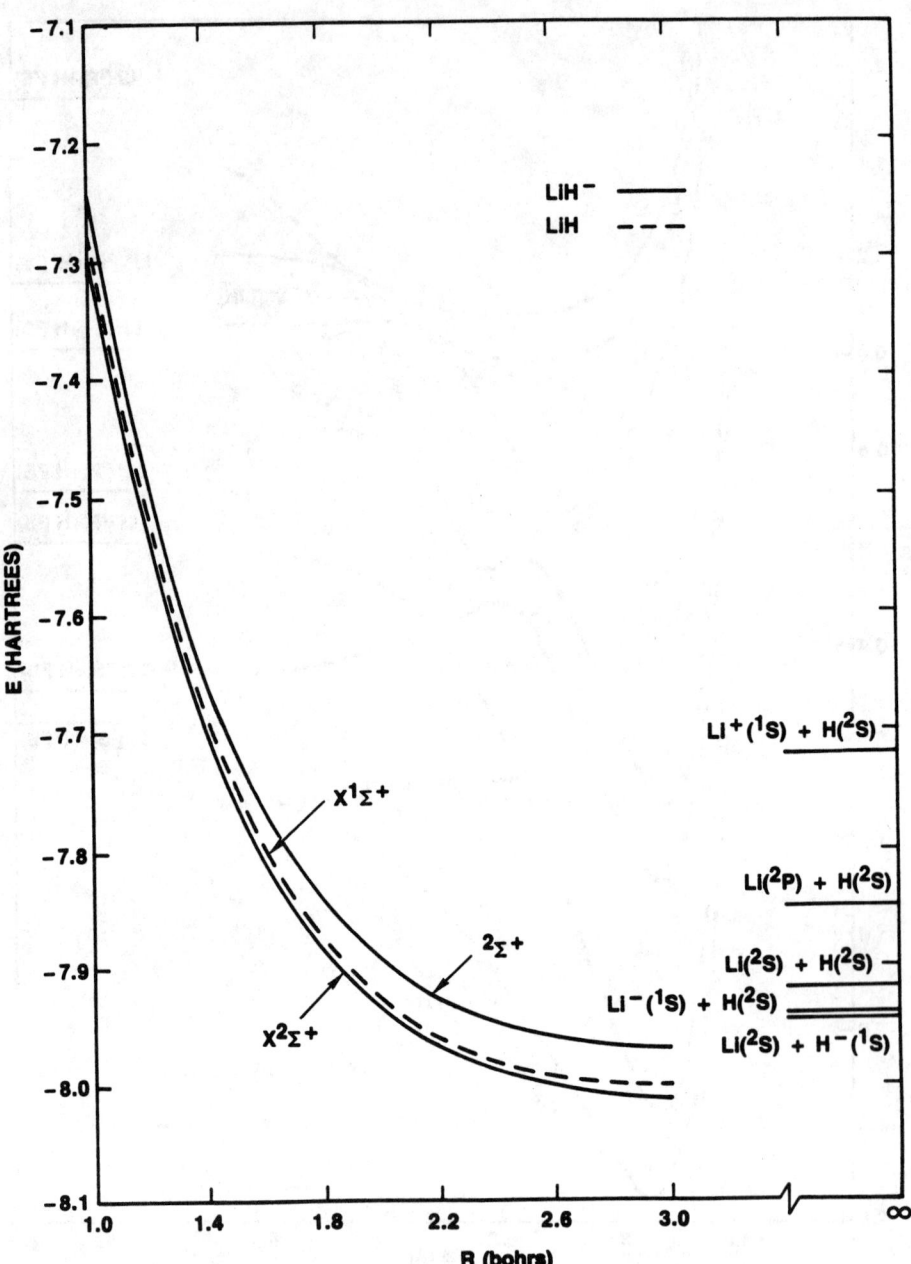

Fig. 2 Short range interaction potentials for LiH/LiH$^-$.

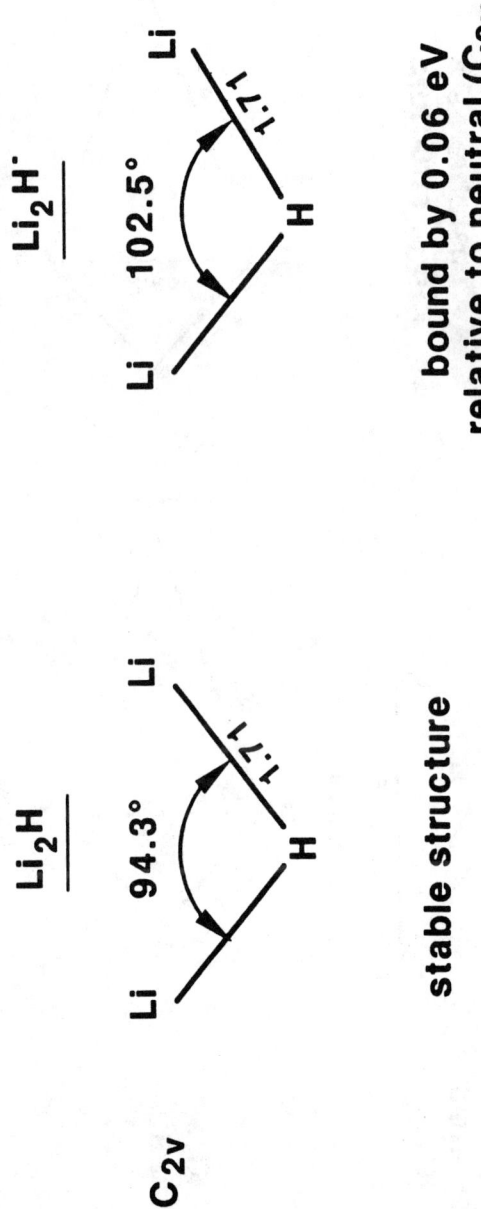

Fig. 3 Li$_2$H/Li$_2$H$^-$ structures.

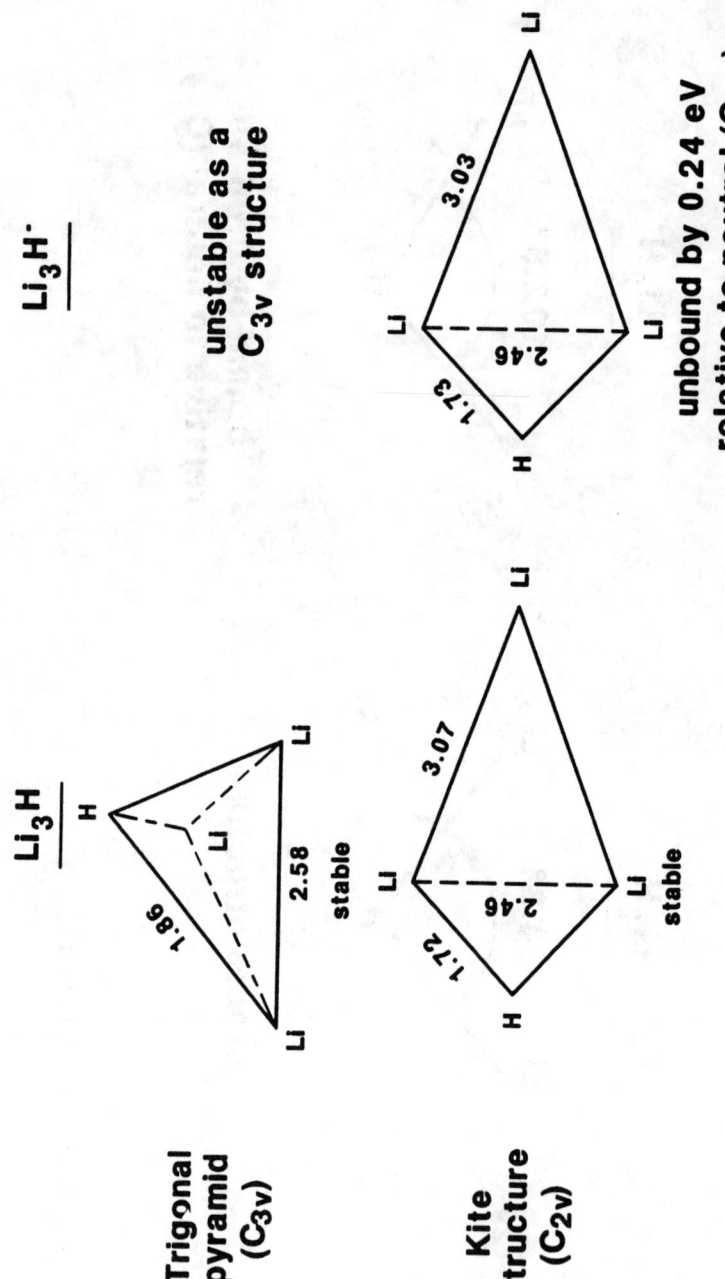

Fig.4 Li$_3$H/Li$_3$H$^-$ structures.

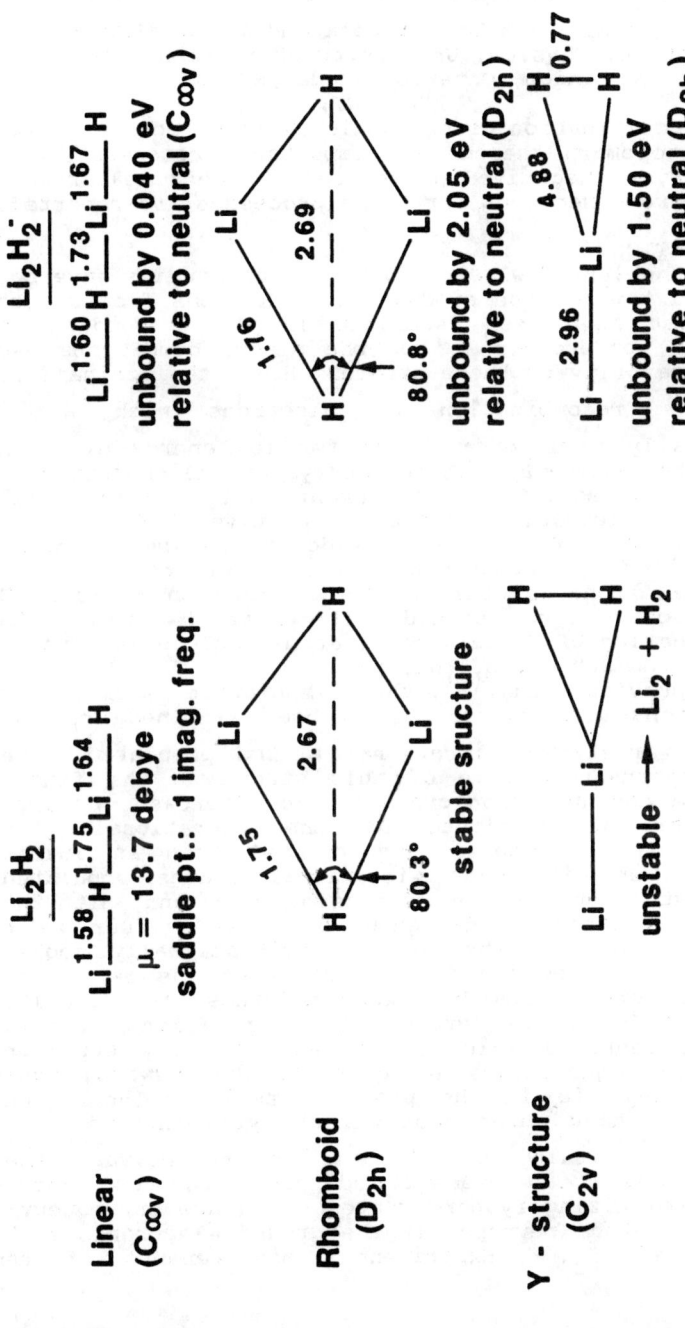

Fig.5 $Li_2H_2/Li_2H_2^-$ structures.

THE DISSOCIATIVE RECOMBINATION OF H_3^+, AND D_3^+.

F.B. Yousif, P. Van der Donk and J.B.A. Mitchell
Dept. of Physics, University of Western Ontario
London, Ontario, Canada, N6A 3K7.

ABSTRACT.

The recombination of triatomic hydrogen molecular ions and isotopomers thereof are important processes in the chemistry of negative ion sources. Recent measurements of the cross sections for these processes are reported.

INTRODUCTION.

Over the last few years, a number of studies have been directed towards the production of H^- ions from sources and these have been described in the proceedings of previous conferences[1]. For the plasma conditions used in the majority of these studies, H_3^+ is the dominant ion and the recombination of electrons with H_3^+ is potentially an important sink for low energy electrons which are believed to be necessary for the production of H^- via dissociative attachment with vibrationally excited molecular hydrogen. Negative hydrogen ion sources are of course considered to be important elements of a thermonuclear reactor and therefore the practical emphasis would be not so much on H^- as on D^- production. One should therefore examine the recombination of deuterated triatomic molecular ions and that is the subject of this paper.

In a previous study[2], the dissociative recombination of H_3^+ ions with electrons has been examined for ions with a variety of vibrational states populated. The most important outcome of this study was the finding that the recombination cross section decreases by about an order of magnitude as the vibrational state distribution is cooled down to v=0 from an initial distribution with many vibrational states populated. This contradicts the conclusions of Adams and Smith that the rate coefficient decreases by three or four orders of magnitude when the ions are vibrationally cooled. Extensive studies aimed at identifying the reason for this discrepancy have been made and these are described elsewhere[3]. Both groups have performed careful remeasurements of this process and both are confident that their conclusions are correct. The answer to this dilemna may lie in the process itself. Theoretical studies[4] have shown that direct recombination of H_3^+ appears to be ruled out but it is not inconceivable that the process could proceed indirectly through capture into autoionizing rydberg states which are subsequently predissociated, perhaps via the ground electronic state of neutral H_3. Experimental measurements of the branching ratio for the product channels of H_3^+ recombination, lend support to this concept as it is found that H_3^* formation accounts for about 8% of the total decay observed 0.1ms after the interaction region[5]. By the time this measurement is performed, many of the H_3^* molecules formed in the recombination must surely have been predissociated to form H+H+H and H_2+H. If indeed the recombination proceeds through the formation of H_3^* molecules, then this may not be apparent in the high pressure environment of a flowing afterglow apparatus, where such species would be collisionally ionized[6].

In a continuing effort to understand this process the dissociative recombination of D_3^+ ions with low internal energies has been studied and the results of this measurement are reported here.

EXPERIMENTAL TECHNIQUE.

The experiment described here was performed using the merged electron-ion beam technique, which has been described in detail elsewhere.[7,8,] The ions are formed in a radiofrequency trap ion source, located in the terminal of a 400 keV Van de Graaff accelerator. It has been demonstrated previously[2,3] that the vibrational excitation of the ions can be controlled by varying the source pressure and the potential applied between the source body and the extraction electrode which is used to draw ions out of the source. If a high potential, (100V) is applied, the ions tend to be excited, if the potential is low, (10V), the ions are vibrationally cold.

The ions are mass analyzed and passed into the interaction following electrostatic cleanup to remove neutrals formed during transit. The electron beam is produced from a barium-oxide indirectly heated cathode and initially travels parallel to the ion beam. It is made to merge with the ions using a trochoidal analyzer which employs an axial magnetic field of 25 gauss and a transverse electric field of 80V to shift the electrons from their initial path to one parallel to, but offset from it. The ions and electrons interact over a distance of 8.6cm before the electrons are demerged again and collected in a faraday cup. The ions and neutrals formed in the interaction region are separated electrostatically, the ions being collected in a second faraday cup while the neutrals are detected using a surface barrier detector. Signals due to background gas collisions are separated from true recombination signals by modulating the electron beam and counting the neutrals in and out of phase with the modulation voltage. Electron currents of 60 microamps and ion currents of 0.1nA were used in this measurement. The electron beam was 2mm in diameter and the ion beam slightly less than this. The overlap of the two beams was measured using a rotating scanner and the effective collision area determined as described in ref.9. Absolute errors in the measurement are estimated to be about 15% and primarily reflect the accuracy in the measurement of the effective collision area.

A new merged beams apparatus, (MEIBE II) was used for this experiment. It is similar in design to the original (MEIBE I) apparatus but it has a better vacuum system and routinely operates at 1×10^{-10} torr. Recent high resolution measurements of H_2^+ recombination have shown that the energy resolution achievable in MEIBE II in the region below 0.1eV of centre of mass energy is 5meV or better.[10]

RESULTS AND DISCUSSION.

Cross sections for the dissociative recombination of D_3^+ ions produced under low extraction conditions, (where a voltage of about 10V is applied to an electrode directly in front of the exit aperture of the source) are shown in figure 1. Results for ions formed at high extraction, (where the applied voltage is 100V) and cross sections previously measured[11] using ions produced in a conventional r.f. source, are shown in figure 2. An estimated 60% of the rf source produced ions were vibrationally excited. Previous experience with the trap source lead us to believe that the low extraction results apply to ions in the ground vibrational state while the high extraction condition produces excited ions, but confirmation of this will require measurements of the thresholds for the dissociative excitation processes:-

$$e + D_3^+ \rightarrow D + D_2^+ + e$$

As in the case for H_3^+ this process should display a threshold at 14.9eV. If the ions are vibrationally excited then the threshold will be shifted to lower energies and so the internal energy of the ions can be measured.

Two window resonances are clearly visible in the results shown in figure 1. The origin of this structure is presumably due to interactions involving rydberg states, the so-called "indirect recombination" mechanism represented by:

$$e + AB^+ \rightarrow AB^R \rightarrow AB' \rightarrow A^* + B$$

Theoretical studies of the effects of rydberg state interactions in the presence of a strong direct capture process have predicted the window resonances although under some circumstances, peaks can also occur.[14] For the case of H_3^+ and its isotopomers, direct capture is apparently ruled out[4] for vibrationally cold ions and so only the indirect mechanism is available for recombination to occur. Interestingly, the resonance structure seen at 0.15eV in the low extraction case is also apparent in the high extraction and rf source results.

Figure 3 shows previously published results for H_3^+ recombination with again cross sections being shown for ions formed in a conventional rf source and for ions formed in the trap source at high and low extraction. It can be seen that the measured cross section for de-excited D_3^+ are about the same magnitude as those for de-excited H_3^+ at low energies although they seem to fall off less steeply above 0.1eV. This is energy dependance is also seen for the vibrationally excited ions formed in a conventional radiofrequency ion source although in this case the cross section for D_3^+ recombination is smaller than for H_3^+ by about a factor of 2 at 0.1eV. Calculations for the dissociative recombination cross section for H_3^+ and D_3^+ have not been performed. It is hoped that these experimental studies will provide encouragement for theoreticians to tackle this difficult but very important problem.

ACKNOWLEDGEMENTS.

This work is supported by the US Air Force Office of Scientific Research and by the Canadian Natural Sciences and Engineering Research Council.

REFERENCES.
1. Production and Neutralization of Negative Ions and Beams (Ed. James G. Alessi) AIP Conf. Proceedings No. 158 1987. and previous proceedings referenced therein.
2. Hus, H., Yousif, F.B., Sen, A. and Mitchell, J.B.A. Phys. Rev. A38, 658, 1988.
3. Dissociative Recombination: Theory, Experiment and Applications (Eds. J.B.A. Mitchell and S.L. Guberman) World Scientific, Singapore, 1989.
4. Michels, H.H. and Hobbs, R.H. Astrophys. J. 286, L27, 1984.
5. Mitchell, J.B.A. and Yousif, F.B. in Microwave and Particle Beam Sources and Directed Energy Concepts, (Ed. H.E.. Brandt) SPIE Proceedings Vol 1061, 1989, p536.
6. Johnsen,R., (Private Communication).
7. Auerbach, D.J.,Cacak, R., Caudano, R., Gaily, T.D., Keyser, C.J., McGowan, J.Wm., Mitchell, J.B.A. and Wilk, S.F.J. J Phys. B. 10, 3797, 1977.
8. Mitchell, J.B.A. in Atomic Processes in Electron-Ion and Ion-Ion Collisions (Ed. F. Brouillard) Plenum Publishing Co., New York, 1986, p185.
9. Keyser, C.J., Froelich, H.R., Mitchell, J.B.A. and McGowan, J.Wm. J. Phys. E. 12, 316, 1979.
10. Van der Donk, P., Yousif, F.B. and Mitchell, J.B.A. in preparation.
11. Mitchell, J.B.A., Ng, C.T., Forand, J.L., Janssen, R. and McGowan, J.Wm. J. Phys. B. 17, L909, 1984.
12. Hus, H., Yousif, F.B., Noren, C., Sen, A. and Mitchell, J.B.A. Mitchell Phys. Rev. Lett. 60, 1006, 1988.
13. Walls, F.L. and Dunn, G.H. J. Geophys. Res. 79, 1911, 1974.
14. See theoretical papers in ref.3.

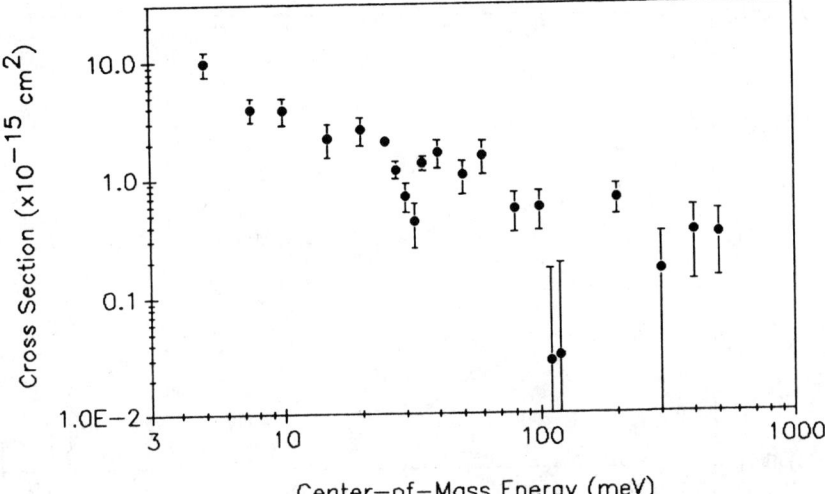

1. Cross sections for the dissociative reombination of D_3^+ produced in a trap source at low extraction potential.

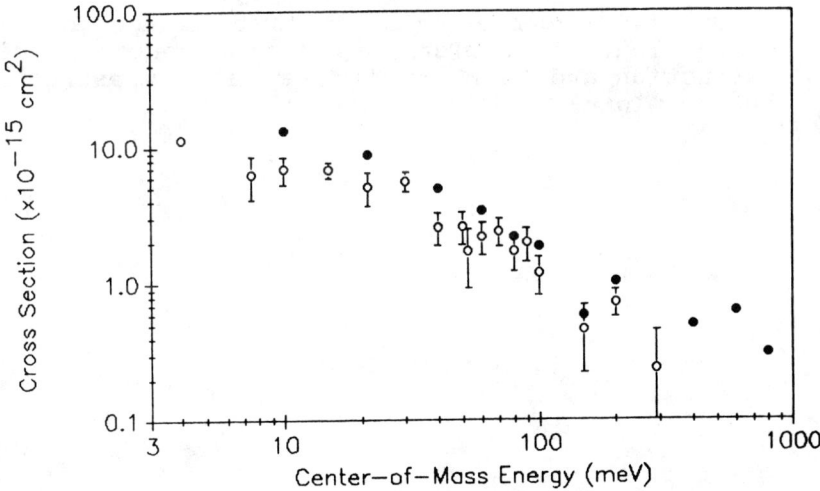

2. Cross sections for the dissociative recombination of D_3^+ ions produced in a trap source with ○ high extraction, and ● formed in a conventional rf source.(11)

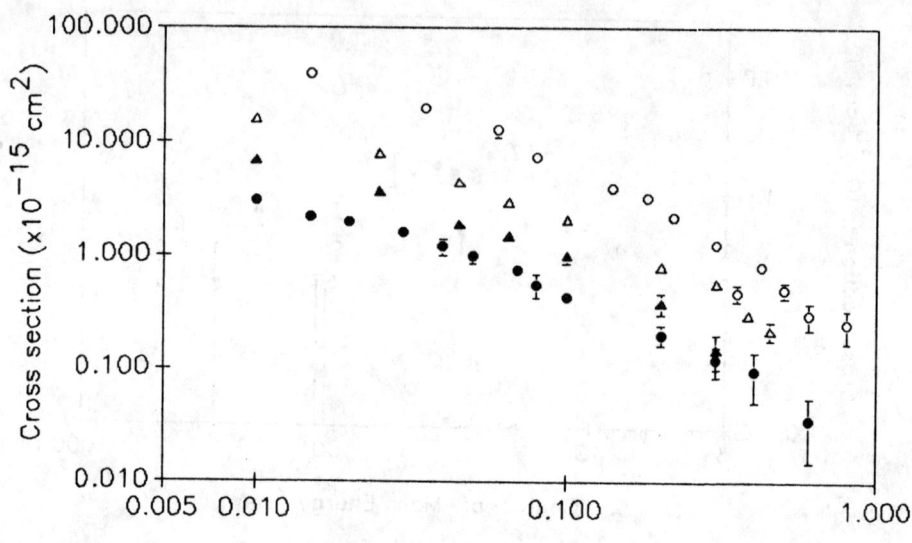

3. Previously measured cross sections for the dissociative recombination of $H_3^+(2)$ produced O in an r.f. source, △ a trap source at high extraction and 10 mTorr, ▲ a trap source at high extraction and 30 mTorr and ● at low extraction and 70 mTorr.

VIBRATIONAL POPULATION OF H_2 PRODUCED BY A DISCHARGE

C. Schermann, R.I. Hall, M. Landau, F. Pichou
Laboratoire de Dynamique Moléculaire et Atomique
Université P. et M. Curie. 75252 Paris Cedex 05. France.
I. Čadež
Institute of Physics
P.O.B. 57, 11000 Belgrade. Yugoslavia.

ABSTRACT

Influence of a discharge (up to 3A) on the vibrationally excited H_2 populations has been studied in a small hot filament source. Varying the intensity of the discharge and H_2 pressure in the gas cell, we observe a large increase of the lower vibrationally excited levels (v = 1 to 4) whereas higher levels only undergo weak enhancement.

INTRODUCTION

It is accepted[1] that high vibrationally excited states of H_2 play an important role in the production of H⁻ by dissociative attachment in negative ion sources. We have recently shown that high vibrationally excited H_2 molecules (v≤9) are produced in a source by simply heating a filament[2] (see accompanying paper[3]). The mechanism for the formation of these $H_2(v)$ molecules has been interpreted by invoking a two step mechanism : atomization of the H_2 molecules on the filament followed by recombination of the H atoms adsorbed on the cold walls of the source. The present work shows the influence of a discharge produced inside the source, on the H_2 (v) populations.

Most experimental techniques employed to detect ro-vibrational molecular state populations, use lasers because of the high resolution and sensitivity of optical methods. However, because of the unique characteristics of its dissociative attachment cross sections, H_2 offers a simple and efficient method for detecting high vibrationally excited levels. Important theoretical efforts have been devoted to the description of dissociative attachment in H_2 [4,5,6] stimulated by the experimental results of Allan and Wong[7]. The dissociative attachment cross sections are characterized by the fact that the magnitude increases rapidly from $10^{-21} cm^2$ for $v = 0$ to a few times $10^{-16} cm^2$ for $v = 6$ and remains roughly constant for higher levels. Moreover they are strongly peaked at threshold.

Our method is based on detection of H^- produced by this reaction viz. :

$$e + H_2(v) \longrightarrow H_2^- \longrightarrow H^- + H$$

The H^- are detected at their respective threshold and their spectrum is observed from 3.7eV (for the $v = 0$ ground state level) down to 0 eV as the incident electron energy is swept (see below).

EXPERIMENTAL SET-UP

The experimental apparatus is schematically shown in figure 1. The electron beam of variable energy is produced by an electron gun comprised of an electrostatic 127° energy filter and its electron optics, and crosses at right angles H_2 molecules effusing from the gas source. The negative ions created by

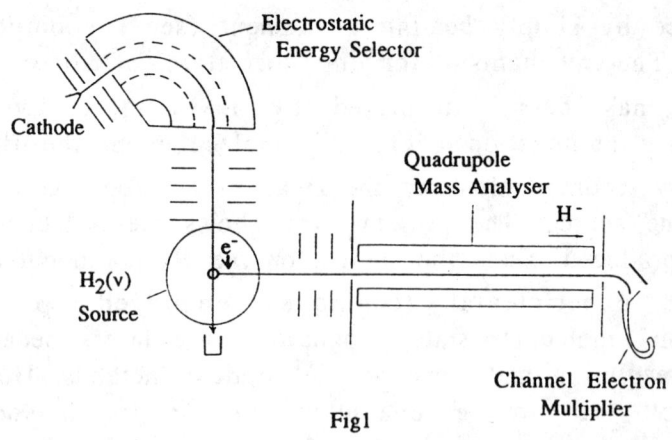

Fig1

dissociative electron attachment in the interaction region are focused by ion optics which also induce a potential well in the interaction center so that the detection of all low energy ions is strongly favoured. H⁻ are then separated from electrons by a quadrupole mass filter and detected by an off-axis channel electron multiplier.

The $H_2(v)$ source is presented in figure 2. The cell, which has an internal diameter of 22mm, a height of 55mm and an orifice of 3mm, is realised in stainless steel, and is shielded by Mumetal to avoid deviation of the incident beam by the magnetic field due to the filament heating. The source wall can be cooled by air or water circulation and its temperature is measured by a thermocouple. We can also determine, the gas pressure in the source, by means of a capacitance gauge, and the filament temperature from its resistivity.

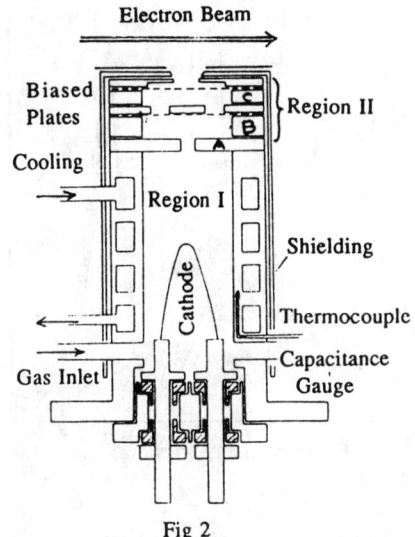

Fig 2

When applying some 50V between the filament and the wall a discharge up to 3A can be produced. In this case, a large number of charged particles are produced in the source which have to be prevented from leaving by biased baffle plates and grids. Otherwise these particles swamp the detection set-up with noise and also perturb the scattering region. Several geometries have been tried for these biased grids which have an influence on the path length of the $H_2(v)$ molecules before their effusion from the source, and thus on their relaxation. This will be discussed in the following.

RESULTS AND DISCUSSION

Figure 3 shows typical spectra obtained with a tungsten filament (heated near 3000 K) with and without discharges (noise substracted).

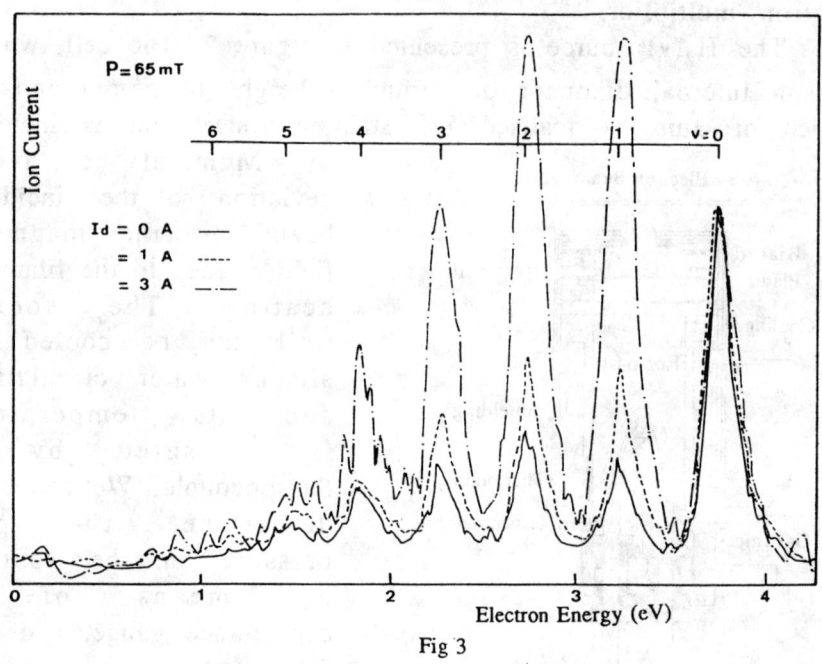

Fig 3

For each spectrum the electron incident energy is swept from 0 to 4.5 eV and the peaks correspond to dissociative attachment to molecules with decreasing v : as the internal energy decreases higher electron energy is required to produce dissociation. In this figure we see clearly that the discharge strongly enhances the lower levels but, as v increases, the peaks are less and less favoured. Owing to the fact that only one half of the v = 0 population is issued from the source - the other part coming from the residual gas in the vacuum chamber - the v = 1 and 2 peaks exceed the ground level by a factor 4.5 for I_d = 3A. This phenomenon is much more important as the discharge current I_d increases; however saturation is observed beyond 2.5 A. Such spectra have also been recorded for different gas source pressures at fixed discharge current (1A) : the v = 1 and 2 levels are slightly enhanced relative to higher levels as pressure increases.

These results are illustrated in figure 4 which gives the ratio of vibrational populations obtained with and without the discharge, for each v level. This ratio can reach 4.5 under favourable conditions for v=1, and rapidly diminishes with v. On the other hand, no important evolution is observed by varying the heating of the filament.

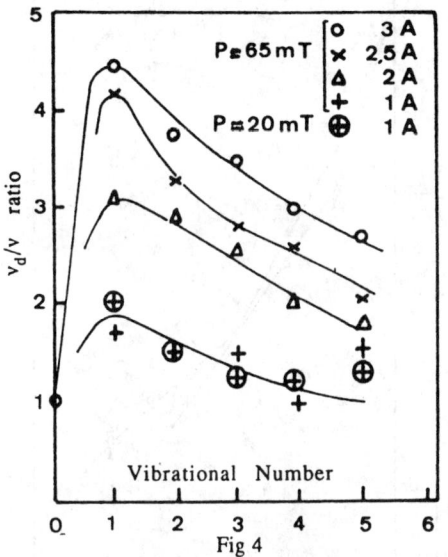

Fig 4

Figure 5 presents, on a semi-log scale, the relative vibrational populations versus the internal energy. These curves are deduced from experimental spectra using theoretical dissociative attachment cross sections of Gauyacq and al.[5] and synthesizes the preceding results.

Without a discharge, the distribution of these populations can be approximated by a straight line between v = 1 and v = 5. This straight line corresponds to a distribution at an equivalent temperature of 2350K. With increasing discharge intensity, the populations deviate more from this line, mainly for v = 1. Then for a strong discharge (beyond 2.5 A) and high pressure (beyond P = 60 mTorr) we observe a straightening of the distribution including v = 0 with an equivalent temperature around 2200K.

Turning back to figure 3, one can notice that in these spectra even without a discharge, only levels up to v = 5 were detected. In previous experiments[2] levels up to v = 9 were observed. The disappearance of the high v populations is due to the presence of the baffle plates at the exit of the source. In order to observe higher vibrational levels, the source cell was modified : the electrodes B and C (fig 3) were suppressed and the thickness of the first exit plate A was reduced to 1mm. However under these conditions the noise is high and observations are difficult and cannot be made above 1A.

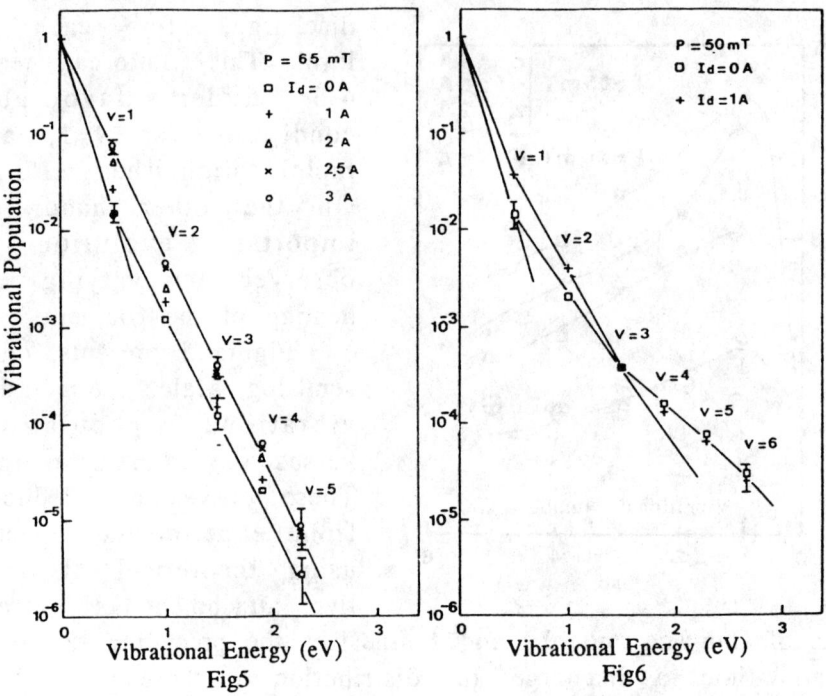

Fig5

Fig6

Figure 6 shows that now, the populations are less relaxed and levels v = 4 to 6 levels are more strongly populated, and the corresponding distribution is not that of Boltzmann. In the first source set-up a molecule undergoes about 30 collisions with the walls in region II (figure 2) after it exits from region I and is more relaxed. For the second set-up, where the path of the molecules is shortened, it undergoes only about 5 collisions. For the first three levels the population variations versus pressure or discharge current are the same for the two set-ups, but they are quite different for higher v. For the second set-up the v = 4 to 6 populations are only weakly dependent on these two parameters, the ratio of these levels to v = 0 varying by a factor between 0.8 and 1.4 . This seems to indicate that, even in the presence of a discharge, the preferential process for exciting molecules to $v \geq 4$ is the atomic H interaction with surfaces. However, we observed that the excitation to $v \geq 4$ tends to weaken during the discharge. When we replaced the tungsten filament by tantalum, higher populations for $v \geq 4$ were obtained. When the discharge is lit, the high level populations drastically diminish and do not reappear when it is turned off. The superficial layer of tantalum evaporated from the filament which covers the source walls and on which the atomic hydrogen recombines must be altered by the discharge and the initial populations are recovered only after a long bake out of the source cell with metal being evaporated from the filament to the walls. Thus the state of the wall surface of the source plays, particularly in the case of tantalum, an important role in determining the vibrational populations above $v = 3$.

CONCLUSION

We have studied the effects of a discharge on the vibrational populations of the H_2 molecules effusing from a small discharge source. The plasma densities obtained would be comparable to those prevailing in a large bucket source but the surface to volume ratio of our source is much greater. The discharge is observed to principally populate the low vibrational levels ($v < 4$) whereas high levels are still populated by the surface mechanism of recombination of atomic hydrogen. Results obtained with a tantalum filament and hence tantalum covered surfaces indicate that these surfaces are irreversibly modified by the discharge and lose much of their ability to populate high vibrational levels. Observations with higher discharge currents required that, in order to reduce the noise in our detection system to acceptable levels, the effusion path length of hot molecules be increased. The observed vibrational distributions were more relaxed and corresponded to a Boltzmann distribution. Again the discharge was seen to populate preferentially the low vibrational levels.

REFERENCES

[1] C. Gorse, M. Capitelli, J. Bretagne and M. Bacal, Chem. Phys. 93,1 (1985).
[2] R.I. Hall, I. Čadež, M. Landau, F. Pichou and C. Schermann, Phys. Rev. Lett. 60, 337 (1988).
[3] R.I. Hall, M. Landau, F. Pichou, C. Schermann and I. Čadež, in this volume.
[4] J.M. Wadhera and J.N. Bardsley, Phys. Rev. Lett. 41, 1795 (1978).
[5] J.P. Gauyacq, J. Phys. B : Atom. Molec. Phys. 18, 1859 (1985).
[6] C. Mundel, M. Berman and W. Domcke, Phys. Rev. A 32,181 (1985).
[7] M. Allan and S.F. Wong, Phys. Rev. Lett. 41, 1791 (1978).

H⁻/D⁻ ION SOURCES

CESIUM MIXING IN THE MULTI-AMPERE VOLUME H⁻ ION SOURCE

Y. Okumura, M. Hanada, T. Inoue, H. Kojima, Y. Matsuda,
Y. Ohara, M. Seki, and K. Watanabe

Japan Atomic Energy Research Institute, Naka-machi,
Naka-gun, Ibaraki-ken, 311-01, Japan

ABSTRACT

A 7.8 A, 50 keV H⁻ ion beam was produced by a cesium seeded volume negative ion source. The source consists of a 25 cm x 46 cm rectangular multicusp plasma generator and a 14 cm x 36 cm multiaperture extractor. Without cesium, the source produced 3.4 A, 75 keV H⁻ ion beams. By seeding a small amount of cesium, we observed a big enhancement of H⁻ production efficiency by a factor of four, and a reduction of optimum operating pressure. Extracted electron current decreased to almost zero when we biased the plasma grid positive with respect to the anode. The effect lasted for more than a week once the cesium was injected for several seconds at an oven temperature of 280-300 °C.

INTRODUCTION

Japan Atomic Energy Research Institute (JAERI) has a plan to construct a negative ion based neutral beam injector for the JT-60 Upgrade tokamak[1]. The H⁻/D⁻ ion source utilized will be a pure volume negative ion source because of its simple structure, good beam performance and good reliability. The present design of the source, which is shown in Fig. 3 in Ref.1, is based on the JAERI multi-ampere H⁻ ion source which has produced 75 keV, 3.4 A H⁻ ion beams[2]. Although the current density achieved in the multi-ampere source was about 25 mA/cm2 at the extraction surface, the design value is chosen to be 12 mA/cm2 (9 mA/cm2) for hydrogen (deuterium). This is because the reduction of operating pressure and extracted electron current is necessary for long pule operation at the expense of the current density. This makes the ion source very large; e.g. the extraction area is 52 cm x 150 cm, and the diameter of the accelerator column is 2.5 meter. Further improvement of the source is preferable to make the ion source smaller and the negative-ion based neutral beam system compact.

In these years, it is reported that the volume source is improved remarkably by mixing cesium vapor in a hydrogen discharge[3,4]. Walther et al. reported that the H⁻ output was enhanced by a factor of 16 by seeding cesium into a small multicusp volume source[5]. At the same time, they observed a substantial decrease of the electron current. It is of great interest to investigate the cesium effect in a large scale multicusp volume source which is directly applicable to neutral beam injectors.

In the present paper, we report that the same effect was

observed in the multi-ampere source. The H⁻ ion current increased to 7.8 A by seeding a small amount of cesium. The electron current and the optimum operating pressure decreased remarkably. In order to study on this effect, cesium ion content, hydrogen ion species ratio and the electron temperature in the plasma were measured by a momentum mass analyzer and a Langumuir probe.

EXPERIMENTAL SETUP

A schematic of the multi-ampere source is shown in Fig.1. The plasma generator is a rectangular water-cooled copper chamber whose dimension is 24 cm x 48 cm in area and 15 cm deep. The chamber is surrounded by Nd magnets to form continuous line cusps perpendicular to the source axis. Two rows of the line cusps at the end of the chamber has a same polarity. The polarity is different on each side wall resulting in a transverse magnetic field, which acts as a magnetic filter. The arc discharge is created by electron emmission from eight tungsten filaments.

The H⁻ extractor consists of five grids called plasma grid, extraction grid, electron suppressor grid, ion suppressor grid, and grounded grid. Each grid has 253 apertures of 11.3 mm diameter within the rectangular area of 14 x 36 cm2. The H⁻ ions are

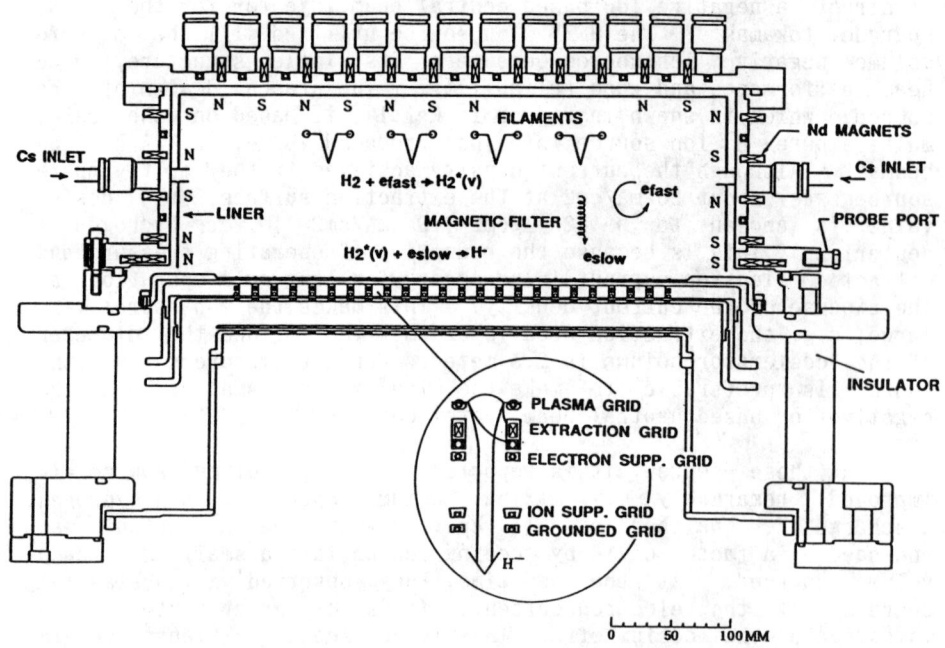

Fig. 1 A schematic of JAERI multi-ampere H- ion source.
Cesium oven is attached in the present experiment.

extracted by a potential between the plasma and the extaction grids, and accelerated up to 75 keV by a potential between the extraction and the grounded grids. The electron suppressor and the ion suppressor grids are connected electrically to the extaction and the grounded grids, respectively. The structure of the original source is given in Refs. 2 and 6.

In the present experiment, an oven containing metallic cesium is attached to the plasma generator via a valve. The cesium vapor was injected into the source from two guiding tubes on the side wall. To prevent condensation of the cesium vapor, the valve and the guiding tubes are heated at the temperature higher than the oven. The inner surface of the plasma generator is covered by a sheet metal liner, which is also heated by the discharge to prevent the condensation.

Two different types of the plasma grids have been tested; one is water cooled copper grid and the other is uncooled molybdenum grid. The molybdenum grid is insulated thermally by ceramic spacers. The temperatures of the plasma grid and the liner were measured by thermocouples during the operation.

The accelerated H^- ion current was measured calorimetrically by a two-dimensional multi-channel calorimeter placed 1.8 m downstream of the ion source. The ion source was typically operated for 0.2 second every 60 seconds. The vacuum was maintained by three turbo-molecular pumps of 2,000 l/s, a cryo-condensation pump of 10,000 l/s and three big cryopumps of 100,000 l/s.

EXPERIMENTAL RESULTS

The ion source was first operated without cesium. Figure 2 shows the dependence of H^- ion current on the arc discharge current. The H^- current increases with the arc current and then tends to saturate, which is a typical tendency of the multicusp volume source. The H^- current achieved was 1.2 A at an arc current of 400 A. The arc efficiency defined by H^- current per unit arc discharge power was 0.043 A/kW. This is about 80 % of the value obtained in the original multi-ampere source. The reduction of the current is partly due to increase of loss area in the liner, and partly due to increase of wall temperature. Indeed, we observed 15% reduction of H^- current when the liner temperature increased to 300°C.

Then the cesium vapor was introduced to the source by opening the valve at an oven temperature of 280°C. The duration of opening was only five seconds. After several tens of shots with electrical breakdown, the H^- current increased by about a factor of two. It is shown in Fig. 2 by the line denoted by "8/8", which means the data was obtained on August 8th. It should be noted that no saturation occured in the H^- ion current; the current increased linearly with the arc current even at a high arc current. The arc efficiency was 0.086 A/kW. Although we operated the source for more than a week

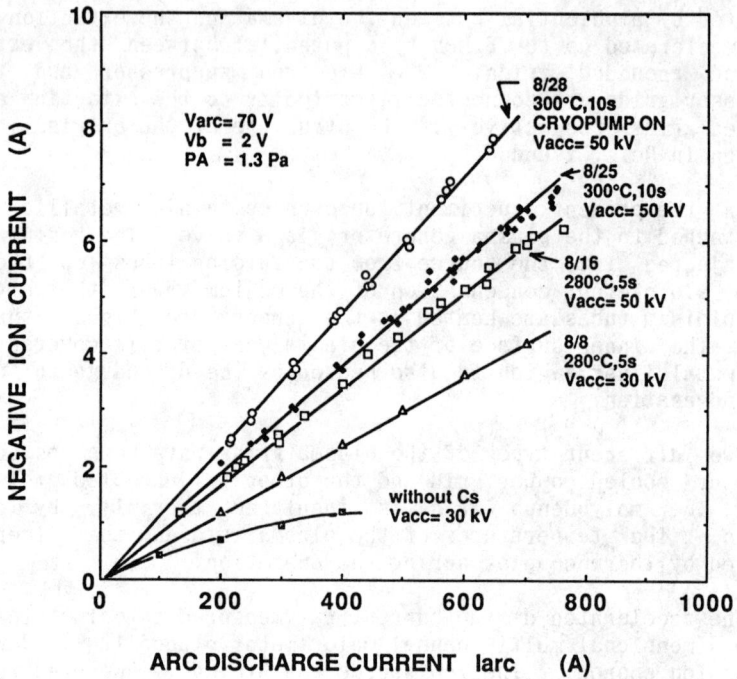

Fig. 2 H- ion current as a function of arc discharge current with and without cesium.

without additional cesium, the efficiency scarecely changed and the data were reproducible.

After eight days' operation, cesium was introduced again at the same oven temperature. The H⁻ current increased further, as shown in Fig. 2 by the line of "8/16". The arc efficiency was 0.13 A/kW. We operated the source for nine days without additional cesium at the same arc efficiency, and then introduced more cesium at 300°C for 10 sec. The arc efficiency improved further, as shown by the line of "8/25". The line denoted by "8/28" in Fig.2 shows the data when the big cryopump was operated to reduce the stripping loss of H⁻ ions in the accelerator column.

The highest H⁻ current of 7.8 A was obtained at beam energy of 50 keV for 0.1 sec with the arc efficiency of 0.17 A/kW. The operating parameters are shown in Table I together with those obtained in the original multi-ampere source. It should be noted that the extraction current, which is the power supply current for extraction, decreased appreciably when the cesium was seeded. The extraction current is a sum of the extracted electron, the extracted

Table I Operating parameters with and without Cesium

	With Cs	Without Cs
H- Ion Current (A)	7.8	3.1
Arc Current (A)	650	1180
Arc Voltage (V)	70	70
Extraction Current (A)	14.2	26.6
Extraction Voltage (kV)	5.0	4.5
Acceleration Current (A)	13.4	5.8
Acceleration Voltage (kV)	50	50
Bias Voltage (V)	2	3.3
Gas Pressure (Pa)	1.3	2.1

H$^-$ ion and backstreaming H$^+$ ion currents. Since the extracted H$^-$ ion current is estimated to be about 12 A taking into acount the stripping loss, the extracted electron current is considered to be almost zero in the cesiated ion source. As reported elsewhere[2], the extraction current or the electron current is a strong function of the bias voltage that is applied to the plasma grid with respect to the anode. Figure 3 shows the extraction current with and without the cesium. The extraction current is smaller in the cesiated source for the same H$^-$ ion current. It decreases more rapidly when the plasma grid biased positively, suggesting that the electron temperature in the cesiated source is lower than that in the uncesiated source.

Another outstanding feature of the cesium effect is a reduction of optimum operating pressure. Figure 4 shows the H$^-$ current as a function of pressure in the arc chamber for various arc currents. This figure should be contrasted with Fig. 5 that was obtained in the original multi-ampere source. In both cases, there are optimum pressures giving highest H$^-$ currents for each arc current. The optimum pressure increases with the arc current in the original multi-ampere source. This is a typical tendency in a multicusp volume source. In the cesiated ion source, however, the optimum pressure is almost constant no matter how the arc current is. In other words, it is not necessary to increase the operating pressure to obtain high H$^-$ current. This is preferable for ion sources for the neutral beam injector, because the stripping loss of negative ions in the accelerator column can be suppressed by decreasing the pressure.

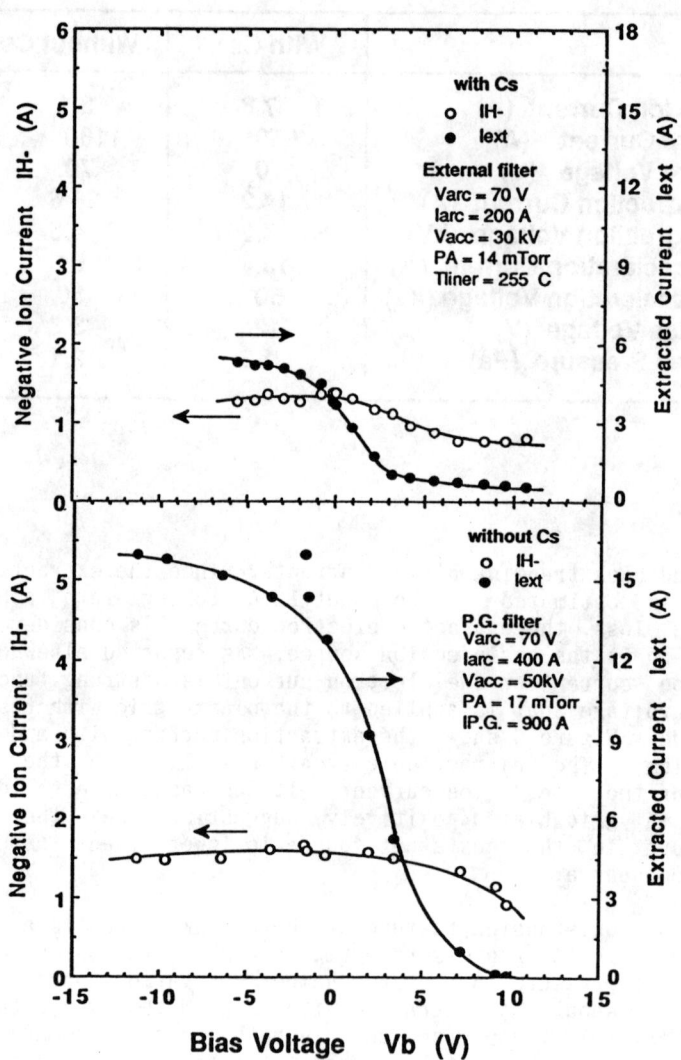

Fig.3 Dependence of H- current and extraction current on bias voltage of the plasma grid, with (upper) and without (lower) cesium.

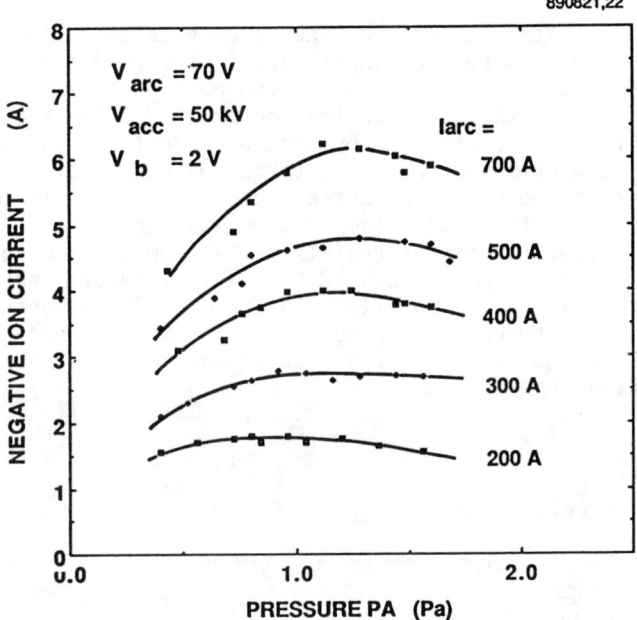

Fig.4
H⁻ current vs. pressure in the plasma generator for various arc discharge currents. The cesium was injected at 300°C, 10 s.

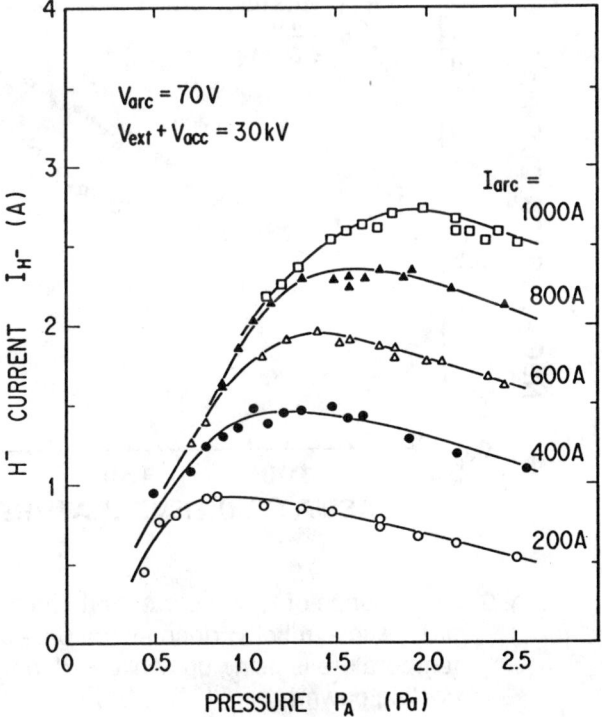

Fig. 5
H⁻ current vs. pressure in the plasma generator in the original multi-ampere source. (Ref.6)

Results described above were obtained in the molybdenum plasma grid, where the tempereture of the grid are considered to be high enough to prevent the cesium condensation. Then we replaced the grid to the water-cooled copper grid. The temperature of this grid can be controlled by turning on/off the water and turning on the filaments during a interval of each shot. Figure 6 shows the dependence of H$^-$ current on the temperature of the plasma grid, where the liner temperature was kept at more than 300°C. The H$^-$ current increases rapidly as the temperature increases above 150 °C. After reaching 220 °C, the temperature was reduced gradually, resulting in a slight increase of H$^-$ current at around 180-200 °C then the current decreased to the initial value. This hysteresis is reproducible. We have tried more than ten times and observed the same hysteresis.

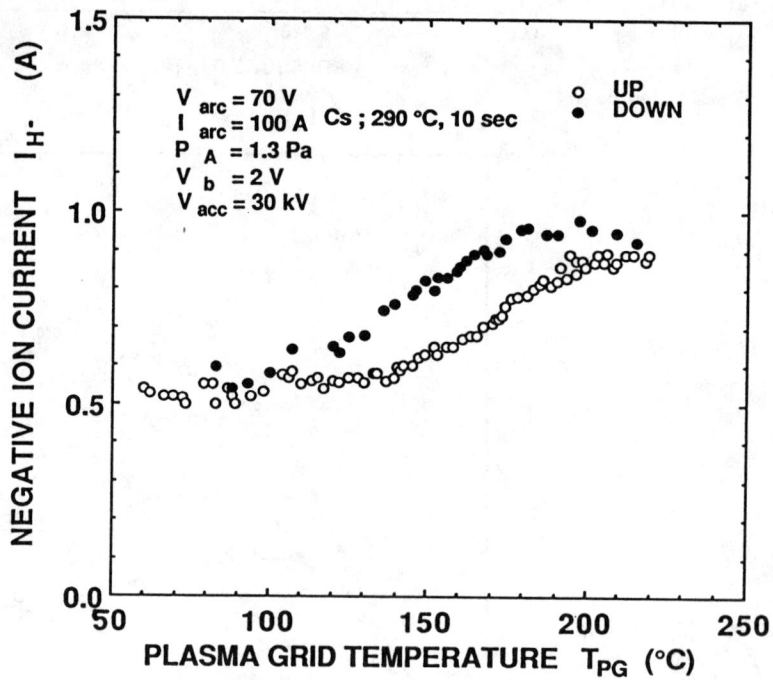

Fig. 6 Dependence of H- ion current on the temperature of plasma grid. The symbol o denotes the data obtained when the temperature is rising up, while ● denotes the data when cooling down.

Figure 7 shows electron temperature measured by a Langumuir probe inserted 20 mm above the plasma grid surface. The electron temperature decreases as the temperature of the plasma grid increases. It is considered that this reduction of the electron temperature is due to the colling effect of cesium. As the plasma grid temperature rises, more cesium evapolates from the grid and it cools the electron via processes of excitation and ionizing collisions. This reduction of electron temperature decreases the H^- loss by electron collisional detachment, and at the same time, may enhance the dissociateve attachment rates.

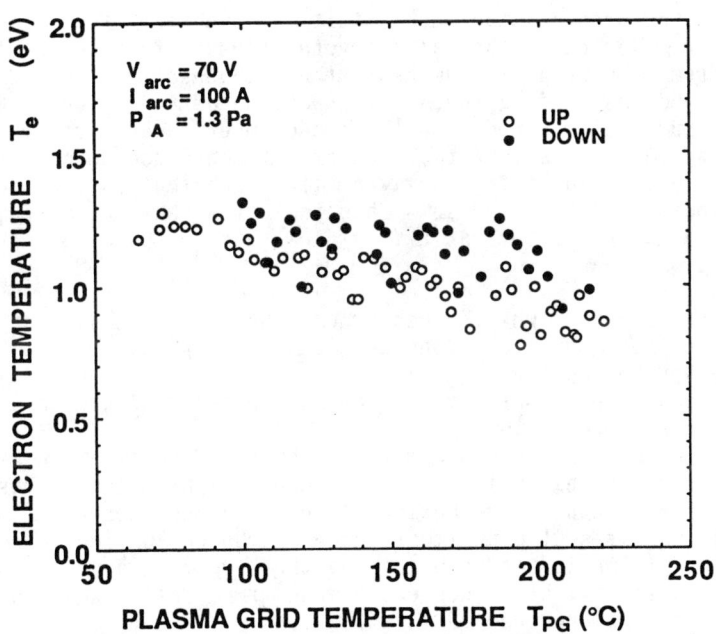

Fig. 7 Variation of electron temperature when the plasma grid temperature was rising up and cooling down.

To estimate the cesium ion concentration in the discharge, positive ions were extracted from the source by reversing the polarity of both the extraction and the acceleration power supplies. The ion species in the extracted beam was analyzed by the momentum mass analyzer with a same manner described in Refs. 7 and 8. Figure 8 shows the cesium ion content in the beam as a function of plasma grid temperature. The cesium ion content increases as the temperature increases, showing a same tendency as the H^- current except for the hysteresis. The cesium ion content in the beam was about 1.2 % at the plasma grid temperature of 200 °C, which corresponds to about 10 % in density ratio in the plasma provided the effective mass of hydrogen ions is 2.

Figure 9 shows the hydrogen ion specie ratio as a function of the plasma grid temperature. There is a tendency that the H_3^+ ion ratio decreases slightly with the temperature.

DISCUSSION

In the present experiment, initial inventory of metalic cesium was 1 g. We opened the valve several times before no cesium came out from the oven. Assuming that no condensation occured in the valve and the guiding tubes, it is estimated that about 100-200 mg of cesium was injected into the arc chamber in each valve opening. Once the cesium was injected, the effect continued more than a week. The H^- production efficiency was almost constant during more than 3000 shots of 0.2 sec pulse. This indicates that the cesium leaves the plasma generator at a very slow rate in spite of the large extraction area.

The reasons why the cesium can enhance the H^- output in the multicusp ion source are considered as follows[5,9];
1) Electron cooling
2) Production of $H_2(v)$ via a reaction between Cs atom and H_3^+
3) Surface production

Walther et al. proposed one more reason that the electron density in the extraction region increases because of ionization of cesium and diffusion through the magnetic filter[6]. We consider this is not the case in the present experiment, because the cesium ion concentration in the plasma is not so high, and the electron density near the plasma grid slightly decreases when the cesium was evapolated because of the presence of negative ions.

Antipov et al. obserbed a strong relationship between the H^- and the H_3^+ densities in their cesiated source[3]. They suggested that the enhancement of H^- current is due to production of highly excited H_2 molecules via process of collisions between cesium atoms and H_3^+ ions. In the present experiment, we obserbed a same tendency that the H_3^+ ion ratio decreases slightly when the H^- current increases with the plasma grid temperature.

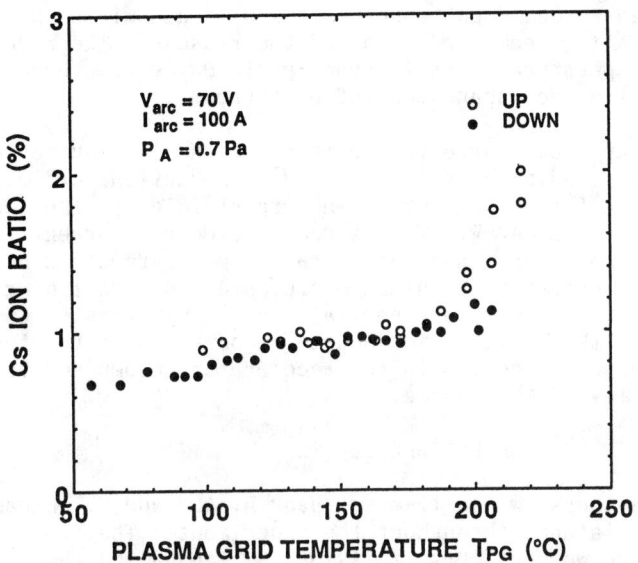

Fig. 8 Cs ion content in the positive ion beam, measured by momentum mass analyzer.

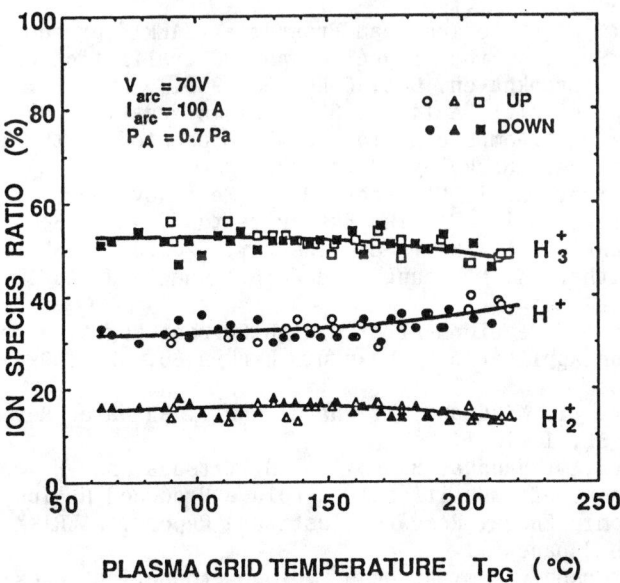

Fig. 9 Dependence of hydrogen ion species ratio on the plasma grid temperature.

Although we could not conclude which is a dominant process, the electron cooling seems to be one of the reasons. The reduction of electron temperature with increasing plasma grid temperature is consistent with the enhancement of H^- current.

In summary, our large multicusp volume H^- ion source was highly improved by seeding a cesium. The H^- production efficiency was enhanced by a factor of four. The arc efficiency increased from 0.043 A/kW to 0.17 A/kW. The extracted electron current could be suppressed almost zero, and the operating pressure could be reduced to 0.5 Pa. These reduction are preferable for designing the negative ion sources for neutral beam injectors. Long pulse operation of the cesiated ion source is now planned to investigate the condensation of cesium in the accelerator column and to confirm the reliability of the source.

ACKNOWLEDGEMENTS

The authors would like to thank H. Abe and M. Kawasaki for technical assistance throughout the experiment. They would like to thank their co-workers, Drs. M. Akiba, S. Tanaka and the members of plasm heating laboratory, for valuable discussion. Thanks are also due to Drs. M. Tanaka and S. Shimamoto for support and encouragement.

REFERENCES

[1] Y. Ohara: "Negative Ion Beam Program at JAERI" presented at the 5th Int. Symp. on the Production and Neutralization of Negative Ions and Beams, Brookhaven, Oct.30-Nov.3, 1989.
[2] M. Hanada, et al.: "A 14cm x 36cm Volume Negative Ion Source Producing Multi-ampere H^- Ion Beam" presented at the Int. Conf. on Ion Sources, Berkeley, June 1989.
[3] S. P. Antipov, L. I. Elizarov, M. I. Martynov, and V. M. Chsnokov: Pribori I Texnika Eksperimenta 4, 42 (1984).
[4] K. N. Leung et al.: Rev. Sci. Instrum. 52, 235 (1987).
[5] S. R. Walther, K. N. Leung, and W. B. Kunkel: J. Appl. Phys. 64, 3424 (1988).
[6] Y. Okumura: " Development of a High Current Negative Ion Source for Fusion Application", Kakuyugo Kenkyu 60, 329 (1988) in Japanese.
[7] Y. Okumura, Y. Mizutani, Y. Ohara, and T. Shibata: Rev. Sci. Instrum. 52, 1 (1981).
[8] Y. Okumura, M. Hanada, H. Kojima, H. Matsuda, and H. Oohara: "Measurement of Impurities in a Volume Produced H^- Ion Beam", Japan Atomic Energy Research Institute Report, JAERI-M 89-090 (1989) in Japanese.
[9] J. R. Peterson: "Comment on H^- Volume Production in Cs-seeded Ion Source", Proc. 4th Symp. on Production and Neutralization of Negative Ions and Beams, Brookhaven, 1989, p.113.

DISCUSSION

Bretagne : I have a question about the improvement of H⁻ density and particularly to this hysteresis effect. You say that the decrease of the electron temperature must increase the dissociative attachment, I agree with this but the other parameter is in fact the electron density. Do you have any values for the electron density.

Okumura: In the region close to plasma grid, the electron density is less than maybe 10^{11} cm^{-3}. But at a position far from the plasma grid, the electron density is very high. More than 10^{12} cm^{-3}.

Dimov : What is the gas efficiency at optimum regimes.

Okumura: The gas efficiency depends on the configuration of the extractor. The conductance of our ion sources is about 3,000 ℓ/s. And operating pressure is 1.3 pascal. So you can calculate the gas flow rate. You can estimate gas efficiency.

Hoffman: Did I understand you correctly, that there is only short puff of cesium admitted to the source and then you worked for a full week, so there is no need to heat your accelerator to prevent coverage of the insulators by cesium, there is essentially no flux of cesium out of your source.

Okumura: That is correct. The consumption of cesium is low and at the beginning of cesium input we observe some breakdown in the accelerator but after 50 shots the ion source is completely conditioned and we have no problem and no breakdowns anymore.

Holmes: What is your operation time, how many seconds per week?

Okumura: More than 3,000 shots with pulse length of .1 to .2 sec.

Hiskes: As you vary from the cesium free to the cesium optimum how does the positive ions species mix change H^+, H^{+2}, and H^{+3}.

Okumura: I will show you the ion species dependance on that on the plasma grid temperature. The H⁻ ion current increases by a factor of 2 but the ion species ratio scarcely changes. That is the answer.

Bacal: Why did you operate always with the same plasma grid potentials? Did you optimize it when you operated with cesium?

Okumura: Sure, we optimized the plasma grid potential. So the optimum point is 2 volts; without cesium we have to bias the plasma grid more than 5 volts to suppress the electron current. But with cesium the electron temperature is lower. Therefore, it is enough to bias the plasma grid to 2 volts to suppress the electron current completely. So we are operating the source at 2 volts.

Seidl: Your surface production negative hydrogen ions could be produced either due to atom bombardment or ion bombardment. What is your estimate of these two fluxes.

Okumura: I have no idea about that estimation, but I think the density of the atomic hydrogen is 10^{11} so the flux is very high. So it might be possible to understand this effect by the atomic bombardment.

Jacquot: I want to understand the real situation concerning the electron extraction. The electron extraction from the source with cesium. You say we have total extraction of 14 amperes with cesium, it's 14 amperes at the final voltage of 50 kV.

Okumura: No, this is a current of the extraction power supply. So the voltage is 5 kV.

Jacquot: And now from this 14 amperes what is the part coming from the plasma and what is the part coming from the stripping in the accelerator gap of 0 to 5 kilovolts.

Okumura: About 12 amperes of this 14.2 ampere is extracted H$^-$ ion current. Electron current is almost 0 or maybe 1 ampere or something like that. Another 1 ampere is backstreaming positive ions.

Jacquot: At the final voltage of 50 kV, what is the part of the electrons, and the part of the ions?

Okumura: The extracted current is 13.4 ampere total current, power supply current. The fully accelerated H$^-$ current is around 8 ampere.

Jacquot: There is no stripping from 0 to 5 kV.

Okumura: No, the H$^-$ current has about 40% ion loss, so it produces electrons. Those electrons are oscillating. So that is a part of this current.

Belchenko: Did you have any change of negative ion current during the pulse length and have you found uniformity of the beam across the beam.

Okumura: We change the pulse length. At first the pulse length is only 50 milli-sec and it is increased to 100 milli-sec, 150 milli-sec, 200 milli-sec and the longest pulse length is 300 milli-sec. And from that experiment the H^- current tends to increase with pulse length.

Belchenko: How much did it increase towards the end of the pulse?

Okumura: Not so much, about 10% or 20%.

PLASMA-VOLUME STATIONARY H^--IONS SOURCE WITH HOLLOW CEZIUM CATHODE

S.P.Antipov, L.I.Elizarov, M.I.Martynov, V.M.Chesnokov
I.V.Kurchatov Institute of Atomic Energy, Moscow, USSR

The main mechanism of negative hydrogen ions formation in modern volume sources is the process dissociative attachment of slow electrons (~ 1 eV) to the vibrationally excited molecules $H_2(v)$, $v > 6$.

Development of the H^--ions source of 1-10 A comes, in first place, to the solving of the following physico-technical problems:
- organization of discharge in such a way that a process of vibrational excitement of H_2 molecules should be intensive in plasma volume;
- method of generation of hydrogen plasma with electron temperature about 1 eV;
- way of electron field creation in plasma volume that provides a preferential movement of H^--ions to the emission surface;
- method of suppression of accompanying electron current.

Moreover, experimentally it was set[1] that the add of cezium steams into the discharge increases significantly the negative hydrogen ions outlet.

To realize these requirements one should put cezium hollow cathode in the source with an axial-symmetric geometry and radial magnetic field[2].

The H^--ion source scheme is presented in Fig.1. The main details are as follows:
- magnetic system 3 that produces radial magnetic field and consists of inner and outer magnet poles;
- cezium hollow cathode 1 ;
- anode 6;
- ion-optical system that consists of electrode 7, intermediate electrode 8 and outlet electrode 9.

The emission electrode 7 is insolated from the discharge chamber. It was sometimes used as the main anode. The magnet poles are hollow cylinders located axially. They are made of magneto-soft steel. Coil 10 is used for the excitation of the magnetic field located in the inner pole. Interpole space is the main discharge region. The H^--ions production takes place here. There are holes in the inner pole for the cezium and electrons outlet into the discharge chamber. The hollow cathode is made of molybdenum tube 6x0,8 mm . There is a diaphragm with an axial hole 0,8 mm in diameter located on the end tube. The cathode is heated with the help of a wire heater 2. The rate of flow cezium vapor control is held at the liquid cezium temperature. Hydrogen is put into the source through the holes in the outer magnetic pole. Its rate of flow is measured with the help of PPK-1 rotameter.

Before start-up the source and cathode are heated with the help of start heater. It switches off after the source outlet is put into working regime. The discharge burns between the anode and the hollow cathode. The ions are extracted in the direction perpendicular to the magnetic field direction through the holes in the emission electrode.

A picture of H^--ions source 140 mm in diameter is shown in

Fig.2.

Local plasma characteristics were studied with the help of a single Langmuir probe. The scheme of the device for probe measurements is shown in Fig.3. The probe moved in plasma volume with the help of an electromagnetic mechanism. The mass-spectrometry measurements results of the accelerated ions beam composition were used for probe characteristics treatment.

Plasma parameters distribution along the discharge chamber axis are shown in Fig.3. As it was expected there is an electric field (~ 4 V/cm) which provides the primary H^--ions movement to the emission electrode. The electric field is located in the region limited by the emission electrode and electrons inlet surface into the discharge chamber.

This electric field must provide the electron temperature increase in the direction to the emission electrode.

Some fall in the electron temperature near the emission electrode is probably connected with cold electrons generation as a result of the negative ions destruction reaction:

$$H_1^- + e \rightarrow H_1^0 + 2e$$

The velocity of the reaction increase quickly as T_e increases. It reaches its maximum at T_e=15 eV ($<\sigma v>_{max}$=7 10^{-7} cm^3/s). The electron concentration growth near the emission electrode can be explained by this process.

The absolute plasma parameters values and the source regime are shown in Fig.3. In Fig.4. one can see the magnetic field distribution in the H^--ions source.

The complicated plasma composition in the negative hydrogen ions source makes it difficult to use the traditional probe method of investigation of the low-temperature plasma.

The most profound and reliable information on the plasma composition one can get with the help of the mass-spectrometry analysis of the ions beam composition.

The extracted beam analysis was held with the help of the small-size magnetic analyzer with a homogeneous magnetic field. Its scheme is presented in Fig.5. It consists of the following elements:

- electromagnet 1 with pole 2, that produce a homogeneous magnetic field in the gap between them and particles receivers;
- electrons 3;
- H^--ions 4;
- heavy ions of any mark 5;
- hydrogen ions 6.

All the analyzer elements are located in a box 7 made of stainless steel. Heat and magnet screens 8 are located in front of the analyzer. The diameter of the analyzer inlet hole is 1,2 mm. The H_1^+, H_2^+, H_3^+ ion beams focusing was performed with the help of the specifically chosen form of the pole magnet outlet boarder. Electron, ion H^- and heavy ion beams were received unfocused. It was possible due to a great difference in the masses of these components. The removal - 12 V was served to the receivers. The analyzer was located in the vacuum chamber on a special device. It allowed one to move the analyzer perpendicular to the beam and to move it up and down in order to combine vertically the beam axis and the analyzer inlet hole. The analyzer allowed one to get the distribution along the deviation angle to the beam axis, not only the relation between the beam components.

The current of any component was measured according to the formulae:

$$I_i = 2\pi r^2 \exp(P\, r\, \sigma_{io}/kT) \int_0^{\varphi_M} j(\varphi) \sin\varphi\, d\varphi$$

where r — distance from source to analyzer;
σ_{io} — exchange cross-section of the corresponding ions into neutrals;
T — temperature and pressure of the gas in vacuum chamber;
k — Boltzmann constant;
$j(\varphi)$ — ion current density distribution, measured experimentally (angularly);
φ_M — analyzer maximal deviation angle from central trajectory.

Experimentally it was set that negative and positive currents angular distributions differ very little. This allowed one to simplify the measurements results treatment process since positive ions full current was also measured.

$$I_i = I_+ \frac{j_i \exp(P\, r\, \sigma_{io}/kT)}{\sum_k j_k \exp(P\, r\, \sigma_{ko}/kT)}$$

where I — full current of positive ions, j_i — current density, measured by the analyzer of any ion component (H_1^+, H_2^+, H_3^+, Cs^+, H^-), — current density measured by analyzer of every positive ion component ($H_1^+, H_2^+, H_3^+, Cs^+$).

The plasma components composition study and its connection with the source working regime is necessary for the clarifying of the main H^--ions generation processes in the plasma of cezium-hydrogen discharge. Note that the results obtained refer to the plasma region close to the emission electrode, not to the whole plasma. The plasma region close to the emission electrode is of great interest since the ion beam extraction takes place in it.

The dependences of the ions and elecrons current densities on the hydrogen rate of flow for three discharge current values: 10 A, 13 A and 17 A are shown in Fig.6. A great number of H^- - ions show to the low electron temperature in the discharge. Note, that under the hydrogen rate of flow change in a wide range of values (from 4 up to 14 cm^3/s) under constant discharge current the electron beam density remains unchanged. Current density and correspondingly, H_3^+-ions concentration grows linearly as the hydrogen expenditure grows, since H_3^+-ions are formated in the discharge as a result of the only reaction:

$$H_2^+ + H_2^0 \rightarrow H_3^+ + H_1^0$$

The H_2^+ concentration remains unchanged. The reaction cross-section is large enough. Under energies 1,2 eV it is 10^{-15} cm^2.

The dependence of H^--ions current density is of complicated character. Its maximum is observed under the expenditures of 6-8 cm^3/s.

The dependence of the plasma composition on discharge current under constant hydrogen rate of flow (14 cm^3/s under normal conditions) is shown in Fig.7.

It is seen that the electron current density grows linearly with the discharge current. H^--ions current density does not grow proportionally to the discharge current, i.e. ions concentration is not connected linearly with the electrons concentration.

It is probably cinnected with the step-by-step mechanism of the negative ions formation:

$$H_2(v=0) + e \rightarrow H_2^- \begin{bmatrix} \rightarrow H^- + H \text{ (single process)} \\ \rightarrow H_2(v>0) + e \end{bmatrix}$$

$$H_2(v>0) + e \rightarrow H_2^- \begin{bmatrix} \rightarrow H^- + H \text{ (cascade process)} \\ \rightarrow H_2(k) + e, \quad k > v \end{bmatrix}$$

since the direct process of the hydrogen negative ions formation does not affect the negative ions generation significantly (see Table 1).

Another peculiarity of the given dependences is the correlated move of the H^- and H_3^--current densities dependences on the discharge current. Such a character of the dependences cannot be by chance. It speaks to the correlation between the processes of these ions formation in the cezium-hydrogen plasmas.

The numerical estimation of the H^--ions current density was 2 10^{-2} mA/cm^2. Factually, the extracted current densities are twice as two orders higher than the numerical ones. The divergence can be explained by the mechanism of H^--ions formation from the vibrational excited H_2 molecules. Using the data on relative populations of the vibrational states in a discharge with parameters close to ours$_4^3$ and the data on H^--ions formation cross-sections from such states one can estimate the contribution of each state and the summary effect of the molecules vibrations on the H^--ions generation. The corresponding values are given in Table 1.

Here v - vibrational level number of the H_2 molecule; $N(v)/N_0$ - part of H_2 molecules on the v-level relatively the number of molecules in the main state; $N(v)<\sigma v>/N_0<\sigma v>_0$ - H^--ions generation velocity from H_2 molecules under excitation relatively the generation velocity from the main state.

It is seen from the Table that the molecules H_2 with the vibrational quantum numbers 6-9 contribute to the H^--ions generation in the discharge mainly. Total H^--ions generation velocity from the vibrationaly excited states of H_2 one can get by summarizing of the relative velocities:

TABLE 1 $T_e = 1$ eV

v	0	1	2	3	4	5	6
$\frac{N(v)}{N_0}$	1	$1.3 \cdot 10^{-2}$	$1.2 \cdot 10^{-3}$	$1.8 \cdot 10^{-4}$	$4.0 \cdot 10^{-5}$	$2.0 \cdot 10^{-5}$	$2.0 \cdot 10^{-5}$
$\langle \sigma v \rangle$	10^{-15}	$6 \cdot 10^{-14}$	$1.5 \cdot 10^{-12}$	$2.0 \cdot 10^{-11}$	$1.4 \cdot 10^{-10}$	$7.2 \cdot 10^{-10}$	$2.9 \cdot 10^{-9}$
$\frac{N(v)\langle\sigma v\rangle}{N_0 \langle\sigma v\rangle_0}$	1	0.78	1.8	3.6	5.6	14.4	58
v	7	8	9	10	11	12	13
$\frac{N(v)}{N_0}$	$2 \cdot 10^{-5}$	$1.5 \cdot 10^{-5}$	$6.1 \cdot 10^{-6}$	$1.4 \cdot 10^{-6}$	$3 \cdot 10^{-7}$	$5 \cdot 10^{-8}$	$1 \cdot 10^{-8}$
$\langle \sigma v \rangle$	$7 \cdot 10^{-9}$	$8.8 \cdot 10^{-9}$	$9.1 \cdot 10^{-9}$	$9.1 \cdot 10^{-9}$	$9.2 \cdot 10^{-9}$	$9.2 \cdot 10^{-9}$	$9.2 \cdot 10^{-9}$
$\frac{N(v)\langle\sigma v\rangle}{N_0 \langle\sigma v\rangle_0}$	140	132	54.6	12.7	2.76	0.46	0.09

$$\sum_{v=0}^{13} \frac{N(v)\langle\sigma v\rangle}{N_0 \langle\sigma v\rangle_0} = 428$$

Highly-excited H_2-molecules take part in the H^--ions generation and, probably, in the H_3^+-ions generation. The cross-section of this reaction grows as the vibrational quantum number grows. Under high hydrogen expenditures the highly-excited H_2 molecules lack is observed. The H^- and H_3^+ ions generation processes start to competated

$$H_2(v) + \begin{bmatrix} + e \rightarrow H_1^- + H_1^o \\ + H_2^+ \rightarrow H_3^+ + H_1^o \end{bmatrix}$$

This means that under the increase of any of the two components (H^- or H_3^+) the current of the other will decrease. The corresponding curvatures will by symmetrical. The picture is shown in Fig.6.

The modernized source composition (Fig.3) allowed us to held a number of experiments on the optimization of some source units. When using a the main (7) or the intermediate (6) anode in the discharge it was set that the anode location does not affect the H^--ions maximal current value. The accompanying electric current decreases one order when using the intermediate anode. The minimal electron-ion correlation decreased from 3,2 to 0,3.

The influence of the accelerating voltage and the emission electrode width on the integral source parameters was studied. Under negative ions movement from the discharge chamber to the emission surface through the holes in the emission electrode across the magnetic field the partial loss of the ions on the hole walls and the corresponding source characteristics deterioration take place. To estimate the characteristics deterioration the beam parameters under different accelerating voltage values were measured. The characteristics of the obtained dependences are shown in Fig.7. A considerable (linear) current growth of negative and positive ions under accelerating voltage growth is seen as well as a considerable (from 9 up to 3,5) reduction of the electron-ion correlation.

Estimations show that the extracted ion current relation to the factual (coming from plasma) one is 0,43 (in the given emission electrode composition: copper, thickness - 3 mm). Other source characteristics also decrease.

The experiments with a thin molybdenum emission electrode (0,5 mm in width) were held to confirm the given estimations. An intermediate anode was used for current drive, the emission electrode was placed under floating potential. The hydrogen negative ions current was 3,3 A; the energetic value of the negative ion was reduced up to 380 eV/ion; the extracted H^--ion current is 0,1 A/cm^2.

So, the emission electrode width in the hydrogen negative ions source is of great importance: when its width reduces from 3 up to 0,5 mm (crack's width - 3,5 mm) the integral characteristics of the source improved three times.

1. Prelec K., Sluyters Th. Formation of Negative Hydrogen Ions in Direct Extraction Sources. - Rev.Sci.Instrum., 1973, v 44, N 10, p.1451-1463.
2. Antipov S.P., Elizarov L.I., Martynov M.I., Chesnokov V.M. Negative Hydrogen Ions Source with Hollow Cathode Working in a Stationary Regime. - PTE, 1984, N 4, p. 42-44.
3. Hiskes J.R., Karo A.M., Bacal M. et al. Hydrogen Vibrational Population Distributions in a Medium Density Hydrogen Discharge. - J.Appl.Phys., 1982, v 53, N 5, p. 3469-3475.
4. Wadehra J.M. Rates of Dissociative Attachment of Electrons to Excited H_2 and D_2. - Appl.Phys.Lett., 1979, 35 (12), p. 917-919.

Plasma-Volume Stationary H$^-$ Ion Source

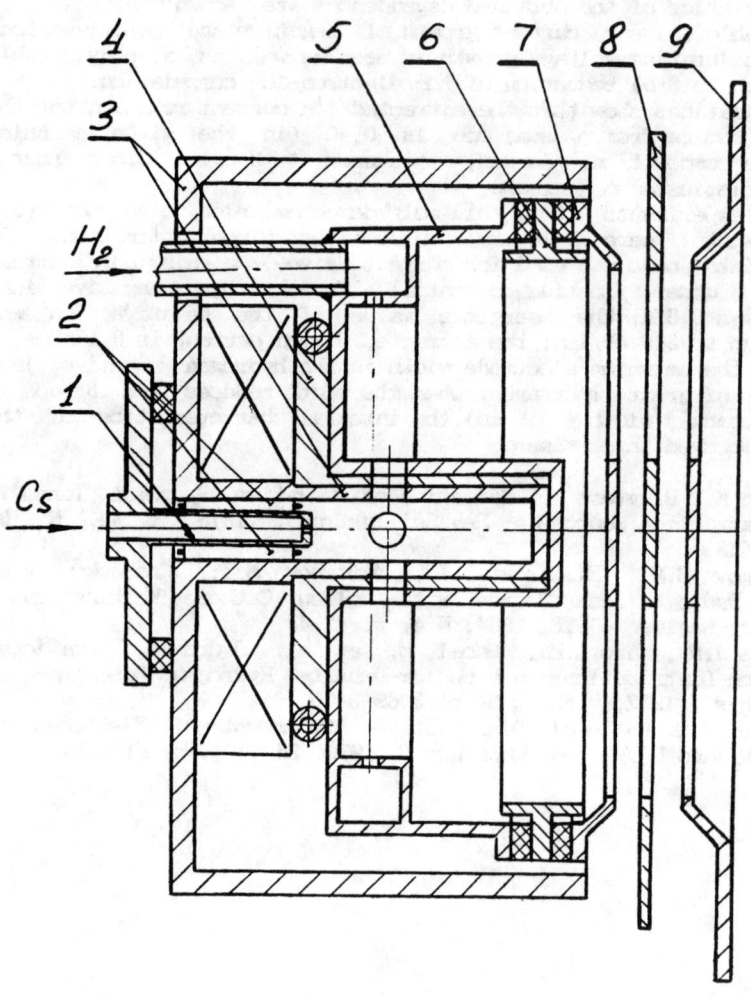

Fig.1. H$^-$-ions source scheme

Fig.2. General view of the ion source

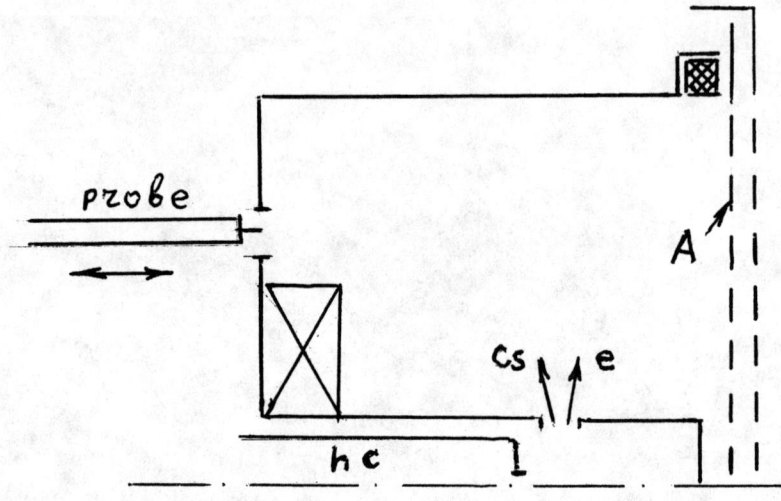

	Parametr value	
	40 cm	43.6 cm
T_e, eV	3.82	2.96
n_e, 10^{-10} cm^{-3}	2.75	5.71
T_i, eV	1.06	1.07
$n(H_1^+)$, 10^{-10} cm^{-3}	4.50	3.86
$n(H_2^+)$, 10^{-10} cm^{-3}	8.45	7.23
$n(H_3^+)$, 10^{-10} cm^{-3}	47.3	40.5
$n(Cs^+)$ 10^{-10} cm^{-3}	152	130
$n(\Sigma^+)$, 10^{-10} cm^{-3}	212	182
$n(H^-)$, 10^{-10} cm^{-3}	209	176

Fig.3. Probe measurements scheme and results

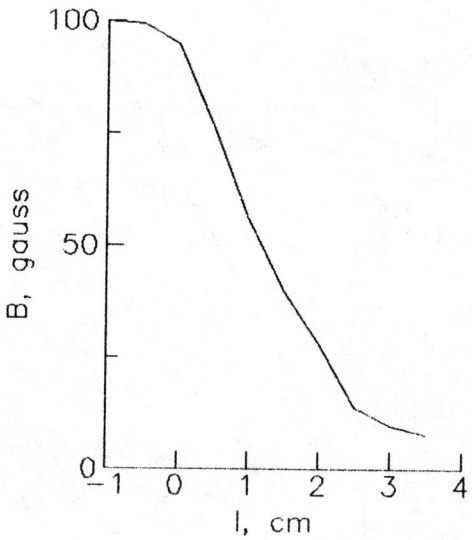

Fig.4 Magnetic field distribution along the source channel axis

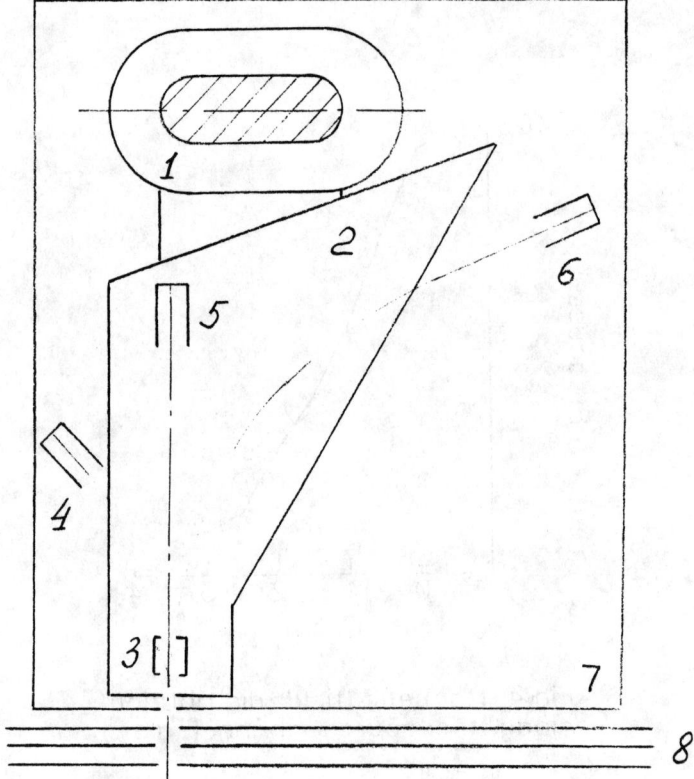

Fig.5. Magnetic analyzer scheme

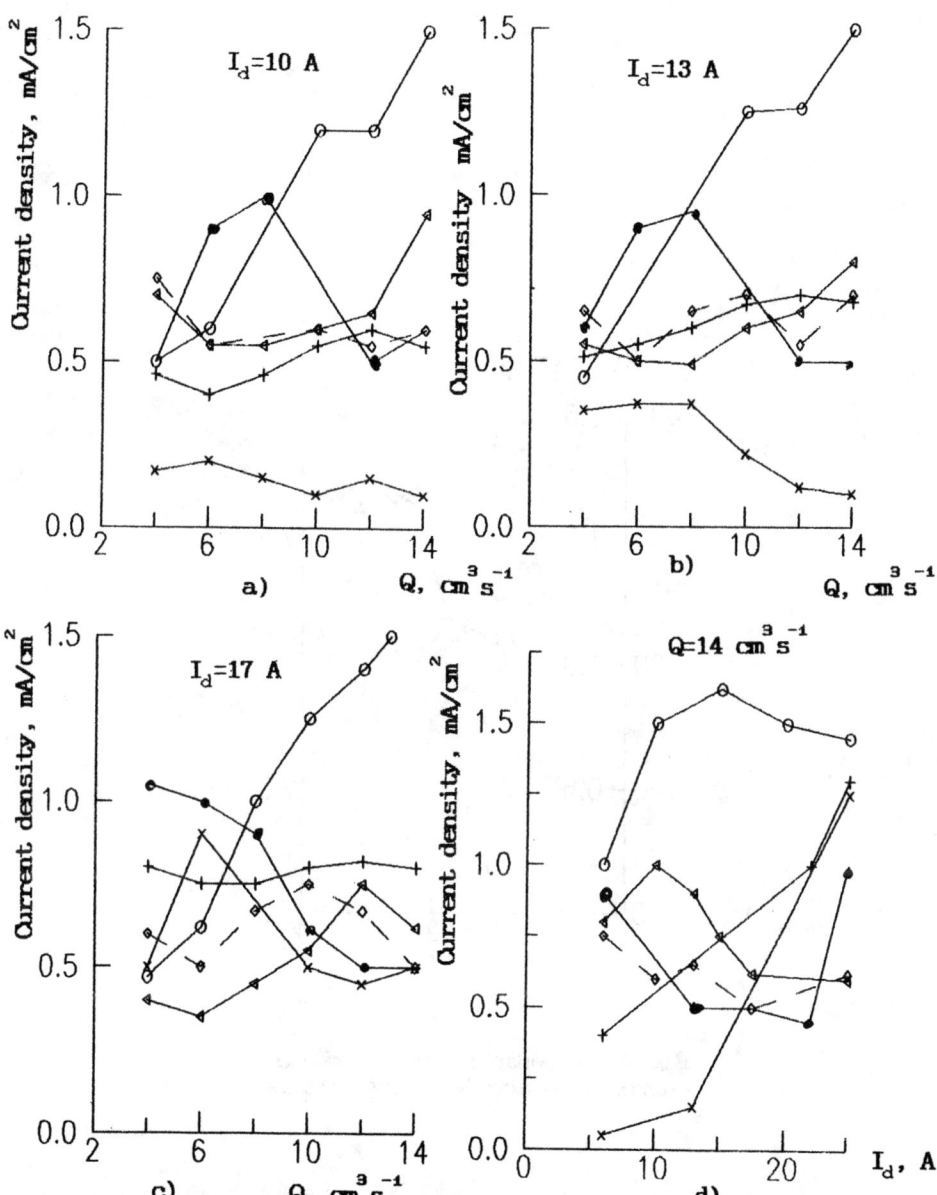

Fig.6. Ions and electrons current density dependence on the rate hydrogen flow (a, b, c) and discharge current (d)
For the both figures △ - current density of H_1^+ ions,
◊ - current density of H_2^+ ions, ○ - current density of H_3^+ ions,
× - current density of Cs^+ ions, ● - 0.1 current density of H^- ions,
+ - 0.01 current density of electrons

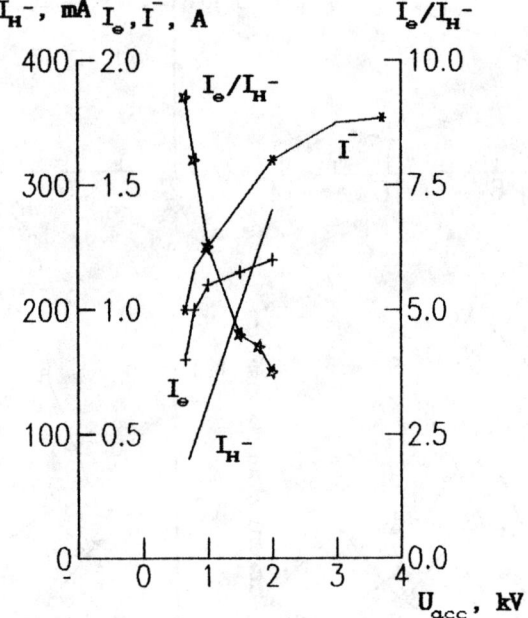

Fig.7. Dependance of extracted currents on accelerating voltage

DISCUSSION

Okumura: What is the material of your extraction grid?

Elizarov: Molybdenum.

Okumura: Have you ever changed the molybdenum to another material?

Elizarov: Yes, we made it from copper, and also molybdenum electrode. Large thermal load. Thickness (of electrode) depended on the beam characteristics.

Okumura: What was the electron to negative ion current ratio?

Elizarov: 0.3

LONG-PULSED SURFACE-PLASMA SOURCES WITH GEOMETRICAL FOCUSING

Yu.I. Belchenko and A.S. Kupriyanov
Institute of Nuclear Physics, 630090, Novosibirsk, USSR

ABSTRACT

We describe models for the long-pulsed surface-plasma multi-aperture quasi steady-state with a hydrogen negative ion output of up to 1 A, with an emission current density of 2 A/cm^2, and with an average current density in a beam of 60 mA/cm^2. The gas efficiency for the production of H^- beams was 12 - 17%. The experimental and the simulated efficiencies of the negative ion production of pure hydrogen and cesiated surface-plasma sources with geometrical focusing are compared.

INTRODUCTION

Powerful neutral beams with energy of atoms up to 1 - 1.5 MeV are necessary for future large tokamaks (NET, ITER) and tandem mirrors. The development of sources of steady-state negative ions for fusion is in progress[1,2]. A steady-state wide aperture surface-plasma source (SPS) with an independent emitter provided[2] H^- beams with average current density of 10 mA/cm^2 and an intensity of more than 1 A. The multiaperture honeycomb SPS provides pulsed H^- beams with a current of up to 11 A and with an average current density of 0.18 A/cm^2 due to Geometric Focusing (GF) of the negative ions[3]. A quasi steady-state model (QM) of honeycomb SPS is described below.

EXPERIMENTAL DEVICE

The QM of the honeycomb SPS was based on similar principles, to the pulsed model[4] (Fig. 1). Overheating of the long pulsed cathode was reduced by increasing the height of the cathode, and by air-cooling the cathode body. The high-current glow discharge with an unclosed electron drift was localized in the gap between the cathode plane and the anode cover. The side projec-tions of the cathode provided electrical oscillations along external magnetic field lines. Hydrogen, and a small amount of cesium, were supplied to the discharge from external containers via channels in the cathode.
Several spherically concave indentations were made on the cathode emitting surface to geometrically focus the cathode-produced NI on to the emission holes. The radius of the indentation concavity was 3mm, and the average depth of the indentation was 0.4 mm. The concavities were arranged on the cathode surface

in an orthogonal matrix with 3 mm shift. The height of the side projections of the QM honeycomb cathode were decreased 0.8 - 0.6 mm to compensate for the enhancement of the discharge in the $\vec{E} \times \vec{B}$ drift[5] direction. For comparison, we tested the QM with the flat cathode-emitting surface and with side projections 1.3 mm in height. Conical emission holes with inner diameter of 0.8 mm were drilled at the NI focusing points on the anode cover (Fig. 1). The emitting cathode surface was approximately 10 cm^2. The total area of the 120 emission holes was 0.6 cm^2. The discharge current in the quasi steady-state mode was up to 90 A. In the pulse mode, the discharge current rose to 700 A.

A steady-state extraction voltage of up to 18 kV was applied to the body of the anode. The multi-slit extractor was made of thick molybdenum wires, springed on their ends (Fig. 2). A set of short high-voltage pulses (up to 1 ms) with a repetition rate of up to 100 Hz, were also used to rapidly observe the model's emission properties. After heating the multi-slit extractor and short-time conditioning, there were no problems in reaching the high-voltage operation of the source.

The NI beam current was measured with the Faraday cap collectors located 20 cm from the extractor; then, the extracted beam was mass-analyzed. The energy spectrum of H$^-$ was investigated with the help of a 90°-electrostatic analyzer (Fig. 3). A "combined" cathode with concaved and flat indentations (Fig. 7) was used to test the efficiency of the NI production. In this case, the anode cover with the test emission hole was shifted by the external driver with respect to the cathode indentations within a distance of 9 mm. A positive ion current was extracted from the QM with the opposite polarity to the extraction voltage.

OPERATION MODES OF THE SOURCE

Several hydrogen-cesium and pure hydrogen modes were observed. The basic parameters and emission properties of these modes are listed in Table 1. NI production was more effective in the hydrogen-cesium mode HC-I with a discharge voltage of 200 - 80 V. Figure 4 shows the oscillograms of the discharge voltage, U_d, and NI output, I$^-$, for 1 s, 45 A, and 0.5 s, 90 A pulses of the HC-I mode. (honeycomb cathode, short-pulse extraction). The typical "steady state" discharge voltage was 90 - 80 V. With a reduced pulse duration or lower discharge current the "steady state" voltage achieved was 100 - 150 V, but it's value had low influence on H$^-$ yield in the HC-I discharge mode.

The main part of the extracted NI beam consisted of H$^-$ ions. It was only in the high discharge voltage start-up of the pulse that the heavy ion component contributed up to 10% of the NI beam. So the initial excess of NI current was caused by this heavy-ion component (Fig. 4). This component decreased to 1 - 3% of the total yield by 0.01 - 0.1 s after the pulse began. This drop was due to the conditioning of the cathode under ion bombardment and also was due to the lowering of the voltage of the bombardment.

MODE	HC-I	HC-II	H-I	H-II
Voltage, V	200-80	60-40	800-700	600-300
H_2 density, Tor	0.03	0.02	0.5	0.2
Magnetic field, T	0.04	0.04	0.1	0.1
H^- yield, A	1.1-0.9	0.6-0.5	0.07	0.15-0.1
H^- origin	cathode 80%, anode + plasma 20%	—	plasma	cathode 25%, anode 50%, plasma 25%

Table 1 QM discharge modes and emission properties. H^- yield is given for honeycomb cathode and discharge current of 90 A.

The output of the H^- was slightly decreased during pulses that caused temporary overheating of electrodes, i.e., discharges with currents up to 50 A, and pulse duration up to 1 s. Higher discharge currents and high starting temperatures of the electrodes caused the efficiency of H^- production to drop towards the end of the pulse. This effect is seen at the 0.35 s point of the 90 A pulse in Fig. 4 or at 0.5 s of the 70 A pulse (at an initial cathode temperature of 500 - 520 C). These points correspond to 3 - 3.5 kJ of discharge energy and to pulse overheating of the cathode surface of up to 800 C. The average temperature of the cathode was increased by 100 C at a discharge pulse of 90 A, 0.5 s. The HC-I mode was stable; H^- emission was stably replicated in hundreds of long pulses.

The increase in the NI yield with the growth of the discharge current is shown in Fig. 5 for honeycomb (GF) and flat cathodes of the HC-I mode and for the H-II pulse mode. The length of the vertical line in Fig. 5 corresponds to the drop in NI current during the pulse's duration. The H^- output for the honeycomb cathode was 3 - 3.5 times higher than that for the flat cathode (Fig. 6), and 6 - 7 times higher than that for the hydrogen mode H-II with activated electrodes. The H^- yield of the honeycomb QM achieved 1.1 - 0.9 A at the discharge pulse 0.35 s, 90 A (extraction with short pulses). Similar dependencies were observed with the steady-state extraction system. Thus, a H^- beam with an intensity of 0.5 - 0.45 A and with pulse duration of 0.25 s was obtained; the total current extracted from the circuit was I < 1.5 A. The NI beams were filtered from any accompanying electrons by a magnetic field transverse to the source. These electrons were effectively dumped at the magnetic poles. Because there was no accumulation and multiplication of electrons, there was a remarkable increase in electrical strength and a rapid recovery of high voltage after nondestructive break-downs.

The H⁻ yield was maximal at minimal hydrogen density in all modes of discharge (see Table 1). Direct measurements of the hydrogen flow from the source for the HC-I mode gave a value 0.5 - 0.7 L · Tor/s · A, i.e., the gas efficiency of H⁻ beam production was 12 - 17%.

GF EFFICIENCY

Figure 7 shows the change of the H⁻ yield I_{H^-} and the extracted positive ion current I_+ for various positions of the emission holes relative to the indentations in the cathode (HC-I). With the emission hole (ϕ 0.5 mm) situated at the GF point, the H⁻ output was 3 - 3.5 times higher than that of the "minimum" point and 2 times higher than that for the flat indentation. The value I_+ was 3 times lower than I_{H^-}, and its magnitude also decreased 2.5 - 3.5 times with the shift of the emission hole from the GF point to the "minimum" point (discharge current of 90 A).

With the emission hole of 0.8 mm in diameter and with the broad anode cavity (Fig. 8), the H⁻ yield in the GF point was twice as large than that in the "minimum" point, while the positive ion current showed little change. With the broad anode cavity or higher side projections on the cathode, the density of the discharge plasma changed slightly across the discharge gap. At the GF point, the heavy negative ion yield was 3 - 3.5 times higher than that at the "minimum" point because of the stronger geometrical focusing of the heavy negative ions.

The relation of the H⁻ yield at the GF point and the H⁻ yield at the "minimum" point, $K = I^{H^-}_{GF}/I^{H^-}_{min}$, was decreased with an increase in the density of the discharge current (Fig. 9). This relation, K, had a value up to 5 at a discharge current density of 5 A/cm².

Figure 10 shows the increase of K with the growth of the discharge voltage. At a discharge voltage of 100 V (HC-I mode), K had a value of 3 - 4. With an increased voltage ≃ 400 - 600 V (mode with a low amount of cesium) K achieved a value of 6 - 7.

This relation was slightly affected by the external magnetic field of the source. The value of K was decreased from 3.2 to 2.7 with an increase of a magnetic field in the 0.07 - 0.2 T range.

An increase in the gas feed had no influence on the value of $I^{H^-}_{GF}/I^{H^-}_{min}$.

ORIGIN OF H⁻ IONS

The energy spectra for ions extracted from the GF point (HC-I) consisted of H⁻, produced on the cathode surface (80 - 85%), while the remaining 15 - 20% was H⁻ produced on anode surfaces and in the near-anode plasma (Fig. 11). In the "minimum" point, the "anode" part of the H⁻ spectra increased by 1.5 - 2.0 times, while the "cathode" part decreased approximately 10 times (Fig. 11c). The "cathode" group of H⁻ ions was approximately equal to the "anode" group at an "average" point, placed between the GF point and the "minimum" point (Fig. 11b). The average energy of the "cathode" H⁻ group was 175 eV for the GF point (discharge voltage - 150 V). This energy increased up to 240 eV during the extraction through the "minimum" point, due to the disappearance of the "sputtered" H⁻ ion fraction from the cathode. The "anode" fraction of the H⁻ beam at the "minimum" point was 2 - 3 times larger than the "cathode" one (Fig. 11).

We observed three groups of ions in H⁻ energy spectra for the hydrogen mode H-II with activated electrodes. These three groups had energies of 0.15 - 0.25 eU_d, 0.5 - 0.9 eU_d, and 1.3 - 1.8 eU_d respectively[6].

The only H⁻ group with "anode" potential was observed for the hydrogen mode H-I.

DISCUSSION

Effective generation of H⁻ is obtained from the cathode surface of the described QM in a long pulse mode with an average discharge current cathode density of up to 9 A/cm². The drop in H⁻ emission at the end of power pulse seems to be associated with an additional increase in cesium in the discharge region caused by the heating of cesium enriched warm gas discharge region in the corners of the chamber. An increase in cesium density in the volume of the gas discharge could cool the plasma electrons and decrease the degree of ionization of the discharged hydrogen, which in turn, reduces the positive hydrogen ion flux incident onto the separate parts of the cathode surface and decrease the secondary emission of the H⁻ ions. Under the "critical" volume for cesium density, the discharge falls into the low-voltage mode HC-II.

The H⁻ output from QM was half that for the same discharge current of the short-pulse honeycomb SPS[4]. A possible reason is that there is a contraction in the discharge near the emission holes in the QM with lower side projections on the cathode. This growth of local plasma target density and thickness exponentially increase the destruction of H⁻ ions on their way to the emission hole. As a consequence, the efficiency of GF in the QM had the value I^-_{GF}/I^-_{flat} = 3 - 4, while for the pulsed honeycomb SPS, the value achieved[6] was 6 - 8 (Fig. 6).

Sixty to Sixty five percent of the extracted H⁻ ions are

sputtered from the hydrogen-saturated cathode surface of honeycomb QM. These "sputtered" ions are strongly focused on the emission hole after acceleration by the near-cathode potential drop due to a low energy spread[6]. These ions fail to hit emission holes that are shifted from the GF point, so they are absent in the energy spectrum at the "minimum" point.

An increase in the "anode" H⁻ fraction at the "minimum" point was probably caused by the low plasma density and the more effective collecting of H⁻ produced on the surfaces of emission holes. On the other hand, the lower electron temperature at the "minimum" point must increase the volume of H⁻ generated by dissociative attachment of electrons.

The decrease in the relation I^{H-}_{GF}/I^{H-}_{min} with the increase of the discharge current was related to the increase and equalization of plasma density along the gas discharge gap.

The efficiency of geometric focusing in the QM compared to the flat cathode had the following values: for the "sputtered" H⁻ fraction 10 - 12, for the "reflected" H⁻ fraction 1.5 - 2, for the "anode" group 0.8 - 0.9, for total H⁻ current 3 - 3.5. For the "reflected" and "sputtered" H⁻ fractions, GF efficiencies were determined with computer simulation data[7].

SUMMARY

For quasistationary SPS models the use of GF led to an increase in total H⁻ ion yield of 3 - 4 times and the increase in output of desorbed H⁻ ions of 10 - 12 times. In models with GF, the yield of H⁻ ions formed on anode surfaces increased up to 20% from the total H⁻ ion current. The gas efficiency of H⁻ ion production was more than 10% for quasistationary models. The average emission current density of H⁻ ions of 75 mA/cm^2 was achieved at the heat loading of electrodes less than 0.6 kW/cm^2.

REFERENCES

1. M. Bacal, Nucl. Instr. and Meth., B37/38, 28 (1989).

2. K.N. Leung and K.W. Ehlers, Rev. Sci. Instrum., 53, 83 (1982).

3. Yu. Belchenko and G. Dimov, Proc. III Intern. Symp. on the Production and Neutralization of Negative Ions, AIP Conference Proceedings, No. 111, p. 363, N.Y. (1984).

4. Yu. Belchenko, Fizika Plazmy, 9, 1219 (1983).

5. Yu. Belchenko, G. Dimov, V. Dudnikov, and A. Kupriyanov, Revue Phys. Appl., 23, 1847 (1988).

6. Yu. Belchenko and A. Kupriyanov, Revue Phys. Appl., 23, 1889 (1988).

7. Yu. Belchenko and A. Kupriyanov, Revue Phys. Appl., 23, 1885 (1988).

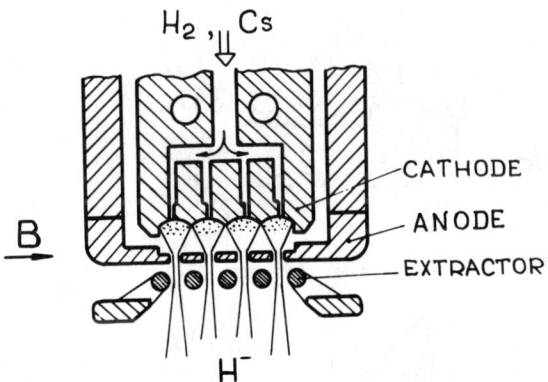

Fig. 1 Layout and view of quasi-steady-state SPS model.
1 - cathode, 2 - anode cover, 3 - hydrogen inlets, 4 - cesium container.

206 Long-Pulsed Surface-Plasma Sources

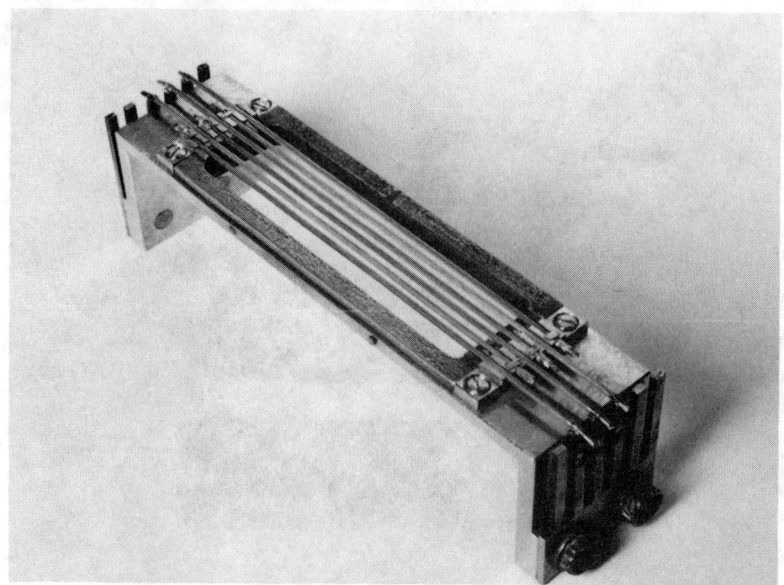

Fig. 2 Photo of the extractor.

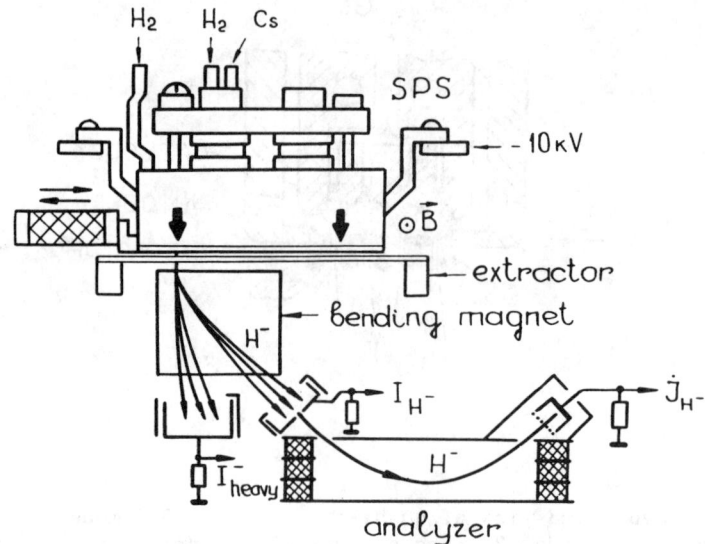

Fig. 3 Scheme of measurements.

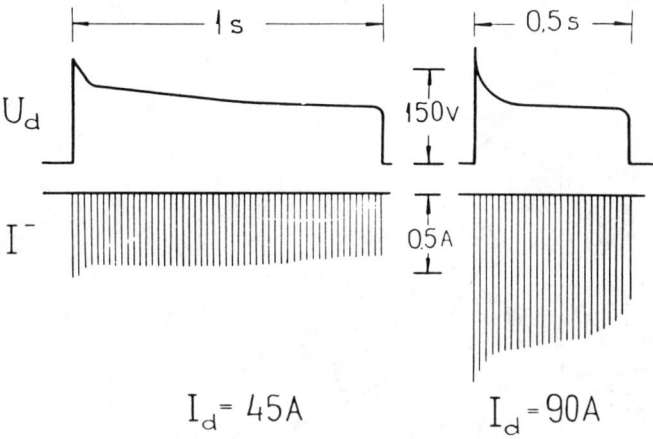

Fig. 4 Oscillograms of discharge voltage U_d and negative ion yield I^- for two discharge currents.

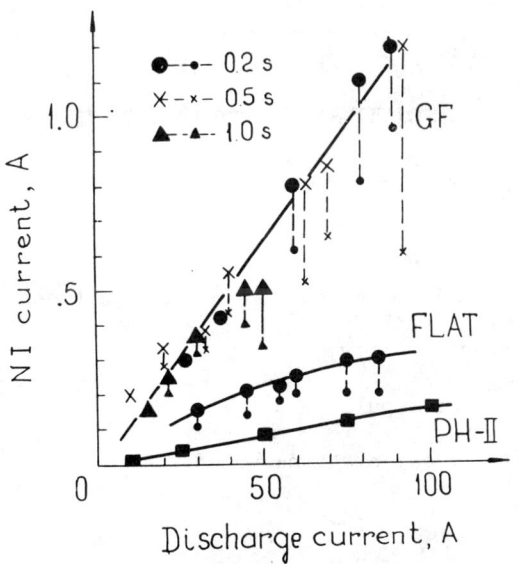

Fig. 5 H^- beam current as a function of discharge current.

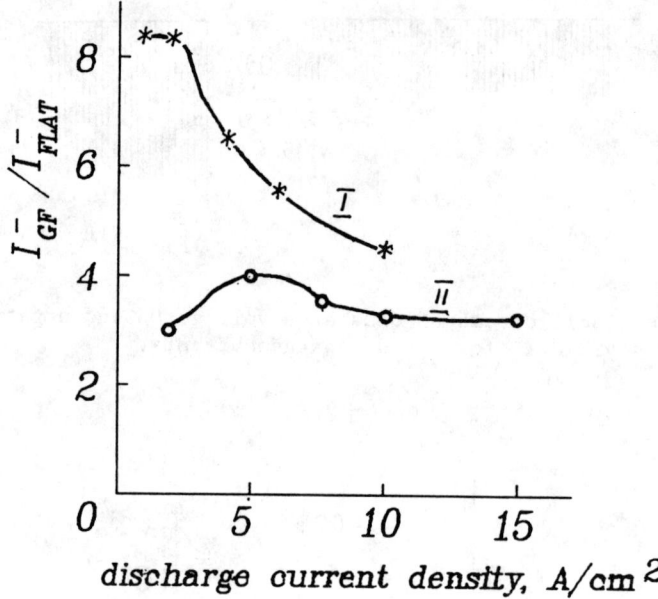

Fig. 6 The ratio of NI current from GF cathode to the that of flat cathode I^-_{GF}/I^-_{FLAT} as a function discharge current density.

I - short pulse model.

II - quasi-stationary model.

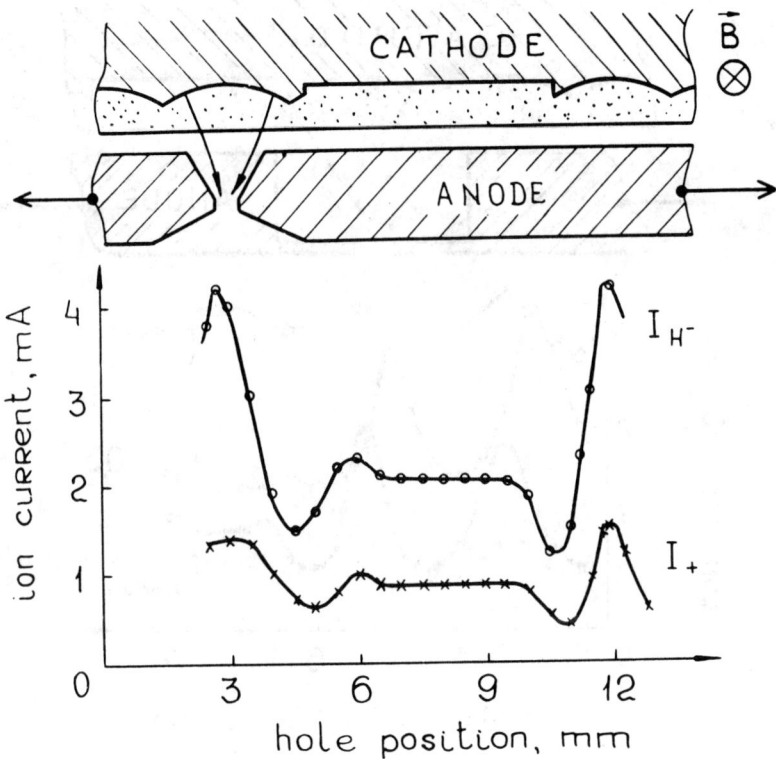

Fig. 7 H⁻ beam current I_{H^-} and positive ion current I_+ as a function of emission hole position. (Conical emission hole).

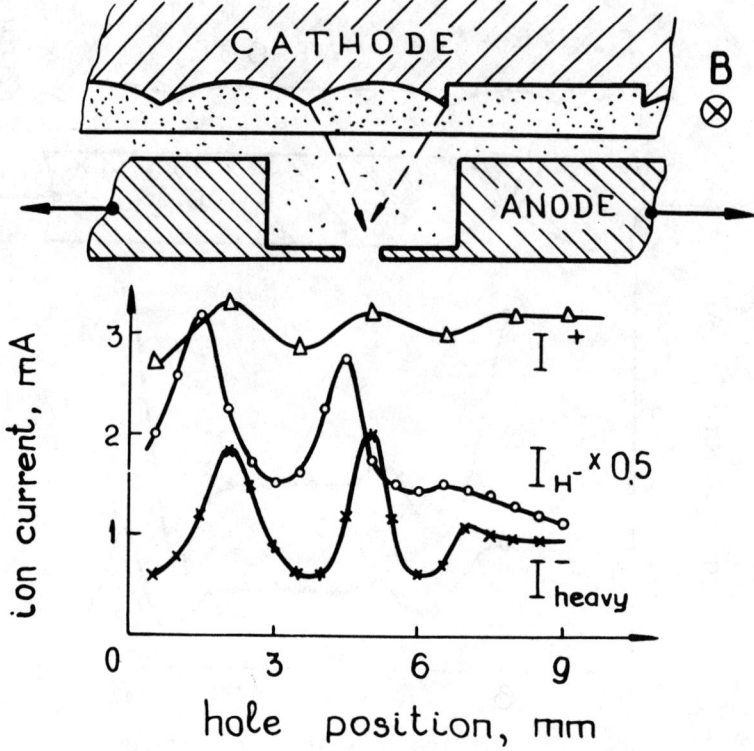

Fig. 8 H⁻ beam current I_{H^-}, heavy NI current I^-_{heavy} and positive ion current I^+, extracted through emission hole with broad anode cavity, as a function of emission hole position.

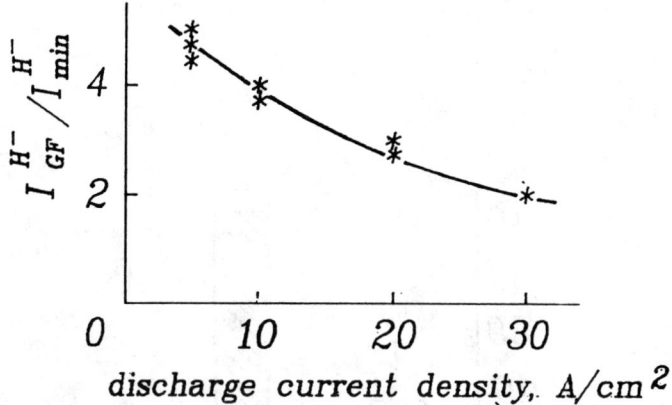

Fig. 9 The relation $I^{H^-}_{GF}/I^{H^-}_{min}$ as a function of discharge current density.

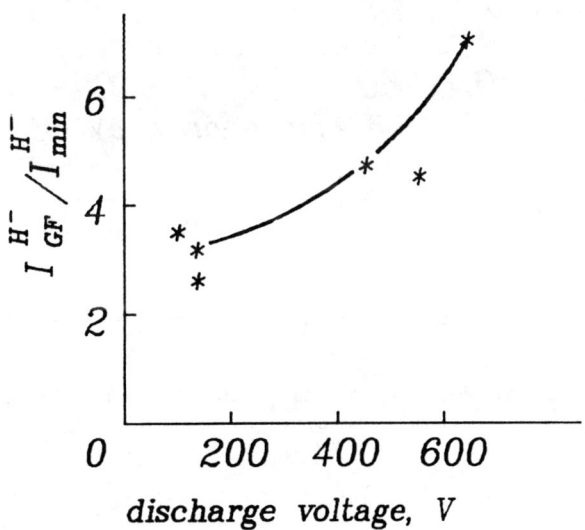

Fig. 10 The relation $I^{H^-}_{GF}/I^{H^-}_{min}$ as a function of discharge voltage.

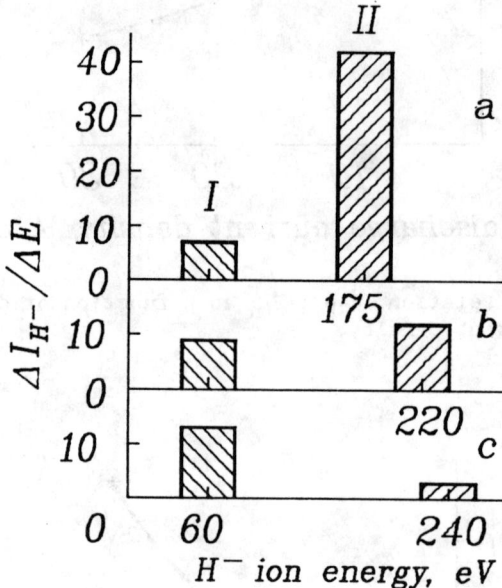

Fig. 11 Average energy and height H⁻ anode and cathode groups extracted.

a - at geometrical focus point; b - at "middle" point; c - at "minimum" point.

DISCUSSION

Hemsworth: Can you tell what the heavy negative ions are and can you classify the fractions that you have in your source.

Belchenko: In the beginning of the pulse, oxygen can be occluded by the cesium on the surfaces. Output of O^- is a little high, approximately 10%. In a few milli-seconds this oxygen is sputtered very quickly. At the plateau of the pulse we have approximately 1% of O^-. But they are geometrically focused very well because they are produced only on the cathode, not on inner surface.

Alessi: As you scan your aperture across the geometrically focused surface, what happens to the ratio of the electrons to H^- current.

Belchenko: There is a very slight change. Inner group of negative ions doubled, but electrons have a slight change, it depends. The positive ion current has no change, it reflects the density of plasma.

Alessi: The electrons aren't constant, coming from the anode rather than from the plasma.

Belchenko: Electrons go from the plasma. In the height of cathode protrusion we have uniform distribution of plasma along the cathode. The same uniform output of electrons, but sometimes with diminished height of cathode protrusion, we can see the discharge contraction in the holes. Distribution of the plasma along the gap became not uniform. Plasma density three times more in the geometrical focus point and low at the intermediate point. So the electrical current was two times less in this case.

NEGATIVE ION PRODUCTION IN AN ION SOURCE OPERATING IN H_2 AND D_2

W.G. Graham
Physics Department, Queen's University, Belfast, N. Ireland

A.A. Mullan
Physical Sciences Department, University of Ulster, Coleraine, N. Ireland

ABSTRACT

The negative ion density in a medium density, $< 5 \times 10^{11} \text{cm}^{-3}$, multicusp ion sources is found to be consistently lower in D_2 than H_2. Under the same source operating conditions the plasma density in D_2 is found always to be higher than in H_2. The effect of the different isotopes on transport processes in the discharge can be understood in terms of a simple model; however, the effect on molecular processes appears more complex.

INTRODUCTION

Negative ion based neutral beam heating of nuclear fusion plasmas requires the production of intense beams of negative deuterium ions. In extracting negative hydrogen ions from the same ion source and under the same nominal operating conditions it has been found that the D^- yields are generally lower than the H^- yields. It is therefore of interest to characterize a discharge operating in both hydrogen and deuterium gas. The range of effects which might be expected to be influenced by the nuclear mass difference can be divided into two categories: transport effects, due to the reduced mobility of deuterium molecules, atoms and ions, and molecular effects due to the different vibrational and rotational energy levels in hydrogen and deuterium and possibly different vibrational and rotational energy distributions in the discharges. Here plasma parameter measurements in a tandem multicusp ion source operating in H_2 and D_2 are compared.

APPARATUS AND EXPERIMENTAL TECHNIQUE

The ion source used for the present measurements is based on a Culham "small" source and has been described in detail elsewhere[1]. Briefly, the source consists of a stainless steel vacuum box 19 x 24 x 19 cm. The discharge is produced by electrons emitted from two hot tantalum filaments which act as cathodes, biased at 60V with respect to the grounded vessel walls. The electrons are confined using permanent magnets on the outer walls in a normal linecusp multipole geometry. A virtual filter is created by breaking the multipole geometry on two opposite sides. The magnetic field lines cross the source preventing fast electrons from the filament—containing "driver" region from reaching the "extractor" region. There are no magnets on the end wall (beam forming electrode, BFE) of the extractor region. The plasma conditions are varied by changing the discharge current (Id) and operating gas pressure.

The gas pressure in the source is measured using a capacitance manometer and maintained by a feedback loop to an automatic valve on the gas inlet. The plasma parameters are measured using cylindrical Langmuir probes. The probe current—voltage characteristic is analysed using a digital technique[1] which allows the measurement of many plasma parameters including the electron energy distribution function (EEDF). Langmuir probes were positioned along the central axis of the source and in the centre of both the driver and extractor region, 12cm and 2cm from the BFE respectively. A movable Langmuir probe could be used to investigate the spatial dependence of the plasma parameters along an axis through the centre of the discharge. A small mass spectrometer capable of analyzing both negative and positive ions was positioned in the centre of the BFE. Its operation is discussed in an accompanying paper[3].

RESULTS

The variation of electron density, n_e; electron temperature, kT_e and plasma potential, V_p with discharge current, measured in both H_2 and D_2, in the extractor region is shown in figure 1. Figure 2 shows the variation of these parameters with gas pressure. Previous measurements[4] have indicated that there is a complex spatial variation of the plasma parameters in this multicusp ion source; however the differences in H_2 and D_2 operation are observed to be similar in the extractor, driver and filter regions of the source. All the plasma parameters measured in D_2 were found to obey the same simple scaling laws with operating conditions found previously in H_2.[4]

In Figure 3 the electron energy distribution function (EEDF) measured in the centre of the driver and extractor regions are shown. In each region the EEDF's measured in H_2 and D_2 are identical. The EEDFs in the extractor region are found to be Maxwellian while those in the driver can be approximated by a bi-Maxwellian distribution and represented by two electron densities and temperatures. At all locations in the source and over a wide range of discharge currents and pressures, no significant differences in the thermal electron temperature, the fast electron temperature or the fast electron density, n_{fe}, measured in H_2 and D_2 is observed.

Differences in n_e and V_p are apparent at all locations. The ratio of $n_e(D_2)/n_e(H_2)$ is approximately the same throughout the source and under various operating conditions, with a value over all readings of 1.20 ± 0.08. Differences in V_p are particularly interesting. At the same discharge current, the plasma potential measured in D_2 is always higher than that in H_2, however when measured at the same electron density (Fig 4) then in the extrator region there is no measureable difference between the values of H_2 and D_2. This is also true for the electron temperature.

As discussed in an accompanying paper[3] the use of mass spectrometry in discharges is not straightforward. In order to extract ions, particularly negative ions which are confined by the plasma potential, a bias must be applied to the mass spectrometer which is in contact with the discharge. This causes a change in the discharge parameters and so for meaningful comparisons between mass spectrometer data and plasma parameters, the discharge must be characterised with the appropriate extraction biases applied to the mass spectrometer. Figure 5 shows that at the same discharge current there are distinct differences in the ion mass distributions in D_2 and H_2; however if the ratios are plotted as a function of the measured electron density, rather than discharge current, then the relative ion mass distributions are almost identical[3].

DISCUSSION

Most of these experimental observations can be interpreted in terms of a simple model of the discharge[4]. The balance equation for the positive ion density, n_+, can be written as

216 Ion Source Operating in H_2 and D_2

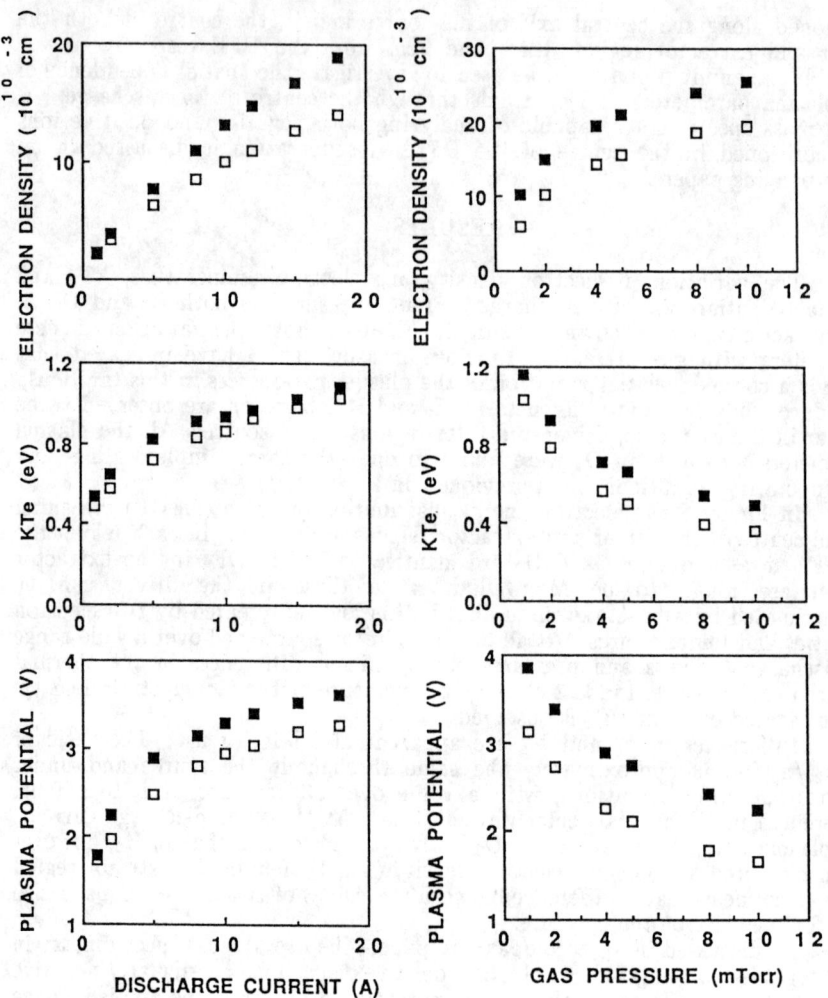

Figure 1

Plasma parameters dependence on discharge current measured in the extractor in ■ D_2 and □ H_2. (Vd = 60V, P = 2 m Torr)

Figure 2

Plasma parameters dependence on gas pressure measured in the extractor in ■ D_2 and □ H_2. (Vd = 60V, P = 2 m Torr)

$$n_+ = n_{fe}n_o(\sigma_i v)\tau^+ = n_e \tag{1}$$

where n_{fe} is the fast electron density, n_o is the neutral gas density, $(\sigma_i v)$ is the rate of ionisation by fast electrons and τ^+ is the ion confinement time.

Under the same operating conditions, ie, with Id, Vd and gas pressure the same, we have shown that n_{fe} remains constant (Figure 3). Therefore only τ^+ in Eq. (1) is isotope sensitive. It can be shown that with the present source and operating conditions, below a plasma density of $5 \times 10^{11} cm^{-3}$, ion loss is dominated by wall collisions. Therefore τ^+ can be approximated by L/v_+, where v_+ is the ion velocity and L is the characteristic length of the source. Assuming the ion temperature in the discharge is the same for D_2 and H_2 then

$$\tau^+(D_2) = (m(D_2)^+/m(H_2)^+)^{1/2} \tau^+ H_2 \tag{2}$$

Therefore, from Eq. (1)

$$n_e(D_2) = 1.4 n_e(H_2) \tag{3}$$

This assumes the ion mass in D_2 is twice that in H_2. In fact as can be seen from Figure 5 the relative ion mass distributions in D_2 and H_2 are different for the same discharge voltage. The effective ion mass ratio is in fact approximately 1.6. Using this value in Eq. (2)

$$n_e(D_2) = 1.26 n_e(H_2)$$

this is in excellent agreement with the experimental value of 1.20 ± 0.08.

In Figure 6 the relative negative ion densities measured in H_2 and D_2 are compared as a function of electron density. The initial increase in H^- and D^- density is found to reach a plateau above an electron density of $4.5 \times 10^{10} cm^{-3}$. It is found that the negative ion density in D_2 is less than that in H_2 ($n_{D^-} \simeq 0.75 n_{H^-}$) in the present operating regime.

It is interesting to note that the lack of an isotope dependence in H^- and D^- production, at electron densities of less than $10^{10} cm^{-3}$, was the initial evidence that dissociative attachment to highly vibrationally ($v > 5$) excited hydrogen molecules was the principle mechanism for negative hydrogen ion production in these discharges[5]. Clearly the effects of nuclear mass on molecular processes, particularly in negative ion production, are quite complex and require further investigation.

CONCLUSION

Measurements of plasma parameters and ionic mass distributions in a low density ion source operating in H_2 and D_2 indicate that the effects of nuclear mass on transport processes in the discharge can be understood in terms of a simple model. The negative ion density in D_2 is found to be consistently less than that in H_2. It would appear that nuclear mass effects on molecular processes are more complex and require further study.

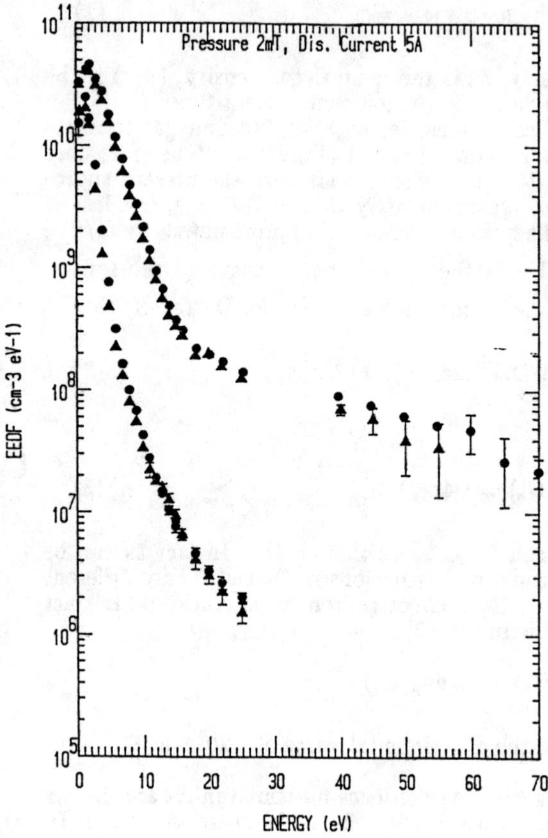

218 Ion Source Operating in H_2 and D_2

Figure 3

Electron energy distribution function measured in the centre of the driver (upper points) and extractor (lower points) in D_2 and H_2.
($Id = 5$ A
 $Vd = 60$ V, $P = 2$ mTorr)

Figure 4

Dependence of the electron temperature and plasma potential on electron density in D_2 (closed symbols) and H_2 (open symbols).
($Vd = 60$V,
 $P = 2$ m Torr)

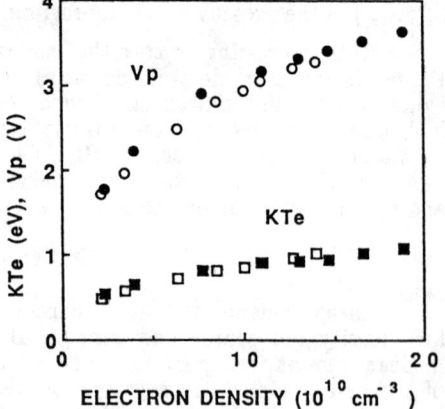

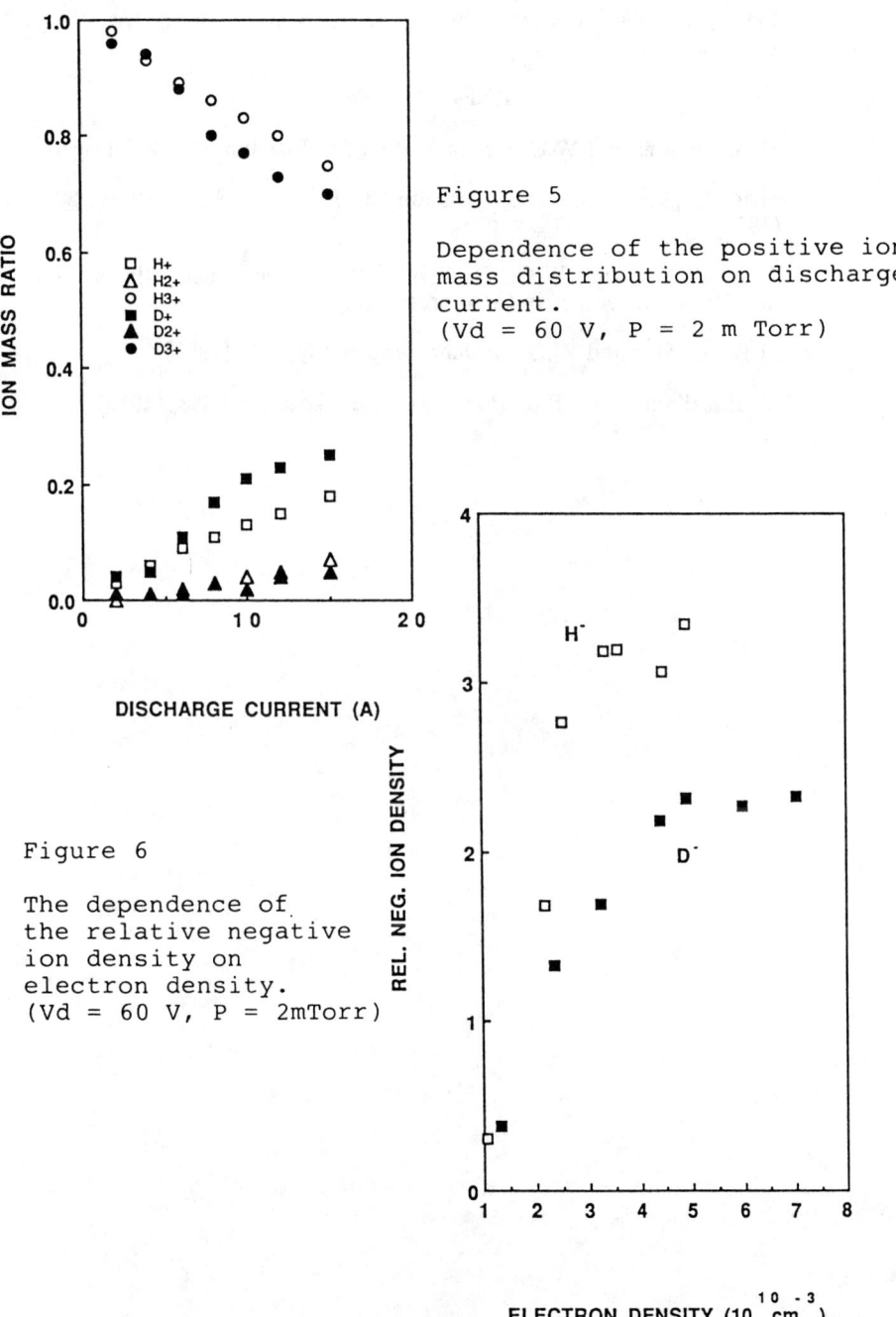

Figure 5

Dependence of the positive ion mass distribution on discharge current.
(V_d = 60 V, P = 2 m Torr)

Figure 6

The dependence of the relative negative ion density on electron density.
(V_d = 60 V, P = 2mTorr)

ACKNOWLEDGEMENTS

This work was supported by an EMR contract from the UKAEA (Culham Laboratory).

REFERENCES

1. M.B. Hopkins and W.G. Graham, Rev. Sci. Instrum. 57, 2210 (1986).

2. M.B. Hopkins and W.G. Graham, J. Phys. D. Appl. Phys. 20, 838 (1987).

3. A.A. Mullan and W.G. Graham, "Mass spectrometry in a tandem multicusp ion source" in these proceedings.

4. M.B. Hopkins and W.G. Graham, Vacuum 36, 873 (1986).

5. M. Bacal and G.W. Hamilton, Phys. Rev. Lett. 42, 1538, (1979).

DISCUSSION

Hiskes: I was interested in your variation of the H_3^+ contribution with the discharge parameters, now it is true that the discharge you are working with is similar to the one that is used at FOM that Eenshuistra used?

Graham: I think so, it's a small source as well.

Hiskes: So your H_3^+ relative population probably is also relevant to that experiment?

Graham: Yes, let's see where this is going to before I to before I commit myself.

Hiskes: With your data, would you speculate on what Eenshuistra's H_3^+ contribution was?

Graham: It is normally pretty dangerous to try to compare one source with another as we note from experience. I'm not prepared to speculate on that until we mix and match. But I think you can almost calculate these, and they work out about right, if we run them on a very simple calculation and we predict those types of ionic mass ratios.

Hiskes: Well, theorists don't believe calculations they would rather see experiments.

Graham: Well I'm glad of that.

Holmes: It's interesting from what Bill has shown for the scaling of the species with the plasma density and the arc current. Some of these have been predicted for a long time. All the beam experiments have been done with real sources and have only had access to the arc current. But when they actually do have probes in there that can measure the electron density, it does behave exactly as predicted.

Hall: I don't know what mass spectrometer you are using but do you have to calibrate the transmission with mass.

Graham: Yes, in fact, it is calibrated with the square root of the mass for each one. And that seems to be fairly consistent with the scaling of the total ion current that we get when we corrected. The measured electron density for example, which is one of the basic tests that we've done. So it is adjusted.

Allison: The space charge effects on the extraction process would lead you to a prediction that the ratio of D^-/H^- is .712. Have you considered that fact?

Graham: No. I don't think we are (space charge limited).

Holmes: I have one question Bill. Your last viewgraph, the one you actually measured the ratios. You are showing there about 75% of the D^-/H^- which is what one would predict. In the standard model, the extraction current density is the part which doesn't change from D to H, multiply by the sound speed which has the isotope effect, but then you show in the last viewgraph that they actually change with pressure.

Graham: Yes, well as I say that is preliminary data, we were surprised to see that and we went back and repeated the mass spectrometry and it still looks pretty good. But of course we are operating at high pressures and we're concerned about stripping effects, and we corrected for the difference between our anticipated difference in stripping for the H^- and D^-. But I really only showed that in the spirit of the meeting: That's something that we are looking at in the future. It won't be in the published proceedings.

OPTIMIZATION OF THE SHEET PLASMA NEGATIVE ION SOURCE

A.Ando, T.Kuroda, Y.Oka, O.Kaneko, Y.Takeiri
T.Kawamato and A.Karita
National Institute for Fusion Science, Nagoya 464-01, Japan

ABSTRACT

Negative hydrogen ions are extracted from a sheet plasma and extracted H^- currents are optimized for several parameters ; a location of the extraction electrodes, a bias voltage, a filling gas pressure and a discharge current. There is an optimum position of the electrodes near the plasma periphery and an optimum pressure for the H^- production. The profiles of the electron density and temperature are measured and the relation between the profiles and the H^- yield is investigated. It is found that the H^- yield is optimized when the electron temperature in front of the extraction grids decreases below 1eV and that this optimum condition can be achieved by changing the position of the electrodes and/or by changing the filling gas pressure.

INTRODUCTION

In the next step of Neutral Beam Injector (NBI) systems, the negative hydrogen/deuterium ion sources capable of generating high ion currents are required, since the beam energy should be, more than several hundreds keV. Recently, the Bucket sources based on the volume production mechanism have been developed because of their simple and reliable operations. In this Bucket sources, the magnetic filter is needed to seperate the plasma into two regions; one is the arc plasma containing fast electrons which are necessary for the vibrational excitations of hydrogen molecules and the other is the diffused plasma containing slow electrons which are dissociatively attached to the excited molecules H_2^*. The need of the magnetic filter, however, makes the sources complicated, and there are other serious problems in the bucket sources such as the large gas load and many electron currents.

The sheet plasma negative ion sources are designed based on the same tandem model and have advantages for its simple geometry to clarify the production mechanism of negative ions. In the sheet plasma the fast electrons are confined in the sheet region by the magnetic field and the plasma are diffused to the front area of the extraction electrode. In this configuration H^- production mechanism can be estimated by one dimensional slab model. We have tested the sheet plasma negative ion source to get data base of the volume production of negative ions and to apply them to the design of the high current negative ion sources for NBI.

Several parameters such as the position of the extraction

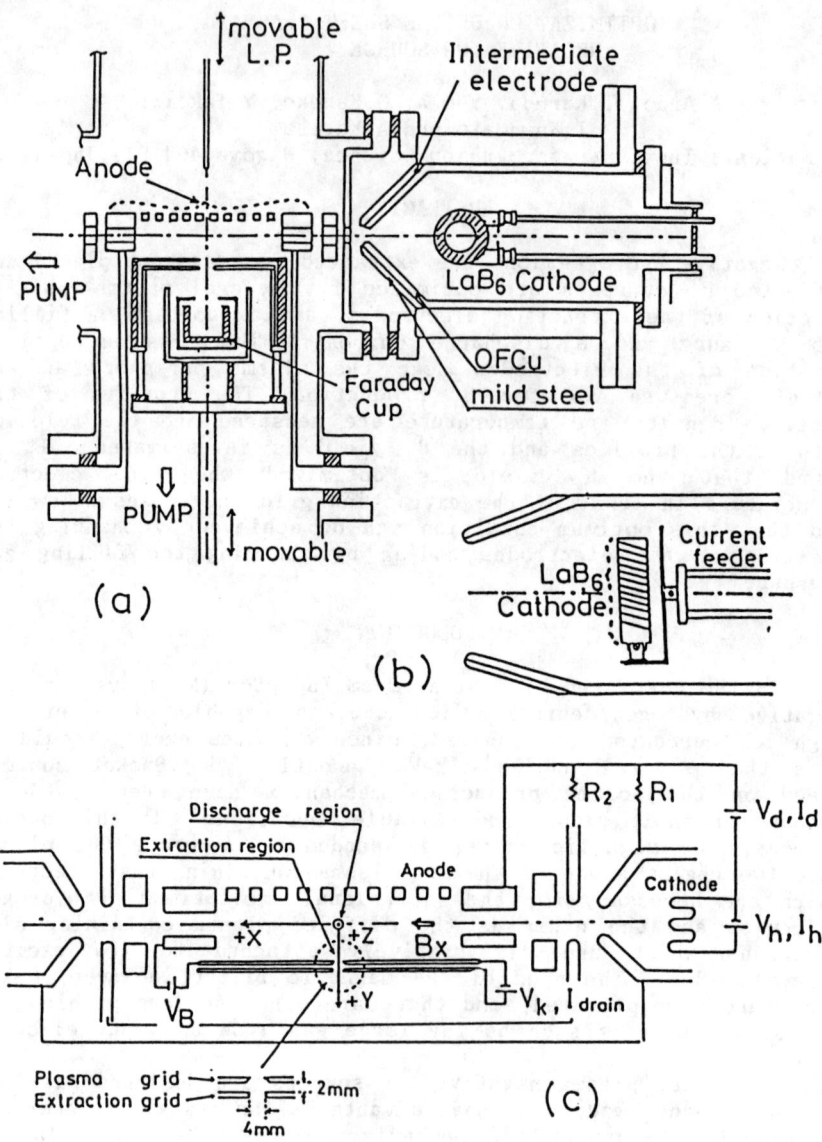

Fig.1 Schematic diagram of experimental apparatus.
(a) experimental arrangement, (b) the shape of the LaB_6 cathode, and (c) an arrangement of power supplies. X, Y, Z coordinates are defined as shown in (c).

electrodes, a bias voltage to the plasma grid, a filling gas pressure and a discharge current are changed, and the H⁻ yield is optimized at each conditions. The electron density n_e and temperature T_e are also measured and the relation between them and the extracted H⁻ current is discussed.

EXPERIMENTAL APPARATUS

A schematic diagram and an arrangement of power supplies of the sheet plasma negative ion source are shown in Figs.1 (a),(b), and (c). The direction of X,Y,Z coordinates is defined as shown in Fig.1(c). Detailed explanation of this device is presented in the previous papers.[1,2,3] Though two plasma sources like a duoplasmatron are provided on the both ends, the source is operated on a single cathode geometry in this experiment. The gas is pumped out to two directions as shown in the figure.

The cathode is a cylindrical non-inductive LaB_6 coil, as shown in Fig.1(b). The intermediate electrode has a slit of 10mm by 50mm, which defines a sheet size, and is made of copper covered with mild steel to guide the magnetic field.

The axial magnetic field is produced by external magnets and guided into the chamber through mild steel cores. The magnetic field strength B_x is almost uniform along the X direction with $\Delta B/B \lesssim 2\%$ in the chamber, and can be changed up to 200G by the coil current. The profiles of B_x along the Y axis is shown in Fig.2.

The electrodes for the H⁻ extraction consist of a plasma grid and a extraction grid, each of which has a single hole of 4mm in diameter on the center. The negative high voltage V_k is applied up to 3keV between the two grids, and the plasma grid can be biased to the anode. The extracted H⁻ beam current is detected by a faraday cup made of carbon. It is placed at 40mm apart from the electrodes. The electrodes and the faraday cup are moved together from Y=15mm to Y=90mm. The extracted electrons can not reach the faraday cup and flow into the extraction grid because of the magnetic field.

The experiments are carried out in the following conditions; the discharge voltage V_d is arranged from 100V to 200V, the discharge current I_d is changed up to 100A, the magnetic field in the sheet center is 190G, the filling gas is hydrogen and its pressure p in the chamber is changed from 2 to 50 mTorr.

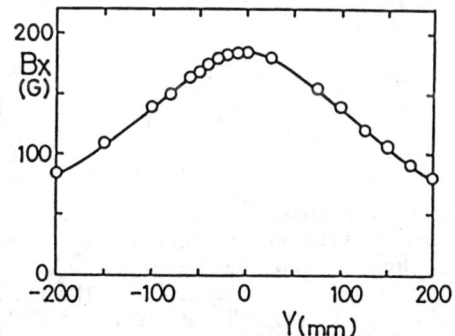

Fig.2 The profile of the magnetic field B_x along the Y axis.

EXPERIMENTAL RESULTS

The plasma parameters are measured by a movable cylindrical Langmuir probe. Fig.3(a) and (b) shows the profiles in the Y direction of the electron saturation current I_{es} and the electron temperature T_e of the sheet plasma at the discharge current I_d = 2A. I_{es} corresponds to the electron density n_e and it is estimated about 1×10^{11} cm^{-3} in the plasma center. The half width of the sheet is about 7mm. The profile of T_e is rather broader than that of n_e. While the density decays exponentially, the temperature is nearly constant ($T_e \lesssim$ 1eV) outside of Y=20mm. Fig.3(c) shows the extracted H$^-$ current I^- measured by the faraday cup and the current flowing into the extraction grid I_e, which is mainly contributed by

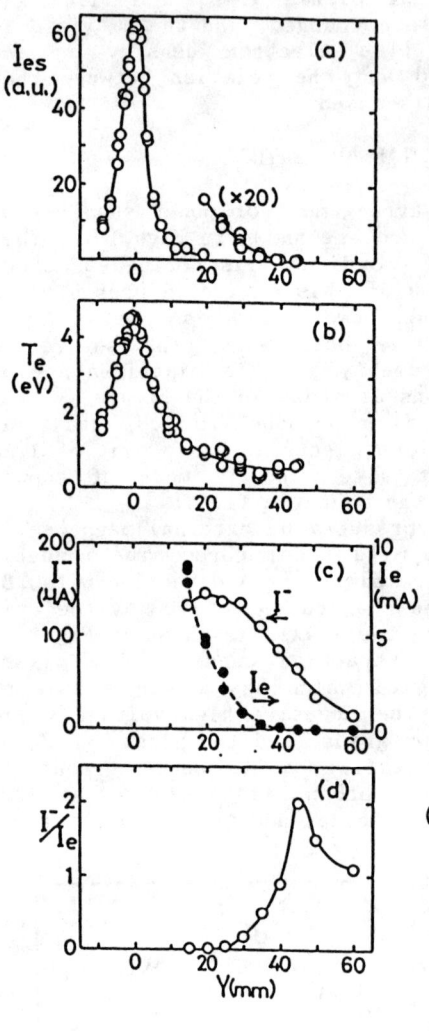

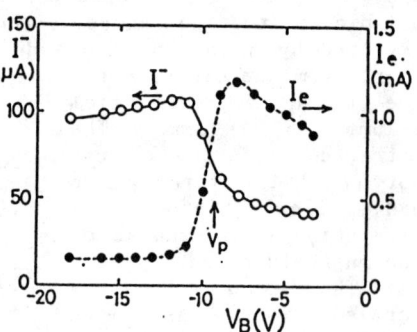

Fig.3 The profiles of (a) the electron saturation current I_{es} (b) the electron temperature T_e in the Y direction. The profiles of (a) the H$^-$ current I^- and (b) electron current I_e when the extraction electrodes is set at each Y position. I_d=2A, p=4mTorr, V_B is optimized

Fig.4 The dependence of I^- (open circles) and I_e (closed circles) on the bias voltage V_B. I_d=2A, p = 7.5mTorr. The electrodes are set at Y=35mm. The arrow shows the plasma potential Vp at Y=35mm.

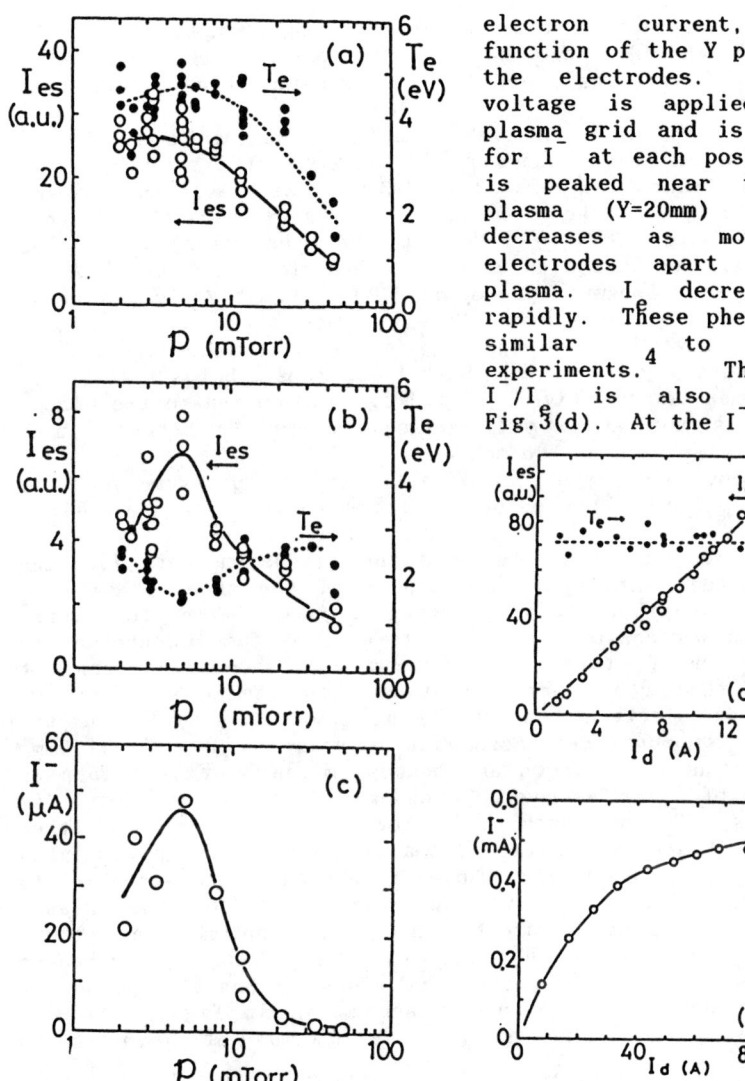

electron current, as a function of the Y position of the electrodes. The bias voltage is applied to the plasma grid and is optimized for I^- at each position. I^- is peaked near the sheet plasma (Y=20mm) and then decreases as moving the electrodes apart from the plasma. I_e decreases more rapidly. These phenomena are similar to other experiments.[4] The ratio I^-/I_e is also shown in Fig.3(d). At the I^- peak

Fig.5 The dependence of I_{es} (open circles) and T_e (closed circles) on the filling gas pressure p, (a) Y=0mm, (b) Y=10mm, and I_d=2A. (c) The dependence of I^- on p. I_d=2A, V_k=3kV. The electrodes are set at Y=15mm, V_B is not optimized.

Fig.6 (a) The dependence of I_{es} (open circles) and T_e (closed circles) at Y=0mm on I_d. (b) The dependence of I^- on I_d. V_k=3kV, p=23mTorr. The electrodes are set at Y=15mm, V_B is not optimized. Right axis is the corresponding current density J^-.

position the ratio is 0.01 and increases abruptly with the increase of the distance between the plasma center and the electrodes. It attains about 2 at Y=45mm and then decreases to 1.

The bias voltage V_B is changed and the effects to the I^- and I_e are measured. Fig.4 shows the dependence of I^- and I_e on V_B when the electrodes are located at Y=35mm. As biasing the plasma grid negatively to the anode, I^- increases slowly and then increases abruptly at $V_B \sim -9V$. On the other hand, I_e behaves oppositely at the same bias voltage. The plasma potential V_p is also measured by Langmuir probe and $V_p \sim -9V$ at Y=35mm as shown in the figure by an arrow. Namely, when the grid is biased slightly negative to the plasma potential, I^- increases and I_e decreases strongly. In the bucket sources, both I^- and I_e increases when the negative bias voltage is applied to the plasma grid. It is not known yet the reason why I_e behaves in a different manner between two sources.

To investigate the effect of a filling gas pressure p to the H^- yield, I_{es} and T_e are measured at two positions and are shown as a function of p in Fig.5 (a)Y=0mm, (b)Y=10mm. The extraction electrodes are set at Y=15mm, and the plasma grid is floating (not optimized). At Y=0mm, the center of the sheet plasma, I_e and T_e increase gradually as p decreases from 50mTorr to 5mTorr, and then becomes constant or slightly decrease as p decreases to 2mTorr. I_e and T_e take their maximum values at p =5mTorr. It seems from the probe characteristics that the plasma has two components of electron temperature at $p \sim$ 5mTorr and that the high energy component decreases with increase of p. At Y=10mm, n_e increases as p decreases and becomes maximum at p =5mTorr as in the case of Y=0mm. However, T_e changes in a different way. As p decreases, T_e decreases to 1.5eV at p =5mTorr and then increases. At $p \sim$ 5mTorr the plasma in front of the plasma grid has enough cold electrons whose temperature is suitable to produce H^-. These observations indicate that the plasma parameters are optimized to the H^- production at the optimum pressure p_{opt} =5mTorr in the case of I_d=2A. The p_{opt} increases as I_d increases. It is expected that many H^- ions are yielded at p_{opt}. The detected H^- currents are shown in Fig.5 (c) as a function of p. H^- current becomes maximum at p=5mTorr as expected.

The dependence of n_e and T_e in the sheet center on I_d at p =23mTorr are shown in Fig.6(a). As I_d increases, n_e increases linearly but T_e remains constant at T_e=5eV. H^- current also increases linearly up to I_d=30A, but it saturates above I_d=40A as shown in Fig.6(b). The bias voltage is not optimized in this case. The increase of I^- in the low I_d range corresponds to the density increase in the center region. The reason why I^- saturates in the high I_d range is not known yet. Further experiments are required. The maximum obtained H^- current density is about 4mA/cm^2 at I_d=100A.

DISCUSSION AND SUMMARY

In the volume production mechanism, the following two processes are important to the H^- production.
1. The vibrational excitation of the hydrogen molecules by collisions with fast electrons. (energy $\varepsilon \gtrsim 40eV$)

$$H_2 + e_f \rightarrow H_2^* + e \qquad (1)$$

2. The dissociative attachment of slow electrons ($\varepsilon \lesssim 1ev$) to the excited molecules.

$$H_2^* + e_s \rightarrow H^- + H \qquad (2)$$

It is important to separate the two regions where these processes take place, since the presence of electrons ($\varepsilon \gtrsim 2eV$) results in the collisional detatchment of H^-. In the sheet plasma, these separation is easily achieved by the magnetic field. The plasma produced in the source flows into a discharge region along the magnetic field lines and is confined. Many fast electrons emitted from the cathode are containing and the excitation process (1) is occured in this region. The thermal plasma is diffused from the discharge region toward the electrodes across the magnetic field. Then, the slow electrons of its energy $\varepsilon \lesssim 1eV$ exist in this diffused region. As shown in the previous section, H^- yield becomes maximum at the diffused region where the electron temperature decreases below 1eV. The reduction of H^- yield near the discharge region is caused by the collisional detachment by hot electrons. As p becomes close to the p_{opt}, n_e and T_e increase and fast electrons exist in the discharge region. In the diffused region, however, T_e decreases to 1eV and n_e increases. The plasma parameters in the two regions becomes suitable for the H^- production and maximum H^- current is obtained in the optimum pressure.

In summary, the H^- yield is optimized in the tandem plasma provided in the sheet plasma negative ion sources. When the extraction electrodes is moved toward the plasma, the optimum position for the H^- yield exists near the plasma edge, where the slow electrons of their energy below 1eV are aboundant. It is found that a small negative bias potential relative to a plasma potential would enhance the H^- yield accompanied by a reduction in electron current. In the optimum gas pressure, the electron density increases and the electron temperature decreased to 1eV in front of the electrodes, which are suitable conditions for the dissociative attachment. The extracted H^- current becomes maximum at this optimum pressure. When the discharge current I_d increases, I^- also increases corresponding to the increase of n_e in the sheet center but it saturates above I_d = 40A. The maximum current density of 4mA/cm^2 is obtained at I_d = 100A.

ACKNOWLEDGEMENTS

This work is partly supported by the Grand-in-Aid for Scientific Researchers of Ministry of Education, Science and Culture, Japan.

REFERENCES

1. T.Kuroda, K.Sakurai, Y.Oka and O.Kaneko,
   in Proceedings of the 4th International Symposium on the Production and Neutralization of Negative Ions and Beams, Brookhaven, 1985, p.289.
2. Y.Oka, Y.Kubota, O.Kaneko, T.Kawamoto and T.Kuroda,
   in Proceedings of the 12th Symposium of Fusion Engineering, Monterey, 12-16 October, 1987, p.468
3. A.Ando, T.Kuroda, Y.Oka, O.Kaneko, A.Karita and T.Kawamoto,
   in Proceedings of the International Conference on Ion Sources, Berkeley, 10-14 July, 1989.
   (to be published in Rev. Sci. Instrum.)
4. A.F.Lietzke and G.Guethlein,
   in Proceedings of the 12th Symposium of Fusion Engineering, Monterey, 12-16 October, 1987, p.1227

DISCUSSION

Hemsworth: I have two questions. One, your target current density for your injector is 30 milliamps per square centimeter, how do you propose to get from 4 milliamps to 30 milliamps per square centimeter?

Ando: The main object of this experiment is to understand production in diffuse plasma. Now we are designing the 30 milliamps source, for which we selected the bucket source.

Hemsworth: My second question is have you any intention of operating with deuterium with this source.

Ando: Deuterium, no experiment (is planned).

Jacquot: It is more or less the same question, but I will add a comment. I worked years ago on an ECR source to produce H^- through the magnetic field, instead of two other discharges with classical filaments. We observed effectively H^- on the edge of the plasma, but the densities are too low that we cannot get these edges to produce large current densities of H^-. You have low temperature and also low density, that means you are not expected to get high current of H^-.

Ando: Yes.

Holmes: I have a comment as well that I would like to make. In your early graphs you showed the electron density decreasing very rapidly as you moved out of the sheet plasma. There is a difference between the bucket source and the sheet plasma source. In the bucket source, that electron density decrease across the field is very much less by a factor of 2 or perhaps 3.

Ando: What you say, is the density profile of the plasma.

Holmes: In the bucket source, this decrease in current as you move across the field is very much less, it looks like something like this.

Ando: In this experiment, transverse magnetic field exists in the extraction grids. The diffusion across the magnetic field is low, in bucket source magnetic field decreases in the front of the extraction grid. The diffusion is large.

Bretagne: I think it is very difficult to compare the sheet plasma and the bucket source, in fact, I think that in one case we are near empty conditions, we have no more molecules and this is not the case, I hope, in the bucket source. So we start with very different conditions, do you have some idea about the mechanism, the processes which lead to the H^- formation?

Ando: The process for H^- production is dissociative attachment.

Belchenko: Your system is very close to system developed by Ken Ehlers approximately 25 years ago, though it may be better to compare the production mechanism by comparing the density of negative ions in the same sources. That is my comment. The question is are you going to introduce a small amount of cesium into this source?

Ando: The main use for the experiment is to get a database for the negative ion bucket source, and we don't have enough data for the best location of the magnetic filter. So we are experimenting with this at present.

Whealton: What's the anticipated gas efficiency of your projected 45 amp?

Ando: Gas efficiency is very bad, I forget what it is exactly, I will inform you afterwards.

OPERATION OF A LARGE NEGATIVE ION SOURCE IN DEUTERIUM

L.M. Lea, A.J.T. Holmes and M.F. Thornton
Culham Laboratory
Euratom/UKAEA Fusion Association
Abingdon, Oxon. OX14 3DB, England

ABSTRACT

Experiments performed in a volume production negative ion source show that the use of a tent or supercusp magnetic filter leads to a more uniform plasma at the normal extraction plane than that obtained with a dipole filter.

For the tent filter geometry the extracted negative ion current density, increased as the accelerator was moved into the source to a maximum value of about 30% greater than that measured at the normal extraction plane.

There is evidence that the primary electrons were well confined by the magnets on the walls of the source, but that significant numbers of positive ions were able to escape from the source by cross-field diffusion, at pressures in excess of 7 mTorr.

INTRODUCTION

For heating and/or current drive in the core of the relatively dense plasmas envisaged within the NET or ITER tokamaks neutral beam energies in excess of 1 MeV will be required. The neutral beam injectors will therefore have to be based on the acceleration and neutralisation of negative ions. This is because the cross-section for the neutralisation of positive ions, and therefore the equilibrium neutral fraction obtainable from them, decreases steeply at energies > 80 keV/nucleon. The efficiency of neutralisation of a negative ion beam is, however, almost independent of the beam energy with a value slightly less than 60% for a simple gas neutraliser, rising to above 80% if a plasma neutraliser is used.

Over the past few years, work at the Culham Laboratory on volume production negative ion sources has produced very encouraging results Lea et al.[1]. The work described here was initiated with the aim of increasing the D$^-$ current density obtainable from a large volume production ion source while maintaining a high uniformity over a large area, so that in principle multi-ampere currents could be provided by a full size extraction system with multiple apertures.

DIAGNOSTICS

Two types of diagnostic have been used on the source test facility. These are a group of sixteen planar Langmuir probes each of area 68 mm^2 set in a cross formation, and a miniature ion accelerator, with an entrance aperture of 1.5 mm diameter, located at the centre of the probe formation. All the Langmuir probes and the miniature accelerator are attached to a drive mechanism which enables them to be moved to any axial position within the source.

To keep the pressure within the accelerator low enough to prevent gas stripping of appreciable numbers of negative ions a separate pumping system is utilised.

PRINCIPLE OF OPERATION AND CALIBRATION PROCEDURE FOR THE ACCELERATOR

The accelerator extraction voltage is typically 10 kV and is applied between the grounded beam-forming electrode and the second electrode. The second electrode contains two sets of permanent magnets. One set is arranged to create a dipole field near the front face of the electrode, and the other creates a reversed dipole field near the rear face. Electrons accelerated from the plasma are deflected into the well in the electrode while only a small deviation is caused to the trajectories of the negative ions.

The Faraday cup is located immediately behind the third electrode and may be biased by up to +100 V with respect to it, to prevent secondary electron emission due to ion impact. The current received by the third electrode is only of the order of 5 percent of the negative ion current reaching the Faraday cup. No Faraday cup signal was seen when a source discharge was run in helium gas; therefore electron leakage past the magnetic traps is considered to be negligible.

By extracting positive deuterium ions and comparing the extracted current density measured by the Faraday cup with the average positive current density on the four adjacent Langmuir probes, it was possible to estimate the gas target in the accelerator for each source pressure. Cross-section data in Barnett et al.[2] for the charge exchange cross-section for H+ and the one electron loss cross-section for H- are in the ratio 1: 1.3 over the energy range of interest (1 to 10 keV), this then enabled the loss of D- within the accelerator to be estimated, Lea et al.[1]. The loss increased with source pressure and reached about 27% at a pressure of 12 mTorr.

COMPARISON OF FILTER GEOMETRIES

Experiments have been performed with two different magnetic filter geometries. The first of these, the dipole filter, is created by re-orientation of the magnets in the second and third rows from the front of the source on both of the long sides, Holmes et al.[3]. On one long side of the source north poles faced into the plasma and on the other side south poles faced, thereby creating a field across the source of peak value 32 Gauss on the centreline. All other magnets on the source walls were arranged in a checkerboard pattern. At a discharge current of 1375 A and a filling pressure of 12 mTorr a maximum D^- current density of 29 mA/cm^2 was obtained after correction for stripping within the accelerator.

The array of Langmuir probes was used to measure the uniformity of the positive ion current density in the normal extraction plane and the results are shown in Fig. 1, for the long direction. The positive ion current density was found to be reasonably uniform in the short direction and therefore it is to be expected that the negative ion current density was also uniform. In the long moved into the source. The plasma potential at first increases from 3.75 V to 5.25 V in the first 40 mm before remaining, on average, near constant to the furthest point of measurement. From the small change in the electron temperature and plasma potential it is clear that at all times the probe and accelerator remained within the extraction volume of the source and did not penetrate into the driver volume. Measurements made with movable probes in a source configured with a dipole filter showed a marked increase in the electron temperature and plasma potential as the probes moved into the filter and then on into the driver volume.

The increase in the positive ion current density up to a point around 90 - 100 mm from the normal extraction plane is not inconsistent with results found on moving into a dipole source. The increase being almost linear as the distance from the filter is decreased, corresponding to an increase in the overall plasma density. Beyond this position both positive ion and electron currents decrease steadily showing a true reduction in plasma density almost certainly due to electrons being trapped on the filter field lines then lost to the back of the source, the positive ions being drawn along to maintain quasineutrality. The fact that the electron current increases on moving from the normal extraction plane to about 40 mm into the source is consistent with the increase in overall plasma density indicated by the positive ion current. The decrease from around 40 mm to 90 mm in, would appear to be primarily due to the fact that there are increasing numbers of negative ions present and so less electrons are required to maintain quasineutrality.

It is of interest to note that the maximum negative ion current is found at a distance into the source which is only slightly less than that for the maximum plasma density indicated by the positive ion current and is at a position close to the filter.

PLASMA UNIFORMITY FOR EXTRACTION FROM WITHIN THE SOURCE

Since the negative ion current measured by the accelerator reached a maximum value at a distance of about 90 mm into the source, the Langmuir probes were used to ascertain the usable extraction area. The positive ion saturation current j_+, the ratio j_e/j_+ and the electron temperature are shown in Fig. 6, for the long source dimension at 90 mm from the normal extraction plane. The positive ion current density is near uniform over most of the distance indicating that the plasma uniformity is reasonable. The ratio j_e/j_+ is unfortunately rather far from uniform with a value increasing from slightly less than 18 near the centreline of the source to around 34 at 220 mm from the centreline. This would imply that the negative ion to positive ion number density ratio decreases by about a factor of three on moving from the centreline to the furthest excursion. This is extremely likely on the basis of the corresponding increase in the electron temperature, which would reduce the dissociative attachment rate and increase the rate of electron detachment leading to a net decrease in negative ion density.

direction, however, the positive ion current density has a severe gradient from one side of the source to the other. This is believed to be due to a $\underline{j} \times \underline{B}$ drift of the charged particles crossing the filter field. Experience with "tent" or "supercusp" filter designs in connection with the JET positive ion sources, encouraged the testing of this geometry for efficiency of negative ion production and uniformity.

The tent filter geometry was created by orientating the magnets in the first two rows from the front of the source on both of the long sides so that south poles faced into the plasma. At the same time orientating the magnets in the centre region of the back plate so that north poles faced into the plasma.

At a pressure of 12 mTorr and a discharge current of 1300 A the D⁻ current density was only 3% less than that measured under similar conditions for the dipole filter geometry. The great advantage of the tent filter can be seen in Fig. 2, where the positive ion uniformity in the long direction is now very good. It can also be seen that the ratio of the electron saturation to positive ion currents is reasonably uniform in the long direction of the source. This would imply that the negative ion density is also approximately uniform since the ratio j_e/j_+ to a probe will decrease almost linearly from the value found in a plasma comprising solely electrons and positive ions as the negative ion density increases. In the short direction the uniformity remains good over the centre region of the source, but the current density and the ratio j_e/j_+ reduce quite rapidly beyond about 70 mm from the centreline as the edge of the tent filter is encountered.

OPTIMISATION OF EXTRACTION PLANE POSITION

Movement of the accelerator into the source for the tent filter configuration produced the results shown in Fig. 3. At around 90 mm into the source the extracted D⁻ current density reaches a peak value 30% greater than that observed in the normal extraction plane, while the electron current reaches a maximum at about 30 mm into the source, before decreasing steadily as the accelerator is moved further in.

Measurements of the electron current density, positive ion current density and the ratio j_e/j_+ obtained by a Langmuir probe adjacent to the accelerator are shown in Fig. 4. As the probe was moved into the source the measured electron current density was found to agree well with the current density measured on the second electrode of the accelerator. The positive ion current density increased steadily as the probe was moved into the source until a maximum value was reached at around 90 - 100 mm from the normal extraction plane. The ratio j_e/j_+ increased by a small amount over the first 10 - 15 mm, but from that point onwards decreased steadily.

The Langmuir probe was also used to measure the electron temperature and local plasma potential as a function of position within the source, the results obtained are shown in Fig. 5. The electron temperature rises from about 1.0 to 1.2 eV as the probe is

In the short direction the positive ion current density increased beyond about 70 mm from the centreline. This is consistent with the fact that the filter field restricts the extraction volume of the source, particularly in the short direction. The increase in the positive ion current is partially due to an increase in the plasma density but also to a clear increase in the electron temperature as the probes pass through the filter into the driver volume. The ratio j_e/j_+ shows a marked increase beyond 90 mm from the centreline, by which point the electron temperature has reached 2 eV and the number of negative ions will be reduced.

PRIMARY ELECTRON AND ION CONFINEMENT

In principle the primary electron confinement time, T_p, may be found from the equation defining the production and loss balance for the primaries, Green et al.[4].

$$\frac{I_{disc}}{eV} = n_p N_0 S_{in} + \frac{n_p}{T_p}$$

V is the source volume and S_{in} the rate for inelastic collisions. It has not been possible to measure primary electron densities directly using Langmuir probes, therefore as an approximation n_p can be eliminated from the equation by use of the equation defining the production and loss balance for positive ions.

$$\frac{I_+}{eV} = n_p N_0 S_i$$

S_i is the rate for ionising collisions of primary electrons on gas molecules, and I_+ is the positive ion current produced.

If it is assumed that practically all positive ions are collected on the metal plate across the front of the source in the normal extraction plane and few reach the source anode due to the magnetic confinement, then $I_+ = k\, j_+$ where j_+ is the positive ion current density at the source extraction plane.

The result is:

$$\frac{I_{disc}}{k\, j_+} = \frac{S_{in}}{S_i} + \frac{1}{N_0 S_i T_p}$$

For the large negative ion source $I_+ = 1000\, j_+$. A plot of I_{disc}/j_+ versus 1/pressure should allow T_p to be estimated for a given primary ionisation rate.

In Fig. 7, I_{disc}/j_+ is plotted against 1/pressure for the source configured with the tent filter. At source filling pressures below 7 mTorr the data set shows an approximately linear dependence of I_{disc}/j_+ on 1/pressure. From the gradient of the fitted line and for a primary ionisation rate S_i of 2×10^{-14} m^3 sec^{-1}, a relatively long primary electron confinement time of 5×10^{-7} seconds is obtained. At pressures above about 7 mTorr the data shows I_{disc}/j_+ increasing with increasing pressure, although this would not be expected from the simple model. It is considered likely that this effect is due to the cross-field diffusion of positive ions to the source anode, a process previously observed at high gas pressures by Goede and Green[5].

The current to the anode is given by:

$$I_{+a} = - A e D_+ \frac{dn}{dx}$$

$$= \frac{A e C_s}{3 N_0 \sigma} \frac{n}{L} \frac{1}{1 + \left(\frac{e B}{M} \frac{1}{N_0 \sigma C_s}\right)^2}$$

where C_s is the ion sound speed, L is the source half width, A is the anode area, B is the magnetic field in the intercusp region and σ is the elastic cross-section.

For high magnetic fields this reduces to a linear dependence of the current to the anode on the product $j_+ N_0$.

$$I_{+a} = \text{constant } j_+ N_0$$

Since the primary electron confinement time is long, $I_{disc}/(kj_+ + I_{+a})$ may be treated as having effectively an almost constant value. I_{disc} is then proportional to kj_+ + constant $j_+ N_0$. A plot of I_{disc}/j_+ against pressure, Fig. 8, will therefore show a linear relationship if cross-field diffusion of positive ions is a major process. This can be seen to be true for pressures in excess of about 10 mTorr. For pressures below 7 mTorr cross-field diffusion is not so significant and the assumption that $I_{disc}/(kj_+ + I_{+a})$ is near constant ceases to be valid.

DISCUSSION

It is clear that if a volume production negative ion source is to be used in a neutral injection system then multi-aperture extraction will be used and the negative ion current density should be as uniform as possible over the full extraction area. Of the two magnetic filter geometries tested the tent filter satisfies this condition best, with the maximum uniform extraction area at the normal extraction plane.

There would appear to be some possibility for trading off the uniform extraction area for higher negative ion current density by using an accelerator which is re-entrant into the source, but without further experimental tests there is a high probability that the large area of the accelerator would perturb the source plasma and thereby reduce the local negative ion current density.

On the basis of the present level of source development it would appear realistic to aim for a conservative current density figure of 15-20 mA/cm² from the source. There are two advantages to working with a current density of this level. First, high D^- current densities require high source discharge currents and suppression of the increased electron flux becomes more difficult. Second, dumping of the power carried by the residual ions, after beam neutralisation, becomes an increasingly difficult problem as the current density is increased for beam energies of order 1 MeV.

Cross-field diffusion of positive ions out of the present source appears to become a problem at source filling pressures in excess of 7 mTorr. In order to obtain an adequate negative ion current density for neutral injection purposes, while keeping the gas stripping within the accelerator to a manageable level, it will be necessary to operate each source at a pressure of around 10 mTorr. The normal checkerboard confinement magnet geometry must therefore be strengthened on future purpose built sources.

REFERENCES

1. Lea L.M., Holmes A.J.T., Naylor G.O.R., Clark D.C. and Newman A.F., Proc. of the Third European Workshop on Production and Application of Light Negative Ions. Amersfoort, The Netherlands, Feb. 17-19 1988, FOM-Institute for Atomic and Molecular Physics (1988).

2. Barnett C.F., Ray J.A. Ricci E., Wilker M.I., McDaniel E.W., Thomas E.W. and Gilbody H.B., Atomic Data for Controlled Fusion Research, Oak Ridge National Laboratory report ORNL-5206, Vol. 1 (1977).

3. Holmes A.J.T., Lea L.M., Newman A.F. and Nightingale M.P.S., Rev. Sci. Instrum., 58(2) p. 223 (1987).

4. Green T.S., Holmes A.J.T. and Nightingale M.P.S., 4th Int. Symp. on the Production Neutralisation of Negative Ions and Beams, Oct. 27-31 1986, Brookhaven National Laboratory.

5. Goede A.P.H. and Green T.S., Physics of Fluids Vol. 25 No. 10 p. 1797 (1982).

240 Operation of a Large Negative Ion Source

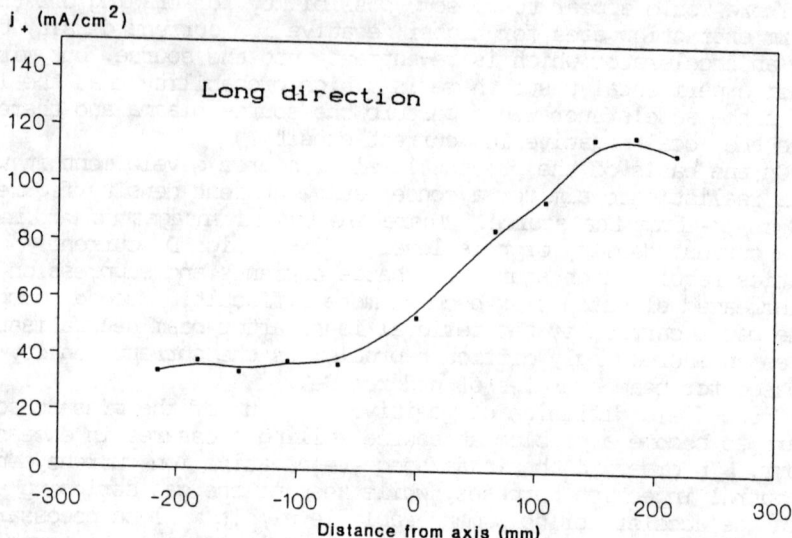

Fig. 1. Plasma uniformity in the normal extraction plane for the dipole filter.

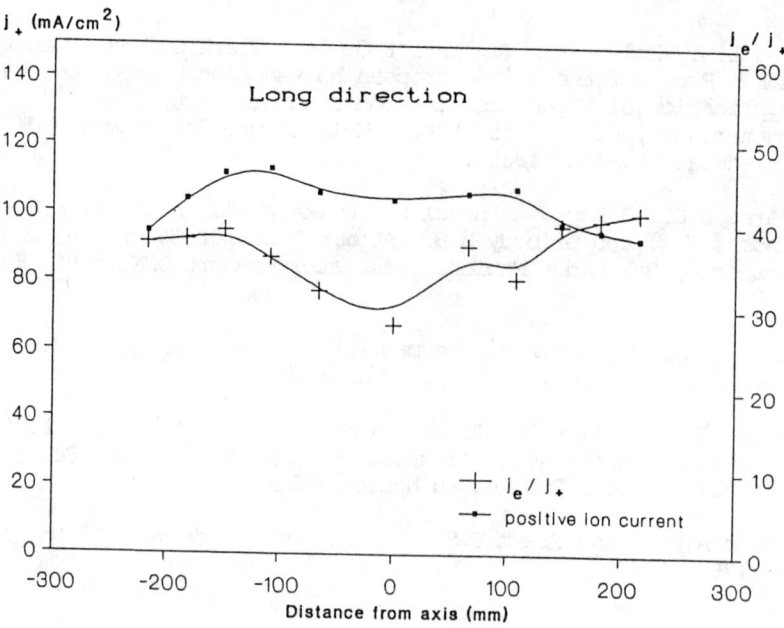

Fig. 2. Plasma uniformity in the normal extraction plane for the tent filter.

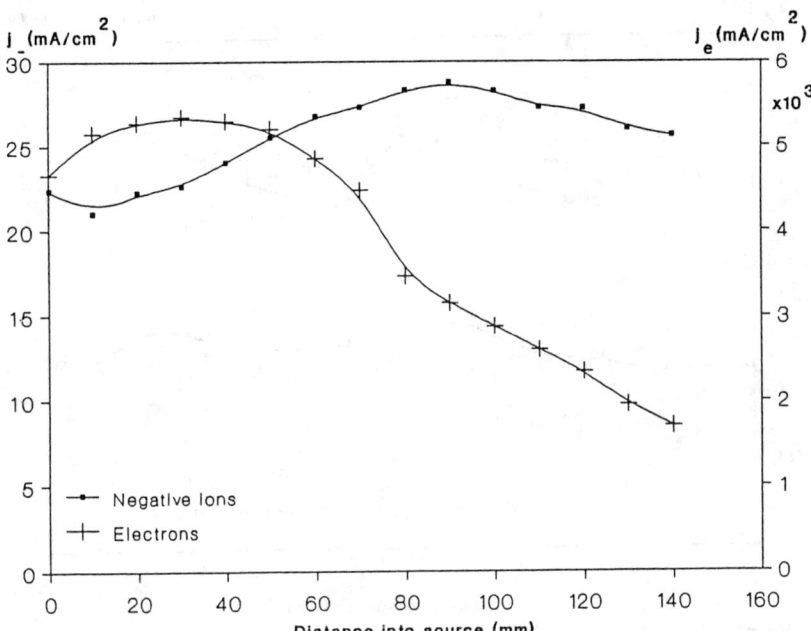

Fig. 3. Extracted negative ion and electron current densities within the source.

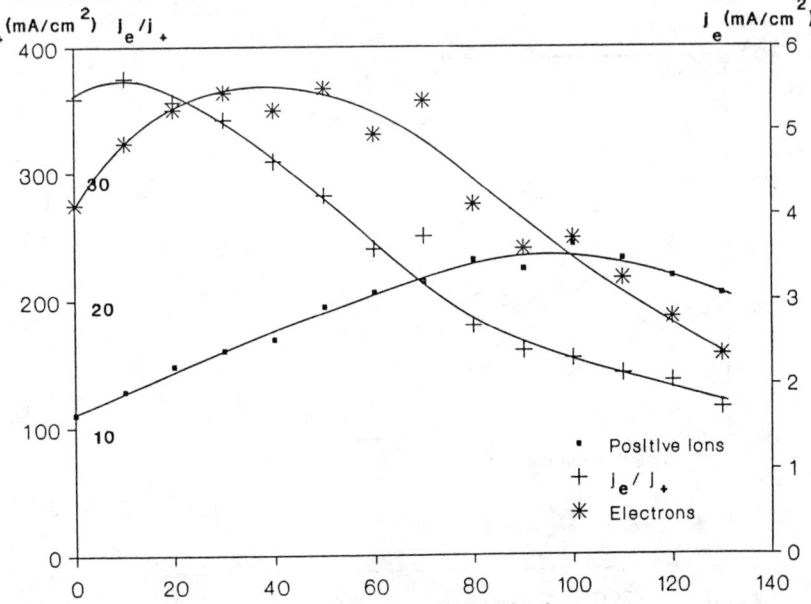

Fig. 4. Positive ion and electron current densities and the ratio j_e/j_+ within the source.

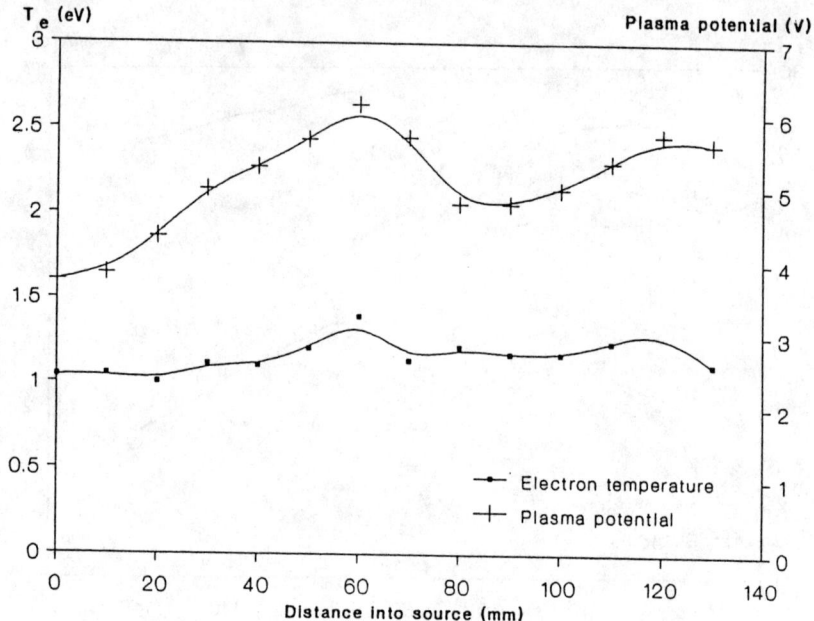

Fig. 5. Electron temperature and plasma potential within the source.

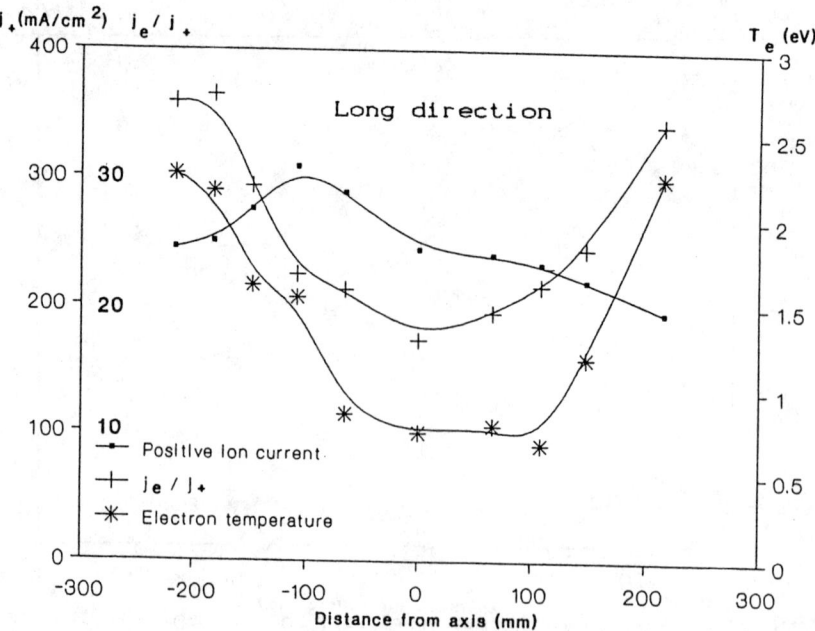

Fig. 6. Plasma uniformity at 90 mm into the source.

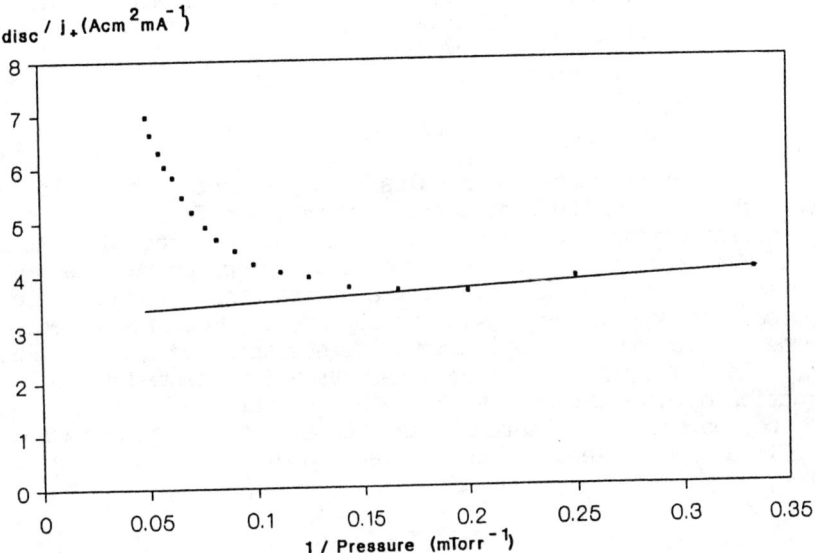

Fig. 7. Primary electron confinement.

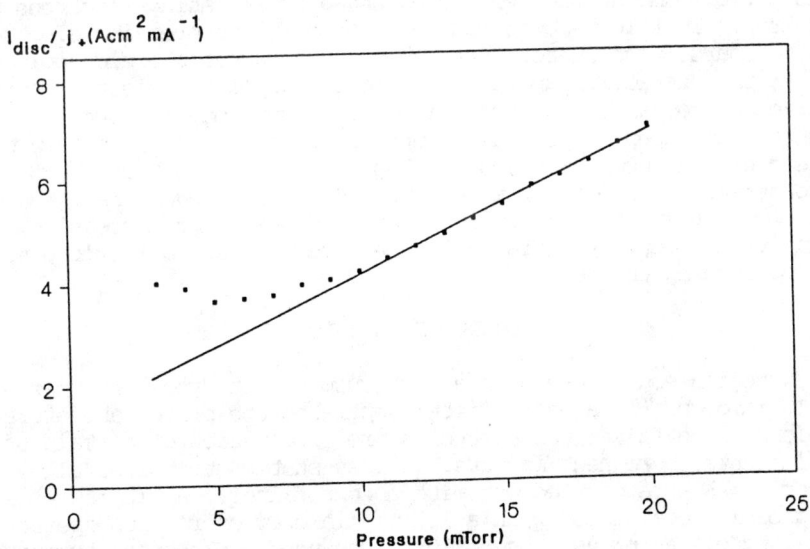

Fig. 8. Positive ion cross-field diffusion.

THE EFFECTS OF ELECTRON SUPPRESSION FIELDS ON D⁻ PRODUCTION

M.F. Thornton, A.J.T. Holmes, L.M. Lea and G.O.R. Naylor
Culham Laboratory
(Euratom/UKAEA Fusion Association)
Abingdon, Oxon. OX14 3DB, England

ABSTRACT

This paper presents the results of experimental studies on a number of different 'current carrying wire' type first grid electron suppressors: or 'inserts', as they shall be called throughout this paper. Extracted negative ion currents are compared for the different inserts over a range of field strengths. Currents are also compared to those extracted with no suppressor present. An attempt is made to relate the negative ion currents obtained for each insert, with field depth and uniformity. Results are presented on the extraction of negative ions with an insert operating some distance into the source. The insert is also tested with a dipole rather than tent field used to separate the driver and extraction regions of the source.

INTRODUCTION

In extracting negative ions from a volume production source, electrons are also removed. With accelerators in use at Culham Laboratory, electrons are dumped in a trap within the second electrode at a fraction of the beam energy. Using this technique a considerable amount of power is expended accelerating electrons to the trap. This lost power can be reduced by 'suppressing electrons' at the source/accelerator interface, via a magnetic field applied across the extraction aperture. A reduction in the electron current extracted from an ion source using a permanent magnet suppressor has been demonstrated by several groups[1-3]. Electron suppression with an insert of the form described in this paper has also been demonstrated[4-5]. Ideally, an electron suppressor would reduce the electron current to zero while keeping constant, or increasing, the negative ion density extracted. The insert offers a flexible method of approaching this goal.

EXPERIMENTAL DETAILS

The ion source used was a multicusp bucket type, 550 mm x 310 mm x 210 mm deep. A magnetic filter separated the driver and extraction regions. The filter was created by re-orientation of certain of the wall-mounted permanent magnets. Unless stated otherwise, all experiments were carried out with a tent filter. A miniature diagnostic accelerator with a 1.5 mm diameter extraction aperture and total accelerating voltage of 10 kV was used. Electrons extracted from the source were magnetically deflected into a trap within the accelerator, while negative ions were collected on a Faraday cup. For all experiments the source was operated at 400 A arc current and 120 V arc voltage. Deuterium gas was used at a pressure of 12 mTorr.

The "insert" device used to suppress the electron current extracted from the source, is shown in diagramatic form in Figure 1. The dipole magnetic field is imposed transverse to the extraction aperture, on the source side. The field lines intersect the insert wall, which may be biased independently of the beam forming electrode. If the coil pitch and depth are kept similar in size, the insert creates a field of comparable thickness to the coil depth. Thus it is possible to produce a relatively intense, local field in front of an aperture, when compared with that produced by a permanent magnet suppressor.

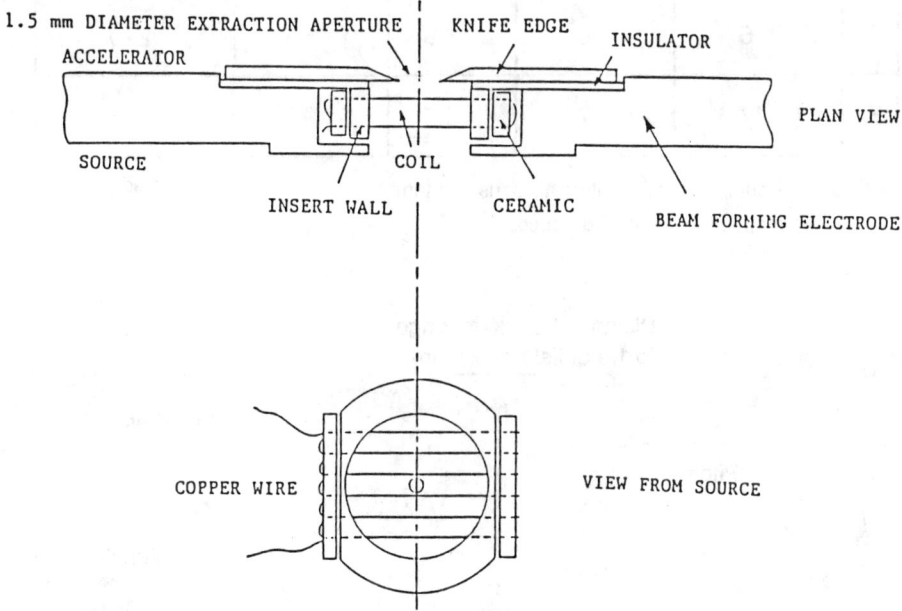

Figure 1 : The electromagnetic insert assembly.

EXPERIMENTAL RESULTS

A number of different insert assemblies have been tested. A summary of the inserts and the magnetic fields they create is given in Table 1. Figure 2 illustrates the various parameters given in Table 1.

	Coil Depth /mm	Coil Pitch /mm	K.E to field max /mm	K.E to 1/e of field max /mm	Variation of B along centre line %
A	4	4	4.0	6.3	20
B	5	5	3.8	6.5	21
C	8	4	6.0	10.0	0
D	5	15	3.8	8.1	240
E	17	17	10.5	19.0	15

Table 1 : A summary of the different inserts and the magnetic fields they produce.

K.E = knife edge

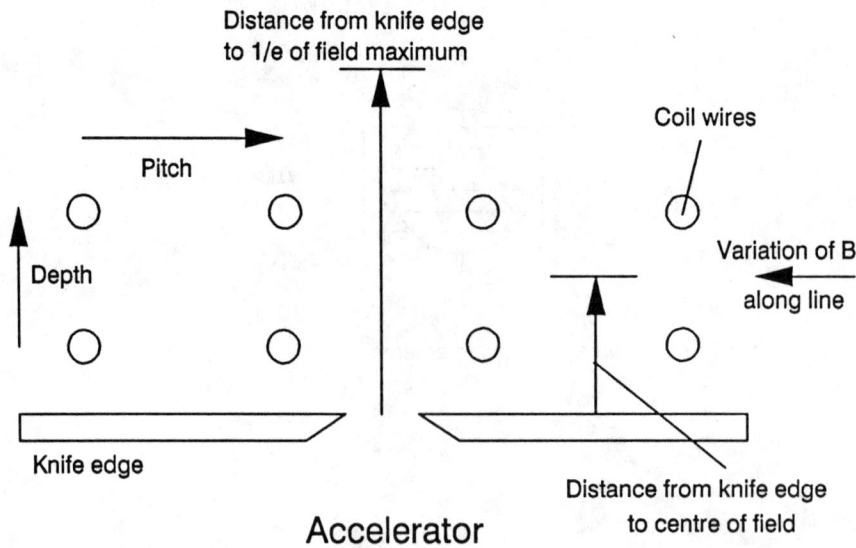

Figure 2 : A cross-section of an insert, indicating those parameters given in table 1.

All the fields given in this table were calculated using a two dimensional electromagnetic analysis package, PE2D[6]. Theoretical predictions were compared with Gauss probe measurements wherever possible and agreement was found to be good (± 15%). The distance from the knife edge, to the point where the magnetic field fell to 1/e times the maximum, was used to gauge the field width. The percentage variation in B from the minimum value, along the centre line of each coil, was used as a measure of field uniformity across the aperture. It can be seen, that inserts A and B (see Table 1) give relatively narrow, uniform fields. Insert C, a broader, uniform field. Insert D, narrow, but highly non-uniform and insert E, a much wider, uniform field.

The extracted negative ion current densities for different inserts at varying field strengths are shown in Figures 3 and 4. The currents are plotted as $J_-/J_-(o)$ where $J_-(o)$ is the negative ion current density obtained under identical source conditions, but with no suppressor present. It is important to distinguish between two factors affecting D⁻ extraction with an insert. One is the enhancement, or the increase in D⁻ yield with the insert field on compared to the field off. The other, is the D⁻ current obtained with no insert field applied. The yields with no insert field lie between 40%-60% of the current extracted with no suppressor. Thus, as in the case of insert C, it is possible for an insert to give poor negative ion yields, while the magnetic field still gives a sizable enhancement. This substantial fall in ion current from the presence of each insert structure can be understood by plasma loss, to the walls of the well and coil wires, in the insert assembly.

Inserts A and B with a short coil pitch and depth, giving narrow and uniform fields, produced easily the best results. The maximum current densities approximately equal the value with no suppressor. Insert A, while giving D⁻ yields equal to those obtained with no suppressor, reduced the extracted electron current by a factor of 27. On the other hand, insert B has been operated to reduce the electron to ion ratio down below 2:1, the negative ion yield was reduced to 60% of the maximum value. Insert C, with its coil depth and therefore field width increased by approximately 4 mm gives a significantly poorer performance. The field still gives an enhancement of negative ion yields but from a lower initial value. Insert D with its slightly narrower, but non-uniform field gives virtually no enhancement. Insert E with a uniform but much deeper field gives a very small enhancement. Note that when insert A was operated, a small reversed field was required to offset the residual field from the electron trap in the accelerator. This offset was not observed with the other inserts.

The next set of results apply only to insert B (pitch = 5 mm, depth = 5 mm). Figure 5 shows the negative ion densities versus insert field, for the extraction aperture at 45 mm and 90 mm into the source. Here $J_-(o)$ remains the D⁻ density extracted with no suppressor present, but with the same aperture position in the source as the experiments carried out with the insert[7]. At 45 mm into the source, one begins to see signs of interaction between the tent filter and the insert field. Negative ion densities were somewhat higher with the coil current in one direction compared to the other, although this was not expected, considering the symmetrical nature of the tent filter field. At 90 mm into the source very little or no increase in D⁻ was observed when the insert field was applied.

Figure 6 shows the extracted D⁻ current densities versus insert voltage for both tent and dipole filter fields in the source. The D⁻ yields with no suppressor present ($J_-(o)$) were very similar for the tent and dipole configurations. When the insert field was applied densities measured with the dipole filter were substantially lower than those with the tent filter. With the dipole filter, no improvement was observed in D⁻ currents when the insert field was reversed. The dipole filter gave a magnetic field strength of approximately 25 gauss in front of the aperture, the tent filter 5 gauss.

DISCUSSION

Table 2 summarizes the results of D⁻ extraction with the five different inserts. Enhancement has been defined as the peak negative ion density obtained with an insert, divided by the density extracted with no field applied. This number is important in deciding which of the insert fields are effective at increasing D⁻ densities extracted from the source.

Insert Type	Field Width /mm	Uniform Field	Peak $J_-/J_-(0)$	Enhancement (peak/zero B value)
A	6.3	Yes	1.03	1.81
B	6.5	Yes	1.00	1.64
C	10.0	Yes	0.68	1.58
D	8.1	No	0.69	1.13
E	19.0	Yes	0.54	1.15

Table 2 : A summary of results on the different inserts.

Increasing the field depth leads to a reduction in the maximum enhancement possible. Similarly, if the magnetic field is highly non-uniform across the aperture, then little enhancement occurs. At the moment this comparison of insert performance with field characteristics is very qualitative in nature, although it may be possible to quantify this analysis after testing a greater number of inserts. At present it has not been possible to operate a successful insert, of this type, with a coil pitch as large as 15 mm. Any attempt to increase pitch while keeping the depth small leads to a non-uniform field: increasing the coil depth and pitch together, leads to a broad field. Both cases give low D⁻ yields. It requires further work to determine the maximum dimensions for coil pitch and depth, while still retaining good D⁻ currents. No work has been carried out on this apparatus with insert wires in front of the aperture: it is to be expected that this shadowing effect will reduce the extracted D⁻ current. This may be important in determining the maximum size of aperture that may be used with this type of suppressor.

None of the inserts tested, gave negative ion currents significantly greater than those obtained with no suppressor present, although there is no reason to believe that this is not possible. It appears from the experiments carried out so far, that the best chance of observing a 'real enhancement', lies in making the magnetic field narrower, but keeping it uniform across the aperture: this could be achieved with an array of more closely spaced wires. However, designing an insert with wires closer than the 4 mm by 4 mm coil already tested, presents technical difficulties. For example, it becomes increasingly difficult to prevent the coil wires from touching and shorting out during operation, while one may be restricted to small aperture sizes to prevent wires blocking the aperture.

Moving the insert into the source, one sees an increasing interaction with the tent filter field. At 45 mm, the enhancement for the positive field is 1.49 and negative field is 1.25, this compares with 1.64 for the same insert back at the beam forming electrode. With the increasing interaction we see a fall in negative ion currents. The dipole filter gives a much stronger field in front of the aperture at the beam forming electrode. This interaction likewise leads to lower D⁻ currents. Thus these two sets of results appear to display essentially the same phenomenon. When the insert was placed in any field stronger than a few tens of gauss very little enhancement took place with the insert field. This interaction is unlikely to be a simple cancelling of the magnetic fields, as the insert fields were much stronger than the source filters. Likewise, any loss due to field cancellation should be removed when the coil current is reversed: this was not observed.

This result may be important if this type of insert is to be used in a smaller source. Here, even the tent filter may adversely affect operation of the insert positioned at the beam forming electrode.

REFERENCES

1. M. Bacal, P. Devynck and F. Hillion. Production and Application of Light Negative Ions. 2nd Euro. Workshop, Ecole Polytechnique, Palaiseau, March 1986, p. 75, Ecole Polytechnique (1986).

2. R. McAdams, A.J.T. Holmes, M.P.S. Nightingale, L.M. Lea, M.D. Hinton, A.F. Newman, and T.S. Green. Production and Neutralisation of Negative Ions and Beams. 4th Symp, Brookhaven, 1986, New York, p. 298. American Institute of Physics (1987).

3. R. McAdams, A.J.T. Holmes, A.F. Newman and R. King. Production and Application of Light Negative Ions. 3rd Euro. Workshop, Amersfoort, February 1988. p. 15, FOM Institute for Atomic and Molecular Physics (1988).

4. L.M. Lea, A.J.T. Holmes, M. Thornton and G.O.R. Naylor. The Suppression of Electrons Extracted from a Negative Ion Source. International Conference on Ion Sources, Berkeley, July 1989. To be published.

5. R. King, R. McAdams, A.F. Newman and A.J.T. Holmes. Physics Test of An Electron Suppressor with Variable Electric and Magnetic Fields. These proceedings.

6. PE2D, Version 8.1, Vector Fields Limited. 24 Bankside, Kidlington, Oxford, OX5 1JE, England.

7. L.M. Lea, A.J.T. Holmes and G.O.R. Naylor. Recent Results from A Large D⁻ Volume Production Ion Source. 3rd Euro. Workshop, Amersfoort, February 1988, FOM Institute for Atomic and Molecular Physics (1988).

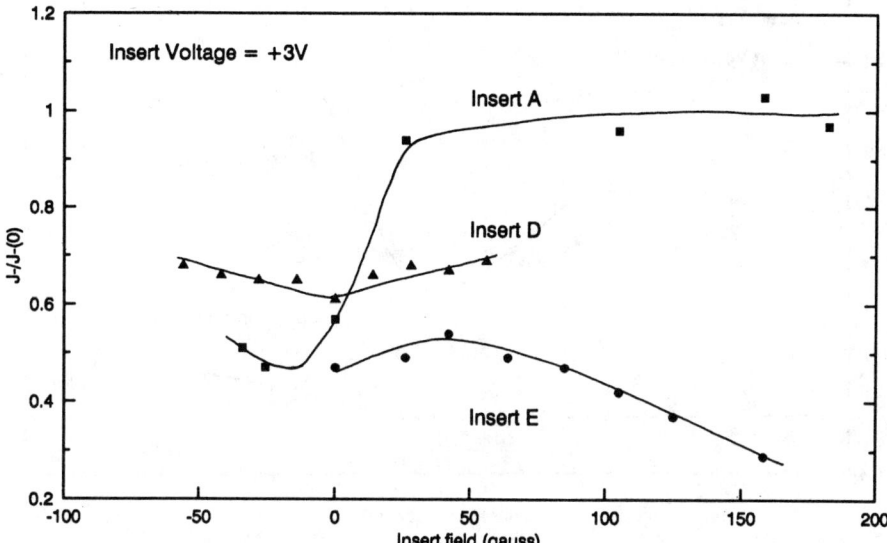

Figure 3 : Negative ion densities with varying magnetic fields, for different inserts.

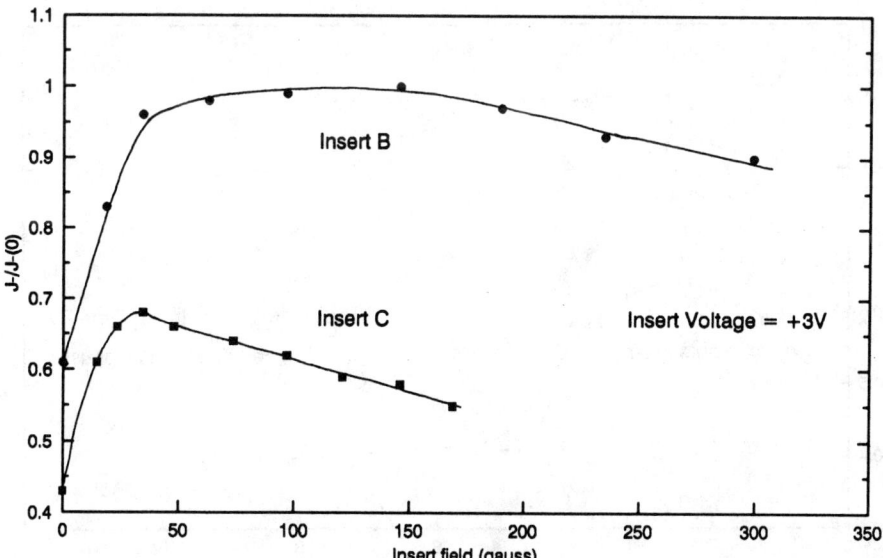

Figure 4 : Negative ion densities with varying magnetic fields, for different inserts.

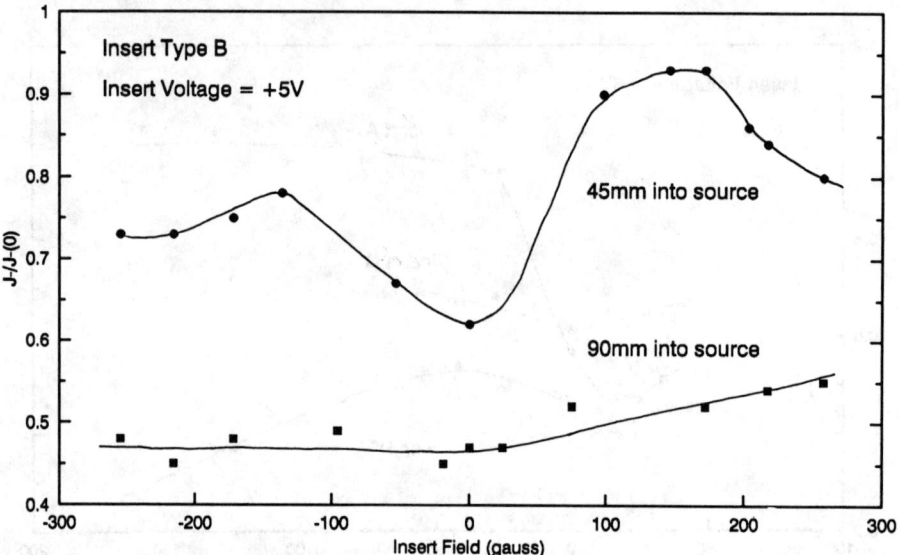

Figure 5 : Negative ion densities with varying insert fields at different positions in the source.

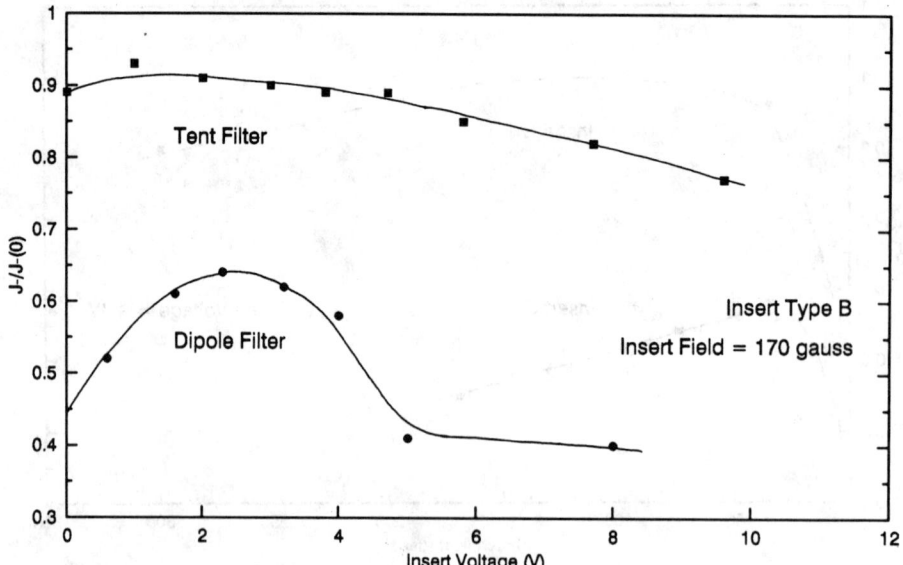

Figure 6 : Negative ion densities with varying insert voltage, for different source filters.

DISCUSSION

Hemsworth: You mentioned what seems to me to be quite important consideration is the shadowing of the aperture by the wires that you expressed to say would limit the dimensions you can use for the aperature. But in a practical system, you are limited to the design of the aperture size by considerations such as beam optics, etc. and if I remember correctly the typical design aperture size is around 30 mm for one MeV accelerators for NET, etc. If you take an aperture that size and use your small pitch filter what is the loss in shadowing of the area and is there a real enhancement as you were defining it?

Thornton: Well, we don't know, we've never tested it on this device. We are limited by our aperture size in this apparatus, so we just don't know what the effect of putting wires in front of an aperture is at the moment.

Holmes: For this reason (shadowing) and plus the engineering problems of building a filter of this type on a large scale, and since its extremely expensive, we decided to move back to the bar magnet type suppressor where we have no shadowing effect. We have actually built a large aperture accelerator operating in D^- with approximately the same pitch as the aperture and that does suppress electrons down to approximately one to one indeed, at high current densities. Though we will design in the field varying ability in the system.

Jongen: When I looked at your first figure showing the beam versus magnet current it seems to be unsymmetrical with current and that was with a tent filter so I was assuming there would be no remaining magnetic field at extraction, could you comment on that.

Thornton: The tent filter gives a field of about 5 Gauss in front of the aperture in the source. But far more important is the straight field from Grid 2, which could be about 15 Gauss, so there you do get nonsymmetry for some of the insets, you don't see it with all the insets but for the one you did.

Jongen: I have a second question. What kind of current density do you have? Would it be possible to use in a dc source?

Thornton: Yes, we used it in a dc source.

Holmes: There's no engineering problem, just money.

Whealton: Have you considering varying simultaneously the bias on the plasma electrode and the tent filter as well, elsewise you might have optimized the other two and leaving you little room for optimization.

Thornton: We haven't tried biasing the beam forming electrode separately, we tried biasing only the insert wall, and the varying the magnetic field, it's possible that things could be improved.

Whealton: I have one other question. Could you tell me the maximum current and current density that you got in the configuration.

Thornton: Well, 400 Amps there that's a positive ion density of about 75 milliamps per centimeters squared, we were getting negative, that is J_o^- of about 18 milliamps per centimeters squared.

PHYSICS TESTS OF AN ELECTRON SUPPRESSOR WITH VARIABLE ELECTRIC AND MAGNETIC FIELDS

R. McAdams, R.F. King and A.F. Newman
UKAEA Culham Laboratory
Abingdon, Oxon. OX14 3DB, England

ABSTRACT

This paper reports tests of an electron suppressor, for volume negative ion sources with variable electric and magnetic fields. The experiments have been carried out with a source and accelerator operating d.c. and with an extraction aperture of 16 mm diameter. The results are compared to those obtained with a permanent magnet suppressor and to a theoretical model of the transport of electrons by diffusion across the magnetic field.

INTRODUCTION

Practical applications of negative ion sources require stringent control of any extracted electrons. For volume negative ion sources operating in hydrogen the electron current can be of the order of 50 times the negative ion current if nothing were done to reduce the extracted electron current. Lea et al.[1] have recently described an electron suppressor with variable magnetic and electric fields which showed some promise. However their experiments used a small probe accelerator with an extraction aperture of 1.5 mm diameter. We have installed such a suppressor on the Culham Ion Source Test Stand using a 16 mm diameter extraction aperture and operating d.c.. The results are compared to those obtained using a suppressor with permanent magnets and to a diffusion model[2] of the electron transport across the magnetic field.

EXPERIMENTAL DETAILS

The volume source and accelerator used in this work have been described elsewhere[3,4]. The plasma generator is a copper magnetic multipole bucket of dimensions 195 x 140 x 85 mm^3. The source filter field was configured in two modes, as shown in Figure 1, the dipole filter has a field strength of ~ 30 Gauss at the extraction aperture whereas the tent filter has only a field of ~ 5 Gauss at the extraction aperture and will produce a more uniform plasma across the extraction plane. The extraction aperture on the source was 16 mm diameter. The accelerator was a triode design which enabled the extracted electrons to be dumped within the accelerator structure at an energy of about 1/6 of the final ion beam energy. The extracted ion beam was measured downstream by a d.c. beam transformer. Figure 2 shows the schematic source and accelerator.

Source Magnetic Filter Configurations

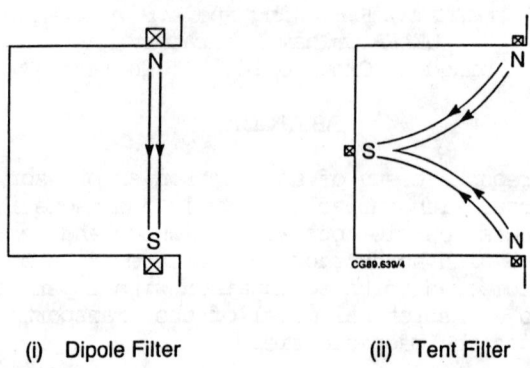

(i) Dipole Filter   (ii) Tent Filter

Figure 1   The source filter field configurations.

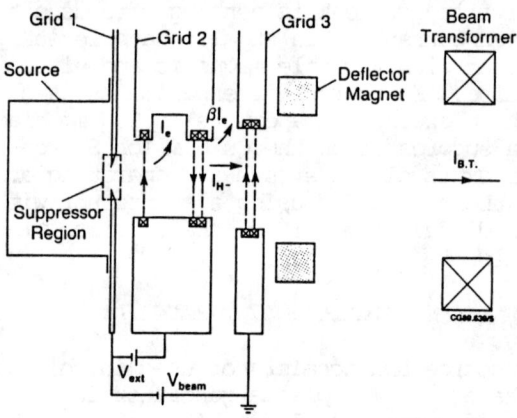

Figure 2   The source and accelerator.

Figure 3 shows a schematic diagram of an electron suppressor. A magnetic field in the region of the extraction aperture magnetises the electrons which might otherwise have been extracted. These electrons move along the field lines and are collected on a biasable electrode threaded by the magnetic field.

Figure 4 shows the permanent magnet suppressor used previously[3,4]. It consists of two 37 x 4 x 4 mm^3 samarium cobalt magnets arranged in a quadrupole configuration. The integral field in the source is ~ 100 Gauss cms. A portion of the extraction electrode forms the electron collector.

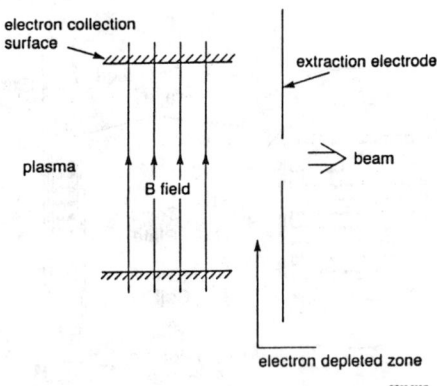

Figure 3   A schematic electron suppressor.

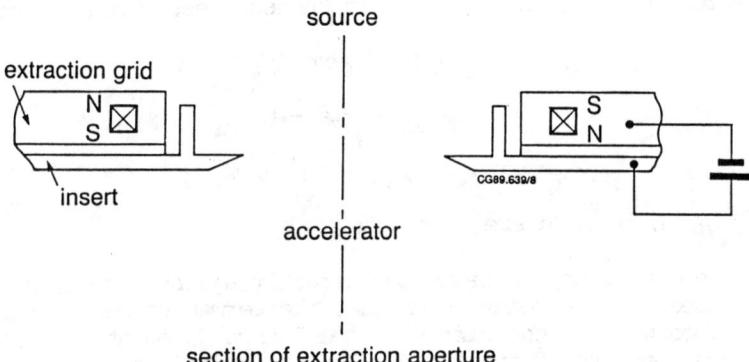

Figure 4   The permanent magnet suppressor.

Figure 5 shows the variable field suppressor. The field is produced by a flat solenoidal coil made of wires space 5 mm apart. The field profile has a FWHM of ~ 5 mm and an peak intensity of ~ 1 Gauss/Ampere. Thermal considerations limit the coil current to 150 A. At each end of the array of wires is the collection electrode.

THEORETICAL MODEL

Green[2] has used a diffusion model to describe the electrons transport across the magnetic field. The electrons diffuse by electron-molecule collisions and are then extracted. Some move along the field lines and are collected by the biasable plate. This model

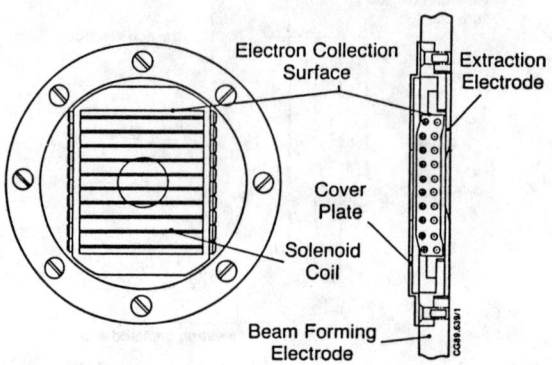

Figure 5 The variable field suppressor.

gives the following relations for the extracted electron current, I_e, the current deposited on the collection electrode, I_{ins}, the magnetic field strength B and the pressure in the suppressor region, p:

$$I_e = (K/B) \exp(-\gamma B) \tag{1}$$

$$\ln(I_e/I_{ins}) = a - bI_{ins} \tag{2}$$

and

$$\ln(I_e/I_{ins}) = c - d/p^{1/2} \tag{3}$$

where K, γ, a, b, c, d are constants.

These equations can be tested directly against the data. The work of Green[2] has already shown that the permanent magnet suppressor is described well by the diffusion model through equations (2) and (3) however because of the fixed value of magnetic field equation (1) could not be tested for this type of suppressor.

Just as important as the effect on the electrons is the effect of the suppressor on the extracted ion current. The diffusion model does not address the possible repercussions for the ion current.

EXPERIMENTAL RESULTS

a) Comparison between the suppressors

Figure 6 shows the electron suppression action of the two devices as a function of the collector bias, V_{ins}.

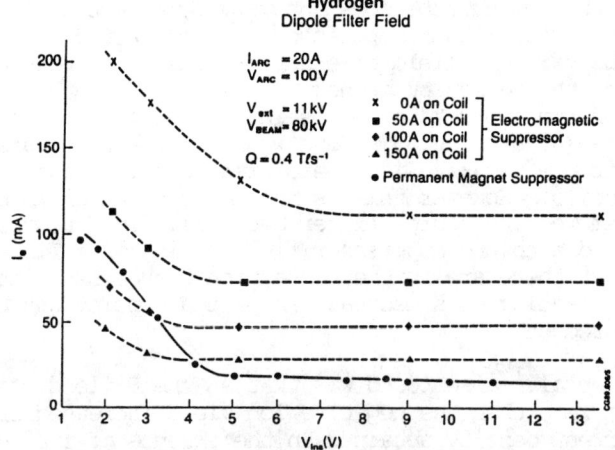

Figure 6 The dependence of extracted electron current on collector voltage for both suppressors.

Both devices do indeed suppress the extracted electron current. The degree of suppression obtained with the variable field device does indeed depend on the strength of the magnetic field. Even at the relatively low arc current used of 20 A, coil currents of > 150 A are needed to reduce the extracted electron current as low as that from permanent field device.

Of course just as important is that the control of the extracted electron flux does not have a drastic effect on the extracted negative ion current. In Figure 7 we show the corresponding data for the ion current, $I_{B.T.}$, as measured at 0.82 m downstream by the d.c. beam transformer.

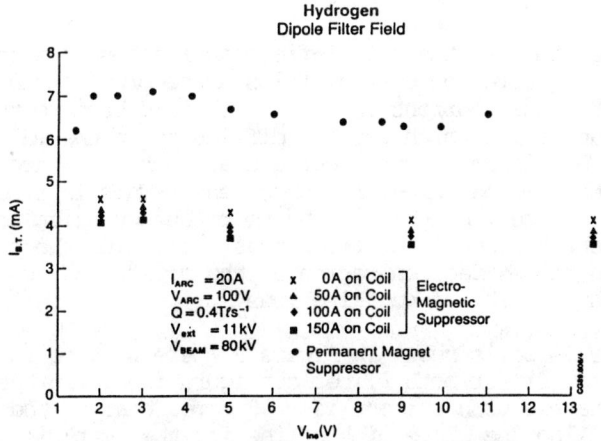

Figure 7 The dependence of extracted negative ion current on collector voltage for both suppressors.

Although the ion current does not vary by more than 10% with the collector voltage for either suppressor, the extracted ion current in the case of the variable field or electro-magnetic suppressor is ~ 60% of that for the permanent magnet suppressor. Now the transparency of the two sets of wires in the variable field suppressor is ~ $(0.75)^2 = 0.55$. This would be the transparency one would expect if the negative ions were created before the suppressor field in the coils. However this is not the case as shown by Lea et al.[1]. They removed the source filter but still found negative ions being produced due to the suppressor field acting as a filter field. It could be that there are further insertion losses associated with the suppressor leading to a decrease in plasma density and hence negative ion production.

What Lea et al.[1] also found was that as the B-field was increased from zero then the negative ion yield increased from 40% to 80% of the current density obtained in the absence of the suppressor and then decreased slowly. This has also be observed by Bacal[5]. The data in Figure 7 does not show this behaviour. The negative ion current decreases with increasing B-field. This experiment was carried out with the source filter field in the dipole configuration. The integrated field from the peak to the extraction aperture due to be dipole field ~ 200 Gauss cms and the field strength at the extraction aperture is ~ 30 Gauss although 1-2 Gauss of this comes from the accelerator fields. Thus even with zero coil current the field is never zero. Thus the results of Lea et al. could not be observed in this filter configuration.

In order to test this hypothesis we reconfigured the source filter field into the tent configuration as shown in Figure 1(b). The field at the extraction aperture then falls to ~ 5 Gauss. In Figure 8 will show a comparison of the dependence of the ion negative current on the coil current (or B-field) for the two filter configurations.

It is seen that as the coil B-field approaches to zero in the tent mode the negative ion current falls hence proving the hypothesis. The coil current is not set to zero because the electron current approaches 1 A which poses a difficulty in thermal management. This large electron current at such a low arc current is associated with the low value of ion current. This is almost certainly due to the fact that in the tent configuration the filaments penetrate the filter field thus increasing the plasma temperature in the extraction region of the source and so producing a decrease in the density of negative ions.

That this is so is shown in Figure 9 where we plot $\ln(I_e)$ versus I_{coil} (or B) for both filter configurations. In the tent filter case the electron current is ~ 10 times that in the dipole filter case. Also the slope of the line for the tent filter data is less than that for the dipole filter. The efficiency of the suppression will decrease with increasing temperature thus we see

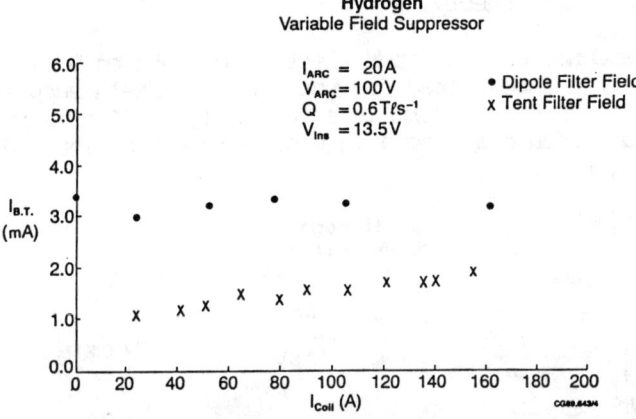

Figure 8 The dependence of extracted negative ion current on magnetic field for both source filter configurations.

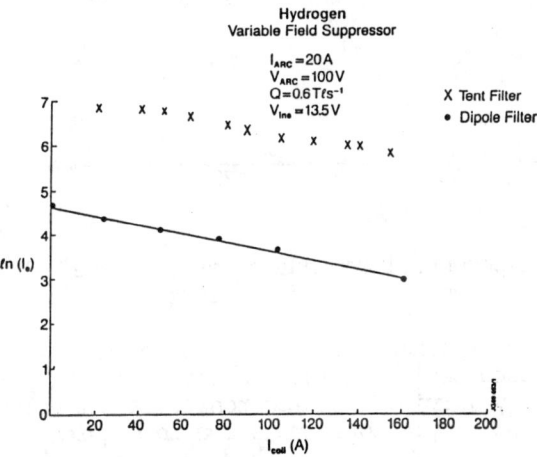

Figure 9 The dependence of extracted electron current on magnetic field for both source filter field configurations.

that the electrons are hotter in the tent filter mode. These hotter electrons will lead to a reduction in negative ion yield through destruction by electron impact.

We are presently redesigning the filaments in the source to avoid the creation of a hot plasma beyond the nominal filter field and so increase the yield of negative ions beyond what we have at present in the tent configuration.

b) Comparison with theory

The dependence of extracted electron current on B-field as given on equation (1) can be tested for the variable field suppressor. Instead of the functional form in equation (1) we find a purely exponential dependence as shown in Figure 10 which plots $\ln(I_e)$ versus I_{coil} (or B).

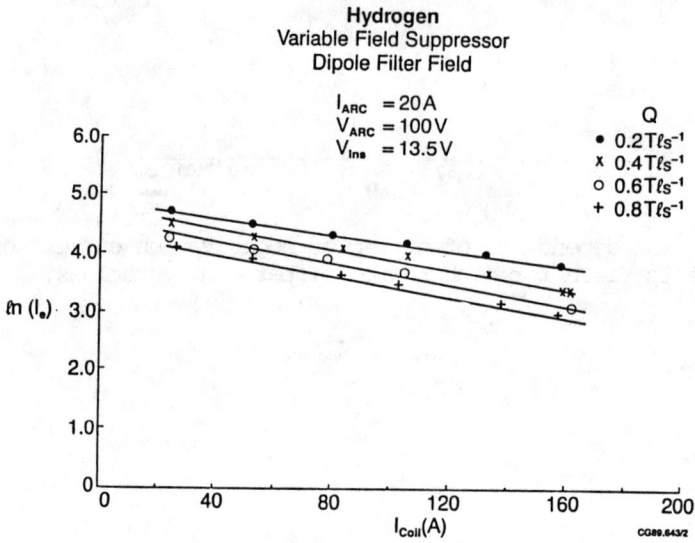

Figure 10 The exponential dependence of extracted electron current on magnetic field strength.

The remaining experimental tests given by the diffusion model can be used for both devices. The dependence of $\ln(I_e/I_{ins})$ versus I_{ins} is obtained by varying the electron collector voltage. The results for the two suppressors are shown in Figures 11 and 12.

For the permanent magnet suppressor the linearity of this curve (and hence agreement with the diffusion model) is much more pronounced than for the variable field suppressor. The deviation from linearity at the lower values of I_{ins} is not entirely due to a failure of the model. At these low values of collection voltage, positive ions are also collected because the plasma potential has not been exceeded thus leading to an apparent decrease in the collected positive current.

The diffusion model is based on electron-molecule scattering in the suppressor region leading to transport across the magnetic field. Thus the pressure dependence of the suppressor action ought to be a very good test of its validity. This dependence is expressed through equation (3). Instead of using pressure we use the gas flow, Q, the two being directly proportional.

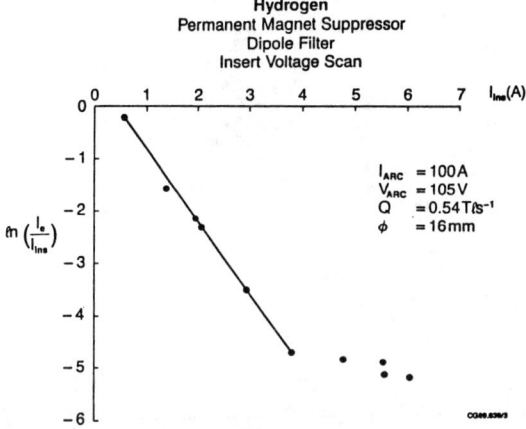

Figure 11 The dependence of $\ln(I_e/I_{ins})$ on I_{ins} for the permanent magnet suppressor.

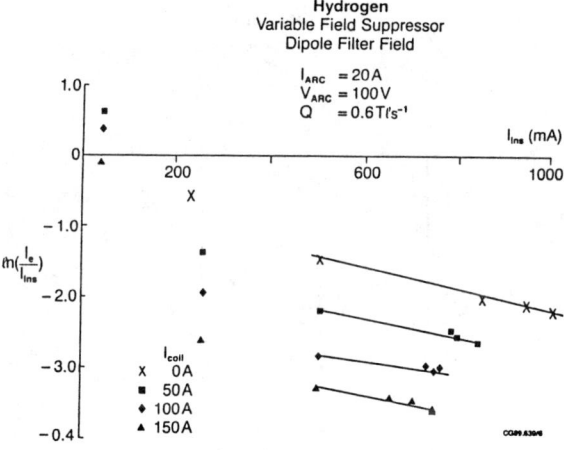

Figure 12 The dependence of $\ln(I_e/I_{ins})$ on I_{ins} for the variable field suppressor.

Figures 13 and 14 show data for both suppressors plotted in the form $\ln(I_e/I_{ins})$ or $\ln(I_e)$ versus $Q^{-½}$ in accordance with equation (3). For the permanent magnet suppressor $\ln(I_{ins})$ only changed by 5% hence the use of $\ln(I_e)$.

It is apparent that the diffusion model is in good agreement with the data from the permanent magnet suppressor but not at all in agreement for the variable field suppressor where it is almost independent of pressure. The deviation from linearity at high pressures (or Q) in the case of the permanent magnet device has been shown previously by McAdams et al[4] to be due to the collection, in

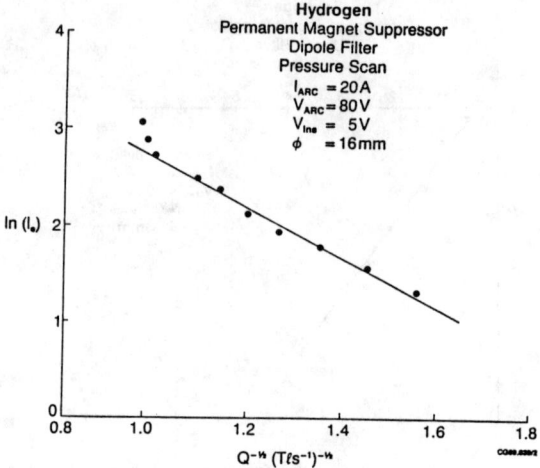

Figure 13 The dependence of $\ln(I_e)$ versus $Q^{-½}$ for the permanent magnet suppressor.

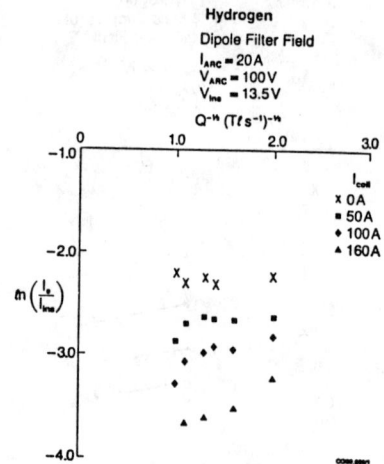

Figure 14 The dependence of $\ln(I_e/I_{ins})$ versus $Q^{-½}$ for the variable field suppressor.

the accelerator, of electron produced by stripping of the H⁻ in collisions with the gas flowing from the source.

CONCLUSIONS

We have described a new electron suppressor with variable electric and magnetic fields based on an extension of previous ideas and guided by previous data[1]. This device worked well with small extraction apertures i.e. with a diameter smaller than the wire spacing but its performance for large extraction aperture diameters is very poor compared to the permanent magnet device. This loss of performance appears to be associated mainly with the transparency of the device. However what we have shown with this variable field device, and Bacal[5], magnetic field close to the extraction aperture leads to an enhancement of the negative ion yield confirming the results of Lea et al.[1].

In our experiments on this enhancement the performance was limited by the filaments penetrating the filter field leading to a hotter plasma and hence lower densities of negative ions. Thus in our continuing work we will optimise the filament/tent filter design in order to see if the enhancement persists of higher negative ion current densities. Until this is done the full potential of the enhancement effect will not be realised.

The characteristics of the suppressor action are well described by a diffusion model in the case of a permanent magnet device. This is not the case for the variable field device. The reasons for this could be in a number of areas. Notably the scale lengths of fields for each suppressor are different as is the detailed magnetic geometry. This problem is being addressed by Holmes and Haas[6] using a hydrodynamic model.

ACKNOWLEDGEMENTS

The authors would like to thank Dr A. Holmes and Dr T. Green for many discussions on electron suppressor action. This work was supported by USAFOSR under contract number F49620-86-C-0064.

REFERENCES

1. L.M. Lea, A.J.T. Holmes, M.F. Thornton and G.O.R. Naylor, Proceedings Int. Conf. Ion Sources (Berkeley 1989), Rev. Sci. Instr. to be published.
2. T.S. Green, SPIE Proceedings Vol. 1061, Microwave and Particle Beam Sources and Directed Energy Concepts, P. 628 (1989).
3. R. McAdams, A.F. Newman, R.F. King and A.J.T. Holmes, Proceedings Int. Conf. Ion sources (Berkeley 1989), Rev. Sci. Instr. to be published.
4. R. McAdams, A.J.T. Holmes, A.F. Newman and R.F. King, Production and Application of Light Negative Ions. 3rd European Workshop Amersfoort p. 15 (1988).
5. M. Bacal unpublished data.
6. A.J.T. Holmes, private communication.

ENHANCEMENT OF NEGATIVE ION EXTRACTION AND ELECTRON SUPPRESSION BY A MAGNETIC FIELD

J. Bruneteau, R. Leroy, M. Bacal
Laboratoire de Physique des Milieux Ionisés
Laboratoire du C.N.R.S.
Ecole Polytechnique, 91128 Palaiseau Cedex, France

J.H. Whealton
Oak Ridge National Laboratory
Oak Ridge, USA

ABSTRACT

We report experiments of extracting negative and positive ions and electrons from a multicusp volume ion source. The effect of a transverse magnetic field in front of the plasma electrode is reported for the three extracted species when the bias of the plasma electrode is changed. Also the effect of the discharge current and of the pressure are reported. The experiments indicate that the effect of the magnetic field is to enhance the negative ion current and to suppress a great part of electrons from the extracted current. An explanation based on electron behaviour in the magnetic field is proposed.

INTRODUCTION

Various experiments have shown the possibility of extracting large negative ion currents from a multipolar hydrogen plasma created by energetic primary electrons [1-3]. The favorable action of a transverse magnetic field near the extractor has been noticed in diminishing the extracted electron current [4-6] and in increasing the negative ion current [4,5,7]. The fact that negative ion currents comparable to positive ion currents are extracted under the same plasma conditions is interesting and deserves a particular attention. Although the density of the negative ions in the center of the plasma is ten times lower than the density of positive ions, the density of the negative ions can be enhanced in the magnetized region.[4]

This work presents experimental results of the extracted currents, (positive and negative ions, and electrons), when parameters such as the bias of the plasma electrode, the magnitude of the magnetic field, the discharge current or the pressure are changed. We do not discuss here the extraction itself, as in reference 8, but the transport of ions and electrons towards the extraction aperture.

Experimental results are discussed and qualitative explanations, invoking potential barriers which can stop a part of the positive or negative ions, are proposed. The spatial potential distribution in front of the extractor is related to the electron flow in front of the extraction aperture.

SOURCE AND EXTRACTION SYSTEM

We performed our study in the hybrid multicusp ion source, extensively described in reference 4. The steady state hydrogen plasma is created by a primary 50eV electron beam and diagnostic facilities for negative ion density and temperature have been used.[9,10]

Ions are extracted through an aperture in one of the end plates of the cylindrical experimental vessel. We used an extraction system consisting essentially of three electrodes, as shown in Fig.1. The plasma electrode (PE) in contact with the plasma has a circular extraction aperture 0.8 cm in diameter. The second electrode called the "separator" is located 0.62 cm from the PE and has also an opening 0.8 cm in diameter. A pair of Sm-Co magnets are located in the separator just behind the opening and create a transverse magnetic field (>300 gauss), strong enough to deflect the accelerated electrons onto the separator. The last electrode collects the ions which are barely affected by the magnetic field. Two small soft iron rods are placed on the PE concentrating the stray magnetic lines to produce a magnetic fied parallel to the PE plane, which penetrates a few centimeters in the plasma. This magnetic field can be varied from 20 gauss to 60 gauss (Fig 2), with nearly the same geometrical distribution. The extraction potential can be reversed in order to collect positive ions. In this experiment the extraction potential is V_{ex} = 2 kV.

INFLUENCE OF THE PE BIAS.

Figure 3 presents the variation of the three extracted charged particle currents when the PE bias, V_b, is changed from -5 V to +5 V. The extracted electron current I_e is reduced and becomes lower than the negative ion current I^- at values of V_b slightly higher than the one corresponding to the optimum V_b for negative ion extraction. In some cases I_e passes through a minimum value.

The negative ion current, after a plateau for negative values of V_b, increases to a maximum, when V_b is nearly equal to the positive plasma potential in that region, and then falls off for more positive V_b.

With the inverted extraction voltage V_{ex}, the positive ion current has a constant value when V_b is negative, decreases monotonously when V_b becomes positive, and finally remains constant at a low value (Fig. 3).

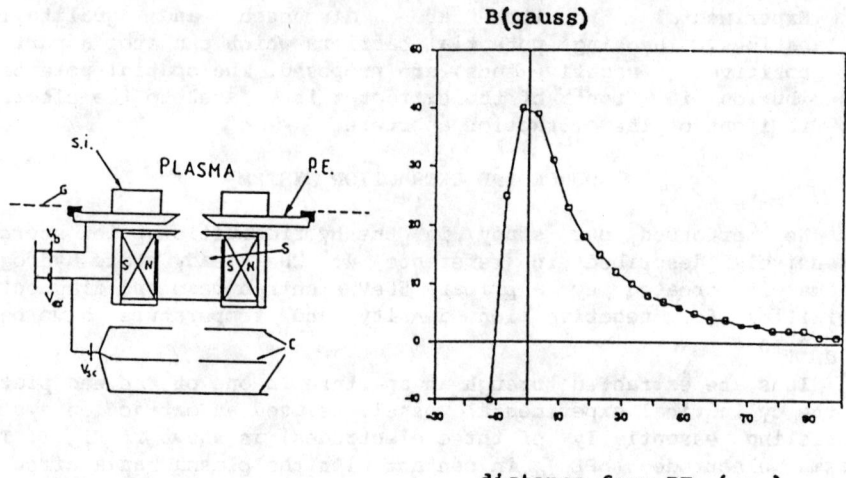

Fig.1. Extraction set up.

Fig.2. Magnetic field distribution near the PE. Case of a 40 gauss magnetic field.

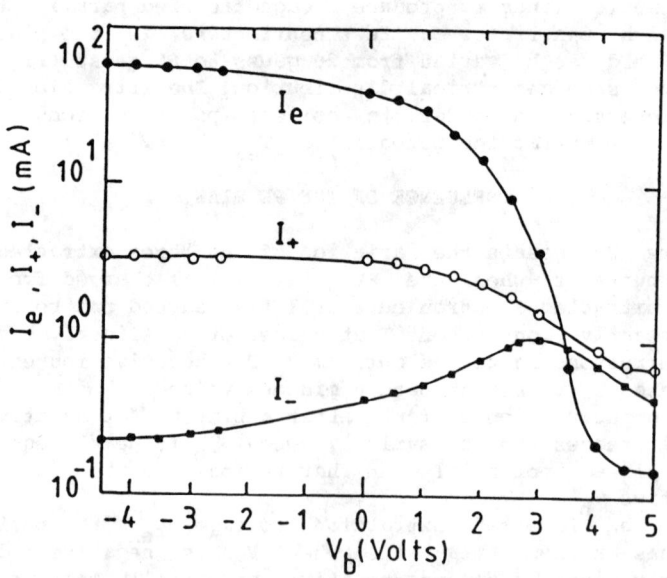

Fig.3. Typical dependence of the extracted currents on V_b. 50 V, 30 A, 3 mTorr discharge. B = 60 gauss.

Fig. 4 shows how potential barriers ΔV_{H+} and ΔV_{H-} can explain the measured positive and negative current variation with V_b, assuming that

$$\Delta V_{H+} = kT_+ \ln(I^+/I^+_{max}) \qquad (1)$$

and

$$\Delta V_{H-} = kT_- \ln(I^-/I^-_{max}) \qquad (2)$$

INFLUENCE OF THE STRENGTH OF THE MAGNETIC FIELD

The variation with the magnetic field of the extracted electron current strongly depends on the PE bias, as illustrated on Fig. 5. With positive PE bias ($V_b = +5$ V, Fig. 5c, or optimum V_b, Fig. 5b), I_e goes down with increasing V_b as B^{-n}, where $n \geq 2$. With negative V_b the variation of I_e with B is much slower, $n \approx 0.7$ (See Fig. 5a).

It will be shown below that according to the theory for electron transverse diffusion I_e scales as B^{-2}. Thus the observed variation of I_e with B at positive V_b can be explained by electron transverse diffusion. As a matter of fact, at positive V_b transverse diffusion dominates over transport along magnetic field lines since electrons can be collected on all the PE surface.

With negative V_b, electrons cannot be collected by the PE. When the positive extraction potential is applied it penetrates into the plasma and allows the escape of a fraction of these electrons. In this case the electron current is largely due to transport along the magnetic field lines.

Figure 6 presents the variation of the extracted currents with V_b for two values of the magnetic field in front of the PE. The increase of B does not largely enhance the ratio of the maximum negative ion to the maximum positive ion currents; this is because with moderate magnetic field the electron density is lower than the ionic densities so the potential distribution does not change appreciably with B.

Figure 7 shows the variation with magnetic field of the extracted negative ion current, with optimum V_b, for two gas pressures, and several discharge currents. After a significant increase at 3 mTorr in the range 0-40 gauss, the negative ion current seems to saturate. However, at 8 mTorr the negative ion current continues to increase with B. Obviously the region above 60 gauss should be further explored at higher pressure.

INFLUENCE OF THE INTENSITY OF PRIMARY ELECTRON CURRENT

Experiments were run varying the intensity (I_d) of the primary electron current between 5 A and 50 A, with a constant discharge voltage of 50 V. Figure 8 presents the dependence of I^- and I_e on the discharge current when $V_b = +5$ V and B = 60 gauss. While I^- goes up linearly with I_d, I_e saturates. Thus the ratio I^-/I_e becomes greater than five at 50 A, while it is only about two at 10 A.

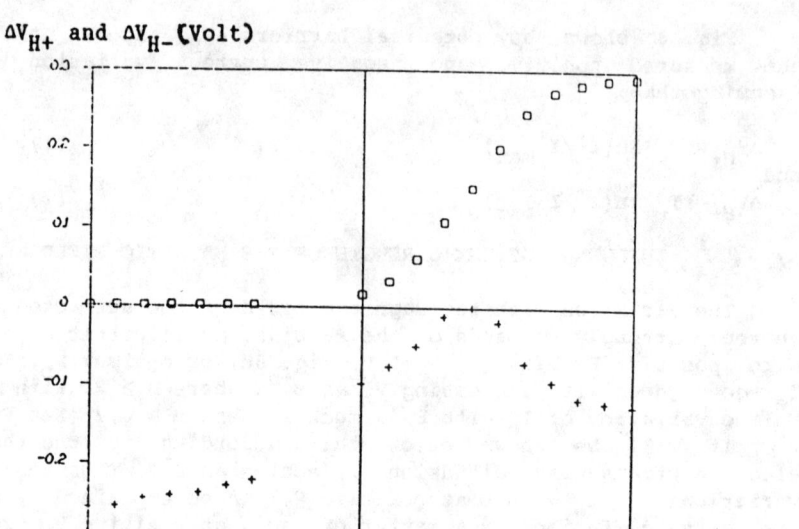

Fig.4. Potential barriers ΔV_{H+} (□) and ΔV_{H-} (+) deduced from positive and negative ion currents. The positive and negative ion temperatures are assumed to be .4eV. Conditions: 50V, 20A, 3mTorr, discharge and B=40gauss.

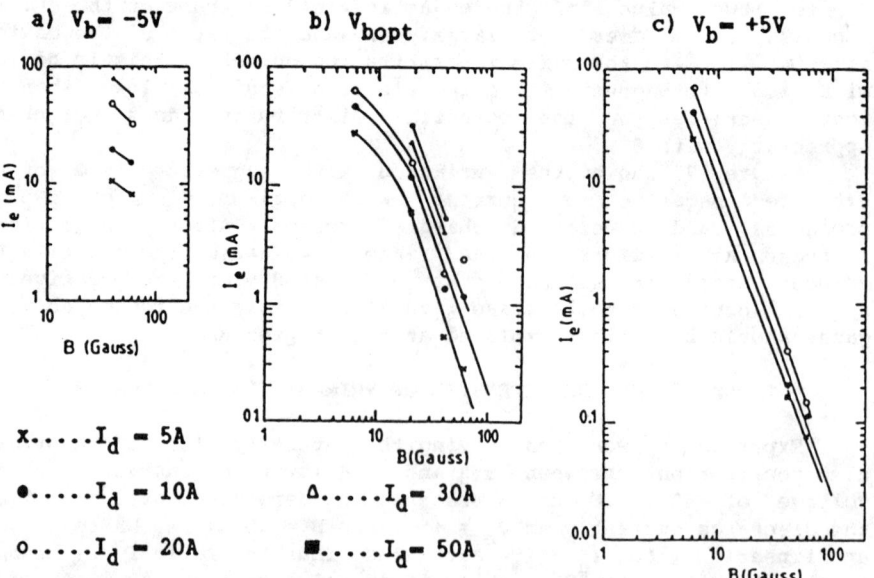

Fig.5. Variations of the extracted electron current for $V_b = -5V$, V_{bopt} and $V_b = +5V$ as a fonction of the transverse magnetic field near the PE. 50V, 3mTorr, discharge.

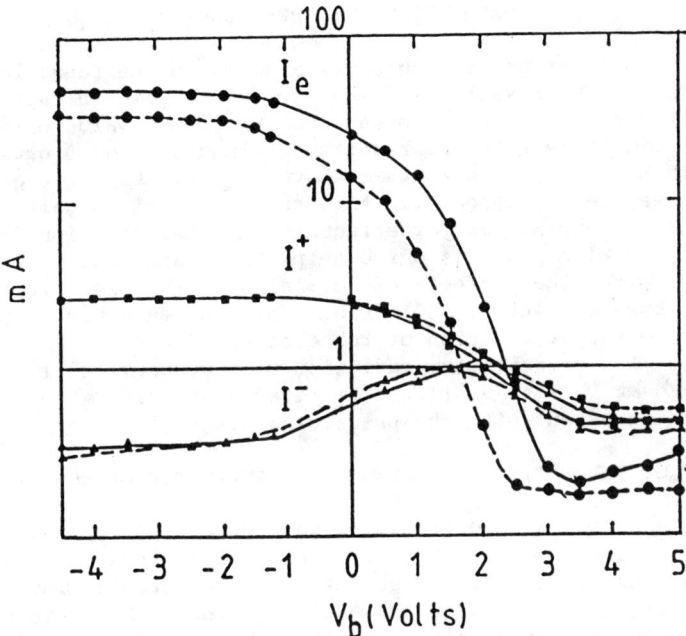

Fig.6 Extracted currents for magnetic field of 40gauss (------) and 60 gauss (———). (discharge 50V 25A 5mTorr)

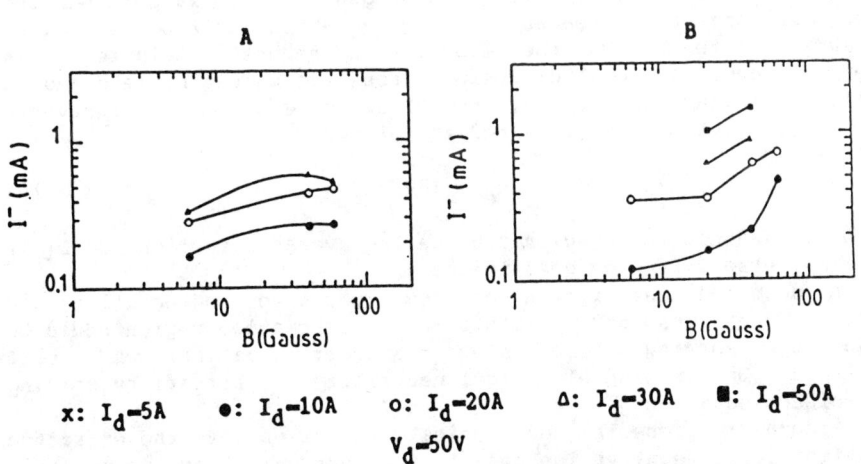

x: I_d=5A   •: I_d=10A   o: I_d=20A   Δ: I_d=30A   ■: I_d=50A
V_d=50V

Fig.7 Extracted negative ion currents for some arc currents as a function of the magnetic field for 2 pressures : A....3mTorr and B....8 mTorr.

INFLUENCE OF THE PRESSURE

The effect of gas pressure was studied in the range between 2 and 8 mTorr. The results are reported as the gain in negative ion current at optimum V_b, with respect to the lowest value observed at $V_b = -5$ V. Fig. 9 reports the results obtained at B = 40 gauss, with a 50 V, 10 A discharge. Note that above 3 mTorr this gain goes down when the pressure increases. (In this case the maximum gain of I^- is 4.) This can be explained by the increase of the collision frequency of electrons with neutrals which helps the transverse diffusion of electrons, with the effect of diminishing the electric field in front of the extractor. Collisions of ions on neutrals can also contribute to the attenuation of the electric field.

Figure 10 presents the variation with pressure of I^-/I_e at V_b optimum and at $V_b = +5$ V. It can be noted that this ratio increases with V_b, in agreement with the preceeding discussion.

ANALYSIS OF THE EXTRACTED CURRENTS AND INFLUENCE OF THE PE BIAS

The negative and positive ion mean free path is longer (about 5 cm for a pressure of 3 mTorr) than the transverse size of the magnetized region in front of the plasma electrode. Their Larmor radius is also very large. In these circumstances the ionic currents depend only on the potential barrier which can reflect them, and the ion velocity distribution function can be described by the collisionless Boltzmann equations, permitting calculation of the ion density for a given potential profile, assuming that the distribution function is known in the center of the plasma.

The mean free path of the electrons, determined mainly by elastic collision with molecular hydrogen, is also larger than the transverse size of the magnetized region, while their Larmor radius is much shorter. Thus the electron transport in this region is dependent on collisional diffusion across the magnetic field and on mobility. Taking in account the Einstein relation the transverse electron current can be described as follows:

$$j_{e\perp} / D_\perp = - dn_e/dz + (n_e/T_e)dV/dz, \qquad (3)$$

where z denotes the abscissa in the transverse direction and $D_\perp$ is the transverse diffusion coefficient.

A detailed description of ion (positive and negative) and electron densities and currents in the extraction region could be given by solving these three transport equations made self consistent by imposing electrical neutrality or, better, by solving the Poisson equation.

Figure 3 presents the typical variation of the extracted positive ion, negative ion and electron currents. For the analysis of these results, the case of a PE without any extraction aperture

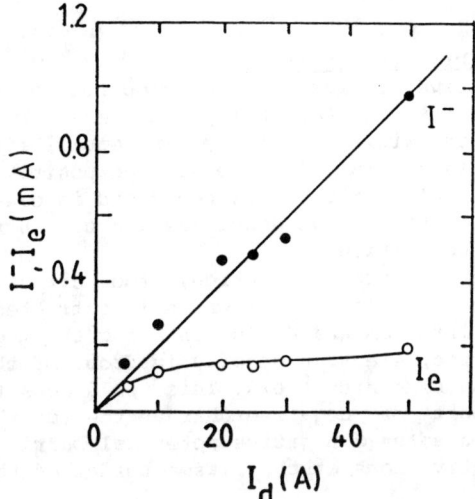

Fig.8. Dependence on the discharge current of the negative ion and electron currents. The magnetic field is 60 gauss and V_b= +5V. 50V, 3m Torr discharge.

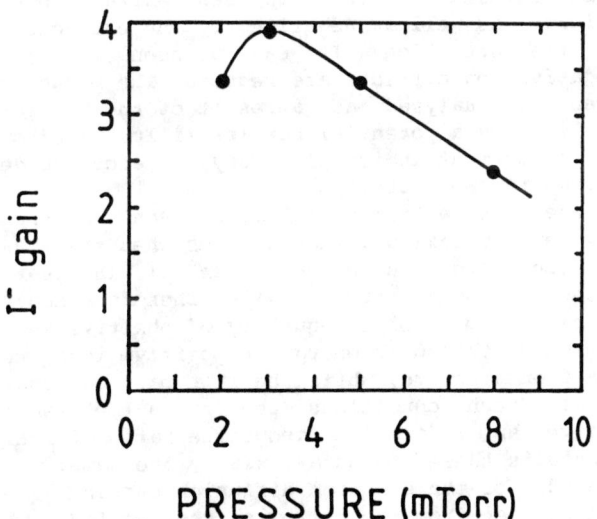

Fig.9. Gain in extracted negative current, $I^-(V_{bopt})/I^-(V_b= -5V)$ for a 40 gauss magnetic field as a function of pressure. 50V, 10A, discharge.

is discussed first.

Case of negative PE bias ($V_b = -5$ V)

For a very negative PE bias, V_b, the electron current collected by the PE acting like a Langmuir plane probe, is very low. The electron distribution will be nearly in equilibrium with the electrons in the plasma. In this case all the positive ions can be collected by the PE. This situation is conserved in case of positive ion extraction: all the charge densities and currents will be the same as in the reference state.

With negative charge extraction, the situation is very different. The large electron current which is extracted is provided by both transverse diffusion and diffusion along the magnetic lines. The depletion of the electron density in front of the extraction aperture causes an electric field. This field acts to facilitate locally the transport of negative charges towards the extraction aperture, but also creates a negative potential barrier which stops a part of the negative ions at the plasma border of the magnetized plasma.

Case of positive PE bias ($V_b = +5V$)

In the opposite case, with very positive bias and no extraction aperture, the electrons can flow onto the PE : an electronic density gradient is formed in front of the PE and also a field to help electron current continuity. In this case extraction of charged particles, positive as well as negative, by the extractor will not notably change the situation. It can be seen (Fig.3) that both positive and negative ion currents are reduced. The reduction of the ion current can be analysed as a result of the reflection of a fraction of the ions by a potential barrier (illustrated on Fig.4). The weak electron current indicates a very low electron density in front of the extraction aperture.

Between these two extremes, it can be seen that the negative ion current presents a maximum. This happens when the bias applied to the PE is close to the plasma potential. The weak electron current indicates a low electron density. Therefore neutrality in front of the extractor involves equality of positive and negative ion densities. Only the more energetic positive ions are able to reach the extraction aperture, while the main part of negative ions are extracted. In such conditions the ratio between the ionic currents can give an indication about the ratio of ionic escape speeds; in our results these speeds are nearly the same.

The dependence of the electron diffusion current j_e upon n_{eo}, T_e, P, and B can be seen from the following scaling laws, where L is the scale length for transverse diffusion, and n_{eo} and n_{+o} the charge densities in the non-magnetized plasma. In the case $V_b = +5$ V, the transverse electron diffusion current is

$$j_e(+5) \propto D_\perp n_{eo}/L, \qquad (4)$$

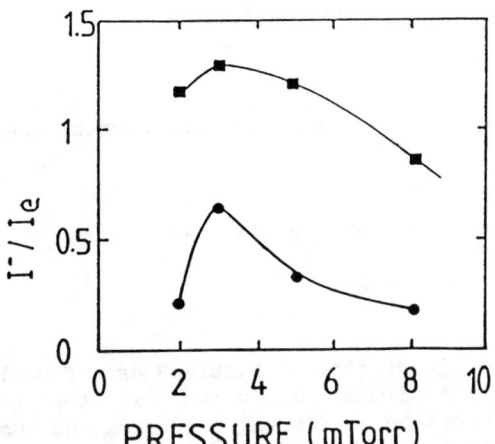

Fig.10. Negative ion to electron extracted currents ratio as a function of the pressure in the case of V_{bopt} (●) and $V_b = +5V$ (■). 50V, 10A discharge.

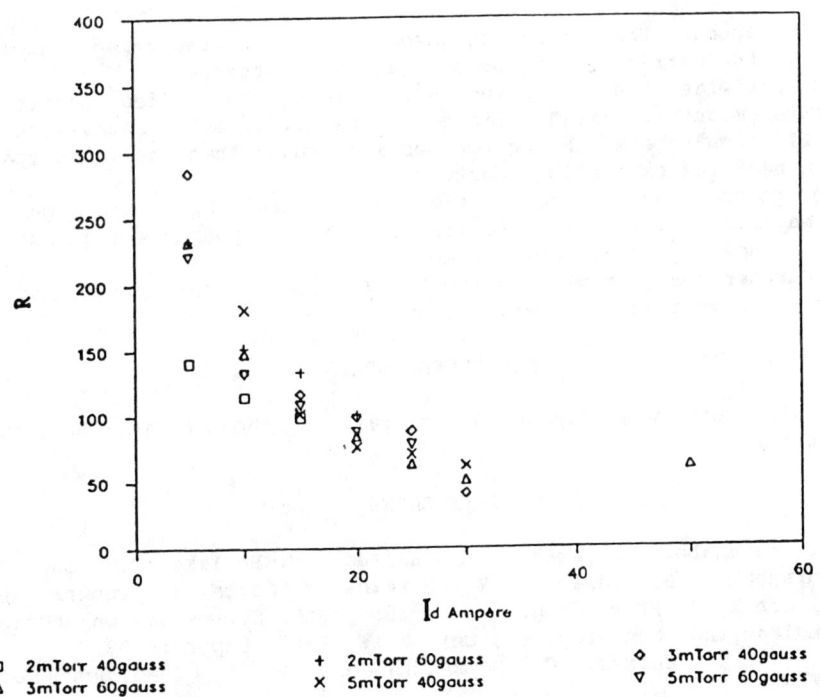

Fig. 11. Ratio R versus the discharge current.

where the transverse diffusion coefficient is

$$D_\perp \propto T_e^{3/2} P/B^2. \tag{5}$$

When $V_b = -5$ V, the extracted positive ion current density is

$$j_+(-5) \propto n_{+o} \approx n_{eo}. \tag{6}$$

From this we find the following scaling law:

$$R = \frac{j_e(+5) B^2}{j_+(-5) P T_e^{3/2}} \approx \text{constant} \tag{7}$$

This ratio R is plotted in Figure 9 as a function of I_d, for the experimental situations, in which all the parameters were available. These include different pressures and magnetic fielde. This ratio is constant within a factor of 2. However it can be noted that the points are distributed around the curve $I_d^{-1/2}$. This disagreemnent between our analysis and experiment could result of our assumption of constant ion temperature.

CONCLUSION

It appears that to extract negative ions a transverse magnetic field in the extraction region has favorable effects:
-to optimize the extracted H⁻ current. This effect occurs with moderate magnetic field, since this result is obtained as soon as the electron space charge becomes much lower than the ionic space charge near the extraction aperture.
-to reduce the extracted electron current. For this purpose, the magnetic field must be as strong as possible and neutral pressure has a prejudiciable effect.

Further research will define the applicabilty of this method to the case of multiaperture extraction.

ACKNOWLEDGEMENTS

This work was supported in part by the Oak Ridge National Laboratory.

REFERENCES

1. K. Watanabe, M. Araki, M. Hanada, H. Horiike, T. Inoue, T. Kurashima, S. Matsuda, Y. Matmida, Y. Ohara, Y. Okumura, and S. Tanaka, Proceedings of the 12th Symposium on Fusion Engineering, Monterey, CA, Oct 12-16, 1987. Paper 10-08.
2. A. J. T. Holmes, G. Dammertz, and T. S. Green, Rev. Sci.

Instrum., 56, 1697 (1985).
3. R. Stevens, R. L. York, K. N. Leung, and K. W. Ehlers, Fourth Intern. Symposium on Production and Neutralization of Negative Ions and Beams, Brookhaven, NY (1986); AIP Conf. Proceedings 158, J. G. Alessi Ed., p. 271.
4. M. Bacal, J. Bruneteau, and P. Devynck, Rev. Sci. Instrum., 59, 2152 (1988).
5. T. Inoue, M. Araki, M. Hanada, T. Kurashima, S. Matsuda, Y. Matsuda, Y. Ohara, Y. Okumura, S. Tanaka, and K. Watanabe, Nuclear Instrum. Methods in Phys. Res., B37/38, 111 (1989).
6. R. McAdams, A. J. T. Holmes, A. F. Newman, and R. King, "Production and Application of Light Negative Ions", Proc. 3rd. European Workshop, Amersfoort, The Netherlands (1988), H. Hopman and W. Amersfort, Eds, p. 15.
7. J. Lea, A. J. T. Holmes, M. F. Thornton, and G. O.R. Naylor, Rev. Sci. Instrum., to be published.
8. J. H. Whealton, P. S. Meszaros, R. J. Raridon, K. E. Rothe, M. Bacal, J. Bruneteau, R. Leroy, and R. A. Stern, Rev. Sci. Instrum., 60, 2873 (1989).
9. P. Devynck, J. Auvray, M. Bacal, P. Berlemont, J. Bruneteau, R. Leroy, and R. A. Stern, Rev. Sci. Instrum. 60, 2873 (1989).
10. M. Bacal, P. Berlemont, J. Bruneteau, P. Devynck, C. Konieczny, R. Leroy, and R. A. Stern, Proceedings of this Symposium.

A LOW FREQUENCY RF DISCHARGE: A POSSIBLE SOURCE OF D^- ?

C. A. Anderson
Applied Physical Science, University of Ulster
Coleraine, BT52 1SA, Northern Ireland.

W. G. Graham
Physics Department, Queen's University,
Belfast, BT7 1NN, Northern Ireland.

ABSTRACT

By using a time-resolved Langmuir probe method, the time dependent plasma parameters of a low frequency (< 500 kHz) rf driven D_2 discharge have been measured. It is found that at certain times during the rf cycle, the electron temperatures measured are very low (0.2 - 0.8 eV). This suggests that rf driven discharges may be suitable as sources of negative ions.

INTRODUCTION

Radio frequency discharges are particularly attractive as sources of ions as they are free from filaments which have limited lifetimes. The high electron temperatures (10-20 eV [1-3]) generally associated with rf discharges would appear to make them unattractive as negative deuterium ion sources where temperatures of ~ 1 eV are considered optimal[4]. Langmuir probes should provide a reliable method for measuring electron temperatures and other plasma parameters. However, it is recognised that the analysis of probe characteristics in rf discharges is difficult due to the presence of rf fluctuations. Langmuir probe techniques employed to date use time averaged I-V characteristics assuming that the rf interference is sinusoidal in nature[1-3]. Plasma induced light emission measurements show that the electron energy distribution is time dependent[5-7], and not at all sinusoidal in nature. This suggests that I-V characteristics will also be time dependent and time average results somewhat questionable. The high electron temperatures generally associated with rf discharges, have been measured using time averaged techniques.
We have recently developed a time dependent Langmuir probe technique[8,9] and measure significantly lower electron temperatures (0.1-0.8 eV) in Ar and N_2 discharges[8-10]. Here we report measurements of the basic plasma parameters in a low frequency (100-500 kHz) rf driven discharge operating in D_2.

APPARATUS

The measurements have been made in a parallel plate type discharge. Two floating electrodes of 53mm diameter, separated by

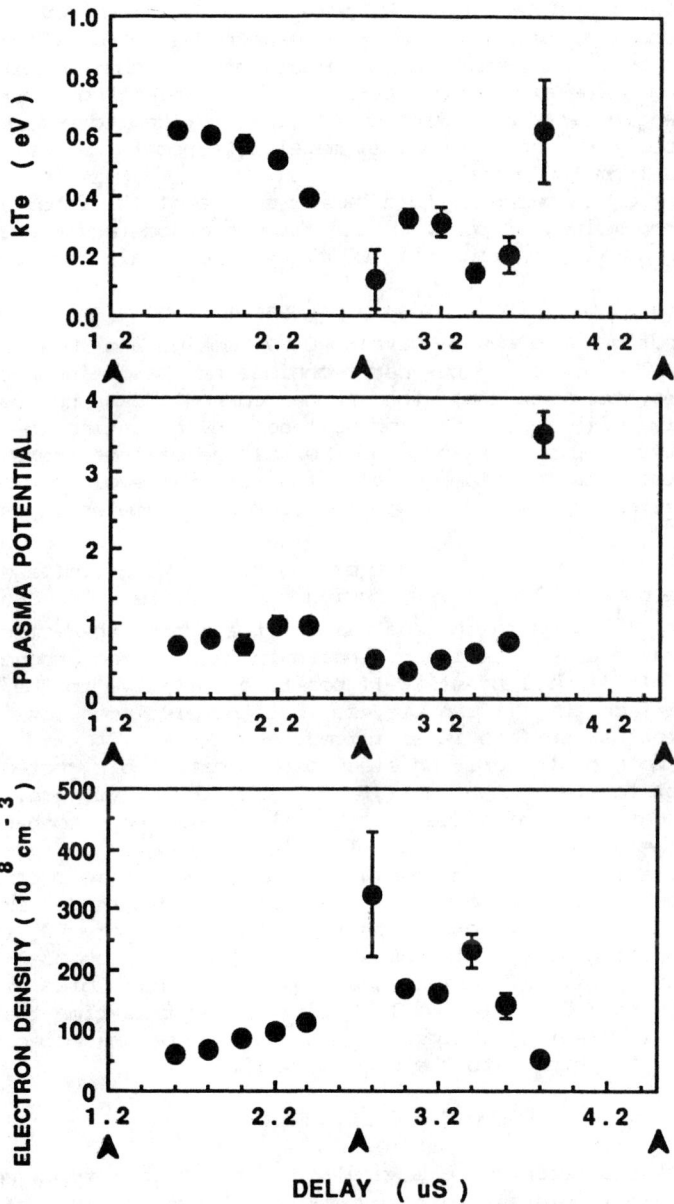

Figure 1. Plasma parameters of a D_2 300 Khz rf discharge as a function of time during one rf cycle. Gas pressure 300 mTorr Rf power 28 Watts. The arrows mark the time at which electrode one and then electrode two are most positive.

30mm, are mounted in a grounded stainless steel vacuum chamber, 106mm diameter and 200mm high. Rf power (25 W at 100, 300 or 500 kHz) is applied between the two electrodes. A cylindrical Langmuir probe constructed of 0.5mm diameter tungsten wire is mounted between the electrodes, parallel to the electrode surface. The probe length is 10 mm. Particular care is taken to ensure that the unexposed wire and any earthed metallic support for the probe are screened from the plasma.

The chamber is pumped (to a base pressure of 10^{-5} mbar) by a baffled turbo-molecular pump. A capacitance manometer is used to measure the working pressure of 300 mTorr used in all the present experiments.

The data acquisition and analysis system used is an extension of the computer based system developed by Hopkins and Graham[11,12] for use in DC plasmas. A box-car technique has been incorporated into this system so that the probe current can be sampled synchronously with the rf driving voltage. By adjusting the time at which the sample is taken, and the voltage of the probe with respect to the grounded chamber wall, the temporal evolution of the probe characteristic, and hence the plasma parameters, can be obtained[8-10].

Due to the periodic nature of the rf driving voltage the plasma-probe potential will vary during the rf cycle. The Langmuir probe theory[11,13] used in the analysis of the probe characteristic requires that a quasi-stationary sheath to have formed around the probe. The plasma will bias itself positive, relative to the most positive surface that it can see, so that the electron loss from the plasma to that surface is minimised.

Since both of the driving electrodes float, their potential, with respect to the chamber wall, along with the rf voltage, will contain a negative DC bias due to the initial electron bombardment from the plasma that is created. The electrodes only go slightly positive, relative to the chamber wall, for a short time during the driving cycle. The peak to peak voltage of both electrodes is 1250 volts and the positive excursion is of the order of 50 volts. During the rest of the cycle the plasma will see the chamber wall as the most positive surface and therefore will bias itself positive with respect to the wall. It is during this time that the probe sheath is stationary since the Langmuir probe characteristics is measured with respect to the chamber wall.

RESULTS AND DISCUSSION

The results reported here were taken in the geometric centre of the parallel plates in the glow discharge region of the discharge. Figure 1 shows the basic plasma parameters of a 300 kHz D_2 discharge as a function of time through one cycle. Note that at certain times (arrowed) no data is presented. This is the part of the cycle where the electrodes go positive, the sheath is no longer quasi-stationary and so the program can no longer analyse the I-V characteristic. Other features to note are the very low

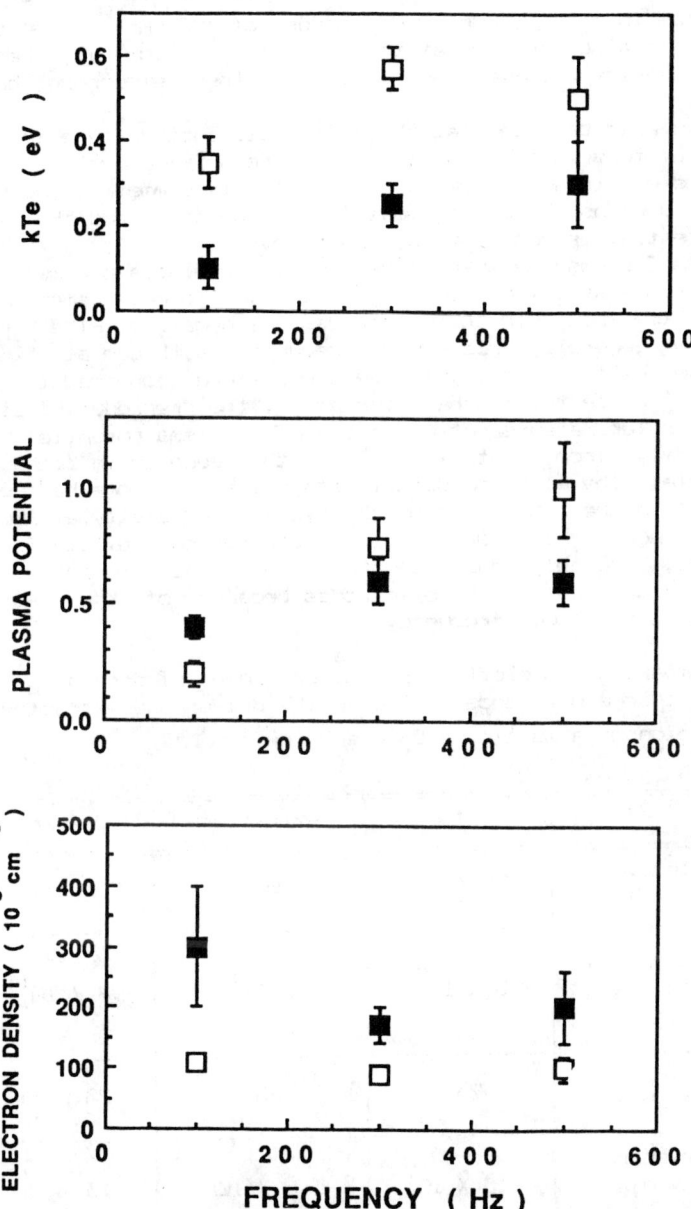

Figure 2. Plasma parameters in a D_2 rf discharge as a function of frequency. Pressure 300 mTorr, power 28 Watts. Measurements are shown for two times during one cycle. □ after electrode one was positive, ■ after electrode two was positive.

electron temperatures of 0.5 eV. We have found these low electron temperatures in both H_2 and Ar discharges at 100 kHz[10]. The plasma potential is also very low at these times, (< 0.5 V). The bulk electron density slowly increases as either electrode becomes negative.

It can also be seen that there is a distinct difference in the plasma parameters in the two periods in the driving cycle. This is because the electrode system is not completely symmetric and so one electrode does not become as positive as the other and hence the plasma potential is not disturbed as much.

Figure 2 shows the average values of the plasma parameters for the two periods as a function of frequency. The measurements were made at a time when both the electrodes are negative. The electron temperature generally rises with frequency, although at 500 kHz during one half of the cycle the temperature appears to drop. However, as reflected in the uncertainty, the reproducibility of this particular measurement was poor. The plasma potential rises quite steadily from 0.4 to 1.0 volts with frequency in both halfs of the cycle. The electron density stays steady in one half of the cycle, but in the other half of the cycle, the density at 100 kHz seems high. Again there were problems with reproducibility for this measurement, probably due to the greater differences in the positive excursions of the electrodes because of the geometric asymmetric at the lower frequency.

TABLE I Comparison of electron parameters in an rf driven D_2 discharge with those found in the driver and extractor region of a multicusp D_2 discharge[15].

	rf	extractor	driver
Bulk temp. kT_e (eV)	0.6	0.8	2.15
Bulk Electron density n_e (cm^{-3})	8.3×10^9	9.5×10^{10}	1.6×10^{11}
Fast temp. kT_{fe} (eV)	23	33	20.5
Fast electron density n_{fe} (cm^{-3})	6.5×10^7	1.8×10^8	10^{10}
n_{fe} as percentage of n_e	0.78 %	0.19 %	6.3 %

It is interesting to compare the plasma parameters measured in the rf discharge with those in a conventional multicusp negative ion source[14,15]. Eedfs taken in the extractor region of such a source can be described as a single Maxwellian and the those taken in driver region as a bi-Maxwellian. The analysis program fits the electron energy distribution function to two temperatures, a low energy bulk temperature and a fast energy temperature. While the eedf measured in rf plasmas[9-10] are neither single or bi-maxwellian, it is interesting to use this type of analysis for comparison.

Table I compares the plasma parameters measured in the rf discharge and the extractor and driver region of a multicusp ion source. The percentage of fast electrons in the rf discharge is four times higher than found in the extractor region of the multicusp source. But it is much lower than that found in the driver region. Too high a density of fast electrons increases the loss of negative ions due to collisions which strip the loosely bound electron from the negative ion.

While plasma conditions in the present rf discharge seem good for D^- production we are at present unable to determine if there will be a sufficient density of $D_2(v^*)$ generated in the discharge by the fast electrons observed in the present measurements, or produced at times in the rf cycle not accessible to the present measurement technique.

CONCLUSION

Electron temperatures, favourable for D^- production, can be produced in low frequency discharges. It is still unclear whether there will be sufficient $D_2(v^*)$ production, thought to be the precursor for D^- production. It would however appear that it is worth looking for D^- ions in low frequency deuterium rf discharges.

1. M.J. Kushner, J.Appl. Phys. 53, 2939 (1982)
2. D. Maundrill, J. Slatter, A.I.Spiers, and C.C. Welsh, J.Phys. D 20, 815 (1987)
3. T.I. Cox, V.G.I. Deshmukh, D.A.O. Hope, A.J. Hydes, N.St.J. Braithwaite, and N.M. Benjamin, J.Phys. D 20, 820 (1987)
4. M. Bacal, A.M. Bruneteau, W.G. Graham, G.W. Hamilton and N. Nachman, J. Appl. Phys 52, 1247 (1981)
5. T. Makabe, and M. Nakaya, J. Phys. D 20, 1243 (1987)
6. C.A. Anderson, W.G. Graham, and M.B. Hopkins. Proceedings of the XVIII International Conference on Phenomena in Ionized Gases, Swansea 13-17 July 1987 (Adam Hilger, Bristol, 1987) p.826
7. G. Rosny, E.R. Mosbur,Jr., J.R. Abelson, G. Devaud, and R.C. Kerns, J.Appl. Phys. 54, 2272 (1983)
8. C.A. Anderson, M.B. Hopkins and W.G. Graham, Appl. Phys. Letts. 52, 783 (1988)

9. M.B. Hopkins, C.A. Anderson and W.G. Graham, Europhys. Lett. **8** 141 (1989)
10. C.A. Anderson, M.B. Hopkins and W.G. Graham, Rev. Sci. Instrum. (to be published Jan. 1990)
11. M.B. Hopkins, and W.G. Graham, Rev. Sci. Instrum. **57**, 2210 (1986)
12. M.B. Hopkins, W.G. Graham, and T.J. Griffin, Rev. Sci. Instrum. **57**, 457 (1987)
13. J.G. Laframboise, University of Toronto, Institute for Arospace studies Report No. 100, (1966)
14. M.B. Hopkins and W.G. Graham, J. Phys. D **20**, 838 (1987)
15. A.A. Mullan and W.G. Graham, (private Communication).

NEGATIVE HYDROGEN VOLUME SOURCE WITH RF MULTIPOLE ION CONTAINMENT

G.Brautti, A.Boggia, A.Rainò and V.Stagno
Dipartimento di Fisica dell'Università e INFN di Bari (Italy)
V.Variale, V.Valentino,
Istituto Nazionale di Fisica Nucleare (INFN) Bari (Italy)

ABSTRACT

A negative hydrogen volume source, whose main feature is the plasma containment in a potential well produced by radiofrequency multipole, is under construction.

In this paper the experimental set-up and preliminary electrical measurements are presented.

INTRODUCTION

By trying to increase the ion production at reduced gas pressure, we increase the depth of the source. This increase gives the additional benefit of an easier separation of the regions of hot from cold electrons; the former electrons excite the vibrational states of H_2, the latter are captured during the molecular dissociation [1].

The increased probability of ion losses at the wall, consequence of the increased depth, is reduced by confining the ions in a RF octupole.
The benefits that we will foresee with radiofrequency confinement are:
- no losses on the walls;
- increased probability of ion production;
- elimination of sputtering, reducing poisoning of the gas and wall corrosion.

RF CONTAINMENT

When we have a particle of charge q in a radiofrequency (RF) rapidly oscillating electric field $E(r,t) = E_0(r) \cos \omega t$, and its motion is "slow", i.e.

$$v \mid grad \mid E_0(r) \mid \mid \ll \omega \mid E_0(r) \mid \qquad (1)$$

where v is the particle velocity and ω RF frequency, the effect of the field is equivalent to a time-indipendent pseudopotential [2]

$$V(r) = \frac{q E_0^2(r)}{4 M \omega^2} n^2 \left(\frac{r}{r_0}\right)^{2(n-1)} \qquad (2)$$

In eq. (1) :
   E_0: electric field due to pseudopotential V;
   M: ion mass;
   q: ion charge;
   n: dipole pair number.

A (two-dimensional) multipole field can be produced by n cylindrical electrodes alternatively connected to the poles of a RF voltage source.

In our present geometry we have built an octupole, because with the same RF voltage it produces a much deeper potential well than a quadrupole as shown in Fig. 1 where potential wells corresponding to some n values. However we foresee the possibility of replacing the electrode assembly with a different multipole.

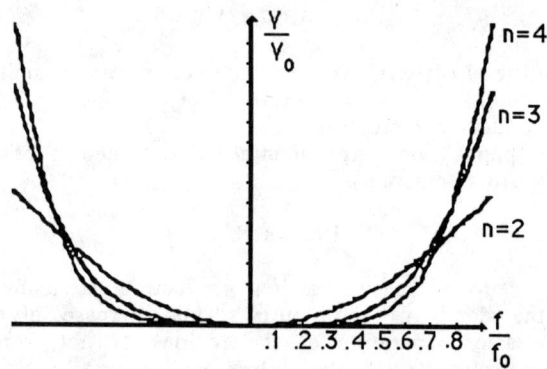

Fig.1 Potential well for different multipoles configurations.

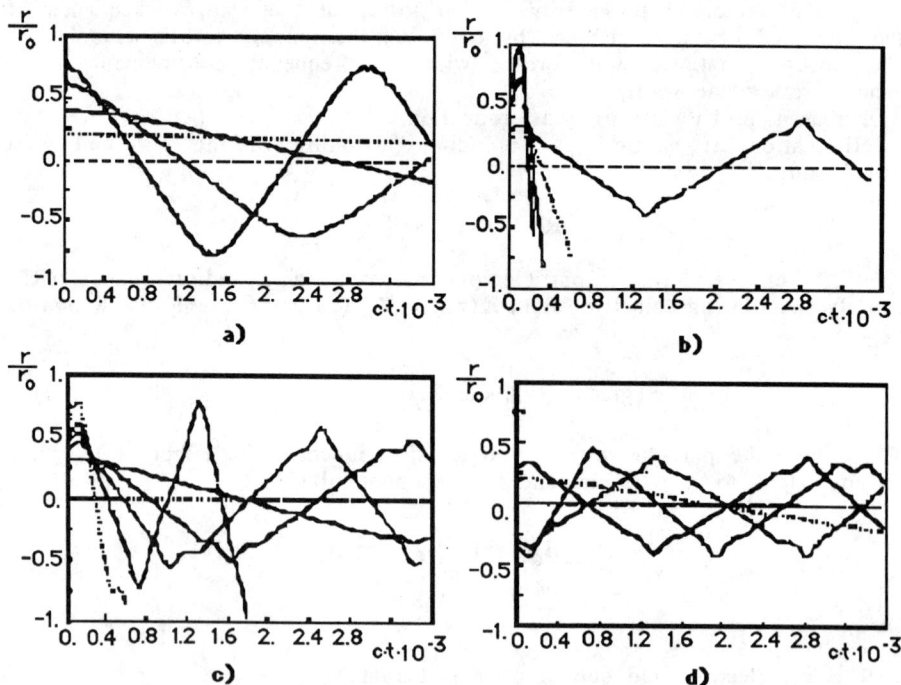

Fig. 2. Radial motion of hydrogen ions in the octupole trap, for several values of the initial radius. Horizontal scale in c·t units. Vertical scale:

ratio of the radial position to the trap dimension. a) RF 5MHz, V=200 volts. b) and d) RF 1 MHz, V=200 volts. c) RF 1MHz, V=100volts.

In some cases the quadrupole may prove particularly advantageous, because it provides an harmonic potential which allows mass-spectrometric selective containment.

Since V(r) is quadratic in the charge, it contains equally well positive and negative ions.It can be shown also that one can obtain stable ion containment at the endcaps. A special extractor must be provided to extract the ion beam.

Single particle motion and containment limits were calculated with a Runge-Kutta method for several radial starting positions (fig.2). As the figures shown, the frequency and voltage values limite the ions confinement region.

We are going to study the space charge effect on the H⁻ motion.

THE PRESENT EXPERIMENTAL SET-UP

<u>Ion Source</u>. A cut - away view of the experimental arrangement is shown in Fig.3.

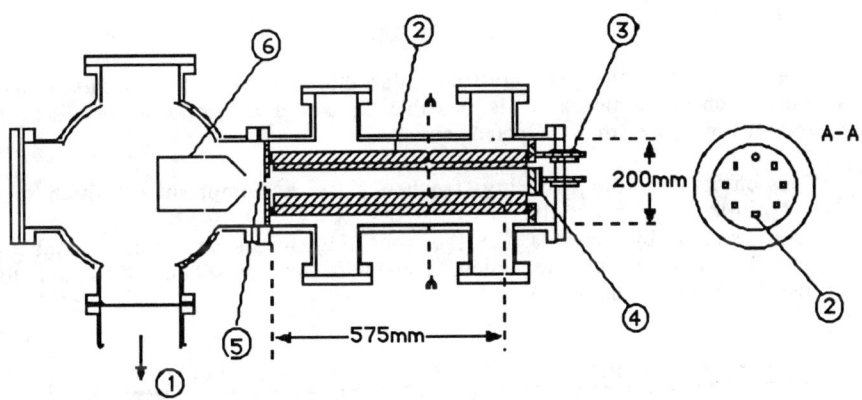

Fig 3. Overall view of the multipole reactor. A-A is a cross section.1) To pump manifold. 2) Multipole electrode. 3) RFinput. 4) Cathode. 5) Extraction slit. 6)Faraday cup.

Presently the vessel contains an RF octupole, with a plasma containment region about 80 cm. long and 11 cm in diameter.

We start with a cold (hollow) cathode, and no magnetic containment of the electrons. Later the addition of a multicusp magnetic field and warm cathodes are foreseen.

In a preliminary experiment with a 20 cm long source, with no cathode at all, the vessel was able to contain stable discharges in nitrogen and argon, while the excitation energy was supplied only by the containing RF

<u>RF circuit</u>. One polarity of the multipole assembly is connected to ground as is the vacuum vessel. The other polarity is connected by an alumina-insulated feedthrough to a tuning-matching external circuit (fig.4).

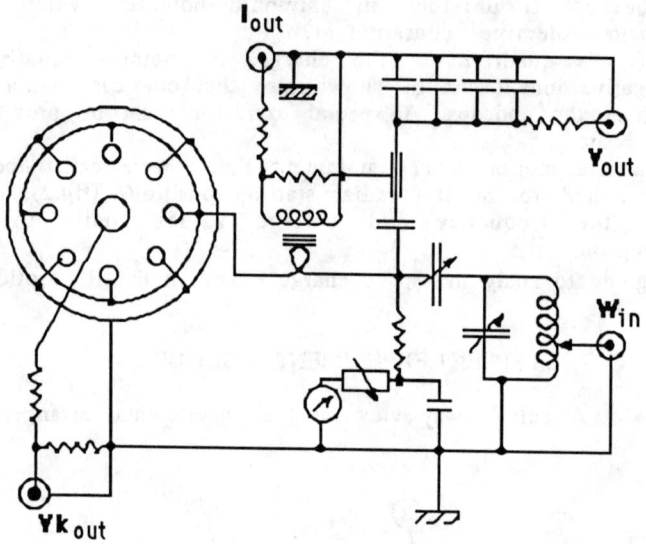

Fig 4. The circuit of the RF multipole. Variable capacitors are used for tuning and matching purposes. It is possible to add a DC bias to one set of poles, and to measure the collected current.

The choice of the working frequency is a compromise which we plan to optimize experimentally.
As shown by formula (2) the confining forces are proportional to the inverse squared of the frequency. However care must be taken not to exceed the limit of formula (1).

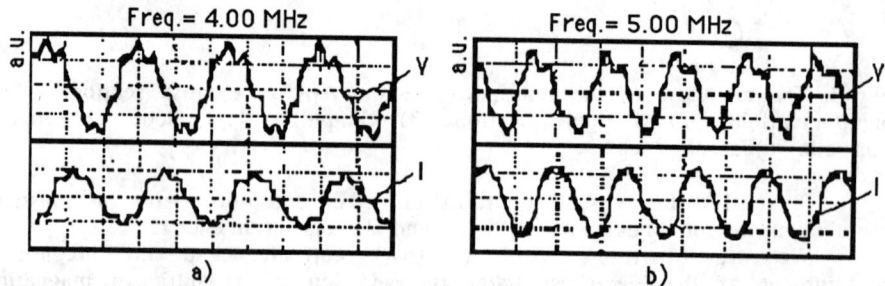

Fig.5. Voltage and Current Waveforms on the octupole electrode for f=4 MHz and f=6 MHz.

We are experienced in trapping nitrogen and argon ions at the fixed frequency of 4.5 MHz.
RF power is delivered by a synthesized RF generator followed by a 200 watts wideband (1-200 MHz) amplifier.

Electrode voltage is monitored by a capacitive divider, while the current is monitored by a self-integrating Rogowsky coil. Voltage and current characteristics are shown in fig.5, where the source was operated with argon gas. There is an effective production of harmonics, which we try to correlate with the electron motion for diagnostic purposes.

Magnetic octupoles will be added next externally to the vessel. Their poles will be azimuthally half-way between the electric ones, to prevent electrons striking the latter.

Besides electron confinement, the field will allow cyclotron acceleration of electrons in an annular region well separated from the attachment region.

We are preparing beam extraction and analysis, and further diagnostics with a Langmuir probe, while we plan to measure the H$^-$ density by laser photodetachement.

REFERENCES

1. J.T. Holmes et al., "Production and Applications of Negative Ions", 2th European Workshop - Palaseau, M. Bacal and Mouttet eds.- EP and CNRS, page 1.
2. L.Landau and E.Lifchitz, Theoretical Physics, Vol.I - Mechanics (Pergamon Press, Oxford, 1960), pag.93.

SELECTION OF CONDITIONS FOR PRODUCTION OF MAXIMUM $\bar{\text{H}}$ BEAM CURRENT DENSITY FROM MULTICUSP SOURCE

A.I.Krylov, V.V.Kuznetsov, D.V.Penkin, N.N.Semashko
I.V.Kurchatov Institute of Atomic Energy, Moscow, USSR

Ion sources with volume generation of $\bar{\text{H}}$ in pure hydrogen discharge[1,2] have been shown to produce beams with significant current densities[3,4] and total ion current[5]. This branch of ion beam investigations have been developed in a number of laboratories for nearly a decade. We joined these experiments two years ago. In our previous article[6] research of $\bar{\text{H}}$ generation in a large filtered multicusp sourse was described. This paper reports about attempts to obtain the conditions for maximum $\bar{\text{H}}$ current density in a small extracted beam.

Two ion sources were tested in our experiments. Schematic view of the them is shown in Fig.1. The first source was rectangular, 34*24 cm, (R-source), the second was cylindrical, with the diameter 20 cm (C-source). The first chamber depth was 10 cm for R-source and 10 or 20 cm for C-source, the second chamber depth was varied from 0 to 10 cm. Magnetic filter in R-source was formed by magnetic field of currents in a system of two rows of copper tubes. Changing the current we could vary the filter strength between 0 and 150 G.cm. The filter of C-source was formed by rows of permanent magnets, its strength was 150 G.cm. Both source chambers were surrounded by permanent magnets to form longitudinal line cusp configuration. The sources chambers were made of stainless steel. Cathodes were tungsten, hairpin type, their number was varied if nessesary. A plasma electrode could be biased relatively anode to optimize $\bar{\text{H}}$ ion extraction. A two-electrode acceleration system was used to extract a beam. Beam pulse duration was 25 ms. The beam was extracted through a hole in the central part of plasma emission surface. The hole was of 10 mm diameter in R-source, in case of C-source the hole was rectangular with dimensions 2*10 mm. Close to grounded electrode a quadrupole magnetic system was located to deflect and measure electrons of the beam. Negative ion beam current was measured electrically at the distance of 10 cm from emission hole with help of ion collector. To obtain ion beam density distribution the ion collector was divided into a number of narrow electrodes. Secondary electron emission was suppressed by negative potential on a grid in front of the ion collector or by the magnetic field along collector surface.

Experiments with C-source were performed to investigate the influence of source parametrs on the negative ion beam current density. The dependence of negative ion beam current density on the second chamber depth with different values of discharge current is represented in Fig.2. Current density slightly grew with the depth decrease. All data below correspond to $z=2$ cm. Fig.3 shows the influence of the distance between cathodes and the filter on $\bar{\text{H}}$ yield. Experiments were performed with the first chamber depth equal to 20 cm. The distance was 13 or 3 cm. One can see that the decrease of cathodes-filter space gave a small increase in $\bar{\text{H}}$ yield. Futher experiments were done with the first chamber depth diminished to 10 cm and cathode-filter distance 3 cm. Fig.4 presents the dependence of negative ion current density on gas pressure in the discharge chamber. Maximum of current density corresponds to pressure 8-10 mTor. Similar dependences were observed with higher and lower discharge currents. The increase in discharge current caused a shift of maximum location to a higher pressure (to 12 mTor). Further measurments were done at 10 mTor (if no special comment).

One should expect strong influence of cusp magnetic field value on negative ion current density. There were three variants of this field. The basic one had BaFe magnets 9-mm pole wide. When the magnets width was doubled the field component perpendicular to chamber surface, Fig.5, measured opposite poles of magnets, did not change (1000 G), but longitudinal component, measured between the poles, grew from 220 to 550 G. Increase in magnetic field along the wall does not change current density of negative ions, curves 2 and 3 in Fig.5. Whereas comparatively small increase in normal field component from 1000 to 1200 G in third variant with SmCo 5 mm-width magnets caused essential increase in current density, curve 1.

A considerable increase in current density was observed with the growth of the discharge voltage. Higher voltage could be realized by the less number of cathodes or/and lower cathode temperature. The increase in voltage from 20 to 100 at constant discharge current gave doubling of H^- current. Futher increase in voltage resulted in a small addition to current. Considerable current gain was obtained when discharge chamber walls were covered with thin copper foil, curves 1 and 2 in Fig.6. But as soon as copper was covered with tungsten from cathodes the effect disappeared. Fig.7 represents the action of different factors: copper wall, cusp field, discharge voltage, pressure$_2$ in diagnostics. Maximum measured H^- current density achieved 20 mA/cm^2. The dependence of pressure in diagnostics on pressure in discharge chamber, curve 1 in Fig.8, permitted to take into account of H^- losses and calculate H^- beam current density on a plasma emission surface, curve 3. Maximum current density in this case is as large as 40 mA/cm^2.

A number of authors reported about admixing of heavy gas in hydrogen discharge in order to enhance H^- yeld[7,8]. We also tested the influence of a small addition of Xe or Ar to hydrogen on yield of negative ions. Ar did not change the yield, Xe gave 20-30% increase, see Fig.9. But addition of heavy gas with relatively higher cross section of ionization led to breakdowns in ion optics. Breakdowns limited the quantity of inert gas admixture in discharge, maximum discharge current and, as a result, maximum value of H^- current in beam.

Distribution of current density in negative ion beam was measured at R-source. Fig.10 shows an alteration of H^- beam shape with extraction voltage. It should be noted that curves of distribution for smaller voltage had high wings. With voltage increase wings disappeared and total H^- current saturated. The width of distribution fell with voltage at first rapidly, then more slowly, almost proportional to reverse ion velocity.

A very important question for H^- ion sources is electron current extracted with ions. Experiments have shown that electron-H^- ion ratio in beam strongly depended on a shape of extracting aperture and transverse magnetic field in this region. So if magnetic field was created only by quadrupole magnets of diagnostics the ratio could achieve 100-1000, see Fig.11 a. When magnetic field was increased by special magnets to 150 G the ratio fell to 20-50, see Fig.11 c. Futher increase to 330 G gives fall to 5-10, and at last with positive bias of emission electrode the ratio becomes equal to 2-3. Unfortunately in the later case H ion current fell too.

The experiments have demonstrated the influence of a number of different parameters on the yield characteristics of H source. Among other methods, wich have been eximined to enhance H current density, higher effect may cause more strong wall magnetic field. It is interesting to develop and use a cathode, that should not cover copper wall of discharge chamber.

1. Bacal M. and Hamilton G.W. //Phys. Rev. Lett., 1979, v.42, p.1538.
2. Bacal M., Bruneteau A.M., Graham W.G. et fl., //J. Appl. Phys., 1981, v.52, p.1247.
3. York R.L., Stevens R.R., Leung K.U.., Ehlers K.W. //Rev. Sci. Instr., 1984, v.55, p.681.
4. McAdams R., Holms A.J.T., Nightingale M.P.S., et al. // Product. and Neutraliz. of Negative Ions and Beams (4th Intern. Symp., Brookhaven, NY 1986), Editor J.G.Alessi New York, 1987, p.298
5. Okamura Y., Horiike H., Inoue T., et al. //Ibid., p.309.
6. Bezverbaja N.K., Krylov A.I., Kuznetsov V.V., et al. //Proceed. of the 3rd European Workshop on Product. and Applic. of Light Negative Ions, edited by H.Hopman and W.van Amersfort, February 17-19, 1988, p.225.
7. Walter S.R., Leung K.N., Kunkel W.B. //Jornal of Applied Physics, 1988, v.64 (7), p.3424.
8. Leung K.N., Ehlers K.N., Pyle R.V. //Proceed. of the 2nd European Workshop on Product. and Applic. of Light Negative Ions, edited by M.Bacal and C.Moutet, March 5-7, 1986, France, p.25.

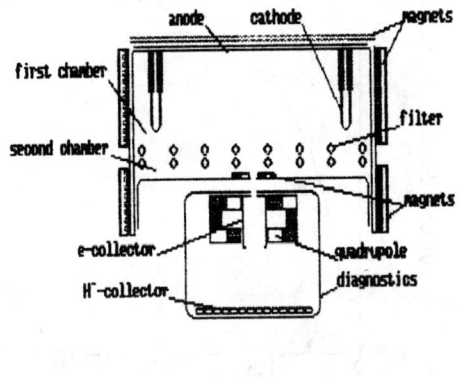

Fig.1 The ion sources scheme

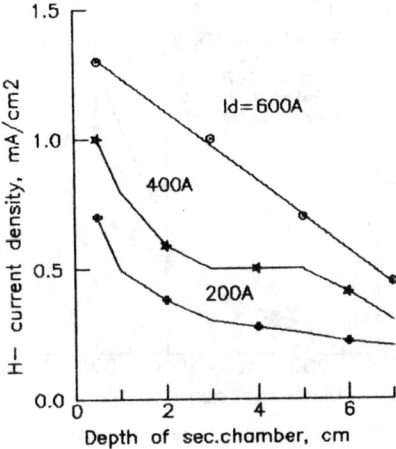

Fig.2 H-yield vs the second chamber depth at different discharge currents

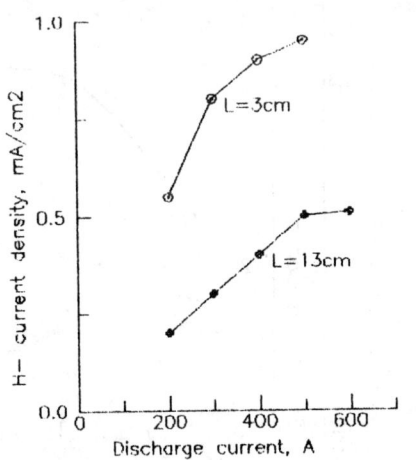

Fig.3 Influence of cathode-filter space L on H- current density
P=6 mTor

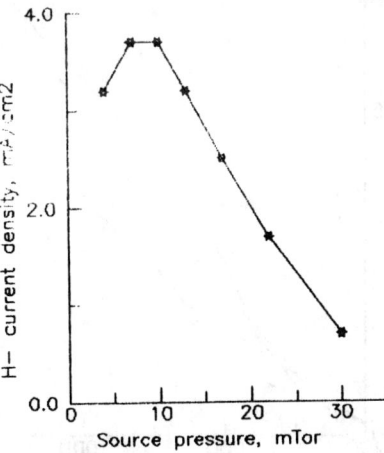

Fig.4. H- current density vs pressure in discharge chamber

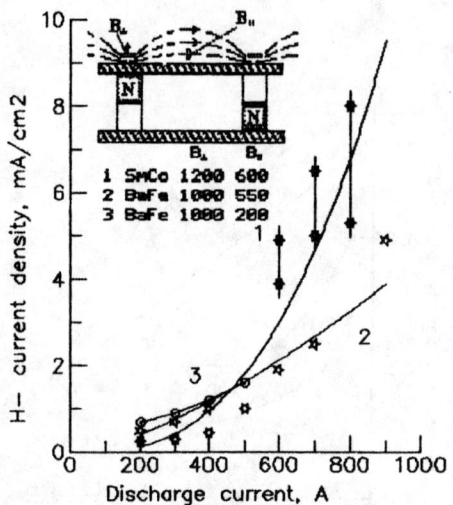

Fig.5 H− yield with different cusp-forming magnets.

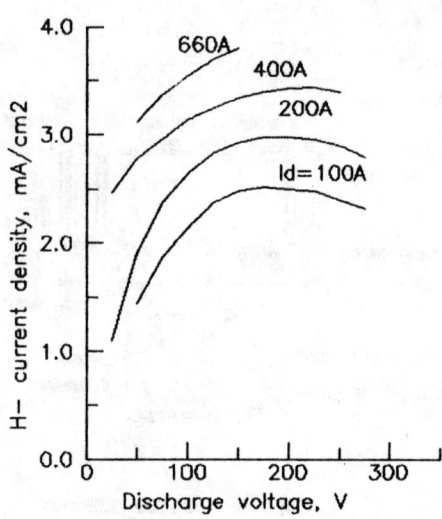

Fig.6 Dependence of H− yield on discharge voltage at different discharge current

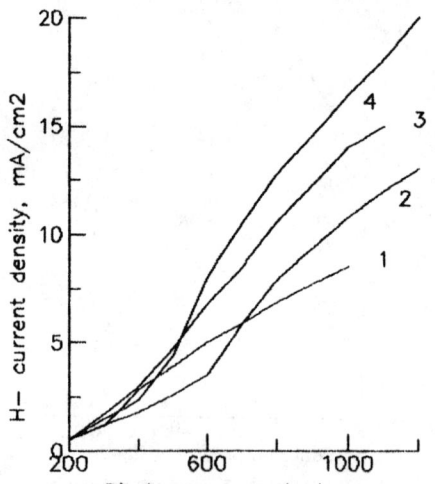

Fig.7 Influence of conditions on H− yield
1. Initial status, 2. Copper foil wall,
3. Increase in discharge voltage,
4. Lower pressure in the diagnostics

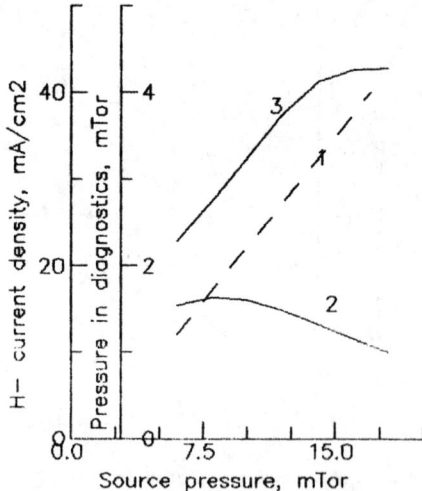

Fig.8 H− yield, measured in diagnostics, 2, and calculated for emission surface, 3, with help pressure in diagnostics, 1

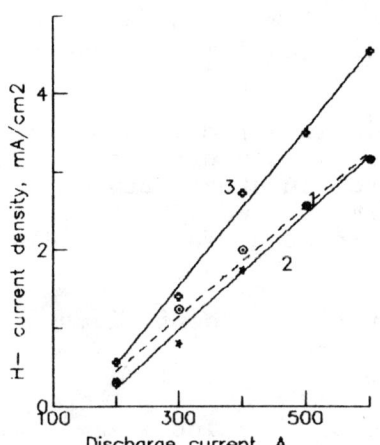

Fig.9 Influence of inert gas admixing on H− current density. 1. Pure hydrogen, 10 mTor, 2. Hydrogen with Ar, 1 mTor, 3. Hydrogen with Xe, 1 mTor

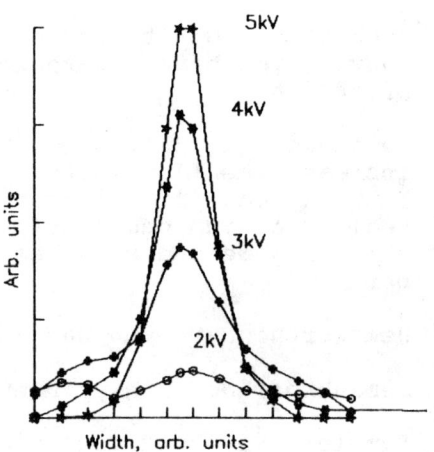

Fig.10 H− beam distribution with different extraction voltage

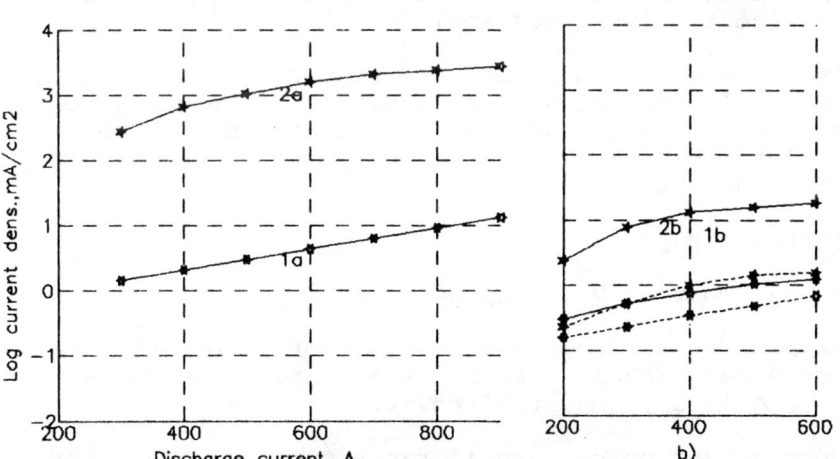

Fig.11 Influence of transverse magnetic field in front of emission hole and positive bias of plasma electrode on extraction of H−ions (curves 1) and electrons (curves 2). Solid − no bias, dashed − optimal bias. a) Field in the gap 80 G, b) 330 G.

DISCUSSION

Jacquot: I don't understand, why with xenon in the discharge, you have breakdown and with cesium you have no breakdown.

Semashko: I don't know, but this is a fact. When we increase the flow in the chamber, we get breakdown. But I think that in Elizarov's source in steady state regime, cesium cools off the chamber. We operate in pulse regime and xenon escapes from the discharge chamber.

Hemsworth: Have you had any operation in deuterium.

Semashko: No. Only in hydrogen.

Pamela: You said your aim is to produce 100 mA/cm^2. Is it with this volume source or by using Elivarov's sources.

Semashko: No, with Elizarov's source. We calculated about 40 mA/cm^2 and we measured 20 mA/cm^2. But, we tried because it is a conceptual design of an ion source for ITER. It is convenient to have 1.1 $Ampere/cm^2$ approximately for accelerating system. I think that it is not possible to accelerate such a current density with a small divergence angle.

Holmes: How do you propose to stop the beam if you have a 100 milliamps per square centimeter at one megavolt with such a small divergence angle of 3 milli rad. Our calculations show its almost unstoppable. We have to reduce the current of density.

Semashko: Maybe.

Holmes: What do you propose?

Semashko: Answers in Russian (Editor's comments) Bacal: He said that Tokamak is such a machine which can swallow and take in any amount of power.

Okumura: You said a considerable current gain occurs, if your sources are covered by a copper foil. I'm very interested in that effect and could you tell me once again how much gain was observed.
Semashko: Initially it was stainless steel, then copper with a gain of approximately 30%. Okumura: But are you sure that the cooper is better than stainless steel?

Semashko: Yes.

Yuan: I've seen that you observed current density increase when the distance between tungsten and filter decreased. Do you have any explanation for that?
Semashko: We need a separate discussion about this since it requires a long explanation.

MAGNETIC FIELD OF A TOROIDAL VOLUME H⁻ SOURCE *

C. R. Meitzler
Department of Physics, Sam Houston State University
Huntsville, TX 77341

ABSTRACT

The magnetic field of the BNL toroidal volume H⁻ source[1] has been calculated using the code PANDIRA[2] for two different configurations. The first configuration was a set of concentric rings of permanent magnets. The second configuration was similar to the first, except that an axial magnet had been added to provide a dipole field on axis. Contour plots of the vector potential and total magnetic field have been provided for each case.

INTRODUCTION

The process of volume production of H⁻ ions was originally reported by Bacal[3] who observed a large H⁻ density in a plasma. This process relies on electrons in two different temperature ranges[4] to produce H⁻: high temperature electrons ($kT > 5$ eV) to produce vibrationally excited hydrogen molecules in the discharge region, and low temperature electrons ($kT < 1$ eV) to cause the excited molecules to undergo dissociative attachment in the extraction region. Magnetic filters have been used to separate the high temperature electrons in the discharge from the low temperature electrons required in the extraction region of the source. Leung et al.[5] have developed a tandem multicusp source, where a planar magnetic filter field produced by lines of permanent magnets, separates the discharge chamber from the extraction chamber.

Prelec[1] has designed a toroidal multicusp volume source that has a cylindrical filter field. The output of this source is sensitive to the magnetic field configuration that is present during operation.[6] In particular, the addition of a Sm-Co disk on the axis of the filament plate has a pronounced effect on the extracted currents. This is believed to be due to the addition of a dipole component to the field close to the axis of the source. In order to understand the sensitivity of this source to its magnetic field configuration, it is first necessary to examine the magnetic field in the region of the axis and extraction aperture. A comparison of the fields with and without the axial magnet will reveal the fundamental differences between the two configurations.

---

* This work was performed at Brookhaven National Laboratory with support by the U.S. Dept. of Energy under contract number DE-AC02-76CH00016.

DESCRIPTION OF THE MODEL

Figure 1 shows a cross-section of the ion source studied by Prelec[1]. Modeling the magnetic fields requires only that the permanent magnets and iron return paths are included in the calculation. The current flowing through the filament was not included in the model since it can be turned off during the source pulse.

The iron sections of the source were modeled using the dimension obtained by scaling directly from a full size drawing of the source. The location of bolt holes and feedthroughs were assumed to be filled with iron. This assumption was made because PANDIRA is a two dimensional code and holes would appear as local perturbations of the field. PANDIRA's internal B-H table was used to provide the magnetic properties of iron: the table corresponds roughly to 1008 steel.

The permanent magnets were Sm-Co rods magnetized along their axes to an energy product of 18 MGOe. The residual induction, B_r, and coercive force, H_c were taken from the manufacturers specifications: $B_r = 8600$ G, $H_c = -8400$ Oe. Each of the toroidal permanent magnet rings was composed of a large number of smaller 6.3 mm diameter, 12.7 mm long Sm-Co rods. To account for the cylindrical shape of the real magnets the width of the model rings was reduced to 5.5 mm. This resulted in the cross-sectional area of the model ring being equal to the total cross-sectional area of the physical magnets comprising the ring. The magnet rings on the outer circumference of the source used the same dimensions as the rings on the filament and extractor flanges.

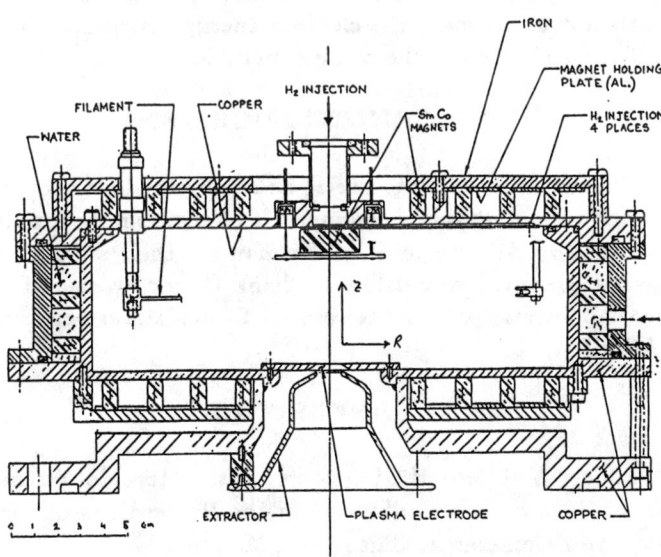

Figure 1. Sketch of the volume source described in Ref. 1.

DISCUSSION

The results of a PANDIRA calculation for the entire source volume without the axial magnet on the filament flange are shown in Figures 2 and 3. The field lines are shown in Figure 2. Figure 3 shows a contour plot of the total magnetic field overlaying the ion source structure. A broad region of low magnetic field can be seen in the center of the plasma chamber. The minimum field is below 100 G, while a large part of the low B region is in the range of 100-200 G.

The next calculation placed a 2.5 cm diameter, 1 cm thick permanent magnet on axis in the plasma chamber. The distance between the plasma electrode and the magnet face was 4.7 cm. The magnet material was assumed to be the same as the magnet rings. Figure 4 shows a TEKDIS display of the magnetic field lines for this configuration, while the total magnetic field contours are shown in Figure 5. A comparison of Figure 5 with Figure 2 shows the presence of dipole-like field close to axis. It is this field that would act like the filter fields found in tandem multicusp sources. The dipole component of the field is seen to weaken as it approaches the plasma electrode.

CONCLUSION

The magnetic field for a toroidal multicusp volume source has been calculated using PANDIRA for two separate cases. The presence of a permanent magnet on axis creates a region close to the axis where there is a large dipole component to the magnetic field. More detailed calculations of the magnetic field and an attempt to estimate the electron energy distribution in the extraction region will be performed in the coming months.

ACKNOWLEDGEMENT

I would like to thank Krsto Prelec for suggesting that I undertake these calculations and providing valuable support and criticism during the initial stages of this work, Jim Alessi and Hovi Kponou for their support while I was at Brookhaven. Finally, I would like to thank Chuck Swenson for several useful discussions about interpreting the results of calculations performed with POISSON/PANDIRA.

REFERENCES

1. K. Prelec, "A volume H$^-$ ion source with a toroidal discharge chamber", BNL-41849 and to be published in the Proceedings of the 1989 Particle Accelerator Conference, Chicago IL, March 1989.
2. Los Alamos Accelerator Code Group, POISSON/SUPERFISH Reference Manual, LA-UR-87-126, (1987).
3. M. Bacal and A.M. Bruneteau, Proc. 3rd Int. Symp. on the Produc-

tion and Neutralization of Negative Ions and Beams (Brookhaven National Laboratory, 1983) ed. Krsto Prelec, AIP Conf. Proc. No. 111 (1984) 31.
4. J.R. Hiskes, Proc. 4th Int. Symp. on the Production and Neutralization of Negative Ions and Beams (Brookhaven National Laboratory, 1986) ed. James G. Alessi, AIP Conf. Proc. No. 158 (1987) 2.
5. K.N. Leung, K.W. Ehlers, and M. Bacal, Rev. Sci. Instrum. 54, 56 (1983).
6. K. Prelec and D. McCafferty, private communication.
7. Thomas and Skinner, Inc., 1120 E. 23rd Street, P.O. Box 150-B Indianapolis, IN 46206.

Figure 2. Magnetic field lines predicted by PANDIRA for the source configuration without a magnet on the axis of the filament flange.

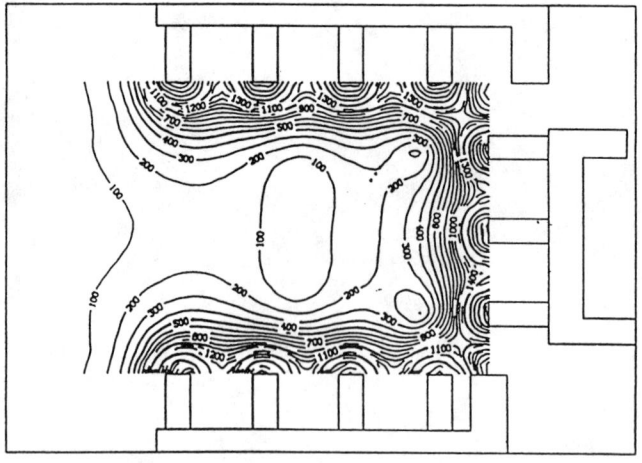

Figure 3. Contour plot of the total magnetic field without a magnet on the axis of the filament flange.

Figure 4. Magnetic field lines predicted by PANDIRA with a magnet on the filament flange axis.

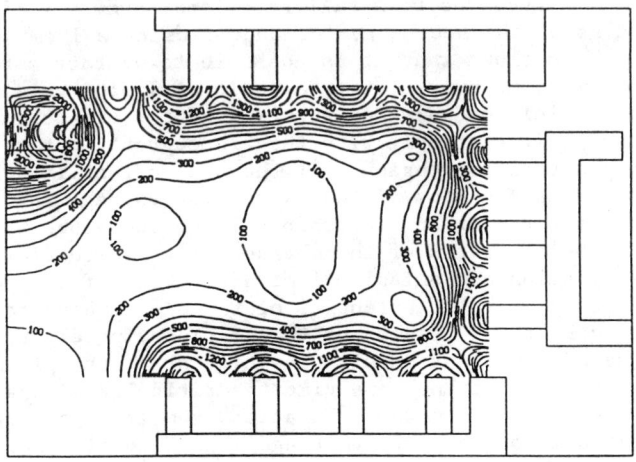

Figure 5. Contour plot of the total magnetic field with a magnet on the filament flange axis.

THE BNL VOLUME H⁻ ION SOURCE*

Krsto Prelec
Brookhaven National Laboratory, Upton, NY 11973

SUMMARY

This paper is a progress report on the studies of the BNL volume H⁻ ion source. We have measured the H⁻ yield, I_H^-, and the ratio I_e/I_H^- as function of the size of the extraction aperture, strength of the conical filter field, size and position of the filament, and of the phase of the filament heating current. The H⁻ current density in the extraction aperture was lower for the largest aperture, while there was a broad maximum when the conical field varied. Position of the filament and the phase of the filament heating current are very important parameters.

INTRODUCTION

The BNL volume H⁻ ion source with a toroidal discharge chamber was described previously[1]; its design evolved from the idea of using a cup-shaped dipole field surrounding the extraction region of a tandem source[2]. There were two benefits expected from such a design: a better utilization of the discharge and a reduction in the beam emittance due to a full rotational symmetry of the source.

Studies of the source performance[3], using a 1 cm² extraction aperture, have shown that it is possible to extract more than 30 mA of H⁻ ions in 1 ms pulses, with a ratio of I_e/I_H^- between 20 and 30. The H⁻ yield measured as function of the arc current increases at first steeply with the arc current, but this is followed by a saturation region. If the arc current is held constant, the H⁻ yield as function of the extraction voltage increases again steeply at first, but at a certain voltage the slope becomes more gradual. The two parts of the characteristics are similar to space charge saturation and emission limited regions of many emitters of charged particles. The potential of the plasma electrode (PE) affected both, the H⁻ yield and the accompanying electron component. The H⁻ yield was usually the highest if the plasma electrode was floating, but at the same time the ratio I_e/I_H^- was also the highest. When the PE potential varied from its floating value through zero into the positive range, the H⁻ yield would decrease somewhat but the electron component would be reduced even more[3].

EXPERIMENTAL ARRANGEMENT

Figure 1 shows a cross section of the source. There are 11 cusp rings assembled from standard Sm Co magnets (6.25 mm diameter, 12.5 mm length) and placed on the inside of the flux return

*Work performed under the auspices of the Department of Energy.

structure (iron). Opposite the extraction aperture, there is a Sm Co disc to create a conical dipole field around the source axis. Another paper[4] presented at this Symposium describes the mapping of the magnetic field. The cathode of the discharge consisted of a single loop of tungsten wire placed outside the conical dipole field. For lower arc currents (up to 150 A) a smaller loop (9 cm diameter) was sufficient, but for arc currents up to 400 A we had to use a loop of 16 cm diameter. Gas injection was pulsed, with the peak pressure in the range of 5 - 15 m Torr.

The H⁻ yield was measured as the voltage drop on a 100 Ω resistor in series with the Faraday cup; a strong dipole field in front of the Faraday cup served to deflect the electrons out of the beam. A standard current transformer served to measure the electron component. For the measurements of the emittance, the Faraday cup was replaced by a slit-and-collector type emittance device; a paper presented at this Symposium[5] describes the results of emittance measurements.

RESULTS

A. Effects of the size of the aperature

Three different apertures have been tried, 0.5 cm^2, 1 cm^2, and 1.87 cm^2. Figure 2 shows the optimized yield as a function of the arc current for the three apertures, at constant arc and extraction voltages and with a 16 cm diameter filament. The H⁻ yield from the largest aperture (1.87 cm^2) increases if a thin tungsten wire cross is mounted across the opening in the plasma electrode; there was no such effect with smaller apertures. The graphs on Fig. 2 show that the H⁻ yield increases linearly with the arc current up to a few tens of amperes; above about 50 A the increase is more gradual and the characteristics approach a saturation above 300 - 400 A. The H⁻ current density does not change much when the aperture changes from 0.5 cm^2 to 1 cm^2; however, there is a substantial reduction for the largest aperture, especially if there is no wire cross on the plasma electrode. Table I is a summary of the best H⁻ yields for the three apertures.

Table I

Aperture	I_{arc}	I_H^-	J_H^-	I_e/I_H^-
0.50 cm^2	300 A	15 mA	30 mA/cm^2	18
1.00 cm^2	400 A	35 mA	35 mA/cm^2	43
1.87 cm^2 (no cross)	300 A	33 mA	18 mA/cm^2	29
1.87 cm^2 (with cross)	400 A	48 mA	25 mA/cm^2	31

B. Comparison of the dipole filter and conical filter

In order to check the effectiveness of the conical filter, we have compared the source performance in the standard configuration (conical filter) with the performance when the Sm Co disc on the filament flange was removed and a linear dipole field established in the vicinity of the extraction aperture by mounting a few small permanent magnets on the plasma electrode. The latter configuration corresponds to those usually existing in tandem H⁻ ion sources. (In the absence of any dipole field, the extracted electron component was extremely high which limited the arc current to 10 - 20 A at most.) Figure 3 shows the optimized yield for the two dipole configurations, once for the larger filament loop and then for the smaller one. It is evident that in either case the conical dipole increases the yield by about 50%, for the same arc current.

C. Size and position of the filament

Figures 3, 4, and 5 show results of studies with filaments having different sizes and different locations with respect to the symmetry plane. First, a smaller loop (Fig. 3) had a better arc current efficiency; however, the arc current was limited to the region below 150 A because of a smaller emitting surface. While the position of the larger filament with respect to the symmetry plane was not critical, the smaller filament was more efficient if placed 1 cm closer to the filament flange (Fig. 4). Finally, even the diameter of the filament wire had an effect on the source performance: there seems to be an optimum for values around 1 mm diameter (Fig. 5). It is not clear what causes this effect, but any explanation would have to include the local magnetic field due to the arc current flowing through the wire (the heating current was interrupted during the pulse).

D. Strength of the conical field

By replacing the Sm Co magnet that produces the conical field with a small coil in the same location, we were able to study the yield as a function of the conical field. Figures 6 and 7 show the H⁻ yield and the ratio I_e/I_{H^-}, as function of the pulsed coil current, for several values of the arc current. The measurement was done both, with the filament heating current on or interrupted during the arc pulse. While there is a broad optimum in the H⁻ yield, shifting slowly toward higher values of the coil current with the arc current increasing, the electron load (or, the ratio I_e/I_{H^-}) depended strongly on the conical field. It should be noted that the best H⁻ yield and the lowest value of the ratio I_e/I_{H^-} do not occur simultaneously.

E. Direction and magnitude of the filament heating current

The effect of the filament heating current was discovered

early in our studies. As the first step, a circuit was added to
bypass the filament during the arc pulse, reducing in this way the
filament heating current close to zero. With an ac heating, we
were able to move the arc pulse over the full ac period and monitor
the effect of the instantaneous value of the filament current.
Figures 8 and 9 show the H⁻ yield and the ratio I_e/I_H^- as a function of the ac phase, for several values of the arc current.
First, for the larger loop (Fig. 8) the H⁻ yield is higher for ac
phases around zero crossings than around peak values of the filament current. There is a symmetry of both sets of curves with
respect to peak values, but an asymmetry with respect to zero values. This observation is in agreement with earlier measurements
when the yield depended on the direction of the direct current for
filament heating. The source with the smaller loop shows a different behavior (Fig. 9). While there is still a symmetry with
respect to the peak instantaneous values of the filament current,
the asymmetry with respect to zero values is much more pronounced.
This agrees again with observations that the H⁻ yield may be even
slightly higher if the filament current (dc) is not interrupted
during the arc pulse. Of course, this is not a general rule
because the behavior can be different if the position of the filament loop with respect to the extraction aperture is changed. As
it was the case with the effect of the wire diameter on the H⁻
yield (Fig. 5), the explanation this time will also have to include
the local magnetic field due to the filament current and its relationship to the local cusp field.

CONCLUSIONS

The work described in this report represents a part of ongoing
studies of the BNL volume H⁻ ion source with a torroidal discharge
chamber. So far only minor variations from the original design[1]
have been investigated, e.g., effects of the size and location of
the filament. The source has performed well, very reliably and
repeatedly, close to the AGS requirements. The data are still not
complete, there are many features to be explained and understood
and more substantial changes of the original design to be explored
(e.g., different cusp configurations; a properly designed extractor; better pumping of the region around the extractor to reduce
stripping losses of H⁻ ions; etc.). We can, however, conclude that
it is not possible to optimize the source performance with respect
to several criteria simultaneously, among them the arc power
efficiency, gas consumption and the ratio I_e/I_H^-. Figure 10
shows, to illustrate this point, the ratio I_e/I_H^- vs I_H^- for many
operating conditions of the source. It is evident from data like
these that there is a broad range of combinations of source parameters and that the selected operating regime will be a certain compromise.

REFERENCES

1. K. Prelec, Proc. European Particle Accelerator Conference, Rome, June 1988, p. 1327.

2. K. Prelec, J.G. Alessi, A. Kponou, A. McNerney, Proc. Second European Workshop on Production and Applications of Light Negative Ions, Palaiseau, France, March 1986, p. 31.

3. K. Prelec, Proc. 1989 Particle Accelerator Conference, Chicago, March 1989 (to be published).

4. C.R. Meitzler, this Symposium.

5. J.G. Alessi, this Symposium.

FIGURE CAPTIONS

Fig. 1   Cross section of the source.

Fig. 2 H^- yield as function of the arc current, for several apertures.

Fig. 3 H^- yield as function of the arc current comparing different filaments and filter field configurations.

Fig. 4 H^- yield as function of the arc current for several positions of the filament.

Fig. 5 H^- yield as function of the arc current for several values of the filament wire diameter.

Fig. 6 H^- yield (full lines) and the ratio I_e/I_{H^-} as function of the conical field strength for several values of the arc current; I_{fil} ON.

Fig. 7 H^- yield and the ratio I_e/I_{H^-} as function of conical field strength; I_{fil} OFF.

Fig. 8 H^- yield (full lines) and the ratio I_e/I_{H^-} as function of the phase of the ac filament current; filament: 16 cm diameter.

Fig. 9 H^- yield (full lines) and the ratio I_e/I_{H^-} as function of the phase of the ac filament current; filament: 9 cm diameter.

Fig. 10 Ratio I_e/I_{H^-} vs I_{H^-} for different configurations and operating conditions.

The BNL Volume H⁻ Ion Source

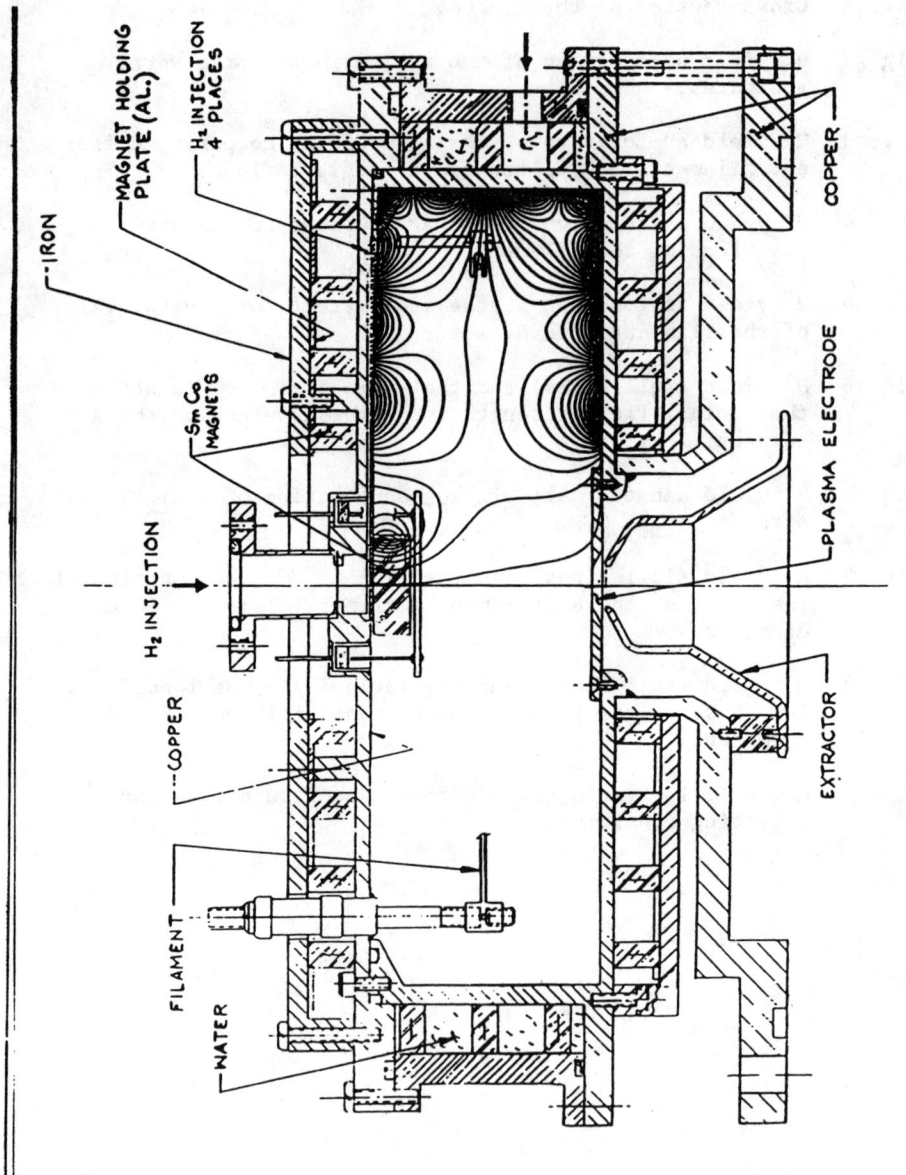

Fig. 1 Cross section of the source.

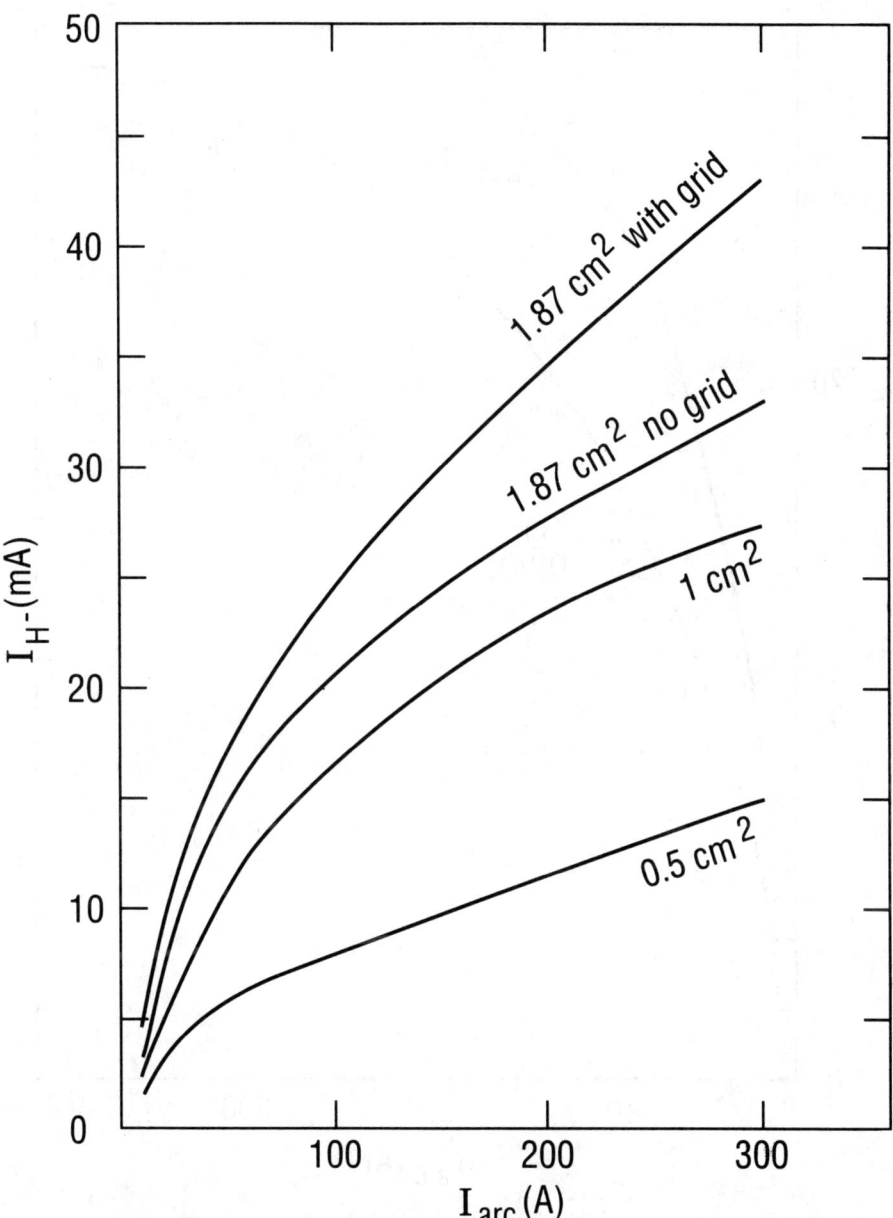

Fig. 2 H⁻ yield as function of the arc current, for several apertures.

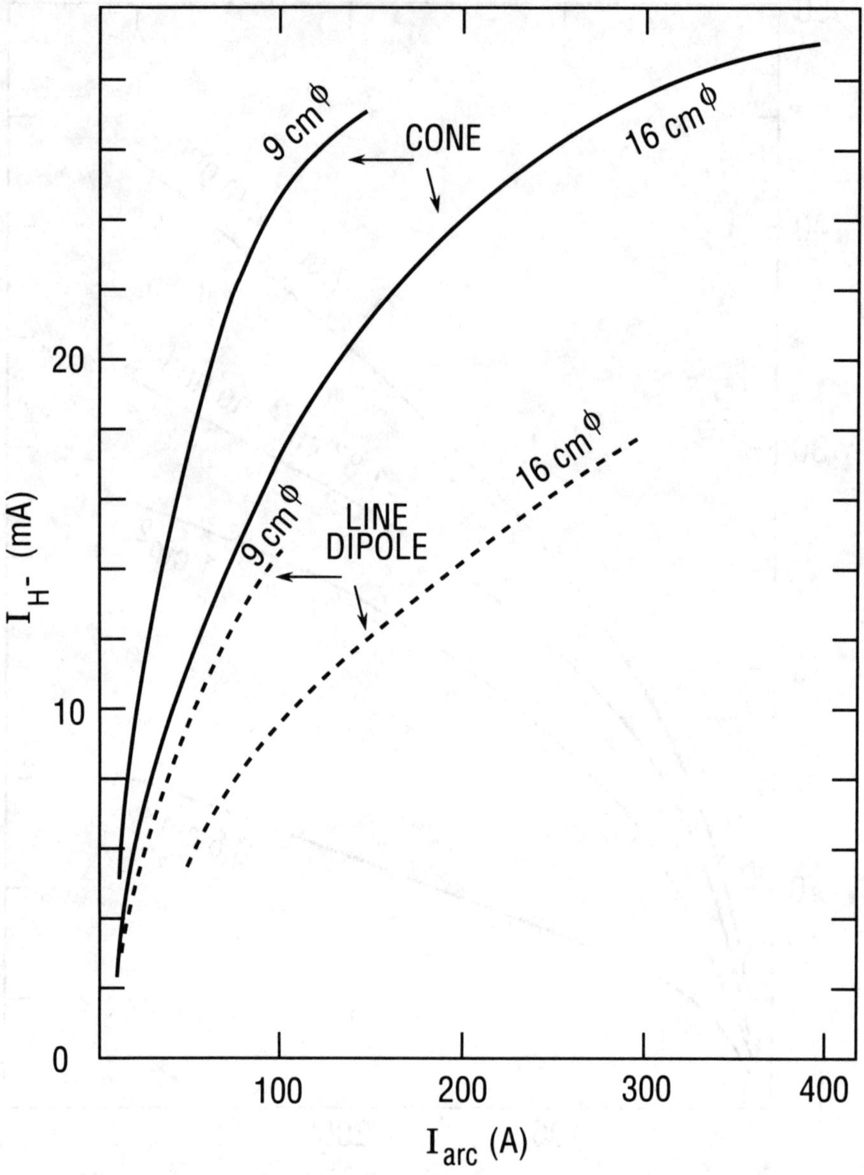

Fig. 3 H⁻ yield as function of the arc current comparing different filaments and filter field configurations.

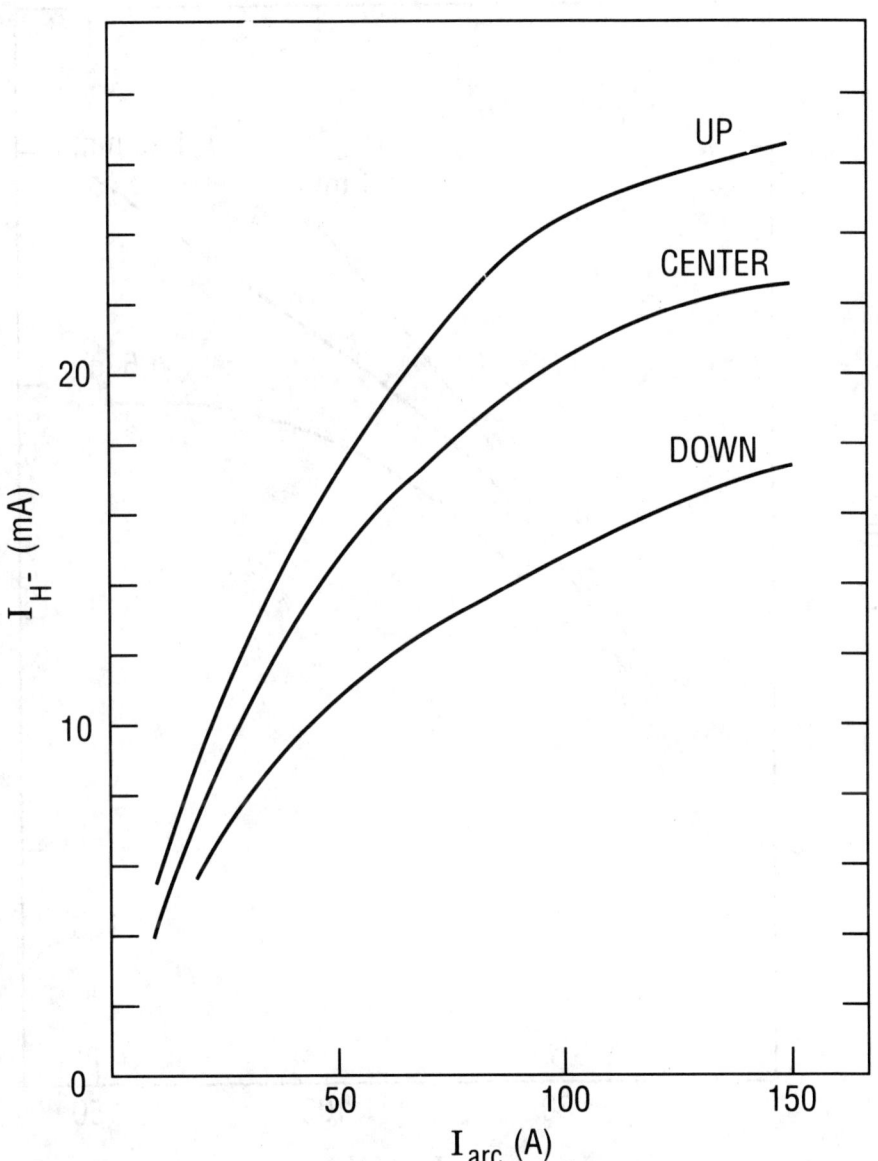

Fig. 4 H⁻ yield as function of the arc current for several positions of the filament.

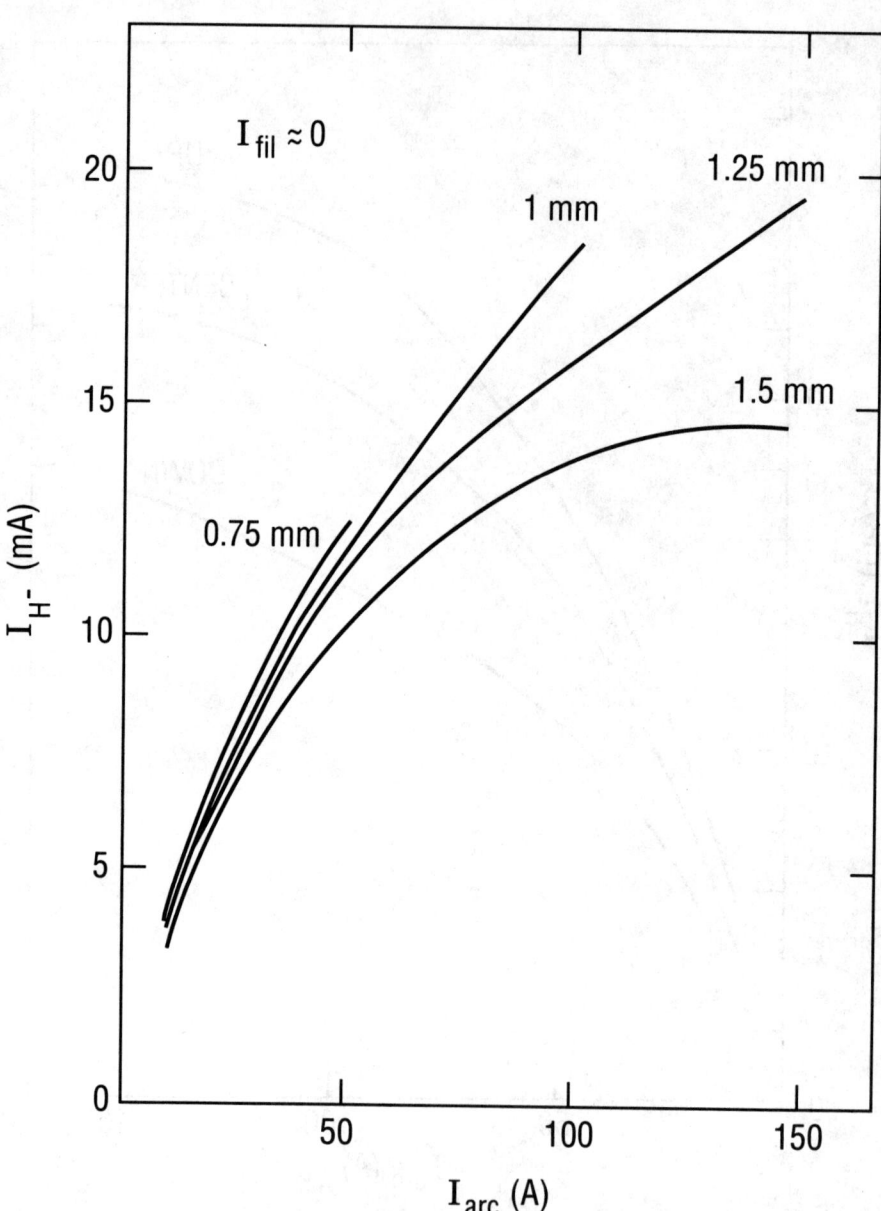

Fig. 5 H⁻ yield as function of the arc current for several values of the filament wire diameter.

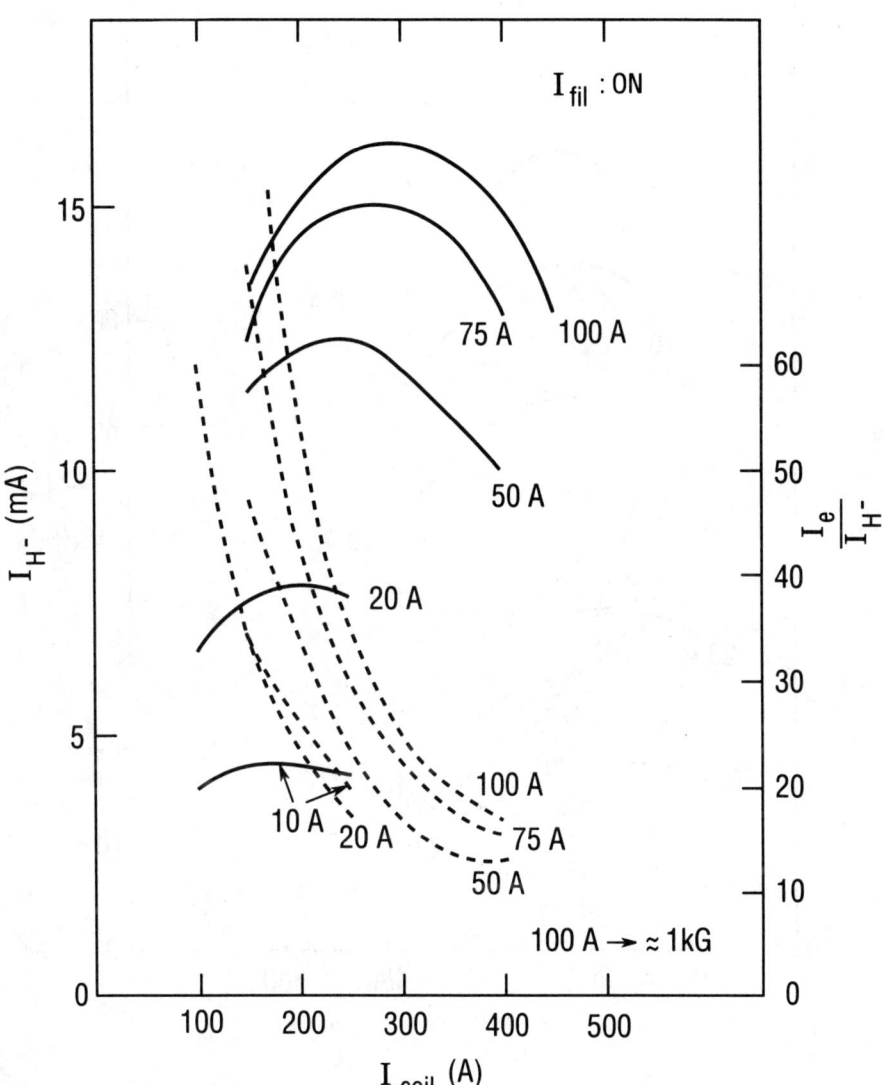

Fig. 6 H⁻ yield (full lines) and the ratio I_e/I_{H^-} as function of the conical field strength for several values of the arc current; I_{fil} ON.

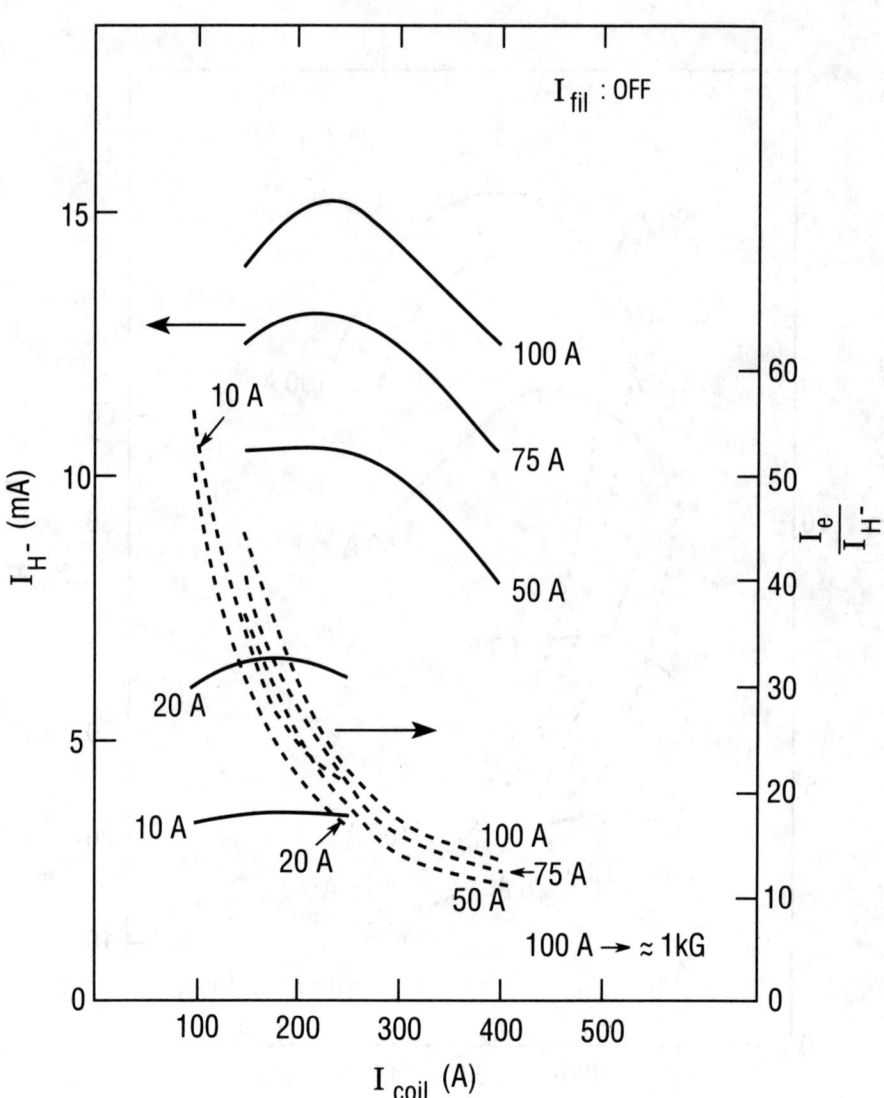

Fig. 7 H⁻ yield and the ratio I_e/I_{H^-} as function of conical field strength; I_{fil} OFF.

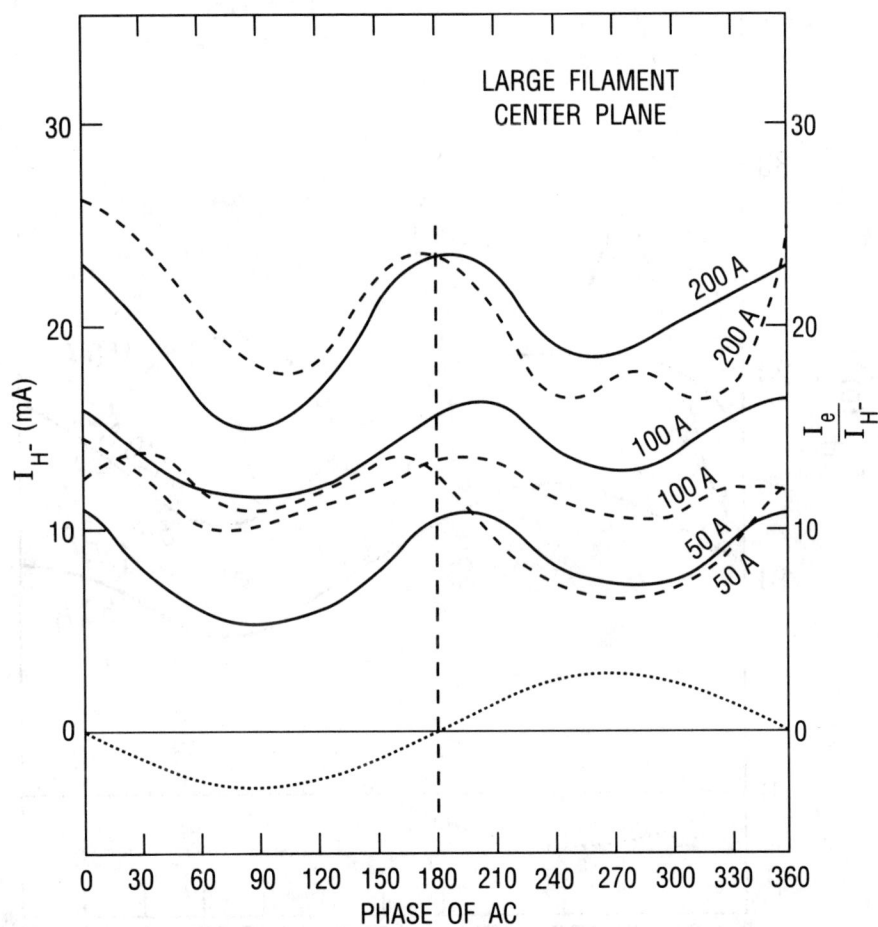

Fig. 8 H⁻ yield (full lines) and the ratio I_e/I_{H^-} as function of the phase of the ac filament current; filament: 16 cm diameter.

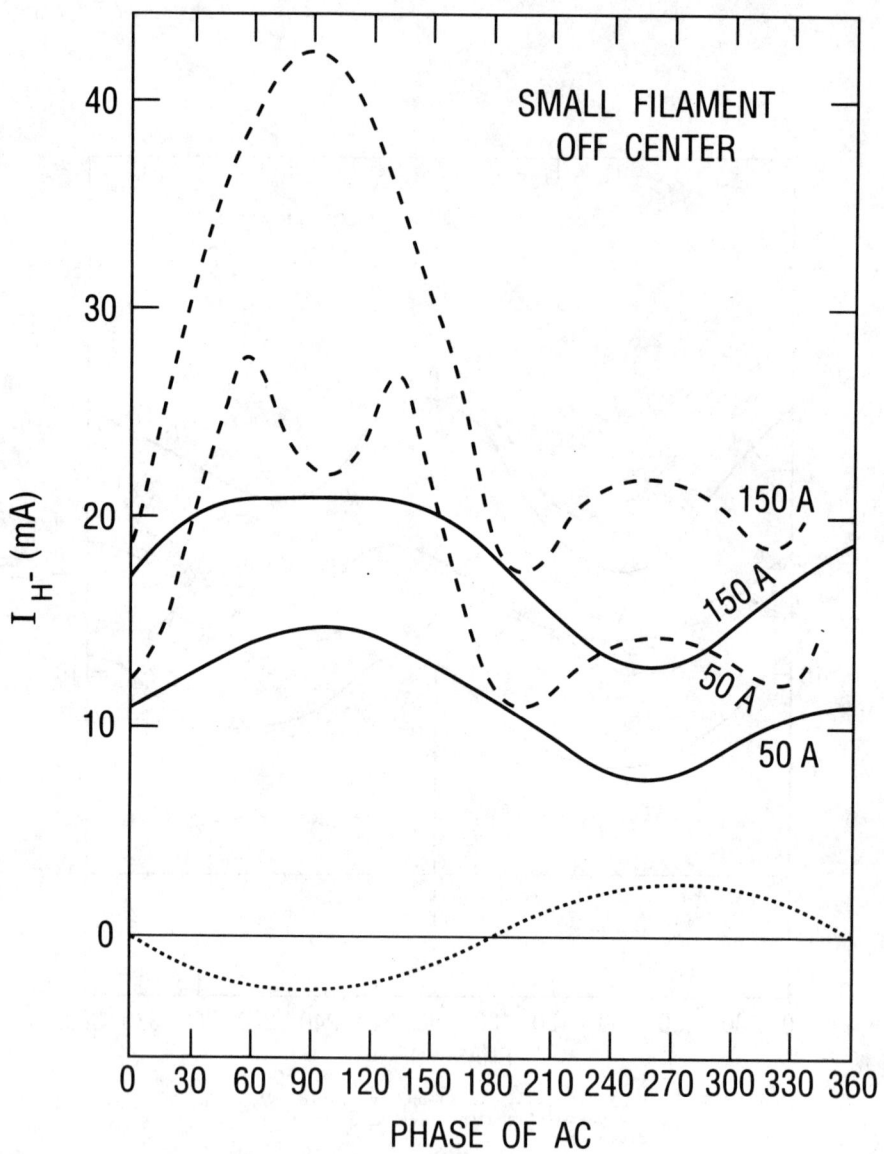

Fig. 9 H⁻ yield (full lines) and the ratio I_e/I_{H^-} as function of the phase of the ac filament current; filament: 9 cm diameter.

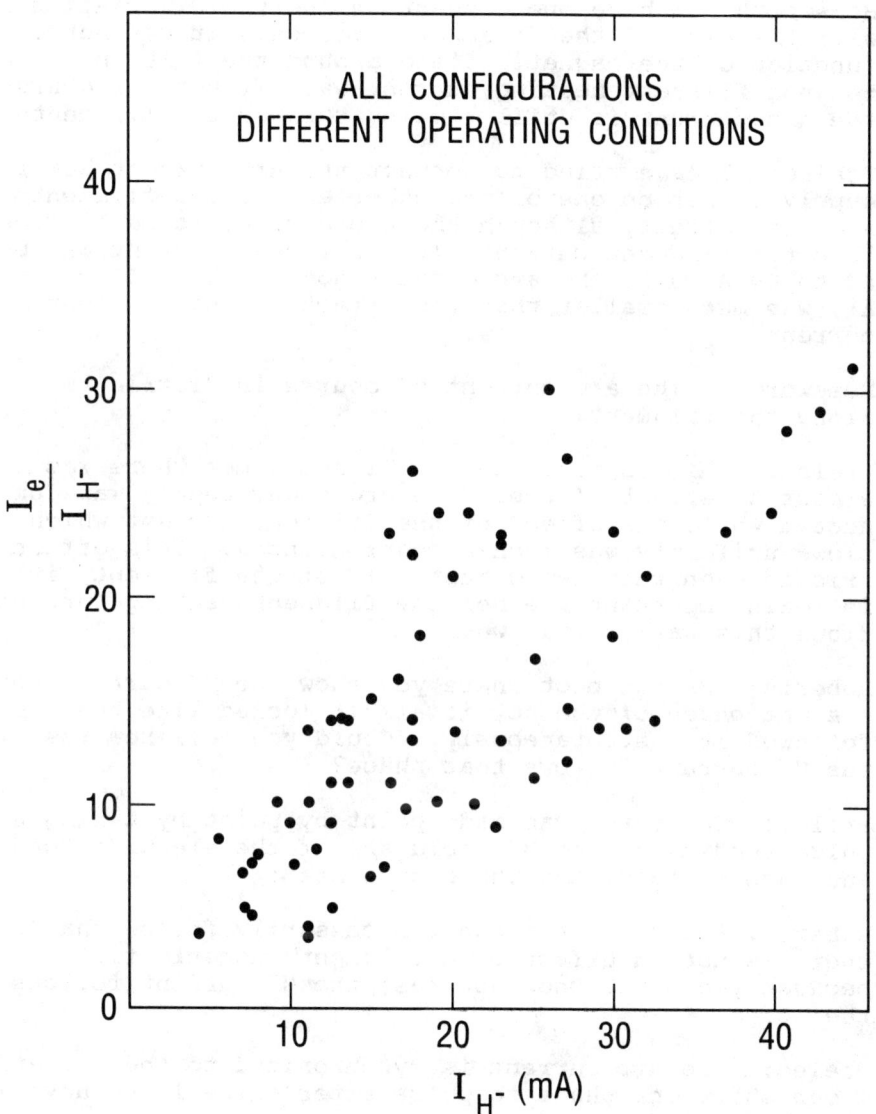

Fig. 10 Ratio I_e/I_{H^-} vs I_{H^-} for different configurations and operating conditions.

DISCUSSION

Hemsworth: I have one comment on the yield variation with the size of the filament. You said it was not a function of the magnetic field around the filament because filament heating current was off but, of course, the arc current is still flowing through the filaments.

Prelec: I have tried to connect the arc current power supply either on one or the other end of the filament and that effect, although the arc current is much higher than the filament current, (the filament current may be 60 to 80 A while the arc current goes up to 200 or 300 A), was much smaller than the effect of the filament current.

Hemsworth: The arc current of course is distributed along the filament.

Prelec: This is true, but still sometimes there was almost no effect of where the arc power supply was connected while the effect of the filament current which flows uniformly was much more pronounced. This effect existed even when using just half of the filament, it is again important whether the filament heating current flows this way or that way.

Roberts: On the plot where you show the H^- current versus the phase of the ac, it almost looked like that it followed it instantaneously. Could you tell how fast the H^- current follows that phase?

Prelec: The curve was made point by point by taking a pulse reading of the H^- yield and of the electron load and then changing the phase by a step.

Roberts: But then it doesn't necessarily follow that's there is not an effect of a filament magnetic field because you don't know how fast the H^- current follows the ac.

Prelec: The arc current is synchronized to the ac, and I can shift the phase so pulse after pulse I can have it at any desired phase.

Roberts: Yes, but it wasn't zero just before or just after the pulse and you don't know how fast the effect follows it because it could have been something to do with the magnetic field plus the current before you made the measurement. I'm not so sure that the magnetic field does not have some effect even though the filament current or the arc current is zero at the moment you made the measurement.

York: A number of years ago, Ralph (Stevens) and I were working with the converter source. We tested ac filaments and we saw the arc efficiency was much better if you went to the zero crossing of ac and we got a 30% increase in arc efficiency when we pulsed the filaments off during the arc pulse. I think my observation of what you are saying is that it is the magnetic field that is affecting the electron emission from the filament and the reason you can see the effect in the volume source is that the drift time from the filament to the extraction geometry is not zero. I mean, you are affecting the plasma, the magnetic field is affecting the plasma in your production region and you are seeing the effect of the time it takes to produce plasma to drift to the extraction geometry.

Roberts: I feel like he said that much better.

Prelec: Let me go back to this. We may have an E x B drift due to the magnetic field of the filament which is azimuthal (around the filament) and due to the radial electric field caused by the cathode voltage drop. The E x B drift would be along the filament and its effect should be the same whether the current flow is this way or that way.

York: The drift could be towards the extraction region. It should be towards the extraction region due to the long term effect based on the magnetic field of the filament.

Prelec: When we moved the filament from a position above the center plane to a position below the center plane, the behavior was just the opposite. So this effect could be due not to the magnetic field of filament current only, but to a coupling between the magnetic field of the filament current and the cusp field.
York: An experiment that might be very interesting is to try something like a LaB_6 filament.

Prelec: I wouldn't like to mention LaB_6 filament because that was the worst disappointment I had. But it was thick, about 2.5 millimeters and could be due to a much larger radius.

Holmes: What went wrong?

Prelec: It just wouldn't perform, I would get half the H^- yield.

Holmes: That's because it only operates in space charge limitation.

Bacal: I wanted to comment on the effect of the filament diameter which is very similar to the effect, which we have observed when I was in JAERI, with the effect of the number of filaments. I mean a filament performs better when it is hotter and that is what you observed: the smaller the diameter the better it was for H^-, which means the filament was hotter to give the same discharge current. To make the long story short, this is an effect of evaporating the filament material, since it is favorable to have fresh filament material on the wall and that is why you had a bad result with the LaB_6 because you don't evaporate tungsten or a metal.

A COMPACT DC CUSP SOURCE

D.H.Yuan, M. Mcdonald, P.W.Schmor

TRIUMF, 4004 Wesbrook Mall, Vancouver, B.C., Canada V6T 2A3

K.Jayamanna

ESE,1151 Kotte Rd. Rajagiriya,Colombo,Sri Lanka

ABSTRACT

This paper presents the results of an experimental investigation of the parameters effecting the quality of the H^- beam extracted from a small source employing multicusp confinement and a magnetic filter to enhance the H^- production. The source was designed to operate in the d.c. mode, to have a long filament lifetime and to provide an intense H^- beam with low emittance. The H^- beam is initially transported about 2 m at 25 keV with a measured 95 % space charge neutralization. The experimental results indicate that the small cusp source is $\sim$ 4 times brighter than the large one[1] for arc current less than 25A. A 7 mA H^- beam with a normalized emittance of 0.34 πmm.mrad at an arc current of 27A and the voltage of 127V is obtained.

The modifications to the extraction system which significantly reduce the electron contamination in the H^- beam and improve beam divergence and emittance, several configurations of magnetic filter and the source design are also described.

INTRODUCTION

During the last few decades, the application of H^- ions in accelerators and producing neutral beams for fusion research has been growing. Several methods were used. Volume production is one of the most reliable and effective methods.

The mechanism of volume production of H^- with arc discharge in multicusp sources can be described in two main stages.[2] First, H_2 molecules collide with high energy electrons to form H_2^+ and H_3^+ ions:

$$H_2 + e^-(\sim 100eV) \longrightarrow H_2^+ + 2e^-$$

$$H_2^+ + H_2 \longrightarrow H_3^+ + H_0$$

These H_2^+ and H_3^+ ions collide with the chamber wall and fast electrons producing vibrationally and rotationally excited molecules inside the first part of the source called driver region.

$$H_2^+ + wall \longrightarrow H_2^*$$

$$H_3^+ + e^-(fast) \longrightarrow H_2^* + H_2$$

These excited molecules move through the magnetic filter into the second part of the source, called extraction region, and collide with low energy electrons producing H^- ions. This can be explained by the resonance model $^2\Sigma_u^+$ temporary ion state calculated by Wadehra and Bardsley.[3] These vibrationally and rotationally states of the molecule are at $\nu = 1$ to 5 and $j = 1$ to 7, respectively. The cross-section of the rotationally exited molecules producing H^- ions is much smaller compared to the cross-section of the vibrationally exited molecules $\nu = 4$ which is $\sigma \sim 10^{-16}$.

$$H_2^* + e^-(\sim 0.5eV) \longrightarrow H_2^-(\sim 10^{-12}sec.) \longrightarrow H^- + H_0(\sigma = 2 \times 10^{-16}cm^2).$$

These H^- beams are characterized by low emittance, high brightness, stable output, low noise. The above properties, along with inherent simplicity of the volume source, and relatively long filament lifetime make it a practical choice for cyclotrons with external H^- injectors. Furthermore, the output beam is axially symmetric, which significantly reduces the complexity of the transporting system while maintaining the low emittance of an intense beam. We describe here a small D.C. cusp source developed at TRIUMF using either a virtual or rod magnetic filter and present some measurements of its performance and a comparison with a larger cusp source previously developed at TRIUMF.

SOURCE DESCRIPTION

An outline drawing of our cusp source and the extraction system is shown in Fig. 1. A cylindrical full line cusp with 10 rows of 3.2 kG $SmCo_5$ magnets located axially on the outside of an all copper water cooled 10 cm diameter × 15 cm deep forms the plasma chamber of the source. A strong virtual magnetic filter formed by flipping the polarity of some end cusp magnets (2.5cm long), creates a magnetic field transverse to the beam axis with $\int Bdl \approx 0.5kG - cm$, dividing the source into two regions. A filament made of tungsten wire (dia.= 1.5mm) is mounted on the back face extending 10 cm into the plasma chamber. The first electrode is insulated from the cusp body such that it can be biased a few volts positive wrt the anode.

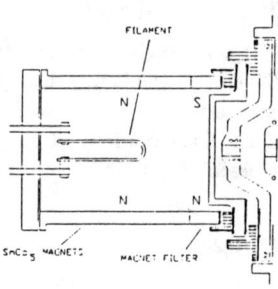

Fig. 1. Schemadic of the source

The extraction system is an axially symmetric four-electrode structure designed to produce an 25 keV beam. The H^- ions are extracted through a small hole by applying a positive $\sim$ 3 kV potential (wrt the cusp body) on the second electrode. The permanent magnets in the second electrode, arranged such that they have a $\int Bdl = 0$, and a peak field of $\sim$ 100 G, serve to sweep the simultaneously extracted electrons from the beam while giving the heavier H^- ions a small net displacement. An additional voltage increase of $\sim$22 kV between the second and third electrodes then brings the ion energy to 25 keV. An positive potential of an order of 100V is applied to the third electrode to prevent backstreaming of low-energy positive ions into the acceleration gap. The gaps of the extraction system can be easily modified by replacing small inserts on each electrode.

The long collimator of the fourth electrode serves to permit a differential pumping to reduce gas stripping in the extracted region to $\sim$ 20% while still allowing the source to run at an optimum pressure. Two turbopumps located on pumping ports in the extracting region of the vacuum jacket are used to evacuate the region between the first and the fourth electrode to an order of 10^{-5} torr. The last aperture in the extraction system limits conductance into the beam line and allows a good vacuum ($2 \times 10^{-6} torr$) to be maintained by two diffusion pumps in the beam line.

Tuning of the source is accomplished with the aid of two sets of steering coils located right after the extraction system correcting the displacement of the beam, a solenoid ensuring a parallel beam, and a graduated Faraday cup measuring the current and ensuring the smallest divergence of the beam shown in Fig. 2. A set of three quartz tubes with 10mm of ID and 10mm long placed 10cm after extraction and 10cm apart from one another serves a self-focusing and steering element, especially useful for low currents.

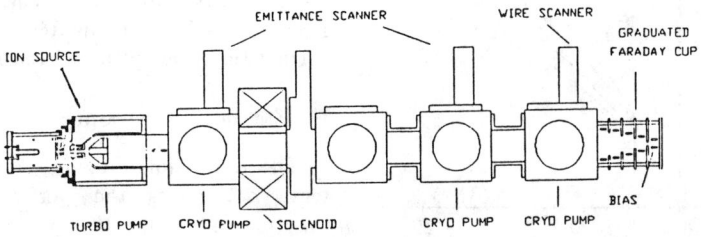

Fig. 2. Schematic diagram of the source, extraction and diagnostics system

The beam emittance is determined by a method developed at LAMPF[4] which employs electrostatic deflecting plates located between two slits and a linear feed screw allows the precise positioning of the slit detector in the beam. A portion of the beam passes through a narrow slit (0.06mm) into a region where two parallel plates (2.8 mm gap, 38 mm long) impose a variable transverse electric field, passes another narrow slit (0.06mm) and is detected by a Faraday cup, then. The plate voltage can be stepped uniformly from -500 V to +500 V and the cup current is digitized for each voltage step. The detector position is moved and a set of curves are generated whose width is proportional to the angular spread of the beam at each position. (see Fig. 3). A computer is then used to contour-plot the beam emittance figure. A X-Y wire scanner is positioned after the solenoid to determine the profile and position of the beam. A second emittance scanner is used to measure the space charge neutralization in the 25 keV region, and the emittance change due to the effects of space charge and lens aberration.

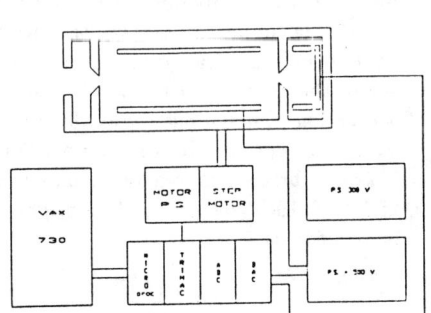

Fig. 3. Block diagram of emittance scanner

EXPERIMENTAL RESULTS

A. Magnetic filter

The magnetic field of the filter separates the source into two regions determining the efficiency of the penetration of ions and low energy electrons into the extraction region, Adjusting the filter field enables us to reduce the electrons out of the source and optimize the H^- current. Varying the strength of virtual filter field is accomplished by setting a permanent magnet bar at each side of the source, creating an extra field parallel to the field of virtual filter, which can be moved while the source is in operation.

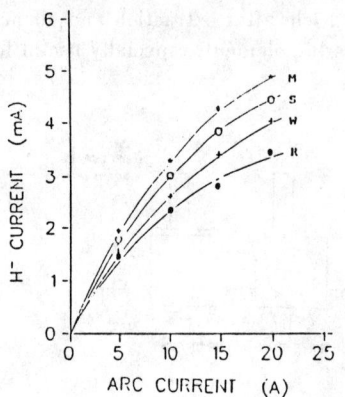

Fig. 4. H^- current versus arc current for four filter strengths

Fig. 4. shows the H^- current versus arc current for four filter strengths. "M" — the optimum case, "W" and "S" represent weaker and stronger field with respect to "M". "R" is a rod filter. The optimum strength yields approximately 30% more current than the optimized rod filter.

B. Extraction system

The extraction system, in some extent, determines the performance of the source. In the extraction region, the vacuum and the beam energy are both relatively low therefore gas stripping becomes significant. In order to reduce gas stripping, a differential pumping technique is used in our system. As a result only 20% of the extracted beam is lost. Fig. 5 shows the comparison between calculated results. Three different vacuum distributions are assumed in the extraction region: 1) linear, 2) exponential, 3) gaussian. The experimental curve is shown in the figure as well, non-differential pumping creates ~ 45% of beam loss[4], which is similar to a gaussian distribution, our system creates a beam loss of ~ 20% close to an exponential distribution. The computer model was tested experimentally by removing differential pumping and measuring the change in current. The result was ~ 30% current lost.

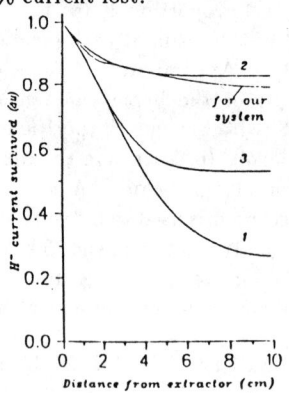

Fig. 5. Gas stripping effects in the extraction system

The gaps between the electrodes have been experimentally obtained by optimizing the beam brightness for the desired beam energy and range of current. Fig. 6 shows the normalized beam brightness as a function of beam energy at three beam currents. For the desired beam (5mA, 25keV) brightness reaches maximum (25A and 100V of arc). The brightness is defined as

$$B_n = \frac{2If^2}{\epsilon_n^2}$$

where

$I \longrightarrow$ the total H^- current measured on the Faraday Cup two meters downsteam of the source

$f \longrightarrow$ appropriate beam fraction

$\epsilon_n \longrightarrow$ normalized beam emittance for appropriate beam fraction

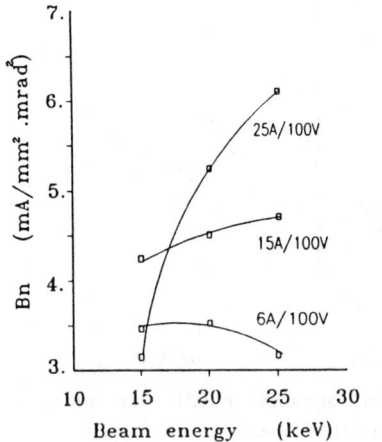

Fig. 6. Normalized beam brightness as a function of beam energy at three beam currents

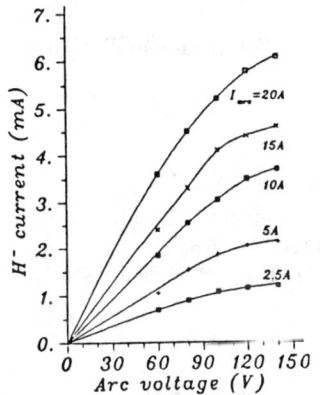

Fig. 7. H^- current as a function of arc voltage for several arc current

C. H^- current

H^- currents from 11 mm aperture as a function of arc voltages for various arc current are shown in Fig. 7. It is found that H^- currents start saturating at arc voltage about 120 V for the small cusp source, where as for the large source, 200 V of arc does not saturate the H_- current[1]. The small cusp source has a smaller volume of plasma, and for same rows of cusp magnet the cusp area is relatively larger than the large one. Therefore, the primary electrons with higher energy (higher arc voltage) have more possibility of colliding to the cusp area or source wall before being thermalized.

Comparison of the extracted H^- current with the big cusp source for 100V of arc voltage and 6.5mm of aperture size is shown in Fig. 8. The H^- current is double and brightness increases by a factor of four for the arc current less than 25A. The results of Langmuir probe measurements indicate that at same arc the e^- temperature is almost the same for both sources. The experimental result confirms that the emittance is almost same (Fig. 9).

Dependence of the H^- current density on aperture sizes for various arc currents was measured. Measured H^- current density was reduced as the size of aperture increased as K.Leung and R.Stevens have seen.[6] The effect of gas stripping along the beam line, especially in the extraction region has been calculated which could be as high as 20% and $\sim$ 2% is found in the beam transporting line, depending on the H_2 flow rate. On the other hand ion emission areas for the aperture sizes were simulated by code AXCEL, the ratios to the areas of πR^2 show also a similar function to the measured depedence of density on the area of aperture. It is believed that any inconstant of current densities versus extraction aperture sizes is partially due to the gas stripping effect and the estimation of the virtual emission areas.

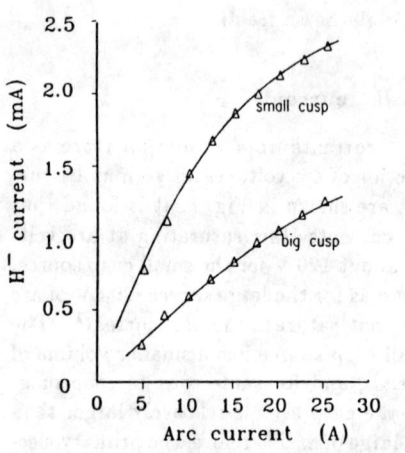 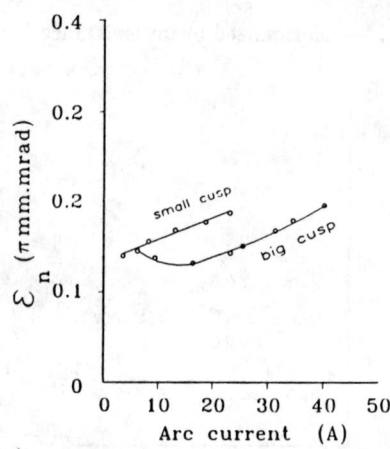

Fig. 8. Comparison of H^- current as a function of arc current between two sources

Fig. 9. Normalized emittance for both sources as a function of arc current

ACKNOWLEDGMENTS

We would like to express our gratitude for the special efforts made by P. Chigmaroff with the electronics, C. Laforge on the apparatus.

REFERENCES

1. K.R.Kendall et al., Rev. Sci. Instrum. **57**,1277(1986)
2. K.N.Leung, K.W.Ehlers and M.Bacal, Rev. Sci. Instrum. **54**,57(1983)
3. J.M.Wadehra and J.N.Barsley, Phys. Rev. Lett., **41**,1795(1978)
4. P.W.Allison, D.B.Holtkamp, and J.D.Sherman, IEEE Tran. Nucl. Sci. NS-30, 2204(1983)
5. J.W.Kwan et al., Volume Produced H^- Beam Experiment at LBL, private communication
6. R.Stevens "Volume Source Studies At LANL", private communication

CONTINUOUSLY OPERATED NEGATIVE ION SURFACE PLASMA SOURCE

A.A. Bashkeev and V.G. Dudnikov
Institute of Nuclear Physics, Siberian Division of the USSR
Academy of Sciences, 630090, Novosibirsk, USSR

Abstract

In this paper we consider negative ion surface-plasma sources with small discharge chambers, designed for long continuous operation. Their important features are a discharge with a cold hollow cathode in a magnetic field, spherical cathode surface for a double geometric focusing of negative ions toward a 1 mm diameter extraction hole, and the plasma drift in crossed fields directed toward the spherical part of the cathode. Regimes with a low noise level have been achieved. H$^-$ beams with currents up to 2.5 mA have been achieved; for a discharge current of 0.7 A and at a voltage of 100 V. The source life is longer than 100 hours. Source operation with mixtures of hydrogen and ethyl alcohol, water, or ammonia vapors has been studied, and mass spectra of extracted ions measured. Heavy ion beams (O$^-$, OH$^-$, NH$^-$) with currents up to 1 mA have been achieved.

Introduction

Negative ion surface plasma source development started in 1971 after the discovery of enhanced negative ion production due to cesium addition to the gas discharge.[1,2] Cesium layer catalyzes negative ion production on the negatively charged electrode surfaces when bombarded by ions and atoms from the discharge. It is important to prevent a destruction of negative ions while they move through the plasma to the emission hole. For that matter, the plasma and gas layer thickness between the electrode surface emitting negative ions and emission hole should be small.

Many modifications of the surface plasma sources, which were optimized for various applications, were developed since 1971. Their description can be found in the proceedings of Brookhaven symposia on the production and neutralization of negative ions and beams (1977, 1980, 1983, and 1986).

Surface plasma sources (SPS) can be divided into three classes:

I. Surface plasma sources using discharges with cold cathodes in crossed electric and magnetic fields. The first versions of such sources were developed in Novosibirsk.[1-3]

II. Surface plasma sources using discharges with hot cathodes in chambers with multicusp magnetic confinement and separate negative ion emitters. The first version of such sources were developed in Berkeley.[4]

III. Surface plasma sources using discharges with hot cathodes with crossed fields are the intermediate case. Such sources were studied in Oak Ridge.[5]

Surface plasma sources with cold cathodes require higher gas densities to initiate the discharge. For that reason, the gap between the cathode-emitter and emission hole should be very small (d < 1 cm), otherwise there will be significant negative ion destruction. Nevertheless, the advantages of SPS-I are: they permit efficient production of negative ions at high discharge current densities (up to hundreds of A/cm^2), they provide record negative ion current densities (up to 8 A/cm^2), and record beam brightness (normalized brightness is up to 10^9 $A/cm^2 \cdot rad^2$). High gas efficiency (up to 30%) can be achieved at high charge current densities only, due to the blocking out of gas and cesium by the discharge plasma. However, the SPS-I operated at very high power densities and such sources were used in a pulsed mode only (pulse duration could be relatively long: 1 to 10 seconds).

In SPS-II, a typical gas density is small and the gap between the electrode-emitter and emission hole is large (approximately 10 cm). However, plasma density and discharge current density cannot be high. SPS-II and SPS-III can operate in a continuous mode in discharge with low power densities. Continuous regimes for SPS-II operation were discussed in Ref. 6. Recently, SPS-II was used successfully for negative heavy ion production.[7]

In this paper various Type I surface plasma sources operated in long, continuous periods at moderate discharge power without forced cooling, will be described.

Experimental Installation

For the SPS under consideration, in the gas flow is reduced due to two-dimensional geometrical focusing of negative ions by the spherical cathode surface toward the emission hole of $\simeq$ 1 mm diameter. A hollow cathode provides discharge ignition at a lower gas density. Source construction provides a tight gas discharge chamber. Ion source schematics are shown in Fig. 1. The gas discharge chamber of the source consists of a Mo cathode 1 and anode 2 with the emission hole. Cathode and anode are separated by ceramic insulator 4. In early versions of the source, the ceramic insulator had copper seals to provide a tight structure (as described in Ref. 8).

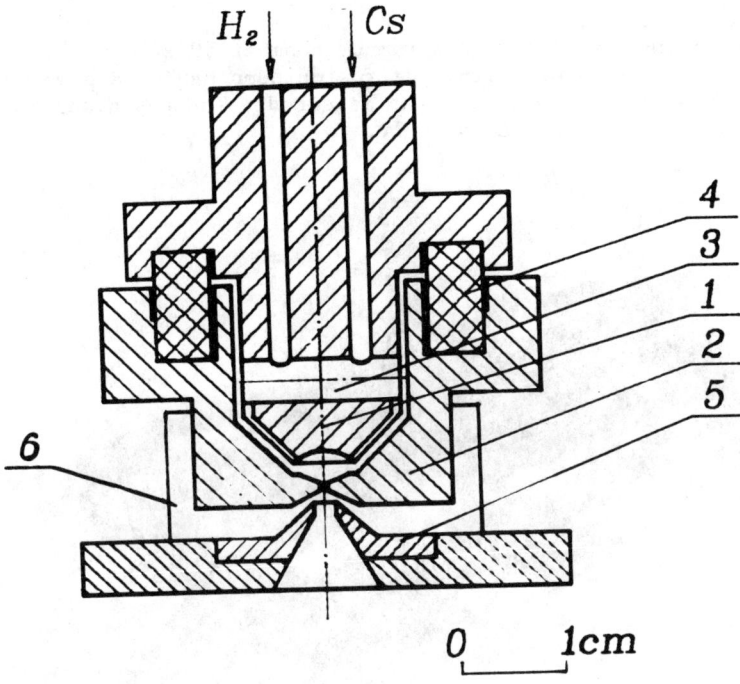

Fig. 1 Design of continuously operated surface plasma source.
1 - cathode 2 - gas-discharge chamber (anode), 3 - hollow cathode, 4 - insulator, 5 - extraction electrode, 6 - magnetic pole.

Hydrogen and cesium are supplied through channels to the cylindrical opening serving as a hollow cathode. Plasma drifts in crossed fields from the hollow cathode 3 along the cathode-canal toward the spherical dimple at the cathode end. Negative ions, which were created on the cathode spherical surface, are then focused to the 1 mm diameter emission aperture. A magnetic field of 0.5 kG defects the 100 eV ions from the center by 1 mm, which is taken into account when the emission hole is manufactured. The gas discharge chamber is mounted between the magnet poles 6 on the plate holding the conical extractor. A magnetic field is generated by the inserts which are ceramic permanent magnets. A general view of the first version of the SPS is shown in Fig. 2. A leak valve or heated palladium leak is used for supplying hydrogen. An additional leak valve is used for supplying vaporous ethanol, water, or ammonia. To control the supply of cesium, there was a heated container filled with tablets with a mixture of cesium chromate with titanium as in Ref. 8.

The source was placed in a vacuum chamber 50 x 50 x 150 cm^3; the vacuum was maintained with a diffusion pump having a pumping speed of 700 ℓ/s. The gas flow was measured. The gas discharge chamber was cooled by radiation only.

Fig. 2   General view of SPS.

The source extraction voltage was up to 20 kV. Discharge and heater voltages were supplied by isolation transformers operating at 20 kHz. The operating parameters of discharge were controlled by optically linked analog bypasses.[9]

The ion mass distribution was registered by the scheme, which is shown in Fig. 3. A simple electrostatic lens 2 focuses the ion beam moving between electromagnet poles 3, then the deflected beam falls through the screen slot onto the collector 4. The relationship of the collector current versus electromagnet current was registered. Unfortunately, the small resolution of such an analyzer did not allow for the separation of close masses of heavy ions.

Experimental Results

In the first experiments, a cathode without cylindrical opening 3 in Fig. 1 but with the canal and spherical dimple was used. To ignite the discharge in this configuration, the required hydrogen density was high which caused strong negative ion destruction even at a minimal radius of 2.5 mm of the spherical surface. When the hollow opening 3 was used, the discharge was ignited and maintained at lower hydrogen density. It is possible to achieve all the discharge characteristics which were observed earlier in the pulsed regimes. Without cesium, the discharge voltage U_d was 500 - 600 V. With cesium, the discharge voltage was reduced a usual. Voltage $U_d \approx$ 100 V corresponded to an optimal negative ion production. The level of oscillations was changing when hydrogen and cesium were supplied. There were regimes with chaotic noise, with coherent oscillations, as well as regimes without oscillations in discharge and beam.

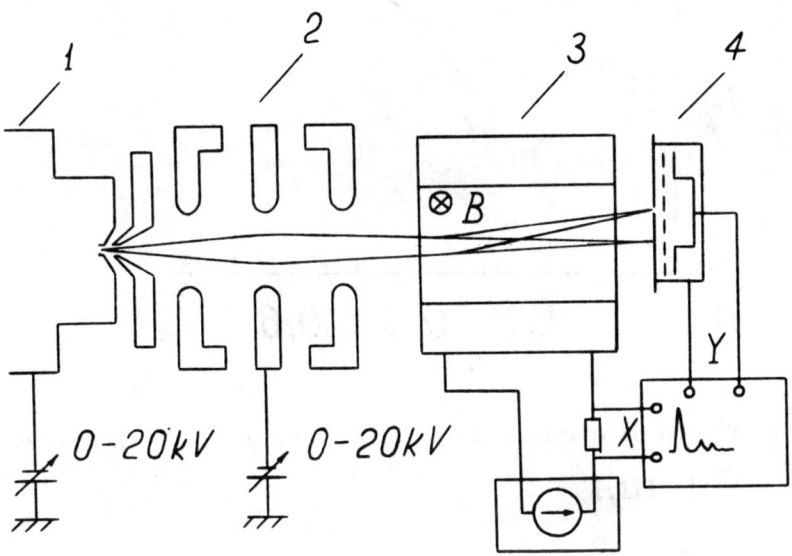

Fig. 3 Schematic drawing of experimental apparatus to study ion beam mass distribution.

The maximum intensity up to 2.5 mA of negative ion beam was achieved when the radius of the focusing sphere was 4 mm (emission aperture diameter 1.2 mm)

Figures 4, 5, and 6 show examples of how beam intensity I^- depends upon discharge current I_d, extraction voltage U_o, and the gas pressure in the chamber. Those relationships are characteristic for surface plasma sources of other types as well. With no forced cooling, the source can work at a discharge power up to 100 Watts. An attainable effectiveness of 5 mA/A of negative ion production is several times lower than in high-current sources. This is due to the use of a small fraction of the cathode surface generating negative ion beam.

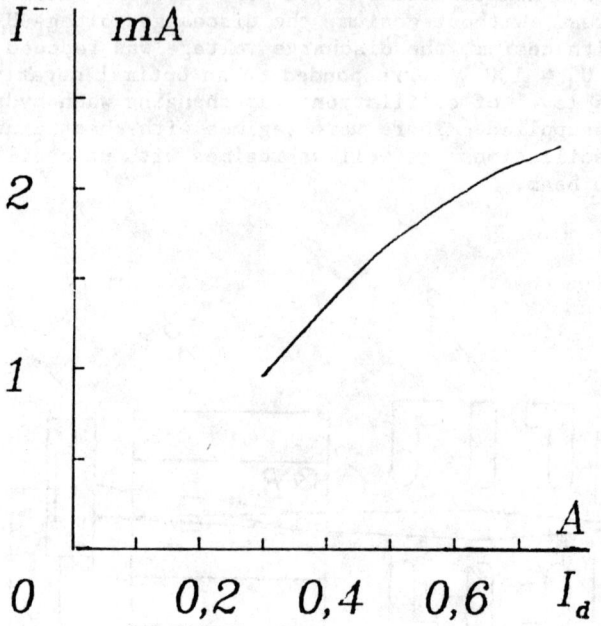

Fig. 4 H⁻ ion beam intensity as a function of discharge current.

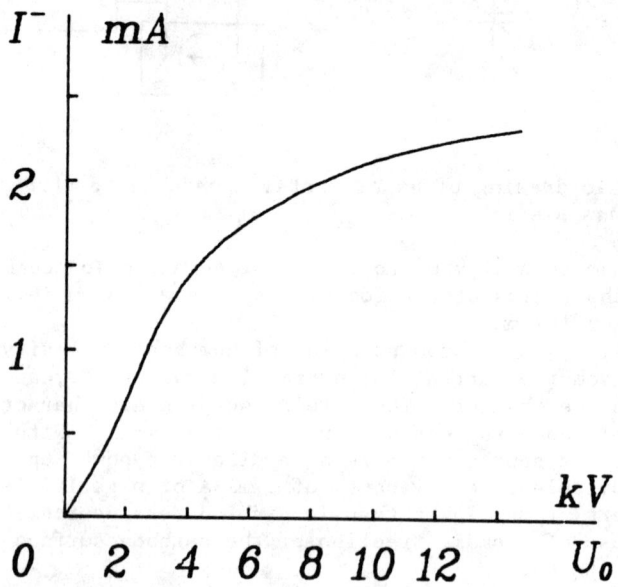

Fig. 5 H⁻ ion beam intensity as a function of the extraction potential.

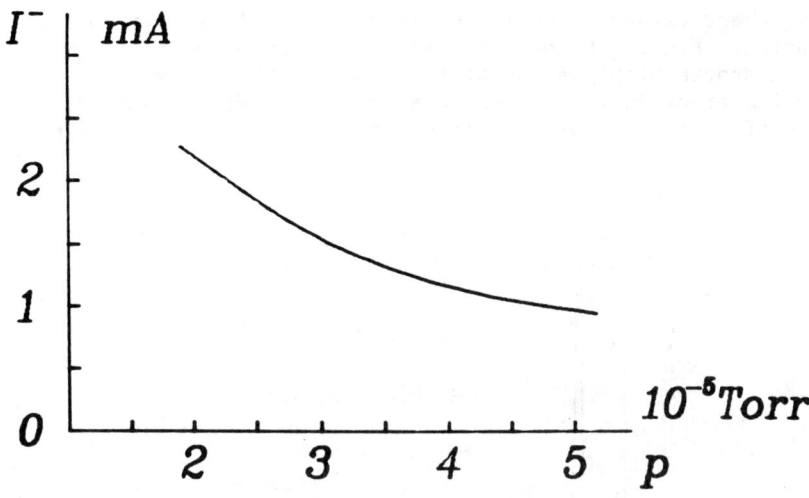

Fig. 6 H⁻ ion beam intensity as a function of the pressure in the vacuum chamber.

With optimal conditions, the full current in the extraction gap was comparable to the beam current, although there were discharge regimes with a large current of extracted electrons. A significant heating of the extractor electrode did not prevent the source from operating in the continuous regime. The optimal adjustment provided the stable operation of the source during several hours with current stability better than 5%.

For long-term testing, the source operated during 10 - 12 sessions, each of 8 - 10 hours. The maximum duration of approximately 120 hours was limited by the amount of cesium in the container (approximately 50 mg). Sometimes there were short circuits between the cathode and the anode due to flaking of deposits from the anode, produced by sputtering of the cathode. However, these shorts were evaporated easily by increasing the current. On the whole, the cathode sputtering was insignificant and occurred mainly at the ignition of the discharge when the voltage is large and cesium does not shield the molybdenum surface from sputtering. In optimized regimes, mainly cesium and hydrogen are sputtered from the surface; both return again on the cathode due to recycling.

Perhaps we can hope that similar type surface plasma sources, including those close in design to the multiaperture SPS,[10] will have a significantly larger life time of operation.

A typical mass spectrum is shown in Fig. 7a, where hydrogen was supplied through a mechanical leak valve. The heavy ion impurity is about 10%. A mass spectrum of heavy ions was not registered when the palladium leak was in use. It was not possible to obtain low voltage discharges while operating with heavy gasses such as air, nitrogen, water, vapor, alcohol, ammonia. However, by adding these gasses to hydrogen, it was possible to provide stable operation. Figure 7b shows a mass spectrum when operation was done using hydrogen with the addition of ammonia. In this case, the heavy ion fraction represents more than 60%. Unfortunately, it was not possible to separate different masses of negative ions in this case.

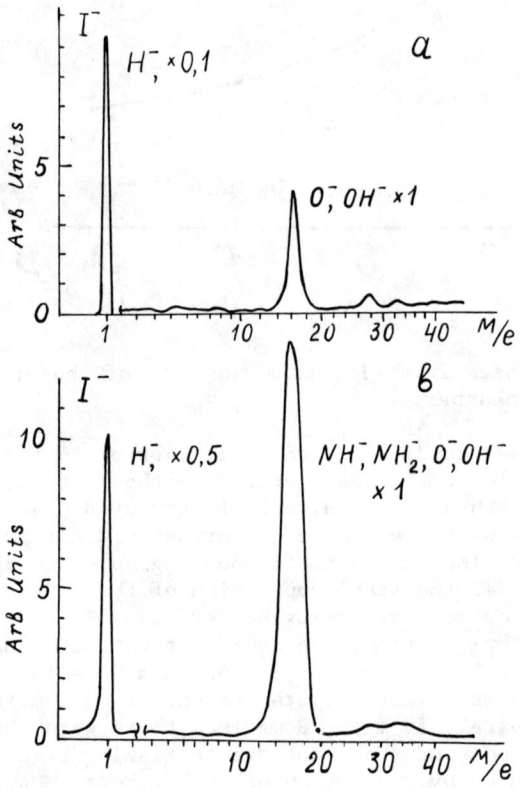

Fig. 7   Mass distribution of ion beams.

   a. Discharge gas is hydrogen.
   b. Discharge gas is hydrogen with addition of ammonia.

When heavy gasses are added, the sputtering rate is increased significantly and the life time decreases down to tens of hours only. It is interesting in this case that a deep hole of about a diameter, 0.2 mm, is sputtered at the cathode center by the beam of fast positive ions.

As it follows from estimates based on Fig. 6 and from gas flow measurements, a minimal gas flux from the source is q ≈ 20 cm^3 Torr/s, which corresponds to ≈ 7 x 10^{17} mol/s. When current I$^-$ = 2.5 mA, the estimated gas efficiency is approximately 1.5%, which is close to the expected value at the small (approximately 0.2 A/cm^2) negative ion current density. It was discovered, however, that the gas went not just through the emission hole, but through other openings in the gas discharge chamber. Because of that, another SPS was manufactured with a more tightly enclosed chamber. The design is shown in Fig. 8. All features of the electrode geometry are preserved and precautions taken to achieve a tighter structure. A schematic of the gas discharge chamber is shown in Fig. 9.

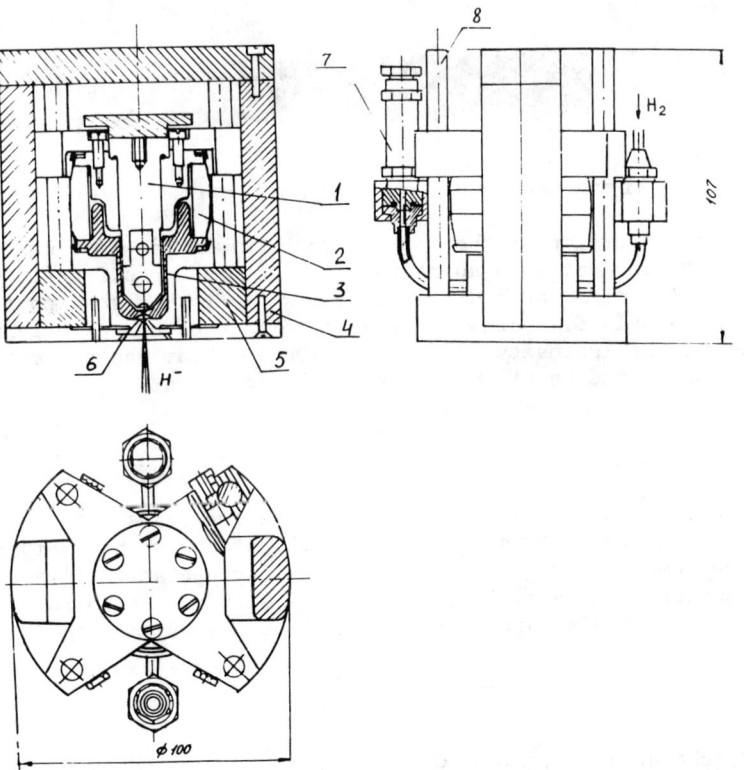

Fig. 8 Design of SPS with vacuum-tight, gas-discharge chamber: 1 - cathode, 2 - metal-ceramic insulator, 3 - gas discharge chamber (anode), 4 - magnet yoke, 5 - permanent magnet, 6 - extraction electrode, 7 - cesium container, 8 - high voltage ceramic insulator.

Fig. 9 View of vacuum-tight, gas-discharge cell.

In this version of the source, the minimum gas flow was proportional to the emission aperture area and was approximately equal to 10^{17} mol/s, at a diameter of 1 mm. At this gas input, there were stable discharge modes at a voltage of approximately 100 V; however, the intensity of the negative ion beam was not more than 0.5 mA. Studies of this source will continue.

Conclusion

The studied sources provided negative ion beams with parameters of interest for practical applications with a very small size, low power, high gas economy, and a long operation life time. For comparison, in Refs. 11 and 12, beams with similar parameters were obtained at discharge currents of hundreds of amperes.

References

1. Belchenko Yu.I., Dimov G.I., Dudnikov V.G., Ivanov A.A. Doklady AN SSSR, 213, 1283 (1973), Preprint INP 81-72, Novosibirsk, 1972.

2. Dudnikov V.G., Author Certificate, Cl HOI 3/04, N 411542, Application 3.III.1972, Bul. N 2, 1974.

3. Belchenko Yu.I., Derevjankin G.E., Dimov G.I., Dudnikov, V.G., Zh. Prikl. Mekh. Tekh. Fiz. 4, 106 (1987); Preprint INP 87-8, Novosibirsk, 1987.

4. **Ehlers K.W.**, Leung K.N., Rev. Sci. Instr., 51, 721 (1980).

5. **Dagenhart W.K.**, Stirling W.L. et al., in Proc. of 2nd Int. Symp. on the Prod. and Neutr. of Neg. Ions and Beams (Brookhaven, 1980), p. 217.

6. **Piosczyk B.**, Dammertz G., Rev. Sci. Instr., 57, 840 (1986).

7. **Mori Y.** et al., in Proc. of Int. Conf. on Ion Sources (Berkeley, 1989) to be published in Rev. Sci. Instr., 1990.

8. Dimov G.I., Derevjankin G.E., Dudnikov V.G., IEEE Trans. Nucl. Sci., NS-24, N 3, 1545 (1977).

9. Belkin V.S., Vibe S.A. Pribory i Tekhnika Eksperimenta, N 6 181 (1984).

10. Belchenko Yu.I., Dimov G.I., Dudnikov V.G., Kuprijanov A.S., Revue Phys. Appl. 23, 1847 (1988).

11. Leung K.N., Ehlers K.W. et al., Rev. Sci. Instr., 59, 453 (1988).

12. Yuan D.H., Baartman R. et al., in Proc. of the 4th Int. Symp. on the Prod. and Neutr. of Neg. Ions and Beams (Brookhaven, 1986), p. 346.

STEADY-STATE PRODUCTION OF H⁻ IONS
BY REFLEX-TYPE Cs FREE ION SOURCE

P.M. Golovinsky, V.P. Goretsky, A.N. Mosijuk, I.A. Soloshenko
A.F. Tarasenko, A.I. Tschedrin
Institute of Physics of Ukr. SSR Academy of Science, Kiev, USSR

ABSTRACT

As early as the 1960's, an ion source based on reflex-type discharge was established which was expected to be highly effective in producing negative ions [1,2]. From the first, considerable discrepancies were observed between the source emissive abilities and numerical estimates of the latter; however, in obtaining such estimates, nothing but the reaction of electron dissociative atachment to non-excited H_2 molecules was taken into consideration. A further attempt to explain the discrepancy assumed that all of the H⁻ ions generated in the discharge column are collected towards the emission slit [2].

In this paper, we present the theoretical and experimental results concerning the source, and discuss the processes of H⁻ production and loss in the light of our most up-to-date knowledge [3,4,5].

EXPERIMENTAL

The sketch of the ion source is presented in Fig. 1. The tungsten cathode (1), heated by 200 A AC, supplies electrons into discharge space. The discharge is radially limited by means of a tantalum diaphragm (2) positioned in front of anode chamber (3). Target-cathode (7) with its potential close to that of cathode acts as a reflector of gas-ionizing electrons. Several of equidistant holes (5) intended to feed working gas into anode chamber. The uniform magnetic field of 2 kGs is oriented axially along the system. The ions are extracted through the emission slit by the electric field of the extractor (4). Mobile anode diaphrams allow adjustment to the gap between the discharge column and anode chamber. The best results obtained correspond to 5 mm and 2.5 mm internal diameters of the anode chamber and diaphragm, respectively.

Figures 2 and 3 show the main emissive characteristics of the source. In Fig. 2, the extracted H⁻ ion current is plotted against the arc discharge current, the pressure of gas in the source chamber being a parameter. As can be seen, the current of extracted ions is at a maximum at a discharge current I_d, ranging from 5 A to 7 A, this effect being independent of gas pressure. The dependences of I_- on gas pressure inside the source with emission slit width as a parameter is given on Fig. 3, where the value of the discharge current is kept optimal ($I_d \approx 6$ A). The curves shows that an increase in slit width leads to the same H⁻ current, while an increase of pressure makes the H⁻ current increase, rising to its maximum and then decreasing. The fact that the H⁻ current increases with widening of the emission slit proves that the effect of ion collection suggested by Gabovich et al. [2] does not actually take place. However, to clear up the issue, several experiments were carried out in which a source with two opposite slits and extracting voltages independently applied to respective extractors were used. Had the effect of H⁻ collection been essential, the extraction of a current through one slit would have resulted into decrease of the current extracted though the other one. Experimentally, this was not observed; H⁻ currents driven from the slits were absolutely independent. This fact finally verifies that all the H⁻ ions generated in the discharge are not collected towards the emission slit.

To analyze the processes occurring in the source in more detail, and to compare the experimental data with the numerical results (the latter will be given below), some plasma characteristics were measured with Langmuir probes sited in the space between the discharge column and anode wall. Figure 4 demonstrates the results of the measurements. As can be seen, peripheral plasma density rises as gas pressure increases while electron temperature falls off, the respective values being $n \simeq 3 \cdot 10^{13}$ cm⁻³, and $T_e = 1.5 - 1.2$ eV in the pressure range where H⁻ emission has its maximum. (It should be noted that a 1 mm-long probe was used, and the values given above are to be considered as average ones referred to 1 mm radial spacing). In this case, the difference between the plasma and anode potentials was about +4 V, thus being in accordance with the results of [6]. Plasma density in the central point of the column was about 10^{14} cm⁻³ proceeding from the magnitude of the target-cathode current. We established that the latter increases linearly as the discharge current rises, and decreases if gas pressure increases. Plasma composition was determined by means of magnetic separation of positive ions after their extraction from the source. Corresponding data referred to different pressures and discharge currents are collected in Table 1 (the data has been obtained with 0.7 x 10 mm emission slit).

TABLE I

P	I_d	I_{H^+}	$I_{H_2^+}$	$I_{H_3^+}$	I^+
Torr	A	mA - %	mA - %	mA - %	mA
5.0 10⁻²	8	3.75 - 89.3%	0.40 - 9.5%	0.05 - 1.2%	4.2
1.1 10⁻¹	8	3.44 - 81.9%	0.56 - 13.3%	0.20 - 4.8%	4.2
1.6 10⁻¹	8	3.10 - 73.8%	0.80 - 19.0%	0.30 - 7.2%	4.2
1.1 10⁻¹	6	2.70 - 77.0%	0.43 - 12.3%	0.32 - 9.2%	3.5
1.1 10⁻¹	3	1.60 - 65.0%	0.46 - 19.0%	0.40 - 16.0%	2.46

We note that within the low-pressure range a rotational instability tends to build up in the discharge column. This phenomenon has a strong influence upon transport processes in the plasma as well as on several characteristics of the extracted beam, such as the depth of the H⁻ current modulation and the effective phase volume. The increase of gas pressure results in suppression of the instability, so that the modulation depth decreases and does not exceed 1 per cent in the vicinity of optimal pressure. The reduced beam emittance has a minimum at $P \simeq 10^{-1}$ Torr, being equal to $3 \cdot 10^{-5}$ cm rad and 10^{-5} cm rad across and along magnetic field respectively, which corresponds to an ion "temperature" or about $T \simeq 0.8$ eV in the region of the emission slit.

Thus, the best characteristics of H⁻ ion beams extracted from this source are observed at a comparatively high pressure ($P \simeq 10^{-1}$ Torr). Particularly, the density of the H⁻ current reaches to $j_- = 80$ mA/cm², the modulation depth of H⁻ current does not exceed 1 per cent, the ion "temperature" in the emission region is close to $T \simeq 0.8$ eV, and the ratio for the H⁻ current density to the electron current density equals 0.2. The results of experiments show that the total value of the H⁻ current depends only upon the area of the emission slit, provided that the discharge conditions are kept constant. Consequently, a further increase in ion current can be achieved by an expansion of the discharge volume, and hence, by elevating the pumping speed. Extrapolating from a pumping speed of $S = 3,500$ liters/s that was required for $I_- = 40$ mA H⁻ current in our experiments, we conclude that for an ion source intended to give $I_- \sim 1$ A, a pumping speed of $S \simeq 10^5$ liters/s would be necessary.

NUMERICAL SIMULATION

The kinetic processes taken into account for numerical simulations of negative H⁻ production in hydrogen plasma are given in Table II. The analysis showed that the dissociative attachment to vibrationally excited molecules $H_2(v)$, i.e.,

$$e + H_2(v) \rightarrow H + H^-, \qquad (1)$$

plays the role of the main production channel for negative ions, in accordance with [7]. Volume negative ions losses are controlled primarily by the following processes:

$$e + H^- \rightarrow 2e + H, \qquad (2)$$

$$H^- + H_2^+ \rightarrow H_2 + H, \qquad (3)$$

$$H^- + H^+ \rightarrow 2H, \qquad (4)$$

$$H^- + H \rightarrow H_2 + e, \qquad (5)$$

$$H^- + H_2 \rightarrow H_2 + H + e. \qquad (6)$$

TABLE II

N	Kinetic Processes	
1	$e + H_2(v=0,..2) \to H_2^+ + 2e$	*
2	$e + H \to H^+ + 2e$	*
3	$e + H_2 \to H_2(v=1,3) + e$	*
4	$e + H_2(v=1,3) \to H_2 + e$	*
5	$e + H_2 \to e + H_2^* (B^1\Sigma_u^+, C^1\Pi_u) \to$ $\to e + H_2 (v=1,14) + h\omega$	*
6	$e + H_2 (v=1,14) \to H^- + H$	**
7	$e + H_2 \to H + H + e$	*
8	$e + H_2^+ \to H + H$	*
9	$e + H^- \to H + e + e$	*
10	$H^- + H_2^+ \to H + H$	**
11	$H^- + H_2 \to H + H_2 + e$	**
12	$H^- + H \to H_2 + e$	**
13	$H^- + H^+ \to H + H$	**
14	$H + H + wall \to H_2$	**
15	$H + H_2 (v=1,9) \to H + H_2$	**
16	$H_2(v=1,14) + wall \to H_2$	**

* Respective rate constants derived from kinetic equation.
** Respective rate constants taken from [7 - 9].

Reaction (of Eq. 2) dominates in the plasma column region, its rate constant depending strongly upon electron temperature. But, in practice, it does not contribute in the region far from the column where electron temperature approaches the detachment threshold energy of the electrons.

H_2 molecules are vibrationally excited on the $v > 5$ level through intermediate electron-excited states H_2^* according to the reaction chain:

$$e + H_2(v = 0) \rightarrow e + H_2^* \; (B^1 \Sigma_u^+, \; C^1 \Pi_u) \rightarrow e + H_2(v') + \hbar\omega. \quad (7)$$

Vibrational kinetics are essentially affected by the processes of dissociative attachment (1), vibrational excitation wall quenching (8), and by scattering on atomic hydrogen (9):

$$H_2(v) + \text{wall} \rightarrow H_2(v'), \quad (8)$$

$$H_2(v) + H \rightarrow H_2(v') + H. \quad (9)$$

To calculate the electron rates constants, the Boltzmann equation describing electron energy distribution function, f_o, was used:

$$\frac{1}{n_e N} \left(\frac{m}{2e}\right)^{1/2} E^{1/2} \frac{\partial (n_e f)}{\partial t} - \frac{\partial}{\partial E} \left[2 \sum_i \frac{m}{M_i} \frac{N_i}{N} Q_{iT} E^2 \left(f_o + T \frac{\partial f_o}{\partial E}\right) \right] =$$

$$= S_{eN} + S_{ee} + A(E) + L(E), \quad (10)$$

where E is electron energy, eV; T is gas temperature, eV; $e = 1.602 \; 10^{-12}$ erg/eV; N_i and Q_{iT} being molecular and atomic densities and their respective transport cross-sections. S_{eN} and S_{ee} are symbols for collisions integrals describing electron-neutral inelastic collisions and e - e scattering respectively, A(E) is the ionization term incorporation the source of primary electrons. The term L is intended to describe the escape of electrons from the column by drift and diffusion.

The shape of the electron energy distribution function is mainly defined by the following inelastic processes:

$$e + H_2(v = 0) \rightarrow e + H_2(1,2,3),$$

$$e + H_2 \rightarrow e + H_2(B^3\Sigma_u^+),$$

$$e + H_2 \rightarrow e + H_2(B^1\Sigma_u^+, C^1 \Pi_u), \quad (11)$$

$$e + H_2 \rightarrow 2e + H_2^+,$$

$$e + H_2^+ \rightarrow 2H.$$

Taking into account inelastic electron-molecule interaction the collision integral was chosen as:

$$S_{eN} = \sum_j \frac{N_j}{N} [(E + E_j) f_o (E + E_j) - EQ_j (E) f_o (E)]$$

$$- EQ_2 \frac{N_{H_2^+}}{N} f_o (E). \qquad (12)$$

while Q_j and Q_2 denote the cross-sections of H_2 molecule excitation with E_j quantum and dissociative recombination, respectively.

The term $A(E)$ was introduced to describe gas ionization by cathode electron beam and cascade electrons:

$$A(E) = \frac{\int_{2E+E_i}^{\infty} E' f_o (E') q_i (E',E) dE' + \int_{E+E_i}^{2E+E_i} E' f_o (E') q_i (E', E' - E_i - E) dE' - E f_o (E) \int_o^{\frac{E-E_i}{2}} q_i (E,E') dE' + S_n \exp\left[\frac{E - E_n}{\Delta}\right]^2}{n_e N \Delta \cdot 1.77 \sqrt{2e/m}} \qquad (13)$$

where E_n, Δ, q_i and S_n are, respectively, the cathode electron beam energy (discharge voltage), FWHM of beam energy spread, the ionization differential cross-section, and the electron beam fluence into the unit volume of discharge column.

Numerical simulation of the processes in the source was performed taking into consideration both the reaction channels of Table II, and the transport equations. It was shown that the volume of the discharge chamber can be physically divided, as first approximation, into two regions which are the discharge column and the column - anode interspace. For practical purposes the plasma parameters out of the column are defined only by the kinetic processes inside the column, the reverse action being less essential. The production of highly excited H_2 levels occurs primarily in the region of the column by high-energy ionization cascade electrons. The production rate of negative ions (i.e., the difference between the numbers originating and their loss) runs to a maximum in the column-anode interspace. This is due to a strong dependence of the rate of detachment (2) on the drop in electron temperature towards the anode wall. Considering the direct vibrational excitation of H_2 to be the main channel of electron energy losses between the column and anode, it is possible to evaluate the radial drop in the electron temperature from equation:

$$\frac{3}{2} v_d \frac{dT}{dx} = E_j \sqrt{\frac{2e}{m}} \, N \int_0^\infty f_o(E) \, Q_j(E) \, dE \qquad (14)$$

where E_j denotes vibrational quantum energy and v_d is the velocity of ambipolar drift.

NUMERICAL RESULTS

Figure 5 shows the shape of electron energy distribution function (EEDF) inside the column. High electron density leads to strong e - e scattering, making EEDF close to maxwellian up to $E \simeq 20$ eV. The high-energy tail of EEDF appears to be smoother than in [7] by virtue of the same causes. The relative percentage of energy losses inside the column for processes (11) is represented in Table III. As stated above, low levels of H_2 vibrational levels is the mechanism primarily responsible for the electrons cooling in the out-of-column region.

The characteristics of hydrogen plasma in the direct current mode are related to the initial density of hydrogen and the discharge gap voltage in a complicated manner. (See Table IV, Fig. 6).

TABLE III

Process	Losses, %
H_2 Ionization	36
Excitation of $H_2(B^1\Sigma_u^+, C^1\Pi_u)$ Levels	24
H_2 Dissociation	19
Dissociative Recombination	12
H_2 Vibrational Excitation Levels ($v = 1 - 3$)	7
Gas Heating	2

TABLE IV

P_o^*, Torr	0.03	0.10	0.20
E_n, eV	120.00	110.00	80.00
n_e, 10^{14} cm^{-3}	7.50	3.70	2.10
T_e, eV	5.00	3.30	2.40
H_2, 10^{15} cm^{-3}	0.23	1.56	4.08
H, 10^{15} cm^{-3}	0.88	1.41	1.34
H^+, 10^{14} cm^{-3}	7.20	2.90	1.10
$H_2^+ + H_3^+$, 10^{14} cm^{-3}	0.32	0.76	1.00
H^-, 10^{11} cm^{-3}	0.36	1.04	2.50

P_o^* - Initial pressure inside the chamber.

The increase of H_2 pressure results in an increase in H^- ion density and a decrease of plasma density inside the column. The first effect is connected with an acceleration of the rate of process (1) due to a drop in electron temperature, the second one being caused by an intensification of dissociative recombination. Vibrationally excited levels occupation by $H_2(v)$ molecules shows a weak dependence on gas pressure (see Fig. 6) because of the inversely proportional relationship between the number of high-energy electrons and N_{H_2}, so that the total excitation rate $\sum_i k_i N_{H_2}$ remains nearly constant.

As mentioned above, the density of negative ions has its maximum in the column-anode wall interspace (see Fig. 7), in which region electron temperature is reduced and, hence, reaction (2) is retarded. As P = 0.1 Torr, I = 8 A near-anode H^- current density approaches closely 50 mA/cm.

CONCLUSION

The experimental and theoretical results described above concerning the source are mutually consistent, both qualitively and quantitively. That is why the experimental and numerically simulated data given in Tables and Figs. 1 - 7 were chosen as mutually complementary.

The following points should be considered separately. An increase in H_2 pressure leads to a drop in electron temperature, which promotes dissociative attachment and causes an abrupt retardation of detachment rate owing to process(2). As H_2 pressure rises, both the negative-ion density in the column region and the density of the near-anode H^- current, I_{H^-}, increase up to a pressure of $P \simeq 0.2$ Torr, I_{H^-} running into 80 - 100 mA/cm. Due to a steep increase in electron cooling and slowing down of the detachment process out of the discharge column, this is the region where H^- density runs into its maximum.

In the pressure range of P - 0.03 - 0.2 Torr, occupation of vibrationally excited levels by $H_2(v)$ molecules varies only weakly, if at all.

REFERENCES

1. Ehlers, K.W., Nucl. Instrum. and Methods, 1965, v. 32, No. 1, pp. 309 - 316.

2. Gabovich, M.D., Naida A.P., Isayev, Ph. M., Ukr, Phys. Zhu. (USSR), 1970, v. 15, No. 4.

3. Allan M., Wong, J.F., Phys. Rev. Lett., 1978, v. 41, No. 26, pp. 1791 - 1794.

4. Kuchinsky, V.V., Mishakov, V.G., Tibilov, A.S., Shukhtin, A.M. Optica i Spektroskopiya (USSR), 1975, v. 39, No. 6, p. 1043 - 1048.

5. Leung, K.N., Ehlers, L.W. Bacal, M. Rev. Sci. Instrum., 1983, v. 54, No. 1, pp. 56 - 61.

6. Jimbo, K., Ehlers, K.W., Leung, K.N., Ryle, R.V., Nucl. Instrum. and Methods, v. A248, Nos. 2,3, pp. 282 - 286.

7. Gorse, C., Capitelli, M., Bacal, M., Bretagne, J., Ladona, A., Chem. Phys., 1987, v. 117, p. 177 - 195.

8. Wadehra, J.M., Appl. Lett., 1979, v. 35, No. 12, pp. 917 - 919.

9. Abroyan, M.A., Golubev, V.P., Komarov, B.L., Chnmyakin, G.V., Negative Ion Sources, preprint OD-4, Leningrad, NIIEFA im. Efremova, D.V., 1974, p. 153.

Steady-State Production of H⁻ Ions

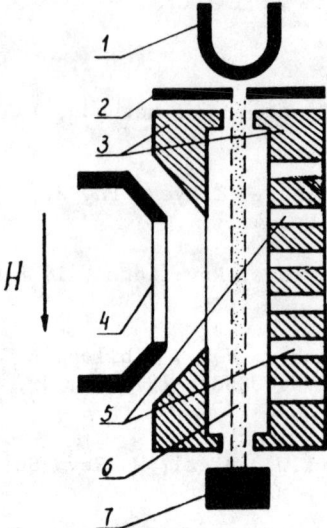

Fig. 1 Sketch of ion source. 1 - directly heated cathode; 2 - tantalum diagram; 3 - anode chamber; 4 - extractor; 5 - gas feed channels; 6 - plasma column; 7 - target cathode.

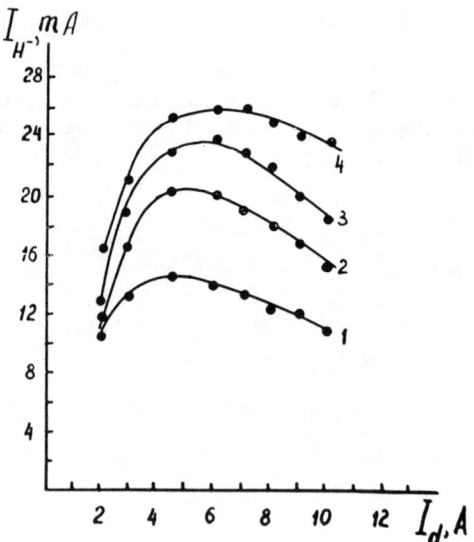

Fig. 2 Extracted H⁻ current against discharge current. Extracting voltage U_o = 14 kV, emission slit of 1 x 40 mm², gas pressure inside the source: 1 - P = 6.5 10^{-2} Torr; 2 - 8.7 10^{-2} Torr; 3 - 1.1 10^{-1} Torr; 4 - 1.4 10^{-1} Torr.

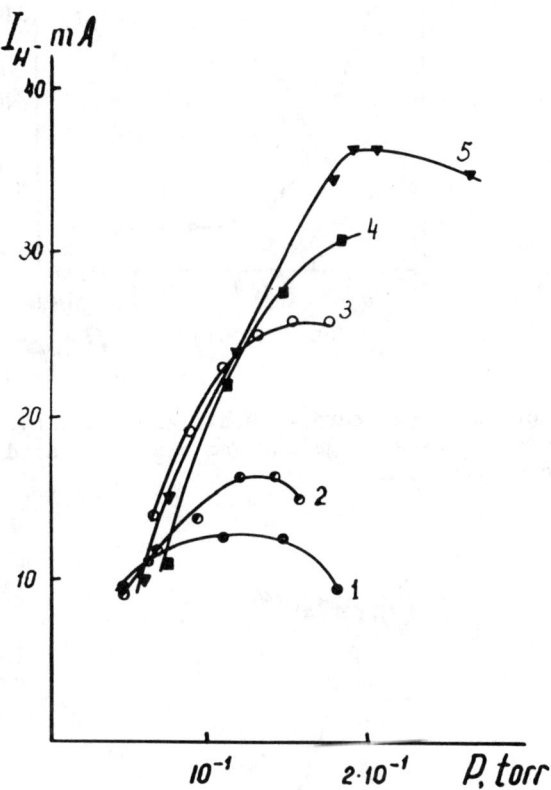

Fig. 3 Extracted H⁻ current against gas pressure. Extracting voltage U_o = 14 kV, optimal discharge current, emission slit 40 mm long; emission slit width: 1 - Δ = 0.6 mm; 2 - 0.7 mm; 3 - 1.0 mm; 4 - 1.2 mm; 5 - 1.5 mm.

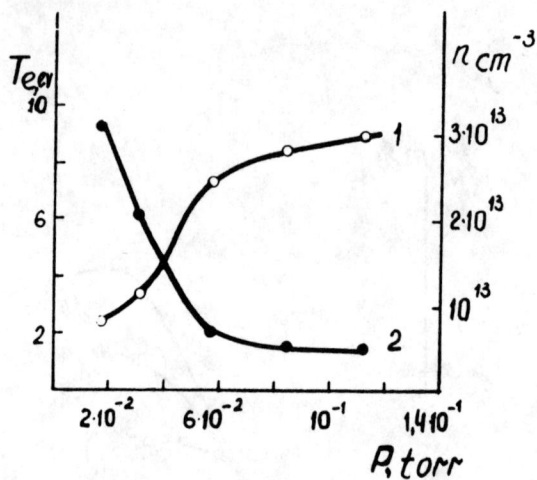

Fig. 4 Electron temperature T_e and plasma density n_e in column - anode interspace against gas pressure inside the source. Arc current 6 A.

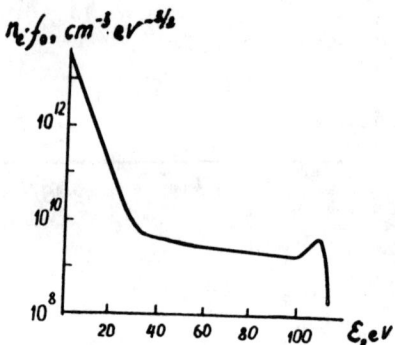

Fig. 5 Electron energy distribution function inside the column. P = 0.1 Torr; I_d = 10 A; E_n = 110 eV.

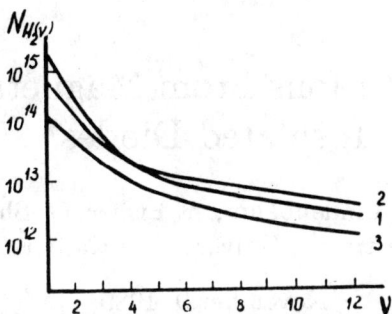

Fig. 6 Vibrationally excited levels occupation for H_2 (v) molecules. I_d = 10 A: 1 - P = 3 10^{-2} Torr, E_n = 120 eV; 2 - 0.1 Torr, 110 eV; 3 - 0.2 Torr, 80 eV.

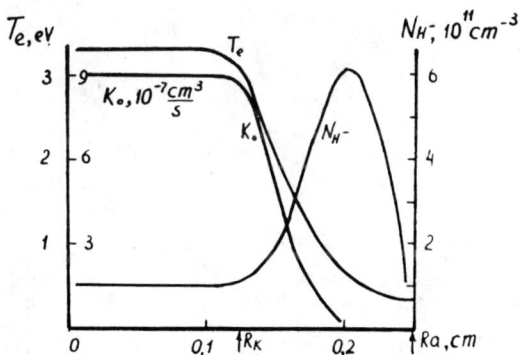

Fig. 7 Radial H^- distribution. T_e - electron temperature, K_o - rate of process (2). I_d = 10 A, P = 0.1 Torr, E_n = 110 eV.

Negative Ions From Magnetically Insulated Diodes*

R. Prohaska, H. Lindenbaum, A. Fisher, G. Sheperd and N. Rostoker
Physics Department, University of California, Irvine, CA 92717

November 9, 1989

Abstract

For the past four years we have studied the production of negative hydrogen ions in magnetically insulated diodes. We have tried to replicate Russian results in annular and racetrack diodes using passive cathodes. Optically triggered active cathodes have been developed. We have studied negative ion producion using either self contained plasma sources or neutral gas puffs with several different gases.

---

*This work was supported by ONR/SDIO

1 INTRODUCTION

The production of neutral particle beams with energy over 100 kV requires starting with weakly bound negative ions which can be accelerated and then stripped to neutrals. Many workers [1] have studied the problem of producing negative ions, usually H^-, in steady state plasmas. We are investigating the use of transient plasmas and gas puffs in conjunction with magnetically insulated ion diodes to explore the possibility of obtaining high (kiloampere) currents of negative ions.

In plasmas H^- is produced most commonly by dissociative attachment in which a slow electron collides with and sticks to a molecule. The molecular ion then breaks into two or more fragments including the desired negative ion. Collisions between negative ions and the seed neutrals from which they are produced naturally eliminate negative ions and the current density obtained is always limited by the relative cross sections for attachment and detatchment.

The probability of dissociative attachment depends strongly on the energy of the collision and is usually expressed as a cross section which has been tabulated [2] for a variety of (simple) molecules. Generally the optimum energy is not more than a few eV. Relativly little attention has been paid to polyatomic molecules of the kind which yield the best results in pulsed diodes.

The dissociative attachment cross section is on the order of 2×10^{-20} for ground state H_2. The ionization cross section is a few orders of magnitude larger. The current density of steady state gas discharge (Penning) sources can be up to 1 A/cm^2 but the total area is small, usually less than 1cm^2. Surface plasma sources, in which negative ions are formed in contact with a heated cesium loaded surface, give on the order of .1 A/cm^2 with a total current of a few amperes. Vibrationally excited hydrogen is reported [3,4] to have several orders of magnitude larger cross secion but the highly collisional conditions needed to cause the excitation result in prompt destruction of any negative ions formed.

The introduction of large density gradients in the cathode plasma minimises the problem of destructive collisions. Usually the density gradients are produced by allowing a dielectric surface to flash over evaporating some of the dielectric into a thin, dense layer of neutrals and plasma. What negative ions form then have a reasonable chance of escaping through the cathode plasma.

2 PASSIVE CATHODES

Polyethelene is the most popular dielectric surface for making proton beams and is the best precursor for making negative ions. Carbon contamination is always a problem. Carbon forms stable negative ions with more binding energy than H^-; we used thin foils to stop the carbon selectivly. A serious complaint is the difficulty in obtaining a uniform cathode plasma which is a prerequisite to obtaining a well collimated beam. Normally a dielectric surface breaks down at a few points on the surface producing wildly diverging beamlets. Drilling arrays of holes in the dielectric surface tends to induce breakdown at each hole but large (megavolt) cathode potentials are required. The result is an array of hot spots which tends to produce a divergent beam.

Our work was inspired by Kolomensky's report [5] of more than 100 A/cm^2 negative hydrogen ion current from passive polyethelene cathodes in a cylindrical diode. We tried without success to confirm the report.

Prompt diagnostics consisted of a pair of Faraday cups biased to collect secondary electrons one of which was covered with a 2 μ mylar foil to strip H^- to H^+ and stop any high Z ions like C^-. The Faraday cups were carefully shielded and we could see currents down to about 1-2 mA/cm^2. Upon getting an ion signature (shown in figure 1 for the foil covered Faraday cup) we installed CR-39 track recording plastic either open (to count particle tracks) or in a pinhole camera to study cathode plasma uniformity and beam divergence. A single layer of 2 μ mylar protected the CR-39 from C^-. A fiber optic and photomultiplier system measured the amount and timing of light emission from the cathode. Cathode voltage for ion extraction was provided by a 700kV, 7 ohm 50 ns pulseline machine with the usual voltage, current and insulation field monitors.

The first experiments were basically a copy of the cylindrical diode experiments done at the Lebedev Institute. There was one important difference: Our machine had a negative prepulse of only about one percent, occurring shortly (250 ns) before the output pulse. The results were null: no H^- was seen.

Upon learning of the huge bipolar prepulse (15-30 percent, 1-2 μs) on the machine used for the early work we attempted to increase ours by removing the prepulse suppression resistor. This had the effect of increasing the prepulse to about 10 percent, but the timing was unchanged and the polarity was negative, in contrast to the positive and negative swing seen on the Russian machine. We observed about 10 A/cm^2 from cathode hot spots, with 1-2 A/cm^2 if we average over the entire cathode. The reproducibility was poor and the divergence varied from 100-300 mr.

These results fell far short of the promise of the initial report, but were not surprising given that we had a much smaller prepulse which occurred only a few hundred nanoseconds before the main pulse. In our experiments, the Faraday cup signals were in the form of short pulses, usually not more than 10 ns in duration at the start of the diode voltage pulse, implying that we were extracting an accumulated population of ions. The Russian workers used nuclear activation methods to monitor ion production, so the time dependence of the ion flow is not known. It seemed likely that we needed to form a cathode plasma, and then wait for an appreciable time (some microseconds) to give negative ions time to form before applying extraction voltage.

3 ACTIVE CATHODES

3.1 PLASMA FLASHBOARDS

It was decided to develop an active (self powered, indpendently triggered) cathode [6] based on plasma sources we developed for other applications. That gave us independent control over the amount of energy invested in the cathode plasma and the timing of its creation.

If a spark is drawn between titanium hydride electrodes a relatively clean hydrogen plasma results, but the usual method is to apply tens of kilovolts to a single electrode which breaks down to form a point source of plasma. We have developed a flashboard plasma source capable of generating surface densities on the order of 10^{16} over 200 cm^2.

Titanium hydride powder is somewhat conductive, and if it is attached to an insulating surface with sodium silicate binder the resulting paint breaks down between grains at the application of only a few kilovolts. The discharge can be made uniform over an area of more than one square centimeter and the low voltage makes it easy to subdivide the flashboard coating into small cells with ballast resistors. The flashboard assembly for the annular diode is shown in figure 2.

The entire system consists of a copperclad circuit board which forms the substrate of the flashboard, a capacitor discharge circuit to deliver the needed energy and an optical trigger circuit (figure 3) to synchronize the flashboard to the beam generator. The flashboard is mounted on the face of the cathode, while the rest of the circuitry resides in the cathode shank. This arrangement was necessitated by the design of the our beam generator, which uses an interface too thin to house isolation inductors or other electrical penetrations. For this reason it was decided to make the active cathode completely self contained and electrically isolated.

A capacitor was charged to 2.7 kV by a DC to DC converter powered by NiCd batteries. A small ceramic metal spark gap switched the capacitor into an isolation transformer, which then drove the individual cells of the flashboard through ballast resistors. The spark gap was triggered by a krytron triggered by an SCR triggered by a FET triggered by a photodiode. The delay from photodiode to flashboard was only a few microseconds, with about half a microsecond jitter.

Two modes of operation were used. In the high current mode the storage capacitor was $2\mu F$ and the discharge current was 3.2 kA. In the low current mode a $.25\mu F$ capacitor was used, with a total discharge current of about 800 A.

The plasma production characteristics were studied using Langmuir probes and light output to determine the temporal and spatial evolution of the plasma. Measurements were made using a double probe in the racetrack diode configuration without insulation or accelerating fields. The plasma had a density of $3 to 13 \times 10^{11}/cm^2$ at a distance of 4-7 cm and an expansion velocity of 8 cm/μs. Application of a .3 T field effected complete confinement to the resolution of the probe (1 cm). Open shutter photographs demonstrated uniform breakdown of each section as shown in figure 4.

The plasma temperature appeared to be somewhere between 1 and 3 eV. The binding energy for H^- is only .6 eV while the ionization energy for H^+ is 13.6 eV. Obviously H^- is not a favored state, but we hoped to find something in the cold end of the distribution.

The active cathode system was tried in both racetrack and annular geometries, and performed about the same in both cases. The best results were 6 A/cm^2 with about 10 mr divergence using the low current mode and 1.5 μs delay between the onset of plasma light and application of accelerating voltage. An illustrative Faraday cup trace is shown in figure 5.

3.2 GAS PUFF CATHODES

Density gradients can also be made using a fast valve to create a gas puff with a risetime short compared to the gas transit time across the anode-cathode gap. The idea was to construct a cathode plasma in which dense cold gas expands through a region occupied by a swarm of trapped, slow electrons in the presence of a strong electric field. The applied field removes the ions as they form, before destructive collisions can occur. It is well known that fiber cathodes ignite uniformly at low fields, and we have incorporated carbon fibers as electron emitters into our ion diode designs. In its present form the gas is emitted from small apertures surrounding carbon fiber bundles which serve as cathodes. Figure 6 gives a sectional view of the cathode assembly and puff valve. Figure 7 shown the electrical circuit which operates the valve.

The choice of gas presents some difficult compromises. Hydrogen would be the cleanest source, but the dissociative attachment cross section in cold gas is only 2×10^{-20}. It has been reported that the dissociative attachment cross section is several orders of magnitude larger for hydrogen which is in a high state of vibrational excitation. The achievement of high vibrational excitation has been attempted [7] by passing the puffed gas through an arc at a few atmoshperes, but it must then be allowed to expand freely: Wall collisions will deexcite it, and unless the density is lowered H$^-$ has no chance for survival.

We made an attempt to measure H$^-$ production from cold H$_2$ puffs but could measure no ions and decided to resort to more complex molecules which had larger published cross sections for dissociative attachment.

Ammonia would seem to be a good candidate with a large cross section (57×10^{-19} cm^2) and no contamination (there is no stable N$^-$) but it gave no measurable ion flux. We also tried loading the cathode fibers with borane ammonia, a complex of BH_3-NH_3, in the hope of seeing H^- derived either from the ammonia or borane. The attachment cross section for borane is not known but the published value[2] for silane SiH_4 is enormous; 2.2×10^3. No H^- was detected. We did not attempt to use silane gas because of its reportedly toxic and pyrophoric nature.

We decided to try a hydrocarbon puff to see if a long molecule worked better. It was speculated that large molecules with many internal vibrational modes could make less elastic collisions with electrons and exhibit higher cross section. Unfortunately we could find no published figures for molecules having more than 5 atoms.

Butane was the longest chain hydrocarbon which we could easily use in the valve. It gave about 10-20 mA/cm^2 of H^-. The resulting beam exhibited very good divergence of about 10 mr. The high molecular weight limited the mobility of the gas and we were able to puff in enough gas to load the diode without shorting it.

That butane worked better than ammonia or borane ammonia leads us to suspect that molecular structure is somehow important. The pressure in the gas puff was lower in the case of butane than any other, so we were working with less starting density. The density gradients were probably smaller than in the case of the borane ammonia loaded cathode fibers.

One of the persistent problems with gas puff diodes has been gas breakdown due either to gas reaching the electrical terminals or strong electric fields due to the rapid rise of the magnetic field. The result of any anode plasma formation in the diode is a short. We experimented briefly with a diode in which the anode was in the form of a solid plate of graphite in the hope that it would keep the gas away from high electric field regions and so avert any breakdown. Ions were to be extracted through a small number of ports in the plate. The result was a significant increase in electron leakage and a complete absence of negative ions.

4 SUMMARY

Three approaches to the production of intense H^- beams have been explored. Passive polyethelene cathodes give the highest current density but suffer from divergence and reproducibility problems. Titanium hydride flashboard cathodes give much better divergence and reproducibility with nearly comparable current density. Gas puff cathodes provide good divergence but the lowest current density. We expect that better results will come from other gases.

We suspect that the quality of the magnetic insulation is important for much more than limiting the flow of leakage electrons. Any dissociative attachment process depends on having collision energies of no more than a few eV. If the magnetic insulation is leaky electrons gain energy temporarily as they gyrate through the cathode field. Energy transfer to the electrons of only a few parts per million is sufficient to ensure that collisions will be too violent to permit the formation or survival of negative ions.

CATHODE TYPE	CURRENT DENSITY A/cm^2	COMMENTS
Polyethelene	5	large divergence poor reproducibility
Active titanium hydride	2	Good divergence and reproducibility
Hydrogen puff	<1 mA	
Butane puff	0.015	Good divergence (10 mr) and reproducibility
Borane Ammonia loaded fibers	<2 mA	
Ammonia puff	<2 mA	
Oxygen puff	0.1	Open Faraday cup

References

[1] Production and Neutralization of Negative Ions and Beams. This is a series of symposia held every two years.

[2] S. K. Srivatava, Present Status of the Measured Dissociative Attachment Cross Sections in Proceedings of the Fourth International Converence on the Production and Neutralization of Negative Ions and Beams, Brookhaven, NY, 1986 p.69.

[3] K. N. Leung and W. B. Kunkel, Phys. Rev. Lett. 59, 787 (1987)

[4] J. M. Wadera and J. N. Bardsley, Phys. Rev. Lett. 41, 1795 (1978)

[5] A. A. Kolomensky, A. N. Lebedev et al. Proceeding of the Fifth International Conference on High Power Particle Beams, San Francisco, 1983 p. 533.

[6] Hayim Lindenbaum, Production of Intense Negative Ion Beams In Magnetically Insulated Diodes, PhD Thesis, University of California, Irvine 1988

[7] R. J. Turnbull, S. R. Walther and J. L. Guttman Generation of Vibrationally Excited Hydrogen for use in a Negative Ion Source in Proceedings of the Third International Symposium on the Production and Neutralization of Negative Ions and Beams, Brookhaven, 1983 p.132.

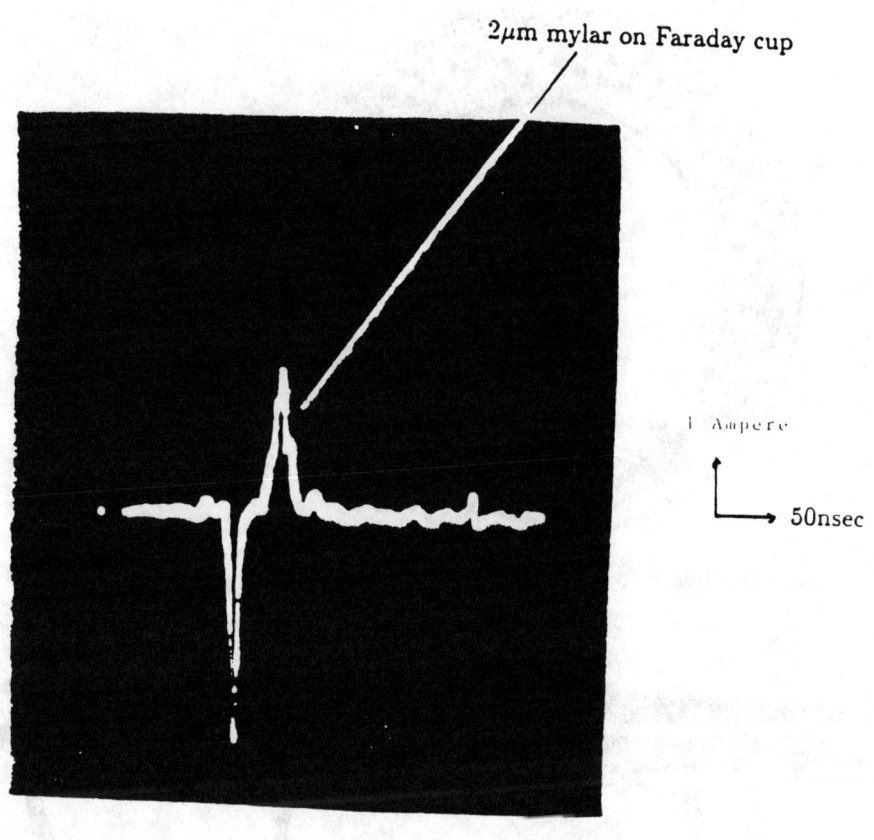

Figure 1: Ion signature on a foil covered Faraday cup from a passive polyethelene cathode

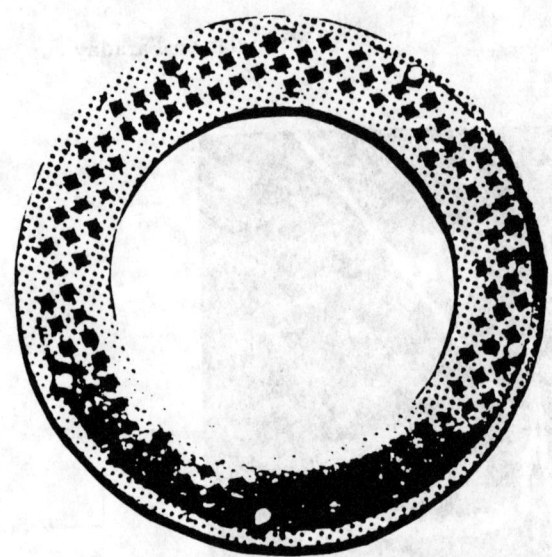

a: Flashboard front wiew.

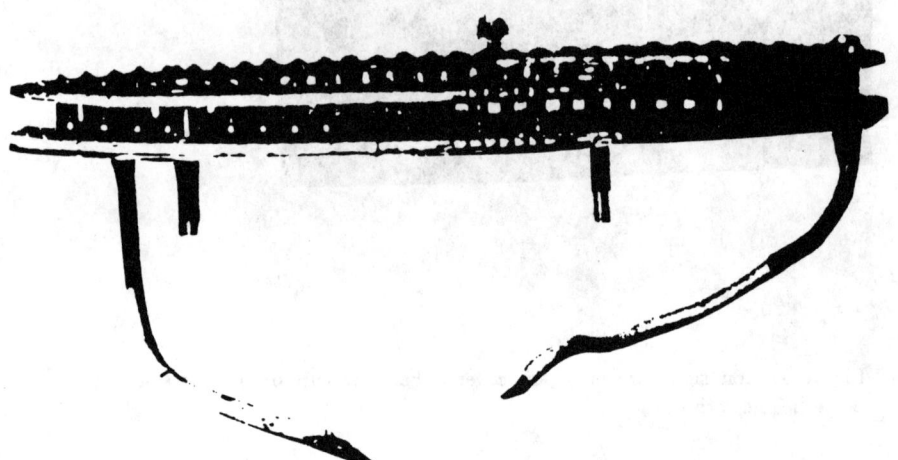

b: Flashboard side view.

Figure 2: Flashboard assembly for the annular diode.

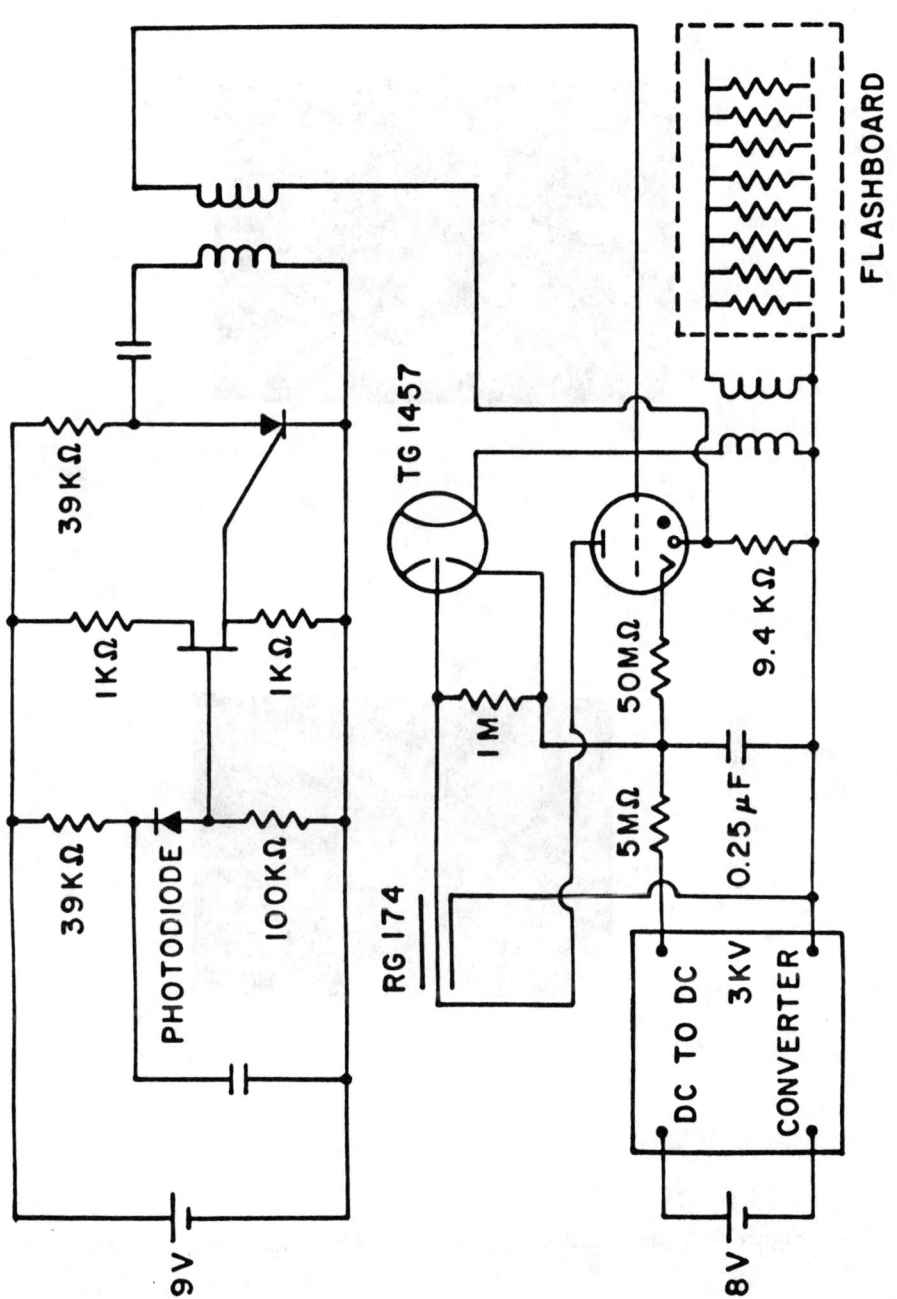

Figure 3: Capacitor discharge and optical trigger circuits for the flashboard cathode.

Figure 4: Open shutter photographs of flashboards in operation.

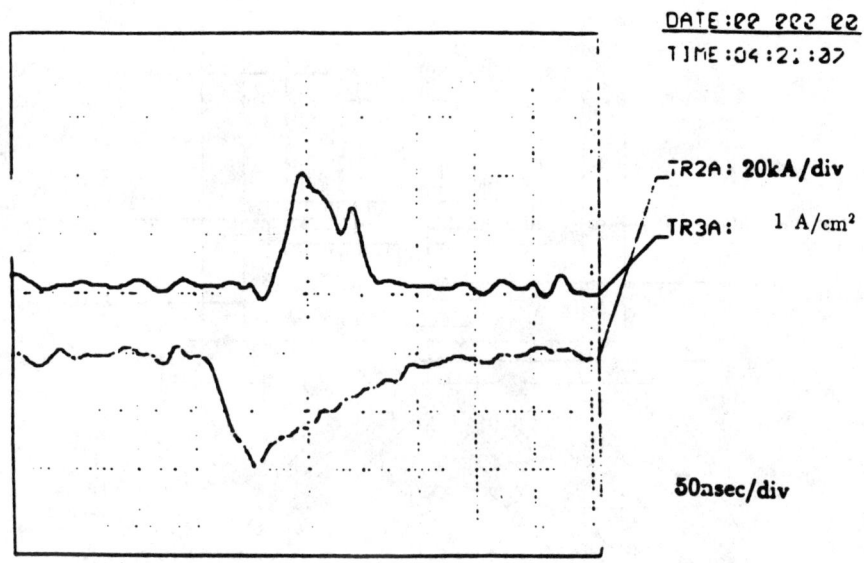

Figure 5: Ion current trace from a flashboard cathode.

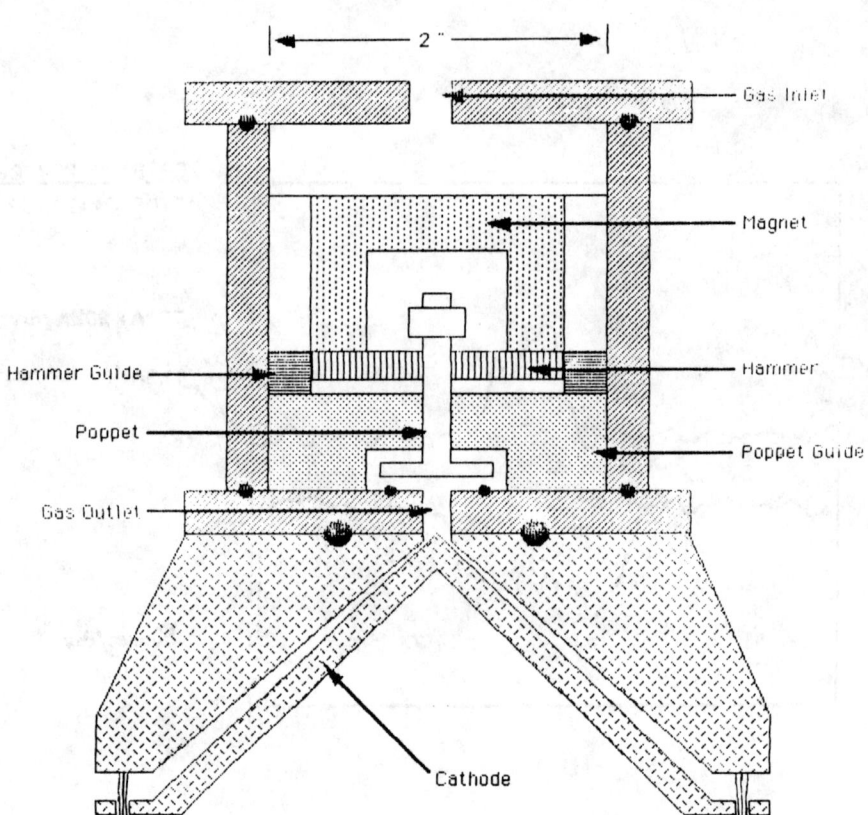

Figure 6: Gas puff cathode and valve.

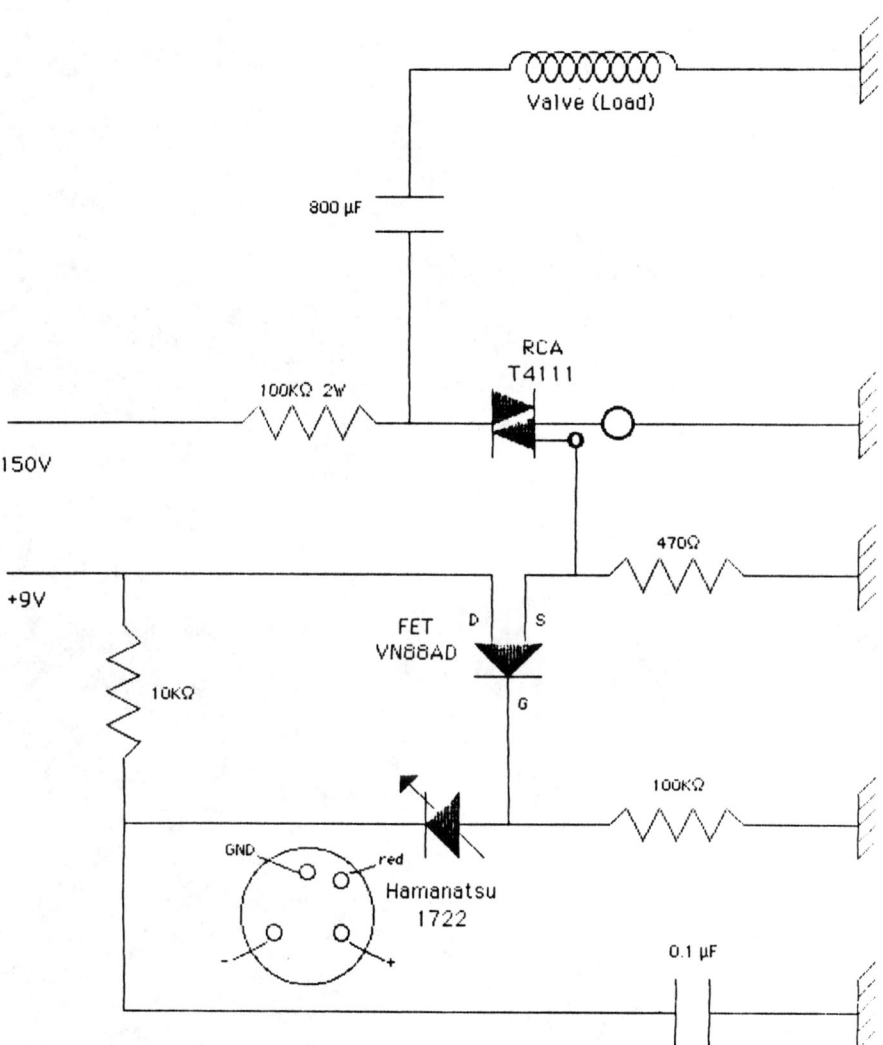

Figure 7: Control circuit for the puff valve.

POLARIZED AND HEAVY
NEGATIVE ION SOURCES

INVESTIGATION OF SPIN-EXCHANGE PROCESSES IN THE OPTICALLY POLARIZED ION SOURCE

A.N.Zelenskii, S.A.Kokhanovskii, V.G.Polushkin,
K.N.Vishnevskii
Institute for Nuclear Research, Moscow, USSR

ABSTRACT

Polarization at spin-exchange collisions of an atomic hydrogen beam with optically pumped sodium vapours has been studied in the optically pumped polarized ions source. The dependence of polarization on beam energy and magnetic field has been measured. At a target thickness of $3 \cdot 10^{14}$ at/cm^2 proton polarization achieves 30-35%.

I. INTRODUCTION

The optically pumped source of polarized ions which was developed in the Moscow Institute for Nuclear Research produces a polarized proton current up to 4 mA and H$^-$ ions current of 0.4 mA with a 65% polarization and the normalized emittance of 0.1 π cm mrad in pulsed mode of operation /1/. Optically pumped sources of polarized H$^-$ ions with the ECR-source of primary protons beams have been developed in KEK, TRIUMF, LAMPF /2, 3/. In these sources the capture of polarized electrons occurs in a high magnetic field of 10-20 kG to decrease the depolarization due to spin-orbital depolarization in excited states. The spin-exchange in a high magnetic field is the main drawback of these sources since the formation of a proton beam in a high magnetic field gives rise to a large increase in angular divergency of the beam and large losses of polarized current. The scheme with a superconducting solenoid and high power RF oscillators makes the source too expensive.

The different original scheme is used in the INR source (see Fig. 1). The proton beam is formed in the external source in a zero magnetic field then it is neutralized in a hydrogen cell and the hydrogen beam is injected into the solenoid where a helium ionizer cell is placed. The cell is actually the proton source inside the magnetic field. Formation of a high brightness primary beam allowed us to produce the record polarized H beam in pulsed mode of operation /1/.

But it is difficult to realize the vacuum pumping of a helium ionizer cell in the case of a high repetition rate or a continuous mode of source operation. Thus, the pick-up polarization technique is very effitient due to a high charge exchange cross-section but requirement of using a high magnetic field leads to limitation of the increase in current and polarization.

II. SPIN-EXCHANGE POLARIZATION

There is the possibility to polarize electrons through spin-exchange collisions of hydrogen atoms with optically pumped alkali atoms. In the case of thermal atoms velocities these processes are well studied and are successively used for optical pumping of different atoms mixtures /4/. Anderson et al./5/ estimated the spin-exchange cross-sections for collisions of fast hydrogen atoms with different alkali targets. The results of calculation are presented in Fig. 2. At beam energy of 2-5 keV, which allows one to get a high current atomic beam with low divergency, the spin-exchange cross-sections can achieve 10^{-15} cm^2. To produce a 95% electron polarization in this case a thickness of optically pumped targets must be $3 \cdot 10^{15}$ atoms/cm^2. A new scheme of the spin-exchange source is very close to that of the INR source (Fig. 1). It is only necessary to remove the helium ionizer cell. The electrons in a hydrogen beam are polarized in spin-exchange collisions, then as usual Sona-transition and ionization in the alkali cell proceed with production of a polarized H$^-$ ions beam.

The spin-exchange polarization technique has many advantages over the electron pick-up scheme.

Pick-up technique	Spin-exchange polarization
1. Pick-up of electrons are most likely in excited states. To prevent depolarization caused by spin-orbital interactions a high magnetic field (10-20 kG) is required.	1. Spin-exchange occurs between ground states of the hydrogen and alkali. The magnetic field as high as 2-4 kG is required for optical pumping of high density vapours.
2. Charge-exchange in a high magnetic field gives rise to emittance growth of the beam. The technique of the ECR source inside the magnetic field has the difficulties of beam formation in a high magnetic field.	2. Spin-exchange occurs between neural hydrogen and neutral alkali atoms. There are no problems with external formation and transport of a high current and high brightness beam.
3. The high current ions beams (which consist of hydrogen and alkali ions) are produced in the charge-exchange processes. The ions collide with the cell's walls and destroy the wall coatings which are used for increasing the spin-relaxation time.	3. In spin-exchange collisions ions are not produced. Thus wall coatings are likely to be used for increasing the spin relaxation time and hence the cell thickness up to $3 \cdot 10^{15}$ at/cm^2.

One can obtain a high proton polarization at a target thickness of $3\text{-}5 \cdot 10^{13}$ at/cm^2 by removing the background caused by neutralization on residual gas. In a spin-exchange technique at least a 10^{15} at/cm^2 thickness of the optical pumping cell is required to produce high proton polarization (higher than 60%).

III. OPTICAL PUMPING OF HIGH DENSITY ($n \geq 10^{13}$ at/cm^3) HIGH THICKNESS ($hl \geq 10^{14}$ at/cm^2) TARGETS

The main factors which determine the limit of a target thickness and polarization are laser power, relaxation time and radiation trapping.

1. Laser power.

The 1W laser power, if properly used, allows one to pump 10^{18} at/s. At present for optical pumping of a sodium cell in cw mode of operation the laser system consisting of a few dye lasers is used, with full power 3-5 W. The laser power in pulsed mode of operation is higher than 100 W. The tunable solid state lasers (alexandrite, titanium-sapphire and others) are used for optical pumping of potassium, rabidium vapours. The cw power of these lasers is higher than 5 W at the 20 W argon laser pumping, which produces 10^{19} at/s.

2. Polarization relaxation time.

The atoms loss polarization in collisions with walls of the cell. These losses depend on the wall material. In the case of stainless steel walls a full depolarization occurs, and relaxation time is equal to atoms time of flight along the diameter of the cell of about 10-20 microseconds. The special dry film coating of the wall increases the relaxation time up to 1-5 ms, the number of bunchers without depolarization achieving 100 ! The dry film is destroyed under the action of alkali vapours. The life time is sufficienly high (about 300 hours). In the OPPIS with the ECR proton source the high intensity sodium ions beam, which was produced in charge-exchange collisions, bombarded the wall and destroyed the wall coating. The life-time of coatings in OPPIS is very short. In spin-exchange collisions ions are not produced, the direct collisions of the well formed hydrogen beam with the cell wall are excluded, so it is reasonable to propose that there will be no strong damage of a dry film in this case. Therefore examination of destruction of different wall coatings in spin-exchange collisions is very important for the spin-exchange source.

3. Radiation trapping.

Radiation trapping imposes limitations on achievable optically pumped target density. These limitations are very strong in the case of a low magnetic field. This was not taken into account in the proposal of "collisional pumping" technique of proton polarization. Actually, it is impossible to pump the vapours of 10^{12} atoms/cm^3 density in the low magnetic field.

The radiation trapping depolarization is much smaller if a magnetic field is sufficiently high for "optical isolation" of transitions with an opposite circular polarization. The results of optical pumping in the presence of radiation trapping are illustrated in Fig.2. This calculation shows that it is possible to optically pump the cell with a (1-3) 10 atoms/cm density for practically available laser power and relaxation time of 100-300 microseconds. The condition of "optical isolation" is fulfilled when Zeeman splitting of opposite helicities transitions is much larger than an efficient linewidth $\Delta \nu$: $\Delta \nu = \Delta \nu_D + \Delta \nu_{SF}$, where $\Delta \nu_D$ is a Doppler broadening and $\Delta \nu_{SF}$ is a width of superfine structure.

For potassium $\Delta \nu_{SF}$ = 770 MGHz and the magnetic field of 2-4 kG is enough to reduce sufficiently the resonance radiation trapping and optically pump the cell with density of 10^{13} atoms/cm^3.

This field does not effect on the beam transportation because both primary and final beams are neutral in the spin-exchange technique. The magnetic field of 5 kG strength increases the relaxation time at collisions with dry wall coating from 0.05 ms up to 0.25 ms as it has been observed at TRIUMF /8/.

IV. MIXTURES OF ALKALI VAPOURS

At optical pumping of sodium mixture with buffer vapours of potassium the spin exchange collisions of optically pumped atoms with buffer atoms produce polarization of the latter ones. The cross-sections of alkali spin-exchange collisions at thermal energies are equal to 10^{-14} cm^2. At a buffer vapours density of 10^{14} at/cm^3 the velocity of motion to the wall decreases and the relaxation time increases at least up to 0.1 ms. The results of optical pumping calculations for sodium-potassium mixture are presented in Fig.3. At laser power of 7W/cm^2 the mixture polarization may be 70% /4/.

In the case of application of the alkali mixture the radiation trapping is not so dangerous because the density of pumped atoms is not too high, and the high full thickness ($\sim 10^{15}$ at/cm^2) is determined by buffer atoms. Furthermore, in this case there are no problems for ionization in the media with a high density of excited states, which was observed in the experiments on optical pumping of high density targets by a high power laser /9/.

The application of alkali mixtures in the spin-exchange source seems very promising. For practical realization the problems of partial pressure stabilization should be solved.

V. EXPERIMENTAL STUDY OF SODIUM-POTASSIUM MIXTURE SPIN-EXCHANGE POLARIZATION IN THE OPTICALLY PUMPED SOURCE

Spin-exchange polarization in the Na-K mixture was experimentally studied at the test-bench of the optically pumped source (see 1, Fig.1). In this experiment potassium vapours were added into the sodium cell from the container with a separate heater and the dependence of polarization on the full vapours thickness was studied.

Polarization was measured by analysing the modulation (M) of a H⁻ ions pulsed current under the action of a pulsed dye laser:

$$M = \frac{I^- - \vec{I^-}}{I^- + \vec{I^-}}$$

Oscillograms of the pulsed current are presented in Fig.4. At the first stage of H^0 production the pick-up of polarized electrons gives rise to hydrogen polarization P(H): $P(H) = \mathcal{E} P_e$, P_e is electron polarization in the Na-K mixture, $\mathcal{E}$ is depolarization factor due to spin-orbital interaction in excited states, for magnetic field of 10 kG, $\mathcal{E} \simeq 0.7$.

The capture probability of the second electrons and hence the probability of H⁻ production depend on P(H) since a single bounded state of a H⁻ ion is $^1S_0$. Thus, the ion current modulation will be proportional to the square of electron polarization: $M \simeq \mathcal{E} P_e^2$. The H ion beam being produced in a high magnetic field has a large divergence and the losses of beam current at transportation are very high. To decrease the current losses the bending magnet was installed at short distance from the solenoid.

Ionization effects under the action of high power laser radiation in the high atoms' density in the cell also result in H current modulation. It was first observed by authors /9/. These effects could be separated since this modulation exists also in a proton current. The results of H⁻ ions yield measurements are presented in Fig.5 as a function of target thickness n(Na) l. The sodium electron polarization is determined from modulation measurements, curve 1 in Fig.6. At a target thickness of 10^{14} atoms/cm² the polarization is equal to 90%. Ionization effects at this density are very small.

In measurements of sodium-potassium mixture polarization this sodium density was fixed $n(Na) \simeq 10^{13}$ at/cm³ and the polarization of mixture was determined in dependence of full target thickness: nl, n = n(Na) + n(K). When potassium density is increased, the polarization drops and then at density higher than 10^{13} at/cm³ it increases due to spin-exchange collisions. The polarization $\sim$ 70% has been measured at a full mixture thickness $7 \cdot 10^{14}$ at/cm².

Stabilization of sodium-potassium components is required for these measurements. The possibility of increasing the cell thickness due to usage of alkali mixtures is extremely useful for application in the spin-exchange source of polarized ions and this work will be continued.

VI. EXPERIMENTAL MEASUREMENTS OF SPIN-EXCHANGE POLARIZATION IN THE OPTICALLY PUMPED H IONS SOURCE.

First spin-exchange polarization measurements were made at the polarized ions source test-bench for conventional geometry with a helium ionizer cell and a sodium cell of 17 cm length (see Fig.1). Polarization was measured by means of a low energy polarimeter. The additional cleaning plates were installed at the entrance of the solenoid for removing ions from the hydrogen beam.

Then the longer length cell (l=35 cm) has been manufactured and installed into the solenoid. The helium ionizer cell was removed. This allowed one to increase the full target thickness up to $(2-3) \cdot 10^{14}$ at/cm^2 at high (80-90%) polarization.

The results of spin-exchange polarization measurements are presented in Fig.7 in dependence of target thickness. The proton polarization of about 30-35% has been measured at hydrogen beam energy of 2 keV, 16% - at 4 keV, 6% - at 8 keV. From these measurements, the cross-section of spin-exchange collisions has been evaluated, at 2 keV $\sigma_{ex} \simeq 1.8 \cdot 10^{-15}$ cm^2. The results of theoretical calculations (see Fig.2) are in a reasonable agreement with the experiment.

The dependence of polarization on magnetic field is presented in Fig.8. The saturation was observed in the field higher than 5 kG, it is in agreement with evaluation of a magnetic field strength which is required for "optical isolation" of transitions (see topic III.4).

CONCLUSION

The spin-exchange collisions in the mixture of sodium and potassium vapours allow one to obtain $7 \cdot 10^{14}$ at/cm^2 in a short length cell of 17 cm, without wall coating increasing relaxation time. The usage of such a target in the source will get more than 10 mA polarized protons and above 1 mA of H$^-$ ions with the emittance acceptable for high energy accelerators.

In the case of monoatomic target it is required to increase the cell length up to 50-100 cm and use the dry film wall coating increasing the relaxation time up to 200-300 microseconds.

REFERENCES

1. A.N. Zelenskii et al., NIM A245 (1986) 223-229
   VII Int. Symp. of High Energy Spin Physics, Protvino (USSR) 1986, p.154-167.
2. Y. Mori et al. NIM 220, p.264, 1984.
3. P. Levy et al., Helvetica Physica Acta, v.59, 1986, p.674.
4. W. Cornelius, Y. Mori, Phys. Rev., A, v.31, p.3718, 1985.
5. D. Swenson et al, Helvetica Physica Acta, v.59, p.662, 1986.
6. D. Swenson et al, Helvetica Physica Acta, v.59, p.652, 1986.
7. D. Tupa, Phys. Rev. A, v.33, p.1045, 1986.

8. P. Levy, J. of Appl. Physics, v.63, p.4819, 1988.
9. A.N. Zelenskii et al., Pis'ma v ZhETF v.44 iss.1, p.21-23, 1986.
10. A.N. Zelenskii et al., Proc. 8-th Int. Symp. of High Energy Spin Physics, AIP Conf. Proc., No 187, p. 1208, 1989.

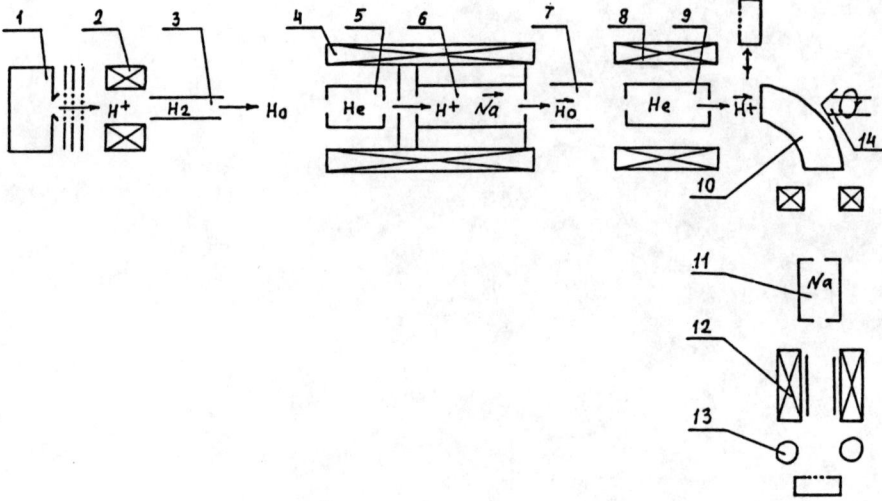

Fig. 1. Schematic lay-out of INR source: 1 - proton source; 2 - focusing lens; 3 - hydrogen neutralizing cell: 4,8 - pulsed solenoids; 5,9 - ionizing helium cells; 6 - optically pumped sodium cell; 10 - bending magnet; 11 - sodium cell of the polarimeter; 12 - spin filter; 13 - detector of H(2S); $CF_{1,2,3}$ - Faraday cylinders.

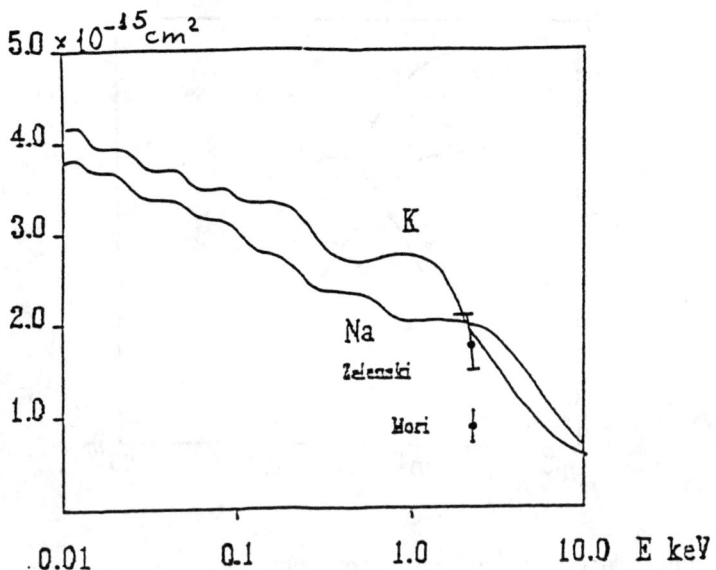

Fig. 2. Calculated spin exchange cross-sections for fast H atoms incident on sodium and potassium target /5/. Experimental results. Y.Mori / AIP Conf. Proc. No 187, p. 1200, 1989 /.

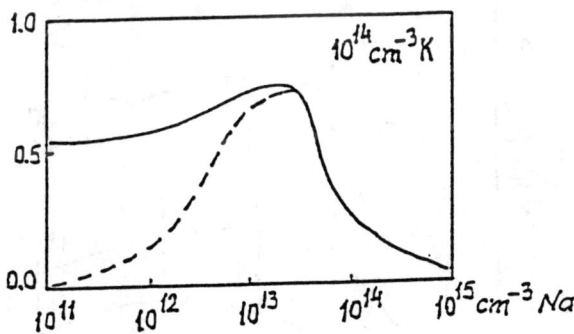

Fig. 3. Calculated polarization of sodium atoms (solid line) and buffer potassium atoms (dashed line) for optical pumping of sodium. The potassium density was fixed at 10^{14} at/cm^3 /4/

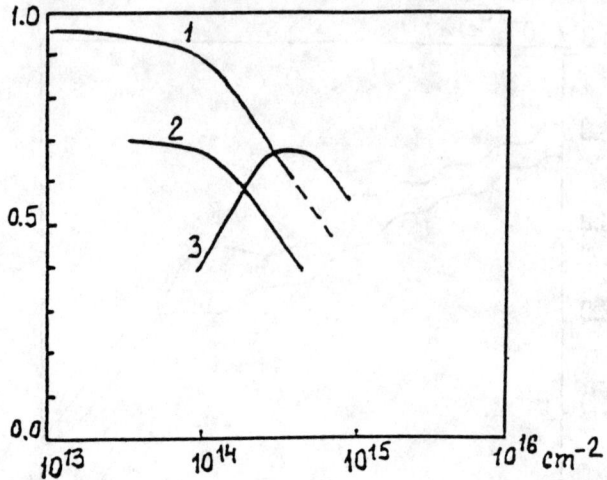

Fig. 6. Dependence of polarization on total target thickness: 1 - sodium electron polarization P_e ; 2 - modulation of H^- from optically pumped target: $M = (I^- - \overline{I^-})/(I^- + \overline{I^-}) = P^2 \mathcal{E}$, $\mathcal{E}$ - factor of spin-orbital depolarization; 3 - electron polarization of sodium potassium mixture determined from H^- yield modulation.

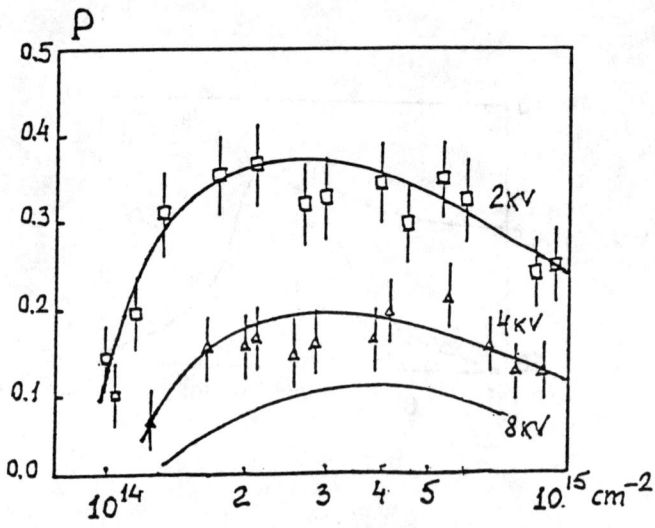

Fig. 7. Dependence of spin-exchange polarization on target thickness at different hydrogen beam energies.

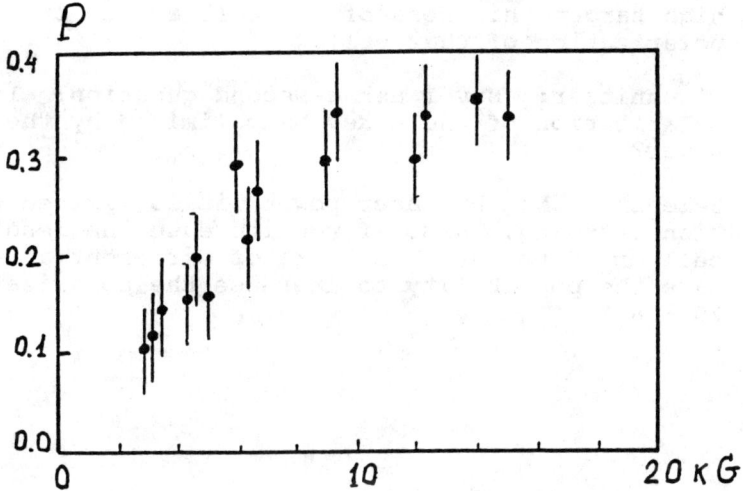

Fig. 8. Dependence of spin-exchange polarization on magnetic field in sodium target.

DISCUSSION

Clausnitzer: What would you propose for the dc operation of your source, spin exchange source?

Zelenski: We have estimated for the dc operation of spin exchange sources especially for the application for TRIUMF cyclotron, an H⁻ current of more than 200 microamperes. Now, it is a very suitable situation for a spin exchange process, the use of high intensity, high power tunable solid state laser available at a power of more than 5 watts. Five watts is enough to produce the high target thickness of the cell and to produce high polarization of that cell.

Clausnitzer: May I ask a second question? Is your 30% polarization of the 8 KeV beam limited by the laser power?

Zelenski: Not by laser power, it is limited by radiation trapping. Also if you increase the length of the cell or if you use a mixture of different alkali you have the possibility to increase the polarization up to 70 - 80%.

RECENT DEVELOPMENTS IN THE BNL INTENSE POLARIZED H⁻ SOURCE PROGRAM*

A. Kponou, A. Hershcovitch, J.G. Alessi, B. DeVito, C.R. Meitzler**
Brookhaven National Laboratory, Upton, NY 11973

ABSTRACT

A program to develop a high intensity polarized H⁻ ion beam for injection into the AGS is under way at this laboratory. The approach we are following is essentially the polarization and ionization of a very cold and intense atomic hydrogen beam. This paper reports on the magnetic focusing of the cold atomic hydrogen beam we have produced.

INTRODUCTION

One aspect of the BNL program to develop an intense source of polarized H⁻ is the production of very cold polarized H° beams. The production of cold **unpolarized** H° beams has been reported[1]. Nuclear polarization of the latter is achieved by a combination of magnetic focusing (the Stern-Gerlach effect), and rf induced transitions. The original plan was to use a superconducting solenoid as the magnetic lens, and beyond it, a set of rf transition units to produce the nuclear polarization. The beam will then be ionized by a ring magnetron ionizer. A review of the entire program was recently given elsewhere[2]. This report deals only with the magnetic focusing of the H° beam.

MAGNETIC FOCUSING

Focusing the neutral hydrogen beam with a superconducting solenoid was not successful at peak beam intensity due, we believe, to intrabeam scattering in the solenoid. This conclusion is based on the observation that focusing decreased with increasing beam density. On the basis of this observation, the H° - H° scattering cross section has been inferred[3] to be $\simeq 100$ Å^2, somewhat higher than values previously reported in the literature.

Modifying the solenoid to give it a more open geometry for improved pumping was not practicable, hence we decided to build a 20 cm long permanent magnet sextupole having a 4 cm bore diameter and a pole-tip field of 7 kG[4]. The individual magnets from which the poles were assembled were made from Nd-Fe and specially coated

*Work Performed under the auspices of the U.S. Depart. of Energy.
**Present address: Physics Department, Sam Houston State University, Huntsville, Texas.

to resist attack by atomic hydrogen. Azimuthally machined slots in the yoke allowed for additional (radial) pumping of the bore.

No significant focusing was observed with this magnet, and our inability to vary the magnetic field was a serious drawback since subsequent simulations showed that the focusing was very sensitive to beam velocity. We have established that the strength of the magnet did not match the velocity spectrum of the beam.

A Two-Magnet System

The permanent magnet has now been reduced to a length of 10 cm. This will be used in conjunction with a conventional electromagnet sextupole which is also 10 cm long, has a 3.6 cm diameter aperture, and is capable of 6.3 kG pole-tip field in d.c. operation (cooling being the limitation) and higher, if it is pulsed. This arrangement is shown in Fig. 1. The permanent magnet will be nearer to the nozzle because its slightly larger bore and pole-tip field give it a larger acceptance. The field of the second magnet will be varied to focus the beam to the detector. The permanent magnet may also be moved axially up to 4 cm, giving us another degree of freedom in optimizing beam focus at the detector.

Simulations

We used computer simulations to determine a suitable configuration of the two magnet system. The simulations involved tracking individual atoms from the nozzle to the detector, which was placed at the position where the ring magnetron ionizer will eventually be located. The Monte Carlo technique was used to launch the atoms. The parameters which were randomly selected are (1) the speed of the atom - according to the measured supersonic velocity distribution, (2) the radial position at the tip of the nozzle - we assumed uniform flux density across the nozzle aperture, (3) the angle of elevation - we assumed a $\cos^5\theta$ distribution but the results are not sensitive to the value of the exponent, and (4) the electron spin state - either 1/2 or -1/2. Azimuthal motion and beam attenuation by scattering were ignored. If the detector was assumed to have a circular aperture, then particles reaching the aperture were weighted with their distance from the axis there. A typical graphical output of the tracking program is shown in Fig. 2. The tracks plotted were also randomly selected and represent 0.025% of 200K starts.

The focusing factor, FF, defined as the ratio

<u>Weighted counts at detector with magnets on</u>
Weighted counts at detector with magnets off

was used as the figure-of-merit to compare the focusing of different sets of operating conditions. In Fig. 3, FF, with typical error bars, is plotted as a function of the pole-tip field of the second magnet. Figure 4 shows FF as a function of velocity for beams assumed to be monochromatic, covering the range of velocities we have measured. We see that with the two-magnet system we should be able to observe focusing over a wide range of beam velocities, by adjusting the field of the variable magnet.

Estimation of beam flux

For a forward H° beam flux density of about 2×10^{20} atoms/s/sr^3, the flux into a 4 mm dia. aperture at the detector plane, 70 cm from the nozzle, is 5×10^{15} atoms/s when the magnets are off. (The permanent magnet can be "turned off" by lowering it out of the beam.) Since the peak value of FF in Fig. 2 is about 7, the expected flux at the detector is about 3.5×10^{16} atoms/s. With a most probable velocity of 575 m/s, the expected beam density at the detector is about 5×10^{12} atoms/cm^3, which can be easily detected by the residual gas analyzer detector. This density is about 20 times greater than the density in our present polarized source. However, scattering will probably prevent us from realizing such a dramatic gain.

PLANS

The modifications to the cold beam source to accommodate the two sextupole magnet system are almost complete. We expect to study the beam focusing early in 1990. The next step will then be to couple the ring magnetron ionizer to the source and study the ionization process.

SUMMARY

Focusing our intense, cold atomic hydrogen beam has, thus far, proved elusive due to scattering effects in the superconducting solenoid, and poor optics in the case of the single permanent sextupole magnet. A two sextupole magnet system, one permanent magnet and one variable strength, has been designed and its installation is almost complete. We feel that this approach solves the problems in our previous attempts to focus the beam with a magnetic lens.

ACKNOWLEDGEMENTS

We gratefully acknowledge the continuing excellent support of this project from our colleagues in the Advanced Source Development Group, W. Hensel and W. Tramm, in particular.

REFERENCES

1. A. Hershcovitch, A. Kponou, and T. Niinikoski, Rev. Sci. Instrum. 58 (4) (1987) p. 547.

2. J.G. Alessi, B. DeVito, A. Hershcovitch, A. Kponou, and C.R. Meitzler, Int. Conf. on Ion Sources, Berkeley, CA. July, 1989, to be published.

3. A. Hershcovitch, Phys. Rev. Lett. Vol. 63, No. 7, 750 (1989).

4. C.R. Meitzler, Private Communication.

FIGURES

Fig. 1 Section through the cold source showing the two sextupole magnets. Some cryopanels were removed to move the dissociator and accommodator forward.

Fig. 2 A randomly selected fraction of the large number of atom tracks used in the computer simulations. The first of the two target planes to the right corresponds to the detector.

Fig. 3 Focusing factor, FF, versus the magnetic field of the variable magnet. The values of the parameters in the velocity distribution were: V_d = 565 m/s, T_b = 0.35 K.

Fig. 4 Focusing factor versus monochromatic beam velocity for three values of the magnetic field of the variable sextupole magnet.

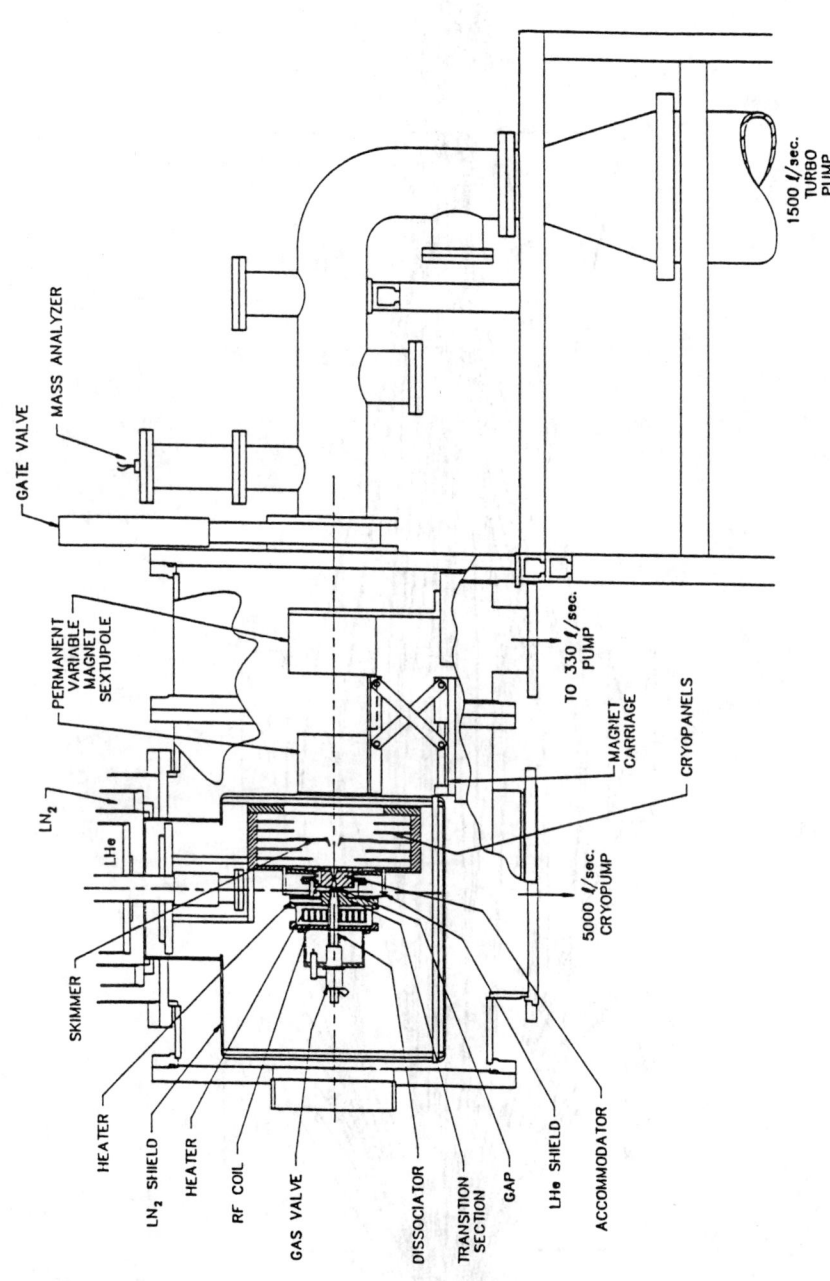

Figure 1

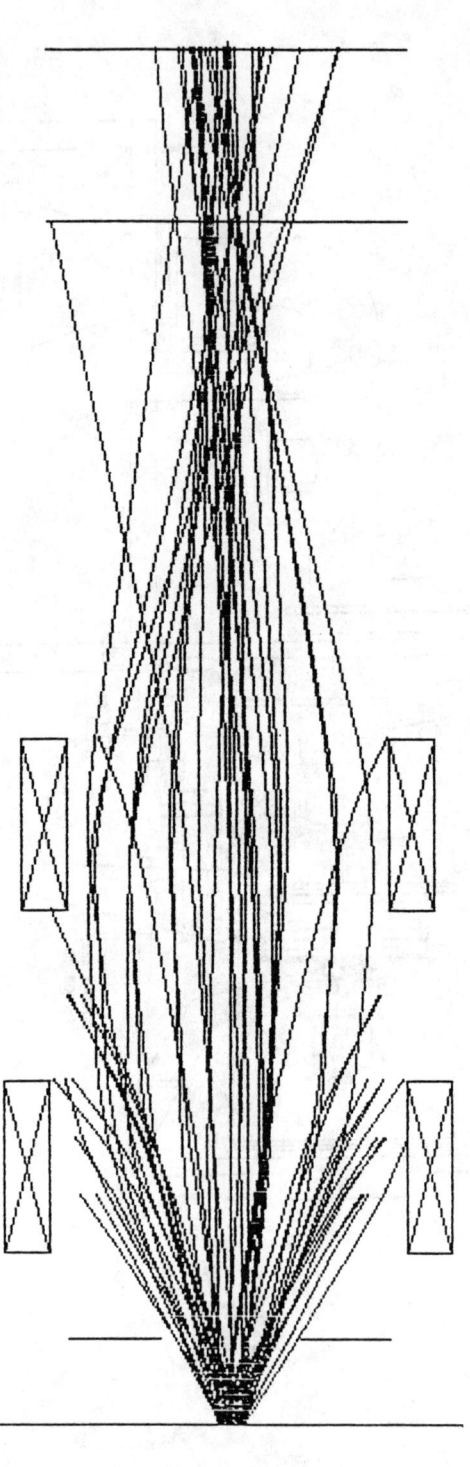

Figure 2

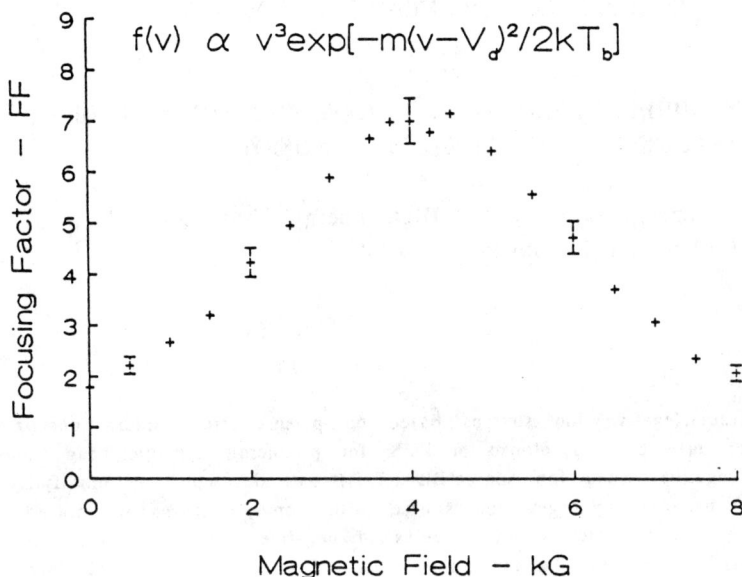

Figure 3

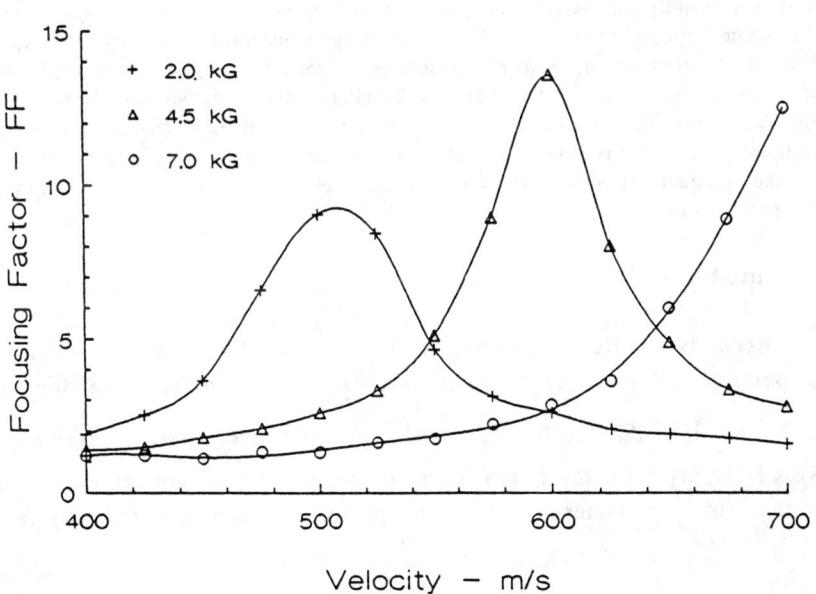

Figure 4

INTENSE NEGATIVE HEAVY ION SOURCES

YOSHIHARU MORI, AKIRA TAKAGI, KIYOSHI IKEGAMI,
AKIRA UENO AND SADAYOSHI FUKUMOTO

National Laboratory for High Energy Physics, Oho 1-1, Tsukuba-shi, Ibaraki-ken 305, JAPAN.

(Abstract) Negative ion sources based on plasma-surface interactions (BLAKE ion source) have been developed at KEK for producing negative heavy ions. The first negative heavy ion source (BLAKE-II) was developed by modifying the ordinary negative hydrogen ion source with converter (BLAKE-I) placed into the plasma. It generates various species of negative heavy ions with intense beam currents. For example, a more than 10mA Au- ion beam was obtained from the ion source. Recently, the large scaled negative heavy ion source (BLAKE-III) has been developed and in the preliminary test experiment, more than 100mA Cu- ion beam has been stably obtained with a 10% duty factor in pulsed operation. The BLAKE-II ion source was attached to the BNL 15MV and Tsukuba University TANDEM accelerators and large current negative heavy ion beams were successfully accelerated in pulsed mode operation. Also, it was found that the space charge effect should be carefully considered for such a large current acceleration in a tandem accelerator, especially at the injection beam line and low energy end. In order to examine the negative ion formation process fundamentally, negative ion production probability related on sputtered particle velocity was measured and the results showed exponential dependence of the production probability on particle velocity as Norskov and Lindquist's theory predicted

1. Introduction

Recently, various new types of negative ion sources which make it possible to generate various species of intense negative ion beams such as $H^-, C^-, Si^-, Cu^-, Ni^-, Au^-$ and so on have been developed at KEK.[1][2][3] In these ion sources, negative ions are produced at the surface of the material which is placed in a hydrogen plasma confined by a cusp magnetic field for producing negative hydrogen ions or a xenon plasma for negative heavy-ions. This type of negative ion source has been originally developed at LBL(Lawrence Berkley Laboratory) for producing an intense negative hydrogen beam for nuclear fusion[4] and then improved for accelerator applications at LANL(Los Alamos National Laboratory)[5] and KEK(National Laboratory for High Energy Physics). Therefore, the ion source has a nickname of BLAKE negative ion source.

The BLAKE-I source was designed as a negative hydrogen ion source. This ion source generates negative hydrogen ion beam of more than 40mA maximum and can be operated stably for more than 2,000 hours with LaB6 filaments. It has been used in the KEK 12-GeV proton synchrotron for more than four years. The BLAKE-II ion source is a modified ion source of BLAKE-I for generating intense negative heavy ions and, for example, it has produced a Au$^-$ ion beam intensity of about 10mA. Recently, a large scaled negative heavy ion source(BLAKE-III) has been developed and more than 100mA of Cu- ion beam has been obtained with high duty factor(10%).

Recently, demand for acceleration of intense pulsed heavy ion beams in an electrostatic tandem accelerator has been increased more rapidly because a tandem accelerator can be used as an efficient injector for heavy ion synchrotrons [6][7][8] and a pioneer work has been already started at BNL with their 15MV tandem accelerator which was operated as an injector for the AGS synchrotron. Also in the field of nuclear physics, such intense pulsed beams are very attractive for some experiments.[9] The BLAKE-II ion source was attached to the BNL 15MV and Tsukuba University 12MV tandem accelerators to accelerate the intense negative heavy ion beams. Relatively higher beam intensities of 0.2 - 1.4 mA for Au$^-$ and Cu- were obtained at the exit of the accelerators. However, in order to accelerate more beam current in the tandem accelerators, it was found that beam emittance degradation due to the space charge effect for such large beam intensities should be eliminated.

In order to examine the fundamental process of negative ion formation on the metallic substrate in the BLAKE ion source experimentally, the velocity dependence of the negative ion formation probability was studied by measuring the negative ion formation yields for C-, Si-, Ag-and Sn- ions sputtered by Xe+ ions. The negative ion formation probability was found to have an exponential dependance on the velocity as predicted by theory and the parameter related the electronic configurations of the negative ion and the metal combination was decided by experimentally.

2. BLAKE-II Ion Source

2-1. Apparatus

Details configuration of the BLAKE negative ion source have already been described in previous papers.[1][2][3] The schematic layout of the ion source is shown in Fig.1. The ion source consists of

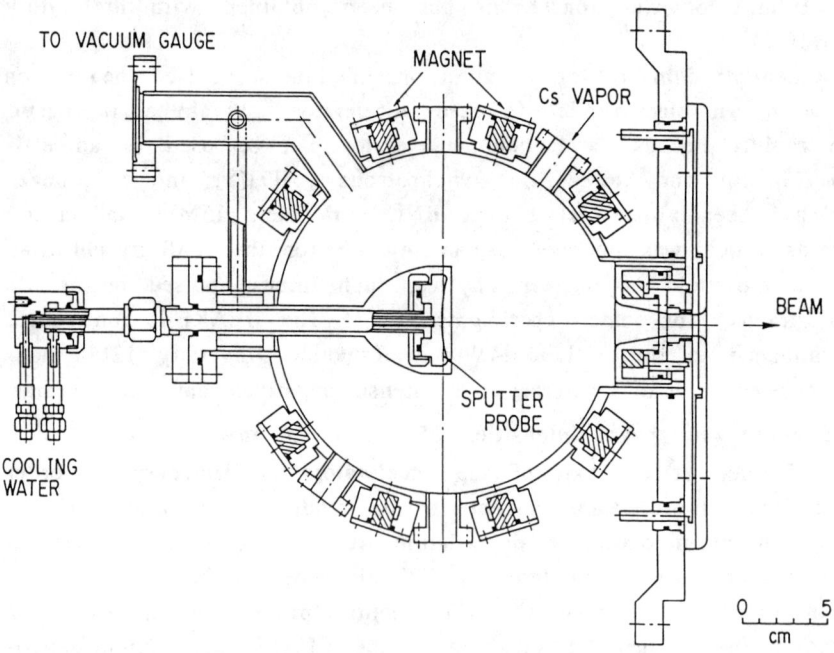

FIg.1 Schematic layout of BLAKE-II ion source.

a cylindrical plasma chamber made of stainless steel, a sputter probe, cesium oven and two sets of filaments. There are eighteen pieces of SmCo permanent magnets surrounding the plasma chamber to make the cusp magnetic field. Two small permanent magnets making a dipole magnetic field of about 100 gauss were also placed at the exit of the anode hole and used to return the extracted electrons back to the anode. The total drain current of the extraction power supply was substantially reduced with these dipole magnets. The sputtering probe was placed at the center of the plasma chamber, which was 12cm from the anode aperture, and biased negatively by a voltage of up to -970V. A quartz glass covered the probe except the surface to the anode hole and helped to prevent the supporting and cooling channel of the probe from sputtering by xenon ions in the plasma.

Two sets of hot filaments made of lanthanum hexa-boride (LaB6)were used for making the arc discharge. The operating temperature of the filaments were about 1400-1500°C, which was almost 1000°C lower to achieve the same plasma condition than that of the tungsten filament . The details of the characteristics and performance of the filaments are described in reference.1

2-2.Pulsed Mode Operation

The ion source was operated in a pulse mode by making a pulsed arc discharge. The arc voltage and current during the normal operation were 30-40 V and 10-20 A, respectively. A current regulated pulsed power supply was used for making the arc discharge. The duration and the repetition rate of the pulsed arc power supply were 100-200msec and 1-20Hz, respectively during the normal operation.

The plasma condition was dramatically changed once cesium was introduced into the plasma chamber. The arc voltages dropped abruptly from 60-80V to 30-40V and the electrons extracted from the ion source were also substantially reduced. It is naturally conceivable that cesium atoms introduced into the plasma would be mostly ionized by the energetic electrons in the plasma because cesium has a low ionization potential. Thus the plasma became a cesium-xenon mixed plasma once cesium vapor was introduced. At the very first stage of the operation with cesium vapor, large numbers of impure negative ions such as oxygen were found in the extracted beam. With the present beam duty factor, several hours were needed to reduce these impurities.

The sputtered probe was normally biased at a voltage of -970 V to the anode wall. Thus, a thin plasma sheath was created at the surface of the probe. Many xenon ions in the plasma were accelerated through the plasma sheath, hit the probe surface and sputtered out a large amount of particles from the surface. Some of sputtered particles formed negative ions on the surface because the work function of the surface was reduced by the cesium coating. The drain current of the probe by these accelerated ions was measured and it was about 300mA at an arc current of 15A.

Beams were extracted at an energy of about 30keV from the ion source and focussed by an einzel lens on a Faraday cup which was placed 120 cm away from the extraction electrode. A large permanent dipole magnet which was 5cm wide, 4cm high and 15 cm long was placed between the einzel lens and Faraday cup to analyze the mass spectrum of the beam extracted form the ion source.

More than 20 species of the negative-heavy ions have been tested and the results of the obtained beam intensities are summarized in Table 1 with the sputtered target configurations for each species. As can be seen from this table, the beam intensities for most of species are almost 50-100 times larger than those obtained

TABLE 1 Negative heavy ion beam intensities from the ion source.

Ion Species	Beam Current(mA)	Ion Species	Beam Current(mA)
Ag	5.4	Ni	6
Al	1.1	P	0.86
Au	10	Pd	6.8
As(As$_2$)	0.67(2.24)	Pt	6.4
Bi	0.13	Si	5.4
C(C$_2$)	3.4(4.6)	Sn	2.6
Co	2.8	Ta	1.4
Cu	10	Ti	0.8
Cr	0.2	V	0.7
Fe	1.7	W	3

from the ordinary cesium sputtered negative-heavy ion source. Beams from the ion source were very stable and reproducible. The measured mass-spectrum for Au ion beam is shown in Fig.2.

The beam emittance was measured by the emittance monitor which was developed for the 750keV H- beam of the 12GeV synchrotron. Figure 3 shows the measured normalized emittance for Ni- ion beam for 2.5 and 6mA respectively. A typical value of the 90% normalized emittance of the beam was about 37 p mm.mrad.(MeV)1/2. This value is about 3-4 times larger than the ordinary sputtered negative ion source. However, the brightness of the beam is relatively large because the beam intensity is 50-100 times larger.

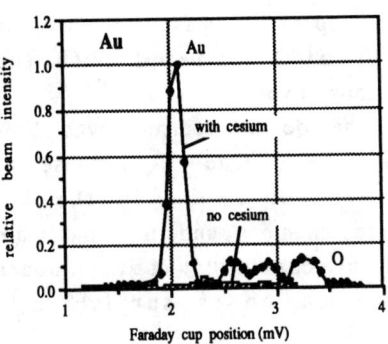

Fig.2 Mass-spectrum of Au- ion beam

2-3. DC-Mode Operation

Compared with pulsed-mode operation, dc-mode operation requires substantially higher cesium flow rates because the cesium coverage on the sputtering.[10] Therefore, a new cesium oven, cesium valve and cesium transport line for feeding more cesium vapor into the ion source were designed. The diameter of the feed line was increased from 6 to 10mm. The distance between the oven and ion source was decreased from 50 to 15cm. The optimum temperature of the new cesium oven for pulsed-

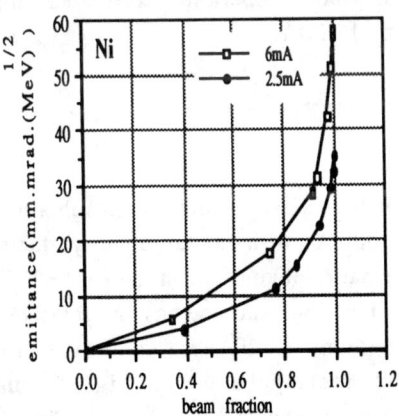

Fig.3 Normalized emittance of Ni ion beam

mode operation was decreased by about 50°C compared with the previous oven.

The dc arc current was limited to less than 5A because of the lack of cooling in the ion source chamber and the maximum current capability of the beam extraction power supply(50kV-10mA).

In the preliminary experiment of dc-mode operation, a spherical geometry copper sputter probe was used.

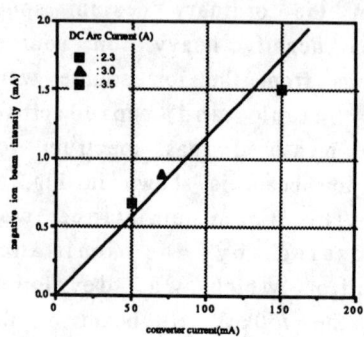

Fig.4 Total beam intensity as a function of probe current in DC mode operation.

At a sputter probe voltage of -610V, the total drain current to the sputter probe was typically 90mA at an arc current of 2A.

The cesium vapor density for an oven temperature of 258°C was estimated to be ~30 times higher than that for pulsed-mode operation(~160°C). Measured total beam intensity as a function of sputter probe current is displayed in Fig.4. The beam intensity is estimated to increase almost linearly with sputter probe current. By linear extrapolation of this data, the beam intensity would reach the same level observed during pulsed-mode operation provided that the arc current could be increased to 15-20A.

3 BLAKE-III Ion Source

A large scaled BLAKE ion source(BLAKE-III) aiming to obtain a relative large current of more than 100mA with a large duty factor has been developed recently. Schematic layout of the BLAKE-III ion source is shown in Fig. 5. This ion source has a relatively large rectangular shape sputter probe(5cm x 20 cm) and the extracted beam shape is not round but rectangular. In Fig. 6, the waveform of the extracted Cu- ion beam is shown. Also in Fig. 7,

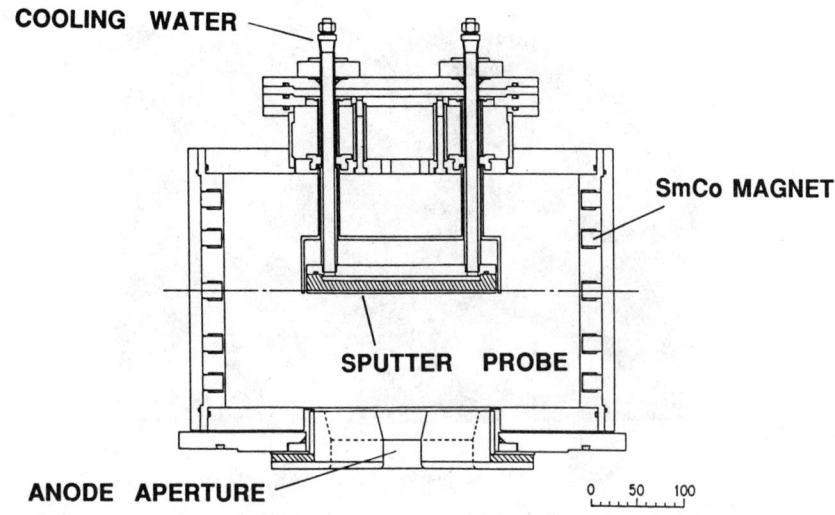

Fig.5 Schematic layout of BLAKE-III ion source.

the photograph of the inside of the ion source is presented. As can be clearly seen from this figure, the negative ion beam was well focused in the plasma by the concave shape of the sputter probe and this means that the space charge neutralization due to the plasma ions works quite nicely.

4 Acceleration Test of Heavy Ion Beams in the BNL 15MV and Tsukuba University 12MV Tandem Accelerators

4-1 BNL 15MV Tandem Accelerator

Intense negative gold and silicon ion beams generated by the cusp negative heavy-ion source in pulsed mode operation,

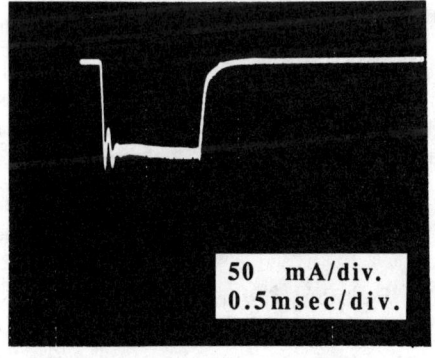

Fig.6 Waveform of 100mA Cu⁻ ion beam.

Fig. 7 Photograph of the inside of the BLAKE-III ion source.

were preliminary accelerated by the BNL 15MV tandem accelerator. The cusp negative ion source was attached to the ordinary injector of the BNL 15MV tandem accelerator. A schematic layout of the injector is shown in Fig.8. The pulse width and the repetition rate of the beam was about 150μsec and 7Hz, respectively, and the peak intensity of more than 3mA was extracted from the ion source for each ion species. Figure 9 shows a waveform of the extracted Au^- ion beam measured at the Faraday cup FC1. The injected beam currents to the tandem accelerator, which were measured at the entrance of the tandem accelerator, were 0.4mA and 0.7mA for the gold and silicon ion beams, respectively and the total accelerated beam current was 1.4mA at the exit of the accelerator. It was understood by computing the beam optics including the non-linear space charge effect of the beam that the poor beam transmission from the ion source to the tandem accelerator was caused by the strong space charge forces in the beams due to the relatively low beam energy in the injection beam line.

4-2 Tsukuba University Tandem Accelerator

Intense pulsed heavy ion beams of silicon and copper beams

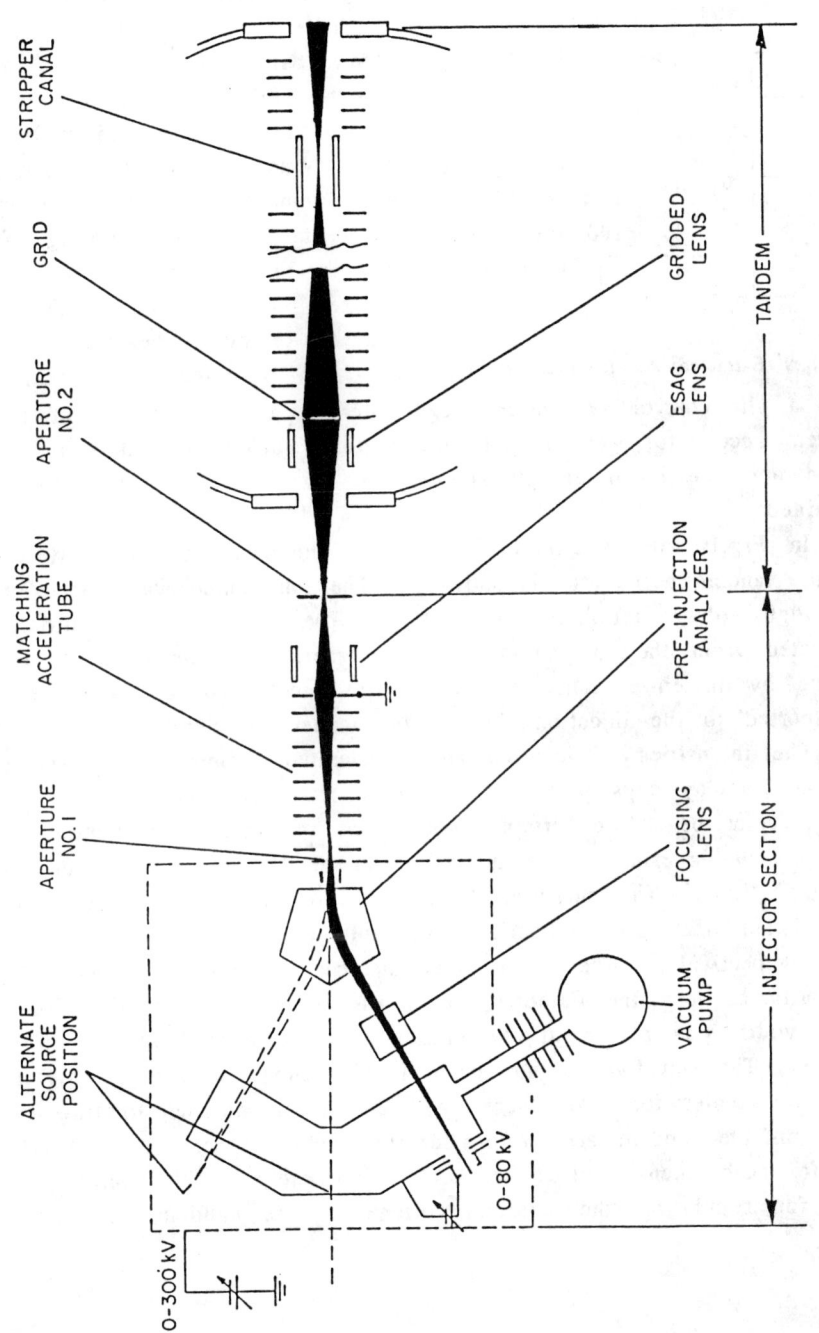

Fig.8　Schematic layout of the injector of the BNL 15 MV tandem accelerator.

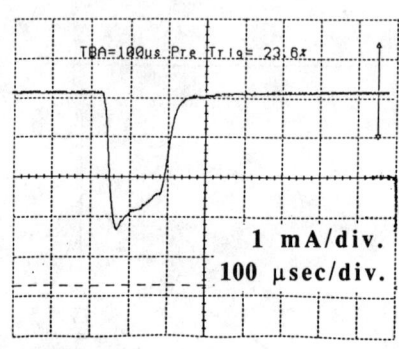

Fig.9 Extracted Au⁻ ion beam at FC1.

generated by the cusp negative heavy-ion source have been successfully accelerated by the 12MV tandem accelerator at Tsukuba Univ. The peak intensity of the accelerated pulsed beam was almost 100 times larger than that of the ordinary DC beam. No deteriorating effect on the acceleration for the intense pulsed beams, such as sparking or heavy beam loading, has been observed. Remarkably, a slit control system for regulating the column voltage of the tandem accelerator worked quite nicely for such a low duty pulsed beam and the beam energy stability of less than 10-4 was easily obtained.

In Fig.10, the schematic diagram of the beam transport system of the tandem accelerator is shown. The ion source was placed on the high voltage terminal of 100kV. The ion beam was initially extracted from the ion source with an energy of 20keV. After focusing by an einzel lens, the beam were accelerated to 100keV and transported to the injection line of the tandem accelerator.

The intensities of injected and accelerated beams were measured by the Faraday cups placed in the beam transport lines of the tandem accelerator. The terminal voltage of the tandem accelerator in the present experiment was normally set at 10 MV and sometimes at 10.5MV. The measured beam intensities at several positions in each beam lines are summarized in Table 2 for silicon and copper ions, respectively. The reason why the beam intensity dropped substantially between the Faraday cup 4 and 5 is that a small slit of 2mm width was placed at the entrance of the each Faraday cup 4 and 5. The waveform of the beam is also shown in Fig.11. During the acceleration test, there have been seen no deteriorating effects on the tandem accelerator due to the beam loading or sparking for such intense pulsed beams. Remarkably, a slit control system for regulating the column voltage of the tandem accelerator

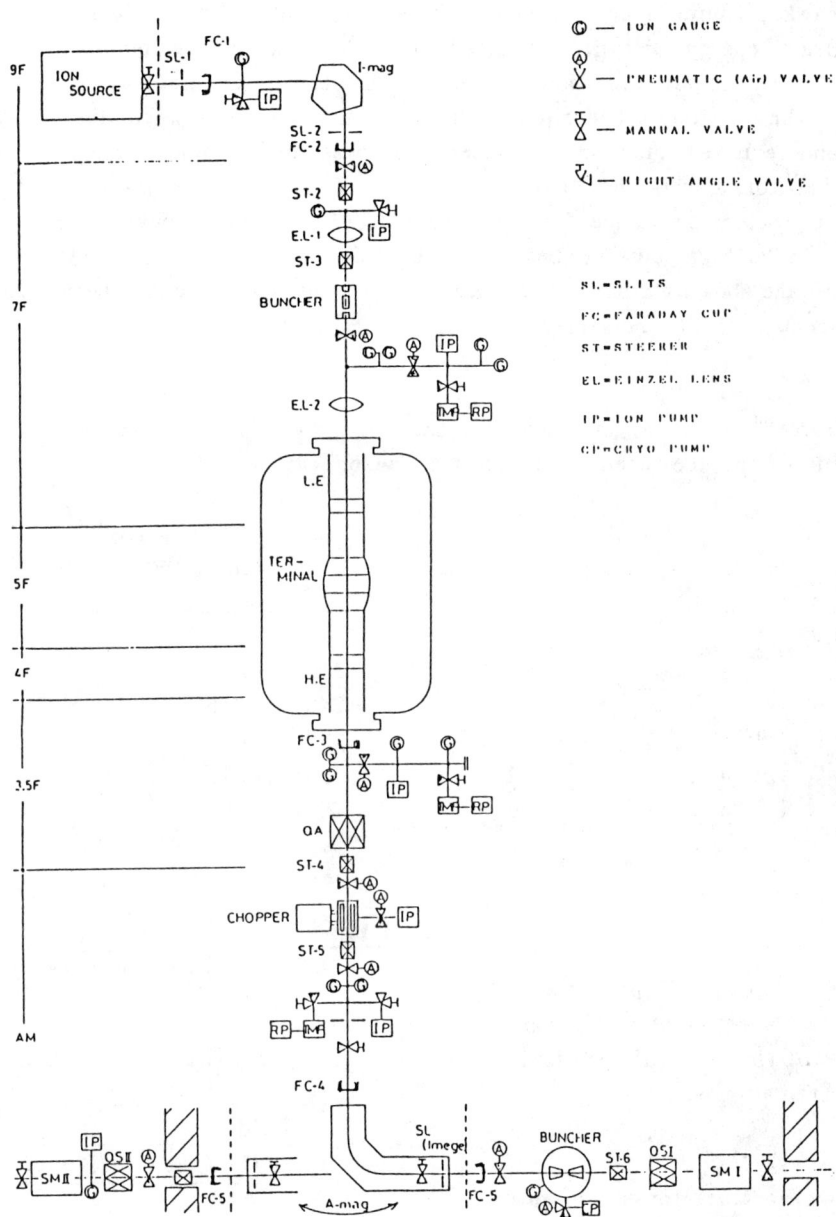

Fig.10 Schematic diagram of the beam transport system.

worked quite nicely for such a low duty pulsed beam and the beam energy stability of less than 10-4 was easily obtained. This is very important when a tandem accelerator is used as an injector of heavy ion synchrotron. Beam transmission between the entrance and exit of the tandem accelerator was rather poor compared with that for a very low current beam. This might be probably due to the beam emittance growth caused by the space charge effect for such a high current beam. In order to improve the transmission of the beam through the accelerator, a higher injection beam energy seems to be necessary.

Table2. Beam intensities measured by Faraday cups which were placed in the beam lines of the tandem accelerator.

Faraday Cup No.	slit width(mm)	Beam Intensity(mA) silicon	copper(^{63}Cu)
1	12	210	195
2	12	240	175
3	-	180	135
4	2	125	140
5	2	23	22

5. Velocity Dependence of the Negative Ion Formation Probabilities of Sputtered Atoms

In the formation process of negative ions sputtered out from the surface

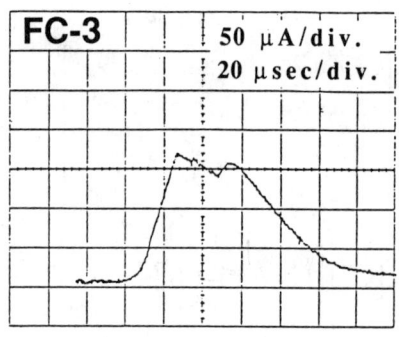

Fig.11 Waveform of the accelerated beam.

of the substrate, the velocity dependence of the negative ion formation probability has been well understood theoretically by considering the time-varying potential experienced by the sputtered particle leaving the surface, and Norskov and Lundquist[11], and Reisser calculated the ionization probability P of negative ions.

According to the theory developed by Norskov and Lundquist, the negative ion formation probability P can be obtained in the following equation.

$$P = 2/\pi \ \exp[-C_1\pi(\phi-A)/h\gamma v] \ \exp[-C_2 \pi / h \gamma v] \ , \tag{1}$$

where ϕ and A are the workfunction of the substrate and the electron affinity of sputtered atom, respectively and v is a velocity of atom leaving from the surface of the substrate. Parameters C_1 and C_2 are related the effective energy difference between the electron affinity and the Fermi energy when the atom goes out of the substrate, γ is a characteristic distance beyond which no further electron exchange between atom and the substrate takes place and v is the perpendicular component of the particle velocity.

In the experimental approach to study the velocity dependence, several experimentss have been done so far. Among them, Yu measured in his excellent work[12] the negative ion formation probability of oxygen ions sputtered out from the chemisorbed oxygen layers on vanadium and niobium by argon ions as a function of the perpendicular component of the velocity emitted from the substrate. He showed clearly a linear velocity dependance of the negative ion formation probability of oxygen ion as predicted by theory and obtained that the coefficient $C_1\pi/h\gamma$ in the exponent of eq.(1) was 4×10^{-5} eV m^{-1}sec in the system of the negative oxygen ion.

When an atom or molecule is sputtered out from the substrate, the most probable energy is almost equal to its sublimation energy Es according to the theory of sputtering. Therefore, the negative ion formation probability on the cesiated substrate can be re-written as,

$$P = 2/\pi \; \exp[-C_1\pi(\phi-A)/h\gamma \; v],\qquad(2)$$

where $v=(2E/m)^{1/2}$ and m is the mass of the particle. According to eq.(2), the negative-ion formation probability for various species of atoms and molecules can be be estimated with the electron affinity and the sublimation energy for various species of atoms and molecules. This is very useful to predict the negative ion beam current from the sputtered type of negative ion sources.

In order to check the validity of eq.(2) experimentally, we made a measurement of negative ion formation probabilities for carbon, silicon, copper, tin and silver atoms and the coefficient $C_1\pi/h\gamma$ obtained in the present experiment was compared with the value taken by Yu.

Measurement of the negative ion formation probabilities for carbon, silicon, copper, tin and silver were made using the cusp negative heavy-ion source which has been developed at KEK recently.

The substrate used in the experiment consisted of carbon, silicon, copper, tin and silver plates of 0.5mm in thickness and it was attached on a mollibudenum metallic base. The measured mass spectrum for the substrate used in the experiment is presented in Fig. 12. For such massive ions as tin and silver, the measured mass peak for each one was not well resolved because the resolving power of the analyzer became poor for heavy mass ions. Therefore, in order to estimate the relative intensities for tin and silver negative ions, the beam intensities data, which were previously measured for pure tin and silver sputtering probes at the same experimental condition, were used. The intensity ratio between tin and silver, ISn/IAg, measured in that experiment was 0.33.

In order to examine the velocity dependence of the negative ion formation probability indicated in eq.(2), the relative intensity for each species has to be normalized by the sputtering rate of each substrate atom. The sputtering rates, S, for those substrates when the incident xenon ion energy is 1 keV have been measured at various institutes so far.

Figure 13 shows the measured negative ion formation probability as a function of the particle velocity ($v = (2E/m)^{1/2}$). This shows an exponential dependence of negative ion formation probability on the particle velocity as predicted by the theory. The gradient obtained from this figure shows that $C_1\pi/h\gamma = 3 \times 10^{-5}$ eV m^{-1}sec and this value agrees well with the value measured by Yu for O^- ion.

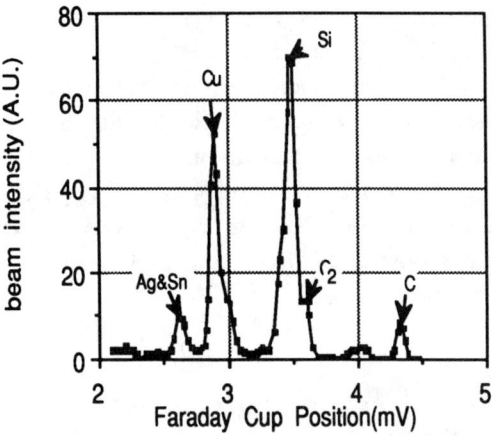

Fig.12 Mass-spectrum of the extracted beam.

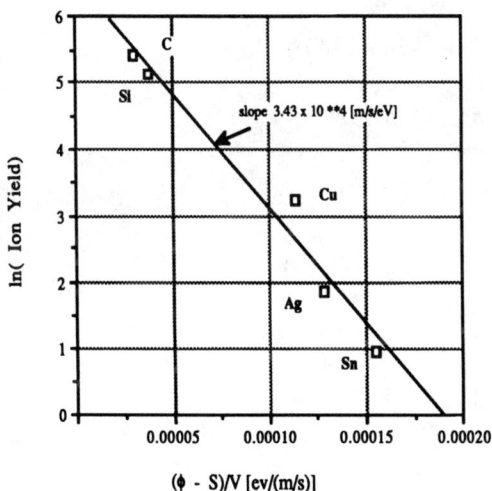

Fig.13 Measured negative ion formation probability as a function of the particle velocity.

6. Conclusion

Characteristics and performance of the newly developed negative ion sources have been described. More than 20 species of negative heavy-ion beams have been obtained so far at the ion source test stand and the beam intensities from the ion source were found to be almost 50-100 times larger than those from the ordinary cesium sputtered negative ion source. Beam emittance was also measured for a Ni beam and the 90% normalized emittance was about 37π mm.mrad.(MeV)1/2.

This ion source might be useful not only for nuclear experiments with a tandem accelerator but also for ion beam applications such as ion implantation.

The authors would like to express their appreciation to Dr. G.D.Alton for valuable discussions. They are also indebted to Profs. T.Nishikawa, S.Ozaki and M.Kihara for encouragement during the experiments.

REFERENCES

1 Y.Mori et al.,;AIP Cof. series,No.158(New York,1987)p.378.
2 G.D.Alton et al.,;Nucl. Instr. Meth.,A270(1988)194.
3 Y.Mori et al.,;Nucl. Instr. Meth.,A273(1988)5.
4 K.W.Ehlers et al.,Rev. Sci. Instrm.,51(1980)721.
5 R.L.York et al.,;AIP Conf. Series,No.111(New York,1984)p.410.
6 P.Thieberger et al., IEEE,NS-30(1983)2749.
7 G.D.Alton et al., Nucl.Instr. Meth.,A244(1986)170.
8 S.Nagamiya,KEK-Internal Report 88-20(1989).
9 G.D.Alton et al.,Nucl. Instr.Meth.,A270(1988)194.
10 Y. Mori et al.,Nucl. Instr. Meth.,
11 J.K.Norskov et al., Phys. Rev. B19(1979)5661.
12 M.L.Yu, Phys.Rev.Let., 47(1981)1325.

DISCUSSION

Jacquot: Have you tried to test another support gas like xenon, have you tried to test a lighter gas.

Mori: Like argon.

Jacquot: For example, do you expect some difference, that is, what is the effect due to the xenon and cesium.

Mori: I have never tried other gases, but I speculate that the test of the beam current depends on the sputtering rate, so a lighter gas, such as argon might reduce the beam current comparatively to the xenon.

Alton: In the species experiment, we had segmented sections so the yields were normalized by sputtering ratios. Did you take into consideration space charge effects which occur at the plasma boundary, the interface boundary, which will modify the yields downstream.

Mori: Actually as you know one of the features of this type of source is that all of ion species have a large initial velocity after leaving the source, and in the source. The space charge is completely neutralized by the xenon ions.

Alton: I would agree in the passage across the double layer but in the double layer, there will be a space charge limit on the current for each particular species in which you modify the beam intensity downstream. The other question is related to high intensity source, did you pulse both the sputter probe voltage.

Mori: No.

Alton: The flat curve is generated by pulsing only the discharge.

Mori: That's right

Alton: Very good.

Holmes: Have you examined the issue of matching the beam to the accelerator? You have talked a lot about the current, how do you get the beam once it leaves the source, how do you transport that beam, because I'm not sure that the space charge can explain all your problems.

Mori: There are two effects of the space charge, one is emittance growth, one is just the beam envelope problem. We can eliminate the envelope effect by compensating the space charge problem. But, even for that case, the emittance growth problem cannot be eliminated because of non-linear effects. So for the very high beam current, we should be very careful about that kind of a program especially for the unneutralized beam. You know this type of source is operated in the very short pulse mode, so it is very difficult to get the space charge neutralization in the beam lines as in a dc beam.

Holmes: But surely you can put a low pressure gas, you must have some gas.

Bejamin: The kind of experiments that Professor Mori did were on a machine that is used for other purposes most of the time. And while he had complete control over the ion source, he didn't have complete control over the acceptance apertures, the gas conductances, the pumping, the type of gas for space charge neutralization between the ion source and the tandem. Furthermore, all the beam handling elements in the tandem are electrostatic in nature: electrostatic quadrupoles, electrostatic steers, which sweep out space charge neutralization effects, so he had a tough fight on his hands there. A completely different type of injection system where space charge neutralization was introduced where magnetic elements rather than electrostatic elements were used, would be greatly beneficial for these high intensities.

Kleyn: You found a remarkable agreement I would say for your negative ion yields for the various species, and my question is did you have any possibility of actually measuring the escape velocity of these pertinent particles.

Mori: Escape velocity, is very difficult.

Kleyn: The velocities that you took, are those from a collision cascade model, or sputter yield or what?

Mori: Yes, that's right, that is one of the issues. But we have just a standard model of the sputtering theory or something like that, so just

Kleyn: Which model?

Mori: I forgot the name of that, sputtering theory, do you remember?

Kleyn: Sigmund?

Mori: Yes, Sigmund.

Yuan: I have two questions about that source. First, what difference did you observe between the dc mode and the pulse mode, at same arc condition. The other question is what is the acceptance of the source you initially designed and what is the agreement between the acceptance and the emittance you measured?

Mori: The acceptance of the adjusted anode hole, you mean that?

Yuan: Yes

Mori: I don't remember that calibration but the emittance is almost the similar to the acceptance, its also different.

Yuan: For a particular ion?

Mori: Well maybe the emittance is limited by that kind of geometry, for every type of different kind of species.

Yuan: What is the difference between dc and pulse modes at the same arc current.

Mori: We have tried this only for copper and it is very similar. At the same arc current we could get similar beam current for both the pulse and dc. But, a different point is the cesium consumption rate, we need lots of cesium for the dc operation. That's a big problem.

DESIGN FEATURES OF AN AXIAL-GEOMETRY, PLASMA-SPUTTER, HEAVY NEGATIVE ION SOURCE

G. D. Alton
Oak Ridge National Laboratory*
P. O. Box 2008
Oak Ridge, Tennessee 37831-6368

ABSTRACT

An axial-geometry, plasma-sputter, negative ion source, which utilizes multi-cusp, magnetic-field, plasma-confinement techniques, is under design at the Oak Ridge National Laboratory (ORNL). The source is based on the principles of operation used in a recently developed radial-geometry source which has demonstrated pulsed-mode peak intensity levels of several mA for a wide spectrum of heavy negative ion species. The pulsed-mode characteristics of the source are well suited for tandem electrostatic accelerator/synchrotron injection applications. Mechanical design features include provisions for fast interchange of sputter samples, ease of maintenance, direct cooling of the discharge chamber, and the use of easily replaced coaxial LaB_6 cathodes. Principal features of the source will be described and the results of computational studies of the ion extraction optics will be presented.

INTRODUCTION

Recent developments[1-4] have demonstrated the utility of a radial-geometry, plasma-sputter, negative ion source for producing high-intensity (several mA) pulsed beams of a wide spectrum of atomic and molecular negative ion species. This source type, as well, has shown promise for dc beam generation at mA-intensity levels.[5] The pulsed-mode performance characteristics of this source type are particularly well suited for use in conjunction with the tandem electrostatic accelerator when used as an injector for a heavy ion synchrotron, while the dc mode of operation is commensurate with stand-alone tandem accelerator operation. For the synchrotron heavy ion accelerator, high-intensity, pulsed beams of widths 50-300 μs at repetition rates of 1-50 Hz for a wide

---

* Research sponsored by the U.S. Department of Energy under contract DE-AC05- 84OR21400 with Martin Marietta Energy Systems, Inc.

"The submitted manuscript has been authored by a contractor of the U.S. Government under contract No. DE-AC05-84OR21400. Accordingly, the U.S. Government retains a nonexclusive, royalty-free license to publish or reproduce the published form of this contribution, or allow others to do so, for U.S. Government purposes."

spectrum of atomic and molecular species are typically required. The feasibility of injecting mA beam intensities with these pulse characteristics into large tandem accelerators without deleterious effects has been recently demonstrated at the Brookhaven National Laboratory and at the University of Tsukuba.[4] The specific need for a high-brightness source for use in conjunction with the Holifield Heavy Ion Research Facility (HHIRF) 25URC tandem electrostatic accelerator at the Oak Ridge National Laboratory (ORNL), which would serve as an injector for the proposed Heavy Ion Storage Ring for Atomic Physics (HISTRAP),[6] was the primary motivating factor which led to the design of the source described in this report.

PRINCIPLES OF NEGATIVE ION FORMATION

The sputter technique has been utilized as a practical means for the generation of negative ion beams for a number of years. In such sources, positive ion beams, usually formed by either direct surface ionization of a Group IA element or in a heavy noble gas (Ar, Kr, or Xe) plasma discharge seeded with alkali metal vapor, are accelerated to energies between a few hundred eV and several keV where they sputter a sample containing the element of interest. The presence of a fractional layer of a highly electropositive adsorbate, such as cesium, on the surface is critically important for the enhancement of negative ion yields during the sputtering process.[7] A fraction of the sputter ejected particles leave the adsorbate covered surface as negative ions and are accelerated through an extraction aperture in the source. The present source, like the source described in Refs. 1-4, will use the plasma sputter technique in forming high-intensity beams (a few to several mA) of a wide spectrum of species.

SOURCE DESIGN FEATURES

The high-intensity, radial-geometry, plasma-sputter, negative ion source, described in Refs. 1-4, has proved to be a reliable, stably operating source with an extremely long lifetime for pulsed-mode operation, which can provide a wide spectrum of negative ion beams suitable for a variety of applications. The intensity levels obtained are often higher by factors of 30-100 than those which can be generated in cesium sputter-type sources such as described in Refs. 8-10 and yet the emittances of the source are comparable for pulsed-mode operation;[11] they also match the calculated acceptances of large tandem accelerators such as the 25URC tandem accelerator at ORNL,[12] and, in principle, ion beams from this source should be transportable through such devices. However, the radial-geometry source is not equipped with provisions for rapid sputter sample interchange such as required for use at a research facility which operates continuously and which must provide beams from a wide spectrum of species on user demand. The radial-geometry source, as well, was designed for exclusive use for generation of H⁻ beams for low-duty-factor synchrotron injection applications which require no direct cooling of the discharge chamber and is, thus, improperly cooled for the high power requirements necessary for dc-mode operation. The LaB_6 filaments used to initiate and sustain the discharge are not designed for fast interchange. When they break, either as a consequence of physical erosion or mechanical stress, an extensive period of time is required for replacement. The source shown in Fig. 1 was designed in an attempt to overcome some of these handicaps and to provide a source which could be used for both pulsed and dc modes of operation.

The operational characteristics of the axial-geometry source are expected to be identical in almost every detail to the radial-geometry source.[1-4] The source is constructed primarily of stainless steel and utilizes metal-to-ceramic bonded high-voltage insulators and low-voltage feedthroughs. Design emphasis has been placed on the ability to rapidly change the source itself and all degradable components. The cesium oven is mounted externally, permitting easy access for servicing, while providing good thermal isolation between the discharge chamber and the oven itself. The main source can be quickly and easily disassembled for cleaning and other maintenance operations. The source assembly is composed of four major independent assemblies: (1) the sputter probe vacuum airlock assembly; (2) a freon-cooled, stainless steel discharge chamber onto which is attached the cesium oven, discharge gas support system, coaxial geometry LaB_6 cathodes, and SmCo and AlNiCo plasma confinement magnets, (3) a ceramic-to-metal bonded alumina (Al_2O_3) insulator to which is attached the high-voltage extraction electrode system, and (4) the coaxial LaB_6 cathode assembly. The power supply arrangement required for operation of the source is shown in Fig. 2.

<u>Sputter Probe Vacuum Airlock Assembly</u>. The sputter probe assembly consists of a thin wall, 12.7-mm-diameter chromium plated copper tube to which is attached the 50-mm-diameter copper sample holder onto which is clamped the material of interest. The sample holder is cooled by continuous freon flow through a concentric tube arrangement. The probe can be inserted into and withdrawn from the source through the airlock valve which is sealed against atmospheric pressure by a conventional elastomer gasket. The vacuum interlock assembly is attached to an externally mounted ceramic-to-metal insulator which in turn is fastened to the back flange of the source. The insulator is used to isolate the probe-vacuum airlock assembly from the source body. Based on experience with sources equipped with similar provisions, the total time required for withdrawing, replacement, and reinsertion of the sputter probe sample material is expected to be the order of 10 minutes.

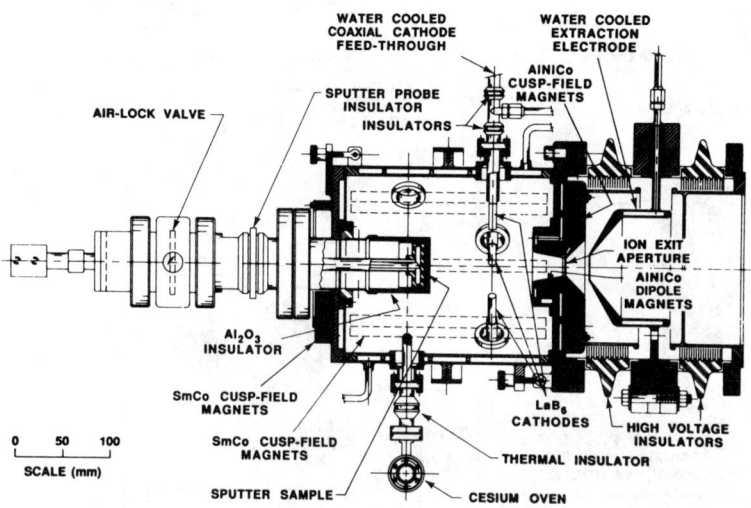

Fig. 1. Axial-geometry, plasma-sputter, negative ion source (top view).

The sputter probe/airlock assembly is insulated from the source housing with an insulator designed to withstand potentials up to ~-10 kV. When fully inserted into the discharge chamber, the sputter probe is surrounded by an Al_2O_3 insulating sleeve which prevents all internal negatively biased components other than the sample material from being bombarded by positive ions extracted from the surrounding plasma. Two geometries of sputter samples will be utilized. In cases where malleable metal sheet material containing the species of interest is readily available, samples 1-1.5 mm in thickness will be pressed by means of a die fixture into a 50-mm-diameter spherical sector probe with radius of curvature $\rho \simeq 210$ mm for focusing the negative ion beam generated in the sputtering process through the exit aperture of the source. Samples which are brittle must be formed from solid materials. Composite sintered compounds or mixtures of compounds will be typically 5 mm in thickness with a spherical radius of 210 mm machined into the face of the material. These samples will be indirectly cooled by clamping the sample to a spherical or flat geometry copper heat sink appropriately contoured to the respective sample geometry. The sputter probe assembly will be cooled by a freon heat exchange unit maintained at 15°C.

Vacuum Discharge Chamber and Plasma Confinement Arrangement. The vacuum/discharge chamber is made of stainless steel equipped with freon coolant passages to protect the ten sets of equally spaced SmCo and ALNiCo plasma discharge confinement magnets from thermal degradation by the radiant power incident on the walls of the

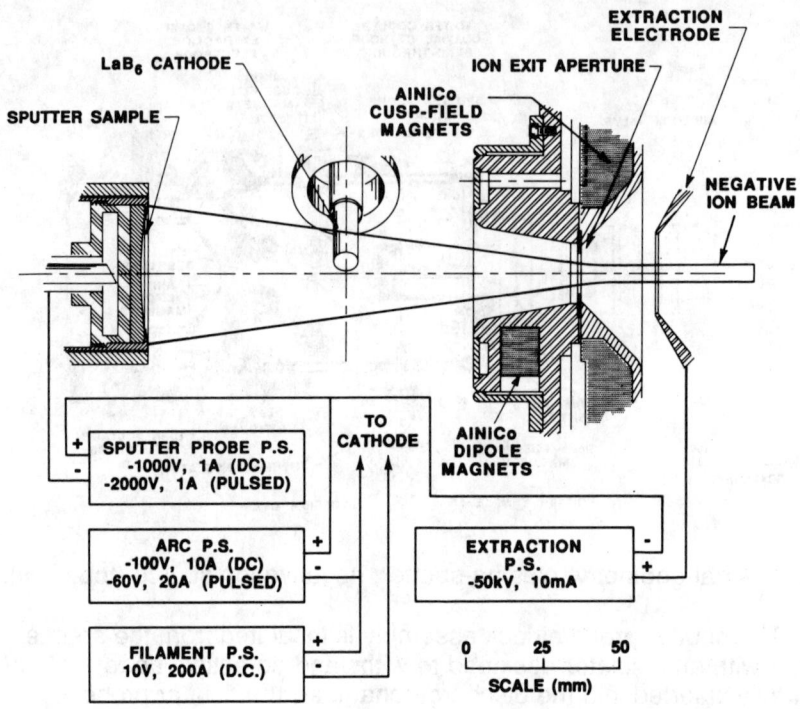

Fig. 2. Power supply arrangement for the axial-geometry, plasma-sputter, negative ion source.

chamber arising from the plasma discharge and high temperature LaB$_6$ cathodes. Plasma confinement is effected by the use of ten rows of SmCo permanent magnets, equally spaced circumferentially around the diameter of the cylindrical chamber. The external and internal flanges of the chamber are equipped with ten equally spaced, azimuthally oriented SmCo and AlNiCo magnets, respectively. The internal flange is equipped with a set of dipole AlNiCo magnets to inhibit electron extraction from the source and thereby reduce loading to the extraction power supply and a replaceable tantalum aperture. Initially, the exit aperture will be 10 mm. The chamber is attached to the rear and front flanges by means of thumb screws for ease in disassembly for cleaning and other maintenance operations.

The Ion Extraction Electrode System and Extraction Optics. One of the advantages of the plasma-type sputter negative ion source lies in the fact that, when operated in a high-density plasma mode, the negatively biased sputter probe containing the material of interest is uniformly sputtered. This characteristic makes it possible to take advantage of the large area spherical- geometry lens system which is formed between the spherical sector sputter probe and the plasma sheath which conforms to the geometry of the probe. Negative ions created in the sputter process are accelerated and focused through the plasma to a common focal point, usually chosen as the ion exit aperture of the source, and then pass into the field region of the extraction electrode system. Within the plasma, the ion beam is free of space charge effects. Thus, the sputtered particle energy and angular distributions, and aberrations in the acceleration plasma lens system determine the beam size at the focal point of the spherical lens system. At high beam intensities, space charge effects come into play when the beam exits the plasma and enters the extraction region of the source. However, because the beam energy is 500-1000 eV upon exit from the plasma region of the source, space charge influences on the beam are reduced. After exiting the source, the beam is further accelerated by a two-stage electrode system insulated by high-quality Al_2O_3 insulators. Typically, the ion beam will be accelerated through a potential difference of 20 kV in the first stage and by an additional 30 kV in the second stage. The optics of the ion extraction aperture and second stage field regions are designed to provide inwardly directed radial restoring forces to the beam to offset, in part, space charge effects which will suddenly appear in the slow moving, intense heavy negative ion beams upon exit from the plasma region of the source.

The ion optics of the ion generation and extraction regions of the source equipped with two different extraction electrode systems have been studied computationally through use of the code described in Ref. 13. Examples of such calculations are shown in Figs. 3 and 4 which display ion trajectories for a 3.5-mA O^- or a 1-mA Au^- ion beam generated at the sputter probe surface and accelerated through the field-free region of the plasma and finally into the extraction lens system at energies up to 81 keV and 51 keV, respectively. The simulation results shown in Fig. 3 are for an extraction system used at LAMPF,[14] while the results shown in Fig. 4 represent the present electrode system. The optics of the latter electrode system are better suited for beam transport into the 25URC tandem electrostatic accelerator injector and, therefore, will be incorporated as the extraction electrode system for the axial-geometry subject source.

Coaxial LaB_6 Cathode Assembly. The source will utilize directly heated coaxial-geometry LaB_6 cathodes to initiate and sustain the plasma discharge. The cathode structure shown schematically in Fig. 5 is very similar to the design described by Leung et al. in Ref. 15. The high melting point, chemical inertness, low work function, and low sputter ratio properties make LaB_6 especially attractive for such applications. The coaxial geometry is desirable because it minimizes the magnetic field surrounding the cathode and thus permits emission and escape of low-energy electrons from the surface. More importantly, the design allows easy and fast interchange of cathodes through a single metal-to-metal vacuum seal feedthrough.

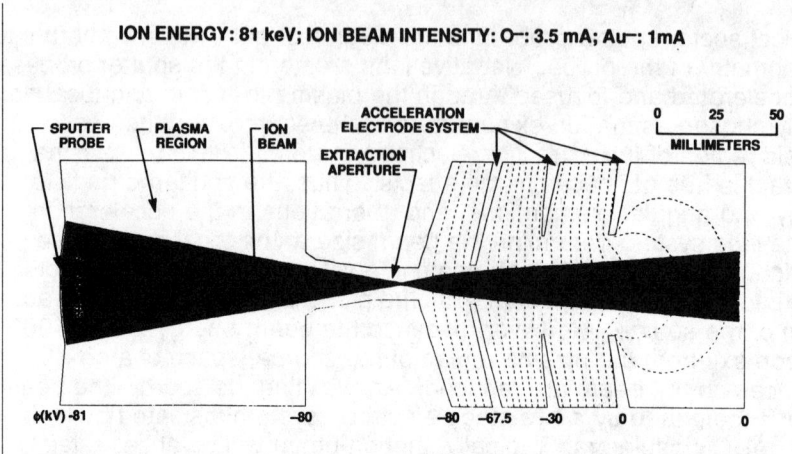

Fig. 3. Negative ion optics of the axial-geometry, sputter heavy negative ion source equipped with the extraction system used at LAMPF.[14]

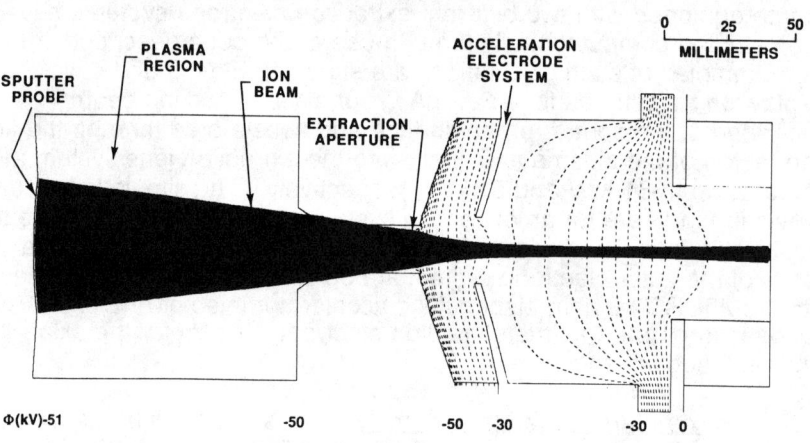

Fig. 4. Negative ion optics of the axial-geometry, sputter heavy negative ion source equipped with a two-stage extraction electrode system shown in Fig. 1.

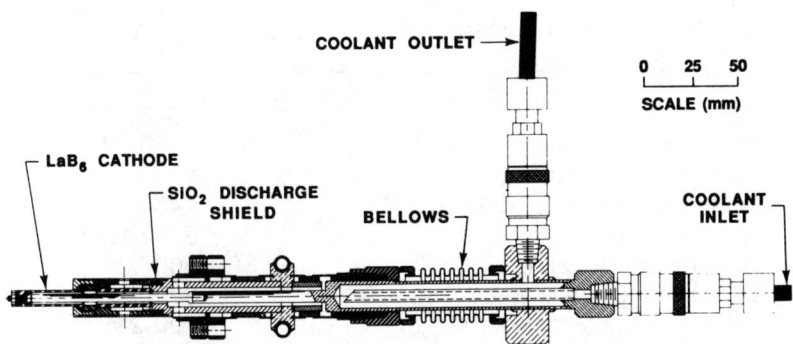

Fig. 5. Schematic representation of the coaxial-geometry LaB_6 cathode assembly.

The cathodes are machined from solid pieces of LaB_6 which have thin hollow cylindrical walls (6.4 mm diameter, 1 mm thick, and 25.4 mm in length). Each cathode assembly is 35 mm in length and has a 2-mm-thick top flange for attachment to the outer conductor and a 5-mm-thick bottom for attachment to the central tantalum conductor as shown in Fig. 5. Rhenium foil is used to provide good electrical contact between the LaB_6 cathode and the top molybedum nut and tantalum outer electrode, and between the bottom molybedum nut and central tantalum conductor.

EXPECTED PERFORMANCE CHARACTERISTICS

The operational parameters and intensity capabilities of the subject source are expected to be similar to those of the radial-geometry source and therefore, the reader is referred to Refs. 1-5 for specific information concerning the species and intensity capabilities, as well as the qualities (emittances), of beams produced in this type of source. The anticipated operational parameters for pulsed-mode operation of the source are shown in Table 1. For example, the source is expected to generate peak beam intensities close to those reported in Ref. 4 for the radial-geometry source, which includes a list of more than 20 negative ion species, including 6 mA C^-, 10 mA Cu^-; 8 mA Pt^-; and 10 mA Au^-. The emittances of beams extracted from the source are also expected to be close to those of the radial-geometry source (typically, 11-17 π mm.mrad $(MeV)^{1/2}$ at the 80% contour level, depending on the beam intensity).

Table 1. Expected pulsed-mode source operating parameters.

Arc current	15-20 A
Arc voltage	30-60 V
LaB_6 cathode current	130 A
LaB_6 cathode temperature	1450°C
Xe gas pressure	2.3×10^{-4} Torr
Sputter probe voltage	500 V
Beam extraction voltage	50 kV
Cesium oven temperature	190-205°C
Beam pulse width	1-5 Hz
Repetition rate	1-50 Hz

ACKNOWLEDGEMENTS

The author is indebted to Dr. K. N. Leung of Lawrence Berkeley Laboratory for supplying design information concerning the coaxial LaB_6 cathode assembly and to Dr. R. L. York of Los Alamos National Laboratory for information on the extraction electrode system used in conjunction with the H⁻ cusp-field source at LAMPF.

REFERENCES

1. G. D. Alton, Y. Mori, A. Takagi, A. Ueno, and S. Fukumoto, Nucl. Instrum. and Meth. A270 (1988) 194.

2. Y. Mori, G. D. Alton, A. Takagi, A. Ueno, and S. Fukumoto, Nucl. Instrum. and Meth. A273 (1988) 5.

3. G. D. Alton, Y. Mori, A. Takagi, A. Ueno, and S. Fukumoto, Nucl. Instrum. and Meth. B40/41 (1989) 1008.

4. G. D. Alton, Y. Mori, A. Takagi, A. Ueno, and S. Fukumoto, to be published in Rev. Sci. Instrum..

5. Y. Mori, A. Takagi, A. Ueno, K. Ikegami, and S. Fukumoto, to be published in Nucl. Instrum. and Meth.

6. D. K. Olsen, G. D. Alton, S. Datz, P. F. Dittner, D. T. Dowling, D. L. Haynes, E. D. Hudson, J. W. Johnson, I. Y. Lee, R. S. Lord, C. A. Ludemann, J. A. Martin, J. B. McGrory, F. W. Meyer, P. D. Miller, W. T. Milner, S. W. Mosko, P. L. Pepmiller, and G. R. Young, The HISTRAP Proposal: Heavy-Ion Storage Ring for Atomic Physics, Nucl. Instrum. and Meth. B24/25 (1987) 26.

7. V. E. Krohn, Jr., Appl. Phys. $\underline{38}$ (1962) 3523.

8. G. D. Alton, Nucl. Instrum. and Meth. $\underline{B37/38}$ (1989) 45.

9. G.D. Alton, IEEE Trans. Nucl. Sci. (in press).

10. G. D. Alton, Nucl. Instrum. and Meth. (in press).

11. G. D. Alton, Phys. Div. Progress Report, ORNL-6420 (1987) p. 222.

12. J. D. Larson and C. M. Jones, Nucl. Instrum. and Meth. $\underline{140}$ (1977) 489.

13. J. H. Whealton, Nucl. Instrum. and Meth. $\underline{189}$ (1981) 55.

14. R. L. York, private communication.

15. K. N. Leung, D. Moussa, and S. B. Wilde, Rev. Sci. Instrum. $\underline{57}$ (1986) 1274.

DISCUSSION

Belchenko: I don't understand why do you prefer to cool the walls but not to cool the sputter probe?

Alton: We do cool the sputter probe, I didn't mention that but it will be chilled as well.

Belchenko: And what was the cooling of the walls?

Alton: The cooling of the walls in the case that we want to go to CW operation, we don't have to do that except for dc operation, for that mode only. In fact, in the radial geometry source, that Yoshi Mori talked about, the walls are chilled enough that it serves as a pump for the cesium. We do not find that a handicap, but we find a serious problem if we didn't cool the walls in the dc operational mode that also serves to protect the samarium cobalt magnets that are just attached, clipped to that side of the chamber. It may be in the future that we find that we have to put an inner liner in there to run the temperature up. We think, though, that if the walls are hot the expulsion of cesium will be greater and the sparking problem will be accentuated (or time before sparking will occur).

Belchenko: Depends on the temperature of the probe. If you cool the probe so...

Alton: Yes, that is a suggestion that I have made to Yoshi (Mori) and we intend to chill that relative to the surrounding so that we can get away with less cesium flow into the source and rely on the discharge to put away the cesium to optimize the cesium coverage.

Belchenko: Another question, what is the initial temperature of the beam included in the calculations of John Whealton?

Alton: I can't answer that but I know it in practice to be of the order of 20 eV. But in his simulation, I suspect it was quite low.

Jacquot: A general question, concerning the surface processes. Normally, when we discuss about H^- prediction, we envision the role played by the H_2^+ and H_3^+ and we produce in this case a lot of negative ions like Au, Ni and so on. Do you have the same processes?

Alton: Actually you are talking about vibrational excitation of that dissociative attachment processes

Jacquot: No, not the vibrational states but the problem of the surface production.

Alton: Probably some of the same mechanisms are taking place. Because I envision even in a H⁻ source that implantation and regurgitation sputtering process have the particles kicked out. And under those conditions, the probability for a negative hydrogen ion formation should be rather high. But certainly the mechanisms for heavier ions where you are ejecting the particle and according to Norskov and Lundqvist's theory which is accepted, (and I believe is experimentally substantiated), it should depend on the velocity. So if you look at the energy distribution of sputtered particles and do a calculation based on that you find that probability is enhanced as one increases the velocity of the sputtered particles. Well you can do that in two ways, you can increase the primary particle energy which accentuates the tail, the other thing that's important is that those distributions are sensitive in shape and magnitude on the work function and different coverages. The probabilities are different for different positions in the energy spectrum. That is predicted by Lundqvist and Norskov, but no experimental verification of that effect has been seen. We hope to do that at Oak Ridge.

DIAGNOSTICS

SPECTROSCOPIC STUDY OF HYDROGEN-CESIUM DISCHARGE PLASMA OF SURFACE-PLASMA ION SOURCES

V.V. Antsiferov, V.V. Beskorovaynyy, A.M. Maximov, P.G. Sova, and L.P. Skripal'
Sukhumi Institute of Physics and Technology
384914, Sukhumi, USSR

Yu.I. Belchenko and G.E. Derevyankin
Institute of Nuclear Physics
630090, Novosibirsk, USSR

ABSTRACT

We made a spectroscopic study of the plasma of a high-current hydrogen-cesium glow discharge with planotron and Penning electrode geometries. The elemental and charge compositions of the plasma were determined. The dynamics of hydrogen, cesium, and molybdenum spectral lines was investigated at different discharge parameters. The changes in the densities of atomic hydrogen, ionized and atomic cesium were observed, along with the sputtering of the electrodes during the discharge pulse.

The H_α- and H_β-line contours were measured with high resolution in a single discharge pulse ($\simeq 10^{-3}$ s). We determined the temperature of atomic hydrogen and the density of the plasma. The radiation energy of the plasma was measured within the 400 - 800 nm spectral range.

INTRODUCTION

A high-current glow discharge in hydrogen with added cesium is widely used in various modifications of Surface-Plasma Sources (SPS) of Negative Ions, developed for injection into accelerators and for fusion devices[1,2]. The physical processes and characteristics of plasma in these discharges have not been well studied. In contrast to dense cesium-hydrogen discharges[3,4], the discharges of SPS are strongly affected by near-electrode nonequilibrium processes and have a more complicated structure due to the external magnetic field. These facts do not permit us to use simple theoretical models of local thermodynamic equilibrium, Maxwellian electron energy distribution in descriptions of the discharge, and modelling.

The small thickness of SPS-discharge plasma layers, the effect of cesium adsorption and the magnetic field hinder the use of probable methods for measuring the parameters of the discharge plasma. Spectroscopy is better for the investigating of high-current discharge plasma. This method has several substantial advantages: high sensitivity, high selectivity and high measurement rate, nonsusceptibility to electromagnetic interference, and the absence of perturbing action on the object. The first spectroscopic measurements of SPS were made in 1977[5]. Later detailed spectroscopic measurements of plasma radiation made it possible to determine the density and temperature of atoms in a hydrogen-cesium discharge with Penning (PIG) electrode geometry[6]. Using vacuum ultraviolet laser-absorption spectroscopy, the $H^°$ density and temperature were measured in the plasma column and drift region of Penning SPS[7].

In this work, we present the results of spectroscopic measurements of a gas discharge with planotron and Penning electrode geometries. The elemental and charge compositions of the plasma were determined. The dynamics of the intensities of the hydrogen, cesium, and molybdenum spectral lines were investigated at different discharge parameters. The H_α- and H_β-line contours were measured with high resolution. We compared the temperature of atomic hydrogen in the planotron and PIG discharges.

1.0 Experimental Arrangements

We investigated the radiation of hydrogen-cesium discharges with planotron electrode geometry[8] and with modified PIG geometry[9]. The arrangement of the planotron is shown in Fig. 1a. A massive cathode made of high-purity molybdenum was placed within an anode chamber, also made of molybdenum. A diagnostic window of 0.6 x 15 mm in area was drilled in the anode's cover to transmit radiation to the diagnostic equipment. The cathode surface, facing the diagnostic window, was shaped as a concave cylindrical groove, with a radius of curvature of 7 mm, as in the planotron SPS with geometrical focusing of negative ions[10,11].

The external magnetic field, whose direction is shown with an arrow, provided electrons oscillations in the region between the opposite sides of the cathode groove and their E x B drift along the groove in crossed electric and magnetic fields. For closing the electron drift around the cathode body, the side projections were put at its ends and its upper part (Fig. 1A - spool-like cathode).

In the modified PIG geometry, electrons oscillate between the opposite cathodes (1a and 1b in Fig. 1b) in E||B fields. The PIG light emission was studied for several discharge regions with four diagnostics slits, each of 0.2 x 10 mm in area (Fig. 1b). Slit I was made in the near-cathode zone, so that it was impossible to see the cathode's emitting surface. The position of slit IV allowed us to see the surface of the cathode 1b within a small part (2°) of the total acceptance angle (12°) of the diagnostic slit. Slit II was situated near the central plane of plasma column, and slit III was at the intermediate point (Fig. 1b). The high-current discharge was supplied with purified hydrogen via channels in the anode wall. Cesium was fed into the discharge from an external heated container of cesiated pyrographite. Both planotron and PIG discharges had the following main parameters:

Discharge voltage, Volts	100 - 600
Discharge current, A	5 - 150
Pulse duration, μs	35,600,850
Pulse repetition rate, Hz	1 - 10
Magnetic field, T	.05 - 0.15
Density of H_2, cm^{-3}	up to 10^{16}
Density of Cs, cm^{-3}	up to 10^{13}
Cathode average temp., °C	400 - 800
Anode average temperature °C	200 - 400
Planotron, cathode area cm^2	5
PIG cathode area, cm^2	2.5

Several stable Modes of the hydrogen-cesium discharge were observed, their voltage and volt-ampere characteristics being different (Fig. 2). Mode IV was realized in a pure-hydrogen discharge without cesium. Hydrogen-cesium Modes I - III differed in the amount of cesium deposited on the electrodes, and were monitored by the electrode temperature and the cesium feed rate to the discharge. Thus, Mode III had only a small of cesium on the electrodes, while for Modes I, and Ia, the electrodes were well activated with cesium ($\theta \simeq 0.5 - 1$).

2.0 Diagnostic Equipment

Plasma radiation passed through the diagnostic slit then through the quartz glass window of the vacuum chamber, and was focused by two lenses on the entrance aperture of the spectroscopic lines. Several methods were used for spectroscopic study.

Spectrograms of the plasma radiation in the 230 - 900 nm range were recorded with an industrial spectrograph STE-1 (line I, Fig. 3); the light was transported with a fiber-optic cable. The element composition of the plasma was determined on the basis of the spectrograms.

Photoelectric measurements of the spectral line intensities were made using an industrial monochromator (MC) MDR-2 and a sensitive photoelectric multiplier (PEM) (line II, Fig. 3). The variation in the intensities of separate spectral lines was traced during the discharge pulse. The absolute intensity of spectral lines was determined by calibration of the spectroscopic line with a standard tungsten lamp, SI-8-200U.

Interferograms of spectral lines H_α and H_β were recorded with a Fabry-Perot interferometer (FPI) (line III in Fig. 3). The specially designed interference filters (IF in Fig. 3) and mirrors of the FPI reduced light losses. The IF transmission bandwidth was 4 nm at the half-amplitude level. The IF transmission coefficient was 80%. The spectral resolution of FPI was 1/50 of the dispersion region. The dispersion region of FPI was 0.1 nm for planotron measurements and 0.2 nm for PIG measurements of the spectral line

H_α. The high transmission of IF and the low absorption of the FPI mirrors allowed us to record interferograms of the H_α and H_β lines in a single discharge pulse (10^{-3} s). The average radiation energy of a discharge plasma in the 400 - 800 nm spectral band was measured by semiconductor detector-convertor (SDC in Fig. 3, line IV). The infrared part of the spectrum was eliminated by a calibrated light filter, while the ultraviolet radiation of the plasma was cut off by the spectral sensitivity of the SDC.

3.0 Experimental Results

3.1 Spectrum and Energy of Discharge Radiation

A large number of spectral lines were recorded in the 230 - 900 nm spectral region. Indentification was made for the spectral lines of atomic hydrogen (H_α, H_β, H_γ, H_δ) and line spectrum of molecular hydrogen ($d^3 \Pi_u \to d^3 \Sigma_g^u$, α - Fulcher system, 580 - 620 nm). Also, electronic-rotational transitions of hydrogen molecules $I^1 \Pi g \to B\ ^1\Sigma_g^+$, $G\ \Sigma_g^+ \to B\ ^1\Sigma_u^+$ were recorded in the 420 - 490 nm region. The spectral lines of cesium atoms CsI, singly-charged cesium ions CsII (300 - 650 nm), and doubly-charged ions CsIII (250 - 290 nm) were identified. At high discharge currents, transitions were recorded from the highly exited levels of cesium atoms with η_{max} to 15 (15 $P_{3/2} \to 6\ S_{1/2}$). Lines of molybdenum MoI, singly-charged molybdenum ions MoII, and oxygen OI were indentified.

The average gas discharge radiation energy of the 400 - 800 nm spectral region, measured with semiconductor detector-converter, had the value $\simeq$ 3 W/cm$^2 \cdot$ ster at planotron discharge power of 10 kW.

3.2 Intensity of Radiation of Separate Spectral Lines

The relative magnitude of separate spectral lines and its variation during the discharge pulse, recorded by the photomultiplier is presented in Figs. 4 - 6. When recording the signals of Figs. 4 - 6, the fluctuations and noise were integrated by RC-filters. Figure 4 is a set of oscillograms for the gas-discharge Mode I of planotron.

The oscillograms in Fig. 5 (Mode III) were obtained in a short pulse (35 μs) of planotron and which allowed us to follow in detail the dynamics of radiation at the pulse fronts.

<u>Hydrogen</u> - The spectral H_α line was the brightest in the studied spectral region of the discharge radiation. In Mode I with reduced hydrogen density in the gas-discharge chamber, the intensity of the Balmer lines was more sensitive to the amount of hydrogen admitted. At low discharge currents (10 - 20 A) in Mode I, the oscillograms of the Balmer line intensities had a "bell-shaped" form, corresponding to the change in the gas density in the discharge chamber. The position of the maximum in the line intensity coincided with the maximum in the gas density, and shifted synchronously with change in the start time of the valve admitting the hydrogen. At a discharge current of 75 - 100 A, the H_α inten-

sity was at a maximum in the initial part of the pulse, especially if the hydrogen was admitted late into the discharge chamber, when a high-voltage step appeared at the beginning of the discharge voltage pulse (Fig. 4).

Inspite of the increase discharge voltage and gas density, in Modes I and II, the intensity of the Balmer lines had approximately the same value as the corresponding discharge currents of Mode I. The oscillogram of the Balmer-line intensities follow the shape of the discharge current pulse; the intensity was directly proportional to the discharge current and varied slightly with the increase in the gas feed to the discharge.

The leading edges of the oscillograms of the Balmer line intensities in Modes II - IV had the same shape and duration as the discharge current pulse fronts (Fig. 5), in contrast to the fronts in the oscillograms of the line intensities of cesium and molybdenum. The oscillograms of the Balmer line intensities for PIG discharge were similar to those of the planotron (Fig. 6).

The change in the glow intensities of the spectral lines of hydrogen as a function of the discharge current is shown in Fig. 7 (Mode I, planotron). The reading of the intensity was made at the "plateau" in the middle part of the pulse. The absolute value of the line intensity was recalculated according to the calibration of spectroscopic path. The data were recorded in the short-time interval so that the thermal conditions of the discharge electrodes did not change. The hydrogen feed was minimized at high discharge currents.

In the 20 - 50 A range of the discharge current, the radiation intensity of the hydrogen lines increased proportionally to the discharge current. At higher currents, the growth rate of the H_α and H_γ intensities decreased, while the intensities of the H_β and H_2 lines were saturated. In the discharge current range we studied, the H_α intensity was 12 - 15 times higher than that of the H_β line, and two orders of magnitude higher than that of H_γ line.

Cesium - The intensity and dynamics of the atomic and ionized cesium spectral lines were strongly dependent on cesium feed into the discharge chamber and on the temperature of the electrodes. In discharge Mode I, the initial intensities of the CsI and the CsII lines were low and increased at the end of the pulse (Fig. 4), especially in the vicinity of the PIG cathode (Fig. 6). This growth rate increased with discharge current and cesium feed. So, for forced cesium feed, (Mode Ia) and for high discharge currents the intensities of the cesium lines increased rapidly during the first 200 μs of the pulse by 5 - 10 times and then saturated (Fig. 6, Mode Ia).

Figure 8 shows the intensity of the cesium lines as a function of discharge current (Mode I of planotron). The readout was made from the intensity maximum and was recalculated according to calibration of spectroscopic path. As in the case of Fig. 7, the measurements were made in the short time interval.

The intensity of the CsII lines was usually higher than for most of the CsI lines. Only the brightest CsI line (852.1 nm) had the higher intensity than did CSII. With an increase in the discharge current, the intensity of the CsI (455.5) line reached saturation, while the intensity of CsI (852.1) and CsII line continued to rise rapidly.

In discharge Mode II and III, the intensity of the cesium lines reached a relatively high steady-state level in 5 - 10 μs after switching on the discharge (Fig. 5). The steady-state level of the CsII lines intensity was proportional to the first power of the discharge current.

The leading edges of the Cs and Mo spectral line pulse were more elongated in duration than the fronts of discharge current pulse and the Balmer lines pulse, while the trailing edges were conversely steeper than those of the discharge current and the Balmer lines (Fig. 5).

<u>Molybdenum</u> - The spectral lines intensities of molybdenum, sputtered from the gas-chamber electrodes had a relatively low value and in Modes I and II and had their maximum at the beginning of the pulse (Figs. 4, 6a). The peak intensity of this maximum was proportional to the discharge current. For Mode I, the intensity of the MoI lines at the "plateau" of the pulse increased linearly with the discharge current. In Mode II, this "plateau" intensity was maximal at a discharge current of 40 A (Fig. 8). The intensity of MoI lines was higher for discharge Modes II - IV with increased discharge voltage and with less coverage by cesium of the electrodes.

<u>Distribution of PIG Radiation</u> - Data on the intensities of spectral lines H_α, CsI, CsII, and MoI recorded for various diagnostic slits of PIG-discharge are shown in Fig. 9. The readings were made at several times during the pulse (at 50 μs, 250 μs, 500 μs, and 750 μs from the start).

The intensity of the Balmer lines was twice as high at diagnostic slits II, III than for slits I, IV. For spectral lines of molybdenum and cesium the distribution of radiation was asymmetrical. Thus, the intensity of the cesium line was 5 - 10 times higher at diagnostic slit IV than at slits I, II, and III. Towards the end of the pulse, the cesium radiation from slit I was 2 - 3 times higher than from slit II and III (Fig. 9).

3.3 Fluctuations in the Glow Intensity of Spectral Lines

The level of the low-frequency fluctuations ($< 10^6$ Hz) and noise (LFN) of the spectral lines' intensity, recorded with a wider band recording system, was increased with discharge current growth. However, the relative level of the LFN, defined as the ratio of the root-mean-square value of fluctuations to the mean value of intensity, generally decreased with the increase in discharge current. The dependence of the relative LFN level for the intensities of the H_β and CsII spectral lines on the external magnetic field, B, is shown in Fig. 10. The figure also shows the dependence of the relative level of discharge current on B. With an increase in magnetic field and with a decrease in hydrogen and cesium feed into the discharge, the LFN level increased. Mode I had much lower level of LFN than did Modes II - III.

3.4 Broadening of Spectral Lines H_α and H_β

Figure 11 presents typical interferograms of the spectral line H_α obtained during one gas-discharge pulse of planotron and PIG discharges. The contours of the spectral lines were determined by microphotometry of the interferograms. The fine structure of the lines is not displayed. The change in the total contour width at the half-height, $\Delta\lambda$, for the H_α line as a function of the discharge current, is shown in Fig. 12 (left scale). With the increase in the discharge current of the planotron from 10 to 100 A, the total width of H_α contour at half-height increased from 0.03 to 0.06 nm (I in Fig. 12) and was slightly dependent on the external magnetic field. The width of H_α contour $\Delta\lambda$ for PIG discharge was 2 - 3 times higher than for the same discharge current of planotron (II and VI in Fig. 12) The data on $\Delta\lambda$ of PIG discharge displayed much higher instability and spread of value.

The total width of the H_β line contour at half-height at a discharge current of 100 A was approximately 4×10^{-2} nm (planotron).

4.0 Discussion of Results

The spectroscopic measurements suggested that in Mode I of the activated electrodes, the density of atomic hydrogen in the discharge was proportional to the discharge current in the 10 - 50 A, range and was sensitive to the hydrogen feed. The initial overshoot in the oscillograms of the Balmer lines at high discharge currents indicates that there is a relatively high contribution from atomic hydrogen desorbed from the surfaces of the gas-discharge chamber electrodes (Figs. 4, 6).

The slowing down in the growth rate of the Balmer lines and the saturation of the glow intensity of the molecular hydrogen lines at high discharge currents from the planotron is evidently caused by the displacement of hydrogen from the discharge zone adjacent to the diagnostic window into the side gaps of the gas-discharge chamber[12].

The experimentally recorded width of the H_α line contour and its change with discharge current were caused mainly by dopler broadening. The apparatus linewidth of the Fabry-Perot internal standard, measured with a single-frequency He-Ne laser (λ = 6328 Å), was 5×10^{-3} nm, with a standard base of 2 mm. The Stark broadening of the H_α line ($\Delta\lambda \simeq 5 \times 10^{-3}$ nm, when $n_e \simeq 10^{13}$ cm^{-3}) can be neglected.

4.1 Temperature of Hydrogen Atoms

The temperature of hydrogen atoms determined by the doppler broadening formula:

$$\kappa T_{[eV]} \simeq 4 \times 10^2 \Delta\lambda^2 \text{ [nm]}$$

had a value of 0.3 eV for a 10 A discharge current: when the discharge current was raised to 100 A, this value increased up to 1.4 eV (for the planotron). We note that these values were obtained without allowing for the fine structure of the H_α line, which is valid if the predominant process of atomic excitation is, for example, excitation during dissociation of hydrogen molecules with electron bombardment, when about 70% of the energy of the H_α line is radiated in the transition $3^2D_{5/2} \to 2^2P_{3/2}$[13]. In the pulsed SPS discharges, a considerable fraction of the hydrogen atoms was formed by desorption from the electrode surfaces, which may lead to broadening of the H_α line contour, in addition to the fine structure. Recalculation, with a correction for the fine structure of the H_α line at planotron discharge currents of 10 - 30 A, yielded a lower atomic temperature of 0.2 - 0.5 eV.

A higher value of atomic temperature for the same discharge current of PIG was caused by the increased density of the PIG discharge current and the corresponding growth in the level of plasma LFN.

4.2 Temperature of the Electrons

From the ratio of the Balmer lines intensities, we can estimate the temperature of the electrons in the discharge from an approximation of local thermodynamic equilibrium and the Maxwellian electron energy distribution. Calculation based on the formula:

$$\kappa T_e = E_j - E_i/\ln[(A_j g_j \lambda_i I_i \kappa_i)/(A_i g_i \lambda_j I_j \kappa_j)]$$

(the designations are standard) yielded a low value of "hydrogen" temperature for electrons $T_e \simeq 0.3 - 0.4$ eV.

The anomalously low electron temperature was obtained when we compared the intensities of different cesium lines and also from the distribution of intensities in the recombination continuum spectrum at the 5d level:

$$Cs^+ = e \to Cs^* + h\nu, \quad \lambda \simeq 550 - 590 \text{ nm}$$

Here the temperature of the low-energy fraction of the energy was calculated from the slope of the line $\ln(I/\nu) = f(h\nu - h\nu_0)$, where I is the continuum intensity, ν is the frequency, and ν_0 is the cutoff frequency[14]. Since the recombination spectrum is inversely proportional to the electron energy $(h\nu - h\nu_0) =$ const, this precludes the need for introducing a correction into the preexponential term when calculating the temperature from the slope of the semilogarithmic lines. The "cesium" electron temperature was 0.12 eV and depended very slightly on the density of cesium. The intensity of the recombination continuum fell off quadratically with a decrease in discharge current.

The anomalously low values for electron "temperature" indicated that the local thermodynamic equilibrium model has limited applicability under the conditions of high-current glowing discharge; this nonequilibrium status of processes is especially evident at the surfaces of the electrodes. In addition to considering the low-energy group of Maxwellian electrons in SPS and related high-current discharges in crossed fields, it is evident that we must also consider the role of the group of fast primary electrons.

4.3 Distribution

The Balmer lines intensity had approximately the same value for PIG diagnostic slits II, III (with a correction for the increased angle of acceptance of slits II and III for plasma radiation). This finding indicates that the distribution of hydrogen and plasma density was uniform along the discharge gap.

The asymmetry of intensities of the cesium and molybdenum spectral lines for slits I and IV (Fig. 9) indicates that the main contribution to this radiation was made by the emission surface of the cathode and the adjacent near-cathode plasma layer, seen only through the slit IV. The asymmetry is apparently caused by accumulation of cesium in the near-cathode region due to rapid ionization of the cesium atoms in the dense plasma and its return to the cathode by discharge electric field[15]. So, the main quantity of cesium is circulated in the near-cathode region, and is accumulated during the discharge pulse. The density of cesium was small in the bulk of the discharge gap (slits II, III).

The high value of the intensity of the CsI (852.1 nm) spectral line and its power-squared dependence on the discharge current (Fig. 8) indicates that there is a large contribution of the recombination processes like

$$Cs^+ + H^- \rightarrow Cs^*(6p) + H_o$$

(with cross-section of 10^{-13} - 10^{-14} cm^2) to the luminescence of spectral line CsI (852.1). The low value of the CsI (455.5 nm) line intensity and its much lower dependence on the discharge current shows the domination of another channel for the Cs(7p) level exitation (for example, electron collisions with cesium atoms).

The slight increase and saturation of the CsI (455.5) line intensity at a discharge current of 50 - 100 A are associated with an increase in the ionization rate of cesium, with a reduction in its density, and with "burning up" in the interelectrode space. The rise in the density of electrons and cesium ions in the discharge caused a rapid power-squared increase in the discharge current in the amplitude of the CsII signal (Fig. 8).

The longer duration (5 - 10 μs) of the leading edges of the CsI and CsII signals than that of the discharge current was due to the additional release of cesium into the discharge caused by bombardment of the electrode. In Modes II and III with hotter electrodes, the initial density of cesium in the gas-discharge chamber space had a higher value, while the cesium coverage of the electrodes was low. The sputtering coefficient was high due to the increased cesium ion content in the current at the cathode and the increased discharge voltage which promotes the CsI and CsII signals arriving rapidly at a relatively high steady-state level (Fig. 5). The linear rise in the intensity of the CsII lines at the plateau of the pulse in Modes II - III indicated a slight change in the bulk density of the cesium ions with variation in discharge current in the range 20 - 100 A.

4.4 Molybdenum

From Figs. 5 and 6, we conclude that in Modes I - II molybdenum was knocked out of the electrodes mainly at the beginning of the gas-discharge pulse. The reduction in molybdenum sputtering and the arrival of the moI signal at the steady-state level in 200 μs after the beginning of the pulse are accounted for by an increase of the cathode coverage with cesium, up to a dynamic equilibrium that is determined by the current of cesium ions toward the cathode and by the coefficient of cesium sputtering.

Molybdenum also is easily ionized and confined by a discharge electric fields in the near-cathode region. Therefore, the molybdenum radiation had higher intensity for PIG slit IV than for slits II - III, especially at the start of the pulse at high discharge currents (Fig. 9).

In Mode Ia, with forced Cs feed into the discharge, the initial increase in the intensity pulse of the MoI spectral line, was weakly detected and the signal "plateau" had a much higher level due to the high growth rate and value of the cesium ion current to the cathode. The "plateau" intensity of MoI signals was slightly increased (Modes I, Ia) or even decreased (Mode II) with the growth of the discharge current in the range 40 - 100 A (Fig. 8) due to an increase in the dynamic-sustained cesium coverage of the cathode.

The dynamic equilibrium of coverage of the cathode with cesium was confirmed by arrival at the steady-state level of discharge voltage (Fig. 5). Fluctuations in the current of cesium ions at the cathode led to local changes in the coverage and in the emissivity of the cathode, which may be a cause of the low-frequency fluctuations in plasma density in the discharge and the fluctuations in the spectral line intensities, and may also lead to heating of the hydrogen atoms, with increase in the discharge current (Fig. 12).

REFERENCES

1. "Proceedings Symp. on the Production and Neutralization of Negative Hydrogen Ions and Beams", ed. Kr. Prelec, New York: BNL-50727, 1977.

2. "Production and Neutralization of Negative Hydrogen Ions and Beams" (Proc. of IV. Intern. Symp., Brookhaven, 1986) ed. J. Alessi: AIP Conf. Proc. No. 158, New York, 1987.

3. Morgulis, N.D. and V.I. Klapchenko, Soviaet Ukranian Journal of Phys., 21, N. 2, 181 (1986).

4. Baksht, F.G. and V.G. Ivanov, Soviet Pis'ma v ZhtF (Letter to Journ. of Tech. Phys.) 12, N. 11, 672 (1986).

5. Grossman, M.W. in [1], p. 105.

6. Smith, H.V., P. Allison, and R. Keller in [2], p. 181.

7. Smith, H.V., P. Allison, E. Pitcher et al. to be published in Rev. Sci. Instrum., January 1990.

8. Belchenko, Yu.I., G.I. Dimov, and V.G. Dudnikov, Nucl. Fus., 14, 113 (1974).

9. Dimov, G.I. G.E. Derevyankin, and V.G. Dudnikov, IEEE Trans. Nucl. Sci., NS-24, N.3, 1545 (1977).

10. Belchenko, Yu.I., G.I. Dimov, V.G. Dudnikov, and A.S. Kupriyanov, to be published in Rev. Sci. Instrum., January 1990.

11. Alessi, J.G., and Th. Sluyters, Rev. Sci. Instrum., 51, 1630 (1980).

12. Belchenko, Yu.I., G.I. Dimov, V.G. Dudnikov and A.S. Kupriyanov, Revue Phys. Appl., 23, 1847 (1988).

13. Polyakova, G.N., A.I. Ranyuk, and V.F. Yerko, Soviet ZhETF, 73, N. 6, 2131 (1977).

14. Moizhes, B. Ya, and G.Ye. Pikus, Thermoemissive Converters and Low-Temperature Plasma, Moskow, Nauka, 1973, p. 480.

15. Belchenko, Yu.I., V.I. Davydenko, G.E. Derevyankin et al. Soviet Pis'ma v ZhTF (Letter to the Journ. of Tech. Phys.) 3, 693 (1977).

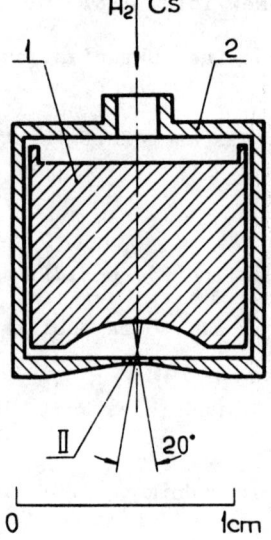 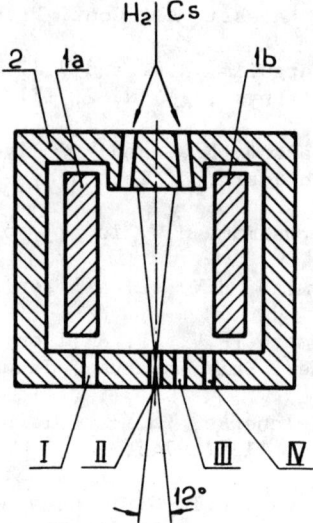

Fig. 1 Layout of the planotron (left) and PIG source (right). 1, 1a, 1b - cathode, 2 - anode. I, II, III, IV - diagnostic slits.

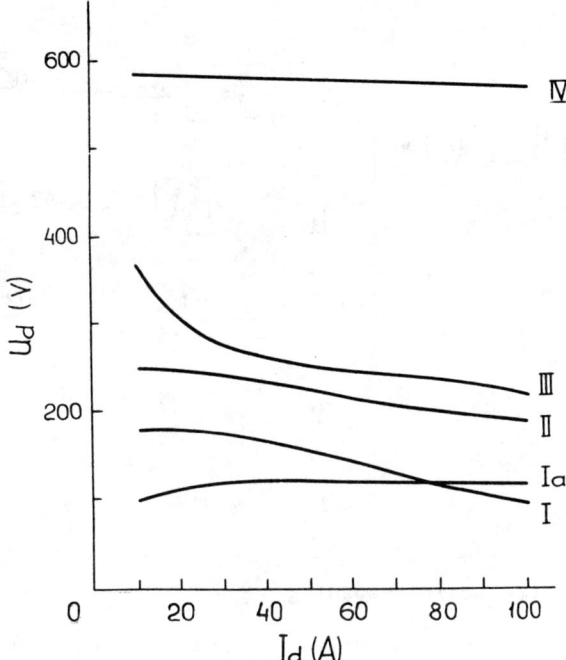

Fig. 2 Volt-Ampere characteristics of several discharge modes of SPS.

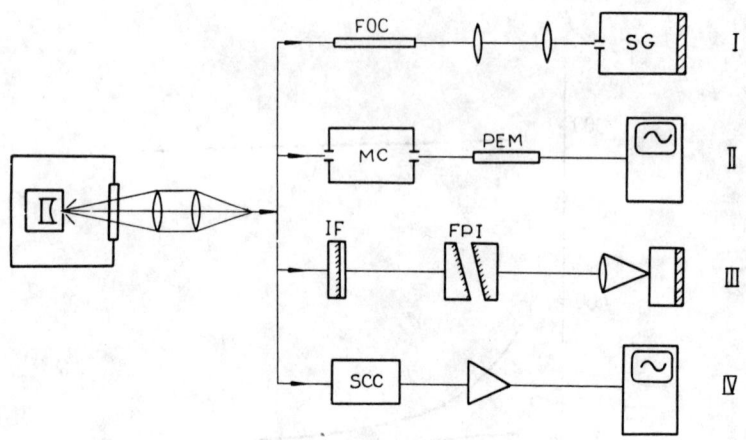

Fig. 3 Spectroscopic recording system

   I - Spectrum measurements: FOC - fiber optic cable, SG - spectrograph

   II - Photoelectric measurements: MC - monochromator, PEM - photoelectric multiplier

   III - Recording of Balmer line contours: IF - interference filter, FPI - Fabry-Perot Interferometer

   IV - Plasma radiation energy measurement: SCC - semiconductor convertor.

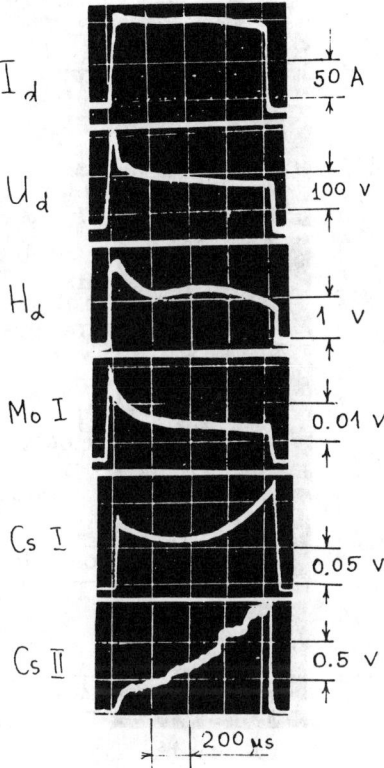

Fig. 4 Oscillograms of discharge current I_d, discharge voltage U_d, intensities of spectral lines H_α, CsI (852.1), CsII (460.4) and MoI (553.3) in Mode I of planotron.

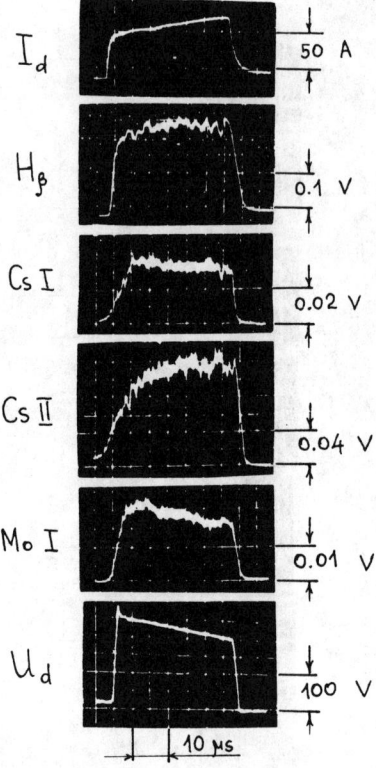

Fig. 5 Oscillograms of discharge current I_d, intensities of spectral lines H_β, CsI (455.5) CsII (460.4), MoI (553.3 nm) and discharge voltage U_d for shortened gas-discharge pulse. Mode III of planotron.

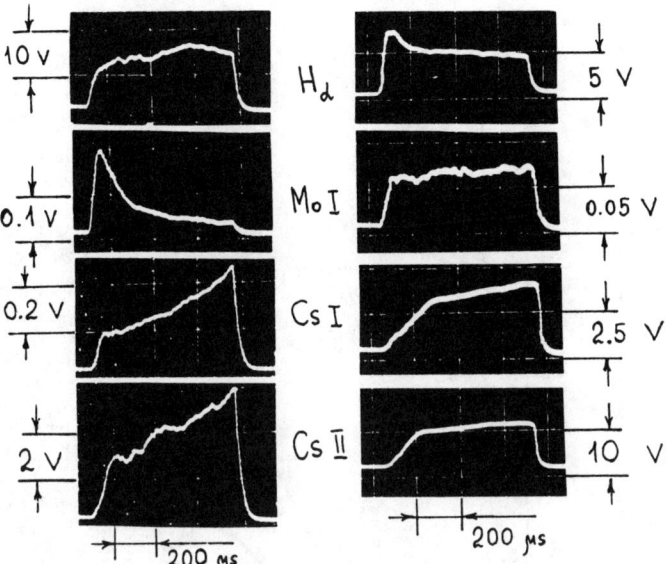

Fig. 6 Oscillograms of intensities of spectral lines H_α, MoI (553.3), CsI (455.5) and CsII (400.4) for PIG Mode I (left) and Ia (right). Diagnostic slit IV.

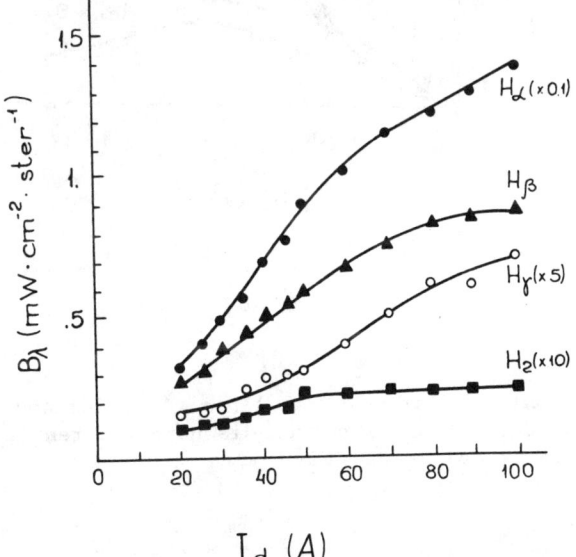

Fig. 7 Luminosity B_λ of Spectral lines H_α, H_β, H_γ, and molecular hydrogen (458.3 nm) as a function of discharge current I_d. Mode I of planotron. Readout made at 400 µs from pulse start.

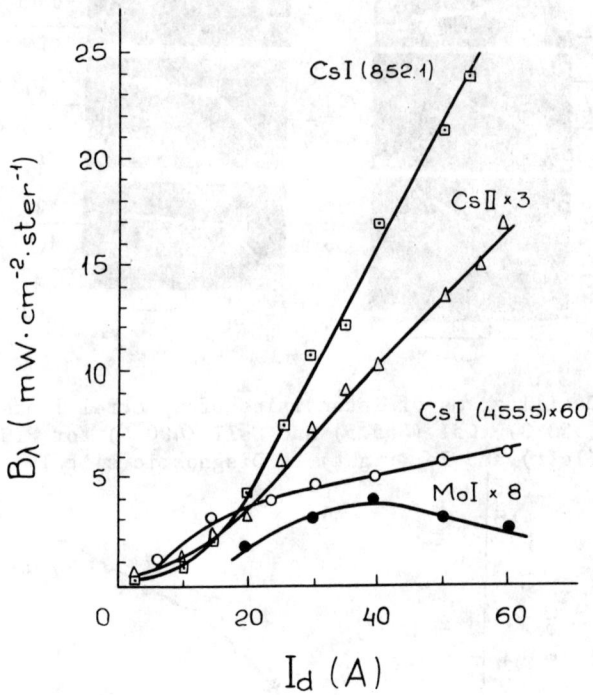

Fig. 8 Luminosity B_λ of spectral lines of cesium and molybdenum as a function of planotron discharge current I_d. Cesium - Mode I, Moly-Mode II.

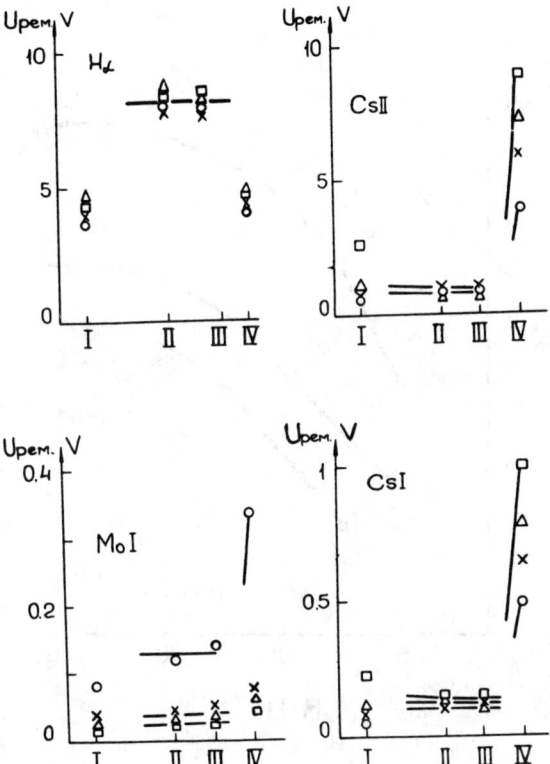

Fig. 9 Intensities of spectral lines H_α, CsII, CsI (455.5) and MoI radiation for various diagnostic slits I - IV of PIG discharge (Mode I, discharge current 80 A).

Circles - readout made at 50 μs after pulse start. Crosses at 250 μs after pulse start. Triangles - 500 μs, squares - 700 μs after pulse start.

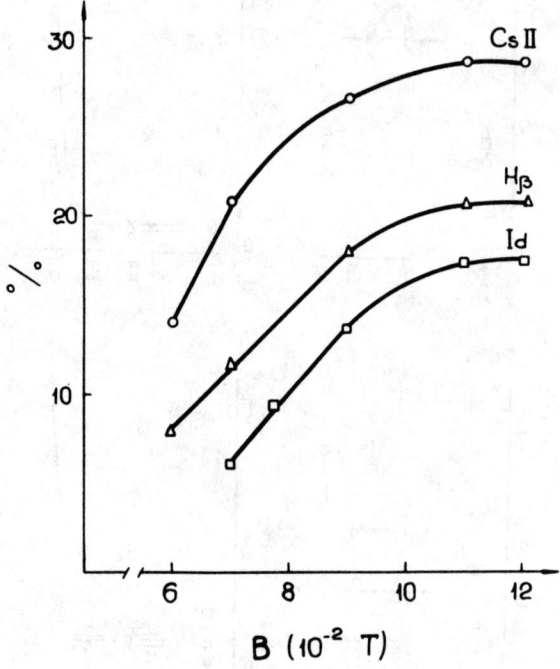

Fig. 10 LFN relative level for spectral lines CsII, H_β intensity and of discharge current I_d as a function of external magnetic field B.

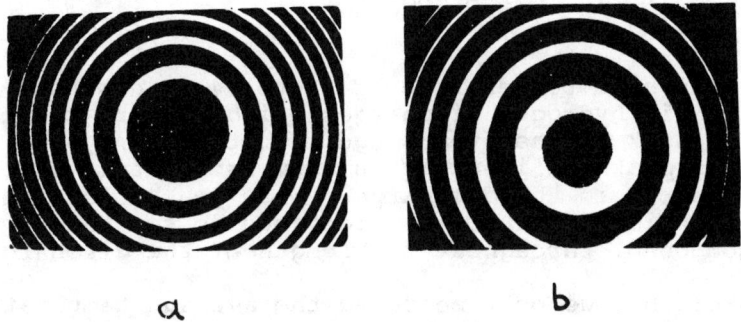

Fig. 11 Interferograms of spectral line H_α for planotron (left) and PIG discharge (right).

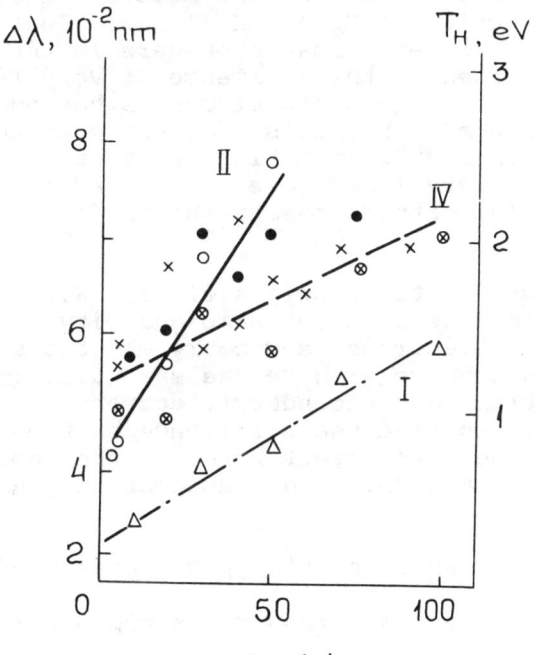

Fig. 12 Dependence of total width of H_α line contour at half-height $\Delta\lambda$ (left scale) and atomic temperature (right scale) on discharge current I_d. Mode - :

I (triangles) - planotron.

II (circles and solid circles) - PIG, slit II.

IV (crosses and crossed circles) - PIG, slit IV.

DISCUSSION

Hiskes: An atom temperature of .3 to .4 eV has been identified in hydrogen discharges going back to the positive ion work fifteen years ago, I noticed in your desorption peak you identify an atom temperature of .3 to .4 eV early in the discharge. Then you follow the H_α during the discharge. Do you continue to see the .3 to .4 eV component throughout the length of the discharge?

Belchenko: No, we only measured the average temperature during the total pulse because the transmission of our filter and interferometer permit us to measure only the average signal for a one millisecond pulse not shorter.

Bretagne: Just a comment on the question about the problem of the profile of H_α and H_β lines. In fact the existence of this very broad shoulders in the lines is directly connected to the existence of very fast electrons in the discharge, with electrons having an energy of more than 30 eV. So, it's a direct mean to control the electron energy distribution function. Probably your plasma is very inhomogeneous. Have you tried to examine near the cathode sheath the profile of the Balmer lines.

Belchenko: No, we tried only 4 slits. With plasma in the middle plane we have the same quantity of primaries and the plasma electrons, secondary electrons. I think that in PIG discharge we have the same distribution of primary electrons and secondary electrons. So it is very difficult to find the difference in the temperature. Maybe the main effect here, is the cooling of atoms near the cathode. Did I understand your question right?

Bretagne: Yes. What is the typical pressure?

Belchenko: The typical pressure is approximately 30 milliTorr.

A-M Bruneteau: In the last slide, how did you obtain atomic temperature?

Belchenko: We usually estimate the temperature of atomic hydrogen by charge exchange of fast positive ions in the near anode region on atomic hydrogen so we can measure the energy spectra of respective negative ions. And when we, for example, change influx of hydrogen or when we change the discharge current, we can see the changes in the output of primary negative ions and secondary negative ions created in the near anode region. So in the near anode region, we have high density of atoms. We know the cross section of such charge exchange processes, so it is possible to evaluate the density of atoms too.

Allison: Would you care to speculate on why the atom temperature goes up with discharge current?

Belchenko: I think that it is due to the reason that already was mentioned by John Hiskes. We have the increased desorption of the atoms in the discharge.

Allison: It certainly is striking that your measurements and ours also show that the atom temperature is apparently higher than the electron temperature.

Belchenko: Yes, somewhat. Because you can desorb an atom with an energy of up to 100 eV. You have a distribution of desorbed atoms. Of course, most of them have a temperature of maybe 5 eV. If you have a very good coverage of cathode with cesium, the cesium condenses many hydrogen atoms, and you begin to sputter hydrogen at high rate so you can have very high temperature atoms.

VUV LASER DIAGNOSTICS OF H⁻ ION SOURCES

A.T. Young, G.C. Stutzin, K.N. Leung, and W.B. Kunkel
Lawrence Berkeley Laboratory, University of California,
Berkeley, California 94720

ABSTRACT

Vacuum ultraviolet laser absorption spectroscopy has been employed to measure the populations and temperatures of ground electronic state H-atoms and vibrationally-excited H_2 molecules in a volume H⁻ ion source. Measurements of both species have been made under a variety of discharge conditions. Vibrational levels to v"=8 have been measured, with the vibrational population distribution well described by a temperature of 4150K.

INTRODUCTION

Ground-electronic state atomic hydrogen, H°, and vibrationally-excited hydrogen molecules, $H_2(v^*)$, are thought to play crucial roles in the generation of H⁻ in multicusp volume sources. In these sources, $H_2(v^*)$ is thought to form H⁻ via collisions with low-energy electrons (dissociative attachment), while H⁻ can be destroyed by H° in the reverse reaction. In order to further the understanding of the physical and chemical processes occurring in these sources, accurate determinations of the density, translational temperature, and state distribution of these species are required. For this reason, vacuum ultraviolet (VUV) laser absorption spectroscopy has been developed to measure these species. This technique allows for the direct, sensitive, and state-specific detection of the ground electronic state H° and $H_2(v^*)$, requiring no assumptions about plasma parameters such as the electron density distribution. The measurement is performed "*in-situ*," in a non-perturbative fashion. The technique gives the average density of the absorbing species along the absorption path, providing a direct measure of the species within the discharge volume.

In this paper, some of the recent measurements of H° and $H_2(v^*)$ using VUV laser absorption spectroscopy are reviewed. These data, obtained on a prototypical multicusp H⁻ source, cover a variety of discharge conditions and will help lead to an increased understanding of the H⁻ production mechanisms in these sources.

EXPERIMENTAL

H° and $H_2(v^*)$ are detected using VUV laser absorption spectroscopy. The experimental apparatus has been described in detail previously,[1] and is shown schematically in Figure 1. Briefly, frequency-tunable 20 ns pulses of VUV are produced by the non-linear optical technique of four-wave sum frequency mixing. For the particular wavelengths of interest, mercury vapor is used as the mixing medium. By using various resonance levels in the mercury, efficient generation of light from 94 to 125 nm has been accomplished. After generation, a portion of the VUV is measured and used for normalization, while the majority of the VUV, the "probe beam," is directed to the plasma chamber where absorption by H° or $H_2(v^*)$ occurs. The measured absorbance $a(\lambda) \equiv \ln[I_{off}(\lambda)/I_{on}(\lambda)]$ is then plotted as a function of VUV wavelength to obtain the absorption spectrum. Here $I(\lambda)$ is the intensity of the normalized probe beam after passage through the plasma chamber and the subscripts refer to the discharge off or on. The narrow bandwidth of the VUV,

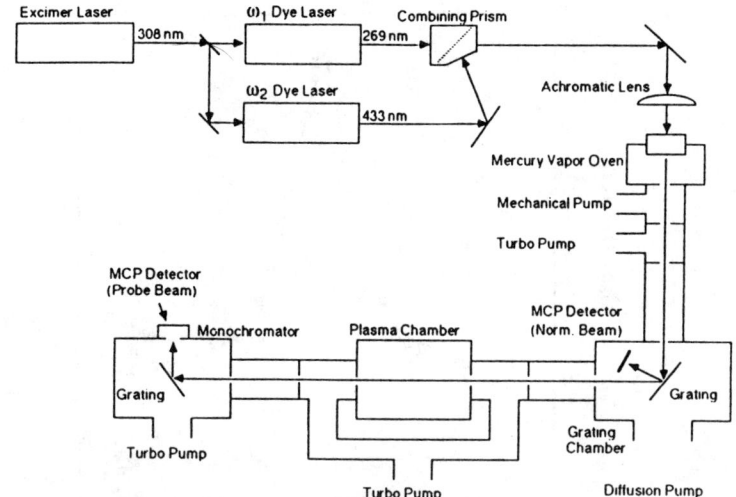

Figure 1. Schematic diagram of VUV absorption spectrometer system. Wavelengths shown are for generation of Lyman-β radiation.

~0.3 cm^{-1} (≈3 x 10^{-4} nm), makes possible the determination of the lineshape of the absorption feature. The total area under the absorption profile is proportional to the density of the absorbing species. The width of the line, assumed to be due to Doppler broadening, gives the translational temperature of the species.

The plasma chamber used for these experiments is a water cooled stainless steel multicusp source, 230 mm long and 200 mm in diameter. Ten rows of permanent magnets along the cylinder walls formed the cusped field. The end flanges of the source are also equipped with magnets. A coaxial LaB$_6$ cathode, 6 mm diameter x 10 mm long, was mounted on a holder of tungsten and tantalum and used as a thermionic electron source in these experiments. The absorption path was longitudinal through the source and slightly offset from, but parallel to, the cylinder axis. Differential pumping apertures defined the pathlength which was measured to be 31 cm. Hydrogen pressures, p_{H_2}, reported here are measured with a capacitance manometer with the discharge off.

Atomic density measurements were obtained using the Lyman β, Lyman γ, or Lyman ε (n=3 ← n=1, n=4 ← n=1, and n=6 ← n=1) transitions of H° at 102.6 nm, 97.3 nm and 93.8 nm, respectively. Individual rotation-vibrational states of $H_2(v^*)$ were measured using the B ← X (Lyman) or C ← X (Werner) bands.[3,4] These lines occur from 110 nm to 126 nm.

RESULTS

A typical absorption spectrum of H° is shown in Figure 2. This spectrum was obtained with the Lyman-γ transition and discharge parameters of 25 A, 150 V at a pressure of 7mTorr. The line through the experimental data points is a least squares fit of the data by the sum of two Gaussians having a common central wavelength but different widths. Such a lineshape is characteristic of a population with a bimodal velocity distribution. The data is consistent with 60% of the atoms characterized by T_{trans} = 700 K and 40% of the atoms at T_{trans} = 7300 K. The total H° density, 9.4 x 10^{12} cm^{-3}, represents 2% of the H_2 initially in the source being dissociated.

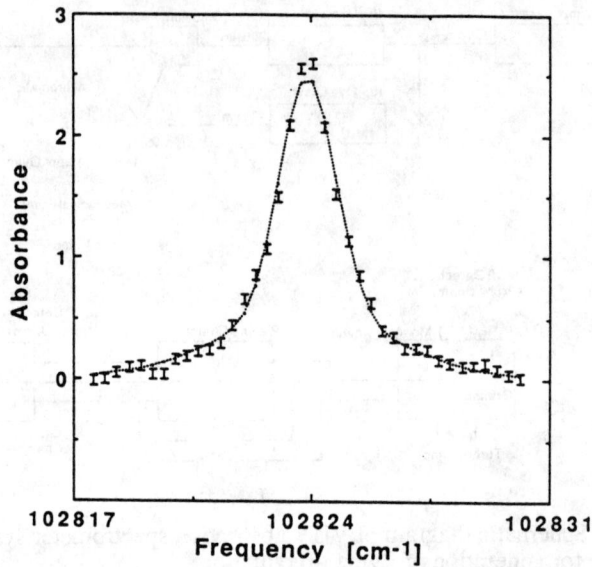

Figure 2. Typical H° absorption spectrum. This spectrum obtained using Lyman-γ Discharge conditions are 25A, 150V, and p_{H_2} = 7 mTorr. [H°] = 9.4 x 10^{12} atoms-cm^{-3}.

The H° density was measured as a function of various discharge parameters, including discharge voltage and current, and H_2 pressure. Figure 3 displays the H° density as a function of H_2 pressure. As can be seen, the H° density rises approximately linearly with H_2 pressure, and represents a constant 5% of the H_2.

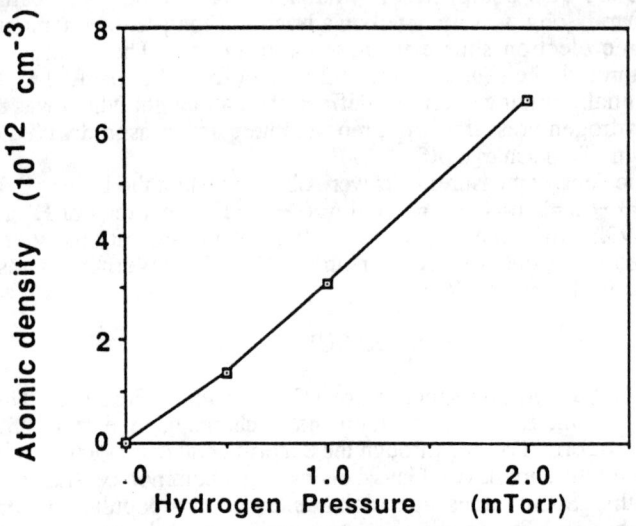

Figure 3. H° density as a function of H_2 pressure. Discharge conditions are 25A and 130V.

Figure 4 shows the results of a study conducted with varying discharge currents. Here the H_2 pressure was 20 mTorr and the discharge voltage was 90V. As can be seen, the H° density rises linearly with current over the range measured. However, the line does not extrapolate to $n_{H°} = 0$ at $I_{discharge} = 0$. Figure 5 shows data obtained at lower H_2 pressure, 1 mTorr. Here, a definite nonlinearitiy in the H° density is observed.

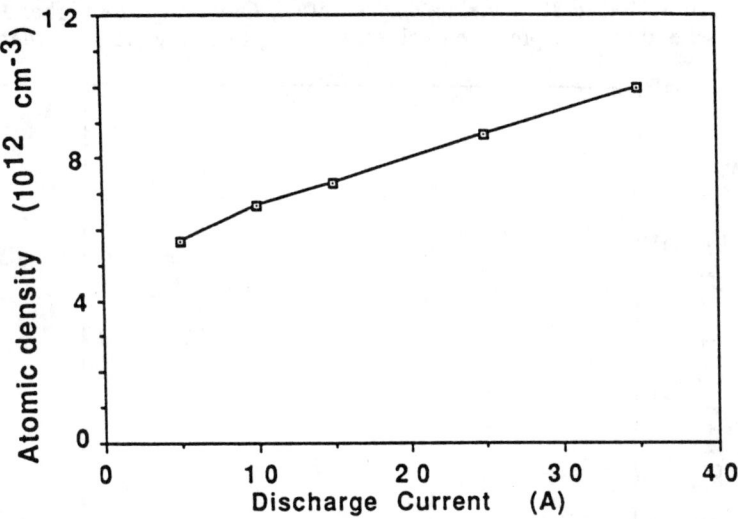

Figure 4. H° density as a function of discharge current. Discharge voltage is 90V and p_{H_2} = 20 mTorr.

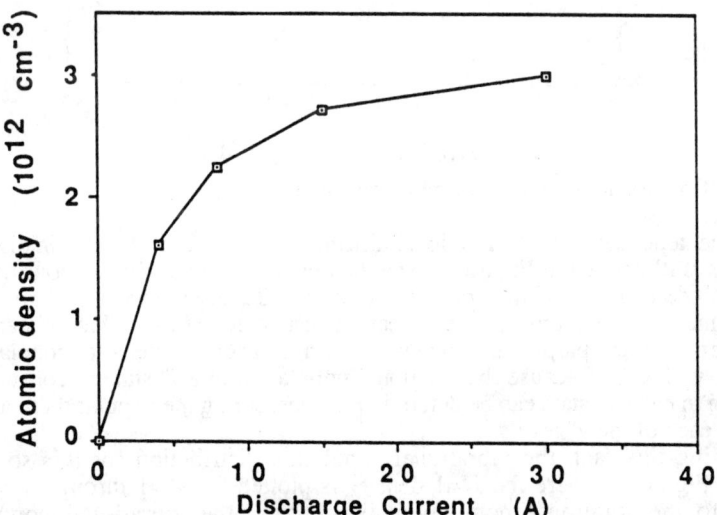

Figure 5. H° density as a function of discharge current, but at a lower H_2 pressure than Fig. 4. p_{H_2} = 1 mTorr, while discharge is at 100 V.

The populations of individual rotation-vibration levels of the H_2 in the discharge have been determined. For v"=1, (where the v" refers to the vibrational level of the ground electronic state) the rotational distribution has been measured to J"=13. This data is shown in Figure 6. A thermal population distribution, when plotted as in Figure 6, would yield data points on a straight line. The slope of the line would then be inversely proportional to the rotational temperature, T_{rot}. As can be seen, this is not the case for our measurements. The first few J levels do lie on a line, however; using J"=0 to J"=3 yields T_{rot} ~ 450K. Obviously, the higher J level populations lie above the predicted values, showing that they are suprathermally populated.

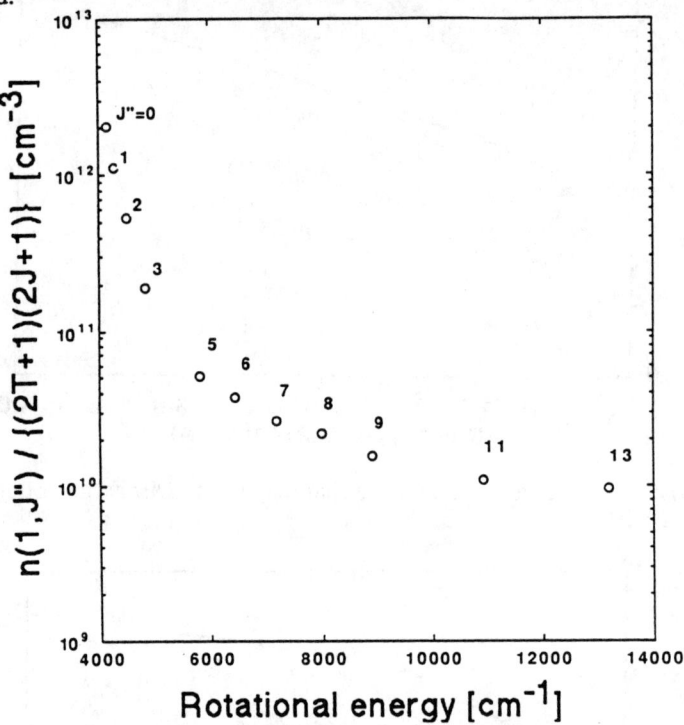

Figure 6. Rotational population distribution for H_2(v"=1).

The dependence of the rotational distribution of v"=1 on changes in discharge conditions is illustrated in Figure 7. Note that over the range of conditions utilized, which include changes of factors of three in the discharge current and two in H_2 pressure, the rotational temperature is nearly constant at ~420 K. This indicates that the population in any particular J state is a constant fraction of the total population in any given v" level. Because the fractional population in a J" state is constant, the population in each v" state can be determined by measuring the population in a single J state for each of the v" levels.

Using this fact, the vibrational population distribution for this source is shown in Figure 8, where the J"=1 density is plotted for v"=1 through v"=8. In contrast to the rotational population distribution, the vibrational population distribution exhibits a Boltzman distribution. The solid line represents a linear least square fit for v"=1-8 with T_{vib} = 4150 K.

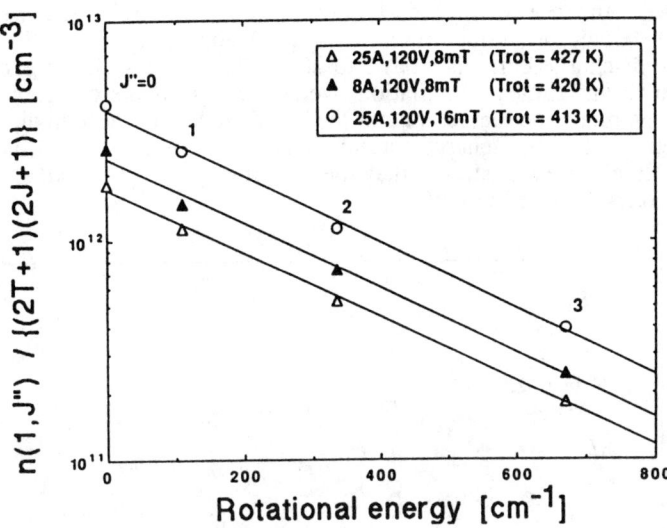

Figure 7. Rotational population of J"=0-3 for $H_2(v"=1)$ under various discharge conditions. Note the invariance of the rotational temperature.

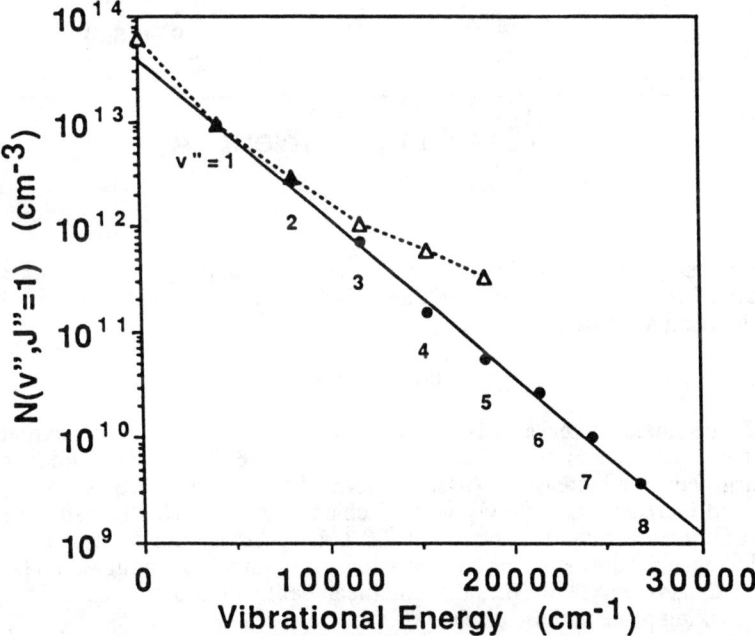

Figure 8. Vibration population distribution for volume source. The experimental points, the closed circles, were obtained at 25A, 120V and 8 mTorr H_2. The solid line corresponds to a vibrational temperature of 4150 K. The triangles represent the calculated values of Hiskes, normalized to the experimental data at v"=1.

The dependence of four high-lying v", J" states on discharge current is shown in Figure 9. Here, the population in v"=4, J"=1 and 3, and v"=6, J"=1 and 3 are shown with discharge currents from 2 to 25A. As can be seen, none of the states displays much variation in population over the range of the currents used. This shows that the rotational temperature of the low J" states for these higher v" states is relatively constant with discharge current, in agreement with the v"=1 data shown in Figure 6. It also demonstrates that the vibrational distribution up to v"=6 is insensitive to discharge current.

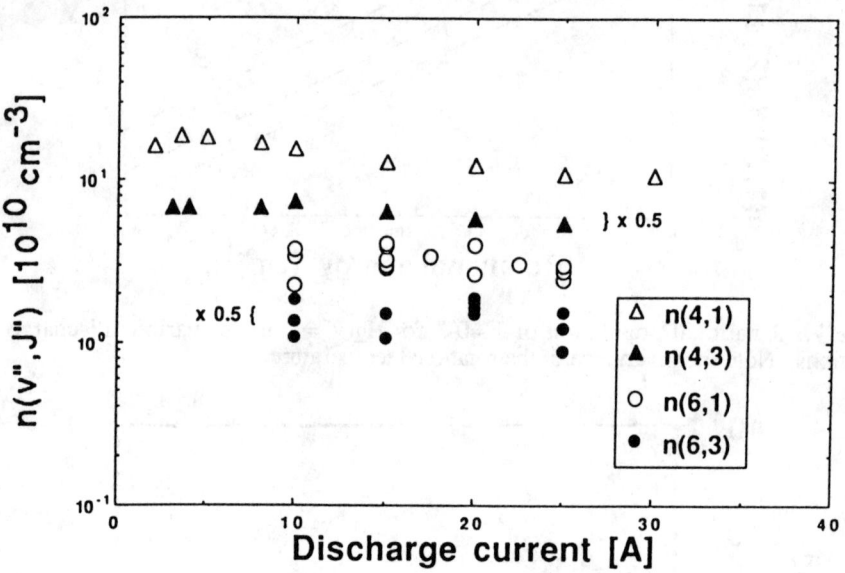

Figure 9. Dependence of the population of several high-lying v", J" states on the discharge current. States measured are v"=4, J"=1 and 3, and v"=6, J"=1, and 3. Note the relative constancy of the populations.

DISCUSSION

The linear increase in H° density, as shown in Figure 4, is qualitatively as expected. That is, at modest discharge powers, more H° would be produced as the discharge current increases. At some point, however, one would expect the H° density to increase more slowly with discharge current. This is also observed, as seen in Figure 5. What is surprising is the modest dissociation fraction at which this slow increase occurs, ~5% for the conditions in Figure 5. Although the H_2 pressure for that data is only 1 mTorr, the nonlinear behavior illustrates the complicated kinetics taking place in these discharges.

The measured rotational-vibrational population distributions show that the H_2 is internally excited. The rotational levels above J"=3 are suprathermally populated, while those below can be characterized by a rotational temperature only slightly above room temperature. This can be explained if rotational relaxation processes occur on a time scale which is short compared to vibrational relaxation. If this is the case, each vibrational level would rotationally equilibrate yielding a thermalized population distribution. The population in the high J" states population may reflect the convolution of the nascent state distribution and the distribution of those H_2 which have only partially relaxed. Alternatively, the high J" states may be the result of V → R energy transfer processes. More work is needed to resolve this question.

It is important to note that the rotational distributions for the low J" levels of v"=1,4 and 6 are very similar, as shown in Figures 7 and 9. This allows the determination of the various v" populations to be made by measuring a single J" state for each v" level, as was done for Figure 8. This is particularly important for the higher v" states as only one or two rotational levels could be observed in these sparsely populated levels. Changes in the rotational distribution for the higher v" levels would cause inaccuracies in the vibrational population distribution; however, no variation in T_{rot} for v" up to 6 was observed. In addition, because a large fraction of the population is in the state observed, J"=1, modest changes in the rotational temperature will not greatly affect the estimated vibrational distribution.

The vibrational distribution shown in Figure 8 exhibits a T_{vib} = 4150 K. Modeling of the source chemistry for these or similar discharge conditions has been performed by Hiskes and Karo[5] and by Skinner, Bacal, et al.[6] The dashed line in Figure 8 is the result of Hiskes for vibrational levels up to v"=5. The calculation and the measurement have been normalized at v"=1. As can be seen, the experiment and model disagree at higher v" levels. The reason for the discrepancy is unknown. Although the Hiskes calculation used an electron density value 50% of the measured 1.6×10^{12} cm^{-3}, increasing the electron density in the calculation would increase the vibrational population, giving a larger discrepancy. The model used by Skinner also overestimates the population in the higher v" states, in spite of using the actual experimental electron density. The large population in v" ≥ 6 in an almost universal feature of models of H⁻ volume sources. This is the so-called "plateau" region, as the vibrational population distribution is predicted to flatten out for v" = 5-9 for a number of calculations.[7,8]

The only other *in situ* observation of $H_2(v")$ within the discharge is by Pealat et al.[9] who used Coherent Antistokes Raman Scattering to observe a high pressure multicusp discharge. That experiment observed v"≤3. The only other experiment to observe the higher v" states is by Eenshuistra et al.,[10] who detected states with v"≤5 effusing from a multicusp source using a multiphoton ionization technique. Under some conditions, this group observed a non-thermal distribution, with a plateau starting to appear at v"=4. As discussed above, the results presented here, which include data at higher v", show no evidence for a plateau. At this time it is not possible to determine if the divergent results are due to different source physics or experimental technique.

The prevailing theory of H⁻ production in volume sources invokes the mechanism of dissociative attachment of low energy electrons to vibrationally excited molecules. This theory requires the presence of H_2 with v"≥6 in relatively large amounts, and is reflected in the models of these sources by the previously discussed plateau. The measurements reported here, the first observations of v">5, exhibit a thermal vibrational distribution out to v"=8, the present sensitivity limit of the technique. It must be pointed out that this source has not been optimized for H⁻ production, and the H⁻ density within the plasma has not been measured. However, the size, shape, and plasma parameters of this discharge chamber are similar to those of optimized tandem multicusp sources, with the exception that the present source does not have a magnetic filter. Instead, additional strong cusp fields are used to increase the plasma confinement. It was anticipated that the plasma processes in this source would be similar to those in the driver region of tandem volume sources. This has not yet been verified, but if true, the formation mechanism of H⁻ in volume sources may need to be re-evaluated.

CONCLUSIONS

Recent results using VUV laser absorption spectroscopy to probe the physics of H⁻ volume sources have been reviewed. The ability to measure quantitativity H* and $H_2(v^*)$ *in situ* in a plasma has been demonstrated. Measurement of H* and $H_2(v^*)$ have been obtained for a variety of discharge conditions. In particular, rotational-vibrational state distributions have been obtained for vibrational states up to $v''=8$. These measurements, in contrast to model predictions, show a thermalized vibrational distribution, with $T_{vib} \sim 4200K$.

ACKNOWLEDGEMENTS

The work presented represents the contributions of many colleagues. These include A.S. Schlachter, J.W. Stearns, G.T. Worth, B. D'Etat, and H.F. Döbele. This work has been supported by the Air Force Office of Scientific Research, Los Alamos National Laboratory, and the Director, Office of Energy Research, Office of Fusion Energy, Development and Technology Division, of the U.S. Department of Energy under contract No. DE-AC03-76SF00098.

REFERENCES

1. G.C. Stutzin, A.T. Young, A.S. Schlachter, J.W. Stearns, K.N. Leung, W.B. Kunkel, G.T. Worth, and R.R. Stevens, Rev. Sci. Instrum. 59, 120 (1988).
2. G.C. Stutzin, A.T. Young, A.S. Schlachter, J.W. Stearns, K.N. Leung, W.B. Kunkel, G.T. Worth, and R.R. Stevens, Rev. Sci. Instrum. 59, 1479 (1988).
3. G.C. Stutzin, A.T. Young, A.S. Schlachter, K.N. Leung, and W.B. Kunkel, Chem. Phys. Lett. 155, 475 (1989).
4. G.C. Stutzin, A.T. Young, H.F. Döbele, A.S. Schlachter, K.N. Leung, and W.B. Kunkel, Rev. Sci. Instrum., in press.
5. J.R. Hiskes and A.M. Karo, Appl. Phys. Lett. 54, 508 (1989).
6. D.A. Skinner and M. Bacal, private communication, and D.A. Skinner, P. Berlemont, and M. Bacal, these proceedings.
7. J.R. Hiskes, A.M. Karo, and P.A. Wilmann, J. Appl. Phys. 58, 1759 (1985).
8. C. Gorse, M. Capitelli, M. Bacal, and J. Bretague, Chem. Phys. 117, 177 (1987).
9. M. Pealot, J.-P.E. Taran, M. Bacal, and A.M. Bruneteau, J. Chem. Phys. 82, 4943 (1985).
10. P.J. Eenshuistra, R.M.A. Heeren, A.W. Kleyn, and H.J. Hopman, Phys. Rev. A 40, 3613 (1989).

DISCUSSION

Hopman: In the comparison between the caculated vibrational distribution and the measured one, I presume there is a great distinction in the two distributions, I think John Hiskes has referred to that. You considered for your vibrational distribution only one rotational level.

Young: That is correct.

Hopman: And if you would be able to measure for all v's up to $J = 10$ rotational distribution, would that sum up the real vibrational distribution. You have not done it yet so I think an honest comparison at this moment is not yet possible.

Young: What you are saying is definitely true and that is why I did make the distinction, I did point out that we were plotting the J equals 1 population for the distribution all the way out to 8. Your comment is particularly well taken if the J distribution is changing shape as a function of vibrational level; up to $v = 5$, we see no evidence for that. And unfortunately, the populations in the higher lying J states for the higher lying v states is too small so that we can't measure it real accurately.

Hiskes: What is the relative population of those above $v = 5$ relative to those below $v = 5$.

Young: Off hand, I don't know. But things are dropping off by a factor of between 3 and 4 for each vibrational level. It's less than a few percent in the upper levels.

Hall: It would be nice now if you could put in tungsten filaments in your discharge to see if that brings up the vibrational populations. Did you see any vibrational excitation without discharge and with the LaB_6 cathode?

Young: We don't see any vibrational excitation without a discharge. What happens when you put in tungsten filaments? The answer is coming out right now as we speak, (well actually it is too early in the morning in California) but we are performing that experiment right now. So, sometime soon you can see the results of that experiment.
Bacal: We do not know that you are the first to meas-

ure such high vibrational levels. We measured before the rotational temperature by CARS, as you know, and we have observed major increases in rotational temperature. I'm astonished that you do not observe them. This doesn't seem reasonable that heating the plasma with a higher discharge current wouldn't lead to any change in temperature. Although, you observe the change in translational temperature. But these two communicate, so I'm astonished and I wonder what is the cause of it, certainly your measurement is correct. I'm thinking also about relavance of these VUV absorptions spectropscopy to what we are interested in: Because on one of the slides it was written that it was spatially resolved. We integrate the absorption on the straight line which crosses both plasma region and empty regions. For example, the regions near the walls were the multicusp magnetic fields keeps the plasma away. So if you have atoms or molecules in there you will measure them but they may be very different from the ones which are inside the plasma. I'm just wondering what would you obtain if you would use a different way of making VUV spectropscopy not absorption spectropscopy but some other measurements? I think it would be a major step in comparing these data, which are extremely interesting, to some enclosed VUV diagnostics.

Young: There is actually a lot of relevance to what you say. First, let me touch on the relative invariance of the rotational energy distributions, and I definitely think that has to do with how rapidly the rotational energy levels relaxed, and that is telling you something about the r to t relaxation rates. The fact is that the higher rotational levels seem to have a higher translational temperature whereas the lower rotational levels have a much more comparable translational temperature seems to indicate that at least the time scales for those two relaxations are somewhat similar. In terms of additional VUV diagnostics, one thing which we would really dearly love to do is laser induced fluorescence. Unfortunately, that is a real problem again because you've got to worry about the intense VUV background that you get from just the plasma. Part of our effort is to create a very intense VUV source so that you would have enough photons to do a direct VUV laser induced flourescence experiment. That work is still continuing, it's a very hard experiment though.

Holmes: The source you described looked very similar to the standard Berkeley negative ion sources. And, if I look at the data which you presented, two things emerge: One, the vibrational density for the high lying states v equals 6 is very, very low. Below that what needed to explain the J⁻ currents that have been reported; and secondly, you are seeing no pressure or temperature variation which we need to explain why the J⁻ goes up with arc current and so forth. So if I take your data as I see it, it looks like you almost destroyed the standard model for negative ion production. Then you will need some other mechanism to explain negative ion yields.

Young: I have a couple of comments: I've almost fooled you in that I didn't tell you exactly what source we were running. This is very similar to the standard Berkeley source. But it's not the one that is conventionally used for these big accelerator experiments although it is very, very similar. This sort of served as the test bed for our work in developing the technology. We are in the process right now of measuring the exact source that we use at Berkeley for accelerator experiments, and we've looked at the H⁻ and the positive ion content coming out of this exact source exactly as it is configured for our VUV experiment. So, I give us about two months and we will be able to give you the ensemble of H⁻, H⁺, H_2^+, H_3^+ atoms and molecules.

Roberts: I would like to say that you have been even more successful than Andrew has indicated because you have also destroyed the marvels for all the electric discharge lasers. The CO_2 lasers where the rotational population is always in translation with the bath: The rotation is in equilibrium with the translation and the rotational distribution is independent of the v levels, they relax much faster at the v levels and if everything is non-thermal it's not the rotational, it's the vibrational. This is necessary for those models to work too. So you've done better than just the volume sources.

H^0 TEMPERATURE AND DENSITY MEASUREMENTS IN A PENNING SURFACE-PLASMA H^- ION SOURCE. II.

H. Vernon Smith, Jr., Paul Allison, E. J. Pitcher,
R. R. Stevens, Jr., and G. T. Worth
Los Alamos National Laboratory,* Los Alamos, NM 87545

G. C. Stutzin, A. T. Young, A. S. Schlachter, K. N. Leung, and W. B. Kunkel
Lawrence Berkeley Laboratory,† Berkeley, CA 94720

ABSTRACT

We recently reported the H^0 density and temperature in the 4X source (a Penning SPS) as a function of the H_2 gas flow and the discharge current, as well as a measurement of the $H_2(v=1)$ vibrational state density. In this paper, we report the H^0 density and temperature in the 4X source as a function of magnetic field for the same source geometry of the previous report and as a function of H_2 gas flow and discharge current for a smaller discharge chamber. When the smaller discharge chamber is used, the plasma column H^0 density increases from 7×10^{14} cm^{-3} to 8×10^{14} cm^{-3} and the plasma column H^0 temperature increases from 1.5 to 1.9 eV.

INTRODUCTION

Surface-plasma sources (SPS) are now being used in pulsed H^- injectors on several different accelerators, including those at Brookhaven National Laboratory,[1] Fermilab,[2] the INR Moscow Meson Factory,[3] Los Alamos National Laboratory,[4,5] and the Rutherford Laboratory.[6] To learn more about the physics of the arc discharge in the Penning-type SPS, we took the 4X source[7] from Los Alamos to Berkeley to measure the H^0 temperature and density as a function of the various source parameters. Knowledge of the H^0 temperature and density could be crucial in constructing a detailed theoretical model of the Penning SPS. The first report on these 4X source measurements is contained in Ref. 8. The rest of these measurements are reported below.

EXPERIMENTAL METHOD

Tunable VUV light is produced by the laser-absorption-spectrometer system[9] at Lawrence Berkeley Laboratory (LBL). The VUV beam is split into a probe beam, which passes through the source plasma, and a normalization beam,

---

*Work supported and funded by the Department of Defense, US Army Strategic Defense Command, under the auspices of the Department of Energy.

†This work was supported by the Air Force Office of Scientific Research and the Director, Office of Energy Research, Office of Fusion Energy of the US Department of Energy under Contract No. DE-AC03-76SF00098.

which monitors the output of the VUV laser pulse to pulse. The transmitted probe beam signal, with and without the discharge, is measured as a function of wavelength over the L_γ absorption peak. The integrated absorbance $A = \int \ln[I_{off}(\lambda)/I_{on}(\lambda)]d\lambda$, where $I(\lambda)$ is the transmitted probe beam signal divided by the normalization beam signal (the subscripts refer to the discharge status): A is proportional to the line density. If the line shape is due to Doppler broadening, then it is a Gaussian with full-width at half-maximum $\Delta\lambda = 2\,(2\,kT_{H^o} \ln 2/Mc^2)^{1/2}\,\lambda_o$ where kT_{H^o} and M are the H-atom temperature and mass and λ_o is the L_γ wavelength.

The 4X source[7] is mounted in the plasma chamber of the VUV absorption system (Fig. 2 of Ref. 9). The source modifications for this experiment are shown in Fig. 1. Note that the 2.5-mm-diam VUV laser beam traverses either the plasma column 6.9 mm from the emitter (extraction aperture) location or the drift region 1.7 mm from the emitter. The emitter is omitted in this experiment because of pumping limitations. The VUV laser beam is directed along the y-axis, perpendicular to the magnetic field. Plots of absorbance vs wavelength are shown in Fig. 2a for the plasma column and in Fig. 2b for the drift region. Two H_2 molecular lines obscure the high-wavelength tail of the plasma column spectrum shown in Fig. 2a. In the drift region, the H^o temperature is low enough that the L_γ peak is almost resolved from the interfering molecular peaks. Obscured portions of the L_γ spectra are omitted from the Gaussian fits (the upper 10 points in Fig. 2a and the upper point in Fig. 2b are deleted).

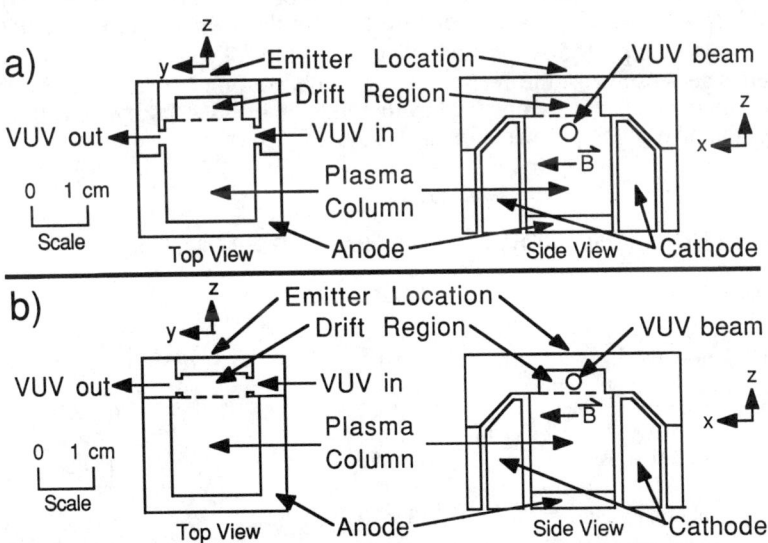

Fig. 1. Schematics showing where the VUV light traverses the 4X source plasma column (a) or the drift region (b).

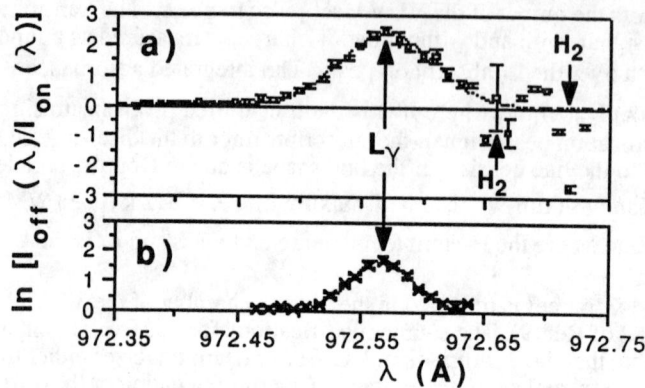

Fig. 2. Sample L_γ absorption spectra for the 4X source plasma column (a) and the drift region (b). Two H_2 molecular lines are present in the plasma column spectrum (see text). The curves are Gaussian fits to the data.

EXPERIMENTAL RESULTS

The H^0 density, n_{H^0}, and kT_{H^0} are measured as a function of discharge current, H_2-gas flow, and magnetic field in both the plasma column and the drift region between the plasma column and the emitter (Fig. 1). The measurements of n_{H^0} and kT_{H^0} vs gas flow and discharge current for the deep slot (17 mm along z) have already been presented in Ref. 8 and will not be given here. Figure 3 shows n_{H^0} and Fig. 4 shows kT_{H^0} vs magnetic field B_d for the deep slot anode for fixed discharge current I_d = 155 A and gas flow Q = 1.2 Tℓ/s. Typical operating parameters and results for the deep slot are given in column 1 of Table I. All the density data in this paper are obtained by dividing the line density by the path length in either the drift or the plasma column, 1.2 cm and 1.6 cm, respectively.

TABLE I

TYPICAL 4X SOURCE OPERATING PARAMETERS AND RESULTS

Parameter	Deep Arc Slot	Shallow Arc Slot
Cathode-cathode gap, mm	17	17
Discharge slot width, mm	17	12
Discharge slot length, mm	16	16
Discharge magnetic field, T	0.07	0.15
Discharge voltage, V	104	98
Discharge current, A	152	155
Gas flow, Tℓ/s	1.2	1.2
Plasma column n_{H_2}, cm^{-3}	3×10^{15}	2×10^{15}
Discharge pulse length, ms	1.0	1.2
Pulse repetition rate, Hz	5	5
Plasma column n_{H^0}, cm^{-3}	7.3×10^{14}	7.9×10^{14}
Plasma column kT_{H^0}, eV	1.5	1.9
Drift region n_{H^0}, cm^{-3}	3.8×10^{14}	Not measured
Drift region kT_{H^0}, eV	0.64	Not measured

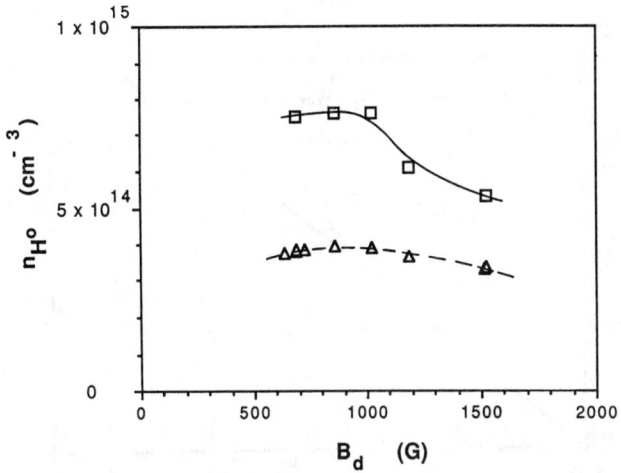

Fig. 3. Plots of n_{H^o} vs B_d for the 4X source plasma column (squares, solid line) and drift region (triangles, dashed line) for the deep slot anode. The lines are guides to the eye.

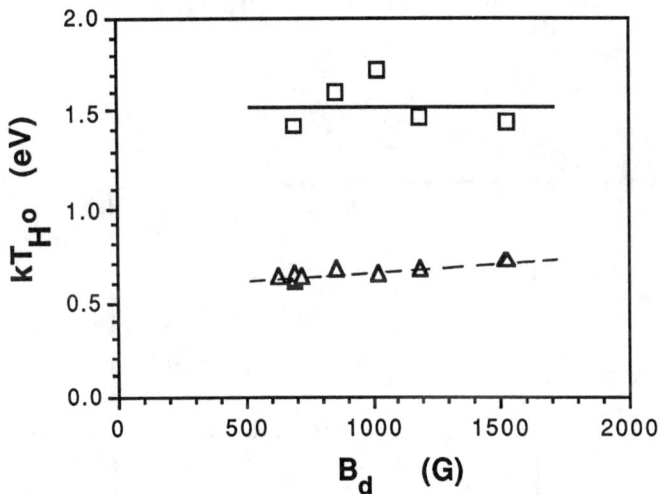

Fig. 4. Plots of kT_{H^o} vs B_d for the 4X source. Otherwise the same as Fig. 3.

Figure 5 shows n_{H^o} and kT_{H^o} vs Q for $I_d = 155$ A and $B_d = 1525$ G, and Fig. 6 shows n_{H^o} and kT_{H^o} vs I_d for $Q = 1.2$ Tℓ/s and $B_d = 1525$ G for the shallow slot anode (12 mm along z). For the shallow slot, we did not measure n_{H^o} and kT_{H^o} vs magnetic field because the discharge would not run at magnetic fields below ~1400 G, nor did we make any measurements in the drift region. Typical operating parameters and results for the shallow slot are given in column 2 of Table I.

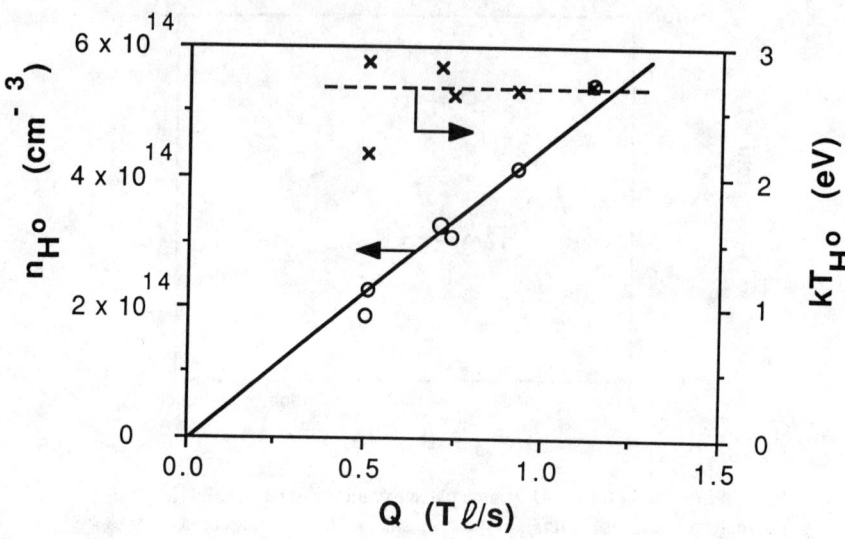

Fig. 5. Plots of n_{H^o} vs Q (circles, solid line) and kT_{H^o} vs Q (crosses, dashed line) for the 4X source plasma column for the shallow slot anode. The lines are guides to the eye.

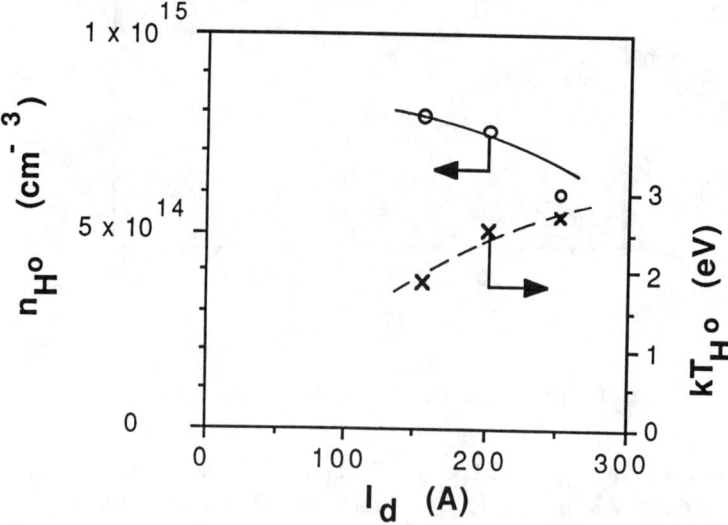

Fig. 6. Plots of n_{H^o} vs I_d (circles, solid line) and kT_{H^o} vs I_d (crosses, dashed line) for the 4X source plasma column for the shallow slot. The lines are guides to the eye.

If n_{H^o} and/or kT_{H^o} vary with time into the arc pulse, a systematic error in these parameters occurs. This is important because for the data shown in Fig. 6, and also in Figs. 5 and 6 of Ref. 8, the time from the beginning of the arc pulse to the laser firing varied from 310 μs for I_d = 325 A to 1250 μs for I_d = 100 A. A test case in which the discharge parameters and the pulse length were fixed and the laser firing time varied shows that kT_{H^o} is constant but n_{H^o} increases with time into the arc pulse. These data are shown in Figs. 7a and 7b. Thus, the kT_{H^o} data in Fig. 6 of this paper and Fig. 6 of Ref. 8 are not adjusted, but a small correction has been made to the n_{H^o} data in Fig. 6 of this paper and Fig. 5 of Ref. 8. The H^o density and temperature just before and just after the discharge was turned off by the arc pulser was also measured. These data are shown in Figs. 8a and 8b for the plasma column (this study was not performed in the drift).

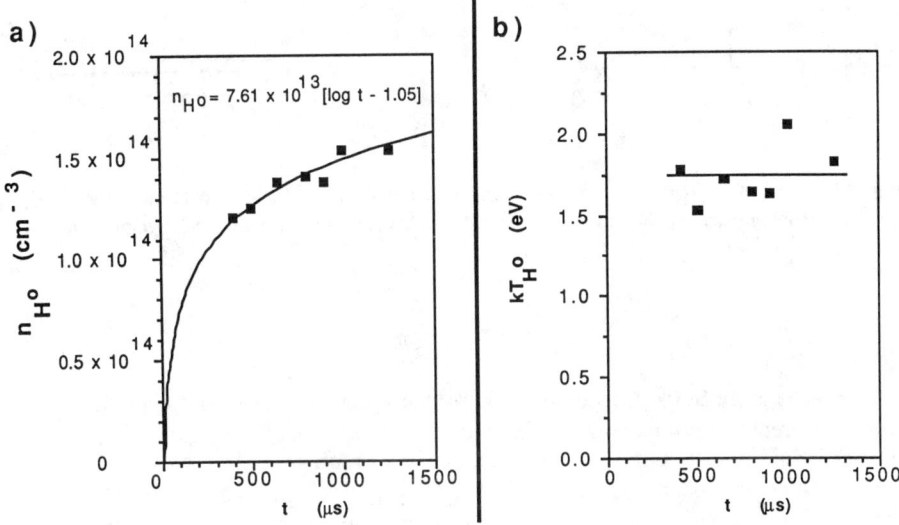

Fig. 7. Plots of (a) n_{H^o} and (b) kT_{H^o} vs time t into the discharge pulse that the measurement is made. These measurements are for the 4X source plasma column with a shallow slot. The discharge pulse length is kept fixed at 1500 μs for these measurements. The calibration curve used to correct the n_{H^o} data in this paper and in Ref. 8 and its equation are given on Fig. 7a. The curve on Fig. 7b is a guide to the eye.

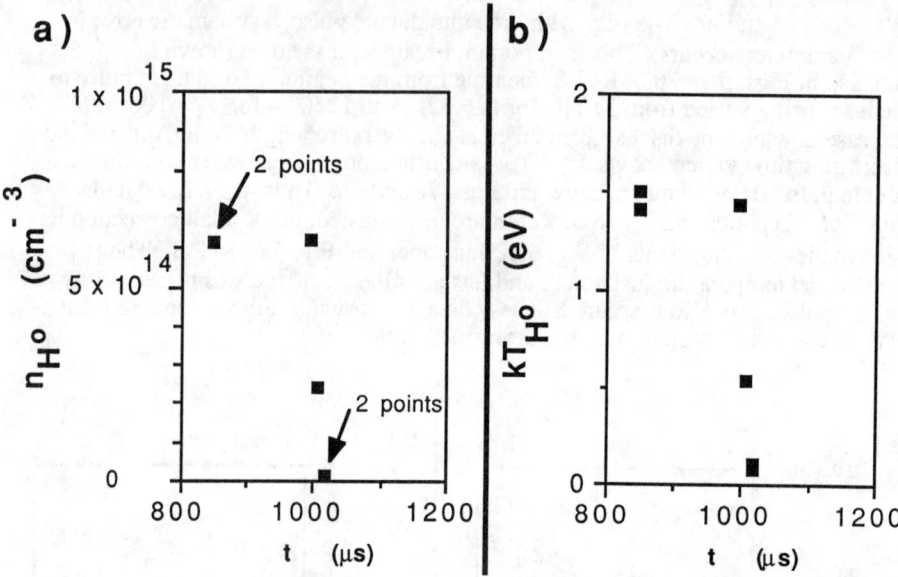

Fig. 8. Plots of (a) n_{H^o} and (b) kT_{H^o} vs time t from the start of the discharge pulse for the 4X source plasma column with the deep slot. This data is for a fixed discharge pulse length of 1000 μs.

DISCUSSION

We ignore Stark broadening of L_γ because the line shapes are Gaussian, with no Lorentzian component admixture, even in the tails of the profiles. In Fig. 2 of Ref. 10, Stark broadening appears as a Lorentzian tail to the Gaussian core of the H_α emission line. The n=3 (n is the principal quantum number) kT_{H^o} measurements in Ref. 10 are for a shallow slot and integrated along the z-direction, from the emitter to the back of the discharge (Fig. 1 in Ref. 10). The n=1 kT_{H^o} measurements in the present work are integrated along the y-direction. Despite these differences, the kT_{H^o} obtained in these two studies compares quite favorably, 2.6 eV at I_d = 200 A in the present absorption work and 2.3 eV at 200 A in the previous emission work (Fig. 3 in Ref. 10).

When the depth of the 4X source arc slot was increased from 1.2 to 1.7 cm, a reduction in the lowest magnetic field at which the source discharge would run was accomplished. In previous experiments,[11,12] when B_d was reduced from ~1500 G to ~600 G, the H⁻ beam noise and H⁻ beam emittance were lowered. With B_d = 1525 G, Q = 1.2 Tℓ/s, and I_d = 155 A, the H⁰ temperature and density are 2 to 2.5 eV and 6 to 8 × 10¹⁴ cm⁻³, respectively, for the shallow slot and 1.5 eV and 5 × 10¹⁴ cm⁻³, respectively, for the deep slot. However, as B_d is lowered with the deep slot, a surprising thing happens — the H⁰ temperature stays constant at about 1.5 eV, but the H⁰ density increases by 50% to ~7.5 × 10¹⁴ cm⁻³. The end result is that with the deep slot at a magnetic field of ~700 G, n_{H^o} is about the same as for the shallow slot at 1525 G, but the H⁰ temperature is reduced from 2 to 2.5 eV to 1.5 eV. This reduction in kT_{H^0} may be related to the reduction in kT_{H^-} observed from the H⁻ beam emittance measurements, because the H⁻ are thought to thermalize in the resonant-charge-exchange reaction H⁰(slow) + H⁻(fast) → H⁰(fast) + H⁻(slow). The reduction in the H⁻ beam fluctuations with lowered B_d is probably related to another phenomenon, such as increased diffusion loss of electrons to the anode.[13] In other measurements, we note reduced H⁻ beam fluctuations with increased Q. From Fig. 5 (plus Figs. 3 and 4 in Ref. 8), we see that increased Q primarily results in increased n_{H^0}, with little change in kT_{H^0}. Thus, increased n_{H^0} is associated with reduced H⁻ beam fluctuations.

We can estimate the wall recombination coefficient for H⁰ if we assume that the decay of n_{H^0} after the discharge pulse (see Fig. 8a) is due solely to recombination on the wall. Careful monitoring of the discharge current I_d showed that it went to zero at 1007 μs after the arc pulse began. Thus, the n_{H^0} and kT_{H^0} measurements made at 1007 and 1017 μs from the start of the arc pulse (Fig. 8) allow an estimate of the wall recombination coefficient to be made without the complication of H⁰ production in the discharge. Assuming an exponential decay of the atomic density gives τ_R = 3.2 μs for the 1/e decay time. We estimate that the time required for H⁰ at 0.53 eV (the H⁰ temperature at 1007 μs) to travel to the walls is dominated by diffusion of H⁰ in H_2. The diffusion coefficient D is given by

$$D = kT_{H^0} / [m_{H^0} n_{H_2} \sigma_{H^0-H_2} v_{H^0}] , \quad (1)$$

where kT_{H^0} = 0.53 eV, n_{H_2} = 3 × 10¹⁵ cm⁻³, and $\sigma_{H^0-H_2}$ = 3.3 × 10⁻¹⁵ cm²; thus, D = 5.1 × 10⁴ cm²/s. The diffusion time τ_D is then given by $\tau_D \approx R^2/D$, where R ≈ 0.4 cm, the weighted distance the H⁰ travel to the walls. The diffusion time τ_D = 3.1 μs is approximately eight times longer than the time for 0.53 eV H⁰ to drift 0.4 cm. Because $\tau_D \approx \tau_R$, it is likely that the wall recombination coefficient is nearly one. We must point out that the state of the cesium and hydrogen coverage of the molybdenum walls for this estimate is unknown. The high H⁰ temperatures in the source discharge indicate that the Frank-Condon energy of dissociation does not relax to the molecular translational temperature. This is possibly due to insufficient H_2 density for the temperature relaxation of the H atoms to occur.

The mean free path λ_{-0} for charge exchange of energetic H⁻ ions (100 eV) from the 4X source cathodes with H⁰ atoms in the plasma column is

$$\lambda_{-0} = \sigma_{-0}^{-1} n_{H^0}^{-1} , \quad (2)$$

where σ_{-o} is the cross section for resonant charge exchange. Using our measured value of $n_{H^o} = 7.5 \times 10^{14}$ cm^{-3}, and $\sigma_{-o} = 7.3 \times 10^{-16}$ cm^2 for 100 eV H$^-$, gives $\lambda_{-o} = 1.8$ cm. The fraction A of fast H$^-$ that charge exchange with H^o is

$$A = 1-e^{-n_{H^o}\sigma_{-o}\ell}, \quad (3)$$

where ℓ is the average distance from the cathode to the emitter, 1.5 cm in the 4X source; $A = 0.56$. Thus, many of the fast H$^-$ ions from the cathode are thermalized in the plasma column and the drift.

The slow H$^-$ ions (kT$_{H^-}$ ~ 1.5 eV) have to reach the emitter to be extracted and formed into a beam. The mean free path λ_{-+} for loss from destruction by mutual neutralization (MN) collisions with H$^+$ ions is given by

$$\lambda_{-+} = \sigma_{-+}^{-1} n_{H^+}^{-1}, \quad (4)$$

where σ_{-+} is the MN cross section. From our previous emission spectroscopy work,[10] the plasma density for $I_d = 150$ A is $n_e = 1.5 \times 10^{14}$ cm^{-3}. Assuming $n_e = n_{H^+}$ and using $\sigma_{-+} = 1 \times 10^{-14}$ cm^2 for 1.5 eV, $\lambda_{-+} = 0.7$ cm. A rough estimate indicates that the loss of H$^-$ from collisions with fast electrons is about the same as the loss from MN. Dudnikov's simple model[14] of the Penning SPS source is that H$^-$ ions from the cathodes are accelerated across the cathode sheaths (~100 V), thermalized in resonant charge exchange collisions, and extracted and formed into a beam. The H^o densities that we measure in the 4X source are consistent with this model.

CONCLUSIONS

When the 4X source arc slot width is decreased from 17 to 12 mm (shallow slot), n_{H^o} and kT$_{H^o}$ increase by 10% and 30%, respectively. When B_d is lowered, the principal change in the 4X source with the deep slot is an increase in n_{H^o}, while kT$_{H^o}$ is relatively unchanged. We note that the H$^-$ beam noise, and hence the H$^-$ beam emittance, is lowest for the conditions that produce the highest H^o density — lowest B_d and higher Q. The present measurements show that H^o temperatures as low as 0.6 eV can be achieved in the 4X source, about 10 times less than the lowest effective H$^-$ temperature at emission. Thus, if relaxation of the H$^-$ temperature to the H^o temperature is the dominant mechanism for determining the effective H$^-$ temperature, there is still room for improvement in the 4X source H$^-$ beam emittance.

ACKNOWLEDGMENTS

The authors thank J. E. Stelzer of Los Alamos National Laboratory and G. J. DeVries and J. W. Stearns of Lawrence Berkeley Laboratory for their assistance.

REFERENCES

(1) J. G. Alessi, J. M. Brennan, A. Kponou, and K. Prelec, "H⁻ Source and Beam Transport Experiments for a New RFQ," Proc. 1987 Particle Accelerator Conf., IEEE Catalog No. 87CH2387-9, pp. 304-306 (1987).
(2) C. W. Schmidt and C. D. Curtis, "Operation of the Fermilab H⁻ Magnetron Source," AIP Conf. Proc. No. 158, 425-9 (1987).
(3) A. M. Anikeichik, et al., "Injectors for High-Intensity Linear Accelerator of INR Moscow Meson Factory," Proc. 1988 Linear Accelerator Conf., CEBAF report no. CEBAF-R-89-001, 660-2 (1989).
(4) P. Allison and J. D. Sherman, "Operating Experience With a 100-keV, 100-mA H⁻ Injector," AIP Conf. Proc. No. 111, 511-8 (1984).
(5) D. Schrage, et al., "A Flight-Qualified RFQ for the BEAR Project," Proc. 1988 Linear Accelerator Conf., CEBAF report no. CEBAF-R-89-001, 54-7 (1989).
(6) P. E. Gear and R. Sidlow, "Present Status of the Rutherford and Appleton Laboratories' H⁻ Source," Rutherford and Appleton Laboratories report RL-81-050 (July 1981).
(7) H. V. Smith, Jr., P. Allison, and J. D. Sherman, IEEE Trans. Nucl. Sci. NS-32, 1797 (1985).
(8) H. V. Smith, Jr., et al., "H⁰ Temperature and Density Measurements in a Penning Surface Plasma H⁻ Ion Source. I.," Proc. Int. Conf. Ion Sources, Berkeley, CA, July 10-14, 1989 (in press). To be published in Rev. Sci. Instrum. (1990).
(9) G. C. Stutzin, A. T. Young, A. S. Schlachter, J. W. Stearns, K. N. Leung, W. B. Kunkel, G. T. Worth, and R. R. Stevens, Jr., Rev. Sci. Instrum. 59, 1363 (1988).
(10) H. V. Smith, P. Allison, and R. Keller, AIP Conf. Proc. No. 158, 181 (1987).
(11) H. V. Smith, Jr., J. D. Sherman, and P. Allison, "Pulsed H⁻ Beams from Penning SPS Sources Equipped with Circular Emitters," Proc. 1988 Linac Conf., CEBAF report no. CEBAF-R-89-001, 164-166 (1989).
(12) H. V. Smith, Jr., N. M. Schnurr, D. H. Whitaker, and K. E. Kalash, "H⁻ Ion Source With High Duty Factor," Proc. 1987 Particle Accelerator Conf., IEEE Catalog No. 87CH2387-9, pp. 301-303 (1987).
(13) G. E. Derevyankin and V. G. Dudnikov, "Shaping of H⁻ Ion Beams from Surface-Plasma Sources for Accelerators," Institute of Nuclear Physics preprint no. 79-17 (Novosibirsk, 1979).
(14) V. G. Dudnikov, "Surface-Plasma Source of Negative Ions With Penning Geometry," Trans. 4th All-Union Conf. on Charged-Particle Accelerators (Nauka Publishers, Moscow, 1975), p. 323.

DISCUSSION

Kleyn: I was surprised that in your source you get an estimated recominbation coefffficient at the walls near unity. I wondered if that could be due to the fact that, as the H atoms diffuse to the walls they undergo many collisions: They are actually cooled before they hit the wall. So that in your case the atoms may be cooler when they hit the wall than in the other sources.

Smith: That is certainly a possible conjecture because as we can see, the atoms continue to cool as they drift towards the wall: their temperature continues to relax. So it is certainly plausible.

Capitelli: When you calculate the recombination on the walls, (on the last slide that you showed), you must put a cross section for the $H - H_2$ and this cross section depends also on the number of atoms because the diffusion coefficient depends on the system. Usually, when you take a cross section, you mean that you consider that you have few atoms in a bath of molecules, but, this is not your case. How do you compare your results with the results presented before by Young which showed that the relaxation of positively charge atomic hydrogen was much lower. Your relaxation cooling time is 1000 μsec, Young's relaxation cooling time is 30,000.

Smith: The discrepancy of the relaxation time is even larger than that. Ours is cooling in 3 microsec and Young's, I believe are relaxing in about 10 millisec. Of course, one of the answers for that is that our plasma chamber is very small. It is only 1.5 centimeters, roughly a cube. The source used by Young, and collaborators at Berkeley is the volume source that is much larger. Secondly, in addition to the increased size, the atom temperature in Young's system is much lower, I believe it is on the order of a tenth of a volt. Our atom temperatures are roughly ten or even higher. So they are going to get to the walls faster.

Hershcovitch: What's the arc voltage?

Smith: The arc voltage is in the vicinity of 100 volts.

Hershcovitch: 100 volts, this means you have about tens of kilowatts of power in the discharge.

Smith: Pulsed power, that's correct.

Hershcovitch: The experience that exists in the community of people doing polarized sources is that when you dump this kind of high power into a discharge, you get very close to a 100 percent dissociation. They do it with a rf discharge, not a dc discharge. They have basically a coil that surrounds a discharge tube with a small volume of gas. I think your power density is even higher since it is probably more efficiently coupled. At this tremendous power, I tend to support Yuri's (Belchenko) assertion that you are probably fully dissociated.

Smith: Well, the assertion that we are fully dissociated certainly explains much of our data too.

DYNAMICS OF NEGATIVE HYDROGEN IONS
IN A VOLUME SOURCE*

R.A. Stern
University of Colorado, Boulder, Colorado 80309

M. Bacal
Laboratoire de Physique des Milieux Ionises
Laboratoire du C.N.R.S.
Ecole Polytechnique, 91128 Palaiseau Cedex, France.

ABSTRACT

A new diagnostic method to measure transport properties of negative Hydrogen ions has been devised. It uses multi-point, multi-time laser pulse photodetachment of H⁻, inducing changes in the electron and ion densities, whose space and time evolution are traced in detail by laser pulses and probes. A kinetic analysis of plasma and self-consistent field evolution in space-time, following localized photodetachment of H⁻, is presented. The fit of measurements to analysis is used to infer H⁻ temperatures.

INTRODUCTION

We have established a program to study the dynamics of negative hydrogen ions (H⁻), with space and time resolution, within volume sources. The program has both analytical and experimental aspects. Its aims are to determine and measure the processes and properties which dominate the transport of H⁻ and the negative-ion beam quality, such as the H⁻ temperature.

Our method is based on an elementary concept : a simple way to measure transport properties is by perturbing the medium in a controlled manner, and measuring the space-time details of its return towards equilibrium. The technique makes use of laser photodetachment. Single-pulse photdetachment, a technique to measure the H⁻ density, assumes that the local, instantaneous (r=0,t=0) electrons released will be equal in number to the H⁻ ions destroyed. The present method goes far beyond this assumption: to measure the H⁻ transport properties, it traces the full space- and time-dependence of the non-local evolution of H⁻ density (all r,t), following a perturbation at some point (r=0,t=0), through the use of specially devised multiple - time, multiple - point laser and probe techniques.

In addition to the experimental procedure, our task therefore includes the construction of realistic models for the full plasma evolution. This paper presents:
 i. the fundamentals of Non-Resonant Optical Tagging (NROT), a diagnostic approach which we have developed for this purpose;
 ii. an analytic description of the kinetic behavior of H⁻ within a source, including the self-consistent electric field; and
 iii. experimental verification through the application of NROT to H⁻ .

NON-RESONANT OPTICAL TAGGING

We adapt for use in H⁻ the most recent approach to plasma ion diagnostics: Optical Tagging (OT)[1], a multi-point, multi-time laser method. In conventional OT, a "pump" laser is beamed across the plasma at a given point in space and time. The laser, tuned to resonance with an optically allowed transition between two long-lived quantum levels of the ion, induces strong local perturbations in the state densities. The subsequent evolution towards equilibrium of the densities is measured using a second, "search" laser[1] or else passively, using the spontaneous fluorescence. By matching the space-time details of the density evolution to a physical model, the transport properties (drift, temperature) of the ions can be inferred.

In analyzing the concepts underlying OT as described above, it is evident that three processes are involved:
- an experimental technique for causing and measuring the changes in a property of the ions, with adequate space/time resolution;
- an analytical model which describes the evolution of the property (e.g. state density), in terms of the transport parameters; and finally
- the fit of model to measurement, in order to unfold actual values of the parameters.

To date, OT has been used in systems where resonant transitions between quantum states of the ion could be induced by means of tuned, resonant laser beams. In these schemes the perturbation is induced in the state density of quantum levels of the "same" ion, with the initial and final states both being bound states of the same ionization state. As a result, the plasma properties of the medium remain unperturbed (other than indirectly, through changes in collisionality).

The variant of OT we now introduce, NROT, avoids the need for tuned radiation and resonant transitions. Instead, a laser is used to photodetach the H⁻ ion, leaving it in a <u>final state lying in the continuum</u>. The final state now consists of two particles: an H atom and a free electron. This provides additional channels through which the "search" function can be performed, since the density of either of the particles comprising the final state can be monitored. As in OT, two schemes are possible: "dark" detection, in which one measures the decrease in the initial-state density, and the "bright" mode, measuring the increase in the final-state density. Unlike OT however, the density of either of the two particles comprising the final state can be used.

In the version of NROT demonstrated experimentally here, the choice of pump laser wavelength results in final states in which the free atom ("photoatom") is generated in its ground level. Interrogating this H atom would require vacuum UV as search radiation. To avoid the technical complications associated with this, the search function in the <u>bright</u> signal mode is performed directly on the photoelectron using positively-biased Langmuir probes, axially and radially movable across the source. For the <u>dark</u> signal mode, the beam searching for the depleted H⁻ density is a second

Nd:YAG pulse delayed in time from the pump, and the detection is carried out on the final-state photoelectron using probes, as above.

Alternatives which may be considered could make use of:

i. UV search laser radiation tuned to one of the Lyman lines, to induce fluorescence from the photoatom generated by the pump laser; or else

ii. a high-energy photon, or a multi-photon scheme, for the pump beam, leaving the final atom in an excited state directly accessible to interrogation by conventional optical methods, e.g. laser-induced fluorescence[2] in the visible, or else passive detection of the spontaneous emission from the excited state.

An elementary measurement and data-analysis cycle in NROT is described below. The scheme concerns negative ions, but may conceptually be extended to photoionization as well.

Let the initial H$^-$ density (in the background plasma) be n$_-$(r,t=0).

1. Pump laser is switched on and off (in a short period) at the initial time t = 0, with the laser beam axis defining the radial coordinate origin, r = 0.
   - Destroys all negative ions within the test volume, a cylinder with r<R, the laser radius.
   i.e. pump causes a "hole" in the negative ions density within r < R.
   - Creates an (equal) excess electron density: $\Delta n_e(r,0) = n_-(r,0)$ within r < R

2. Electron-collecting Langmuir probe positioned at r registers an immediate[3] pulsed signal increase proportional to $\Delta n_e(r,0) = n_-(r,0)$.

3. Excess negative ions from surrounding region r > R flow into test volume r < R.
   - Ion density in r < R increases (from 0), reaching a value n$_-$(r,t) at a time t.

4. Search laser is switched on and off (in a short period) at a delayed time t = τ, at the position r = 0 (overlaps pulse laser radius).
   - Destroys all n- (r,τ).
   - Creates an (equal) excess electron density: $\Delta n_e(r,\tau) = n_-(r,\tau)$.

5. Electron-collecting Langmuir probe registers an immediate pulsed signal increase proportional to $\Delta n_e(r,\tau) = n_-(r,\tau)$.

By varying the delay τ from 0 to a large value (a few μsec), the Langmuir probe signal increase in the last step is caused to trace out the fractional time-recovery of the negative ion density, yielding n$_-$(r,t)/n$_-$(r,0) . Matching these measurement to a physical model which describes this ratio in terms of the background plasma unfolds the plasma properties (density, temperature, velocity, etc)

Note that the instantaneous equality of the excess electron density with the negative ions density, required in steps 1 and 4, implies that the laser photodetachment is thoroughly selective. It has been fully demonstrated[3] that in Hydrogen gas-discharges, with the type of laser radiation and under the conditions used in the experiments described here, only H⁻ is photodetached so that the equality is ensured.

Unlike OT, the splitting of the final state into two particles, whose charge/mass ratio is unlike that of the initial state, will cause a direct change in the plasma properties of the medium. The model which will be used to analyze the data must therefore describe the full dynamics of a three-species plasma, consisting of H, electrons and positive ions, taking into account also the evolution of the self-consistent fields and the effect they have on the properties to be measured (e.g. the ion velocity distribution). Thus the task of modelling the data is more complex.

DYNAMICS OF PHOTODETACHED PLASMAS

Since the background plasma is neutral, and the laser photodetachment pulse perturbation produces no net charge, the plasma is initially stationary and electric-field free everywhere. This is a relatively unusual initial condition, unlike the common situations, in which the plasma is initially rendered non-neutral by an applied field or injected charge (e.g. a beam). Instead, we are dealing with the evolution of two concentric, equipotential (field-free) neutral plasmas with the same total charge density, stationary (i.e. with no relative fluid speed), and differing only through the fraction of heavy (ion) to light (electron) negative charge densities which they contain.

The perturbation which starts the process is the gradient in electron and H⁻ densities on either side of the interface between two regions r>r and r<R, where r is the radial coordinate centered on the pump laser beam axis, and R is the beam radius. The kinetic motion (finite temperature) of the species begins to smear out the gradient, immediately following the pump pulse. Since the densities and thermal velocities differ, the fluxes of electrons and ions can be unequal, causing net charge accumulation and the build-up of a self-consistent field.

The field will tend to reduce the fluxes, may oscillate and eventually lead to a stationary state. As the field builds up, the positive charge density also begins to change. In principle, a complete description of the density evolution for all species can be generated. It is useful however from an experimental viewpoint to isolate the early-time regime in which the field is still weak, so that the simplest possible model can be developed to interpret the measurements.

Basic Equations.

The problem is treated kinetically using the coupled Boltzmann - Poisson equations. We limit ourselves to early times in the evolution, i.e. a period short in comparison with collisional times. The equations consist of the set:

$$\Delta \cdot E = -\Delta^2 \phi = 4\pi \Sigma \, en$$

$$\frac{\partial f}{\partial t} + v \cdot \frac{\partial f}{\partial r} + \frac{e}{m} E \cdot \frac{\partial f}{\partial v} \approx 0$$

with a separate equation for the velocity distribution function f of each species. The macroscopic observables we seek are the densities, obtained as integrals:

$$n(r,t) = \int dv\, f(r,v,t).$$

We linearize and apply the method of characteristics. This yields the formal solution of the Boltzmann equations, as the sum of a ballistic and a collective term:

$$f(v,r,t) = f(v',r',t')|_{t'=0} + \frac{e}{m}\int_0^t dt'\, E(r',t')\, \frac{\partial f_o(v,r')}{\partial v}$$

Here the regular and primed coordinates are connected by the zeroth-order orbit equations:

$$d^2 r'/d^2 t' = 0\, ;\ dr'/dt' = v'\, ;\ r'(t'=t) = r\, ;\ \text{and}\ v'(t'=t) = v.$$

<u>Early-Time Approximation.</u>

Initially, the field is negligibly small, but there are strong perturbations in the H⁻ and electron densities (not in the positive ions). For early times therefore, the ballistic terms will dominate the H⁻ and electron densities. These densities are given by:

$$n_B(r,t) = \int dv\, f(v',r',t')|_{t'=0}.$$

Because of the light mass, the ballistic period is quickly completed for the electrons. Therefore the approximation must include also the collective term for the electrons.

The calculation[4] consists of the following steps:

i. transformation of the integral defining the density n_B. Since r' depends on v, $n_B = \int dv\,(\ )$ can be transformed into $\int dr'\,(\ )'$, using the Jacobian evaluated from the solutions of the orbit equations.

ii. introduction of the background plasma state. We consider here for all species Maxwellian velocity distributions $f_o(v) = \pi^{-3/2} v_{th}^{-3}\, e^{-v^2/v_{th}^2}$, and piecewise-constant initial density distributions $g(r')$, with values $n_{o-,+,e}$ within cylindrical regions defined generally by radii $R_{min} < r < R_{max}$, $z_l < z < z_u$, and zero elsewhere.

iii. use of the orbit equations to express the velocity v in f_o in terms of r', and introduction into the expression for n_B. This yields:

$$n_B = \pi^{-3/2} v_{th}^{-3} \int_{-\infty}^{\infty} dr'\, t^{-3}\, e^{-[(r-r')/(v_{th}t)]^2}\, g(r')$$

Because of the symmetry, the integration is most conveniently carried out in cylindrical coordinates. The angular integral is simple to carry out, in view of the angular independence of the initial conditions. When the aspect ratio (length $z_{u,l}$ of chamber to laser diameter) is large, the axial integral reduces to a constant. The radial integral can be shown to be:

$$n_B(r,z=0,t)/n_0 \cong J([R_{min}/v_{tht}]^2,[r/v_{tht}]^2) - J([R_{max}/v_{tht}]^2, [r/v_{tht}]^2), \text{ where}[5]:$$

$$J(x,y) \equiv 1 - 2e^{-\rho^2} \int_0^\beta te^{-t^2} I_0(2\rho t)dt \int_0^b t^{-t^2} I_0(2rt)dt ;$$

$$x = \beta^2 = (R_{min}/v_{tht})^2 \text{ or } (R_{max}/v_{tht})^2; \; y = \rho^2 = (r/v_{tht})^2$$

The function J is shown plotted in Fig. 1 as a function of t in time units R/v_{th}, for values of r/v_{th} ranging from nearly 0 to 3, i.e. at points r both < and > R. As seen, the function changes form drastically on either side of $r/R = 1$. For r<R is is monotonically growing, with the slowest growth rate at the center $r = 0$. For $r/R > 1$, it is non-monotonic, and exhibits a local minimum which propagates nearly with the constant velocity v_{th} over a large range. This model therefore predicts the ballistic density recovery both within and without the laser beam.

<u>Results of the Analysis.</u>

Applying the appropriate initial conditions for each species, one obtains:

for the H⁻ density: $n_{-B}(r,z=0,t)/n_{-0} \cong J([R/v_{th-}t]^2, [r/v_{th-}t]^2) \equiv J_-$
the positive ion density : $n_{+B}(r,z=0,t)/n_{+0} = 1$ (unchanged in the approximation)
the excess electrons in the ballistic limit:

$$\Delta n_{eB}(r,z=0,t)/\Delta n_{e0} = 1 - J([R/v_{thet}]^2,[r/v_{thet}]^2) \equiv 1 - J_e$$

At large times, $v_{thet} = \infty$ so that $J_e = 0$. For the collective part of the electron density, the conventional[6] linear approximation is used : $n_{e\,coll} = n_{eo}(1 + e\phi/kT)$.

Combining these results and inserting them into the Poisson equation yields the expression determining the early-time evolution of the field :

$$\Delta^2\phi \cong -4\pi e\{n_{+B} - n_{-B} - n_{e0}(1 + e\phi/kT_e)\}$$

$$= 4\pi e\, n_{+0} [\varepsilon J_-(R/v_{th-}t, r/v_{th-}t) + \frac{e\phi}{kT_e}]$$

This is the inhomogeneous (driven) Helmholtz equation, the solution of which has well-known treatments in terms of integrals of Bessel functions.

We first estimate the magnitude of ϕ at early times, and therefore the size of the field-dependent correction to the negative ion ballistic density, by means of the "plasma approximation". This assumes that the static screening is sufficiently effective that the term $\Delta^2\phi$ be negligible in comparison with ϕ/λ_D^2 (the last term on the RHS above). The remaining terms then yield the value of the normalized potential:

$$\frac{e}{kT_e}\phi \approx -\varepsilon J_-(R/v_{th-}t, r/v_{th-}t)$$

where ε is the ratio of initial H⁻ density to the total negative charge n_{+o}. Since this ratio is <1, as is J at early times, it is straightforward to define the period within which the potential is weak. In this region therefore the field has a negligible effect on the H⁻ velocity, and the ballistic term dominates the density recovery. The process described here is analogous to ambipolar diffusion, except that it is the negatively-charged H⁻ ions whose <u>in-flow</u> into the laser-beam volume slows down the outflow of the electrons. We therefore characterize it as "monopolar transport".

EXPERIMENTAL VERIFICATION: H- THERMAL VELOCITY

The NROT technique was applied to an H⁻ multipole hybrid volume source[7,8]. We use two synchronized Nd:YAG lasers, firing pulses of energy > 30 mJ and 15 nsec duration, with delay between pulses variable from 0 to a few μsec. and jitter < 1 nsec. The photodetachment cross-section at these wavelength is > 10^{-17} cm², so that 10 mJ ≈ 6 x 10^{16} photons in a 1 cm dia. beam suffice to completely photodetach the H⁻ within the beam. Ballistic times for ions are of order R = 0.3 cm / (v_{th-} = 10^6 cm sec⁻¹) ≈ 300 nsec. In Hydrogen discharges at 3 mT, mean free paths are typically 3 cm, >> R (0.1 to 0.6 cm), so that the collisionless approximation is applicable.

Fig. 2 shows a schematic of the diagnostic instrumentation and geometry, together with typical probe current pulses at r = 0. The traces describe the response to 1. pump pulse only; 2. pump pulse followed by delayed search pulse; 3. computer-calculated difference between traces 1 and 2, showing the negative - ion density recovery at the time of the search pulse. For convenience, the base lines are displaced vertically. The abcissa is time, 200 nsec/division. The ordinate shows the probe electron current, linear scale (more light = signal down). The marks ^ indicate laser switch-on times (t = 0 and τ).

Typical results obtained using the "dark-mode" version of NROT are shown in Fig. 3. The data points trace the recovery of the H⁻ density as a function of time, at r = 0. The curves are data fits to the ballistic approximation [J(r=0)]. The best fit yields v_{th-} = 8 +/- 1 x 10^5 cm sec⁻¹, in agreement with indirect estimates[8].

The "bright-mode" version of NROT is illustrated in Fig. 4. Here the pump laser is pulsed on at r = 0, causing a delayed <u>increase</u> in the electron density at r>R. The signal shape peaks at a delayed time Δτ, as predicted by the ballistic approximation. As the probe is scanned along a diametral chord through r = 0, the value of Δτ varies as shown. Note that for r<R, it has the value 0, as it should. The slope of the r - Δτ variation yields a value for v_{th-} within the range found in the dark mode.

COMMENTS AND CONCLUSIONS

The agreement between the bright and dark modes of NROT, and the possiblility of carrying out non-local measurements (probe outside laser beam), suggest that the more complicated two-laser scheme may be used as a calibration technique, primarily. Future directions include the systematic measurement of H^- temperatures in a variety of source conditions and configurations, including the extraction region, as well as the development of models which will enable late-time (strong electric field) and collisional conditions to be analyzed and measured.

In summary, methods have been developed which yield both the analytical (kinetic) and the experimental descriptions of the behavior and transport properties of H^-, with space and time resolution, within volume sources. A solution of the Boltzmann-Poisson equations, including large ballistic terms, is developed to predict the evolution of H^- for an initial-value problem. A new measurement technique, Non-Resonant Optical Tagging, has been worked out and shown to be an effective diagnostic of H^- properties, in several modes of operation. It is found that H^- transport is governed by the negative ion inertia -with a secondary role assumed by the electrons and positive ions- for a period following the onset of an intense density perturbation. The experimentally determined H^- thermal velocities are consistent with theoretical models and independent estimates.

The preceding indicates that our program may provide the physical understanding and control techniques needed for negative-ion source and beam design and operation.

FOOTNOTES AND REFERENCES

*Supported by Direction des Recherches, Etudes et Techniques (France), ORNL and NATO grant RG85/0452. R.A.S. also acknowledges support under NSF grant PHY-8707338.

1) R.A. Stern, D.N. Hill and N. Rynn, Phys. Lett. 93A, 127 (1983); R.A. Stern, Europhysics News 15, 2 (1984).
2) R.A. Stern and J.A. Johnson III, Phys. Rev. Lett. 34, 1584 (1975); R.A. Stern, Phys. Fluids 21, 1287 (1978).
3) M. Bacal, G.W. Hamilton, A.M. Bruneteau, H.J. Doucet and J. Taillet, Rev. Sci. Instr. 50, 719 (1979).
4) R.A.Stern, P. Devynck, M. Bacal, P. Berlemont and F. Hillion, to be published.
5) "Integrals of Bessel Functions", Yudell L. Luke , McGraw-Hill, New York 1962.
6) S.G. Tagare and R.V. Reddy, Plasma Phys. and Contr.Fusion 29, 671(1987) ; A.V Gurevich, L.V. Pariiskaya and L.P. Pitaevskii, Soviet Physics JETP 22, 449(1966)
7) M. Bacal and G.W. Hamilton, Phys. Rev. Lett. 42, 1538 (1979); M. Bacal, A.M. Bruneteau and M. Nachman, J. Appl. Phys. 95, 15 (1984).
8) P. Devynck, J. Auvray, M. Bacal, P. Berlemont, J. Bruneteau, R. Leroy and R. A. Stern, Rev. Sci. Instr. 60, 2873 (1989).

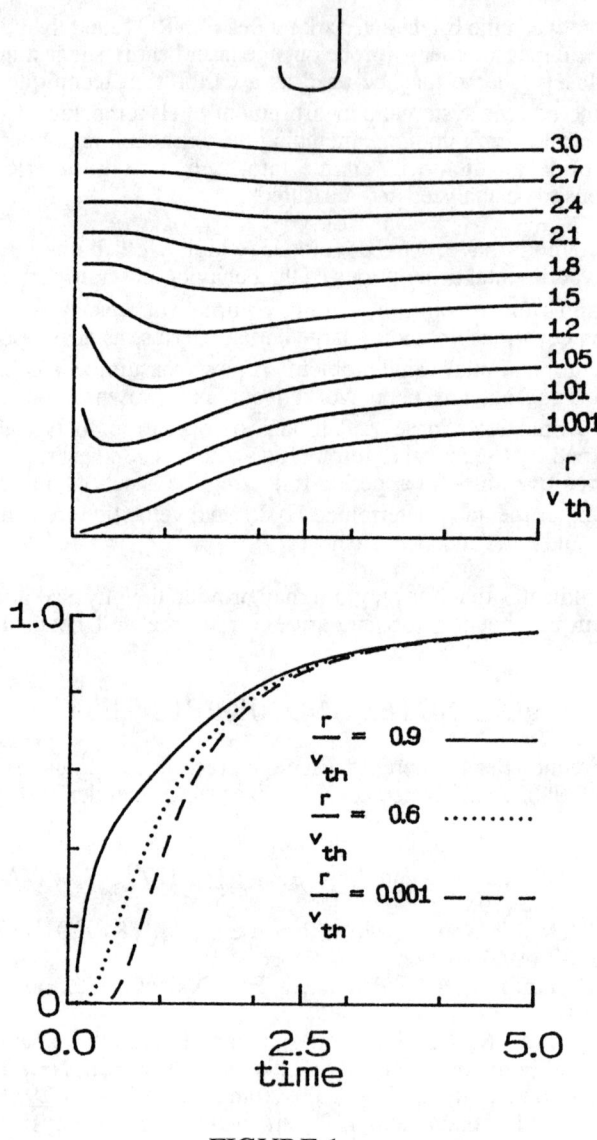

FIGURE 1

Function $J(1/t, r/v_{th}t)$ vs t
Top : $r/v_{th} > 1$, curves displaced vertically, to avoid overlap

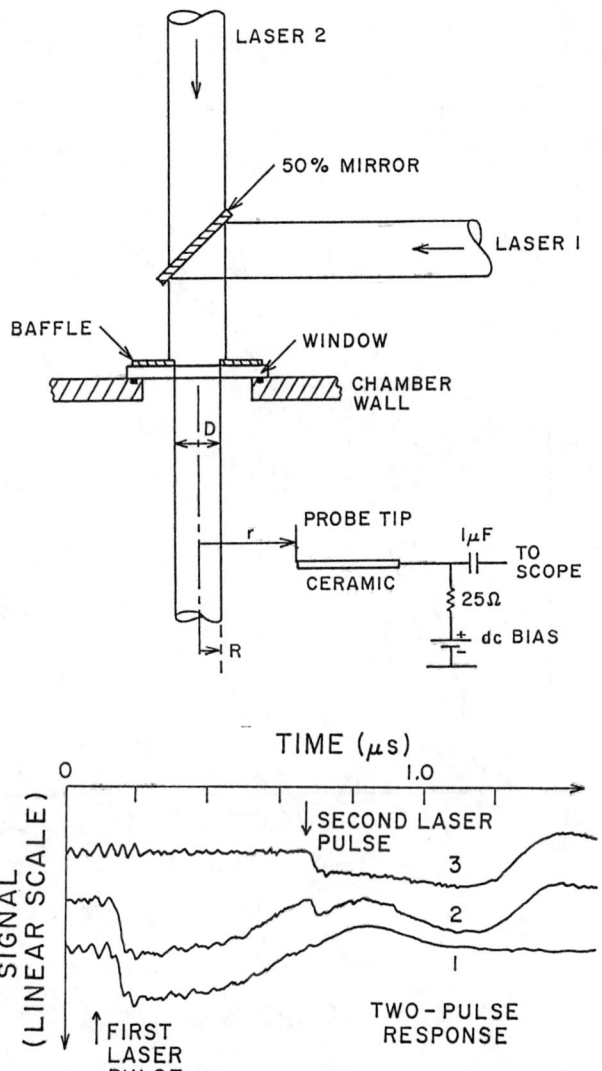

FIGURE 2

TOP: Diagnostic Configuration

BOTTOM: Typical Signals (Raw and processed data)

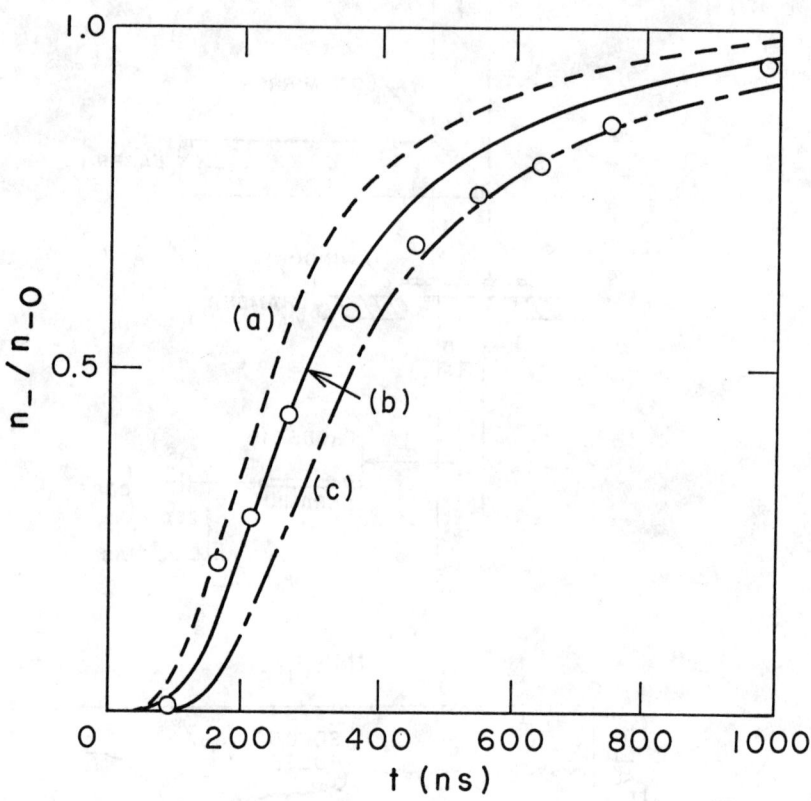

FIGURE 3

Recovery of H⁻ density at r = 0, for R = 0.2 cm.,

Data fits using the ballistic approximation. Parameter values R/v_{th}:

a. 2×10^{-7} sec, b. 2.5×10^7 sec., c. 3×10^{-7} sec.

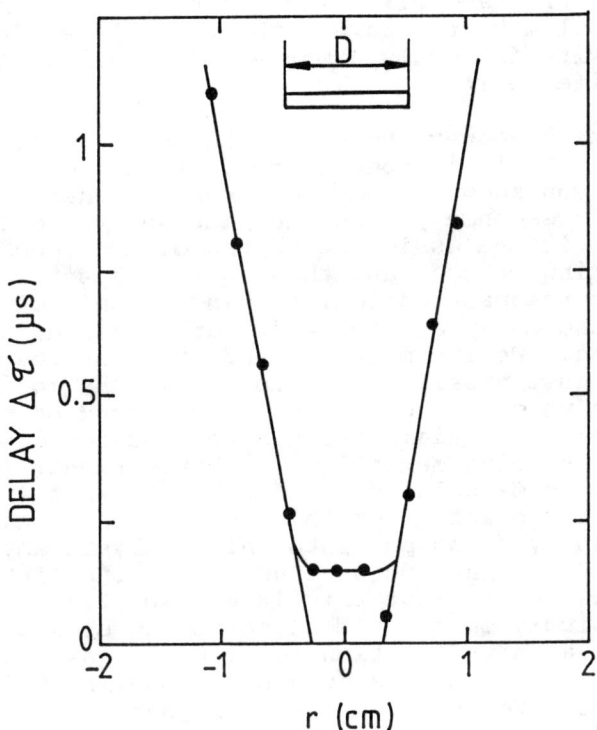

FIGURE 4

Variation of Peak Delay $\Delta\tau$ with Probe Position r along Diametral Chord

Insert shows radial location of laser beam (test volume)

DISCUSSION

Hershcovitch: First, this is model dependent, and you assume that the electric field at the very start is zero. Now if you have low frequency turbulence this will not be true and will affect the model. Also, another problem that I have with it a little bit is that you start with f_0 equals a Maxwellian and I wouldn't stake my life on it.

Stern: Let me answer those two things, two very good points. First, it is model dependent but you know everything you know is model dependent. When in laser induced fluorescence you scan a line and you get a line and look at it and assign it to one or two functions, you are making a model and this is the reason why I showed this resonant optical tagging technique where basically the delay and the width are modeled in terms of something. We are model dependent ourselves and so I will not excuse myself except to say if the model is correct, we've done a good job. The second part about it is the electric field being zero, and the distribution function being Maxwellian. That's right. It does not have to be Maxwellian, we don't know what it is but of course, the point about this is it doesn't have to be a Maxwellian, you can put into this analysis any distribution function that you want and it is the fit to the data that tells you what this is all about. I took the Boltzmann simply because it gives us the closed solution in terms of a known function. Our colleague, Pascal Devynck, for instance has done some modeling using general algebraic formula for this. It can all be done. We certainly have no time to discuss this. The turbulence part of this, yes and no, remember that this is a background electric field, which basically exists there all the time. You can measure this using the probe and essentially deduct it out from the equation. The electric field that we are concerned with is the growth of the electric field during the period of only nanoseconds and which you see the great advantage is, it is synchronous and in coincidence with the laser pulse. We can extract it because at no matter what phase of the turbulent background fluctuations, our rise in the electric field will always be the same. This is the equivalent of a coherent as opposed to a turbulent field and therefore, it has a completely different characteristic.

Hershcovitch: Can I make an extra remark regarding laser induced fluorescence? One can try to photo-detach H^- using the shape resonance and let it drift. The L_α

will decay within two nanoseconds or less. There is a
component in the 2s level that can be passed through an
electric field to induce its decay.

Stern: That's a very good suggestion and we have
thought about it in great detail. Of course, the kind
of lasers that you need for that are beyond our means
but this technique exactly is a very hopeful one and I
have on occasion in fact spoken to people at both Berkeley and Los Alamos. When Fred Schlachter started his
vacuum UV laser, he suggested that once it is built, we
can combine these techniques because building too many
laser systems in the same place is impossible. But the
suggestion is certainly a very good one and I would
dearly love to have corroboration of our results by
directive laser induced fluorescence using the shape
resonance.

Hershcovitch: That has its own problem since you have
to measure everything very close to the Lyman alpha
line. In a plasma there is too much L_α background.

Stern: Yes, what you are saying is very true, but there
is a way beyond that too. You see there is as you say a
lot of vacuum UV in the background. The enemy, the
noise, the radiation, however, those of us who know a
little bit about atomic radiation know that there is a
difference between the photon arrival rate, from laser
induced fluorescence than from spontaneous background or
collisional plasma because one is the radiation that
occurs in the presence of an electric field or strong rf
laser electric field, and the other one is spontaneous.
So using the equivalent of photon counting, one can
deduce the two and we have given some thought to the
matter, but we are far from having the funds or the
staff for such an ambitious project.

Bretagne: Could you include kinetic effects in your
model. By this I mean a combination of neutralization.
This is very important due to the fact that you consider
very fast times.

Stern: Let me answer the very second part first. In
late times we have no problem with: instead of solving
the abbreviated Helmholtz equation, we solve the full
Helmholtz equation which also has full known solutions,
although more complex than the ones that I have here.
So that can be done. The other part that you asked
about, how about the recombination and other kinetic
processes, the answer is in principle, yes, practically
I don't know how to do it. For the following reason.
Some of these processes have cross sections which are

velocity dependent. That means on the right hand side of the Boltzmann equation, you are going to come up with rather complicated integrals. Instead of a differential equation, it will be an integral differential equation, and here is where we rely very heavily on skills such as your own. I know that Marthe relies very heavily on your expertise in here, and we look forward to any input that you can give us in how to handle that.

MEASUREMENT OF THE H⁻ THERMAL ENERGY BY TWO LASER PULSE PHOTODETACHMENT

M. Bacal, P. Berlemont, J. Bruneteau, P. Devynck, C. Konieczny,
R. Leroy and R.A. Stern*
Laboratoire de Physique des Milieux Ionisés
Laboratoire du C.N.R.S.
Ecole Polytechnique, 91128 Palaiseau Cedex, France

ABSTRACT

The H⁻ negative ion thermal energy measured using the two laser-pulse photodetachment technique is reported to be in the range from 0.1 to 0.7 eV for various conditions of volume ion source operation (pressure - from 2 to 7 mTorr, discharge current - from 1.5 to 20 A). The hydrogen pressure has a significant effect in lowering the negative ion temperature, while the increase of the discharge current leads to a raise in T_-. It is found that T_- represents a fraction of the electron temperature, T_e, which is approximately independent of the discharge current, but strongly depends on the gas pressure. T_- scales linearly with the electron temperature and exceeds the highest values predicted by the theory of dissociative attachment for T_- at formation. The possible mechanisms for H⁻ heating are discussed.

INTRODUCTION.

The negative ion temperature is an important characteristic of the negative ion source, since it determines the emittance of the negative ion beam. Only indirect determinations of the negative ion temperature have been reported.[1-3] Recently MacAdams et al[3] measured the emittance of the negative ion beam extracted from a volume source and deduced from this measurement that the negative ion temperature ranged from 0.5 to 12.1 eV over the complete data set, with average values in the range 1 - 3 eV.

We have recently developed a method of measuring the negative ion temperature. Our diagnostic scheme is based on laser induced photodetachment of the H⁻ ion into an atom and a free electron, followed by detection of the free electron using a Langmuir probe. This scheme has been used earlier for determining the relative H⁻ ion/electron density.[4] Two techniques have been developed and tested. A direct determination of the H⁻ temperature is obtained using two laser pulses delayed in time. The first pulse depletes the local H⁻ density, while the second pulse releases free electrons from the H⁻ ions which have flowed into the laser-illuminated volume during the delay. A probe located within the illuminated volume measures the electron density raise in time and enables the transport speed of H⁻ to be evaluated. An indirect determination using a single laser pulse is based on the space-time history of the free electrons released in a single photodetachment pulse.[5,6] It turns out that the rate of outflow of electrons from the laser illuminated volume is governed, because of plasma neutrality requirements, by the much slower in-flow rate of H^- from the surrounding volume. In this paper we will describe the direct determination of H^- thermal energy in the two laser-pulse experiments.

EXPERIMENTAL SETUP

These experiments were performed in the hybrid multicusp volume H^- ion source, under study at Ecole Polytechnique. This volume source has been described in detail earlier[7]; the energetic primary electrons are confined near the sidewall, in the multicusp magnetic field produced by the wall magnets. Therefore, the center of the source is a low electron temperature plasma, suitable for volume production of H^- ions.

The plasma parameters (electron density and temperature) were measured using a microcomputer-controlled electrostatic cylindrical probe (0.5 mm in diameter, 1 cm long). This diagnostics technique uses the numerical results of Laframboise, as described by Hopkins and Graham.[8]

Figure 1 shows the experimental setup used in the photodetachment experiments. A Nd-YAG laser generates a cylindrical beam of 1.2 eV energy photons. The radius of this beam can be changed by interposing discs of different diameters in its path. The illuminated H^- ions release electrons with an energy equal to the energy of the photons (1.2 eV) minus the electron affinity of the hydrogen atom (0.745 eV), i.e. 0.455 eV. A cylindrical probe is located on the axis of the laser beam. The signal collected by the probe after the laser pulse is decoupled from the background current by a capacitor and recorded across a 50 Ω resistor. A transient recorder, consisting of a digitizer (LeCroy 6880A) and its accessories, is used to sample the signal, with a digitizing rate of 1.35 Gigasample/sec i.e. in steps of 0.742 nsec, with a total of 10,016 samples.

In the present experiments, two Nd-YAG lasers are synchronized, to obtain two laser pulses with a variable delay. The jitter is less than 1 nsec. The axis and the diameters of the two lasers are identical. The first laser beam detaches all the negative ions in the illuminated region. The second laser is fired after a delay Δt. It will detach those H^- ions which flowed into the laser beam volume during Δt. This will produce an excess electron signal, proportional to the integrated H^- flux. As the delay is changed, we can reconstitute the entire recovery of the H^- density after the initial photodetachment and thereby infer the thermal speed of the H^- ions.

Figure 2 shows the signal Δi_e collected by the positively biased probe in the two laser beam experiment. In the case illustrated on Figure 2a the second laser pulse is fired after a delay of only 0.59 μsec following the first laser shot, and

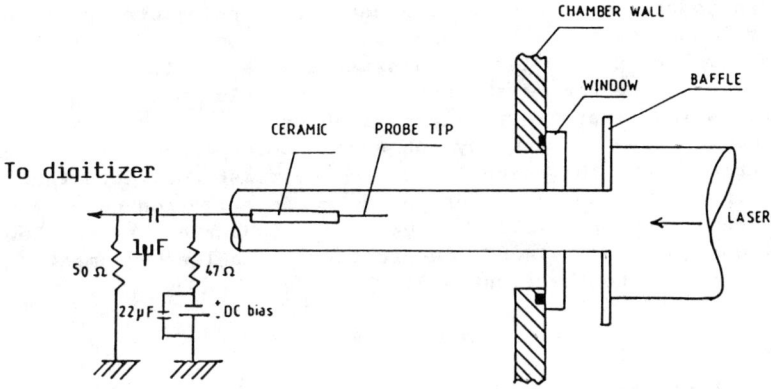

Fig. 1. Diagram of the two laser beam photodetachment experiment. The cylindrical probe movable along the source radius is coaxial with the laser beams.

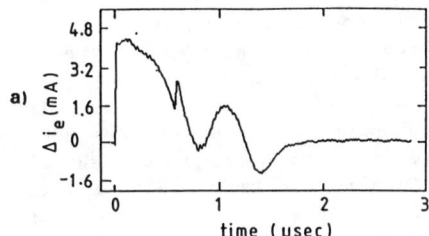

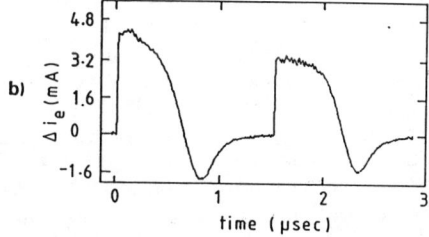

Fig. 2. Photodetachment signal collected by the positively biased cylindrical probe located on the axis of the laser beams, in the two laser beam experiment. The laser beam diameter is 0.8 cm, the probe bias is +20 V. 3 mTorr, 50 V, 5 A discharge. The delay between the two laser pulses 0.59 μsec in (a) and 1.55 μsec in (b).

the photodetachment signal it generates is superposed upon that due to the first laser. In the case shown on Figure 2b the second laser pulse is fired after a delay of 1.55 µsec, i.e. just after the end of the probe current pulse due to the first laser (which lasts in this case 1.50 µsec). Now the signal due to the photodetachment by the second laser appears completely separated from the signal due to the first one. Note that at the time of the second laser pulse the negative ion density has recovered only partially ($\sim$80%), as indicated by the lower amplitude of the signal due to the second laser, compared to the signal due to the first one.

EFFECT OF LASER FLUX

We have recorded the signal due to a single laser pulse and studied the changes introduced by varying the laser flux. In particular we have measured the length of the plateau, Δt_1, and the amplitude of the signal, Δi_{e1}. Figure 3 shows the recorded values of the plateau duration, Δt_1, and the amplitude of the photodetachment signal, Δi_{e1}, versus the laser flux, measured with a calorimeter. It can be noted that Δi_{e1} attains saturation for a laser flux of approximately 12 mJoule/cm². The duration of the plateau slowly decreases when the laser flux is reduced. This result is unexpected. We had anticipated that the duration of the plateau might go up when the flux is reduced below the saturation value but in actual fact it went down. The reason why we thought that this duration might go up was related to the development of an electric field, accelerating the flow of negative ions from the background plasma into the illuminated region. Such an electric field will develop because of the higher flux of outgoing electrons in comparison with the flux of inflowing H⁻ ions.

One possible explanation for this behaviour is the presence of reflected laser light outside the cylinder illuminated by the incident laser beam. If some negative ions are photodetached by the reflected light outside the cylinder defined by the incident laser beam, the recovery process will take longer, since the influx of negative ions (product of density and velocity) is reduced.

We have verified this hypothesis by measuring the photodetachment signal with a probe located outside the region illuminated by the directly incident laser light. The result of this measurement is shown on Figure 4. The photodetachment signal has in this case two distinct parts: (a) a fast rise at the time of the laser pulse followed by a plateau, and (b) a slow signal rise, followed by an overshot, occurring after a delay, which increases with the distance from the laser beam.

We investigated the change of this signal when the laser flux is increased above the saturation value. It was found that the plateau value following the fast rise increased in proportion to the laser flux, while the delayed, slow rise signal was saturated. The latter feature, as well as its nearly

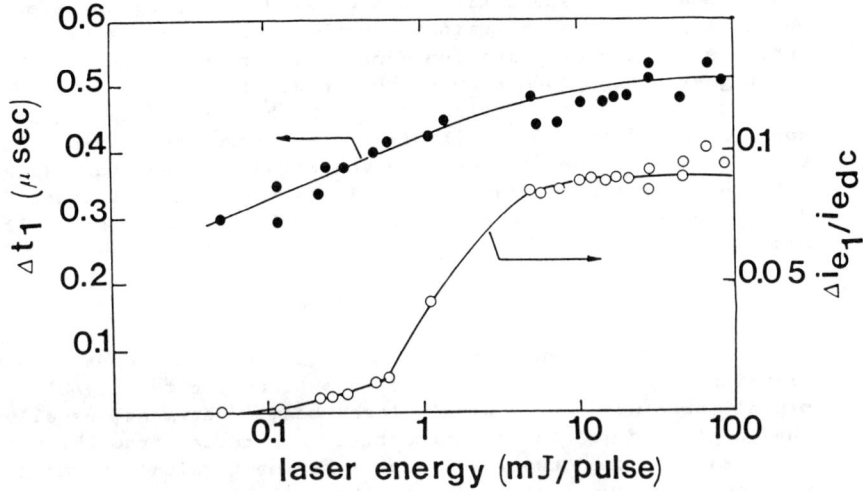

Fig. 3. Dependence of the photodetachment signal plateau duration and amplitude versus laser beam intensity, in single laser beam experiment. 50 V, 5 A, 3 mTorr discharge. Laser beam diameter 0.8 cm. Probe bias: + 20 V.

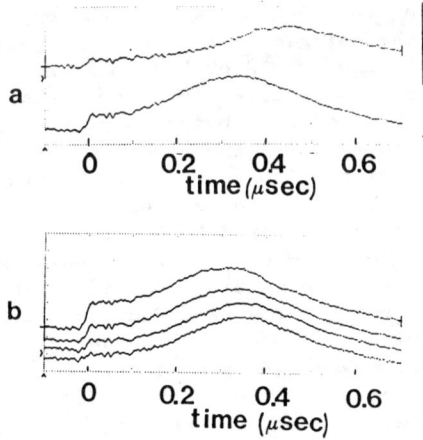

Fig. 4. Photodetachment signal from probe located outside the laser beam. a) Signals taken at two different distances from the center of the laser beam. b) Signals taken with different laser fluxes, corresponding to attenuations 1, 2, 3 and 4. 50 V, 5 A, 3 mTorr discharge. Probe bias : + 20 V.

linear delay with increasing distance from the center of the laser beam, r=0, identified this delayed signal as the electron burst leaving the illuminated region. Figure 5 shows that the increase of the negative ion density destroyed by the reflected light goes up in proportion to the laser flux.

We reduced the reflected light by installing a graphite cone on the flange which is hit by the laser beam. This reduced the level of the plateau in the signal of the probe located outside the laser beam, as well as the effect of laser flux upon the overshot region of the probe located inside the laser beam.

DATA REDUCTION

The information about the negative ion recovery in the illuminated region is obtained by subtracting two signals: the signal obtained in a single laser beam experiment, usually at the beginning of the experiment, is subtracted from the signal obtained in the two laser beam experiment, with a given delay between the two lasers. This subtraction is carried out by the computer associated with the LeCroy digitizer. We usually start off by averaging each of the two signals over a certain number of laser shots. The difference between the two averages at the time of the second laser shot is then read. This difference is used to plot the time dependence of the negative ion density (see Figure 6).

Several difficulties appeared to us in relation with this procedure. The first problem is the slow change in plasma conditions between the beginning of the experiment, when we usually measure the single laser beam signal, used as a reference for all the two laser beam measurements. It appears that a small difference can appear between the two signals to be compared simply because of the change in plasma conditions.

In an attempt to increase the precision of our analysis, we read directly the abrupt rise on the signal obtained in the two laser beam experiment at the instant of the second laser pulse. Discrepancies appeared compared to the results of the analysis using the difference between the two signals. A more detailed analysis indicated that these discrepancies were related to the finite duration of the laser pulse, and thus to the rise in the photodetachment signal ($\sim$10 nsec); during this time a significant change may occur in the reference signal, at times when the signal variation is fast (after the end of the plateau). Therefore, the direct reading of the fast change in the two laser beam signal does not account correctly for the negative ion density at that time. It appears that the initially proposed method of subtracting two signals is correct but has to be improved, by avoiding a large time lapse between the measurement of the reference signal and the measurement of the two laser beam signal. This could be done for example by subtracting two sequential two-laser-beam signals measured with a finite time delay.

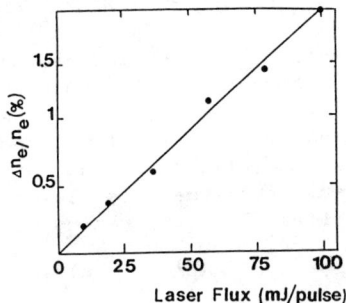

Fig. 5. Dependence of the negative ion density destroyed outside the laser beam upon the laser flux. 50 V, 5 A, 3 mTorr discharge.

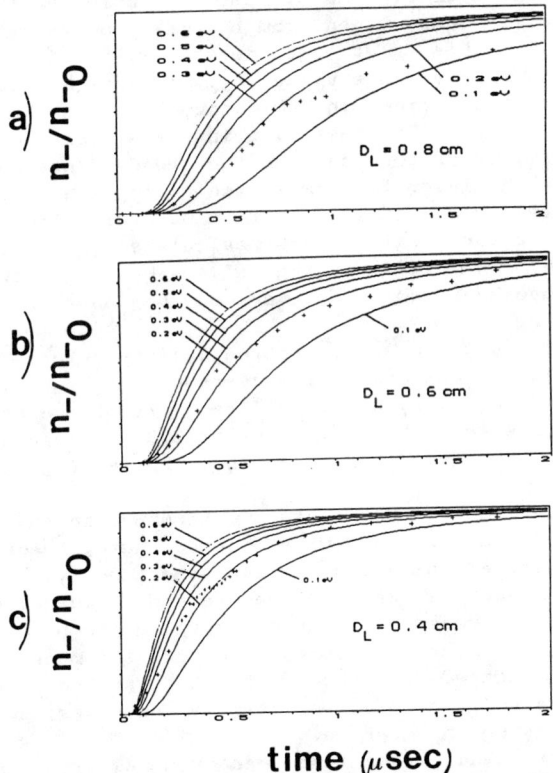

Fig. 6. Normalized H⁻ density versus the time delay after laser induced photodetachment. 50 V, 5 A, 3 mTorr discharge. Laser beams diameters: a) 0.8 cm; b) 0.6 cm; c) 0.4 cm. Probe bias: + 20 V. Theoretical curves calculated using Eq. 1 are also plotted, with the H⁻ temperature as a parameter.

MEASUREMENT OF THE H⁻ TEMPERATURE

Our theoretical model[9] of laser induced photodetachment of negative ions in plasma relies on the idea that the departure of excess electrons from the illuminated plasma column occurs when they are replaced by negative ions from the surrounding plasma. In this model the time dependence of the negative ion density at the center, after the first laser shot, is, in the ballistic limit, as follows (R is the radius of the laser beam):

$$n_- = n_{-o} \exp[-R^2/(v_{th} \times t)^2] \quad (1)$$

The measured time variation of the negative ion density after the first laser shot is plotted on Figure 6a, for the case of a laser beam 0.8 cm in diameter, and a 3 mTorr, 50 V, 5 A discharge. Similar data are plotted on Figure 6b and 6c for laser beam diameters of 0.6 and 0.4 cm, respectively. The time on the abscissa is determined from the delay between the two laser shots. In this plot the negative ion density is normalized to its steady state value, n_{-o}. This is accomplished by dividing the fast rise in the signal due to the second laser, Δi_{e2}, by Δi_{e1}. The fast rise in the signal due to the first laser, Δi_{e1}, is proportional to the steady state negative ion density, if the laser flux is sufficiently high to destroy all the negative ions. This condition was fulfilled in this experiment. The theoretical curves calculated for different negative ion temperatures from Eq. (1) are also plotted on Fig. 6. Two important observations can be made from the analysis of Fig. 6:

1. The time at which the electron density returns to its steady state value at the end of the overshot, Δt_e, corresponds to a partial (although very high, 80%) recovery of the negative ion density. With a laser beam diameter of 0.8 cm (Fig. 6a) the full recovery is observed at 5.5 μsec, while Δt_e is only 1.2 μsec.

2. The negative ion temperature determined in any of the three measurements by fitting to the corresponding theoretical curves shown on Fig. 6 a,b, and c, is the same, namely 0.2eV.

Our first objective was to measure the influence of the discharge current increase upon the negative ion temperature in the center of the hybrid multicusp source. The measurements tend to be complicated by the fact that, as the discharge current is increased, the electron density increases, too, and the fluctuations go up in meantime.

On the other hand, when the measurements are made at constant probe bias (+20 V in our initial measurements), the shape of the photodetachment signal also changes, as the discharge current is increased. We noticed at a discharge current of 20 A that the plateau, usually observed in the initial part of the photodetachment signal, was replaced by a maximum (a 'bump'). It was found that this maximum disappeared and the usual shape with a plateau was again observed when the probe bias was increased to +40 V. Thus, we report now the T_- values measured with a suitably chosen probe voltage. In future measurements we should maintain a constant, higher value of the probe bias through all the measurements.

Effect of discharge current and pressure. Figure 7 shows the results of a series of measurements on the effect of the discharge current I_d, on the negative ion temperature T_-, at constant pressure (2, 3 and 7 mTorr). We also show on Fig. 7 the corresponding variation of the electron temperature T_e with discharge current. Note that at 3 mTorr T_- increases from 0.2 eV to 0.4 eV when the discharge current is enhanced from 1.5 to 20 A. The corresponding increase of the electron temperature is from 0.43 eV to 0.86 eV. It can be noted that in this case T_- increases in proportion to T_e, and its value is approximately $0.5\,T_e$. The ratio T_-/T_e is almost independent of the discharge current, but is strongly affected by the gas pressure. T_-/T_e decreases from 0.7 at 2 mTorr to 0.2 at 7 mTorr (see Fig. 8).

Since the negative ion temperature appears to be related to the electron temperature, it can be concluded from Figures 7 and 8 that T_- goes down with pressure for two jointly acting reasons: (a) the reduction of the electron temperature when the pressure increases; (b) at higher pressure T_- represents a lower fraction of T_e.

Figure 9 shows the results of an experiment in which the pressure was varied, while the other discharge conditions remained unchanged. In the example of a 5A discharge the increase of the pressure from 2 to 7 mTorr leads to a reduction of T_- by a factor of five (from 0.45 to 0.1 eV), while the electron temperature only goes down by a factor of two (from 0.78 to 0.43eV).

Effect of the electron temperature. Figure 10 presents the dependence of T_- versus the electron temperature T_e using all the results obtained at different pressures and discharge currents. These measurements were made with two laser beam diameters (0.4 and 0.8 cm). It can be noted that in the studied range of T_e, T_- increases with the electron temperature following the equation (where both kT_- and kT_e are in eV):

$$kT_- = 0.94\,(\,kT_e - 0.3) \qquad (2)$$

Wadehra[10] calculated the negative ion average energy at formation by dissociative attachment from a particular vibrationally excited molecular population, when the electron energy distribution is maxwellian. This quantity depends on the internal energy of the molecule and on the electron temperature.[10] The average negative ion energy can be related to the H^- ion temperature (if the H^- energy distribution is also maxwellian) as follows:

$$E = (3/2)\,kT_- \qquad (3)$$

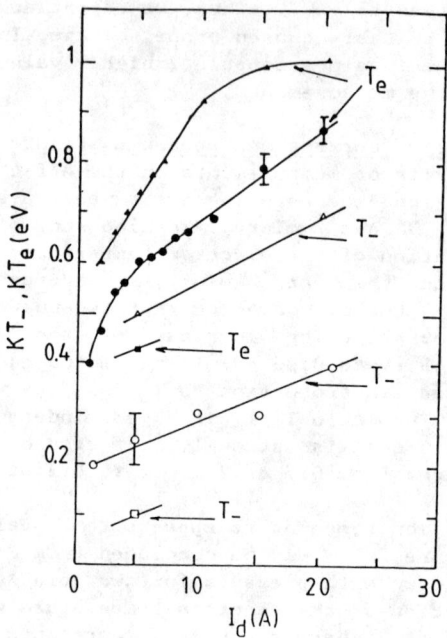

Fig. 7. Negative ion and electron temperatures versus discharge current, at three gas pressures (2,3 and 7 mTorr). Triangle: 2 mTorr; circle: 3 mTorr; rectangle: 7 mTorr. The open symbols indicate the negative ion temperature; the full symbols - the electron temperature. Laser beam diameters used: 0.4 and 0.8 cm.

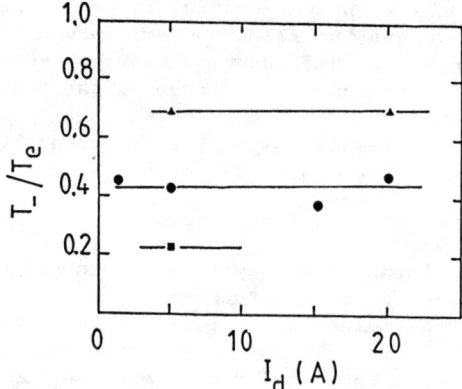

Fig. 8. Variation of T_-/T_e versus the discharge current at three different pressures. Triangle: 2mTorr; circle: 3mTorr; rectangle: 7mTorr.

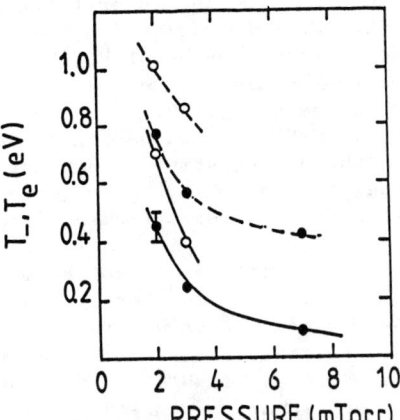

Fig. 9. Negative ion and electron temperatures versus pressure, at constant discharge current. The discharge voltage was 50 V. Open circles: 20 A; full circles: 5 A; full line: T_-; dotted line: T_e.

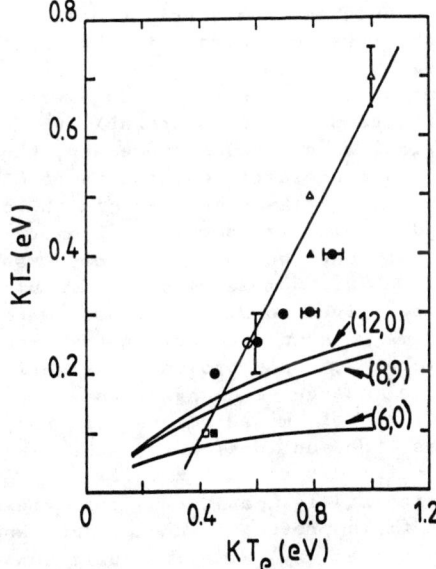

Fig. 10. Variation of the negative ion temperature with the electron temperature. The different symbols indicate different pressure. Rectangle: 7 mTorr; circle: 3 mTorr; triangle: 2 mTorr. The open symbols indicate that the measurement was done with a laser beam diameter of 0.8 cm; the full symbols correspond to a laser beam diameter of 0.4 cm. The three full curves represent theoretical T_- from Ref. 10, for the rovibrational states indicated.

We plotted in Figure 10 the theoretical values of kT_- for three levels of the H_2 molecule: (v"=6, J=0), (v"=8, J=9) and (v"=12, J=0). The internal energy of H_2(v"=8, J=9) is 3.967 eV and is very close to the limiting energy when the dissociative attachment becomes exoergic (3.994 eV). The internal energy of the state H_2(v"=12, J=0) is even higher than this limiting energy, and thus the dissociative attachment to this state is exoergic. According to Wadehra's calculations the highest H^- temperature corresponds to ions formed from these molecular states.

It can be noted that the T_- values measured at 2 and 3 mTorr exceed the highest values of T_- at formation, which follow from theory. The T_- value observed at 7 mTorr is the only one close to the theoretical characteristic for H_2(v"=6, J=0). Let us remind that it is from v"=6, 7 and 8 that modelling predicts the formation of negative ions.[11]

Two factors can modify the energy of the H^- ions acquired at formation: elastic collisions and the electric field in the plasma.

The Coulomb collisions of H^- ions with electrons could be invoked to explain the correlation of the negative ion and electron temperatures. However the H^+-H^- relaxation times for energy transfer are much shorter than the H^--electron relaxation times, because the energy transfer between charged particles is much more efficient if the particle masses are approximately equal. Therefore (a) the H^- energy exchange with positive ions dominates the energy exchange with electrons, and (b) the positive ion temperature is probably very close to T_-.

Among the elastic collision processes, the shortest mean free path (MFP) is for elastic collisions of H^- with hydrogen molecules. The energy relaxation MFP in H^- collisions with positive ions and atoms is somewhat longer. Since the H_2 molecules are cold, one could reasonably assume that H^- is cooled by collision with molecules. H^- could be heated in collisions with positive ions, if those were hot. Charge exchange with atoms, known to have a temperature well above that of the molecules, can provide energetic H^- ions. The uncertainty in the charge exchange cross section of H^- with atoms, and the lack of measurements of the positive ion temperature, makes the definition of the major collisional heating mechanism difficult at this time.

Another possibility is that the H^- ions are accelerated by the weak electric field present in the plasma, and the acquired velocity is randomized by elastic collisions. This could contribute to the heating of the H^- ions only if their lifetime is longer than the characteristic time between two elastic collisions. This condition is fulfilled in the conditions of this experiment.

Our study of the plasma potential profiles along the axis and the radius of the hybrid multipole[12,13] indicate that the potential is extremely flat at pressures above 5 mTorr, while a plasma potential variation can be observed at lower pressures.

This may explain our observation of higher T_- values at 2 and 3 mTorr. The plasma potential and its gradient are related to the electron temperature and the observed corelation of T_- and T_e may be the consequence of the dependence of T_- on the plasma potential gradient.

If the gradients in plasma potential were the cause of H^- heating, a very flat plasma potential profile in the extraction region would be essential for low emittance sources. In meantime, the low negative ion temperatures observed in this work are compatible with ion optics at high energy.

Acknowledgements. This work was supported in part by the Oak Ridge National Laboratory and by Direction des Recherches, Etudes et Techniques (France). We acknowledge useful discussions with A.M. Bruneteau, G.W. Hamilton and D.A. Skinner.

REFERENCES

* Permanent address: University of Colorado, Boulder.
1. H. Vernon Smith Jr. and P. Allison, Rev. Sci. Instrum., 53, 405 (1982)
2. R. Keller and H. Vernon Smith Jr., 1985 Particle Accelerator Conference, Vancouver, British Columbia, May 13-16, 1985. Preprint LA-UR- 85-1559.
3. R. MacAdams, A.J.T. Holmes and M.P.S. Nightingale, Rev. Sci. Instrum., 59, 895 (1988)
4. M. Bacal, G.W. Hamilton, A.M. Bruneteau, H.J. Doucet and J. Taillet, Rev. Sci. Instrum., 50, 719 (1979)
5. P. Devynck, Rev. Phys. Appl. (Paris), 24, 207 (1989)
6. P. Devynck, J. Auvray, M. Bacal, P. Berlemont, J. Bruneteau, R. Leroy and R.A. Stern, Rev. Sci. Instrum. 60, 2873 (1989)
7. M. Bacal, F. Hillion and M. Nachman, Rev. Sci. Instrum., 56, 649 (1985)
8. M.B. Hopkins and W.G. Graham, Rev. Sci. Instrum., 57, 2210 (1986)
9. R.A. Stern and M. Bacal, Proc. of this Symposium.
10. J.M. Wadehra, Phys. Rev., A29, 106 (1984)
11. M. Bacal, D.A. Skinner and P. Berlemont, SPIE Proceedings Series, vol. 1084, "Microwave and Particle Beam Sources and Directed Energy Concepts", Editor H.E. Brandt, 1989, p. 528.
12. P. Devynck, M. Bacal, J. Bruneteau and F. Hillion, Rev. Phys. Appl. (Paris), 22, 753 (1987)
13. P. Devynck, M.B. Hopkins, J. Bruneteau, J.P. Stephan and M. Bacal, "Production and Application of Light Negative Ions", Proc. 3rd European Workshop, Amersfoort, The Netherlands, (1988), p. 47.

DISCUSSION

Hershcovitch: I noticed you were trying to look at various scattering by ions and by atoms. A few years ago, I examined the possibility of making very cold H^- with Jim Alessi's magnetron. It is an inverted magnetron where very cold H^0 (with $T \approx 2$ K) would go through the center of the magnetron and undergo charge exchange with D^-. But the D^- attract some D^+, and one has a little plasma. Although the magnetron is only two centimeters long, I calculated that in no time the H^- will thermalize, not by single particle collision but rather by the effect of the whole plasma on the H^-. Using Trubnikov's model with slow test particles, one finds indeed that one thermalizes very fast with the plasma.

Bacal: Okay, so could you conclude from that that we have to make a new experiment in a very flat plasma. Our plasma is very flat anyway.

Hershcovitch: No, if you produce the H^- at an energy lower than the temperature of the ions and electrons in the plasma. This H^- ion will thermalize very fast with the plasma particles.

Bacal: Okay, with which particle?

Hershcovitch: The ions and electrons in the plasma. There is a formula for calculating collisions due to many small angles scattering. This effect is usually shown in elementary plasma physics books to be a hundred times bigger than the effect of single particle scattering.

Bacal: Well so you say that thermalization goes even faster than we can think about.

Hershcovitch: That is correct, I am trying to propose another mechanism for thermalization.

Jacquot: I have a general comment and a question. Because we are working in plasma physics we know that the behavior of particles and heating and temperature does not depend on collisional process but on turbulence or the electric field. For example, in this paper you have some discrepancy between the H^- temperature and the temperature of the atoms. This can be explained by electric fields or turbulence. I have a question containing that. When you increase the pressure in any negative ion source, could you get a better agreement between the H⁻ temperatue correlated to any other temperature of molecule of atoms. Normally, when you increase the pressure, the electrostatic turbulence, or magnetic turbulence, should be disappear more or less. This means you move closer to an agreement between the temperature of molecules, atoms, and H⁻.

Bacal: We have not finished this comparison unfortunately. We have to finish analyzing our data on atoms at least. So I can't answer that question in a general way. However, if we come back to the paper given by Vernon Smith, I realized that I have to find out what are fields in the source. I understood that H⁻ ions were hotter than atoms, so I wandered whether the explanation by the field in the source wasn't more relevant. But we have yet to analyze this.

Kleyn: We have measured the drift velocity of the H⁻ ions as they are diffusing towards the acceleration gap. I just looked at it and it seems the scaling that you observed, for the ion temperature seems to be so comparable to what we see. Did you actually compare the numbers?

Bacal: Well we also measured the drift velocities. They were in that range. They were certainly of that order of magnitude. However, what we did not do recently, is to repeat these measurements in the extraction region and compare the drift velocities as deduced from extraction and densities and with this velocities. I have a feeling they will be comparable, but however, this has to be done since there may be some other features in the extraction region which are not observed here.

ATOMIC TEMPERATURE AND DENSITY IN MULTICUSP H⁻ VOLUME SOURCES

A.M. Bruneteau, G. Hollos[*], R. Leroy, P. Berlemont, M. Bacal
Laboratoire de Physique des Milieux Ionisés
Laboratoire du C.N.R.S.
Ecole Polytechnique, 91128 Palaiseau Cedex, France

J. Bretagne
Laboratoire de Physique des Gaz et des Plasmas, LA73 du CNRS,
Université de Paris-Sud, 91405 Orsay, France

ABSTRACT

The Balmer ß and γ line shapes have been analyzed to determine the relative density and the temperature of hydrogen atoms in magnetic multicusp plasma generators. Results for a 90 V, 4-40 mTorr, 1-18 A conventional multicusp plasma generator and a 50 V, 4 mTorr, 1-15 A hybrid multicusp plasma generator are presented. The relative number density of hydrogen atoms increases smoothly with pressure and discharge current but never exceeds 10 %. The absolute atomic number density in a 90 V - 10A discharge varies in proportion with pressure. The atomic temperature (in the 0.1 - 0.4 eV range) decreases with pressure and slowly increases with the discharge current. The role of atoms in the processes determining the H⁻ temperature and the H_2 vibrational and rotational temperatures is discussed. The results confirm that in multicusp negative ion sources collisional excitation of ground-state atoms and molecules by energetic electrons is the dominant process in Balmer ß and γ light emission.

INTRODUCTION

Atomic hydrogen plays an important role in the processes determining the negative ion fraction in multicusp negative ion sources. The experimental and theoretical investigation of these sources indicates that the dominant H⁻ formation mechanism is dissociative electron attachment of slow electrons to highly vibrationally-excited molecules[1]. The rate of destruction of H⁻ by the inverse reaction clearly depends on the atomic hydrogen density. Collisions[2,3] with atoms can also vibrationally relax the excited molecules and reshape the H⁻ ion energy distribution. On the other hand, wall recombination[4] of atoms can generate vibrationally excited molecules. The knowledge of the density and temperature of the atomic hydrogen fraction in these sources is, therefore, of utmost interest.

Péalat et al[5] proposed a method of determining the ratio of atomic and molecular hydrogen densities from the line shape of Balmer light emitted from the source. The method is based on the discovery by Freund et al[6] of large, characteristic wings in the Balmer lines produced by electron impact dissociative excitation of low pressure molecular hydrogen. Later work on the subject showed that in the dissociative excitation (DE) of hydrogen molecules by fast electrons two groups of excited atoms are produced : a fast group, which is responsible for the appearance of the wings, and a group of slow excited atoms whose contribution is included in the central peak. The main contribution to the central peak comes from atoms of the atomic hydrogen fraction in the plasma excited by fast electron collisions (HE). The analysis of the line shape thus permits to determine the relative proportion of atoms and molecules in a low pressure hydrogen plasma.

Recently, the validity of this method has been questioned[7,8] as a result of experiments performed in tandem plasma sources. Two observations appeared to contradict the basic assumption in the method of Péalat et al : a) light emitted from the excitation chamber (driver) which contains a large density of energetic electrons [7] did not display the characteristic feature of the "fast atom wings"; b) light observed from the extraction chamber, known to contain only electrons with a Maxwellian energy distribution corresponding to a temperature of the order of 1 eV, did nevertheless contain Balmer α and β lines[7,8]. The results of these observations lead Nightingale and Forrest[7] to conclude that processes other than collisions with energetic electrons are at the origin of Balmer light emitted from multipole plasma generators and that, therefore, the analysis of the Balmer line shapes cannot be used neither for the determination of the density of the atomic hydrogen population nor for the calculation of its temperature.

In the present work we have investigated the Balmer β and γ light emitted from a single chamber conventional plasma generator, as well as from different parts of a tandem multipole source. The results bring further evidence for the validity of the model underlying the method proposed in Ref.5. Some possible causes of the discrepancies with the experiments reported by Nigthingale and Forrest[7] and Graham and Hopkins[8] will be discussed. New data on the atomic/molecular density ratio and the atomic temperature in the investigated two sources will be presented. Relying on the density of the molecular hydrogen measured by CARS[5], we have also determined the absolute atomic density and its pressure dependance in particular operating conditions of the single chamber multicusp plasma generator.

EXPERIMENTAL SET-UP

Two magnetic multicusp plasma generators have been studied. The first one, described in Ref.5, was a single chamber conventional multicusp plasma generator, 16 cm in diameter and 20 cm high, which is similar to the excitation chamber of the tandem source. Two thoriated tungsten filaments were placed in the magnetic field-free region at about 4 cm from the optical axis. The present work was done at two gas pressures (4 and 40 mTorr) and discharge currents up to 18 A. In these conditions, the electron temperature and density were, respectively, lower than 1 eV and 10^{12} cm^{-3}.

The second plasma generator described in detail elsewhere[9], was a tandem negative ion source denoted as a hybrid multipole. Here the excitation region containing the energetic electrons was located at the periphery, around ten filaments placed in the multicusp magnetic field, which acts as a magnetic filter and prevents the primary electrons from entering the central part of the plasma. Therefore the entire central region of the generator is characterized by a very low electron temperature favourable for H$^-$ ion formation. In our operating conditions, the electron temperature and density were lower than, respectively, 1 eV and 1.3×10^{11} cm^{-3}. The emitted radiation was analyzed both along the radial (4 cm from the filament) and axial directions, for a 4 mTorr, 50 V discharge, and for discharge currents up to 15 A. In measurements made in the axial direction, light was collected only from the central region which contains no fast electrons. The plasma region analyzed in the radial direction contained energetic electrons in the neighbourhood of the filaments.

The light emitted by the plasma was collimated by a lens or brought by an optical fiber on the slit of a 2 m spectrometer equipped with a 1200 lines/mm grating. The resolution of the spectrometer was determined by measuring the broadening of the He-Ne laser line at 6328 Å and found to be 126000 when using 20 μm wide, 1 cm high curved slits (which corresponds to an instrumental bandwidth of FWHM = 50 mÅ).

RESULTS AND DISCUSSION

It can be shown[10] that, in Balmer α line emission from a volume negative ion source plasma, disssociative recombination of H_2^+ with electrons may play a significant role, in addition to direct electron excitation. On the contrary in the production of H (n=4,5) the contribution of all the examined concurrent channels (dissociative recombination of H_2^+ with electrons, DE of vibrationnally excited molecules, excitation of metastable levels, dissociative excitation and wall collisions of molecular ions) were found to be negligible. Because of this, and also because the detection sensitivity for Hα of our apparatus was very low, we limited our investigation to the Balmer β and γ lines.

A. Line shape of Balmer β emission from different regions of the hybrid multipole.

Figure 1 shows an example of Balmer β line shapes from a 50 V, 10A, 4mTorr discharge observed along a radial direction (Fig.1a) and along the axis of the plasma (Fig 1b). The emission observed along the axial direction was about ten times weaker. The common feature is that in both cases we see the peaks with wings. The 4861.7 Å molecular line was also detected, but not the filament continuum. The appearance of the wings in the lines observed in the radial direction is natural if we take into account that the wings are the result of the interaction of molecules with energetic electrons and that these electrons are concentrated at the periphery, around the filaments. The fact that

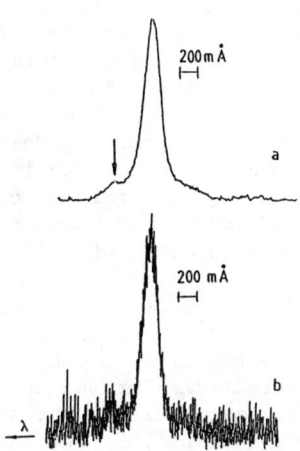

Fig. 1 : Balmer β line emission of a 50 V - 10 A - 4 mTorr hybrid multipole plasma along radial (a) and axial (b) directions. The sensitivity of light detection was set 10 times higher in (b) than in (a). The arrow indicates the position of the 4861.7 Å molecular line.

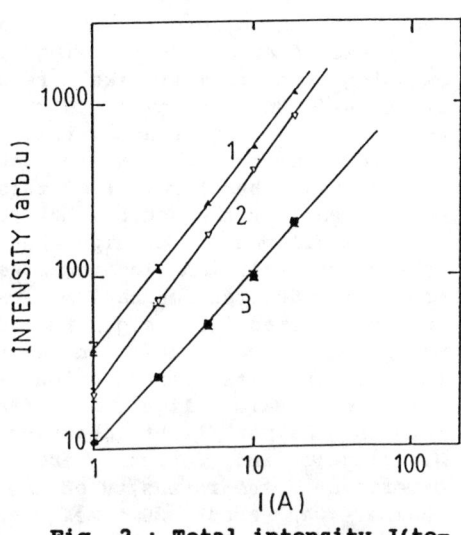

Fig. 2 : Total intensity J(total) (1), and fractional intensities J(H), (2) and Jf(H_2), (3), in H_β line shape versus the discharge current I for a 90 V - 40 mTorr conventional multicusp discharge.

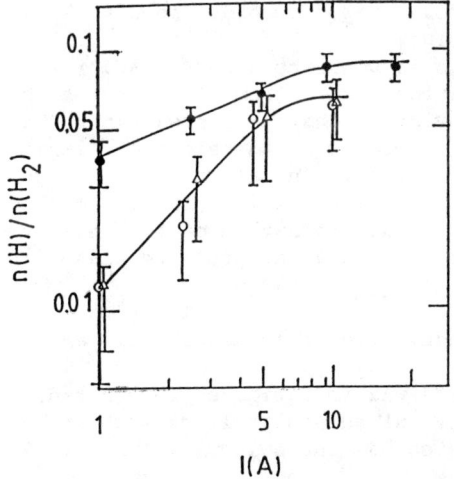

Fig. 3 : Relative atomic density n(H)/n(H_2) versus the discharge current I, obtained from the H_β line shape at 40 mTorr (●), at 4 mTorr (o), and from the H_γ line shape at 4 mTorr (Δ) in a conventional multicusp discharge.

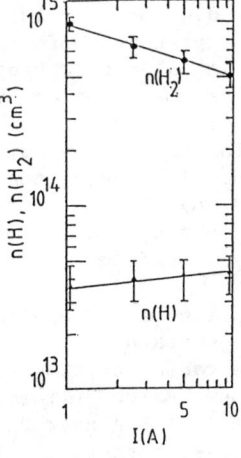

Fig. 4 : Atomic n(H) and molecular n(H_2) densities versus the discharge current in a 90 V - 40 mTorr conventional multicusp discharge.

the line shapes detected in the axial direction (fig 1b.) display the same characteristic wings leads to the conclusion thas this emission too is most likely related to the presence of energetic electrons. But in axial sighting, only light coming from the central part of the plasma, that is from a region characterized by very low electron temperature, can reach the spectrometer. We have to assume, therefore, that this light does not originate in the central part of the source, but comes from the same regions, close to the filaments, of high electron temperature, and reaches the spectrometer in an indirect way, by some process of radiation transfer. Several mechanisms can be envisaged to explain this radiation transfer, e.g. reflection of the visible light by the walls, or Lyman radiation emission in the excitation chamber followed by its reabsorption in the extraction chamber and emission of Balmer light. A similar radiation transfer can account for the Balmer light observed in the extraction chamber by Nightingale[7] and Forrest and Graham and Hopkins[8]. In the latter experiment the intensity of the light coming from the extraction chamber was about 20 times weaker than that of the light coming from the source chamber (against a factor of 10 in our case), which shows that in their device the conditions for the radiation to penetrate from one part of the discharge into the other were even less favourable. That such a transfer can really take place is supported by the observation by Graham and Hopkins[8] of the fact that the atomic Balmer lines from the extraction chamber were superposed upon the filament continuum, even though the filaments were located in the excitation chamber. The lack of the characteristic wings in the line shapes recorded in Ref. 7 was probably due to insufficient resolution.

The presence of the wings in both the Balmer β and γ line shapes in all operating conditions and from all parts of the source can be taken as further evidence that processes other than dissociative excitation of molecules and electron impact excitation have a negligible role, if any, in Balmer β and γ light emission. In these conditions the application of the method proposed by Péalat et al for the determination of the relative atomic density and atomic temperature in a low pressure plasma is entirely justified.

B. Light emission from the conventional multicusp plasma generator.

The radiation from this source was investigated in the radial direction. Balmer lines and several molecular lines such as the line at 4856.5 Å near H_β were detected. The molecular line was 70 times weaker than H_β.

Investigation of the effect of the discharge voltage on the Balmer β line shape showed that the ratio of the intensities due to HE and DE (fast atoms) increases as the discharge voltage is decreased from 90 V to 30 V, in good agreement with the data of Refs. 11 and 12. This confirms the proposed excitation mechanism involving fast electrons.

The H_β intensity was in all conditions by a factor of 5 to 10 higher than the intensity of H_γ. The fast atom component was relatively stronger in H_β than in H_γ.

Examination of the H_β and H_γ line shapes at different discharge currents (1 to 15A) and pressures (4 and 40 mTorr) reveals that :
- the relative fraction of fast excited atoms decreases when the pressure increases both for H_β and H_γ; this can be attributed to the energy degradation of the fast electrons ;
- the relative fraction of fast excited atoms decreases when the discharge current is increased ; this means that the relative atomic hydrogen density increases with the discharge current.

Fig. 2 reports the total intensity J(total) and the partial intensities J(H) and Jf(H_2) of the H_β line versus the discharge current I for a 90 V - 40 mTorr discharge. The partial intensities have been calculated using the method of Péalat et al. We find that the intensities scale with the discharge current as follows :

$$J(total) \sim I^{1.2} \qquad Jf(H_2) \sim I^{1} \qquad J(H) \sim I^{1.3} \qquad (1)$$

The total line intensity, essentially due to HE, increases more rapidly than I_d as has been also shown by the experiments of Graham and Hopkins[8]. Again in agreement with their results we found that the height h of the molecular lines varies slower than the discharge current :

$$h(\lambda = 4856.5 \text{ Å}) \sim I^{0.7}$$
$$h(\lambda = 4634 \text{ Å}) \sim I^{0.85} \qquad (2)$$

C. Atomic hydrogen density in the conventional multipole source.

Knowing the spectral intensities $J(H)_5$ and $Jf(H_2)$ we can calculate the relative atomic hydrogen density :

$$n(H)/n(H_2) = (J(H)/Jf(H_2)) \times (Nf(H_2)/N(H)) \qquad (3)$$

Here N(H) and Nf(H_2) are the pumping rates to produce a given excited state (e.g. n=4 when we deal with the Balmer β line) through the two concurrent channels HE and DE (fast component). The pumping rates were calculated (Table 1) using the known cross-sections for HE[13] and[11,12] for the DE channel leading to the production of fast atoms[11,12], and the electron[14] energy distribution function calculated by Bretagne et al[14]. This function has[15] been previously found to be in good agreement with experiments[15].

Fig. 3 shows the dependence of the relative atomic density upon the discharge current at 4 mTorr and 40 mTorr. The relative atomic density increases with the discharge current up to a certain level, and is always higher at elevated pressure but never exceeds about 10 %. Pressure rise degrades the electron energy thus favoring molecular dissociation. The agreement between data

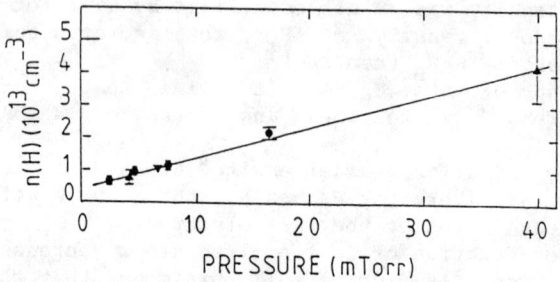

Fig. 5 : Atomic density versus pressure for a 90 V - 10 A conventional multicusp discharge (▲). For comparison, we also plot the results (●) reported in arbitrary units in Ref. 16, after fitting them to our experimental value at 4 mTorr. (▼) is from Ref.17.

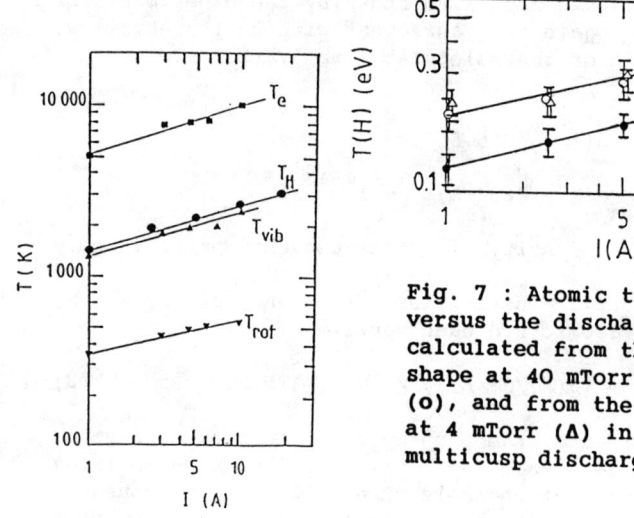

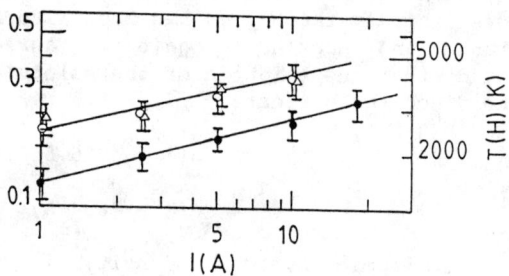

Fig. 7 : Atomic temperature T(H) versus the discharge current I, calculated from the H_β line shape at 40 mTorr (●), at 4mTorr (o), and from the H_γ lineshape at 4 mTorr (Δ) in a conventional multicusp discharge.

Fig. 6 : Electronic (■), atomic (●), vibrational (▲) and rotational (▼) temperatures versus the discharge current in a 90 V - 40 mTorr conventional multicusp discharge.

obtained from the analysis of H_β and H_γ is good and confirms the validity of the model ; in particular it shows that the e-H_2 dissociative recombination is negligible in our operating conditions.

Table 1

Pumping rates in a 90 V - 10 A discharge in s^{-1}

n	40 mTorr			4 mTorr		
	N(H)	Nf(H_2)	Ns(H_2)	N(H)	Nf(H_2)	Ns(H_2)
3	12.3	0.33	0.54	21	0.75	0.99
4	5.05	0.10	0.055	8.78	0.24	0.095
5	2.44	0.031	0.013	4.31	0.07	0.022

Given the n(H)/n(H_2) ratios, we can determine n(H)$_5$ by using earlier data on T(H_2) measured by CARS in the same source[5]. Fig. 4 shows the dependence upon discharge current of n(H_2) and n(H), at 40 mTorr. Note that at constant pressure n(H_2) decreases as I increases due to the higher degree of gas dissociation and higher temperature, while n(H) is practically constant. The dependence upon pressure at I = 10 A is shown in Fig. 5, where we also plotted the data measured (in arbitrary units) by Bonnie et al[16] using a multiphoton ionization technique. Note that n(H) increases linearly with pressure up to 40 mTorr. The functional dependence of n(H) with pressure, found by Bonnie et al[16] is in good agreement with our data. The value found by Schlachter et al.[1] is also represented in fig.5. Though it was measured in a 140V-25A discharge; it compares well in order of magnitude with our data at 90V-10A.

D. Atomic temperature in the conventional multicusp plasma generator.

The residual profile of the central peak of the experimental Balmer line shapes (fig.1), after substracting the contribution of the slow excited atoms, represents a convolution of the Doppler broadening caused by the motion of the radiating atoms, the instrument profile, and the fine structure splitting of the sublevels. At moderate resolution the component lines in H_β and H_γ merge into two fine structure lines. Over about 90 % of its height the resulting shapes approximated very closely (to better than 2 %) a Gaussian shape. This is attributed to the fact that both the instrumental bandwidth (FWHM = 50 mÅ) and the peak separation of the fine structure components (77 mÅ for H_β and 64 mÅ for $H\gamma$) are small compared with the Doppler width (150 to 200 mÅ in our measurements). In these conditions the Doppler contribution (and the atomic temperature) can be calculated by equating the square of the observed width with the sum of the squares of the contributions.

Fig. 6 shows the variation versus I of the atomic temperature in a 40 mTorr plasma. For comparison we also represent the electron temperature T_e, and the vibrational T_v and rotational T_r temperatures measured by CARS spectroscopy[5]. T_v has been derived from the population of the first four levels while T_r was found from J = 0 - 3.[6] The functional dependence of all these temperatures is $I^{0.25}$.

It is remarkable that the vibrational temperature of the molecules and the atomic temperature are practically equal. This occurs because the atoms effectively exchange energy with the molecules. The relevant energy exchange process is T-V (translation-vibration) transfer $H + H_2(v=0) \rightarrow H_{18} + H_2(v=1)$.

Indeed, model calculations of this plasma show that, at 40 mTorr, this process is as effective in producing $H_2(v=1)$ molecules as the e-V process $e + H_2(v=0) \rightarrow e + H_2(v=1)$.

The atomic temperatures at 4 and 40 mTorr are plotted in Fig. 7 versus the discharge current. Note that the temperature varies in proportion to $I^{0.25}$ and that its values at 4 mTorr are higher than at 40 mTorr since at lower pressure the atoms are cooled less by molecules.

At 4 mTorr we used the line shapes of both H_β and $H\gamma$ to determine the atomic temperature. The two sets of values, as shown in Fig. 7 are in very good agreement. They compare well also with the temperature determined by Launay et al[19] in a different magnetic multicusp plasma generator from the Doppler broadening of Lyβ lines.

It can be inferred from the curves in fig. 3 that the atomic density at 4 mTorr is lower than at 40 mTorr in the whole range of discharge currents. It follows that at low pressure T-V collisions become less effective than e-V[18] collisions (by a factor of 3 according to model calculations). Therefore one can expect that the vibrational temperature would be lower than T(H). Indeed, Lefebvre et al[20] obtained T_{vib} = 2096 K for Id = 10A in a discharge similar to ours, for which we determined T(H) = 4060 K.

Rotational excitation of H_2 by collision with atoms is also possible[21]. The higher atomic density is, therefore, one of the possible causes of the higher rotational temperature observed at 40 mTorr (550 K at 40 mTorr in this measurement, compared to 355 K at 4 mTorr[5,20], for I = 10 A).

E. Atomic temperature and relative atomic density in the hybrid multicusp generator.

The study of the atomic temperature and relative atomic

density in this device is of special interest because most of its volume represents an extraction chamber characterized by a low electronic temperature and the absence of fast electrons. . The variation with the discharge current of the relative atomic density and of the atomic temperature is plotted, respectively, in Figs. 8 and 9 for a 50 V - 4 mTorr discharge. Note that the trends observed are similar to those found for the conventional multicusp plasma generator (Figs. 3 and 7).

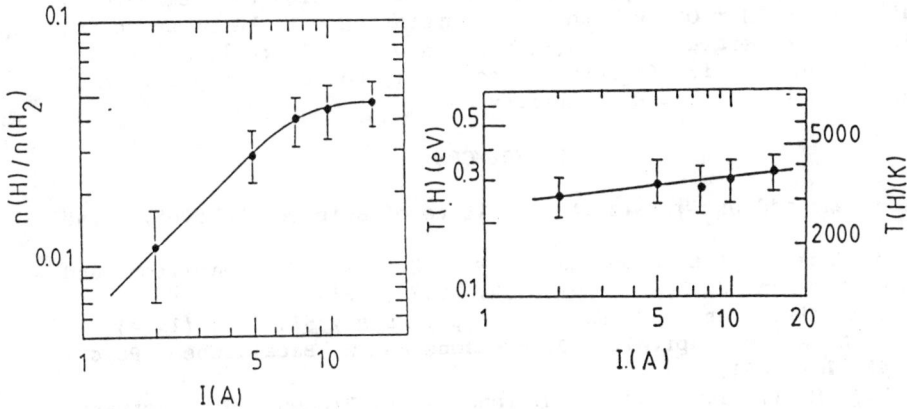

Fig. 8 : Relative atomic density $n(H)/n(H_2)$ versus the discharge current I in a 50 V - 4 mTorr hybrid multicusp discharge.

Fig. 9: Atomic temperature T(H) versus the discharge current I in a 50 V - 4 mTorr hybrid multicusp discharge.

F. Negative ion temperature.

Wadehra[22] had calculated the average energy carried by H^- ions at their formation by dissociative attachment and showed that this energy is proportional to the average electron energy. For an average electron energy 0.65 - 2 eV the average energy carried by H^- ions is 0.2 -0.4 eV. The electron temperature decreases when the pressure goes up and thus a pressure increase leads to a reduction of the H^- energy at formation.

Since the plasma represents a potential well for negative hydrogen ions, the lifetime for H^- loss to the wall is much longer than the lifetime for H^- destruction by plasma volume processes Before their destruction by mutual neutralization, associative detachment and electron detachment collisions, the ions exchange energy in elastic scattering and charge transfer collisions with atoms, in elastic collisions with molecules, and in $H_3^+ - H^-$ Coulomb collisions. On this basis it was estimated[23] that in the absence of atomic hydrogen the negative ion temperature should be close to that of molecular hydrogen.

Taking into account the atomic hydrogen population and its temperature determined in this work $(n(H)/n(H_2)) = 0.03$, $T(H) = 0.3$ eV for a 50 V - 5 A - 4 mTorr hybrid multipole plasma, Fig. 8 et 9), we find that elastic collisions between H and H^- are slightly more effective than H_2-H^- collisions and $H_3^+ - H^-$ Coulomb collisions except for low discharge currents. The H^- temperature should therefore depend on the atomic hydrogen temperature. This was in fact observed in direct measurements of H^- velocities$_{24}$.

In a general way, one can estimate, using earlier data[23], that for $T(H) = 0.2$ eV the H^- temperature will be close to the molecular temperature when $n(H)/n(H_2) \leq 0.025$. It follows that low atomic density is favourable not only for achieving a high H^- fraction but also low H^- temperatures.

REFERENCES

*Permanent address : Weizmann Institute of Science, Rehovot, Israël.

1. M. Bacal, A.M. Bruneteau, W.G. Graham, G.W. Hamilton and M. Nachman, J. Appl. Phys., 52, 1247 (1981).
2. J.R. Hiskes and A.M. Karo, J. Appl. Phys., 56, 1927 (1984).
3. C. Gorse, M. Capitelli, J. Bretagne and M. Bacal, Chem. Phys., 93, 1 (1985).
4. R.I. Hall, I. Cadez, M. Landau, F. Pichou, C. Schermann, Phys. Rev. Lett., 60, 337 (1988).
5. M. Péalat, J.P.E. Taran, M. Bacal and F. Hillion, J. Chem. Phys., 82, 4943 (1985).
6. R.S. Freund, J.A. Schiavone and D.F. Brader, J. Chem. Phys., 64, 1122 (1976).
7. M.P.S. Nightingale and M.J. Forrest, Production and Neutralization of Negative Ions and Beams, Fourth Intern. Symp., Brookhaven NY 1986, edited by J.G. Alessi, AIP Conference Proceedings No 158 (1986), p. 154.
8. W.G. Graham and M.B. Hopkins, Production and Neutralization of Negative Ions and Beams, Fourth Intern. Symp., Brookhaven NY 1986, edited by J.G. Alessi, AIP Conference Proceedings No 158 (1986), p. 145.
9. M. Bacal and F. Hillion, Rev. Sci. Instrum., 56, 2274 (1985).
10. A.M. Bruneteau, G. Hollos, M. Bacal and J. Bretagne, PMI Report n$_0$ 2006. Ecole Polytechnique. Palaiseau, France
11. G.R. Möhlmann, F.J. de Heer, and J. Los, Chem. Phys., 25, 103 (1977)
12. M. Higo, S. Kamata and T. Ogawa, Chem. Phys., 73, 99 (1982).
13. K. Sur and N.S. Sil, Phys. Rev. A23, 715 (1981).
14. J. Bretagne, G. Delouya, C. Gorse, M. Capitelli, and M. Bacal, J. Phys. D : Appl. Phys., 18, 811 (1985).
15. M. B. Hopkins, Experimental Measurements in a Multipole Discharge : Application to H^- Production. Thesis for degree of Doctor of Philosophy, University of Ulster (1987).

16. J.H.M. Bonnie, P.J. Eenshuistra and H.J. Hopman, Phys. Rev. A37, 1121 (1988).
17. A.S. Schlachter, A.T. Young, G.C. Stutzin, J.W. Stearns, H.F. Döbele, K.N. Leung and W.B. Kunkel, SPIE Proceedings Series, Vol. 1061, "Microwave and Particle Beam Sources and Directed Energy Concepts", Edited by H.E. Brandt (1989), p. 610.
18. P. Berlemont, Private Communication.
19. F. Launay, M. Bacal, A.M. Bruneteau and F. Hillion, Production and Applications of Light Negative Ions, Proceed. of the 2nd European Workshop, March 5-7, 1986, Ecole Polytechnique, Palaiseau, France, edited by M. Bacal and Ch. Mouttet, p. 129.
20. M. Lefebvre, M. Pealat, J.P.E. Taran, F. Hillion and M. Bacal, Production and Applications of Light Negative Ions, Proceed. of the 2nd European Workshop, March 5-7, 1986, Ecole Polytechnique, Palaiseau, France, edited by M. Bacal and Ch. Mouttet, p. 107.
21. H.S.W. Massey, E.H.S. Burhop and H.B. Gilbody, in Electronic and Ionic Impact Phenomena, Vol. III, Oxford University Press, 1971.
22. J.M. Wadehra, Phys. Rev. A, 29, 106 (1984).
23. A.M. Bruneteau and M. Bacal, J. Appl. Phys., 57, 4342 (1985).
24. P. Devynck, J. Auvray, M. Bacal, P. Berlemont, J. Bruneteau, R. Leroy, R.A. Stern, Rev. Sci. Instrum., 60, 2873, (1989).

MASS SPECTROMETRY IN A MULTICUSP ION SOURCE

A. A. Mullan
Applied Physical Science, University of Ulster,
Coleraine, BT52 1SA, Northern Ireland.

W. G. Graham
Physics Department, Queen's University,
Belfast, BT7 1NN, Northern Ireland.

ABSTRACT

Mass spectrometry has been used for the detection of positive and negative ions in a multicusp ion source operating with both hydrogen and deuterium gas. The mass spectrometer operation has been optimized and it is shown that applying ion extraction voltages can disturb the discharge. Using this technique combined with a Langmuir probe technique we are able to study the positive ionic fractions present when operating with both gases (and the negative ion densities.)

INTRODUCTION

A full understanding of ion source operation requires knowledge of the positive and negative ion species present in the discharge. This is particularly important when comparing sources operating with hydrogen and deuterium gas. The interpretation of positive and negative mass spectra from ion sources is difficult since extraction voltages are required to remove the ions, particularly for negative ions which are generally confined in the discharge by the positive plasma potential. This extraction voltage is found to change the plasma parameters, especially the electron density. Here we have measured the positive and negative densities in both a H_2 and D_2 plasma with particular attention to the effect of biasing the mass spectrometer on the plasma parameters.

APPARATUS and EXPERIMENTAL TECHNIQUE

The ion source used is based on a Culham "small source[1,2] and has been described in detail elsewhere[2] The source consists of a stainless steel vacuum box 19*24*19cm. The primary electrons are emitted from two hot tantalum filaments which are biased negatively with respect to the walls. The electrons are confined using rows of permanent magnets on five walls in a line cusp geometry. A virtual filter is created by breaking the multipole geometry on two opposite sides. The magnetic field lines cross the source dividing it into two regions: the "driver" region, where primary electrons produce the plasma and the "extractor" region, kept free of primary electrons by the filter field. There are no magnets on the end wall of the extractor (beam forming electrode). The centre portion of the end wall has been replaced by a small mass spectrometer.

The plasma parameters are measured using cylindrical Langmuir probes. The probe current - voltage characteristic is analysed using a digitial technique which allows the accurate measurement of many plasma parameters including the electron energy distribution function (EEDF)[2].

The mass spectrometer is based on a design by Leung et al[3] and consists of a stainless steel pill box which sits in the gap between two solenodial coils. The coils are mounted inside an open ended mild steel box. The walls of the box form a return yoke to confine the magnetic flux. Thus, very little flux leakage occurs outside the spectrometer. The mild steel box, in turn, is housed inside a copper case which is water cooled. The copper case has dimensions of $5 * 5 * 5$ cm^3.

When extracting ions, the copper and the mild steel boxes are biased with a small voltage (Vspec). Since the stainless steel pill box is isolated from the magnet and the mild steel box by thin mica foils, it can be biased at a higher potential (Vacc). Charged particles, accelerated by the potential difference created between the plasma and mass spectrometer pass through the entrance aperture (1mm in diam) and enter the pill box. The beam is first collimated and then deflected 90 degrees by the magnetic field before being collected by a small Faraday cup. The cup is normally maintained at the same bias as the pill box.

Fig 1 shows typical mass spectra for positive and negative ions in both H_2 and D_2. The current collected j_i on the Faraday cup is related to the density of ions (n_i) in the discharge by

$$j_i = An_i ev_i/4 \qquad (1)$$

$$= (An_i e(2E_i/m_i)^{.5})/4 \qquad (1a)$$

where

A is the collector area
e is the electronic charge
v_i is the ion velocity
E_i is the ion energy
m_i is the ion mass

There is always uncertainty about the values which should be used for A and particularly v_i, in this paper we report relative measurements of n_i where

$$n_i \propto j_i(m_i)^{.5} \qquad (2)$$

The total collected current is found by integrating the peaks for each mass seperately. Each is corrected for background noise. The total current is then multiplied by the square root of the mass of the particular ion, to give a relative density for that ion. The total ion density is obtained by summing the densities for all the ions. Ionic fractions are obtained by expressing the densities of each species as a fraction of the total.

EXPERIMENTAL CHECKS

A series of experimental checks were performed to determine the optimum operating conditions for the mass spectrometer. Fig 2 shows the effect of different spectrometer biases (Vspec) on the discharge. The plasma parameters were measured in the extraction region, using a Langmuir probe approximately 2 cms from the spectrometer. It is clear that for mass spectra to be related to plasma operating conditions, plasma parameters must be measured under ion extraction conditions. This is particularly important when comparing positive and negative ion species. It was confirmed that the plasma parameters were independent of the magnitude and sign of the accelerating voltage (Vacc). Fig 3 shows that while as expected, the positive ion current increases with Vspec, the ion ratios are essentially independent of the ion extraction voltage. The positive ion current is also found (Fig 4) to be dependent on the voltage used to accelerate the ions in the spectrometer, Vacc. Fig 5 shows that the negative ion density dependence on Vspec and on Vacc is similar to that seen with positive ions. The negative ion current is much more sensitive to Vspec, as expected since they are confined by the positive plasma potential. These experimental checks confirmed that if plasma parameters were measured, under the same extraction conditions, it was possible to obtain reliable, relative measurements of the ionic mass densities.

RESULTS and DISCUSSION

The mass spectrometer was used to analyse the ion species in the extractor region of the ion source, operating in H_2 and D_2. For the measurements reported here the pressure was held constant at 2 mTorr. The discharge current could be varied from 1 Amp to 15 Amps and the discharge voltage was 60 Volts. The magnitude of the accelerating voltage, Vacc, was 40 Volts for both positive and negative ion detection, with Vspec −5 Volts for positive ion analysis and +10 Volts for negative ions.

As shown in Fig 6 the total positive ion density is linear with electron density, as expected from charge neutrality. A comparsion of the ionic fractions in both H_2 and D_2 as a function of electron density (Fig 7) indicates that these fractions are similar in both H_2 and D_2 discharges. In Fig 6 the negative ion density in H_2 and in D_2 are compared. Both are shown as a function of electron density and it is found that the negative ion density in D_2 is consistently less than that for H_2; the D^- density being about 70% of the H^- density. It is also shown that both signals tend to saturate at an electron density of 3.5 and $5*10^{10}$ cm^{-3} respectively. The relationship between negative ion density and electron density is more complex than for the positive case since it depends on the plasma chemistry. The details of ion source operation in H_2 and D_2 is discussed in more detail in an accompanying paper[4].

CONCLUSIONS

Mass spectrometry has been shown to be a useful diagnostic tool for examining the positive and negative ion densities in plasmas, however due to the disturbance of the plasma, caused when biasing the mass spectrometer the plasma parameters must be measured under ion extraction conditions.

Using mass spectrometry we have shown that under the same discharge conditions the ionic fractions in both a H_2 and D_2 discharge are the same and that the negative ion density in H_2 is greater than that in D_2. The present mass spectra measurements when combined with accurate plasma parameter measurements can provide a useful diagnostic technique for ion source discharges.

ACKNOWLEDGMENT

This work was supported by an EMR contract from the UKAEA Culham Laboratory.

REFERENCES

1. A J T Holmes, G Dammertz, T S Green and A R Walker, Proceedings of the International Ion Engineering Congress, Kyoto, Japan, 12-16 September 1983 (unpublished) p71.

2. M B Hopkins and W G Graham, Rev Sci Instrum $\underline{57}$, 2210 (1986)

3. K N Leung and G Guethliein, Rev Sci Instrum $\underline{56}$, 1480 (1985)

4. W G Graham and A A Mullan "Negative ion production in an ion source operating in H_2 and D_2 (These proceedings)

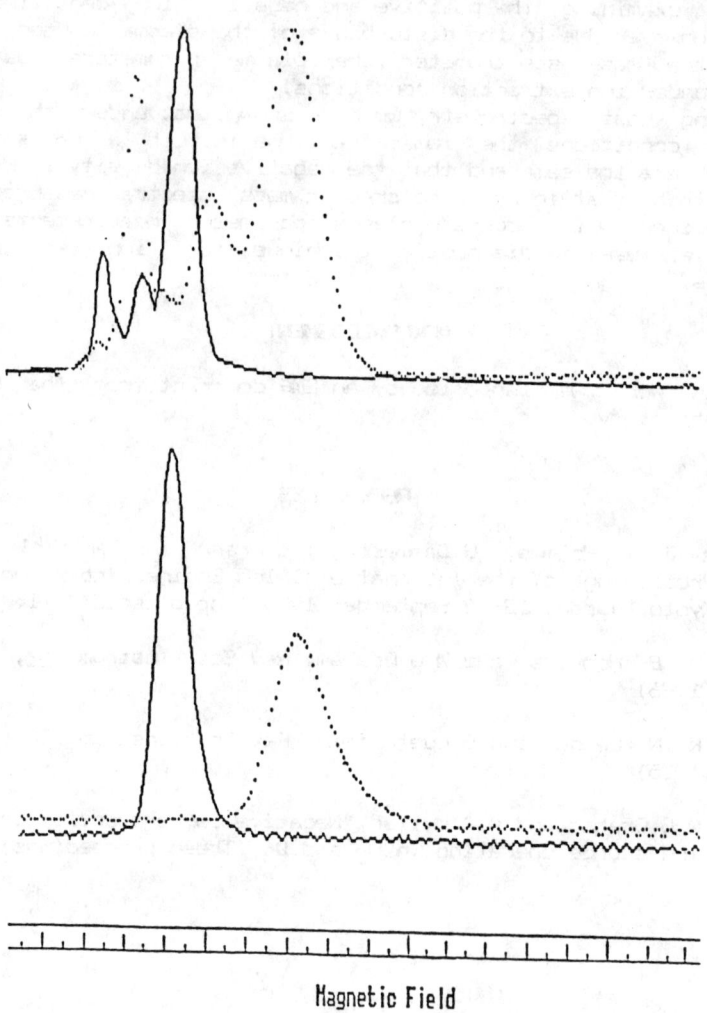

FIG 1 Typical mass spectra for positive and negative ions in both H_2 (line spectra) and D_2 (dot spectra)

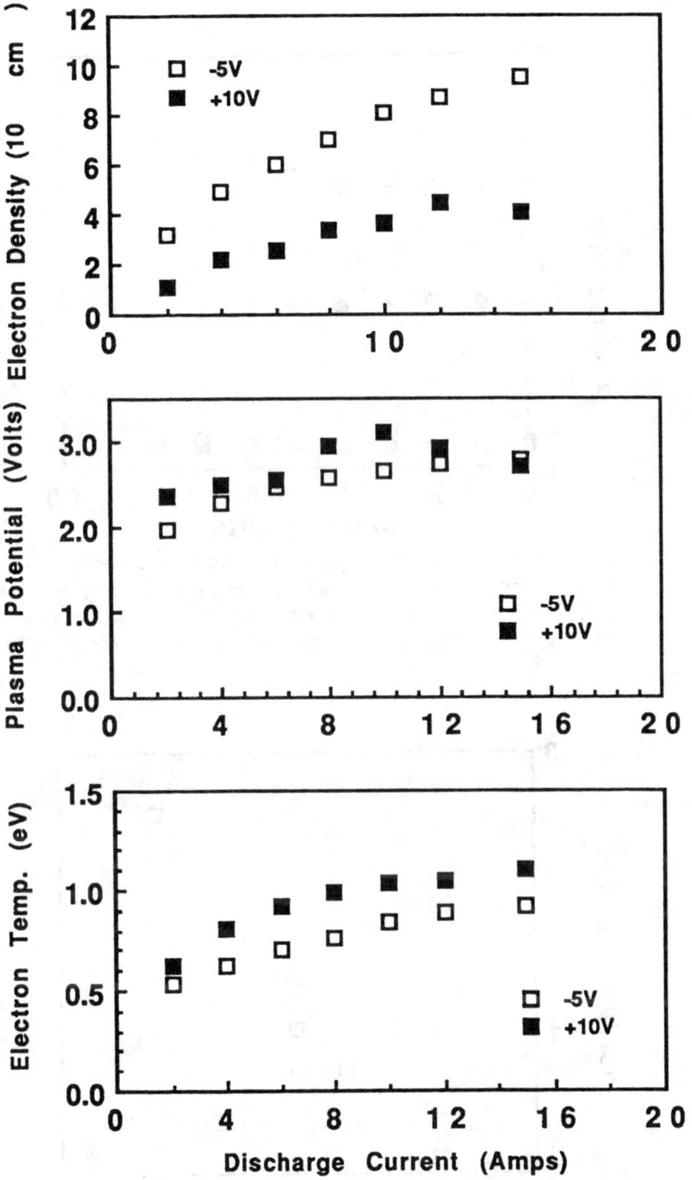

FIG 2 Dependence of electron density, electron temperature and plasma potential on discharge current in H_2 gas at 2 mTorr and Vd = 60V.

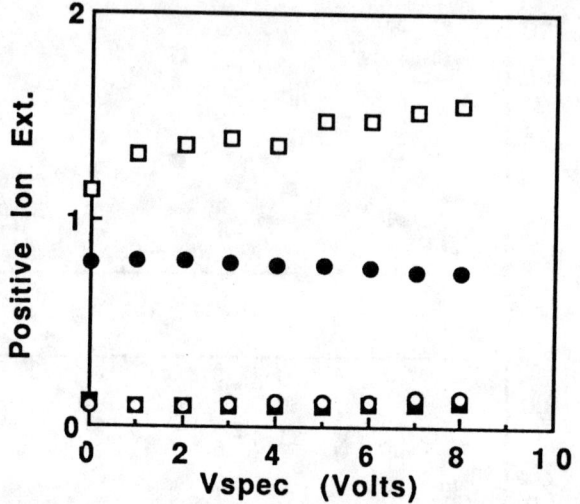

FIG 3 Dependence of the positive ion current, and the ionic fractions, H^+ ■, H_2^+ ○, H_3^+ ● on the extraction voltage, Vspec, in H_2 gas at a pressure of 1mT and discharge current of 5A and Vacc = 60V.

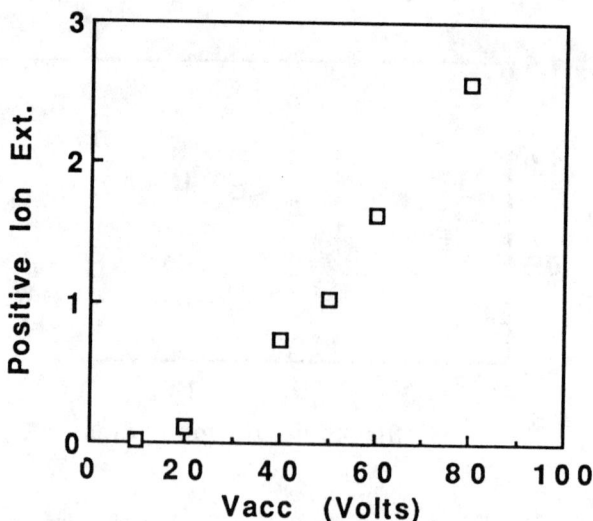

FIG 4 Dependence of the positive ion current on the accelerating voltage, Vacc, in H_2 gas at a pressure of 1mT, discharge current 5A and Vspec = -4V.

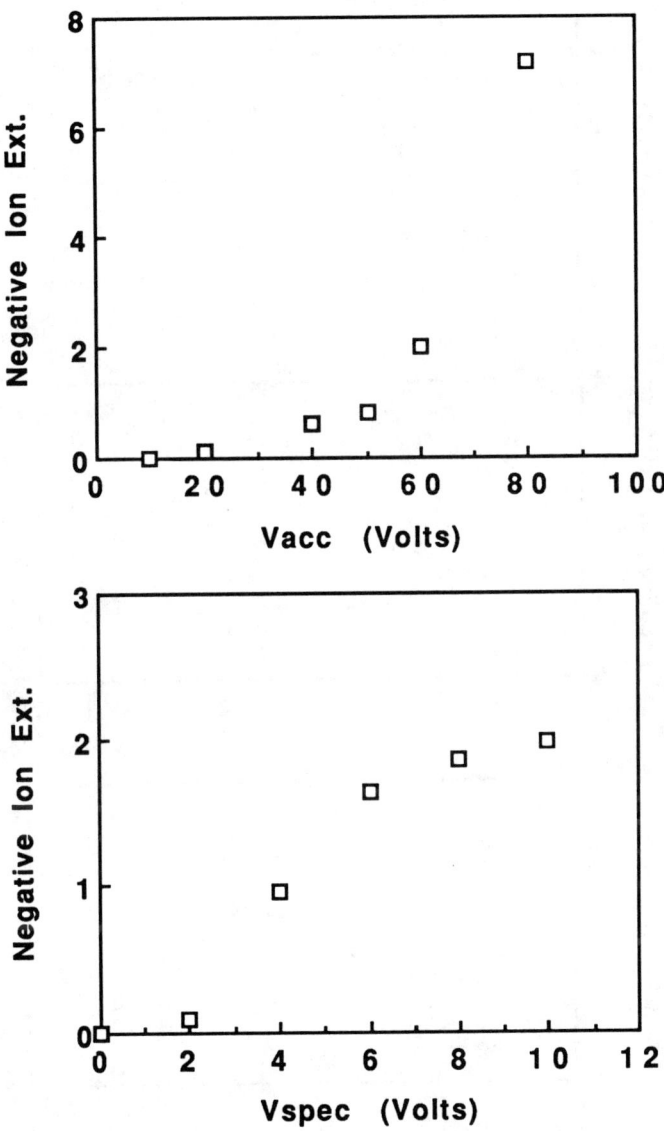

FIG 5 Dependence of negative ion signal on Vspec at a pressure of 1mT, Vacc = +60V and discharge current = 5A and on Vacc with pressure at 1mT, Vspec = +10V and discharge current = 5A.

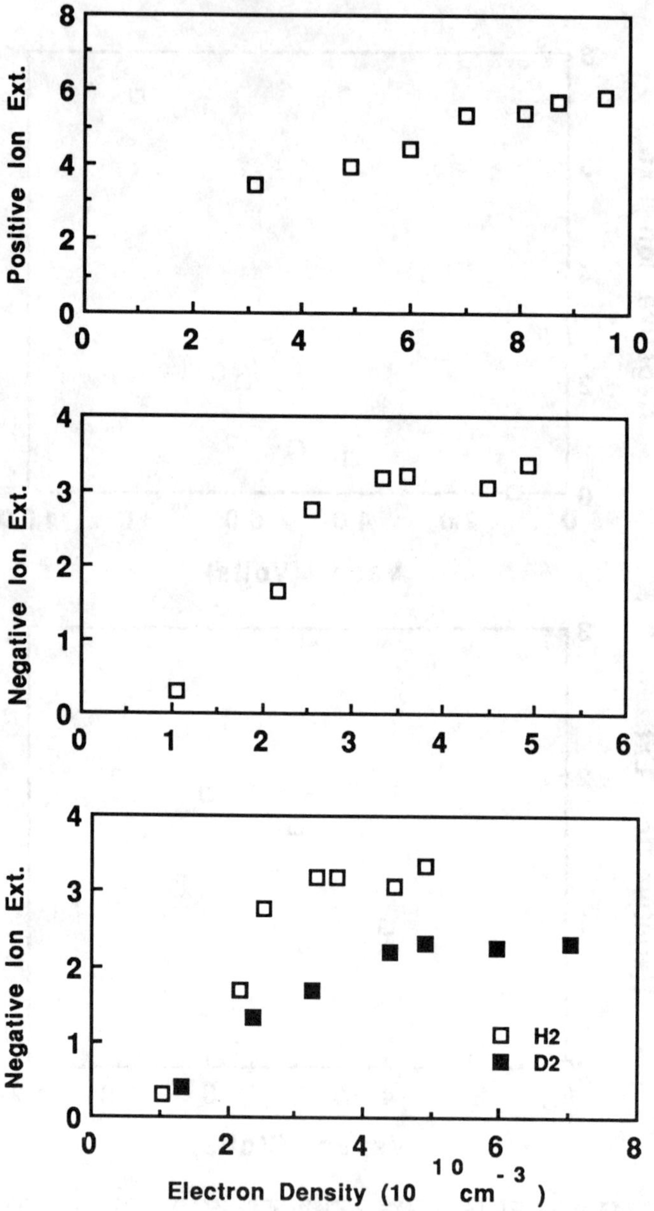

FIG 6 Dependence of positive ion current in H_2 and negative ion current in both H_2 and D_2 gas on electron density at a pressure of 2mT and Vacc = +/- 40V and Vspec = -5V for positive ion extraction and +10V for negative ion extraction.

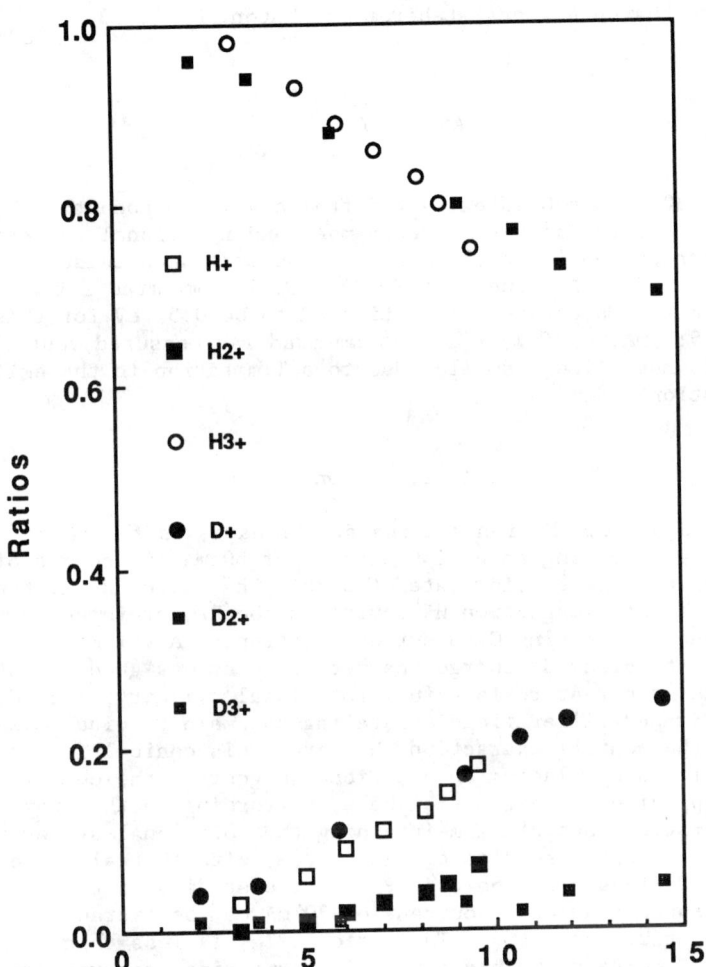

FIG 7 Dependence of the ionic fractions in both H_2 and D_2 gas on electron density at a pressure of 2mT, Vacc = 40V and Vspec = -5V.

EMITTANCE MEASUREMENTS ON A VOLUME H⁻ ION SOURCE

J.G. Alessi
Brookhaven National Laboratory, Upton, NY 11973

ABSTRACT

A current of 30 mA has been extracted from a volume production H⁻ source having a toroidal discharge chamber and rotational symmetry. This is a current density of 30 mA/cm^2. The emittance measurement gave a normalized, 90% value of $\epsilon_N(90\%) = 0.32$ π mm-mrad for a 13 mA beam. The ion temperature is estimated to be 0.57 eV for this case. For 25.5 mA, $\epsilon_N(90\%) = 1.11$ π mm-mrad was measured, but the true value is most likely smaller due to a limitation in the emittance resolution.

I. Introduction

Studies of volume H⁻ ion sources at BNL have, as the objective, a source producing an H⁻ ion current of 50 mA, in pulses of 1 ms duration at a repetition rate of 5 Hz. The source could then be used instead of a magnetron H⁻ source on the RFQ preinjector at the Brookhaven Alternating Gradient Synchrotron.[1] A volume H⁻ ion source with a toroidal discharge chamber has been designed and studied.[2,3] Its main feature is a full rotational symmetry, including a conically shaped filter field separating the main toroidal discharge from the central extraction chamber. This conical filter field is achieved by placing an additional magnet in the center of the flange opposite the extraction hole, perturbing in this way the cusp configuration, but still maintaining the rotational symmetry. Figure 1 shows a cross section of the source, with the calculated magnetic field lines also shown.[4] Parametric studies of this source have shown that an H⁻ current of 30 mA can be extracted through an aperture of 1 cm^2. The ratio I_e/I_{H^-} is less than 30 at the highest H⁻ currents. This paper will summarize measurements of the H⁻ emittance, including a comparison with the case where the source was reconfigured to have a standard dipole filter field.

II. Experimental Arrangement

The ion source, described in Refs. 2 and 3, was mounted on a vacuum box and pumped by a ≈ 400 l/s turbomolecular pump and an 1800 l/s oil diffusion pump. The source was typically operated with a 1.2 ms discharge pulse width at a 0.5 - 1.3 Hz repetition rate. All measurements were done with pulsed gas injection, and
*Work performed under the auspices of the U.S. Depart. of Energy.

the range of peak pressures in the source chamber was 5 - 15 mTorr, depending on the arc current. The peak pressure in the vacuum box was ≈ 2 x 10^{-5} Torr during the gas pulse. The plasma electrode was isolated from the chamber, and was floating. The anode and extractor apertures were 1.13 cm diameter, and the extraction gap was 0.97 cm. The extraction geometry was in no way optimized, and the extracted beam was quite divergent. The source was isolated and connected to a negative HV power supply (dc extraction voltage), and the extraction electrode was at ground potential. The extracted beam has a large electron component, so a strong dipole field (≈ 1.0 x 10^{-3} T-m) was placed near the source exit to remove electrons from the beam soon after the extractor, while deflecting the H$^-$ only slightly. The emittance device was located 9.7 cm from the extractor, and a Faraday cup was also mounted at the emittance head location to monitor the beam current.

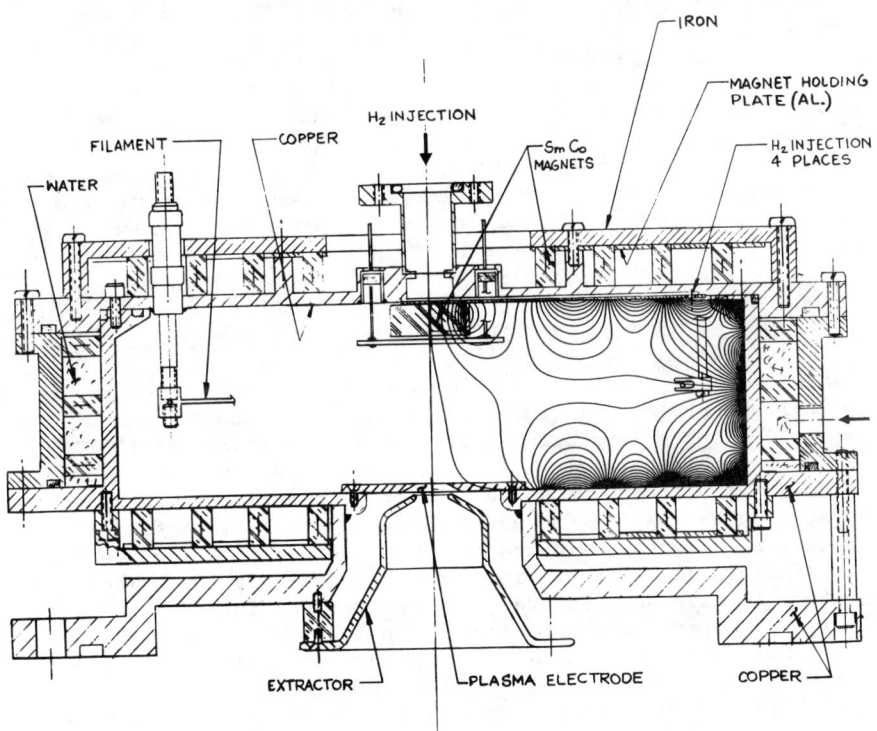

Fig. 1: Cross section of the H$^-$ source. The calculated magnetic field in the source from Reference 3, is also shown.

The emittance measurements were done with a slit-and-collector type emittance head which was stepped through the beam in the vertical direction. The head has a 0.1 mm wide x ≈ 80 mm long entrance slit, and, to detect the angular spread in the beam at that position, an array of alternating collector foils and insulating strips, parallel to the slit. The 30 collector foils have a spacing of 0.26 mm, center-to-center. There is a grid in front of the collector array which can be biased, either negative, to suppress secondary electron emission when reading the H$^-$ current, or positive to read instead the secondary electron current. The distance from slit to collectors can be adjusted, when the head is removed, to give the desired angular range for the measurement. The dimension chosen is a compromise between total angular spread that can be measured, and the resolution of the individual channels. Initial measurements were done with this dimension at 25.4 mm, which gave a total angular resolution of ± 150 mrad (10 mrad/collector). This was required in order to measure the full beam under a variety of operating conditions. Following these measurements, the slit-to-collector separation was increased to 50.8 mm (5 mrad/collector) in order to improve the resolution in cases where the total divergence was ≤ ± 75 mrad. The emittance head was stepped across the beam in 100 steps (1 step per beam pulse) over a total range of 2.5 - 5 cm. The current on the 30 collectors was sampled and held during a flat portion of each beam pulse (usually 0.6 ms into the pulse). The data was stored and analyzed via computer. Emittances in the two source planes were measured by rotating the source by 90° on the vacuum box.

III. Results

Based on our previous experience with a negative bias on the grid in front of the emittance collectors, one cannot prevent secondary electrons coming off a collector from hitting neighboring collectors. This then gives an emittance larger than the true beam emittance. Therefore, we normally choose to operate with a positive bias on this grid, and detect instead the current from the secondary electron emission caused when the beam hits the collectors. This gives a larger signal of positive polarity. When operating in this way, however, any neutral particles in the beam are also detected. With the positive grid bias, we were able to measure two emittances in the beam, separated in angle. Two examples of this, for different source parameters, are shown in Fig. 2. The lower emittance comes from the H$^-$ beam, deflected by the electron-sweeping dipole field at the source exit. The upper emittance is primarily H^0, coming from H$^-$ stripped in the extraction region and undeflected by the dipole. To verify that this was the case, we took some emittances with a negative grid bias. In this case only a very small component was seen at the location of the upper emittance, this being some heavy negative impurity in the beam (ex. O$^-$). The current in this emittance is < 4% of the H$^-$ current. Thus, the great majority of the upper peak seen with a positive bias is due to neutrals. By comparing the integrated counts in the two emittances, we estimate that the intensity of the H⁰ beam is 25 - 50% of the H⁻ beam intensity under various conditions. (This component is not detected in the Faraday cup current). Fortunately, with this separation between the H⁻ and H⁰ beams, one could easily do an analysis of the H⁻ emittance alone.

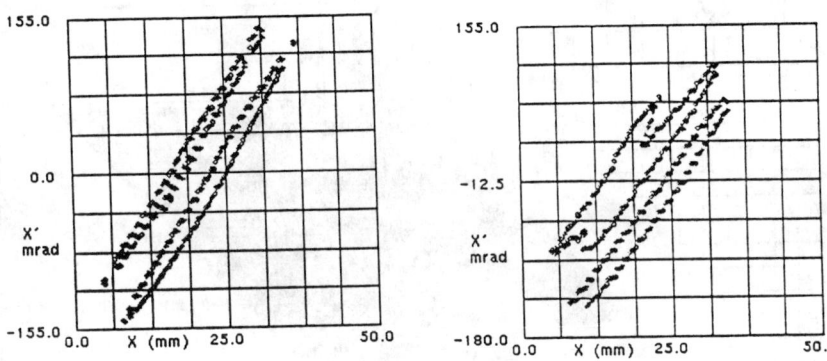

Fig. 2: Emittances with a positive grid bias, showing the H⁻ (lower) and H⁰ (upper) emittances.

Figure 3 shows an emittance measurement for a 14 kV, 18.6 mA beam (arc current = 100 A). The H⁰ beam emittance has been removed. This measurement gave a normalized emittance for 90% of the beam of $\epsilon_N(90\%) = 0.55 \pi$ mm-mrad. The choppiness of the emittance in the 3-D plot was typical. This is due to the fact that the angular spread in the beam is less than the resolution of one collector, so the beam is essentially hitting only one collector at a time. Therefore, for many of the measurements the actual beam emittance is very likely smaller that what was measured with the 10 mm/channel resolution. As mentioned previously, the emittance head was readjusted at one point to have a finer resolution. For several cases a comparison could be made between measurements with 10 mrad/collector and the finer resolution 5 mrad/collector, under the same beam conditions. As expected, the total angular divergence of the beam and the spot size remained the same, but the width of the emittance was less. The choppiness on the emittance was reduced, but still present, indicating that the resolution still was not good enough (i.e., the beam was still hitting only 1 or 2 foils). In spite of this, the 90% emittance was reduced by 57 - 66%. A scaling by this ratio for the emittances measured with the coarse head setup would probably be valid in many cases. Figure 4 shows emittances taken under the same source conditions for 10 and 5 mm/channel resolution.

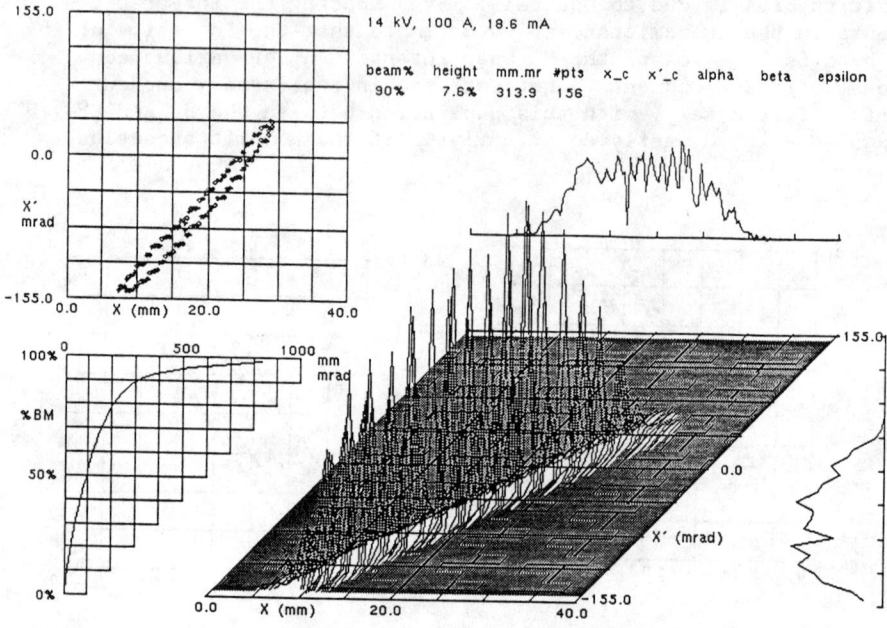

Fig. 3: Emittance measurement for an 18.6 mA, 14 keV beam. $\epsilon_N(90\%) = 0.55$ π mm-mrad (10 mrad/channel).

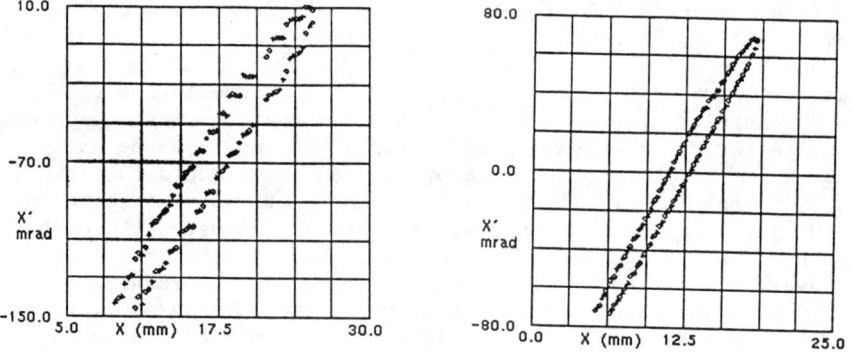

Fig. 4: Emittances for a 13 mA, 12 kV beam with a) 10 mrad/channel and b) 5 mrad/channel emittance head resolution. Except for an offset, both emittances are shown on the same scale. a) $\epsilon_N(90\%) = 0.53$ π mm-mrad b) $\epsilon_N(90\%) = 0.32$ π mm-mrad.

Table I

Filter Geometry	H⁻ (mA)	V (Ext) (kV)	Arc Current (A)	Emittance Resolution (mrad/channel)	Normalized 90% Emittance (π mm mrad)	RMS Emittance (π mm mrad)	H⁻ Ion Temp (eV)
Conical	13.0	12	50	5	.32	.070	0.57
Conical	12.7	12	50	10	.53	.110	1.41
Conical	19.0	14	100	5	≤ .44	.115	1.54
Conical	18.6	14	100	10	.55		
Conical	25.5	18	190	10	1.11	.238	6.61
Dipole	12.0	12	100	5	.38	.080	0.75
Dipole	15.0	12	100	10	.67		
Dipole	16.5	14	150	5	.49	.100	1.17
Dipole	19.5	14	150	10	.74		

A detailed analysis of the emittance as a function of various source parameters has not been carried out. Measurements have been done for arc currents in the 50 - 200 A range and extraction voltages from 10-18 kV. The maximum current from the source with the conical filter field was 30 mA, while with the dipole filter field only 20 mA could be obtained. Table 1 shows some emittance values measured for both the conical and dipole filter field configurations. At higher currents than shown in the Table, the divergence of the beam was greater than the maximum angles that could be detected.

Emittances in the other plane were measured by rotating the entire source assembly by 90°. The emittances were larger by ≈ 70% for both the conical and dipole filter configurations. Presumably, this is due to the fact that in this plane the H^o and H^- emittances were not separated. The electron sweeping magnet which produced the separation rotated with the source, so the deflection was now in the other plane. Except for this sweep magnet, the conical configuration is perfectly symmetric.

If one plots the normalized emittance as a function of $\ln[1/(1-F)]$, where F is the beam fraction, then if the beam has a Gaussian distribution one will see a linear dependence from which the RMS emittance can be determined[5,6]. This was done for several of the above measurements, and this dependence was always linear up to > 80% beam fraction. The departure from Gaussian at higher beam fractions is presumably due to extraction optics effects. The results of this analysis is also shown in Table 1 for several cases. If one assumes that the emittance is determined by a Maxwellian energy distribution of the ions of temperature kT, then the emittance $\epsilon_{4RMS} = 2r(kT/Mc^2)^{1/2}$ [6], where r is the anode aperture radius. Estimates of the ion temperature based on this are also given in Table 1. The lowest ion temperature measured was 0.57 eV at 13 mA of beam current.

IV. Conclusion

Emittances were measured for a toroidal volume H^- source with both a conical and standard dipole filter field. A value $\epsilon_N(90\%) = 0.32\ \pi$ mm-mrad was measured at 13 mA, or $\epsilon_N(RMS) = 0.07\ \pi$ mm-mrad. This corresponded to an ion temperature of 0.57 eV. The maximum H^- current obtained with the conical filter field is higher than the standard dipole, and emittances for the conical filter were somewhat lower than those for the dipole filter, for approximately equal H^- currents. We are, at this point, limited by the resolution of the slit-and-collector emittance device. An electric-sweep scanner type device is probably required to improve the measurement.

Acknowledgements

The support of the BNL Advanced Source Development Group is greatly appreciated. Special thanks go to Krsto Prelec, who designed the source, for helpful discussions, and to Dan McCafferty, who skillfully operated the source and made the many modifications required for these measurements.

References

1. J.G. Alessi, J.M. Brennan, and A. Kponou, "H⁻ source and low energy transport for the BNL RFQ preinjector", these Proceedings.

2. K. Prelec, Proc. 1989 Particle Accelerator Conference, Chicago, March 1989.

3. K. Prelec, "The BNL volume H⁻ ion source", these Proceedings.

4. C.R. Meitzler, "Magnetic field of a toroidal volume H⁻ source", these Proceedings.

5. P. Allison, Proc. Fourth International Symposium on Production and Neutralization of Negative Ions and Beams, Brookhaven, 1986, AIP Conf. Proc. <u>158</u> (1987) 465.

6. R.R. Stevens, Jr., R.L. York, K.N. Leung, and K.W. Ehlers, ibid, p. 271.

NEUTRAL BEAM DETECTORS*

U. von Wimmersperg
Brookhaven National Laboratory, Upton, NY 11973

ABSTRACT

A technique is described which utilizes the electrons stripped off fast neutral atoms and negative ions to monitor beam profiles with sub-micron position accuracy and with 1500 GHz bandwidth.

Fast neutral atoms or negative ions can be detected via electrons stripped off the parent nucleus. Consider the example of 200 MeV H^o atoms or H^- ions which carry "companion" electrons with a laboratory energy of about 109 keV as a result of their velocity ($\beta = 0.567$). By intercepting such particles with a thin carbon foil, these companion electrons can be liberated and detected remotely after being guided out of the beam with the aid of a weak magnetic field ($\sim 10^{-2}$ T). Beam position and profile monitoring can be obtained to micrometer accuracy by intercepting the beam with a thin carbon fiber. A single fiber with a thickness of a few microns can be stretched on a frame and traversed across the beam to monitor position with a resolution down to about 1/10 of the fiber diameter. Instantaneous beam profiles with a resolution equal to the fiber diameter can be obtained using a "harp" of several fibers spaced in parallel.

Companion electrons offer several advantages over secondary emission electrons commonly used in beam instrumentation. The stripping yield of companion electrons is 100% and 200% respectively for H^o and H^- beams as compared to about 4% for secondary emission electrons. Thermal loading at high beam current densities leads to thermionic emission of low energy electrons which are indistinguishable from secondary emission and thus destroy the linearity of such measurements. No such confusion arises with high energy companion electrons. Carbon fibers of 5 μm thickness can withstand thermal loads from beam densities of up to 10 mA/cm^2 CW of 200 MeV H^o, extending the upper limit in current density for companion electron monitors by at least a factor 10 over what is possible with secondary emission monitors. At the other extreme of current density, the detection of high energy companion electrons with a scintillator and photomultiplier offers single particle efficiency in a simple way. Further advantages stem from the geometrically simple and non-invasive external magnet arrangement of transporting companion electrons from the beam to a point where the electrons are detected. The high kinetic energy of these electrons also ensures immunity to space-charge effects within the beam.

An elegant application of companion electron detection applies to the monitoring of the longitudinal structure of H^o or H^- particle bunches. A single thin carbon fiber, intercepting a small fraction of beam, provides sufficient signal to monitor bunch structure. The stream of companion electrons emerging from the

---

*Work performed under the auspices of the USA Strategic Defense Command.

fiber accurately reflects the longitudinal density profile of the beam bunch. Such an electron bunch, stripped from 200 MeV hydrogen, has a momentum which is low enough to allow manipulation by small magnetic and electric fields. After emerging from the bending magnet at 90° to the beam direction, the companion electron bunch is passed through a transverse plane-wave electric field deflector, triggered by the approaching beam bunch. The entire electron bunch thereby simultaneously (from head to tail) acquires a transverse velocity, immersing it abruptly into a strong transverse DC field. The longitudinal density profile is then obtained by landing the electron bunch broadside onto a position-sensitive MCP (multi-channel plate electron multiplier) detector. Assuming a spacial resolution of 0.1 mm for the MCP, this implies an equivalent bandwidth of 1500 GHz for beam bunches with β of order 0.5.

A simpler version of longitudinal bunch monitors using companion electron stripping can be constructed by passing the H^0 or H^- beam through the walls of a plane parallel transmission line constructed of thin metal foil just thick enough to stop and capture the electrons while the resulting protons pass through. Typical dimensions for the stripline would be 0.1 mm thick foil 10 mm wide and spaced 1.3 mm to provide a 50 Ω signal path for the stopped electrons. Such a stripline can be supported on a conventional 50 Ω vacuum feedthrough. The bandwidth is then limited by the monitoring oscilloscope. When high frequency response is not required, H^0 or H^- beam intensities can be monitored to high accuracy by passing the entire beam through an insulated metal foil thick enough to stop the companion electrons and integrating the electron current from this "Faraday cup."

EXTRACTION LOSSES IN AND MODELING OF VOLUME SOURCES

Extraction induced emittance growth for negative ion sources*

J. H. Whealton, P. S. Meszaros, R. J. Raridon, K. E. Rothe
Oak Ridge National Laboratory, Oak Ridge, Tennessee 37831

Abstract

Nonlinear emittance growth produced by ion extraction is considered by a 3-D analysis in a Vlasov-Poisson-Boltzmann formulation. Phenomena considered include: (1), presheath effects, including electron depletion, (2), electron sheath accumulation (for large transverse magnetic fields), (3), nonlinear sheath fields (obtained by a self-consistent solution with an assumed quasi-equilibrium positive ion distribution and at least one Vlasov distribution), (4), nonlinear fringe fields produced by the accelerator-extractor itself (obtained self-consistently with item 3 above, (5), nonlinear space charge of the beam itself, (6), beam in conjunction with extracted electrons, and (7), ion acoustic waves. For specific volume negative ion source configurations, an investigation of the contribution of aberrations caused by an electron trap and electron accumulation in the extraction sheath are studied. Either of these effects can contribute significantly to the beam emittance, possibly dominating the contribution of the negative ion temperature in the source.

---

*Research sponsored by the Exploratory Studies Program of the Oak Ridge National Laboratory, operated for the U.S. Department of Energy by Martin Marietta Energy Systems, Inc., under contract DE-AC05-84OR21400.

Negative ion sources are the basic ingredient of neutral beam injectors of energy greater than 200 kV. There are many applications of this technology: e.g., (1) injection of neutral beams into high density plasma confinement devices to provide auxiliary heating and, in the case of tokamaks, current drive, and (2) injection of energy from a distance into targets in a vacuum. Previously, the state-of-the-art tools for negative ion accelerator design are like those for positive ion sources 20 years ago. As with positive ion sources, improvement is possible if a validated tool can be developed to actually design extractors and preaccelerators.

We have proposed a phenomenological description of space-charge imbalance in a negative ion source presheath. This is identified as a principal source of difference between the apparent inscrutability[1] of negative ion source extraction optics compared to positive ions. For positive ion extraction sheaths, particularly at low magnetic fields, the plasma electrons are available to instantly and completely (on the order of kT/e) cancel any excesses of positive ion space charge that might occur just before extraction. The properties of this nonlinear sheath in many cases are well known. However, for negative ion sources the situation is different. Electrons are extracted or hit a wall before being affected by space-charge fluctuations. Only the positive ions are present to cancel the space-charge imbalance.

This description is coupled with a full multi-dimensional, nonlinear Vlasov-Poisson analysis which includes nonlinear field effects, space-charge effects, and image charges. Preliminary results of this modeling have appeared in Ref. 2. We have modeled intense volume negative ion sources with respect to transverse space charge limits and RMS emittance production. We compare the theoretical treatment and the experimental data taken at the Los Alamos National Laboratory[3] (LANL). We also make some theoretical observations on an injector proposed by Lawrence Berkeley Laboratory (LBL).[4]

We consider in detail the geometric configuration of a negative ion source at LBL.[4] If we compute the total output current as a function of plasma density, then we arrive at Fig. 1. We see that for low values of the current density, j (cases A and B), there is a linear increase in output current. For extremely low values of the current (case A), we see a slight rise above this linear value. The prototypical case for low values of current density is illustrated in Fig. 2 (b), where only the ions inside the geometrical shadow of the extraction electrode are extracted and pass through theaccelerator. For the case of extremely low densities, we have the situation as typified in Fig. 2 (a). In this situation, some of the ions outside of the shadow of the extraction plasma electrode enter the extraction region due to the presheath fields, which occur as fringe fields of the applied potentials.

For sufficiently high current density, as shown in case C of Fig. 2, we have a transverse space-charge limitation, where the ions are impinging upon the extraction electrode. This is denoted as the first transverse space-charge limit. For yet higher values of the current density, shown by Fig. 2 (d), the onset of the second space-charge limit is reached. Here the ions now intercept the electron filter electrode. For an increase in plasma density, the net current decreases. This is what is meant by a transverse space-charge limit. In this case, the transverse space charge is determined principally by the small bore of the extraction electrode. The space-charge limit of this accelerator is approximately 130 mA (H^-). This value might exceed the current density capability of the ion source, which would indicate that the space-charge limit is not a practical inhibition upon the performance of the accelerator. However, the aberrations present below the space-charge limit may be. For positive ion extraction, the space-charge limit appears at approximately 270 mA, as shown in Fig. 1.

The RMS transverse emittance of the beam is examined and the results shown in Fig. 3. Here we see a wild behavior: we begin with a rather high emittance value at very low densities, then we go to a very low value at slightly higher plasma densities, and then steadily increasing to a much higher value and finally sharply decreasing again, but not to as low a value as before. This erratic behavior is fully explained and can be understood by reference to Figs. 2 (a)-(d). Very low densities, indicated by point (a) in Fig. 3, are shown in Fig. 2 (a). Here the ions have a very nonlinear force applied to them which produces significant beam halo or aberrations. These ions were tricked out of hitting the plasma electrode by the presheath fringe fields, due to the applied electrostatic accelerator potentials. As shown in the figure, the outside trajectories are crossing over neighboring interior trajectories. This produces a relatively high transverse RMS emittance, such as indicated in Fig. 3. However, for case (b) in Fig. 3, where the emittance is minimum, the sheath for negative ion extraction produces relatively small aberrations of the beam, as seen in Fig. 2 (b). For yet higher densities near the first space-charge limit, the emittance rises, and the outside trajectories, due to the fringe fields at the electron trap, cross over significantly the penultimate trajectories, producing a highly aberrated beam. The design of the electron trap has not been made with respect to minimizing the beam transverse emittance at maximum achievable current, but only with respect to trapping electrons. Also, the electron trap was designed assuming that the sheath was at the same plane as the exit aperture. There is no evidence to support this conjecture. If we increase the density still further, then we see a decrease in the emittance, as in case 3 (d). Even though we are beyond the first and second space-charge limits, the electron trap is vignetting the worst trajectories on the outside, thereby eliminating them from the emittance measurement.

Next, we analyzed the experimental data[3] of Ralph Stevens and colleagues of LANL using an LBL-type negative ion source. This two-dimensional analysis for the nonlinear negative ion extraction sheath shows promising results, at least in the zero temperature limit. Examples of the agreement between the subject analysis and the experiments are indicated in Fig. 4. Here, we show negative ion current as a function of inverse perveance for a space-charge dominated case. The analysis is indicated by the solid line and the experimental-based determinations by the symbols. Using this analysis, ion trajectories and equipotential contours are shown by solid and dashed curves, respectively, in Figs. 5-7; the four cases shown correspond to the cases A-D indicated in Fig. 4. In Fig. 5 are shown four different perveance beams in the Stevens negative ion source (labeled a-d). The desirable operating regime is most closely represented by the one labeled (b). This perveance is near, but slightly less than,, the transverse space-charge limit of this accelerator. Notice that even near the space-charge limit, there is still significant field penetration in the plasma at the sheath for ion extraction. For lower perveance, as shown in the top part, Fig. 5 (a) the beam crosses over and intercepts the extraction electrode. The field penetration is very significant, and the beam is extremely aberrated. This is not a useful mode of operation. Figure 5 (c) shows a perveance which is significantly higher than the transverse space-charge limit. Here the ions are shown to be intercepting the extraction electrode. If one were to increase the plasma density in the source, the number of ions that would hit the electrodes would increase faster than those that would actually exit the ion source. Thus the higher the source plasma density, the lower the extracted current which gets accelerated. This is illustrated in Fig. 5 (d). In Fig. 6, a blowup of the sheath region is featured. Here, we see a significant amount of field penetration [Fig. 6 (a)] as well as space-charge limited perveance [Fig. 6 (d)]. So, we have for the same geometric configuration both field penetration and space-charge limit.

Emittance, ε^N, measurements[3] on this source are shown in Fig. 7 by the curve labeled S. As a function of voltage for a fixed source plasma density (arc current), a transverse component y of emittance was considered. Calculations in 2-D (ε_r^N) and

3-D are also shown. The behavior at low perveance (high voltage) is similar to the observations, but at high perveance (low voltage) a significant difference obtains. This region of disagreement corresponds to a region of perveance higher than the transverse space charge limit (like cases C and D of Fig. 4). The computations show, if anything, a drop at this point corresponding to the fact that some of the most aberrant beams intercept the electrode and are no longer counted. In the experiment, these ions may instead be reflected and still contribute to the emittance.

A preliminary three-dimensional analysis of this same source follows. We studied the RMS transverse emittance for extraction from the Stevens[3] negative ion source at LANL. In Fig. 8, we show the negative ion source extraction sheath featured in Fig. 5, but in full three-dimensions. A calculation of the entire geometry is shown in Fig. 9, where the region of Fig. 8 (a) is indicated. In Fig. 8 (a), the extraction sheath is symmetric; however, in Fig. 8 (b), we have included the effect of the electron space charge of those electrons which are in an $E \times B$ guiding-center drift across the sheath, on their way out of the ion source. Here we see the sheath as asymmetric with potential contours as shown. This oblique sheath and the nonlinear aberrations associated with it give rise to an additional RMS emittance of the beam. However, the degree of contributionis less than 0.003 for reasonable parameters, and no explanation of the high perveance anomaly is suggested. Ion acoustic activity, such as shown in Fig. 52 (compared with Fig. 49) of Ref. 1B, would be a resonable expectation for a case like 86 or 7d (compared to 8A or 7a-b).

References

1. (a) J. H. Whealton, P.S. Meszaros, R. J. Raridon, K. E. Rothe, M. Bacal, J. Bruneteau, and P. Devynck, Revue. Phys. Appl. **24**, 945 (1989); (b) J. H. Whealton, M. A. Bell, R. J. Raridon, K. E. Rothe, and P. M. Ryan, J. Appl. Phys. **64**, 6210 (1988).

2. R. McAdams, A. J. T. Holmes, M. P. S. Nightingale, L. M. Lea, M. D. Hinton, A. F. Newman, and T. S. Green, "Production and Neutralization of Negative Ions and Beams," 4th International Symposium, Brookhaven National Laboratory (1986), Ed., T. G. Alessi, AIP Conf. Proc. **158**, p. 298ff (1967).

3. R. Stevens, private communication 1987.

4. K. Leung, private communication 1988.

FIGURE CAPTIONS

Fig. 1. Output current vs. plasma density for an LBL source showing that the space charge limits are above the anticipated operating range.

Fig. 2. Ion extraction for cases A-D shown in Figs. 1 and 3.

Fig. 3. RMS traverse emittance for the case considered in Figs. 1 and 2.

Fig. 4. Negative ion current vs. inverse perveance for the Stevens source when the symbols denote observations, and the solid line the subject negative ion extraction theory and the standard positive ion theory.

Fig. 5. LANL/LBL beam optics for cases A-D shown in Fig. 4.

Fig. 6. Blowup of LANL/LBL beam optics for cases A-D shown in Figs. 4 and 5.

Fig. 7. RMS normalized emittance for the source considered in Figs. 4 through 6. The curve, S, stands for observations,3 and the curves labeled 2 and 3 refer to 2-D (ε_r) and 3-D (ε_y) theories.

Fig. 8. End and side view showing the effect of oblique sheath due to ExB electron drift for the source considered in Figs. 4 through 7.

Fig. 9. 3-D calculation of the source considered in Fig. 8.

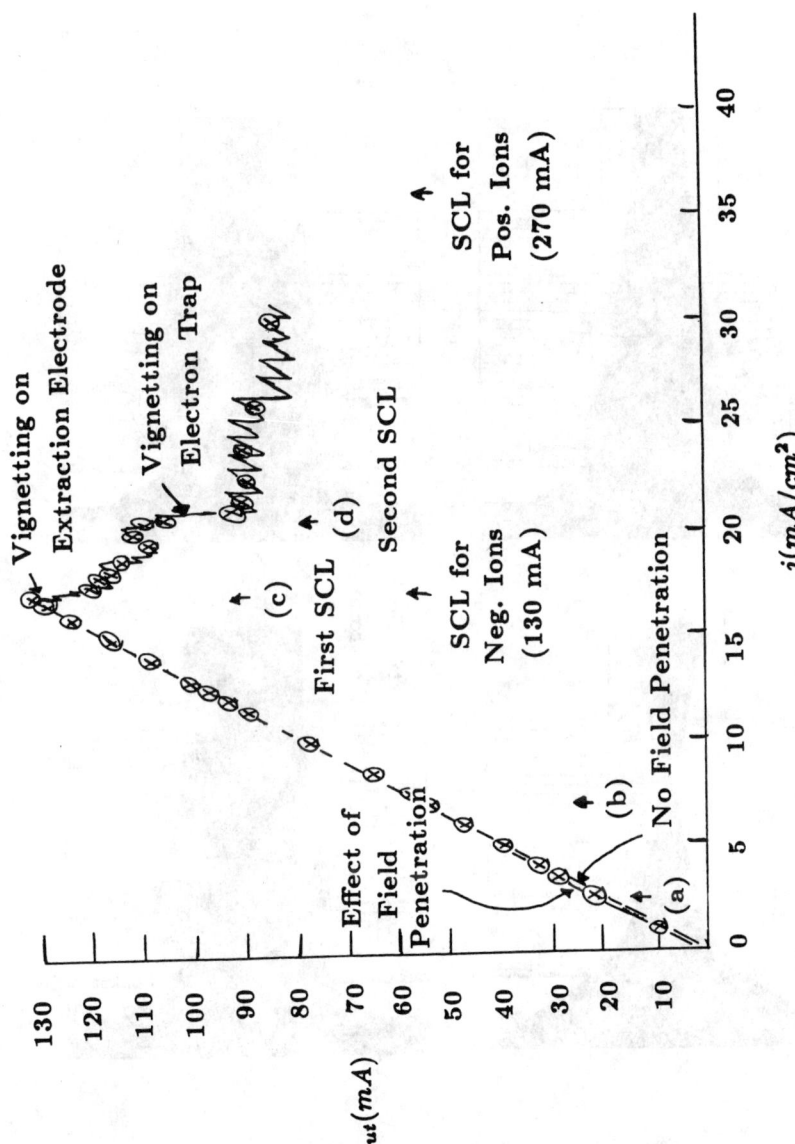

Figure 1. Output current vs. plasma density

546 Extraction Induced Emittance Growth

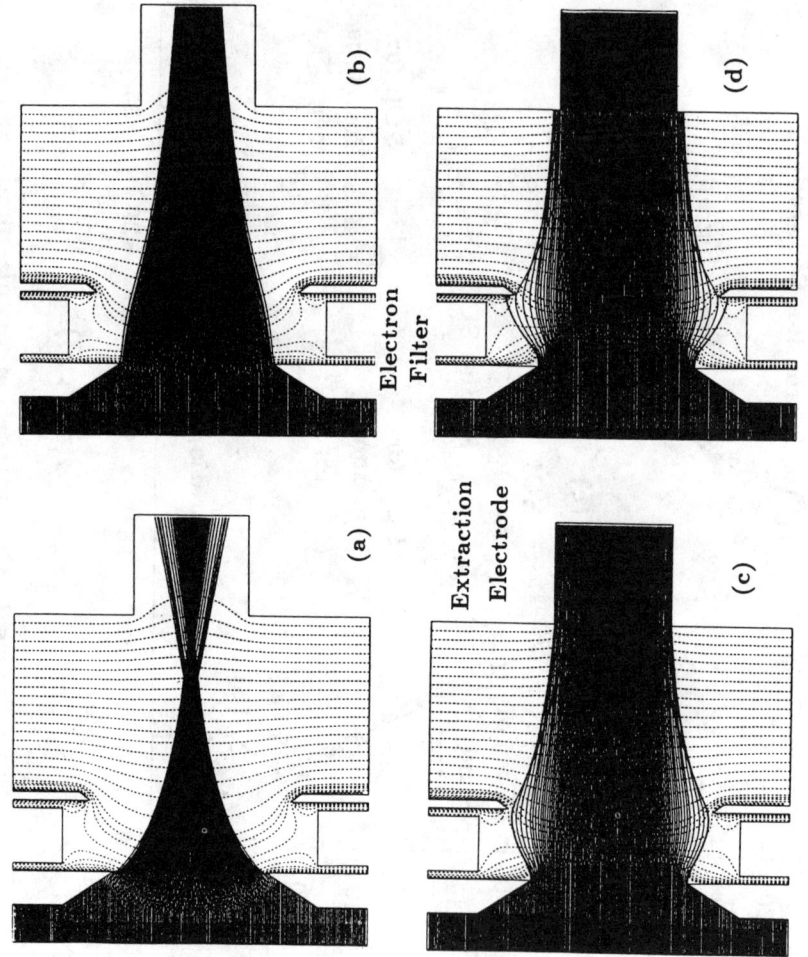

Figure 2. Ion extraction.

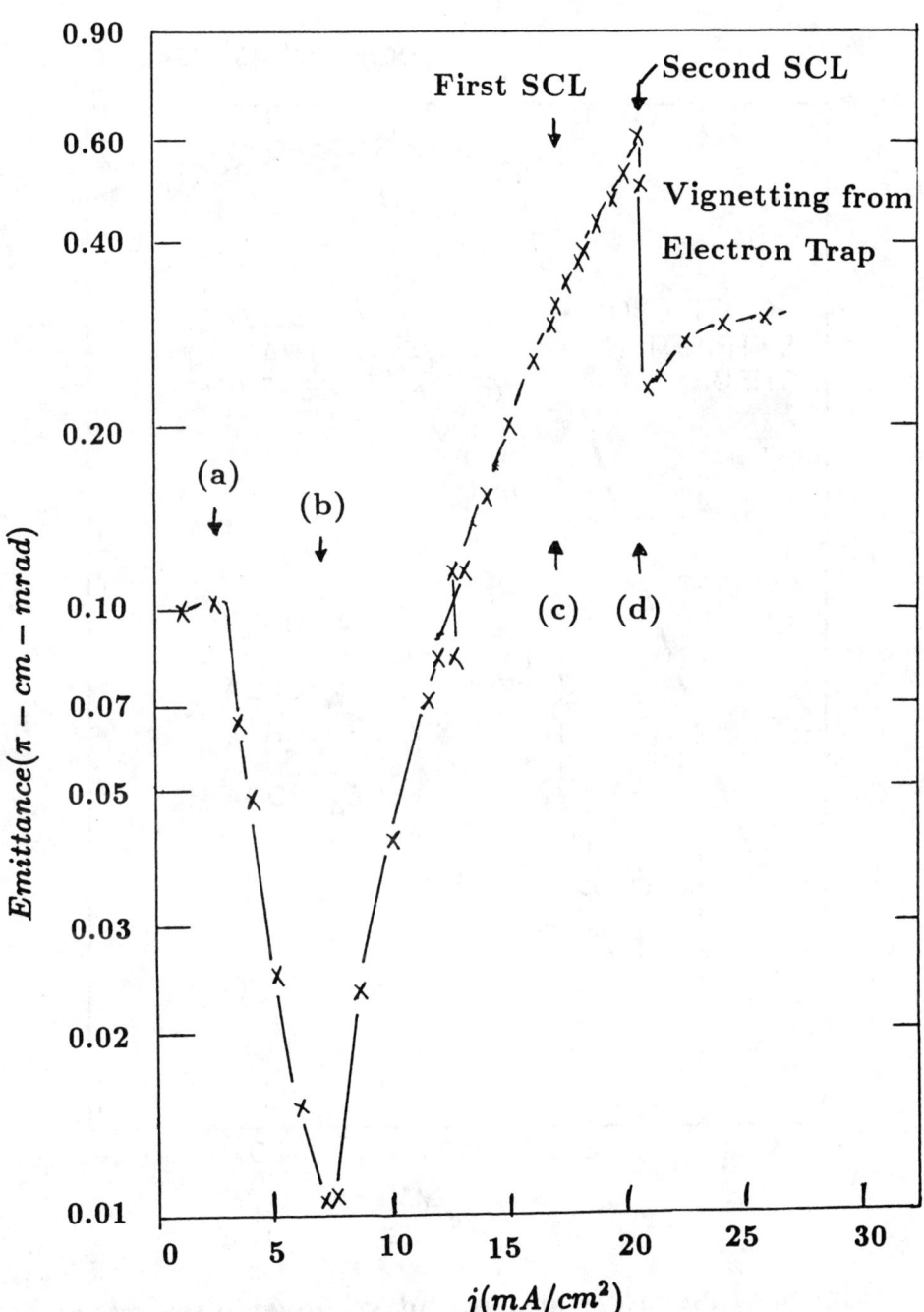

Figure 3. RMS transverse emittance.

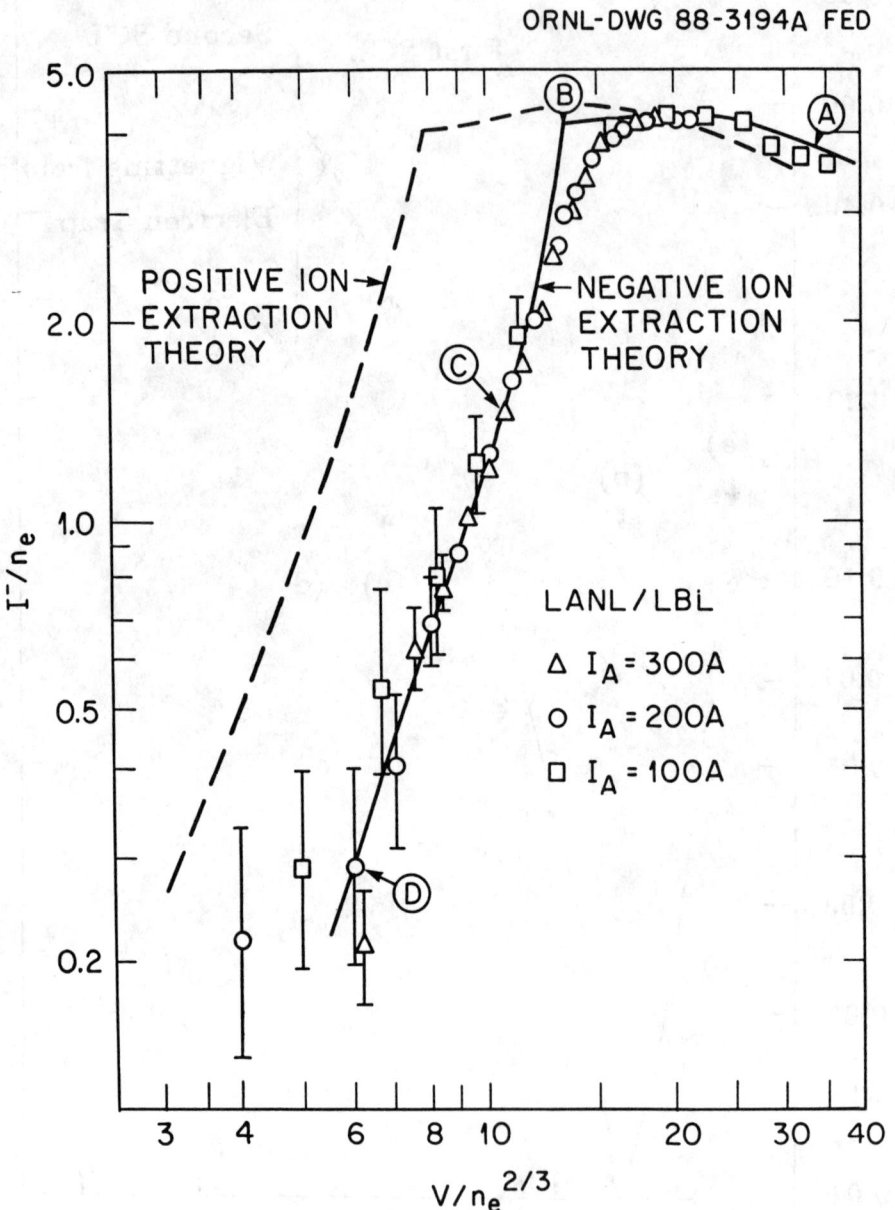

Figure 4. Negative ion current vs. inverse perveance.

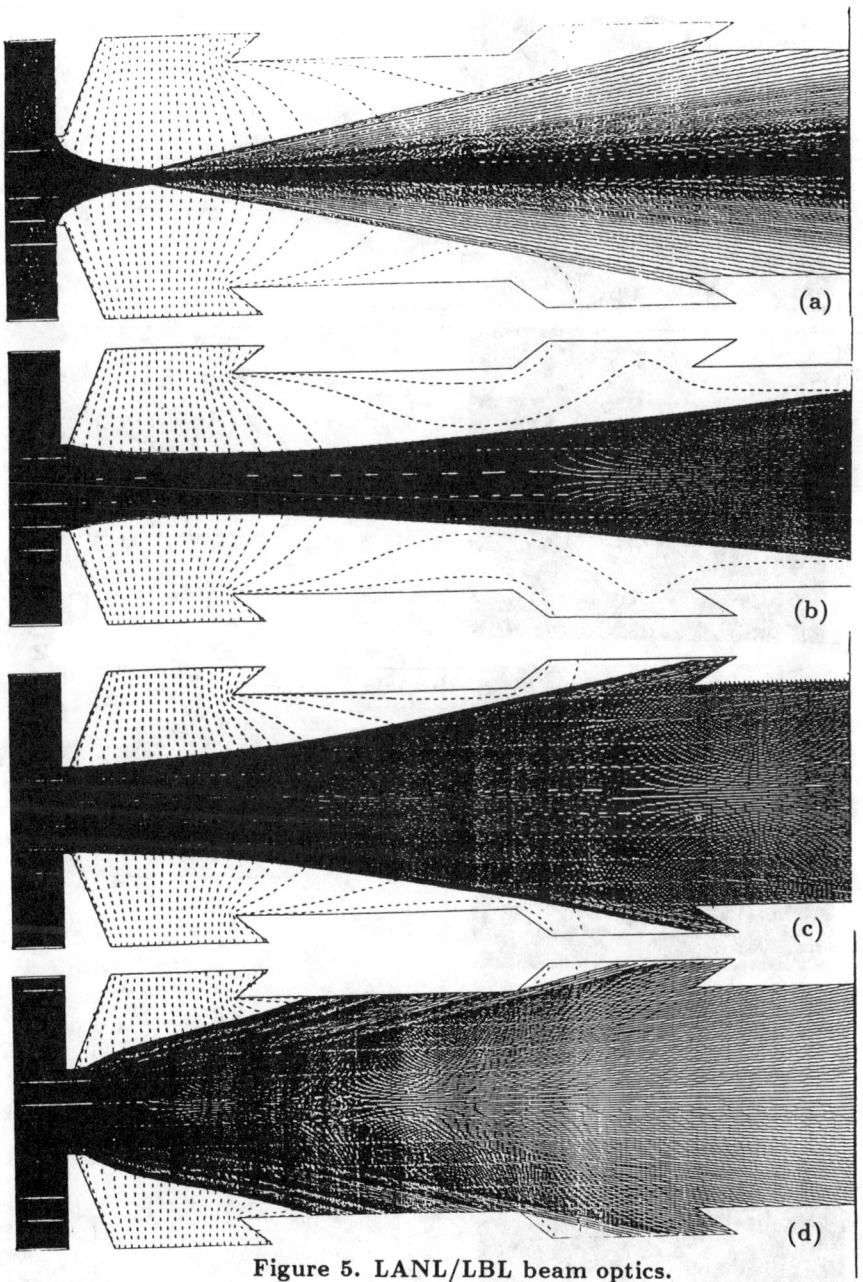

Figure 5. LANL/LBL beam optics.

550 Extraction Induced Emittance Growth

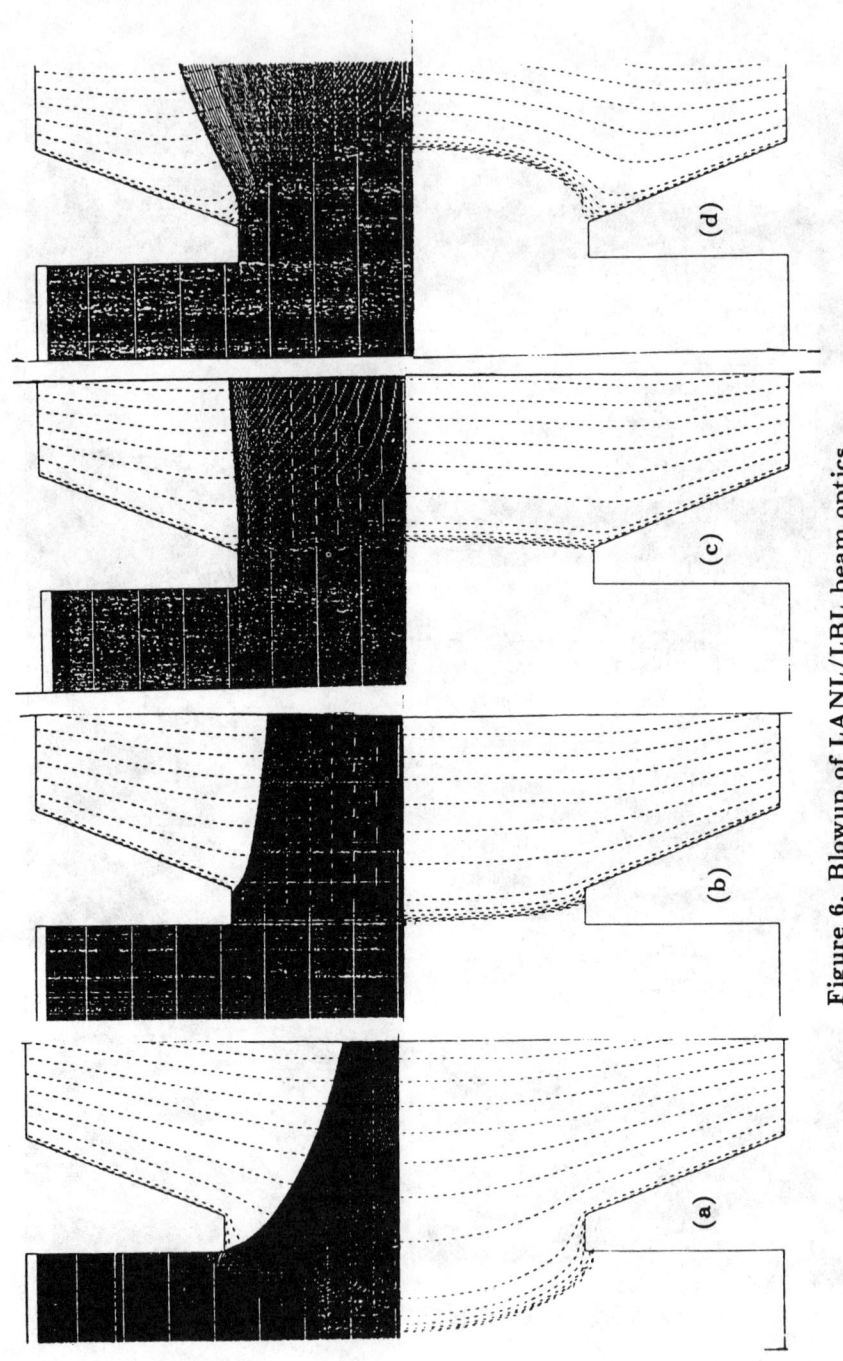

Figure 6. Blowup of LANL/LBL beam optics.

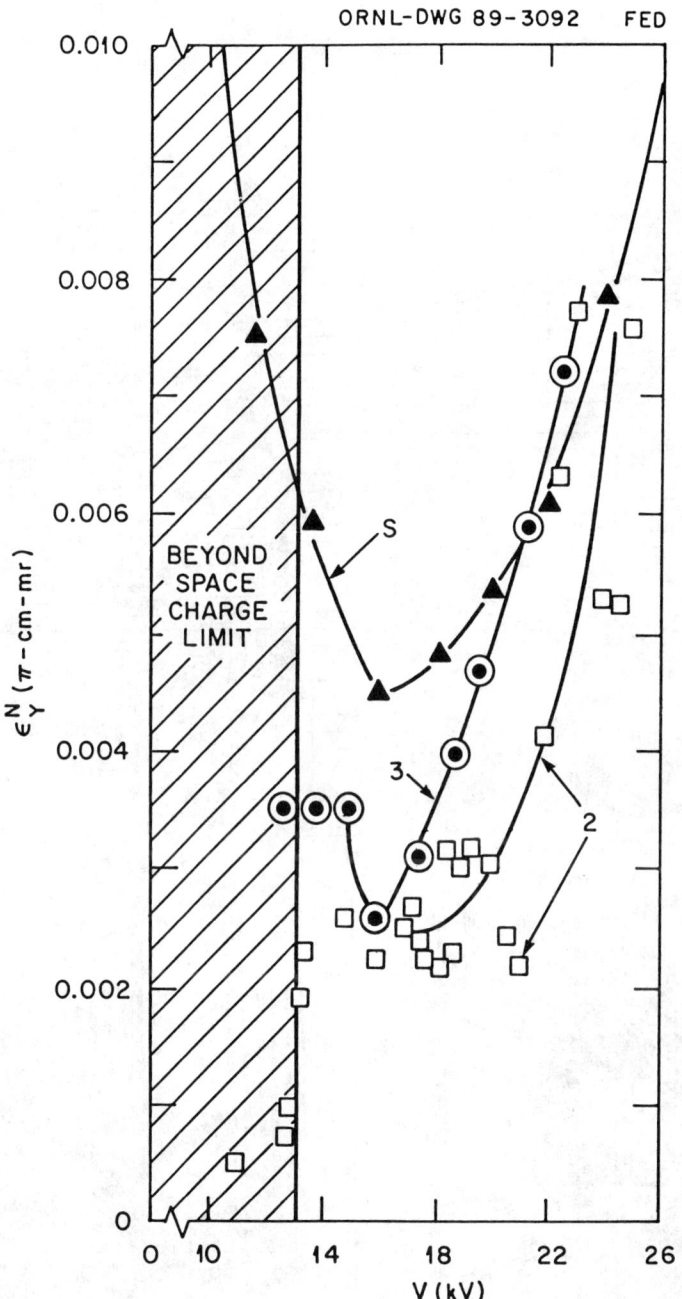

Figure 7

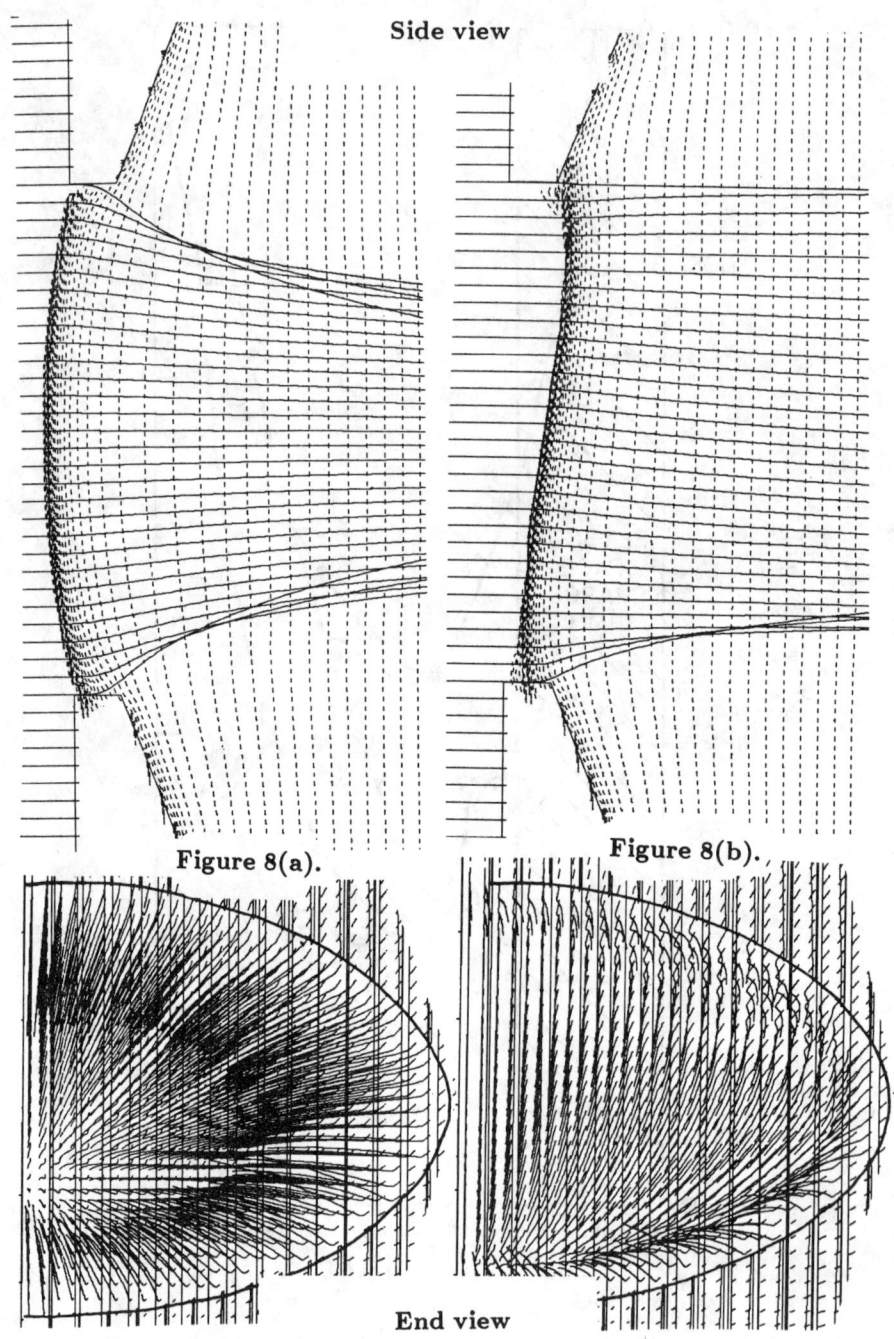

Figure 8(a). Figure 8(b).

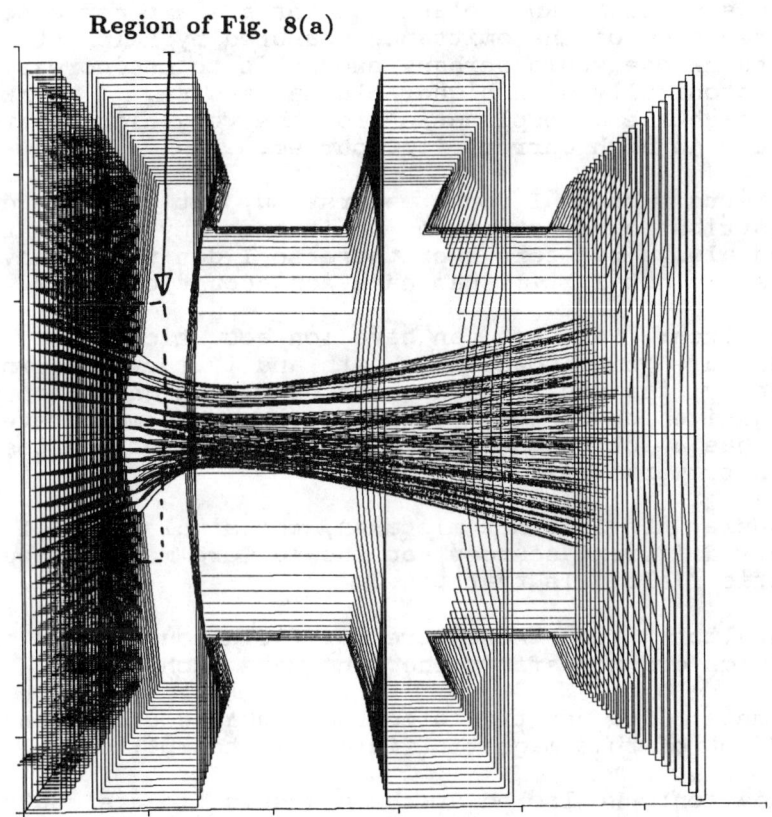

Figure 9. 3-D calculation of LBL source.

DISCUSSION

Roberts: Could you explain why we would have ion acoustic waves on both sides of the minimum and not at the minimum?

Whealton: No, I did not mean to imply that ion acoustic waves would be an explanation for a sharp perveance dependence of the emittance measured by Ralph Stevens. I think one would perhaps expect it to be increasing monotonically with higher plasma density. I think that it might be an explanation for the very high emittances found in high current ion sources.

Roberts: Maybe I don't understand, but when you do the distribution function you don't have any driving term, you always put zero over there so I don't have any collisions but I still can have ion acoustic waves?

Whealton: Yes, you can have ion acoustic waves. The ions are going down the sheath and if they go down too slowly, the positive ions charge gets higher than the negative ion charge. It's the same phenomena reversed, it has a lot of the implications for effective positive ion temperatures.

Pamela: If I am not mistaken, the simulations of the four different sources you showed were made without magnetic field. Is that true?

Whealton: Yes, there were low field sources and the fields did not effect the ions very much.

Pamela: Did you try to estimate what could be the effect of this magnetic field. What amplitude?

Whealton: We looked at the magnetic field effects with respect to the Steven's source in investigating the source of emittance. And we concluded that to the extent examined there were no very large magnetic field effects. It was mostly due to aberrations in the nonlinear sheath. Even for the various three phenomena we considered.

Jongen: I can see quite clearly the physical origin of this effect in a positive ion source since electrons are being repelled by the acceleration field. Is this the identical phenemon, but replacing the electrons by the positive ions in a negative ion source?

Whealton: I don't think that it is identical I think that there would still be a sheath, positive ions go so far, no further. However the ability of the positive ions to cancel out and mitigate any changes in the space charge such as the electrons do for positive ion sources, is far less. So I think that the sheath in negative ion sources is likely to be noisier to the extent that the electrons don't see the extraction fields. Part of the sheath where the electrons see the extraction field, I expect it to be noisier since the positive ions would do a worse job.

Roberts: Is noisier and less well defined the same thing? Do you mean there is time dependence in there? What do you mean by noisier.

Whealton: Yes, a time dependence. If the ion space charge is not being compensated correctly locally, there are going to be fluctuations.

Roberts: Does that show up in your movies?

Whealton: Yes, it did show up in that.

Roberts: Did it kind of sum up the whole thing?

Whealton: In that particular case, that was a symmetric positive ion extracton and there were three nodes that appeared. We haven't made any deeper investigation of that. It's not necessarily uniform over the sheath, even for that simple case.

Holmes: When you simulated Ralph Steven's data, what was your input data? Did you pick a temperature for the ions and the electrons when you did the emittance simulation?

Whealton: For the negative ions, we assumed zero temperature and that's why we got these very sharp transitions to space charge limit. However, for the electrons the issue is to what extent is there complete space charge neurality in the sheath region. And so the space charge neutrality was not as strong as it is for the typical positive ion source.

Holmes: The reason I am asking is that we see in sources dissimilar to Ralph's U shaped emittance graph with beam current: The emittance goes up for too low or too high beam current. So it is very similar to what Ralph's data shows.

Whealton: At low perveance, we see aberrations increase. At the high end, if we didn't consider plasma oscillations, we would find that emittance did not go up on the high end and so I think that these types of phemonema are perhaps very important for understanding emittance at high perveances.

MODELING OF VOLUME HYDROGEN NEGATIVE ION SOURCES

D.A. Skinner, P. Berlemont, M. Bacal
Laboratoire de Physique des Milieux Ionisés, Laboratoire du CNRS
Ecole Polytechnique, 91128 Palaiseau, France

ABSTRACT

Two negative ion sources for which $H_2(v'')$ vibrational spectra have been reported are studied with a numerical model. The model gives reasonably good agreement with available experimental measurements, with the notable exception of the vibrational spectra. There are significant quantitative differences at the higher v'' levels, but also some intriguing qualitative similarities.

A scaling study of high power sources is also reported. Gas dissociation and possibly vibrational cooling by thermal electrons appear to explain the saturation and in negative ion production with increasing input power. The surface-to-volume ratio becomes an important parameter for high power sources, since it affects the atomic recombination rate and the plasma density.

INTRODUCTION

It is generally understood that volume production of negative ions is a two-step process, involving (a) the formation of vibrationally excited molecules and (b) dissociative attachment of low energy electrons to these vibrationally excited molecules.[1] But numerous collisional processes occur when a current is struck in a gas, and computer modeling can be very useful in understanding the results.[2]

Recently measurements have been made of the $H_2(v'')$ spectrum. In Sec. 2 we report our efforts to present the sources in which these measurements were made. In Sec. 3 we also report a modeling study relevant to high power operation of sources which was motivated by the observed saturation of negative ion production with discharge current. In the final section we discuss some implications of these studies for high yield tandem negative ion sources.

1. THE COMPUTER MODEL

A simplified model of negative ion production in a multipole bucket source and the corresponding computer code have been described by Gorse et al.[3,4] The code simultaneously solves the Boltzmann equation for the electron energy distribution function and the vibrational master equation for the $H_2(v'')$ spectrum, as well as solving for the densities of atoms, molecules, electrons, and the ion species H^+, H_2^+, H_3^+, and H^-. Due to the complicated chemistry which is included in the model, only a point model is solved—i.e., there are no spatial dependencies and the source is assumed isotropic and homogeneous. The model may be interpreted as a spatial average of a real source—this is most accurate for sources in which mean free paths are large in comparison to device size and for single chamber rather than tandem sources.

Simulating a negative ion source with our model requires a choice for input parameters, which we will review. The physical configuration is represented by several surface and volume specifications. V_{gas} and S_{gas} denote the total volume of gas and its bounding surface area. V_{plas} and S_{plas} represent the volume of the unmagnetized plasma-confining region and its surface area. S_{elec} is the electron loss area, along line cusps plus to any magnetically unshielded surfaces. The plasma potential V_p governs electron losses to S_{elec}. Losses to magnetically unshielded surfaces biased at other potentials may also be modeled. Dissociation on the filament surface S_{fil} is significant only in low power discharges. The ratios $V_{\mathrm{gas}}/S_{\mathrm{gas}}$ and $V_{\mathrm{plas}}/S_{\mathrm{plas}}$ are also used as scale lengths for particle flight to boundaries for gas particles and ions respectively. The temperatures T_{H} and T_{H_2} determine the average speeds of gas particles, and also affect V-T cross sections. We take $T_{\mathrm{H}_2} = 500°K$ and $T_{\mathrm{H}} = 4000°K$ in all of the work presented here (save for a slightly different choice for T_{H} for the LBL source). In some cases ion loss velocities are determined by the ion temperatures, which we take to be $T_{\mathrm{H}_2^+} = T_{\mathrm{H}_3^+} = T_{\mathrm{H}_2}$ and $T_{\mathrm{H}^+} = T_{\mathrm{H}}$. Input power is determined by the discharge voltage V_d and discharge current I_d. Gas density is governed by the initial filling pressure P_{fill} along with a normalization condition which will be discussed below. The plasma potential V_p is chosen to yield a neutral plasma in the final steady state condition, and the recombination coefficient γ_{H} is adjusted to give the desired atomic density. The last adjustable parameter is b in the wall relaxation term [5], where $b = 1$ corresponds to the strongest relaxation and $b = 4$ or 2 is our usual choice. This parameter has little effect on high power discharges in which volume reactions dominate surface reactions.

The plasma potential V_p has a strong influence on the temperature and density of the low-energy thermal part of the electron distribution function, since it controls the losses of the lowest energy electrons to the surface S_{elec}. We now determine the sheath potential V_p which balances electron against ion losses to produce a neutral plasma, by using a simple model for positive ion production and assuming that ions are lost and neutralized when they reach the boundary of the unmagnetized plasma region. Following the suggestion of M. Hopkins [6], the average ion velocity is taken to be

$$\bar{v}_{\mathrm{H}_n^+} = \frac{S_{\mathrm{elec}}}{V_{\mathrm{plas}}}(0.6)\sqrt{\frac{kT_e}{m_{\mathrm{H}_n^+}}} + \frac{S_{\mathrm{plas}}}{V_{\mathrm{plas}}}\sqrt{\frac{kT_{\mathrm{H}_n^+}}{m_{\mathrm{H}_n^+}}}. \tag{1}$$

When there are magnetically unshielded surfaces in the source and S_{elec} is of the order of magnitude of S_{plas}, then the first term on the r.h.s. dominates the second. In this case the ions move with the ion acoustic velocity in the Bohm pre-sheath to the unshielded surface. But if all the walls of a source are magnetically shielded then S_{elec} is small, in which case the second term dominates and the ions move with their thermal velocity. This view is supported by measurements showing that the plasma potential can be relatively flat in the unmagnetized region of a source with completely shielded walls. [7]

It is customary to use the filling pressure in specifying source conditions. This is not necessarily the most meaningful parameter since it is stripping loss during extraction which imposes low pressure operation, and the stripping loss depends on the molecular and atomic densities during source operation. If the stripping loss Δn_{H^-} varies linearly with gas densities in the source, $\Delta n_{\mathrm{H}^-}/n_{\mathrm{H}^-} = S_1 n_{\mathrm{H}} + S_2 n_{\mathrm{H}_2}$, where S_1 and S_2 are constants of proportionality, then

$$n_{\mathrm{H}_2} + \frac{S_1}{S_2} n_{\mathrm{H}} = \mathrm{const} \tag{2}$$

is the condition for constant fractional stripping loss. In a fuller analysis, S_1/S_2 depends on the stripping loss over the negative ion trajectory from the point of extraction, $\ell = 0$, through extraction and acceleration,

$$\frac{S_1}{S_2} = \frac{n_{H_2}(\ell=0)}{n_H(\ell=0)} \frac{\int_0^n n_H(\ell) \; n_{H^-}(\ell) \; \sigma_1(v) \; d\ell}{\int_0^n n_{H_2}(\ell) \; n_{H^-}(\ell) \; \sigma_2(v) \; d\ell}, \qquad (3)$$

where σ_1 and σ_2 are electron detachment cross sections for collisions with H and H_2 respectively, and $v(\ell)$ is the negative ion velocity during extraction and acceleration. For extraction energies below 1 kev where σ_1 is an order of magnitude larger than σ_2,[8] atoms have a much worse effect than molecules. The smallest value that S_1/S_2 could take is $\frac{1}{2}$, which would occur if negative ions were quickly accelerated during extraction so that most stripping losses occur at energies above 70 kev where $\sigma_1/\sigma_2 \approx \frac{1}{2}$. In the work presented here, we have assumed $S_1/S_2 = \frac{1}{2}$, in which case Eq. (2) reduces to a simple "closed box" condition for the gas. We have thus taken the most optimistic view of the effect of atomic hydrogen on stripping loss. The model can work equally well with other S_1/S_2 values determined by a stripping loss analysis, in which case dissociation could only be more detrimental to negative ion production in our model than indicated in the results of Sec 3.

2. SIMULATION OF FOM AND LBL SOURCES

We have modeled two negative ion sources in which the $H_2(v'')$ vibrational spectrum has recently been measured: the source studied using REMPI (Resonance-Enhanced MultiPhoton Ionization) at FOM Institute [9,10], and the source studied using VUV absorption spectroscopy at LBL. [11,12] We will henceforth refer to them as the FOM and LBL sources. Both sources are single chamber multipole devices for which our simulation code is well suited.

Extensive measurements are reported for the FOM source over variations in both filling pressure P_{fill} and discharge current I_d. Experimental measurements and simulation results for the negative ion density are compared in Fig. 1 over variations in P_{fill} and I_d. A similar comparison is made in Fig. 2 for the atomic density. Given the uncertainties and approximations of the model, the qualitative and even quantitative agreement between experiment and model is remarkably good. The only clear disagreement is for the highest pressure case. We have noticed in previous work that the model becomes less reliable at higher pressure, presumably because reduced mean free paths in the system make the model less suitable.

Given the good agreement between experimental and modeled H^- density, one might expect that the experimental measurements of the $H_2(v'')$ spectrum would confirm the model calculations. But as shown in Fig. 3-a, the measured and calculated spectra are far apart. Table 1 summarizes this case—all calculated results are close to experimental values, except for the vibrational distribution.

Fig. 4 shows that the difference between experimental and modeled vibrational spectra exists at all discharge currents. Here both measured and calculated spectra are displayed normalized by the density $H_2(v''=0)$ so that variations in the spectral shapes are evident. The experimental spectra show little variation in response to increasing discharge current over the range $1 \leq v'' \leq 4$, in striking contrast to the calculated spectra. Indeed, were it not for the measurements at $v''=5$, it might appear that the spectral shape depends weakly on discharge current and saturates as I_d increases past 10 Amps. But the experimental $v''=5$ densities show more variation and are amazingly proportional to the calculated values even though they are an order of magnitude smaller—see Fig. 5. The quartet of densities $n_{H_2(v''=5)}$ for $I_d = (5, 10, 20, 30)$ amps may be normalized to the $I_d = 10$ amp value to exhibit the proportionality and then take the form $1.60 \cdot 10^{10} \times (0.56, 1, 1.63, 2.13)$ cm^{-3} for the density measurements and $1.43 \cdot 10^{11} \times (0.53, 1, 1.70, 2.14)$ cm^{-3} for model results.

Turning now to the LBL source, Fig. 3-b compares the experimental and calculated $H_2(v'')$ spectrum for a relatively high power case. The two spectra are both nearly linear over the measured region, although characterized by different temperatures. The model predicts a plateau-like distribution at low power, but at high power predicts a nearly linear distribution up to $v''=9$. With no experimental data available for the total molecular density or the density of the ground state $H_2(v''=0)$, it is unclear how much gas should be specified in the simulation and thus at which v'' the theoretical spectrum should be normalized to the experimental spectrum. We have chosen to normalize it at $v''=0$ by using a linear extension of experimental data for the experimental ground state density. With this normalization, there is again a significant quantitative difference between theory and experiment at those vibrational levels $v''=5$ through 8 which should be responsible for negative ion production. Taking

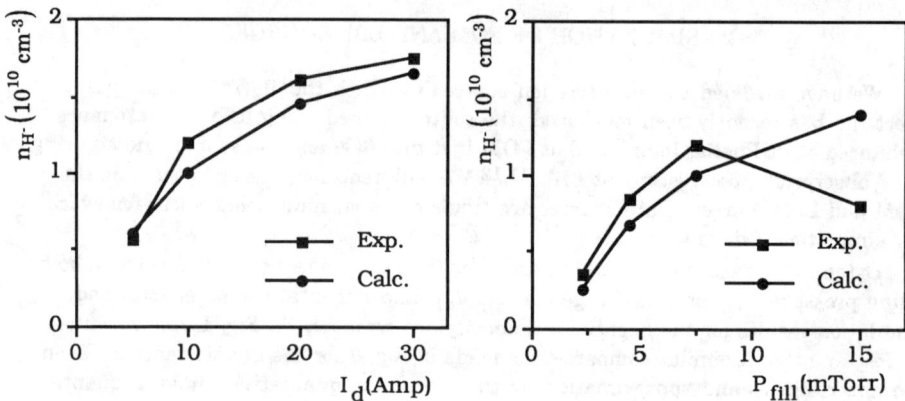

Fig. 1. Negative ion density n_{H^-} in the FOM source.
(a) $n_{H^-}(I_d)$ at $P_{\text{fill}} = 7.5$ mTorr. (b) $n_{H^-}(P_{\text{fill}})$ at $I_d = 10$ Amp.

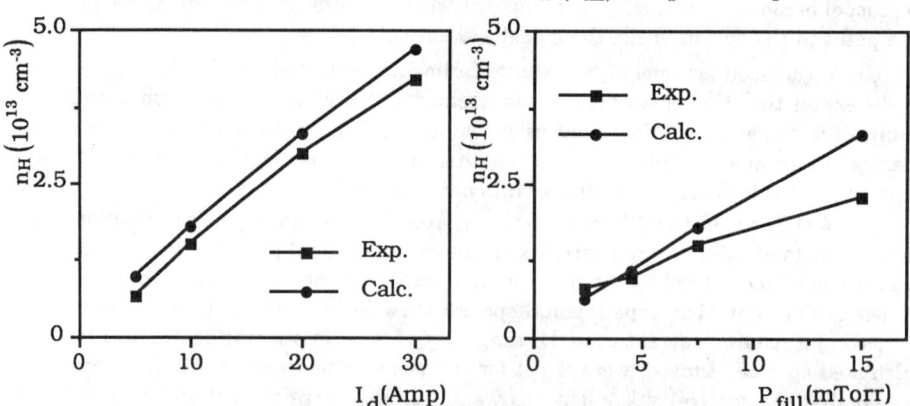

Fig. 2. Atomic density n_H in the FOM source.
(a) $n_H(I_d)$ at $P_{\text{fill}} = 7.5$ mTorr. (b) $n_H(P_{\text{fill}})$ at $I_d = 10$ Amp.

$\gamma_H = 0.4$ in the model then reproduces the measured atomic density. The result for $\gamma_H = 0.1$ in Fig. 3-b indicates the sensitivity of the spectrum via V-T vibrational cooling to the atomic density, which there is quadrupled over that of the $\gamma_H = 0.4$ case. Table 1 summarizes the parameters and other results for this case, which are otherwise in reasonable agreement. Measurements are not available for several important quantities including the molecular density and the negative ion density.

We have used the model to study the effect of the filling pressure. Fig. 6 shows the variation of the $n_{H_2(v''=5)}$ density with P_{fill}. It is qualitatively quite similar to the measured variation of $n_{H_2(v''=5,J''=1)}$ with filling pressure,[12] which is also indicated in Fig. 6 on a different scale. The saturation and decrease with pressure occurs for all vibrational levels in the modeling and is due to gas cooling of energetic electrons.

A relatively high value of the recombination coefficient $\gamma_H = 0.4$ was required in the simulation to reproduce the measured atomic density. In earlier work we have usually found γ_H to be of the order 0.1. The LBL source used LaB_6 cathodes rather

		FOM		LBL	
		Exp.	Calc.	Exp.	Calc.
P_{fill}	(mTorr)	4.5		8	
I_d	(Amps)	10	10	25	25
V_d	(V)	105/115	115	120	120
T_{mol}	(°K)	390	500		500
T_{atom}	(°K)		4000	700/7000*	5400
b(v"=1)			2		2
γ_H		0.06 †	0.05	0.025 †	0.1
n_e	(cm^{-3})	2.6 10^{11} §	3.9 10^{11}	1.6 10^{12}	2.0 10^{12}
$n_{fe}(\varepsilon > V_{dis}/3)$ (%)		1.0 §	0.85		0.4
T_e	(eV)	1.3 §	1.6	2.5	2.0
V_{plas}	(V)	2.9	4.6		1.8
H$^+$	(cm^{-3})		1.1 10^{10}		4.6 10^{10}
H$_2^+$	(cm^{-3})		1.0 10^{11}		3.1 10^{11}
H$_3^+$	(cm^{-3})		2.8 10^{11}		1.7 10^{12}
H$^-$	(cm^{-3})	8.5 10^9	6.8 10^9		5.2 10^{10}
H	(cm^{-3})	1.0 10^{13}	1.1 10^{13}	1.0 10^{13}	1.0 10^{13}
H$_2$	(cm^{-3})	8.3 10^{13}	8.2 10^{13}		1.5 10^{14}
T_{vib} (v"=0:1)(°K)		2360	2500		4671
H$_2$(v"=0)(cm^{-3})		7.5 10^{13}	7.3 10^{13}		1.0 10^{14}
H$_2$(v"=1)(cm^{-3})		6.9 10^{12}	6.7 10^{12}	2.4 10^{13}	2.8 10^{13}
H$_2$(v"=2)(cm^{-3})		7.3 10^{11}	1.5 10^{12}	6.5 10^{12}	1.1 10^{13}
H$_2$(v"=3)(cm^{-3})		1.6 10^{11}	4.4 10^{11}	1.8 10^{12}	4.7 10^{12}
H$_2$(v"=4)(cm^{-3})		4.0 10^{10}	1.7 10^{11}	3.8 10^{11}	2.2 10^{12}
H$_2$(v"=5)(cm^{-3})		1.1 10^{10}	8.5 10^{10}	1.4 10^{11}	9.7 10^{11}
H$_2$(v"=6)(cm^{-3})				6.7 10^{10}	3.9 10^{11}
H$_2$(v"=7)(cm^{-3})				2.5 10^{10}	1.6 10^{11}
H$_2$(v"=8)(cm^{-3})				9.1 10^9	9.5 10^{10}

Table 1. Comparison of experiment and calculation for FOM and LBL sources. Notes: (*) two temperature distribution; (†) post-discharge γ_H; (§) electron measurements in a lengthened source.

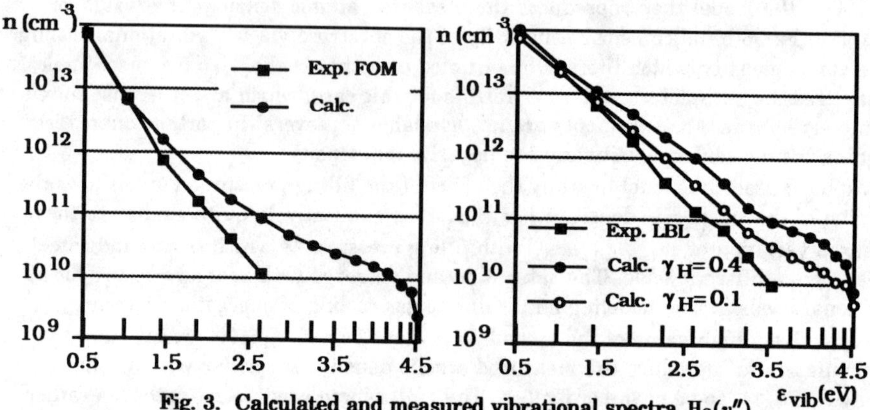

Fig. 3. Calculated and measured vibrational spectra $H_2(v'')$.
(a) FOM, $I_d = 10$ Amp, $P_{fill} = 4.5$ mTorr.
(b) LBL, $I_d = 25$ Amp, $P_{fill} = 8$ mTorr.

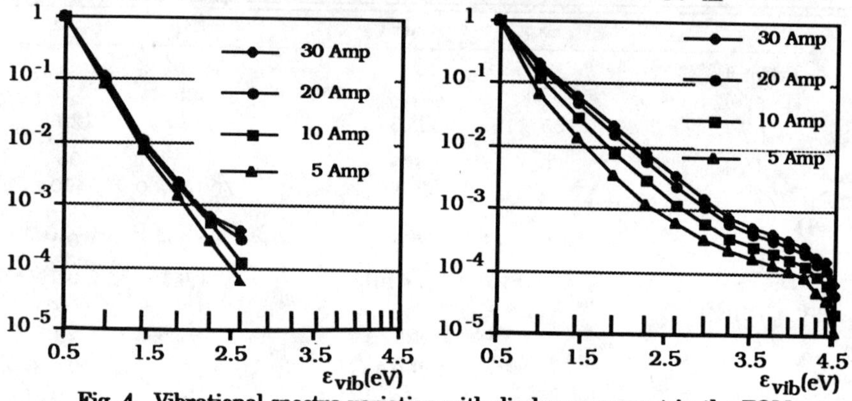

Fig. 4. Vibrational spectra variation with discharge current in the FOM source for $P_{fill} = 7.5$ mTorr, densities normalized at $v'' = 0$, for (a) measurements and (b) calculations.

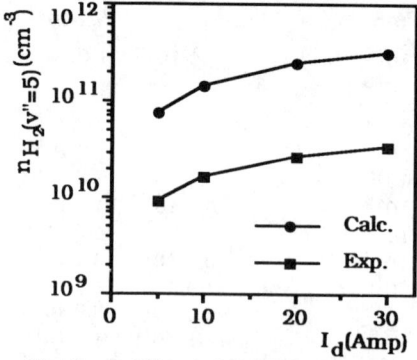

Fig. 5. $H_2(v'' = 5)$ density as function of discharge current in FOM source, calculated and measured.

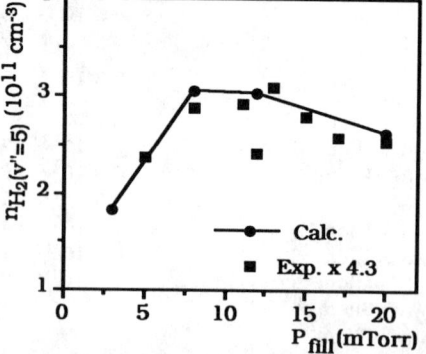

Fig. 6. $H_2(v'' = 5)$ density as function of filling pressure in LBL source, measurements multiplied by 4.3.

than the tungsten filaments used on most other sources, which might explain the different γ_H. This value of γ_H is however an order of magnitude larger than the 0.025 indicated by the post-discharge decay of atomic density.[13] This may indicate that the surface conditions change drastically as the discharge ends.

A recombination coefficient γ_H may be estimated from post-discharge measurements of atomic density, but γ_H during the discharge would be of more direct interest. We note that γ_H might be estimated *during* some discharges by measuring the growth of atomic hydrogen after the discharge is struck. In time-dependent simulations we have noticed that dissociation proceeds on a slower time scale τ_H than the production of vibrationally excited molecules or ions. Soon after the beginning of the discharge the system settles into a quasi-equilibrium whose further evolution is governed by dissociation. So long as the dissociation level remains low (and unless the atomic hydrogen modifies wall properties), there should be little further evolution in the system, and specifically the production rate of H should remain constant. The rate equation, $dn_H/dt = S_H - \gamma_H (S_{gas}/V_{gas}) \bar{v}_H n_H(t)$ then has the solution $n_H(t) = n_\infty (1 - exp(-t/\tau_H))$, where $\bar{v}_H$ is the average atomic velocity, n_∞ is the longterm density $n_H(t \gg \tau_H)$, and S_H is the constant source term. In the event that τ_H and n_∞ can be identified from experimental data, $S_H = n_\infty/\tau_H$ and $\gamma_H^{-1} = \tau_H \bar{v}_H (S_{gas}/V_{gas})$.

To summarize this section, there is reasonably good agreement between our model and experimental measurements, to within a factor of two for nearly all measured quantities and much better for most, including: plasma density, electron temperature, fraction of primary electrons, dissociation fraction and even the negative ion density. However, two experiments have reported $H_2(v'')$ profiles which are significantly colder than predicted by our model. It is a challenge to relate the measurements to our theoretical understanding of negative ion production. Some qualitative agreements between theory and experiment were noted, specifically the variation of $n_{H_2(v''=5)}$ with P_{fill} at LBL and with I_d at FOM.

3. PARAMETER STUDY OF A HIGH POWER SOURCE

We have studied a model system which approximates the source region of the bucket source with virtual magnetic filter at Culham Laboratory.[14] This is a scaling exercise intended to elucidate the observed saturation and degradation of negative ion production with increased power, not a detailed simulation as in Sec. 2—the Culham source with its broad filter field and the associated continuous gradients in the electron temperature is not the best candidate for our zero-dimensional model. Calculations were made for about 25 sets of conditions which span variations in filling pressure $P_{fill} = 1-15$ mTorr and discharge current $I_d = 10-1000$ amps at $V_d = 100$ V, for a system with total gas volume V_{gas} of 17 liters and a surface-to-volume ratio of 0.3 cm^{-1}. The dissociation coefficient was assumed constant, $\gamma_H = 0.1$.

Fig. 7 shows the calculated electron energy distribution function for three values of I_d. The oscillatory structure most evident in the $I_d = 10$ amp e.d.f. occurs because injected primary electrons lose energy predominately through reactions with molecules. As I_d is raised, electron-electron collisions become more important for energy transfer and the oscillatory structure is smoothed out. This structure may not actually exist in sources, since the beam instabilities which should occur as electrons leave the filament are not included in our model. Plasma density increases with discharge current, as shown in Fig. 8, but varies little with filling pressure over the pressure range examined.

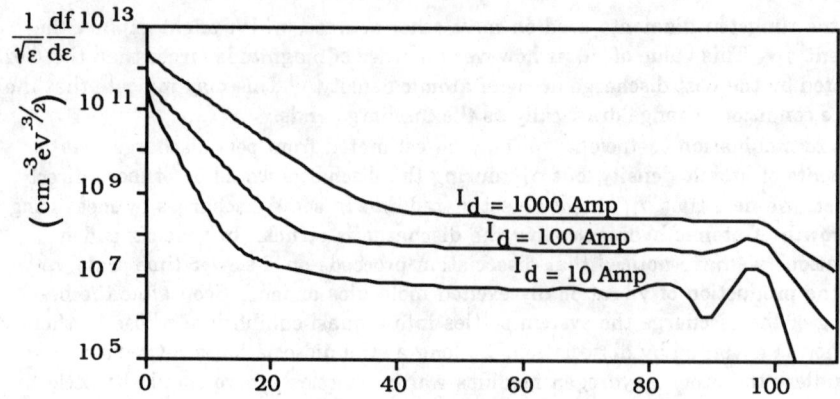

Fig. 7. Electronic energy distribution for three values of discharge current, $I_d = 10$, 100, and 1000 Amp, in high power study.

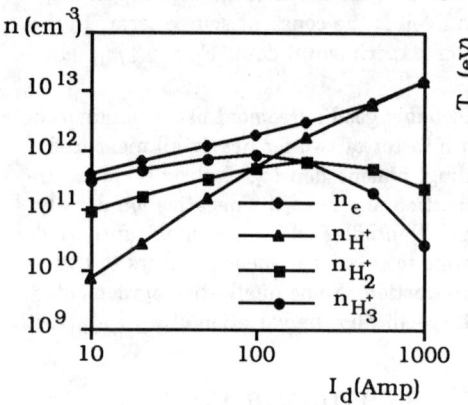

Fig. 8. Plasma and ion densities as functions of discharge currrent, $P_{fill} = 3$ mTorr.

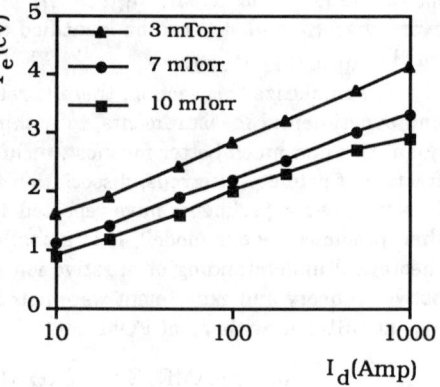

Fig. 9. Electron temperature as function of discharge currrent.

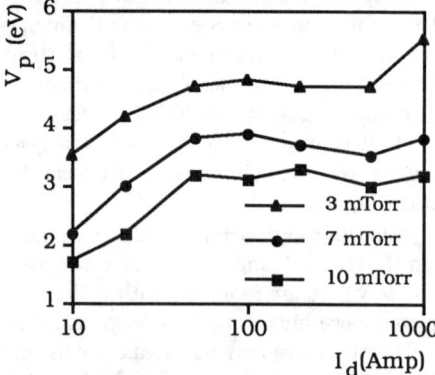

Fig. 10. Plasma potential as function of discharge current.

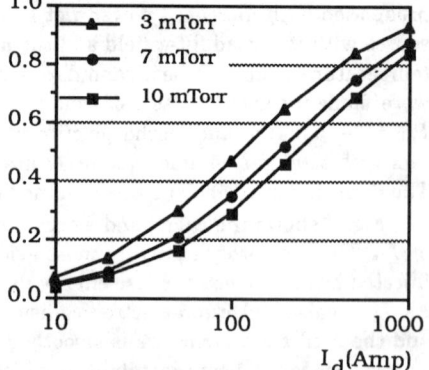

Fig. 11. Atomic fraction $n_H/(n_H + n_{H_2})$ as function of discharge current.

Fig. 8 also shows how the positive ion fraction varies with I_d. The temperature of the low energy thermal part of the e.d.f. is plotted in Fig. 9, and the plasma potential in Fig. 10.

As power input is increased, both the plasma density and the primary electron density increase. Volume reactions generally dominate surface reactions. The surface reactions which remain important in our model are the recombination of atomic hydrogen (there being no volume reaction which disposes of atomic hydrogen) and neutralization of positive ions.

The atomic fraction $n_H/(n_{H_2}+n_H)$ attains high values in the simulation, as shown in Fig. 11. Note how a similarly shaped curve is traced at each filling pressure, the curve being displaced in the direction of higher discharge current as the pressure increases. The atomic fraction depends on γ_H, which is held constant in this simulation. We adopted the value $\gamma_H = 0.1$ based on our experience simulating low power discharges and on measurements.[15] These measurements were surely made on contaminated surfaces, since it is known from careful ultra high vacuum work that γ_H is essentially one for most truly clean metals.[16,17] Sputter cleaning of the surface by energetic atoms or ions might cause γ_H to vary with discharge power (possibly depending on pumping rates, wall and filament materials, and other details of source construction and operation).

In this simulation, molecular dissociation is the principal cause for poor negative ion production at high power. With increasing power the molecular vibrational distribution becomes hotter while the total molecular density falls—the maximum negative ion production occurs at an intermediate power. As is evident in Fig. 12, the $I_d = 10$ amps and $I_d = 1000$ amps discharges produce about the same vibrationally excited populations in the range most important for negative ion production $v'' = 5-9$, well below those of the $I_d = 100$ amps case. With increasing input power the molecular vibrational distribution becomes hotter while the total molecular density falls. Maximum negative ion production occurs at a power which optimizes the contradictory trends. As indicated in the plots of $n_{H_2(v''=7)}$ in Fig. 13, peak production occurs at a higher discharge current for a higher filling pressure.

We note that the competing trends of dissociation and vibrational excitation also explain the peaking and subsequent decay of negative ion production in pulsed sources.[18] Because the time scale is slower for dissociation than for vibrational excitation, negative ion production peaks as the vibrational distribution is established then decays more slowly as the molecular density drops. This raises the interesting possibility of comparing a negative ion source's steady state production to its peak production when pulsed as a indication of how detrimental dissociation is in steady state operation.

If dissociation becomes so important at high power, then the most obvious way to increase negative ion production is increasing the atomic recombination rate. This would permit effective operation at higher power. For the single case $P_{\text{fill}} = 3$ mTorr and $I_d = 1000$ amps we have investigated the effect of (a) $\gamma_H = 1.0$ rather than 0.1 and (b) a four-fold increase in surface area S_{gas} and S_{plas} for the same volume. As is evident from Fig. 14, the two changes produce similar dramatic increases in the $v'' = 5-9$ vibrational populations important for negative ion production. These new values of $n_{H_2(v''=7)}$ densities are marked in Fig. 13 and hint at how much higher the optimal discharge current and negative ion yield could be.

A second potential reason for poor negative ion production at high power is the growing role of thermal electrons in establishing the vibrational distribution. In our 100 and 1000 amp cases, the e-V process (thermal electron) dominates others in establishing the low v'' distribution. In our simulations, the vibrational spectrum characteristically becomes nearly Boltzmann over the range where the e-V process is dominant. The e-V process appears to be associated with a particular vibrational temperature, which presumably depends on the thermal part of the e.d.f. In Fig. 15 we compare the 3 mTorr, 100 amp vibrational distribution with the distributions that would be formed either without the e-V process or without the E-V (energetic electrons) process. In the absence of E-V, the thermal electrons nonetheless promote molecules far up the vibrational ladder via e-V. When e-V is omitted from the calculation, the resulting distribution is more plateau-like, and it is evident that e-V actually cools the

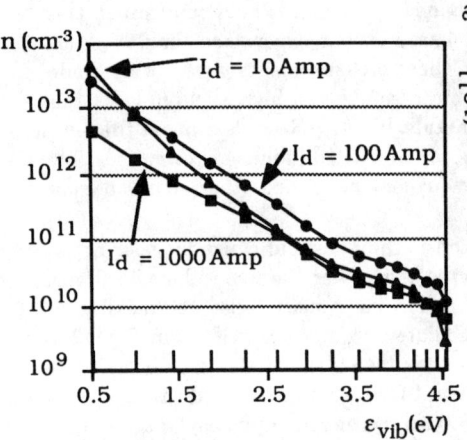

Fig. 12. Vibrational spectra for $I_d = 10$, 100, and 1000 Amp at $P_{fill} = 3$ mTorr.

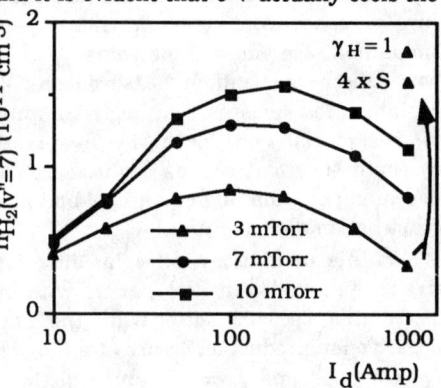

Fig. 13. $H_2(v''=5)$ density as function of discharge current. Isolated points show the $P_{fill} = 3$ mTorr case for γ_H raised from 0.1 to 1 and for quadrupled surface area.

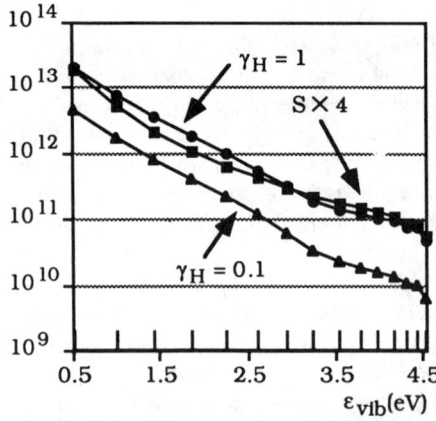

Fig. 14. Vibrational spectra $H_2(v'')$ at $I_d = 1000$ Amp, $P_{fill} = 3$ mTorr showing the effect of quadrupling surface area and of raising γ_H from 0.1 to 1.

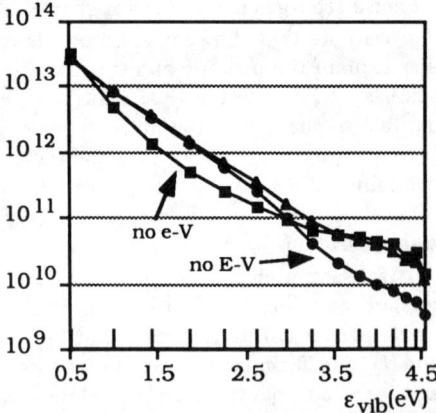

Fig. 15. Vibrational spectra $H_2(v'')$ at $I_d = 100$ Amp, $P_{fill} = 3$ mTorr showing the effect of neglecting e-V or E-V in the calculation.

highest vibrational states. In this case e-V has a net positive effect on negative ion production. But if dissociation were reduced or controlled so that sources could be operated at higher power and plasma density, the e-V would become a net loss at lower and lower v'' levels and limit the benefits of higher current at the vibrational levels of interest.

4. IMPLICATIONS FOR HIGH YIELD NEGATIVE ION SOURCES

Several mechanisms are capable of producing vibrationally excited hydrogen molecules: the E-V process (collisions with energetic electrons); the e-V process (collisions with low energy electrons); the recently discovered production of highly vibrationally excited molecules by recombinative desorption; the four step mechanism proposed by Hiskes and Karo [20] for H_2^+ and H_3^+ wall neutralization. How a negative ion source should be optimized depends on which $H_2(v'')$ producing mechanism is dominant. There are indications that recombinative desorption does not produce $H_2(v'')$ while the walls are exposed to a discharge,[10,21] so a negative ion source based on this interesting mechanism would be quite unlike current volume sources. A source which optimizes $H_2(v'')$ production by positive ion neutralization at the walls would also probably be unlike current volume sources. Both the e-V and E-V processes can be important mechanisms in volume sources. The best source may be one in which one or the other is dominant, i.e., one in which either the thermal electrons or energetic beam electrons play the dominant role—the two mechanisms are not optimized in the same fashion.

In our study of high power sources reported in Sec. 3, we found that at high plasma density the thermal electrons and the e-V reaction play an important role in establishing the vibrational distribution. The vibrational distribution of Fig. 15 calculated without E-V indicates that thermal electrons alone can produce significant densities of highly vibrationally excited molecules. Could a negative ion source operate with a high density of thermal electrons and a much reduced density of energetic electrons? Such an approach would have the advantage of reducing dissociation by energetic electrons. The ideal device for such an approach might be quite different from current multipole sources. would be required to produce and confine such a plasma.

The vibrational distribution for the source with increased surface area in Fig. 14 indicates how a source can be optimized for the E-V reaction. Increasing the source surface area addresses both of the reasons identified in Sec. 2 for poor source performance at high power: gas dissociation and possible cooling by thermal electrons. A smaller fraction of the gas would be dissociated because the larger surface increases the recombination rate. The plasma density also falls because the ion loss rate increases and thus thermal electrons become less important. Making smaller sources (reducing S/V while maintaining the same input power density) should have the same results. For a given power volume density, reducing source size would also reduce power input and wall loading.

The S/V ratio plays an important role in negative ion source operation. For S/V too large (sources too small), vibrationally excited molecules are quenched by the walls. For S/V too small (sources too large) and given sufficient power density, the gas is dissociated. The optimal size lies somewhere between the two extremes, and depends on the input power density. The two extremes are determined by surface conditions which are poorly understood and which may be modified by the discharge. Under some conditions vibrationally excited molecules survive hundreds of wall collisions,[19] while they are thought to survive, at the most, tens of collisions in volume sources. In Sec. 2 we noted that there appears to be an order of magnitude difference between the discharge and the post-discharge recombination coefficient for the LBL source. It has also been shown that the filament and wall material affect negative ion yields, presumably because of varying wall properties.[1]

Modeling studies can be no more valid than the model employed. Comparisons between experiments and our model in sec. 2 show a need for more work on the basic production mechanism of negative ion sources. The modeling and the comparison between theory and experiment is much easier for a simple single chamber source than for sources with more complicated geometry. Measurements of the $H_2(v'')$ spectrum, the e.d.f., gas and ion densities are all important in modeling a source–all these quantities have been measured, but never simultaneously nor even in the same source geometry. It is also evident that surface reactions can play an important role in source operation, but little is yet known about how surface conditions influence these reactions and are in turn influenced by details of the source construction or by operating conditions.

We acknowledge the computational resources granted by the Conseil Scientifique du Centre de Calcule Vectoriel pour la Recherche, France. We also acknowledge the support of the Oak Ridge National Laboratory and the European Atomic Energy Community.

REFERENCES

1. M. Bacal and D.A. Skinner, Comm. in Atomic and Mol. Phys., to be published.
2. M. Bacal, P. Berlemont and D.A. Skinner, SPIE Vol. 1061 Microwave and Particle Beam Sources and Direct Energy Concepts (1989), p. 528.
3. C. Gorse, M. Capitelli, J. Bretagne and M. Bacal, Chem. Phys. 93, 1 (1985).
4. C. Gorse, M. Capitelli, M. Bacal, J. Bretagne and A. Laganá, Chem. Phys. 117, 177 (1987).
5. J.R. Hiskes, A.M. Karo and P.A. Willman, J. Vac. Sci. Technol. A3, 1229 (1984).
6. M. Hopkins, Private Communication.
7. R. Limpaecher and K.R. MacKenzie, Rev. Sci. Instr. 44, 726, (1973)
8. M.W. Gealy and B. Van Zyl, Phys.Rev. A36, 3091 (1987).
9. P.J. Eenshuistra, A.W. Kleyn and J.H. Hopman, Europhys. Lett. 8, 423 (1989).
10. P.J. Eenshuistra, R.M.A. Heeren, A.W. Kleyn and H.J. Hopman, Phys. Rev. A40, 3613 (1989).
11. G.C. Stutzin, A.T. Young, A.S. Schlachter, K.N. Leung and W.B. Kunkel, Chem. Phys. Lett, 155, 475 (1989).
12. G.C. Stutzin, A.T. Young, H.F. Döbele, A.S. Schlachter, K.N. Leung and W.B. Kunkel, Proc. Int. Conf. on Ion Sources, Berkeley, 1989, Rev. Sci. Instr. Special Issue, to be published.
13. A.T. Young, Private Communication.
14. A.J.T. Holmes, L.M. Lea, A.F. Newman and M.P.S. Nightingale, Rev. Sci. Instr., 58, 223 (1987), and Private Communication.
15. H. Wise and B.J. Wood, in Advances in At. and Mol. Phys., 3, 291 (1976), edited by D.R. Bates and I. Estermann, Academic Press, N.Y. and London.
16. D.E. Rosner, Annual Review of Materials Science 2, 573 (1972).
17. G.A. Beitel, Recombination of Atomic Hydrogen on Metal Surfaces in Ultra High Vacuum, thesis presented at the Physics Dept., University of Wisconsin–Madison, Madison, Wisconsin 53706, U.S.A. (1969)
18. M. Bacal, Nucl. Instr. and Meth. in Physics Research, B 37/38, 28 (1989)

19. R.I. Hall, I. Cadez, M. Landau, F. Pichou and C. Shermann, Phys. Rev. Lett. **66**, 337 (1987)
20. J.R. Hiskes and A.M. Karo, SPIE Vol. 1061 Microwave and Particle Beam Sources and Direct Energy Concepts (1989), p. 542.
21. R.I. Hall, I.Cadez, M. Landau, F. Pichou and C. Schermann, Fifth International Symposium on the Production and Neutralization of Negative Ions and Beams, (1989), to be published.

DISCUSSION

Kleyn: I have two questions: First of all in your sort of parameter space that have you varied b and the deactivation per wall collison. Second, have you varied the atomic temperature? This morning we learned that the atomic temperature apparently in the Berkeley source is very low. I found out to my surprise and what would happen if you do that. And finally a remark, what we measure outside the source in terms of vibrational distribution might be something different than you calculated inside the source.

Skinner: Yes, we varied b some. It depends on which regime you are in. If you are in a high current regime and vary b, the relaxation won't change things very much. We've moved down a little bit; we can't make the LBL result look any better by varying b. It's not very sensitive. From the very lowest discharge from FOM that would be $b = 2$ and we can't go much further than $b = 1$. Regarding atomic temperature, LBL measured two temperature distributions: half of it was cold, and half was hot. We did modify the temperature in a sense that we took a new velocity for the mean temperature for the atomic hydrogen that would give us the mean flow rate to the wall, which you would get from their temperature distribution. And that has an effect for the LBL result. Our distribution is somewhat sensitive to the density of atomic hydrogen. If you were to change the cross sections as well because of the temperature, we would see dependence, I don't know how it would vary. We are sensitive to that for the high power cases.

Holmes: You show a proton yield of 94%. We had a problem in the early days that we couldn't get in this kind of source, much beyond 60% proton yields, the molecular ions just existed regardless of what you did. You could turn the pressure down, you could go to currents beyond a 1000 amps, 1500 amps, it made no difference, it was just saturated, which suggests that there are mechanisms in the discharge which clamp the atomic fraction to not more than a certain fraction of the total gas. You can't get to these almost perfect permanent dissociated situations.

Skinner: The only process we have that turns atoms into molecules is recombination on a wall. Are you saying there was another process?

Holmes: We had all the discharge parameters you're describing. But we could not get the molecular, fractions down. That's when we had to introduce the filter. All together, even with the filter we still didn't get much higher than 90%. I have a feeling that you may be over estimating the role of atomic hydrogen in your model, and there is another process which clamps the atomic fraction to not more than a certain value, and that would be nice to get lower still. Which you could do by enhancing the surface area of the wall, or the recombination rate, I don't think it is quite as bad as you are making out.

MODELING JAERI 1 AND 5 A TANDEM H⁻ VOLUME SOURCES*

Joel H. Fink
Negion, Inc.
Hayward, CA 94542

ABSTRACT

From approximations based on Hiskes' theories of Tandem-Volume Sources, expressions are derived for the H⁻ currents obtained from JAERI's 1 and 5 A sources, and estimates made of the temperature of the neutral gas molecules in the source.

The calculated currents, normalized with respect to their maxima at optimum source pressures, are compared with data extrapolated from JAERI's graphs, showing H⁻ ion currents vs. source pressures at several discharge currents. At pressures less than or equal to optimum, the observed and calculated currents are very close.

INTRODUCTION

The purpose of this work is to devise empirical relationships by which the performance of tandem volume H⁻ sources could be estimated under conditions not previously tested. In the following, I assume the shape of the characteristic curve of gaseous negative ion sources, i.e. the H⁻ beam current versus pressure at a specified discharge current and voltage, is the consequence of negative ion losses resulting from collisions with the background gas. Accordingly, Hiskes' tandem volume source theories[1] are used to derive an approximate relationship between the H⁻ beam current, the discharge current, and the gas pressure in the first chamber of the source. I then determine the conditions which make the H⁻ current maximum.

TANDEM VOLUME H⁻ SOURCES

The first chamber, as shown in fig. 1, houses a gas discharge that generates electrons with energies in excess of 20 eV. Collisions, between these high temperature electrons and the background gas, raise the neutral molecules to vibrationally excited states[2]. Some fraction of the newly excited molecules, however, are subsequently de-excited as the result of additional collisions[3]. The magnetic filter blocks the passage of the hot electrons, while permitting neutrals, positive ions and low

*This work was done for the Lawrence Livermore National Laboratory under U.S. Air Force Contract: ISSA-89-0039.

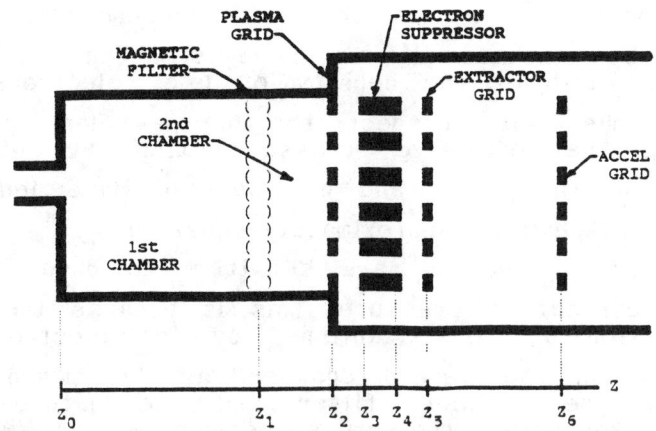

Fig. 1. Schematic of an H⁻ Tandem Volume Source

temperature electrons to escape from the discharge region and enter the second chamber. Assuming a specific source design and given electrode potentials, the density of excited molecules in the second chamber, between Z_1 and Z_2, is such that:

$$n^* \propto n_{o1} \cdot I_D , \qquad [1]$$

in which n_{o1} is the density of the gas in the first chamber, and I_D the discharge current.

In comparatively high pressure sources, the principle source of negative hydrogen ions results from dissociate attachment of low temperature (roughly 1 eV) electrons with excited molecules. At the same time some fraction of the negative ions are lost by the combination of negative and positive ions, as well as negative ions and neutral atoms which, upon the release of their extra electron, form neutral molecules.

Neglecting wall losses, the fraction of excited molecules that form negative ions that ultimately arrive at the plasma grid, can be described by:

$$F_N = \left[1 - \exp\left\{ -n_{o1} \int_{Z_1}^{Z_2} \sigma_{da} \cdot n_e / n_{o1} \cdot dz \right\} \right] \cdot \exp\left\{ -n_{o1} \int_{Z_1}^{Z_2} \sigma_1 \cdot n_+ / n_{o1} \cdot dz \right\} \qquad [2]$$

in which the first term, in large brackets, represents the fraction of the excited molecules that generates negative ions and the second term, i.e. the exponential, is the fraction of negative ions not lost to either to ion-ion neutralization or associative detachment of electrons from the combination of negative ions and neutral molecules.

In the above, σ_{da} is the effective cross section for dissociative attachment by which negative ions are formed, n_e/n_{o1} the ratio of the density of 1 eV electrons in the second chamber with respect to the gas density in the first, σ_1 the effective cross section by which the negative ions are lost, and n_+ the positive ion density in the second chamber, approximately equal to n_e.

Neither σ_{da} nor σ_1 are sensitive to changes in the discharge current or pressure. This is because the average cross section σ_{da} is established by the electron energy distribution in the second chamber, as determined by the design of the magnetic filter and the nature of the discharge. Meanwhile, the cross section σ_1, by which the negative ions are lost, depends upon the ion velocity distribution, and that is determined by the energy available when the combination of low energy electrons and excited neutrals split into negative ions and neutrals. Because the high energy electrons in the first chamber establish the degree of excitation of the molecules, they also determine the energy distribution of the newly formed negative ions.

Only a small percentage of the excited molecules actually form negative ions. Thus the fraction of excited molecules that do so, the first term in Eq. 2, is roughly proportional to n_{o1}. As a result, Eq. 2 becomes:

$$F_N \propto n_{o1} \cdot \exp\{- K_1 \cdot n_{o1}\} \qquad [3]$$

in which the exponential in Eq. 1 has been modified to include K_1, as defined by:

$$K_1 = \int_{Z_1}^{Z_2} \sigma_1 \cdot n_+/n_{o1} \cdot dz \ . \qquad [4]$$

From Eqs. 1 and 3, the negative ion density in the vicinity of the plasma grid is:

$$n_2 \propto I_D \cdot n_{o1}^2 \cdot \exp\{- K_1 \cdot n_{o1}\} \ . \qquad [5]$$

To evaluate the negative ion current which leaves the accel grid at Z_σ, it is first necessary to determine the current that passes through the plasma grid at Z_2, i.e.

$$I_2 = (n_2 \cdot e \cdot v) \cdot A_a \qquad [6]$$

in which e is the negative ion charge, v its velocity and A_a the total area of all the apertures in the plasma grid.

Beyond the accel grid at Z_6, the H^- current leaving the source equals the ion current I_2 times F_s, the fraction of negative ions not neutralized by collisions with the background gas in the Z_2-Z_6 region.

Thus, at Z_6:

$$I_N = K_N \cdot I_D \cdot n_{o1}^2 \cdot \exp\{- K_1 \cdot n_{o1}\} \cdot F_s \qquad [7]$$

in which K_N, a proportionality term which contains all the undefined variables by which the negative ions were formed, is not a constant, i.e. is not independent of the pressure or discharge voltage.

Consider the un-neutralized fraction of the negative ions traveling between Z_i and Z_{i+1}:

$$F_s \Big|_{Z_i}^{Z_{i+1}} = \exp\left\{- n_{o1} \cdot \int_{Z_i}^{Z_{i+1}} \sigma_i \cdot n_i/n_{o1}) \cdot dz \right\} \qquad [8]$$

where σ_i, the stripping cross section, is a function of the average negative ion velocity between Z_i and Z_{i+1}, and n_i/n_{o1} the ratio of the average gas density, in the region of interest, with respect to the gas density in the first chamber.

The values of $(Z_{i+1} - Z_i)$ and σ_i are the same for identical sources with the same grid voltages, while the ratio n_i/n_{o1} is a function of the grid geometry. This is because a string of grids, as shown in Fig. 1, divides the pressure across them in a manner similar to a series of resistors which act as a voltage divider. For this to be accurate, however, the diameter of each grid aperture must be small enough to make the line density of the plume of gas, streaming out of each aperture, negligible in contrast to that of the background gas between the grids.

As a consequence, the integral in Eq. 8, as a function of the grid geometry and bias, is constant in a given source and it can be rewritten as:

$$F_s \Big|_{Z_i}^{Z_{i+1}} = \exp\{- K_i \cdot n_{o1}\} \quad . \qquad [9]$$

Thus for the extractor and accelerator regions, extending from Z_2 to Z_6:

$$F_s = F_s \Big|_{Z_2}^{Z_6} = \exp\{- (K_2 + K_3 + K_4 + K_5) \cdot n_{o1}\} , \qquad [10]$$

and by adding the constants in the exponents of Eqs. 7 and 10 and letting:

$$K_E = K_1 + K_2 + K_3 + K_4 + K_5 , \qquad [11]$$

those equations can be combined to give:

$$I_N = K_N \cdot I_D \cdot n_{o1}^2 \cdot \exp\{- K_E \cdot n_{o1}\} . \qquad [12]$$

MAXIMUM H$^-$ CURRENT

A source, operating with a fixed discharge current I_D and voltage V_D, delivers a maximum ion current I_M when the first chamber is at its optimum gas density n_M. The optimum density is found by setting the derivative of I_N equal to zero. As a result, from Eq. 12:

$$n_M = 2/K_E , \qquad [13]$$

whereby the maximum current becomes:

$$I_M = 4 \cdot \exp\{- 2\} \cdot \left(K_N/K_E^2\right) \cdot I_D \qquad [14]$$

and, from Eqs. 12 and 14, the ratio of beam current at a density n_{o1} to its maximum at the optimum density n_M is:

$$I_N/I_M = \left(n_{o1}/n_M\right)^2 \cdot \exp\{2 \cdot \left(1 - n_{o1}/n_M\right)\} . \qquad [15]$$

Although K_N and K_E have been cancelled out of Eq. 15, the result is accurate only to the extent that these terms do not vary with the gas density, n_{o1}.

According to Eq. 13, the optimum gas density is independent of the discharge current, while weakly dependent on the discharge voltage, via K_1 of Eq. 4. The ratio I_M/I_D in Eq. 14, meanwhile, is directly proportional to K_N and, because K_N is not a constant, the ratio is not truly independent of the discharge voltage or pressure.

GAS FLOW AND TEMPERATURE

The pressure $\mathbb{P}_{o1}$ (Pa), cited in the JAERI[1] papers, corresponds to the fill pressure in the first chamber prior to the ignition of the discharge. The gas flowing out of a high pressure tank, via a needle valve, sets up an almost constant flow that is independent of the operating conditions of the source. But once the discharge starts, the electrodes, continuously bombarded by charged particles, assume a temperature that balances the power density of the bombardment with the effectiveness of the electrode cooling.

The pressure in these sources, however, is such that the mean-free-path of the neutral molecules is larger than, or at least of the same order as the average inter-electrode spacing. As a consequence, the molecules are heated by their frequent collisions with the hot electrodes, causing the gas temperature to approach that of the electrodes and the temperature increase to be some function of the power of the discharge, $I_D \cdot V_D$.

Once the needle valve is set, the product of the molecular gas density and velocity in the first chamber becomes a constant, as a consequence of the sustained gas flow mentioned above. Thus, for any discharge voltage or current:

$$\left(\frac{\mathbb{P}_{o1}}{T_c^{1/2}} \right) = \left(\frac{\mathbb{P}_{o1}}{T_{o1}^{1/2}} \right) \qquad [16]$$

in which the density of the gas molecules is proportional to the pressure divided by their temperature, i.e.

$$n_{o1} = P_{o1}/k \cdot T_{o1} \ , \qquad [17]$$

and the velocity of the gas molecules is proportional to the square root of their temperature.

Introducing Eq. 17 into 16 results in:

$$\mathbb{P}_{o1} = n_{o1} \cdot k \cdot (T_c \cdot T_{o1})^{1/2} \ , \qquad [18]$$

while from Eq. 18, assuming the gas temperature T_{O1} does not change significantly with the pressure, the ratio of the fill pressure P_{O1}, with respect to the optimum fill pressure P_M is approximately:

$$P_{O1}/P_M = n_{O1}/n_M .\qquad [19]$$

Thus, from Eqs. 15 and 19, the ratio of the negative ion current, at any pressure, with respect to the maximum ion current at the same discharge voltage and current, is:

$$I_N/I_M = \left[P_{O1}/P_M\right]^2 \cdot \exp\left\{2 \cdot \left(1 - P_{O1}/P_M\right)\right\} . \qquad [20]$$

Notice, that Eq. 20 was derived on the assumption that, at any given discharge voltage and current, neither K_N, K_E nor T_{O1} would vary significantly with changes in the gas density n_{O1}.

A rough estimate of the temperature of the molecules in the JAERI sources can be gotten from measurements made in other types of discharges operated at similar pressures. For example, a 15 kW discharge in a Penning source of negative ions[1], with a volume of just a few cm³, has a gas temperature that is less than 1000 K; while a 3.5 kW discharge in a chamber of roughly comparable size, using a coaxial Lanthanum Hexaboride cathode[2], has a temperature of 450 K. Thus it seems reasonable that a 70 kW discharge, with I_D = 1000 A and V_D = 70 volts, in a 5 A JAERI source of disproportionately large volume and electrode area, should have a gas temperature not greater than 600 K.

EXTRAPOLATED DATA

H⁻ beam currents, obtained from the JAERI 1 A tandem volume source[3] at various discharge currents, are shown in Fig 2 as functions of the first chamber pressure P_{O1} prior to starting the discharge. The ratios of the observed ion currents with respect to their peak values:

$$R_O = I_N/I_M , \qquad [21]$$

as extrapolated from Fig. 2, are shown in Fig. 3 with respect to the initial fill pressure for discharge currents of 140, 300 and 500 A. For comparison, the figures also show the ratios:

$$R_C = I_N/I_M , \qquad [22]$$

as calculated from Eq. 20. The ratios of the ratios R_O/R_C are also given to illustrate the quality of the match between the observed and calculated data.

H^- beam currents vs. P_{O1} obtained by JAERI from their 5 A source[1], are shown in Fig. 4 and values of R_O, R_C, and R_O/R_C, are plotted in Fig. 5 for discharge currents of 200, 600, and 1000 A.

In all of these curves, at pressures equal to or less than the optimum, the agreement between the observations and calculations is very good. But at higher pressures, the observed currents are greater than the calculated values, while at even higher pressures and discharge currents the difference is markedly reduced.

Extrapolated values of V_D, I_D, P_M and I_M, taken from Figs. 2 and 4, are shown in Table I. The ratios of I_M/I_D for each of the sources tend to go down as the discharge current and pressure go up. Of course the limited data and inaccuracy of the extrapolations make it impossible to tell if these tendencies are real. But as briefly noted below Eq. 7, the ratio is proportional to K_N, and K_N is a function of the pressure.

The ratio P_M^2/I_D, also shown in the table, appears to be a constant at a given discharge voltage in both the 1 A and 5 A sources.

Table I
Maximum H^- currents and Optimum Pressures in
JAERI 1 and 5 A Tandem-Volume Sources

V_D	I_D	P_M	P_M^2/I_D	I_M	I_M/I_D
(V)	(A)	(Pa)	(Pa²/A)	(A)	(A/A)
1 A Source					
70	140	0.58	.0025	0.21	.0015
70	300	0.84	.0024	0.39	.0013
70	500	1.06	.0025	0.70	.0014
5 A Source					
70	200	0.88	.0039	0.94	.0047
70	400	1.25	.0039	1.46	.0037
70	600	1.44	.0035	1.95	.0033
70	800	1.63	.0033	2.37	.0030
70	1000	1.92	.0037	2.76	.0028

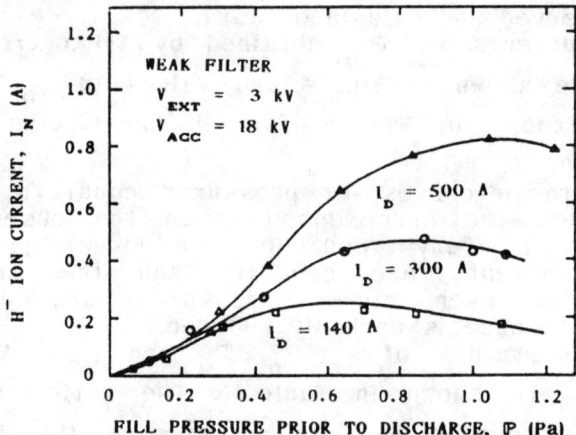

Fig. 2. H⁻ current vs. fill pressure for 140, 300 and 500 A discharges in a 1 A JAERI source.

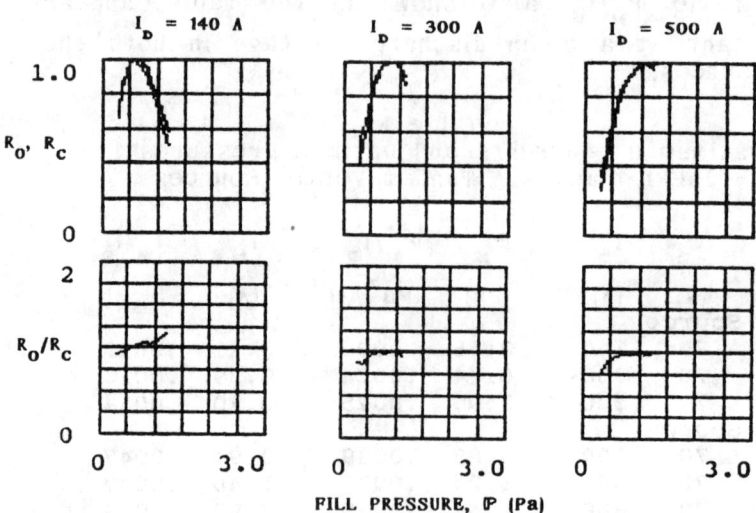

Fig. 3. The ratio R_0 of H⁻ currents with respect to their maximum at discharge currents of 140, 300 and 500 A as shown in Fig. 2, the ratio R_c as calculated from Eq. 20, and the ratio of the ratios R_0/R_c as a function of the fill pressure P.

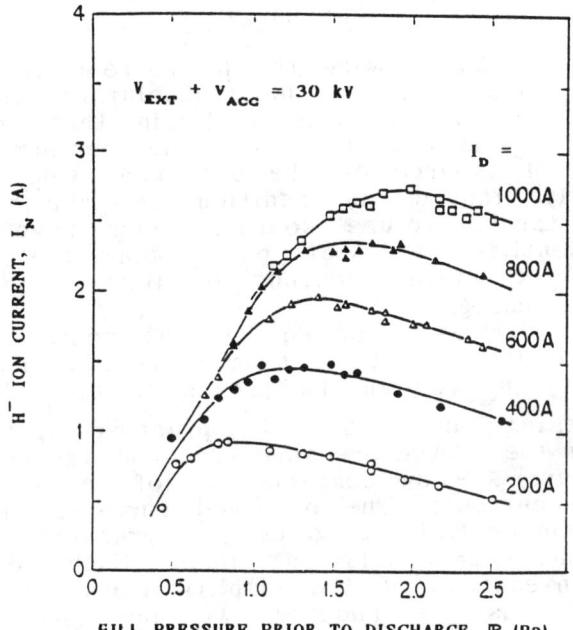

Fig. 4. H⁻ current vs. fill pressure for 200, 400, 600, 800 and 1000 A discharges in a 5 A JAERI source.

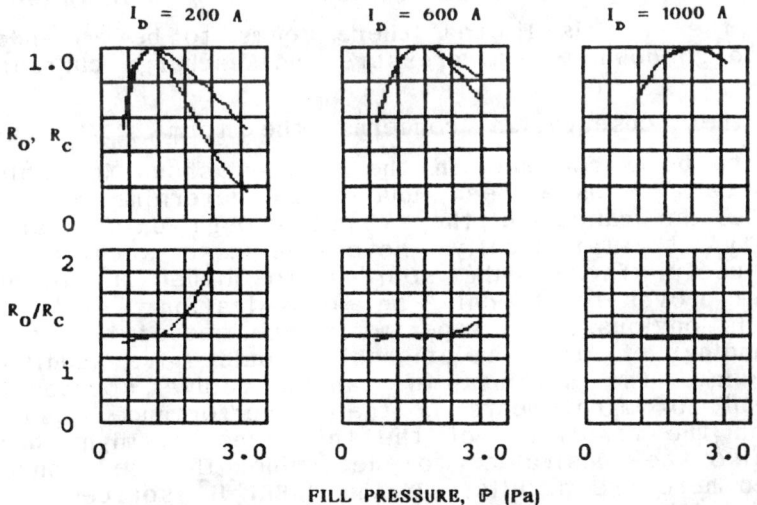

Fig. 5. The ratio R_o of H⁻ currents with respect to their maximum at discharge currents of 200, 600 and 1000 A as shown in Fig. 4, the ratio R_c as calculated from Eq. 20, and the ratio of the ratios R_o/R_c as a function of the fill pressure P.

CONCLUSION

Although Eq. 20, showing the H^- current as a function of the fill pressure, is not absolutely correct, it fulfills the objective of this work, in that it provides an empirical relationship by which the performance of a tandem volume H^- source can be estimated under conditions not previously tested. In addition it illustrates some features of tandem volume sources which have not been previously identified and may be common to all gaseous sources of ions whose output is limited by collisions with the background gas.

From the derivative of Eq. 12, the source density is found at which the output H^- beam current is maximum. The optimum density, n_M per Eq. 13, proves to be a function of the grid design and, for all practical purposes, is independent of the source pressure and discharge current.

Figs 3 and 5 show comparisons of the observed and calculated H^- currents. The observed currents tend to be larger than the calculated values at pressures in excess of the optimum, particularly at lower discharge currents. This is a consequence of the simplifications used in the derivation of Eqs. 15 and 19. At any given discharge current, both K_N and the gas temperature T_{O1} were assumed to be independent of the pressure.

A measure of the variability of K_N as a function of the discharge current can be obtained, in accordance with Eq. 14, from the observed values of I_M/I_D in Table I. Although the data is limited, there seems to be a tendency for K_N to go down as the pressure and discharge current go up.

Another observation concerns the term P_M^2/I_D, which appears to be a constant in the sources shown in Table I. Unfortunately, I have been unable to determine why this should be the case and there isn't enough data available to verify it statistically. Never-the-less, such a term can be useful for scaling sources to higher H^- outputs, even if it proves to be only an approximation.

It is obvious, more experiments are needed because an understanding of this relationship would give additional insights into the workings of tandem volume sources and might lead to improvements in their performance. But as useful as the resolution of the this question might be, it would also be desirable to determine if the concepts presented here are peculiar to the JAERI H^- sources, or of general significance in all gaseous negative ion sources.

REFERENCES

[1] J.R. Hiskes, A.M. Karo, P.A. Willmann, J. Appl. Phys. 58 (5), p1759, (1985).

[2] J.R. Hiskes, A.F. Lietzke, "Electronic Excitation and Ion Source Optimization," IAEA Technical Committee Meeting on Negative Ion beam Heating, Culham Laboratory, U.K., July 15-17, 1987.

[3] J.R. Hiskes, A.F. Lietzke, C. Hauk, 4th Int'l Symp. on the Production and Neutralization of Negative Ions and Beams, Brookhaven, p.231, 1986.

[4] T. Inoue, of Japan Atomic Energy research Institute, Naka-machi, Naka-gun, Ibaraki-ken, 311-02, Japan, Personal Communication (June, 1989).

[5] H. V. Smith, Jr., et. al., International Ion Source Conference, Berkeley, CA, July 10-14, 1989, Los Alamos Report LA-UR-89-1876, Los Alamos National Laboratory, Los Alamos, New Mexico, 87545.

[6] G.C. Stutzin, et. al., Chem. Phys. Letters, 155, 475, 1989.

[7] Y. Okumura, H. Horiike, T. Inoue, T. Kurashima, S. Matsuda, Y. Ohara, S. Tanaka, Fourth International Symposium on the Production and neutralization of Negative Ions and Beams, p309, Brookhaven, NY, AIP Conference Proceedings No. 158, New York, 1987.

[8] Y. Okumura, "Negative Ion Beam Development at JAERI," Informal Meeting at Lawrence Berkeley Laboratory, Feb. 22, 1989.

DISCUSSION

Bacal: Did you understand that these currents are measured calorimetrically?

Fink: Yes, I took them as nomimal currents.

Bacal: Yes, but they are measured calorimetrically so even the stripped ions are measured. Is that correct Dr. Okumura?

Okumura: We measured the current calorimetrically, so the neutral particles which are stripped also hit the calorimeter but these neutral particles have a low energy, because almost all stripping occurs in the extractor. But at that region particle energy is very low. So I think that the calorimetric measurement is almost equal to the electrical current at the exit of the source.

Fink: You still have to explain the fit. I mean even if we put any condition in, there is some physics buried in here. Although I say the statistics isn't right but it's unbelievable, I am almost embarrased to present curves like this. I just look at them and scratch my head: I did cancel the constant, I did ratio, and I did bury what might be different by means of ratio, but I'm still left with scaling laws.

GENERALIZED MULTIBODY COMPUTER SIMULATIONS OF PLASMA-WALL DESORPTION AND ENERGY-TRANSFER PROCESSES

A. M. Karo, J. R. Hiskes, and T. M. DeBoni
Lawrence Livermore National Laboratory
Livermore, California 94550
and
J. R. Hardy
University of Nebraska
Lincoln, Nebraska 68588

ABSTRACT

We present calculations that describe the kinematic factors influencing the recombinant desorption of hydrogen from metal surfaces and a preliminary study of the sputtering and possible recombination of subsurface hydrogen following surface impact of a flux of ions. A new multi-body computer molecular dynamics method used in these studies is discussed. In the first instance we are able to describe both a scattering channel in which backscattering is the dominant result and also a reactive channel leading to molecular formation with the molecules leaving the surface in vibrationally-excited states. The importance of these processes to negative ion formation is discussed.

INTRODUCTION

We have discussed elsewhere[1] in detail the techniques and codes that have been developed for simulating atomic and molecular processes that can occur in the region forming the boundary between a fusion plasma and the surface of the containment vessel. Recently our work has been focussed on carrying out simulations that attempt to describe those factors that influence the recombinant desorption of hydrogen molecules from metal surfaces containing a partial monolayer of hydrogen atoms, when a flux of energetic (1 to 20 eV) protons is directed toward and impacts the surface. A preliminary study has also been initiated that describes the sputtering and possible recombination of hydrogen atoms placed in equilibrium sites just below the surface of a lattice of metal atoms when a flux of ions hits the surface. Such simulations can be expected to shed light on two additional mechanisms that have been proposed for negative ion

---

*Work performed under the auspices of the U.S. Department of Energy by the Lawrence Livermore National Laboratory under contract No. W-7405-ENG-48 and Air Force Office of Scientific Research contract No. AFOSR-ISSA-89-0039.

production: the direct generation of H^- ions by electron capture on the H atoms ejected or back-scattered from the surface, and the formation of H^- ions by resonant electron capture from the surface to the desorbing $H_2(v')$ molecules. The energetics of this latter process has been discussed in some detail in another communication.[2]

PLASMA-WALL DESORPTION AND ATOMIC/MOLECULAR SPUTTERING

In previous work[1,3] we examined both the de-excitation and the re-excitation of vibrationally-excited molecules undergoing wall collisions, from thermal (500K) velocities up to translational velocities in the range of 50-100 eV. The results showed that wall recombination of H_2^+, H_3^+ to form $H_2(v'>6)$ could make an important contribution to the total vibrationally excited population and thus to the negative ion yield, comparable to that from the fast-electron process.

We note that although we are dealing with light molecular systems, several factors make it possible to justify the simplifying approximations associated with a classical Newtonian description of the equations of motion in treating the interparticle forces. Primarily this simplification is due to the fact that we are dealing with the combination of highly-vibrationally-excited states and/or large translational kinetic energies. In addition, the nature of the problem has allowed one the use of the sum-of-pair-potentials approximation wherein the interaction between pairs of atoms are central and are not influenced, to first order, by changes in the surrounding force field. Additionally, parameters defining the strength and range of the interaction can be varied over a wide range to ensure that the qualitative statistical results obtained are not sensitive to these factors. Subsequent trajectory studies in which these restrictions were lifted have verified that our original assumptions are a reasonable approximation.

However, in treating recombinant desorption and sputtering, it is necessary to develop a more general treatment of the forces acting on the particles in order that the simulations be carried out in a realistic fashion. Therefore, multi-body force terms, derived from an approximation to relevant features of the multi-dimensional potential energy surface, are incorporated into the general molecular dynamics codes. The development of multi-body molecular dynamics algorithms thus allows us to include more realistic simulations of bond breaking and bond formation, taking us beyond the sum-of-pair-potentials approximation.

Formally, the potential between any two particles in an ensemble will be influenced by the positions of all other particles. That is, the force acting on each particle will result from spatial derivatives of the general n-body potential energy surface. Since the effective range of the forces depends on the nature of the atomic particles, interaction distances can be truncated so that the residual error remains insignificant for each case. Although we must allow for modification of the force between any pair of particles due to the ever-changing environment, to carry out our simulations in a tractable manner we have restricted ourselves to an effective two-body representation of the interactions. Thus, n-body effects are incorporated into a time-dependent effective potential acting between each pair of particles. Inclusion in the dynamics of the true intrinsic n-body forces by fitting a numerical n-dimensional potential energy surface via algebraic expressions offers an alternative route that is possible in principle, but is in fact precluded due to the enormous expense of generating the n-dimensional surface for an assembly of many interacting atoms.

Methods for incorporating important features of surfaces that describe reactive channels have received increased attention in recent years as computational power has increased. Developments have proceeded from "simple" trajectory studies of non-reactive and reactive atomic and molecular scattering (e.g., H + HH; H + Li_2, H + LiH),[4,5] where "brute-force" n-body potential energy surfaces were fitted analytically, to simulations where angle-dependent (non-central) terms or terms simulating "chemical" processes (i.e., endothermic bond breaking and exothermic bond forming events) are explicitly included in the forces (e.g. Tersoff and embedded atom potentials).[6,7] These latter approaches are conceptually based on time-dependent switching of the forces describing the interactions as the electron-nuclear environment changes. A simple example would be the co-linear reaction A + BC = AB + C wherein if the strength of the bond AB > BC, we would have an exothermic event. The influence of nearby atoms/molecules can affect this reaction and must be taken into account, e.g., A + BC--D, where the species D overlaps the wave function describing the molecule BC.

In the work we are presenting here, we have introduced directly into the Newtonian equations of motion algorithms that permit controlled switching to and from different portions of the underlying potential energy surface. This is the surface incorporating all the many-body interactions that would describe in complete detail our n-body ensemble. Thus we control the forces directly as they relate both to the nuclear configuration and to the underlying chemistry that is associated with changes in the electronic coordinates (i.e., we take into account appropriate reactive or non-reactive channels or pathways from "reactants" to "products."

Figures 1 and 2 show the two configurations we have used in the studies reported here. Figure 1 shows the placement of hydrogen atoms in equilibrium positions on the surface. The surface and bulk atoms that are explicitly included in the n-body equations of motion are shown within the dashed region. Additional bulk atoms outside the dashed region interact with the rest of the

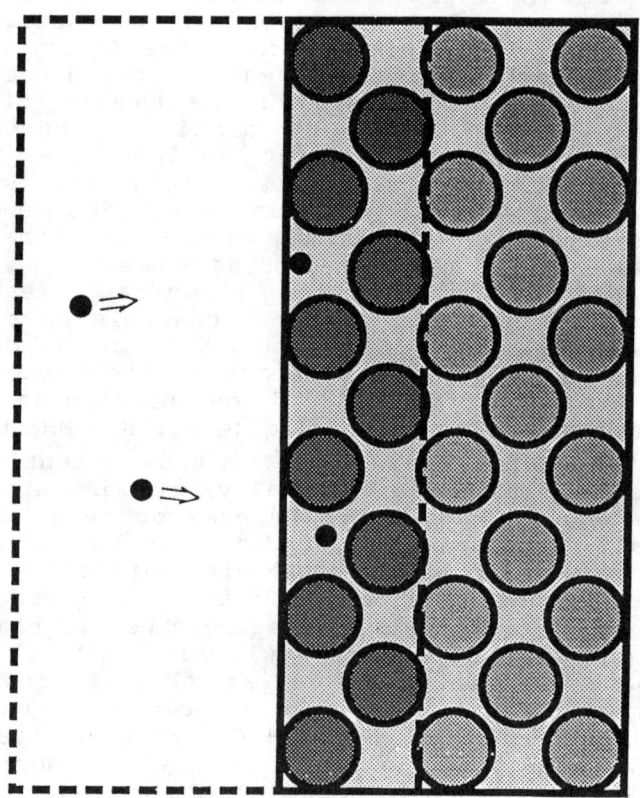

Figure 1. The schematic representation of the assembly of particles used to study recombinant desorption of hydrogen from a metal surface. Multi-body forces are included in the dynamics of the cluster within the dashed region. The remainder of the bulk atoms interact via two-body central potentials.

ensemble via central two-body forces. In Fig. 2 we show
the initial configuration for the assembly of atoms used

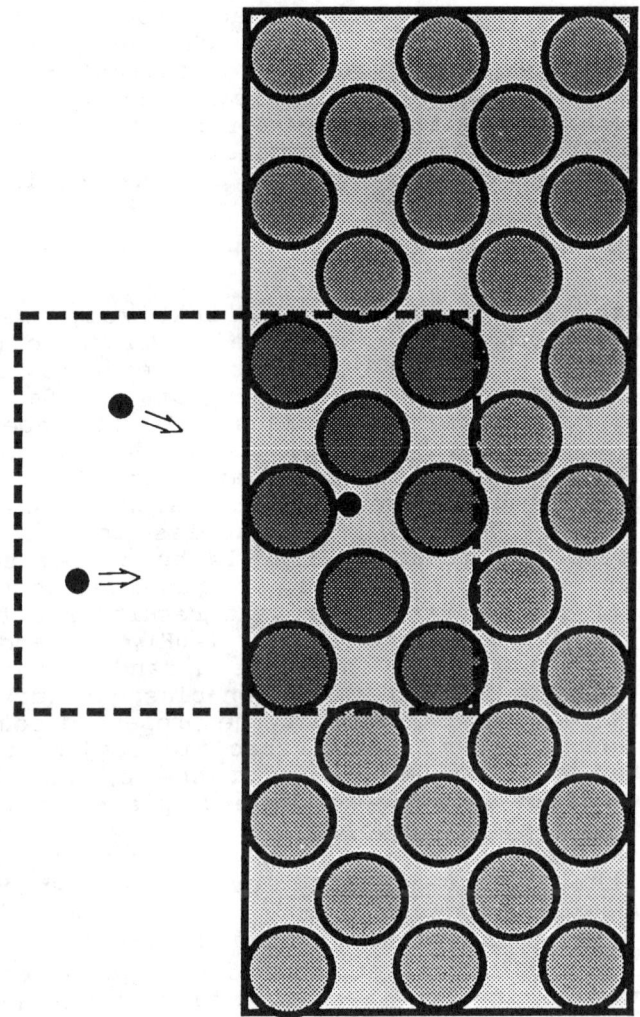

Figure 2. The schematic representation of the assembly of particles used to study sputtering and possible recombination of hydrogen atoms placed just below the surface. Multi-body forces are included in the dynamics of the cluster within the dashed region. The remainder of the bulk atoms interact via two-body central potentials.

to study sputtering and possible recombination of hydrogen atoms placed in equilibrium positions just below the surface. As described above, those atoms explicity included in the n-body equations of motion are shown inside the dashed region, with the additional bulk atoms outside this region interacting through central two-body potentials. Therefore, we could characterize our simulations as including both chemical dynamics when the H atoms interact with the atoms forming the multi-body cluster, and conventional molecular dynamics in the treatment of the multi-body cluster and the remainder of the wall.

RESULTS

Recombinant Desorption. Qualitatively we can describe the results of a large number of trajectories as shown in Fig. 3. For normal incidence at energies around 1 eV the flux directed in the vicinity of a surface atom is reflected back from the surface. As we enter the channel shown by cross-hatching, the trajectory of the incoming atom is deflected in a way that alows for a closer interaction with the surface H atom as shown in trajectory (b) of Fig. 3; trajectories in which the impact point of the incoming H would be closer to the surface metal yield the sort of results shown by trajectory (a). Although our present results are for an initially quiescent situation (i.e., T=0K), the addition of a wall temperature will introduce a random character that will be important in forming non-biased statistical averages. However, for purposes of the present discussion this is not necessary since any specific trajectory can be correlated to some extent with one of our basic trajectories, after allowing for some displacement of the wall atoms. Statistical studies are underway for several options that describe qualitative differences in the topology of the potential energy surface from which the switching dynamics are derived.

For the 1 eV flux, we have found H_2 molecules to be readily formed within the recombinant desorption channel and vibrational excitation in the vicinity of the surface to be the dominant result. Obviously many more trajectories must be run to provide a statistical basis for the formation of $H_2(v')$ and for the relative populations of the different vibrational states. As indicated above this work is in progress. Additionally, we are examining incoming fluxes at other energies, principally at 4, 10, and 20 eV, energies corresponding to those we have selected in our studies on $H_2(v')$ de-excitation.[1,8] It does not appear necessary at this time to consider fluxes at other than at the normal angle of incidence, although computationally this would not be difficult.

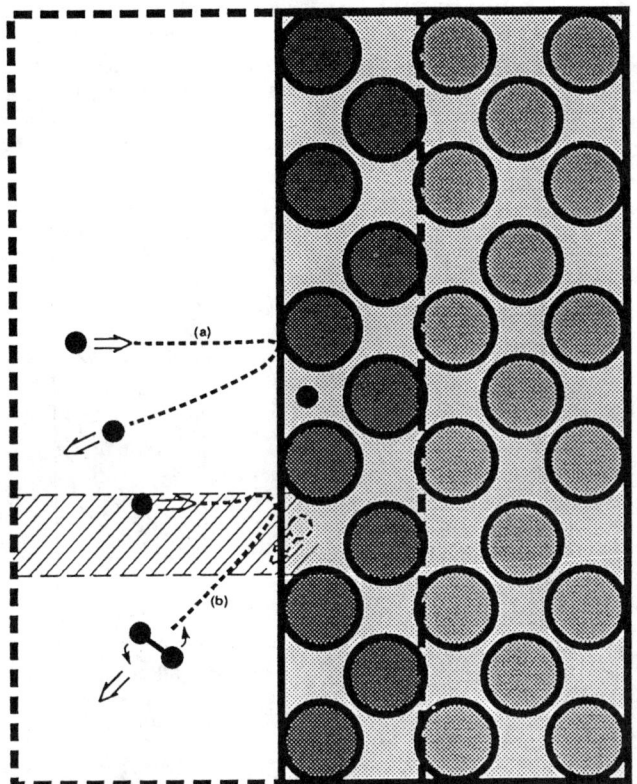

Figure 3. The schematic representation of the assembly described in Figure 1. Trajectory (a) shows the path of an incoming atom that is backscattered by one of the heavy metal surface atoms. The reactive channel is shown by cross-hatching. Trajectory (b) shows the trajectory of an incoming atom within the reactive channel. It is deflected in a manner that allows for the formation of a vibrationally excited H_2 molecule. The original position of the H atom on the wall is depicted by a dashed open circle.

In Fig. 4 we show some preliminary results for the $H_2(v')$ molecules that are formed near the surface, as a function of the surface impact parameter b, (b = [y(H) - y(impact)]). In most instances, because the small translational energy (1 eV) of the incoming H atom is partially equipartitioned among the vibrational and rotational degrees of freedom, the molecules formed remain near the surface and are eventually recaptured. As the incoming energy is increased, the molecules will have enough translational energy to escape.

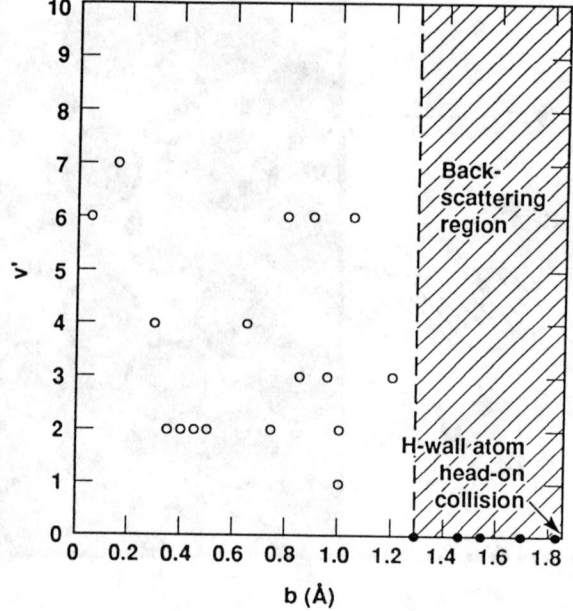

Figure 4. Vibrational excitation v' for molecules leaving the surface shown in Fig. 1 following impact from 1 eV hydrogen particles. Open circles denote the values of v' and closed circles where scattering occurred without molecule formation.

Overall, we can summarize our results as follows:

1) Recombinant desorption is a function of the surface impact parameter and is most favorable when it is within a certain critical range, whereas an impact in the vicinity of a surface atom leads to backscattering and will not result in a recombinant process leading to desorption.

2) There exists an incoming reactive channel wherein the hydrogen atoms can interact, leading to an enhanced probability of desorption for a molecule which, in addition, can be in a vibrationally-excited state. Motion of the surface atoms due to temperature will blur the

boundaries of the reactive channel, but we would expect the qualitative picture to hold.

Sputtering. It is known that hydrogen atoms lie in equilibrium configurations below the surface of the bulk material.[9] We have begun to extend our multi-body codes toward an investigation of the sputtering of this occluded hydrogen (cf. Fig. 2), a process which could affect the concentration of atomic species in the plasma as well as provide an additional mechanism for the creation of excited molecular hydrogen. Although certain incoming trajectories can result in an outgoing molecule and a distribution of outgoing molecular states, it will require a large number of trajectories to provide meaningful statistical distributions. Dependence on the incoming flux energy is expected to be important, and this study is in progress.

SUMMARY

A realistic simulation of surface processes such as recombinant desorption and sputtering of partially-occluded atoms and/or molecules requires effective potentials that go beyond the conventional sum-of-pair-potentials approximation. We have applied a new multi-body computer molecular dynamics method that incorporates the essential features of a reactive potential energy surface to investigate the particle dynamics underlying these phenomena. It is seen that there exists a scattering channel in which the incoming flux is merely backscattered to the plasma and a reactive channel leading to molecular formation wherein the molecules leave the surface in vibrationally-excited states. The importance of this to negative ion formation has already been emphasized. Further work to provide a solid statistical foundation is in progress and will examine the influence of the flux energy on the results, as well as a possible dependence of the results with respect to the direction of the flux relative to the surface. The possibility of the importance of processes involving subsurface hydrogen interacting with an incoming proton flux is discussed, and this is currently being investigated.

References

1. A. M. Karo, J. R. Hiskes and R. J. Hardy, J. Vac. Sci. Technol. **A3(3)**, 1222 (1985).

2. J. R Hiskes and A. M. Karo, *Proceedings of the NATO Advanced Study Institute on Non-Equilibrium Processes in Partially Ionized Gases*, Aquafredda di Maratea, Italy, June 4-17, 1989.

3. A. M. Karo, J. R. Hiskes, and T. M. DeBoni, *Proceedings of the Fourth International Symposium on the Production and Neutralization of Negative Ions and Beams*, edited by J. G. Alessi (AIP Conference Proceedings No. 158, American Institute of Physics, New York, 1987), p. 97.

4. D. G. Truhlar and C. J. Horowitz, J. Chem. Phys. **68**, 2466 (1978); **71**, 1514 (1979); P. Siegbahn and B. Liu, J. Chem. Phys. **68**, 2457 (1978). (H + HH).

5. W. B. England et al., J. Phys. Chem. **81**, 772 (1977). (H + Li_2; Li + LiH).

6. J. Tersoff, Phys. Rev. Lett. **56**, 632 (1986); Phys. Rev. **B37**, 6991 (1988).

7. M. S. Daw and M. I. Baskes, Phys. Rev. Lett. **50**, 1285 (1983); Phys. Rev. **B29**, 6443 (1984).

8. J. R. Hiskes and A. M. Karo, J. Appl. Phys. (submitted for publication).

9. J. R. Hiskes and A. M. Karo, *Proceedings of the Symposium on the Production and Neutralization of Negative Hydrogen ions and Beams*, edited by K. Prelec (Brookhaven National Laboratory Report BNL 50727, January 1978), p. 42.

DISCUSSION

Kleyn: What do yo think is the barrier against dissociation when you bring the molecule up very close to the surface.

Karo: The barrier for exit molecule from the surface would

Kleyn: I mean along the vibrational degree of freedom the molecule . . .

Karo: If it is along the vibrational degree of freedom, then we would actually introduce a potential barrier which would be taken from best guess or best approximation. So if it wants to be 1 volt potential we could put that barrier in. If indeed we want to use something lower 0.7, 0.5 that would go in. That comes from the Abnecio calculation so we would take those results and parametrize them into our potential.

TRANSPORT PROCESSES THROUGH MAGNETIC FILTER IN NEGATIVE ION SOURCE

M. Ogasawara, T. Yamakawa, F. Sato
Faculty of Science and Technology
KEIO University, Yokohama 223, JAPAN

Y. Okumura
Japan Atomic Energy Research Institute
Nakamachi, Nakagun, Ibaraki 311-02, JAPAN

ABSTRACT

Decrease of electron number density and temperature through the magnetic filter is invetigated theoretically. Expression of electron flux and heat flux that include interference of two irreversible processes, i.e., diffusion and thermal conduction, are employed. Results are in near agreement with experimental ones. Interference is shown to be as important as or more important than the diffusion and thermal conduction.

INTRODUCTION

Recent negative ion source has magneic filter within the source chamber. The magnetic filter impedes the flow of energetic electrons from the plasma source (first) chamber to the extraction (second) chamber. Then the temperature is also reduced in the second chamber as compared with that in the first chamber. Since the filter region has a finite width, there exist gradients in electron number density and temperature. Holmes[1] measured the electron number density and temperature in each chamber and showed that the measured values agree well with the results of theoretical model. However, in Holmes' theory interference between diffusion and thermal conduction has not been considered. Then in this paper we will derive the electron number density and temperature based on the expressions of the electron flux and heat flux that include the effect of interference of the two irreversible processes.

BASIC EQUATION

In a filter region with a strong magnetic field such as $(\omega\tau)^2 \gg 1$, ω and τ being electron cyclotro frequency and mean collision time of electron with ion, electron flux Γ and heat flux q are generally expressed in terms of gradients of number density dn/dx, and of temperature dT/dx and the electric field E as[2]

$$\Gamma = -\frac{\sigma}{e}E - D\frac{dn}{dx} + \frac{1}{2}D_T\frac{dT}{dx}, \tag{1}$$

$$q = -\mu E - TD\frac{dn}{dx} - 1.92TD_T\frac{dT}{dx} \tag{2}$$

The coefficients are[2] $\sigma = ne^2A$, $\left[A = (\tau/m)(\omega\tau)^{-2}\right]$ the electrical conductivity, $D = TA$ the diffusion coefficient, $D_T = nTA$ the thermal diffusion coefficient and $\mu = neTA$. The third term on the right of (1) and the second term of (2) express the interference effect. Onsager's reciprocity relation is satisfied among the coefficients of (1) and (2).

Since we are interested in the stationary state, Γ and q are taken to be constant in space and time. Giving Γ and q, we will solve (1) and (2) to obtain $n(x)$ and $T(x)$ under suitable boundary condition.

BOUNDARY CONDITION

Region of magnetic filter is bounded from both sides by two chambers. In each chamber, electron number density and temperature are constant in space and time. They depend on the filter strength and have higher values in the first chamber compared with those in the second chamber.

RELATION BETWEEN n AND T

We will solve $n(x)$ and $T(x)$ from (1) and (2). After solving dn/dx and dT/dx from the coupled equations (1) and (2), we have

$$\frac{d}{dx}\ln n = \frac{1}{2}\frac{d}{dx}\ln T + 2.42\frac{\left(1+\frac{\sigma E}{e\Gamma}\right)\Gamma}{q-\Gamma T}\frac{dT}{dx} \tag{3}$$

As will be shown later, we can neglect the effect of electric field, i.e., $\sigma E/e\Gamma \ll 1$. Integration of (3) gives

$$\ln \frac{n_1}{n_2} = \frac{1}{2}\ln \frac{T_1}{T_2} - 2.42 \ln \left(1 - \frac{1}{\beta-1}\frac{T_1-T_2}{T_2}\right), \qquad (4)$$

where subscripts correspond to the chamber 1 and 2, and as was used by Holmes we have put

$$q = \beta T_2 \Gamma. \qquad (5)$$

This comes from the energy balance in the second chamber. Incoming heat flux is lost by convection by plasma electron of which outgoing flux is Γ. Then outgoing heat flux can be expressed as $\beta T_2 \Gamma$, where βT_2 is a transported energy.

Figure 1 shows the relation (4) together with the experimental results by Holmes[1] and JAERI, where β is taken as a parameter. Holmes' results are on the line of $\beta = 9 \sim 10$ and JAERI's data are along a curve of $\beta \simeq 7$ in Fig.1. With use of the expressions of Γ and q that do not include the interference terms, Holmes[1] treated the same problem. His data are on the line of $\beta = 2.5$ in his formulation.

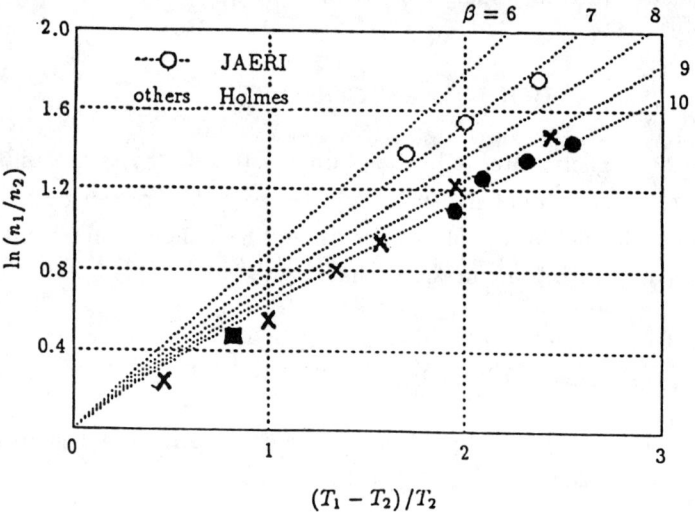

Fig.1. Relation between $\ln(n_1/n_2)$ and the temperature ratio $(T_1 - T_2)/T_2$. Parameter β is the coefficient of transported energy.

MAGNETIC FLUX DEPENDENCE OF T AND n

From (3) we have a relation between n and T

$$\frac{n}{n_2} = \left(\frac{T}{T_2}\right)^{\frac{1}{2}} \left[\frac{(q - T_2\Gamma)}{(q - T\Gamma)}\right]^{2.42} \quad (6)$$

With use of this expression and (1) and (2), we can eliminate n and obtain the differential equation for $T(x)$. If we take the magnetic field as $B(x) = B_0 exp(-x^2/a^2)$, temperature ratio $\tilde{T}_1 = T_1/T_2$ is determined by the following equation as a function of magnetic flux

$$\tilde{T}_1^{\frac{3}{2}} \int_1^{\tilde{T}_1} \frac{\tilde{T}^{\frac{1}{2}} d\tilde{T}}{(\beta - \tilde{T})^{5.84}} = \frac{1}{2.42\gamma} \sqrt{\frac{\pi}{2}} \frac{\tilde{T}_1^{\frac{3}{2}}}{(\beta - 1)^{4.84}} \frac{m}{M} \frac{(\omega_0 a)^2}{\Gamma a} \quad , \quad (7)$$

where $\gamma = \nu / \left(nT^{-3/2}\right)$, ν is the electron-ion collision frequency, $\omega_0 = eB_0/mc$ and a is the width of the region of magnetic filter. Although experimental results[1] show that T_1 increases slightly with the magnetic flux, we will neglect this small increase and take $T_1 = 3eV$. For Γ we use the measured electron current $j_e = -e\Gamma$ [1], which depends strongly on the magnetic flux.

After solving T_1/T_2 from (7) as a function of the magnetic flux, n_1/n_2 is obtained from (6). Results are given in Figs.2 and 3. Both experimental data

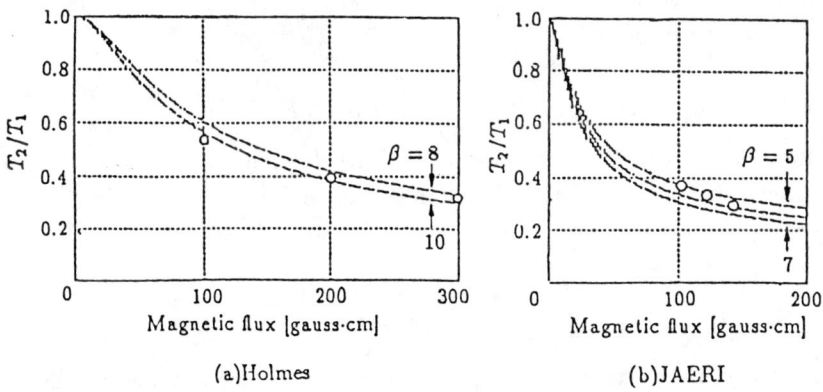

(a)Holmes (b)JAERI

Fig.2. Dependence of the temperature ratio T_2/T_1 on the magnetic flux

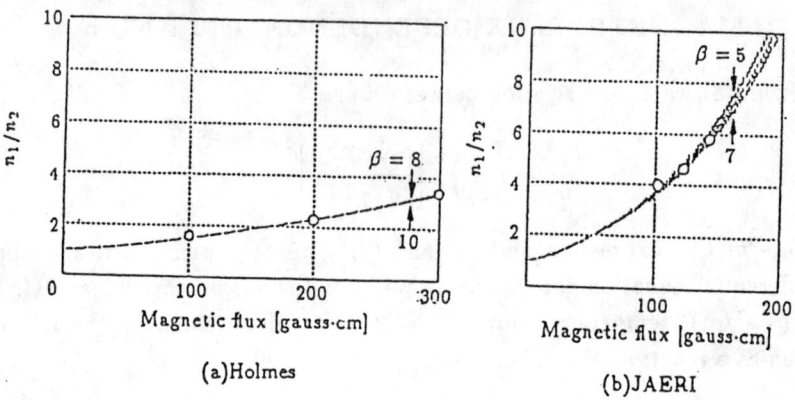

Fig.3. Dependence of the ratio of number density n_2/n_1 on the magnetic flux

nearly agree with the semi-theoretical curves. Here the word semi comes from the fact that we have used experimentally observed electron current density to evaluate the value of the flux Γ in (7). From Figs. 2 and 3, $\beta = 9 \sim 10$ and $5 \sim 6$ for Holmes and JAERI's data respectively. These β's are consistent with those obtained from Fig.1.

DISCUSSION

Effect of electric field Since the Larmor radius and the mean free path of the ions are of the order of or larger than the width of region of magnetic filter, ion flux Γ_i can be given as

$$\Gamma_i = (n_1 - n_2) v_i - (n_1 + n_2) \mu_i E \quad . \tag{8}$$

Experimentally ion current is less than 1/10 of the electron current. Then we will evaluate the value of E by putting $\Gamma_i = 0$ in (8). Note that thus determined value of E is an overestimated value. This E gives

$$\frac{\sigma E}{\Gamma e} \approx \frac{n_1 - n_2}{2n_2} \sqrt{\frac{M}{m}} \sqrt{\frac{T_i}{T_e}} \frac{1}{(\omega \tau_{ei})^2} \quad . \tag{9}$$

For typical values of electron densities, and magnetic field in ion source, $\sigma E/\Gamma e \ll 1$ is well satisfied and the effect of E is negligible in (3).

Coefficient β of transported energy From fitting the experimental data of Holmes[1] and JAERI in Fig.1, values of β are 10 and 7 respectively. These values are rather high. For neutral particles having Maxwellian velocity distribution, it is known that $\beta = 2$. But in the case of charged particles flowing through the sheath of the extraction grid, β is given by[3]

$$\beta = 2 + \ln\sqrt{\frac{M}{m}} \ . \tag{10}$$

Since there are ions of H_3^+, H_2^+ and H^+, $M = 2M_p$ (M_p = proton mass) is a reasonable value for an effective mass of the ions, which gives $\beta = 6.1$. This value nearly agrees with the experimental ones. But there still exist difference in Holmes' case.

Relation to Holmes' formulation If we neglect the interference terms in (1) and (2), we are in the same situation as Holmes based. Based on (1) and (2), we have

$$\ln\frac{n_1}{n_2} = \frac{1.92}{\beta}\frac{T_2 - T_1}{T_2} \ , \tag{11}$$

which gives $\beta = 3.13$ by fitting Holmes' experimental data. Holmes claimed $\beta = 2.5$. Difference comes from the numerical factor 1.92 of the thermal conductivity, while it is 1.5 in Holmes' formulation. This shows that the numerical factor in the transport coefficients has great influence on determining the value of β. Judging from β obtained with and without the effect of the interference terms, the interference phenomena are as important as the diffusion and thermal conduction.

CONCLUSION

0. Electron number density and temperture across the magnetic filter are investigated on the basic equations that includes diffusion, thermal conduction and the interference between them.

1. Results nearly agree with Holmes' and JAERI's experimental data.

2. In obtaining the good agreement with the experimental data, the numerical factors of the transport coefficients have delicate influence.

3. Effect of the interference between diffusion and thermal condution is as large as the effect of diffusion and thermal conduction in JAERI's experiment and larger in Holmes' experiment.

ACKNOWLEDGEMENT

The authors are grateful to Mr. M. Hanada and Mr. T. Inoue of JAERI for their useful discussion.

REFERENCES

1. A. J. T. Holmes, Rev.Sci. Instrum. $\underline{53}$ 1517, 1523 (1982).

2. I. P. Shkarofsky, T. W. Johnston and M. P. Bachynski, The Particle Kinetics of Plasma (Addison-Wesley, Massachusetts, 1966), p.361.

3. V. E. Golant, A. P. Zhilinsky and I. E. Sakharov, Fundamentals of Plasma Physics (John Willy & Sons, N.Y.,1977), p.207.

COMPUTERIZED ANALYSIS OF HYDROGEN PLASMA IN A COMPACT H$^-$ CUSP SOURCE

D.H. Yuan, K. Jayamanna and P.W. Schmor
TRIUMF, 4004 Wesbrook Mall, Vancouver, B.C. V6T 2A3

ABSTRACT

A cylindrical Langmuir probe with diam of 0.5 mm, length of 5 mm and the Laframboise theory are used to give an analysis of the plasma parameters in a H$^-$ cusp source including temperatures and densities of slow, fast electrons and positive ions in a median density plasma. The iteration technique overcomes the problems of conventional Langmuir probe analysis. A VAX based program is used to control the motion and analyze data from the probe. In this paper, we briefly describe the program and present initial results obtained from a compact H$^-$ volume multicusp source.

INTRODUCTION

To date, the Langmuir probe is still one of best techniques to analyze and understand the process in a hydrogen plasma. In order to take a spatial measurement over the entire region of the plasma a computerized technique is required, which gives a fast measurement and analysis of the plasma parameters. TRIUMF has developed two H$^-$ dc cusp sources which have shown some advantages over the other sources. The sources are divided into two regions (1) driver region, (2) extraction region separated by a magnetic filter, which allows cold electrons ($\sim$ 1 eV) to penetrate with positive ions into the extraction region from the driver region, but stops fast e^- from doing so. Consequently, H$^-$ ions are formed in the extraction region and extracted. In this paper the development of a VAX computer controlled Langmuir probe as a plasma diagnostic is described. The technique provides a rapid measurement by a modified program initially developed for emittance measurements and an iterative analysis using the Laframboise theory[1], which shows a good agreement of positive ion current with experiment to a highly negative biased probe and also allows the measurement of fast electrons.

EXPERIMENTAL SETUP

A compact cusp source[2] is used in the experiment. A cylindrical Langmuir probe with a tungsten wire (0.5 mm in diameter and 5 mm in length) is driven by a stepping motor along the entire axis of the source and measures an I-V curve at each axial position. The block diagram of the data acquisition system is shown in Fig. 1. A stepping motor is controlled by a VAX to bring the probe to the requested positions in the source through CAMAC. A step voltage is applied at each position to the probe from a -V to a +V. At each step of the voltage the current is collected by the biased probe to form the data of I–V curve stored in the VAX. In order to have a good resolution there is a natural scaling parameter $\chi = e(V - V_p)/kT_e$. The accuracy of fitted curve depends on the number of the points per χ rather than per volt. In the extraction region $kTe \leq 1$ eV 700 points are acquired for the I–V curve.

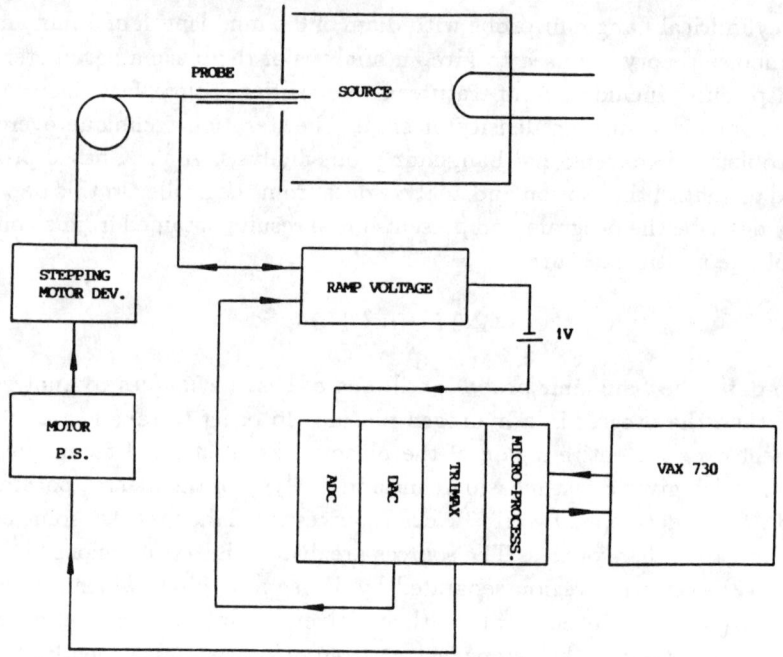

Fig. 1. Scematic of probe system.

PROBE THEORY

An ideal I–V curve is shown in Fig. 2. After a cylindrical Langmuir probe has been inserted into the plasma, a plasma sheath is formed around the probe. The thickness of the sheath depends on the voltage applied to the probe with respect to the anode of the plasma chamber (see Fig. 3). Here the following symbols are used: V, the potential difference between the probe surface and the anode which is composed of two parts; namely the potential drop between the anode and the sheath boundary, V_p, and the potential difference of the probe surface with respect to the sheath edge V_{sh}. If the probe potential V_{sh} is sufficiently negative i.e. $-eV_{sh} \gg kT_e$, which is achieved by making V more negative with respect to V_f, the floating potential, almost no low e^- will reach the probe while all positive ions passing the sheath edge in the direction of the

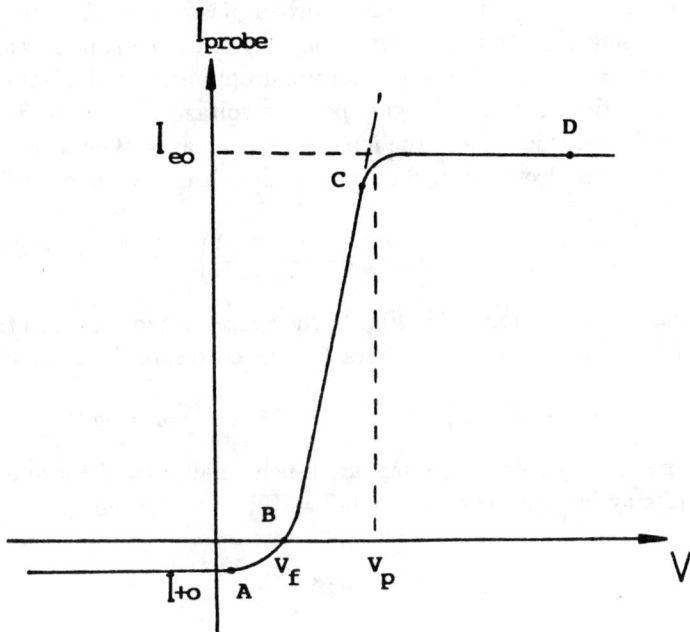

Fig. 2. I–V curve of an idealized probe.

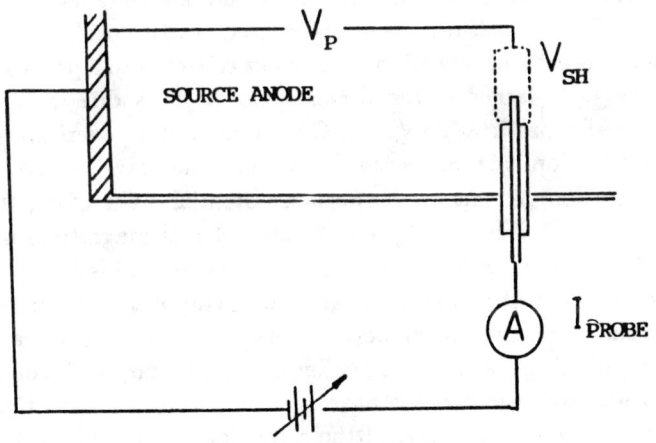

Fig. 3. Probe circuit.

probe are absorbed. Some higher energy e^- still can penetrate the sheath to the probe. As V is made more and more negative the high e^- become less and eventually no e^- are observed. The probe current then is equal to the ion saturation current I_{+0}:

$$I_{+0} = \frac{1}{4} e n_+ A \left(\frac{8kT_e}{\pi m_+} \right)^{\frac{1}{2}} \tag{1}$$

where n_+ is the ion density in the undisturbed plasma and A is the probe surface area. Raising V in the other direction, higher than V_f, more and more e^- are able to overcome the retarding potential drop; initially, the faster e^- of the velocity distribution and at still higher positive voltage of V_{sh} also the slower e^-. Simultaneously the thickness of the ion sheath decreases. The density distribution of Maxwellian electrons in the positive sheath is governed by Boltzman's law ($V < V_{sh}$):

$$n_e = n_0 \exp\left(\frac{e(V - V_{sh})}{kT_e}\right) \qquad (2)$$

As V approaches V_p, at point C in Fig. 2, the space charge sheath in front of the probe vanishes. This point of the characteristic corresponds to the condition:

$$V_{sh} = 0; \qquad V = V_p; \qquad I_{probe} = I_{+0} + I_{e0}.$$

Since the mean velocities of the ions are much smaller than those of the electrons, I_{probe} may be approximated by ($V = V_p$):

$$I_{probe} \cong -\frac{1}{4}en_e A \left(\frac{8kT_e}{\pi m_e}\right)^{\frac{1}{2}} \qquad (3)$$

which is the e^- saturation current. In practice, in the medium density plasma (10^{10} to 10^{12}cm^{-3}), the I–V characteristic does not saturate moving to the right from point C. The nonsaturation of the current to the probe is explained by considering in the extraction region the current collected to a probe of increasing effective area (even larger in the driver region). It is due to additional ion production in the presheath region. Customarily it is overcome by linearly extrapolating the ion current measured at highly negative voltage, to give the expected saturation current at the plasma potential. Therefore, It was found that the value of n_+ can vary by almost an order of magnitude in the driver region and by a factor of 2 in the extraction region. This indicates that the expansion of ion current is nonlinear, and the extrapolation technique does not provide a accurate result. Laframboise theory provides a numerical solution to the equations which govern ion collection in a stationary collisionless plasma, and a set of ion expansion curves have been deduced from this theory. These curves are a function of the ratio of the radius of the probe, R_p, to the Debye length, λ, which is given by

$$\lambda = \left(\frac{\epsilon_0 k T_e}{e^2 n_e}\right)^{\frac{1}{2}} \qquad (4)$$

where ϵ_0 is the permittivity of free space. Laframboise used the plasma potential as a reference. A natural scaling parameter $\chi = e(V - V_p)/kT_e$ is introduced (Fig. 4). We assume in the calculation that the dominant ion species is H_3^+;

the other ion densities are neglected. From the theory the expansion of the ion current to the probe can be written as a function of applied voltage χ:

$$\frac{I_+(\chi)}{I_{0+}} = f\left(\frac{R_p}{\lambda}, \chi\right) \qquad \chi < 0 \qquad (5)$$

where $I_+(\chi)$ is the ion current to the probe as a function of χ or probe voltage. I_{0+} is the ion saturation current at plasma potential. $f(R_p/\lambda, \chi)$ is the ion current expansion factor as calculated in the theory in the zero ion temperature limit (see Fig. 4). R_p is the radius of probe.

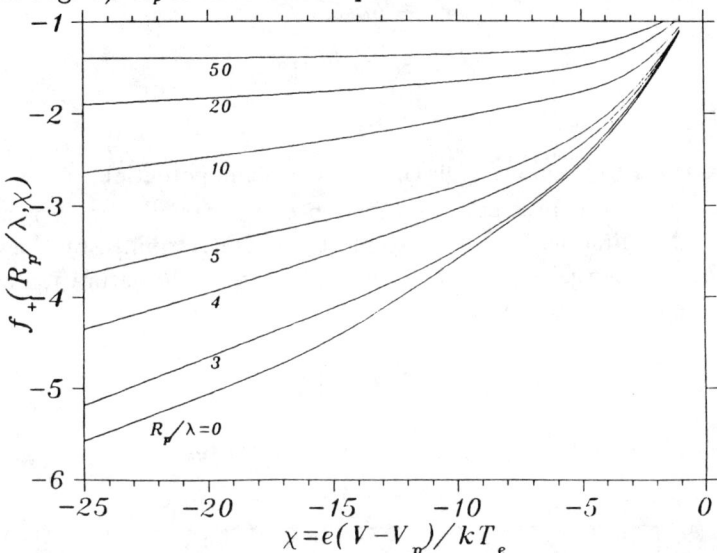

Fig. 4. Ion current expansion factors as a function of natural scaling parameters, χ for various values of R_p/λ.

PROGRAM DESCRIPTION

In order to use Laframboise theory to analyze the I–V curve, one has two difficulties. First, the theory uses the plasma potential V_p as a reference value, which is not yet known. Secondly, one needs to select a particular theoretical curve from Fig. 4, which depends on the ratio of the probe radius to the electron Debye length, R_p/λ. This ratio, in turn, depends on kT_e and n_e, which have to be determined. A iteration technique is used here to solve the problems and give accurate results of the plasma parameters.

Initial V_p, kT_e, and n_e

A derivative of the equation (2), $dI(V)/dV$, gives a maximum value at the plasma potential, V_p, and

$$\frac{I(V)}{dI(V)/dV} = \frac{kT_e}{e} \qquad (6)$$

That is, in the exponential region, the ratio of current to the first derivative is equal to the electron temperature, kT_e in units of eV. When the first derivative passes through the maximum, the current and voltage at maximum, $I_{maxderiv}$, $V_{maxderiv}$ and the value of derivative itself, I'_{max} are stored in the computer. We assume when the first derivative reaches 1/10 of I'_{max}, the probe current is a reasonable estimate of saturation e^- current, I_{sat}.

To a first approximation we then have

$$V_p = V_{maxderiv},$$
$$kT_e = \frac{I_{maxderiv}}{I'_{max}},$$
$$I_{0e} = I_{sat}$$

where I_{0e} is the maximum electron current at plasma potential.

V_p and χ can be expressed as a function of voltage. Figure 5 shows the I–V curve and its first derivative. Customarily, V_p is determined by the voltage at which the exponential region of I–V curve intersects the saturation current. Here it can be approximated by

$$V_p = V_{max} + kT_e/e \ln\left(\frac{I_{sat}}{I_{maxderiv}}\right) \tag{7}$$

Substituting kT_e and I_{0e} into equations (3) and (4) gives the initial values for n_e and λ.

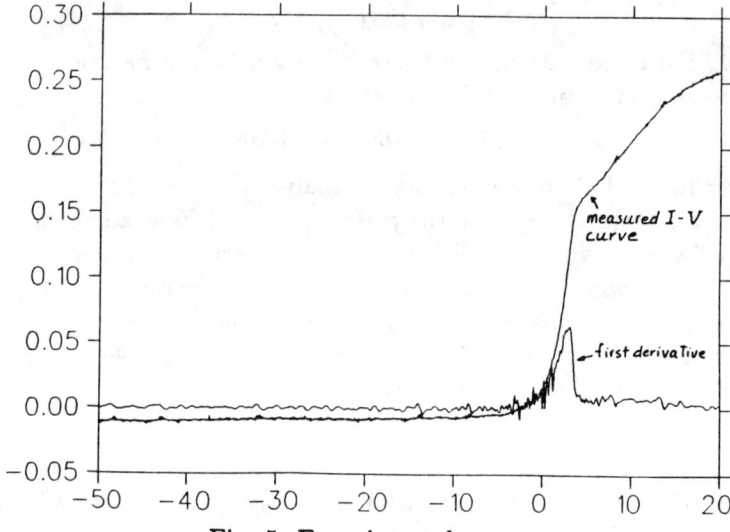

Fig. 5. Experimental curve.

Calculation of n_+

The first approximation of kT_e, V_p, and n_e are used to calculate the parameter R_p/λ and the function χ, and to select a curve in Fig. 4, which is fitted and stored in the computer. From Eqs. (1) and (5) we get

$$n_+ = \frac{4I_+(\chi)}{eA(8kT_e/\pi m_+)^{1/2} f(R_p/\lambda, \chi)} \tag{8}$$

where, again, $I_+(\chi)$ is the current to the probe at a negative voltage wrt the plasma potential.

The measurement of $I_+(\chi)$ is complicated due to the presence of both the tail of Maxwellian distribution of the thermal e^- in the form of $I_{0e}\exp(\chi)$ and fast e^-. This is overcome by measuring $I_+(\chi)$ at a sufficiently negative voltage to remove contributions from all electrons.

Calculation of n_{fe}, kT_{fe}

For $\chi < -10$, the bulk thermal e^- contribution is effectively zero. Then, the current to the probe due to the collection of fast electrons $I_{fe}(\chi)$ is calculated as follows:

$$I(\chi) = I_+(\chi) + I_{fe}(\chi)$$

where $I(\chi)$ is the measured current to the probe and $I_+(\chi)$ is the ion contribution. Thus we have:

$$I_{fe}(\chi) = I(\chi) - \frac{1}{4} e n_+ A \left(\frac{8kT_e}{\pi m_e}\right)^{\frac{1}{2}} f\left(\frac{R_p}{\lambda}, \chi\right) \qquad \chi < 10$$

$I_{fe}(\chi)$ is fitted by an exponential regression to:

$$I_{fe} = I_{0fe} \exp\left(\chi \frac{kT_e}{kT_{fe}}\right)$$

where I_{0fe} is the fast electron current to the probe at plasma potential, V_p, and kT_{fe} is the fast electron temperature.

Figure 6 shows an exponential fit to the experimental curve after positive ion contribution subtracted.

Iteration

Now, we have the initial values of $kT_e, n_e, V_p, n_+, kT_{fe}$, and n_{fe}. The contribution to the probe current due to fast electrons and the ion current are subtracted from the probe current in the region $-2 < \chi < 0$ (see Sec.5 discussion). The remaining current is due to the low energy electrons only:

$$I_e(\chi) = I(\chi) - I_+(\chi) - I_{fe}(\chi) \qquad -2 < \chi < 0$$

$\ln[I_e(\chi)]$ is then fitted by a linear regression to V. The inverse of the slope gives

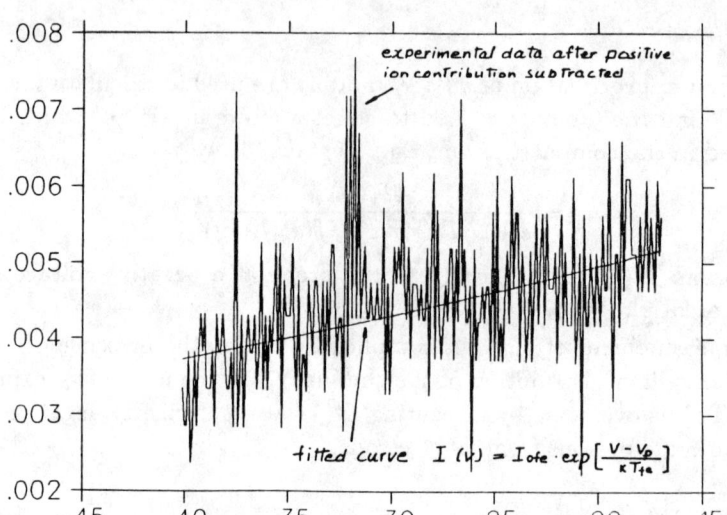

Fig. 6. Experimental curve fit for fast electron.

new values of kT_e and n_e with the ion and fast electron current removed. The plasma potential is corrected in a similar manner (Eq. (7)) by:

$$V_p(new) = V_p(old) + kT_e/e \ln\left(\frac{I_{0e}(new)}{I_{0e}(old)}\right)$$

These new values define a new χ as a function of the probe voltage and a new value of R_p/λ. Now the program goes back and repeats the calculations of all the plasma parameters until a self consistent solution is reached. We continue the calculation until V_p converges to within 5% of the previous estimate. In calculation of V_p, I_{0e} is replaced by $I_{0e} + I_{0fe}$.

Figure 7 shows an exponential fit to the experimental curve after contributions of positive ion and fast electrons subtracted.

Calculation of V_f

The floating potential, V_f, is calculated as the I–V curve passes through zero.

The results of plasma parameters of TRIUMF small dc cusp source calculated using the program are shown in Fig. 8.

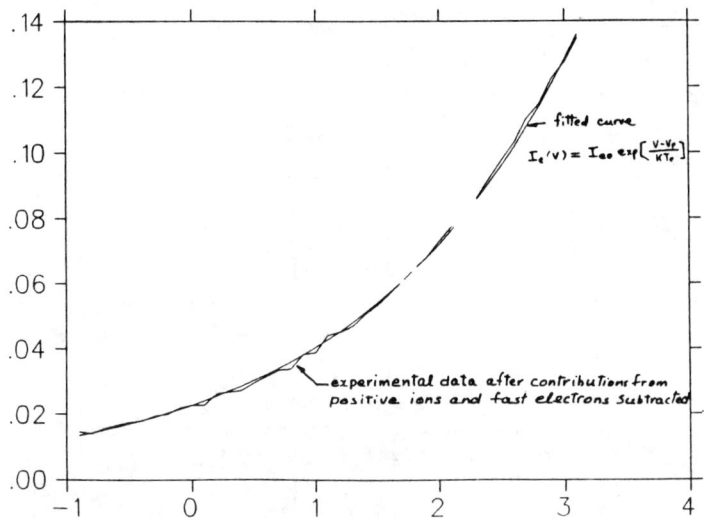

Fig. 7. Experimental curve fit for low electron.

DISCUSSION

It is believed that a probe biased near or higher than the plasma potential may seriously disturb the local plasma and lead to an error in the measurement of the saturation electron current. There are three possible sources of error. First, the probe collects charged particles from the plasma. This problem is partially overcome by making the area of the probe as small as possible. The probe area we used was 0.08 cm^2, which takes a current of the order of 4 mA which is negligible comparing to the arc current. Secondly, there is a possible error resulting from an effect observed by Kunkel[3] due to a depletion in the number of electrons with energy less than or equal to $kT_e/2$. This results in an electron saturation current lower by a factor of two or more. It is assumed that the depletion of low electron occurs. In order to compensate for this effect, the plasma potential is calculated by extrapolation of the points lying lower than the plasma potential which, in tern, was calculated during the previous iteration cycle, and the points above that are not used. Thirdly, there is a depletion due to the magnetic field, which in the small cusp source is 5G in the driver region and up to 150G in the filter region. In order to use the probe in a magnetic field, it is important to satisfy the relationship $R_p/R_{\text{larmor}} \ll 1$, where R_p is the radius of the probe and R_{larmor}, the Larmor radius. Customarily, $R_p/R_{\text{larmor}} = 0.25$ is the criterion, below which the perturbation is small. With this condition, in our case, the magnetic field has to satisfy:

$$B(G) < 20\sqrt{kT_e(eV)}$$

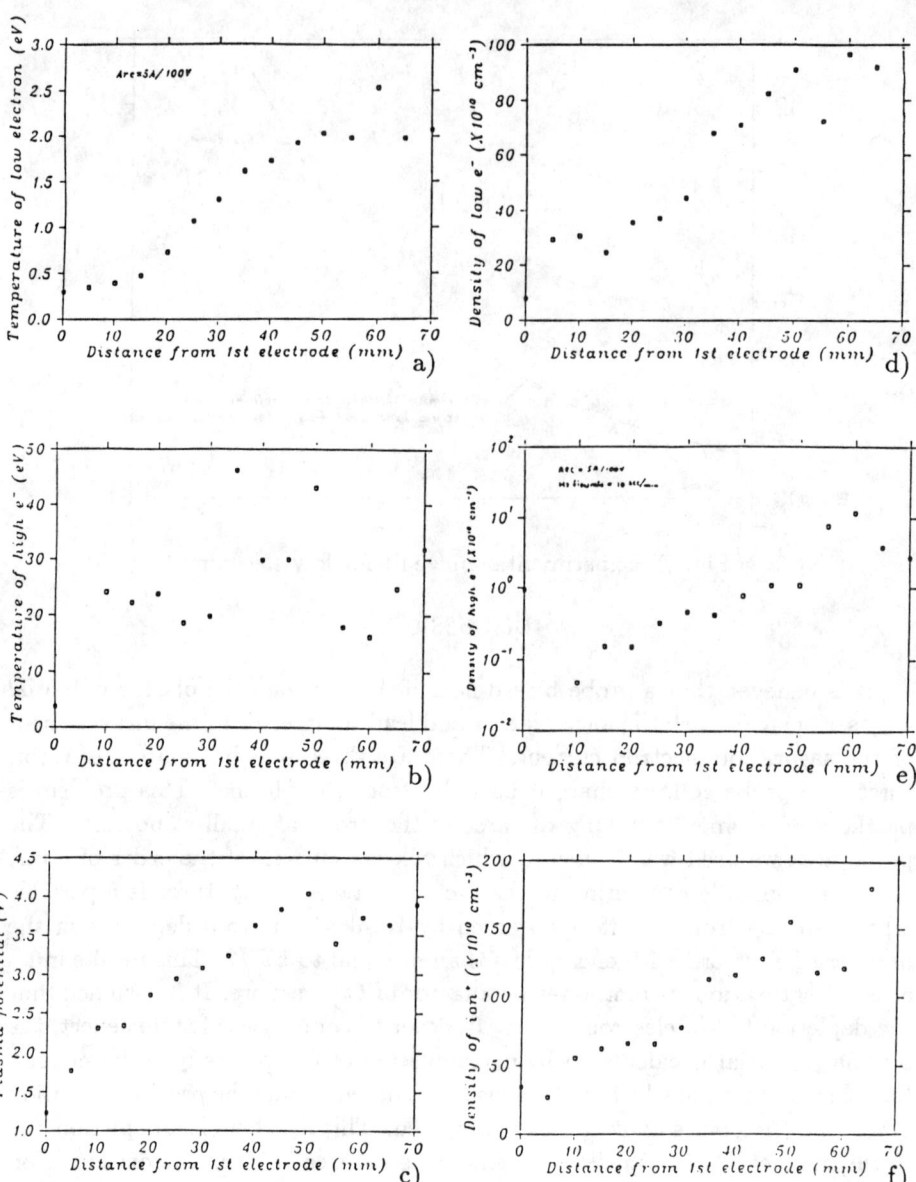

Fig. 8. Plasma parameters at positions from the extractor. a) Low energy electron temperature, KT_e; b) high energy electron temperature, kT_{fe}; c) plasma potential, v_p; d) low energy electron density, n_e; e) high energy electron density, n_{fe}; f) positive ion density, n_+.

It is a problem in the extraction region of the H⁻ cusp sources where the electron temperature is low ($< 1eV$) and the B field is relatively high. In order to meet the criterion a very small diameter probe has to be used. In our case, the probe dimensions lead to a significant perturbation in the extraction region.

In the calculation of kT_{fe} and n_{fe}, a Maxwellian distribution was used. In order to validate this, the second derivative of the fast electron current could be taken by the computer. This has not yet been done, however, it has been established that a bi-Maxwellian energy distribution is a good approximation by the method of Druyvesteyn[4], for an isotropic distribution. In a low pressure, the primary electrons rebounding many times from the magnetic field ensure an isotropic distribution.

ACKNOWLEDGMENTS

We would like to express our gratitude for the special efforts made by P. Chigmaroff with the electronics, H. Wyngaarden on the apparatus.

REFERENCES

1. J.G. Laframboise, University of Toronto, Institute for Aerospace Studies Report No. 100 (1966):and in Rerefield Gas Dynamics, edited by J.H.de Leeuw (Academic, New York, 1965), Vol. II, p.22.

2. D.H. Yuan, K. Jayamanna and P.W. Schmor "A Small Intense dc H⁻ Cusp Source in TRIUMF"; to be published.

3. K.F. Schoenberg and W.B. Kunkel, J. Appl. Phys. **50**, 4685 (1979)

4. M.J. Druyvesteyn, Z. Phys. **64**, 781 (1930).

SIMULATION OF CHARGED PARTICLE DYNAMICS IN GAS MEDIUM OF A FUSION REACTOR INJECTOR

V.P.Sidorov, S.Yu.Udovichenko
Sukhumi Institute of Physics and Technology,
Sukhumi, USSR

A.M.Astapkovich, P.N.Afanasyev,
Yu.V.Zuev, Yu.A.Svistunov
D.V.Efremov Scientific Research Institute
of Electrophysical Apparatus, Leningrad, USSR

ABSTRACT

The present paper deals with the results of a number of researches on numerical simulation of phase dynamics of a fast negative ion beam in an ambipolar electrical field of a weakly ionized plasma which is produced by the beam itself in a rarefied gas just behind the ion source and in a dense gas of the neutralizer as well. The problem is solved for ensuring the phase characteristics for neutral atoms at the neutralizer exit suitable for future successful transportations of beams over the distances exceeding 40-50 metres and for their injection into a tokamak reactor.

INTRODUCTION

Negative ion beams are widely used for heating plasmas up to fusion temperatures in tokamak-type closed magnetic systems. While producing stationary beams of charged particles, and later, those of neutral particles, there arises a problem of their transportation over long distances. Considering both the necessity of heating the central part of a toroidal plasma and the possibility of fast particle losses to metal walls of the transport channel, it is desirable that the beam radius should not essentially change in comparison with the initial one.

Within the injector, the negative ion beam propagates in the atomic or molecular gases of sufficiently high pressure: except for the region of electrostatic or high-frequency acceleration, the gas pressure is $P \gtrsim 10^{-5}$ Torr. Within the region of the ion beam formation and precondictioning which is done to supply the beam into the accelerator, there is a rather high drop in the pressure of the gas which is delivered from the plasma ion source. The gas neutral atom density near the source is such that the plasma produced as a result of ionization yields a complete compensation of the beam volume charge. Meanwhile, the plasma particle density may essentially exceed that of the beam particles. Since the plasma ion and electron mobilities differ, in the systems limited in their radial direction (the injector length is much greater than its transverse dimensions), there arises a plasma ambipolar electric field. As the distance from the source increases, the neutral gas density becomes lower and when its value is less than the critical one, the beam becomes partially compensated and the sign of the electric field inside the beam is opposite to the sign of the plasma ambipolar field[1].

THE BEAM DYNAMICS IN THE RAREFIED GAS

In Ref.[2], using the numerical simulation, the transverse dynamics of the negative ion beam just behind the source are investigated under the assumption of homogeneity of the rarefied gas density along the injector and that of full compensation of the fast particle charge by a quasineutral plasma. While solving this problem, the Poisson equation has not been applied and the electric field of the beam-plasma system has been analytically derived using the steady-state hydrodynamic equations of motion and continuity of plasma components. Having made the optimum choice of parameters for a quasineutral collisionless ion beam plasma, it is shown that its ambipolar electric field may have a considerable effect upon the transverse motion of the fast beam ions. Fig.1 shows the phase volumes of ions in the beam prior to (a) and after (b,c) the drift of 100 cm in the gas from the Dudnikov source [3], respectively. The energy W of the longitudinal motion of particles is 20keV; the cross-sectional distribution of particles in the beam is nearly Gaussian; the maximum angle of ion departure from the axis is $\alpha_{xm} = \alpha_{ym} = 4\cdot 10^{-3}$ rad; maximum transverse dimensions of the beam, X_b and Y_b are 0.7cm and 2.4cm, respectively; the gas pressure P is $5\cdot 10^{-4}$ Torr; the gas molecular density n_g is $1.78\cdot 10^{13}$ cm^{-3}; the plasma ion and electron densities $n_i(Y_b)$ and $n_e(Y_b)$ are $6.14\cdot n_b$ and $5.14\cdot n_b$, respectively, where n_b is the beam ion density; the electron temperature T_e is 10eV and the residual gas is nitrogen N_2. At further distances in the gas the phase volume becomes distorted and the effective beam emittance increases. There exists some optimum length for which (the phase volume being unchanged) the maximum angle of departure from the beam axis in one or both planes may become less and even change its sign.

PLASMA DYNAMICS WITHOUT EXTERNAL FIELDS

Based on nonstationary one-dimensional equations of two-fluid hydrodynamics and Poisson equations for the self-consistent field, involving the parametrical dependence of variables upon the second coordinate, we have further [4] obtained a radial distribution of the resulting electric field of the ion beam and plasma which fully or partially compensates the volume charge of fast ions in the gas medium inhomogeneous along the injector. Apart from plasma oscillations, the nonstationary problem makes it possible to take transient processes into account, such as plasma build-up when the beam enters the gas and the plasma dynamics for small gas densities. It should be noted that during the extraction, additional acceleration and transportation processes, the beam moves on within the electric field of the accelerating electrodes and magnetic focusing lenses. Techniques for calculating the electric fields within

the system of electrodes and magnetic fields of solenoidal
and quadripole lenses are well developed. The authors aim at
determining the sum of all the fields within the injector,
both matched with the beam and influencing its dynamics; however, this task requires having a sufficiently reliable me -
thod for calculating the fields within the beam-plasma system.

The equations of plasma dynamics mentioned above are presented in a one-dimensional form taking into account the fact
that the injector is much greater in length that in width, i.e.
$\partial/\partial z \gg \partial/\partial t$; there exists a small inhomogeneity in plasma
and gas parameters along the injector. It has been also assumed that the beam produces a single ionization of the gas,
the gas ionization by secondary electrons is negligible, and
elastic collisions take place in the plasma itself. In terms
of these assumptions, the beam-plasma system has been simulated by five integro-differential equations:

$$\int_V \frac{\partial n_\alpha}{\partial t} dV = \int_V \nu_H n_g dV - \oint n_\alpha u_\alpha dS , \qquad (1)$$

$$\frac{\partial u_\alpha}{\partial t} = -u_\alpha \frac{\partial u_\alpha}{\partial z} \pm \frac{eE_z}{m_\alpha} - \nu_H \frac{n_g}{n_\alpha} u_\alpha - \frac{T_\alpha}{m_\alpha n_\alpha} \frac{\partial n_\alpha}{\partial z} - \nu_{\alpha o} \frac{m_o}{m_\alpha + m_o} \cdot u_\alpha , \qquad (2)$$

$$\oint_S E_z dS = 4\pi e \int_V (n_i - n_g - n_e) dV . \qquad (3)$$

Here, the index α acquires the meaning of i,e (ions, elect -
rons); n_α and u_α are the densities and directed velocities of
plasma ions and electrons; n_b and v_b are the density and the
velocity of beam particles, respectively; ν_H and $\nu_{\alpha o}$ are collision frequencies which are respectively equal to:
$\nu_H = n_g(z) \sigma_i v_b$ and $\nu_{\alpha o} = n_g(z) \sigma_{\alpha o} v_\alpha$, where $n_g(z)$ is a slightly
nonuniform density of the gas, σ_i is the cross-section of the
gas ionization by the beam ions, $\sigma_{\alpha o}$ is the cross-section of
elastic collisions of plasma components with gas neutrals; T_α,
u_α, m_α are the temperature, the thermal velocity and the mass
of plasma particles respectively, m_o is the mass of neutrals in
the gas, V and S are the integration volume and area, respective-

ly. With no violations in integrity, it is assumed that the directed velocity of the gas neutrals is equal to zero. Boundary conditions (not all of them being independent) are

$$u_\alpha |_{z=0} = (\partial u_\alpha / \partial z)|_{z=0} = (\partial n_\alpha / \partial z)|_{z=0} = 0, \quad (4)$$

$$\partial(\partial n_\alpha / \partial z)/\partial z |_{z=R} = 0, \quad \partial(\partial u_\alpha / \partial z)/\partial z |_{z=R} = 0. \quad (5)$$

Conditions (4) are well-known, while conditions (5) are substitutions for the equality of electron and ion currents onto the wall and imply the particle flux conservation while proceeding to the last cell of integration from the last but one. Zero values of densities and directed velocities of plasma components correspond to the initial conditions of the problem. Note that the equation system studied does not include the equation of motion of the beam ions. It is assumed that negative ions of hydrogen have Gaussian distribution over the beam cross-section and it remains unchanged during the calculations.

Integration of the system of Eqs.(1)-(3) has been carried out according to the explicit scheme of finite differences. In order to stabilize the plasma oscillations, for the first step of integration, the term including the elastic plasma friction in the residual gas differs from zero. Further out, this term has been neglected since moderate gas densities have been taken while the plasma has been considered collisionless. Figs.2-3 show the numerical simulation results for the beam-plasma system for the residual gas densities of $n_{g1} = 3.6 \cdot 10^{13} cm^{-3}$ and $n_{g2} = 3.6 \cdot 10^{12} cm^{-3}$ for the time moment $t = 8.43 \cdot 10^{-6}$ s after "switching" the beam on. The peak density of particles in the beam, n_b, at the system axis is $10^7 cm^{-3}$; the effective radius of the beam, r_b, is 3cm; the electron temperature, T_i, is 0.03 eV; the velocity of fast ions, v_b, is $2 \cdot 10^8$ cm/s, and the casing radius, R, is 20 cm. Fig.2 illustrates the case when the condition of full compensation for the beam charge is met: $n_i = n_e + n_b$; n_i, $n_e \gtrsim n_b$. Under such conditions, the beam-plasma system is ideally matched, and the electric field potential versus radius dependence corresponds to that observed experimentally[5]:

$$\varphi = -\varphi_R \left\{ 1 - \exp[-(\beta z/z_6)^2] \right\}, \quad \beta \simeq 0.39. \quad (6)$$

The system plasma-beam has been investigated under conditions n_e, $n_i \gg n_b$ too. In this case the resulting field is determined by plasma ambipolar field. The results presented correspond to the calculations defining the stationary ambipolar field in a quasi-neutral beam plasma from Ref.[6].

Fig.3 represents the case when undercompensation of the volume charge of the negative ion beam takes place. Under such conditions, the outer electron density exceeds the electron density within the beam and this fact is consistent with the experimental data [7]. It is seen that the moment in time indicated corresponds to a certain point with a coordinate which is less than the beam radius and where the electric field sign change takes place. When the beam is undercompensated, as it is considered above, the fields of the beam and those of the plasma are not matched, therefore, some nonmonotonicity is observed in the resulting field close to the beam boundary. These perturbations may be avoided if to take the beam dynamics (i.e., the change in its effective radius) into account.

THE BEAM DYNAMICS IN A DENSE GAS

In Ref.[8], the ion beam motion is considered in a dense gas in the neutralizer, and using numerical simulations, the fast ion dynamics are studied in the electric field of a quasi-neutral (n_i, $n_e \gg n_b$, $n_i = n_b + n_e$) collisional plasma produced as a result of beam ionization of the gas. It is shown that in the chamber with the dense gas, apart from the formation of fast neutral particles, a "plasma lens " is generated, and its ambipolar electric field may have a considerable effect upon the transverse dynamics of the beam.

For rather high densities in a weakly ionized plasma, collisions between slow ions and neutrals of the gas become prominent. In such a case, when the condition $u_i \ll R \nu_{io}$ is met, the electric field value for the beam with the uniform density of particles is found from Eqs. (1) and (2):

$$E_z(z \leq z_\delta) = \frac{T_e}{e z_\delta} \cdot \frac{z}{z_\delta} \left[\frac{2T_e}{m_i \nu_{io} R \nu_s} + \ln \frac{R}{z_\delta} + \frac{1}{2}\left(1 - \frac{z^2}{z_\delta^2}\right) \right]^{-1}, \quad (7)$$

$$E_z(z > z_\delta) = \frac{T_e}{e z_\delta} \cdot \frac{z_\delta}{z} \left[\frac{2T_e}{m_i \nu_{io} R \nu_s} + \ln \frac{R}{z} \right]^{-1}, \quad (8)$$

where v_s is the sound velocity. The calculation results have shown that, depending on the choice of parameters (gas composition, density, temperature, etc.), the "plasma lens" may enlarge or diminish the regular constituent of the angular deviation of particles from the beam axis, i.e., the angle of inclination of the phase ellipse with respect to the coordinate axis of the phase plane.

INFLUENCE OF THE CONVERSION OF THE BEAM CHARGE COMPONENT ON PARTICLE DYNAMICS

First of all, it is of interest to consider the phase volume of the beam of fast neutral atoms since they are used for plasma heating in a reactor. As the conclusions in Ref.[8] for the neutral part of the beam are valid only in a rather approximate way, the phase dynamics problem for a two-component beam has been further investigated [9].

While passing through the gas target, the fast ion beam composition is continuously changed. The efforts are undertaken to ensure the maximum factor for converting the negative hydrogen ions into neutrals and to obtain such values of neutral atom phase characteristics at the target exit which will allow to transport the beam over the distance of $\gtrsim 40$ m towards the reactor.

The simulations were carried out in the EVM EC-1055 M computer via the PL/1 programme describing the three-dimensional particle dynamics. Since it has been conjectured that the injected beam is continuous we have considered the motion of particles available over a single cross-section. Initial particle distributions in coordinate and phase spaces have been given by means of a sensor of random numbers, the distributions being nearly homogeneous. The phase portrait of the beam in a certain chosen phase plane X-X' (the problem is symmetrical one), at the entrance into the "plasma lens" is given in Fig.4. There were 1200 noninteracting particles over the initial cross-section. While moving further into the target, the densities of ions and neutral atoms undergo changes according to the following expressions:

$$n_b^-(z) = n_b^-(0) \cdot \exp(-\sigma_{-10} \cdot n_g \cdot z), \qquad (9)$$

$$n_b^o(z) = n_b^-(0) \cdot \frac{\sigma_{-10}}{\sigma_{-10} - \sigma_{01}} \left[\exp(-\sigma_{01} n_g z) - \exp(-\sigma_{-10} n_g z) \right],$$

where σ_{-10} is the cross-section of the electron detachment from a negative ion, σ_{01} is the fast neutral ionization cross-section, n_g is the homogeneous gas density. In order to illustrate the plasma target performance, the results of calculations for the 1 MeV ion dynamics are given. The first case shows the motion of a single component beam of charged particles. Fig.5 shows the beam emittance at the "plasma lens" exit, the "lens" length, L, being 50cm, which is twice as little as the optimum length, L_{opt}, in terms of maximum conversion for H^- into H^o. It should be noted that for 80% of particles the regular constituent of the particle departure from the axis is compensated by an ambipolar force. The lens parameters defining the electric field which influences the beam are: T_e=12eV, T_i=0.3eV (it is assumed that atomic hydrogen is preliminarily heated up to 3500°C), $R=r_b$=1cm, n_g=5.2·10^{14}cm^{-3}, σ_{io}=(7 - 10)· ·10^{-15}cm^2, v_s=8.4·10^6cm/s. The second case illustrates the motion of a two-component beam under the same conditions. The neutral beam at the target exit is divergent, having a large mean angle of particle departure from the beam axis. In the third case, the target length is increased and amounts to L_{opt}. Fig.6 shows the phase portraits of charged and neutral components at the exit. With the target length being extended, the effective emittance of negative particles increases as a result of nonlinear distortions in the phase volume while the mean angle of departure from the axis of the neutral component becomes smaller.

CONCLUSIONS

Based on the results obtained, a number of conclusions may be drawn:

1. By means of a "plasma lens", it is much easier to form the required phase volume of charged particles than that of neutrals. This is due to the fact that the ambipolar force continuously acts upon the charged particles over the whole length of the lens, while with respect to the neutrals the effective force appears weaker as a result of the unsimultaneous conversion of fast ions into neutral atoms.

2. The optimum "plasma lens" length for ensuring the minimum mean angle of the neutral component departure from the system axis coincides with the neutralizer length, $L=L_{opt}$, for which case, nearly 80% of particles have maximum angles of departure not exceeding $3 \cdot 10^{-4}$ rad (the maximum angle at the entrance is nearly 10^{-3} rad).

3. It is shown that the stationary ambipolar electric field of the weakly ionized plasma produced by the beam itself in the gaseous medium of the injector can have a considerable effect upon the transverse dynamics of the beam. The phase volume is defined for the neutral beam component which is formed as a result of fast ion conversion in the gas target.

4. The numerical code developed gives the opportunity to optimize the injector parameters in order to obtain such values of the phase characteristics of the neutral atom beam as to ensure its transport and injection into the tokamak reactor.

REFERENCES

1. M.D.Gabovich, L.S.Simonenko, I.A.Soloshenko. Compensation of the Volume Charge of the Intense Beam of Negative Ions, Zh.Tekh.Fiz.(Sov.Phys.-Tech.Phys.), vol.48,p.1389 (1978).
2. P.N.Afanasyev, Yu.A.Svistunov, V.P.Sidorov, S.Yu.Udovichenko. Studies on Charged Particle Dynamics in Rarefied Gas Media in the Injector of a Fusion Reactor, Preprint issued by Scientific Research Institute of Electrophysical Apparatus, P-V-0759, Moscow (1987).
3. G.E.Derevyankin, V.G.Dudnikov. Production of High Brightness H^- Beams in Surface Plasma Sources: ALP Conference Proc. N 111. Production and neutralization of negative ions and beams, Brookhaven, p.376 (1983).
4. A.M.Astapkovich, Yu.V.Zuev, Yu.A.Svistunov, V.P.Sidorov, S.Yu.Udovichenko. The Dynamics of Plasma, Fully or Partially Compensating Charge of the Ion Beam (to be published).
5. A.J.T.Holmes. A Theoretical and Experimental Study of Space Charge in Intense Ion Beams. Preprint CLM-PS, Abingdon, Oxfordshire (1978).
6. E.B.Hooper, Jr.O.Anderson, P.A.Willman. Production and Flow of Plasma in Ion Beams. Phys.Fluids, vol.22, p.2334 (1979).
7. V.P.Goretsky, A.P.Nayada, L.S.Simonenko. Screen Formation Around the Quasi-Neutral Beam of Negative Ions. Zh.Tekh.Fiz. (Sov.Phys.-Tech.Phys.), vol.54, p.1362 (1984).
8. P.N.Afanasyev, Yu.A.Svistunov, V.P.Sidorov, S.Yu.Udovichenko. Numerical Simulations of Fast Particle Dynamics in a Dense Gas Medium in the Injector of a Fusion Reactor. VANT, Nuclear Fusion series, N 1, p.56 (1987).
9. P.N.Afanasyev, Yu.A.Svistunov, V.P.Sidorov, S.Yu.Udovichenko. Studies on Fast Particle Dynamics in a Dense Gas of the Injector of a Fusion Reactor. VANT, Nuclear Fusion series, N 1, p.28 (1989); Preprint issued by Scientific Research Institute of Electrophysical Apparatus, P-V-0787, Moscow (1988).

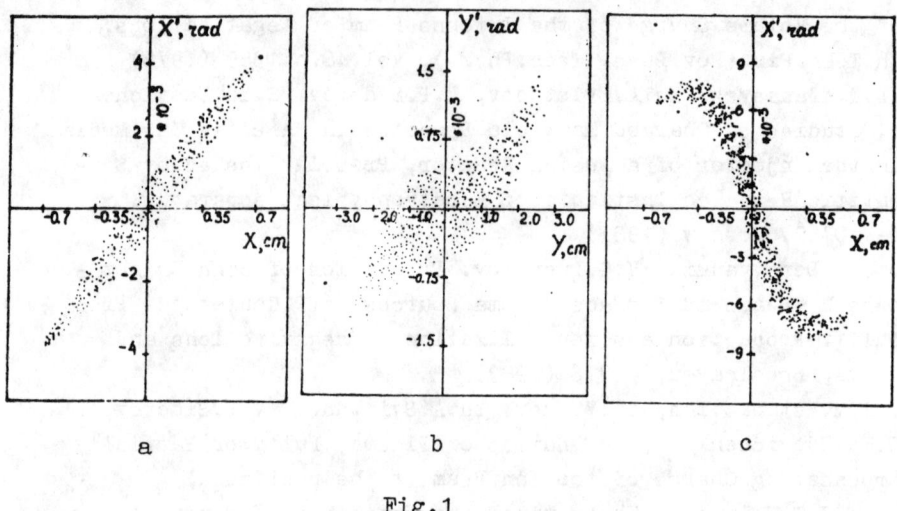

a b c

Fig. 1

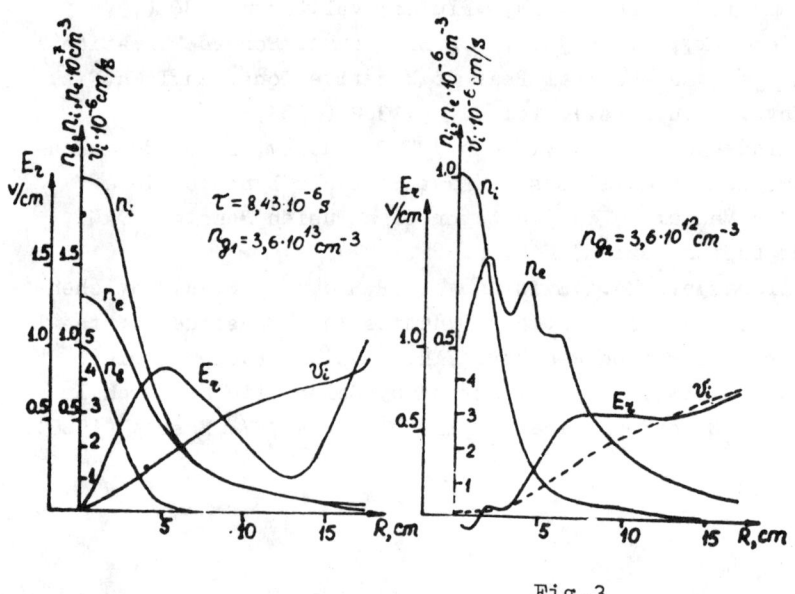

Fig. 3

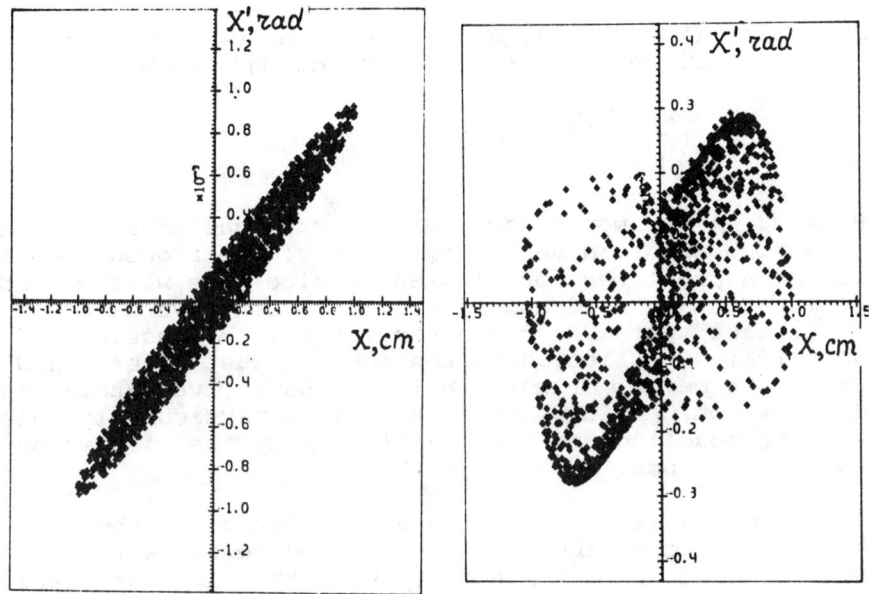

Fig.4 Fig.5

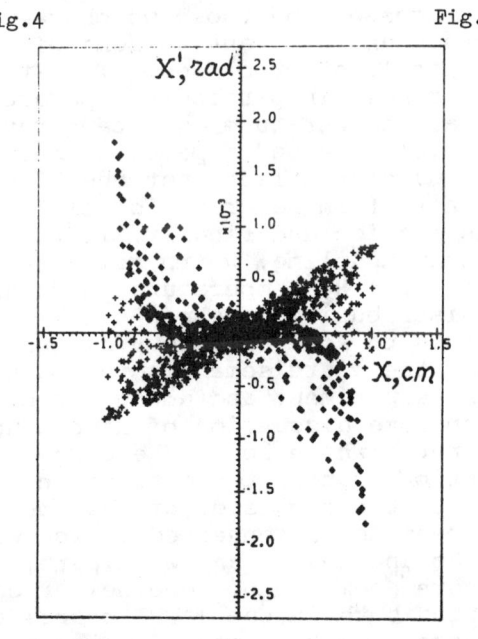

Fig.6

PANEL SESSION: EXTRACTION LOSS AND MODELING OF VOLUME SOURCES

J. Hiskes (Moderator), J. Bretagne, M. Capitelli, L. Elizarov, L.M. Lea, and M. Ogasawara

Hiskes: A few months ago our chairman called me on the telephone and said he thought it would be a good idea if we had a panel session. I wasn't quite sure what he had in mind but he said modeling is very controversial so we'll take that as our charter for this discussion. What I'll do is introduce the subject just briefly and then each member of our panel will have five minutes to discuss some topic which he thinks is particularly relevant to modeling and then we'll open up the discussion to the audience.

Modeling usually begins with a calculation of the electron energy distribution function and we had a few papers on that at this meeting. With the electron energy distribution function in hand, one can calculate the atomic rate processes and those which seem to be particularly uncertain and relevant to some of the distributions of the little eV processes. The atom recombination on the surface, in particular, parametrizes a function of the wall temperature, gas temperature, and vibrational level. We had a paper on neutralization and that reminded me of the fact that the H^- - H_3^+ mutual neutralization to form neutrals is still a completely unknown cross section and finally, if we move on to spacially dependent modeling, we're going to need to know something about the H_2 temperature in detail. The vibrational distributions are now available both experimentally, thanks to some excellent diagnostic work, and calculated and there are some discrepancies between the theory and the experiment and among the calculations. There has been some discussion of H^- production at this meeting, but not a whole lot. There was a paper on, not the H^- production, but on the extraction and the perveance, i.e., problems outside of the source. New mechanisms which seem to be connected to low work function surfaces barium and cesium and with particular reference to cesium (there seems to be some new processes) are being manifested both in the JAERI source and in the Kurchatov source, and the specific processes have not yet been identified. And then, just a general comment, modeling is a function of electron density, provided the electron density is less than 10^{12}. The problem is approximated by four formation processes: little eV, big eV, surface vibration, and atom-wall recombination, (the atom vibration process and the relaxation of the distribution is dominated almost entirely by the wall relaxation process). As we go beyond 10^{12} towards 10^{13}, which is really the density at which the high power discharges are going to operate (high power negative ion sources are going to operate) there is a qualitative change in what is going on, namely; much of the discharge becomes uncoupled from the wall and this will bring us into a new modeling regime in which there is almost no information available, in particular it introduces atom - H_2 processes, and the destruction process becomes the dissociation and ionization by electrons, H^+, H, plus a related set of processes that have been discussed by Capitelli in some detail. So with that, let me call upon our first speaker Dr. Bretagne: five minutes.

Bretagne: I think that I need no more than five minutes. First of all a general remark about this conference. I think that the attention of the papers is drawn from the volume processes to surface processes. I think it corresponds to a question we have already raised. Referring to the problem of modeling, I want to draw your attention about two points. The first one is the need for comparison between modeling and experiments. In my opinion, speaking about modeling without any comparison with experiments is really nonsense. So we also have to use in the model experimental parameters. First of all, we must be sure that we speak about the same type of discharge, in fact, various discharges would lead to various results and various modeling. Also, one point is the geometry of a discharge. We have to take into account (when we speak, for instance, about multicusp discharge) effective plasma volume, and the loss surfaces. Another very important point is the nature and the state of the surface. As we know if you work with a copper surface, very rapidly the surface, due to filaments in the discharge, is covered with tungsten so the nature of the walls is changing during the experiment. Also I must draw your attention on one particular aspect. Sometimes we compare the results obtained with a given discharge voltage or current. So I think that the most important point to take into account is the power density in the discharge. I think the thing to do is to use, for checking of the model against experiments, cross diagnostics. For instance, when we are interested in the measurement of H density, we can use a VUV laser absorption multiphoton ionization, and resonantly enhanced to multiphoton ionization or classical emission spectroscopy. A suggestion that

I could make is to implement different methods of diagnostics on the same source, and at the same time use a different model for the source in order to test both diagnostics and the models. The second point is on the relative importance of volume processes and surface processes. We know, in fact, that walls are very important in many processes. From the results of Richard Hall's group in Paris and the FOM group, we know that walls and the nature of the walls are very important for the formation of vibrational excited states. So, we have to control the walls; also, we need more reliable data. I think there is also a great effort to continue to obtain these data. For instance a problem I just mentioned, is the recombination coefficient of the walls, and a question for a given experiment is what is really the wall? We have to determine precisely the products of the various processes, both for volume and wall processes. For instance, what is a product of a recombination of H_3^+? This problem is quite unsolved in volume recombination and in wall recombination it is just a very short line [H_3^+ + e (wall → H + H_2 (V˜= 0)] but it contains the very last problem of cross section. Thank you.

Hiskes: Thank you. That's five minutes right on the nose. The next speaker is Professor Capitelli.

Capitelli: First of all I want to summarize what has been happening in the last few years for the input data for models, and on the status of electronic cross section. Now we have a better set of cross sections for the associative attachment and the small eV processes presented by Wadehra at this conference. We have a good big eV cross section and now we have a complete set of H_2 - H_2 cross sections. We have a better cross section, a complete set of cross sections for H_2 (v) with atomic hydrogen, and here is a big problem. The big problem is about translational temperature of atomic hydrogen and I want to show you what happens in the very simple scheme, the dissociation into two atomic hydrogens, one in the ground state and the other one in an excited state, it is the dissociation for the repulsive part of the bound state or the complete repulsive state. If you assume a Frank-Condon distribution for translational energy of atoms you can obtain a kind of a distribution so that one cannot speak about a translational temperature. Of course, this distribution will be thermalized in some cases as you have seen this morning. During the conference you have seen that the translational temperature of atomic hydrogen is very hot, in some cases 2 eV and the same, for example, if you look at the translational distribution of protons coming from the dissociative ionization in two typical multigas conditions, in different currents you can see that the translational distribution of protons is far from the equilibrium conditions. You have the cold protons and hot protons coming from a repulsive state, and a repulsive state or a bound state, and you can see that this is an open problem in this kind of an audience. For the problem of H_2 on the walls, we have seen at this meeting two papers, one from Karo and Hiskes on iron, the other from Cacciatore on copper. This kind of result is far from being a coincidence. Then we have a very good paper by John (Hiskes) about the recombination. This is a new process that we must insert in the modeling. It is dissociative ionization and I have already spoken about that. Now in the programs under development at Bari and at Ecole Polytechnique, we can insert all this new important data and allow discussion about the deuterium discharge because of the goal of our modeling. So, well, what about the Lyman β distribution found by Young this morning? I think that this is not new and since modeling can model all things this is no problem. The paper by C. Grose, myself, Bretagne, and Bacal in 1985 already saw this kind of distribution, so no problem. The problem now is a question for Young, if he is present, about the convolution of the experimental data because we spoke about the input data to the modeling but never about the experimental results. Well, and then a proposal (a small proposal) in order to avoid repeating in each meeting the same things, one possibility is to build a so called H^- reference cell which is similar to rf reference cell. We must build several (3 - 5) multicusp cells, for example, to be built by the same company and distributed to different people to do the same diagnostics. A few diagnostics must be the same and other diagnostics cannot be the same, in order that our models have the same cross-sections so that we can compare things. This also leads to the things said before by Jean Bretagne.

Hiskes: Thank you Mario. I hope there are some representatives from the funding agencies here to back up your proposal. The next speaker is Dr. Elizarov.

Elizarov: The modern volume source, in my opinion, is to pass to a new regime with source. It is a low arc voltage regime of approximately less than 10 volts in cesium-hydrogen plasma. This low voltage discharge regime of less than 10 volts exists. Now the H^- current density is in volume sources, in my opinion, limited due to a large amount of positive ions because of recombination processes. I think that future volume ion source will work at arc voltage of less than 10 volts. Thank you.

Hiskes: All right, thank you. The next speaker is Dr. Lea.

Lea: I'd like to talk very briefly about an aspect of modeling which I consider to be very important. This is the one dimensional plasma model of the ion source. I think that it's very important because we talk about the driving region and the extraction region of a volume production source, but, the plasma has to get from one to the other so the filter field within the source is a very important part of it. So the model I am going to describe, very briefly, is being developed at the Culham Laboratory and it's based on the fluid equations, which are moments of the Boltzmann equation, and reference for fluid equation derivation really goes back to Braginskii. The three moments the Boltzmann equation have been solved in time independent form, initially for electrons and positive ions but the aim is to go and solve that for negative ions as well. I will run very quickly through the equations and some initial results which have come out very recently. In fact, the exact form of the fluid equations is described in Phys. Fluid, $\underline{29}$, 463 (1986) with $dq/dx = 0$ and Braginskii type collision terms, and that computation work is being done at Culham Laboratory by Diane Mynor, who is working for her Ph.D. at Oxford University. The assumption she's making at present is that the electron and positive ion densities are equal so that there are no negative ions yet and that different velocities of electrons and positive ions are common. The equation of the zeroth moment is the continuity equation for particles that applies to both electrons and positive ions. Momentum equations describing the gain and loss of momentum, the ions and the electrons, and then the second moment is the energy gain and loss equations again for electrons and positive ions. Within the momentum equation you have friction between the different species and in the energy equation there's equipartition of energy between the different species. Now the work that's been done so far in solving these equations is to use a shooting integration technique, so we start in the driver region and then integrate the equations in small steps from the driver region across the filter and then into the extraction region of the plasma. This does cause some problems because we need a consistent set of starting conditions in order that the integration doesn't crash before we get to the extraction aperature. We have very recently been out to achieve the integration and that's working quite well. Ken Smith and Alvin Glasser, Los Alamos, have done some work on this in the past using time

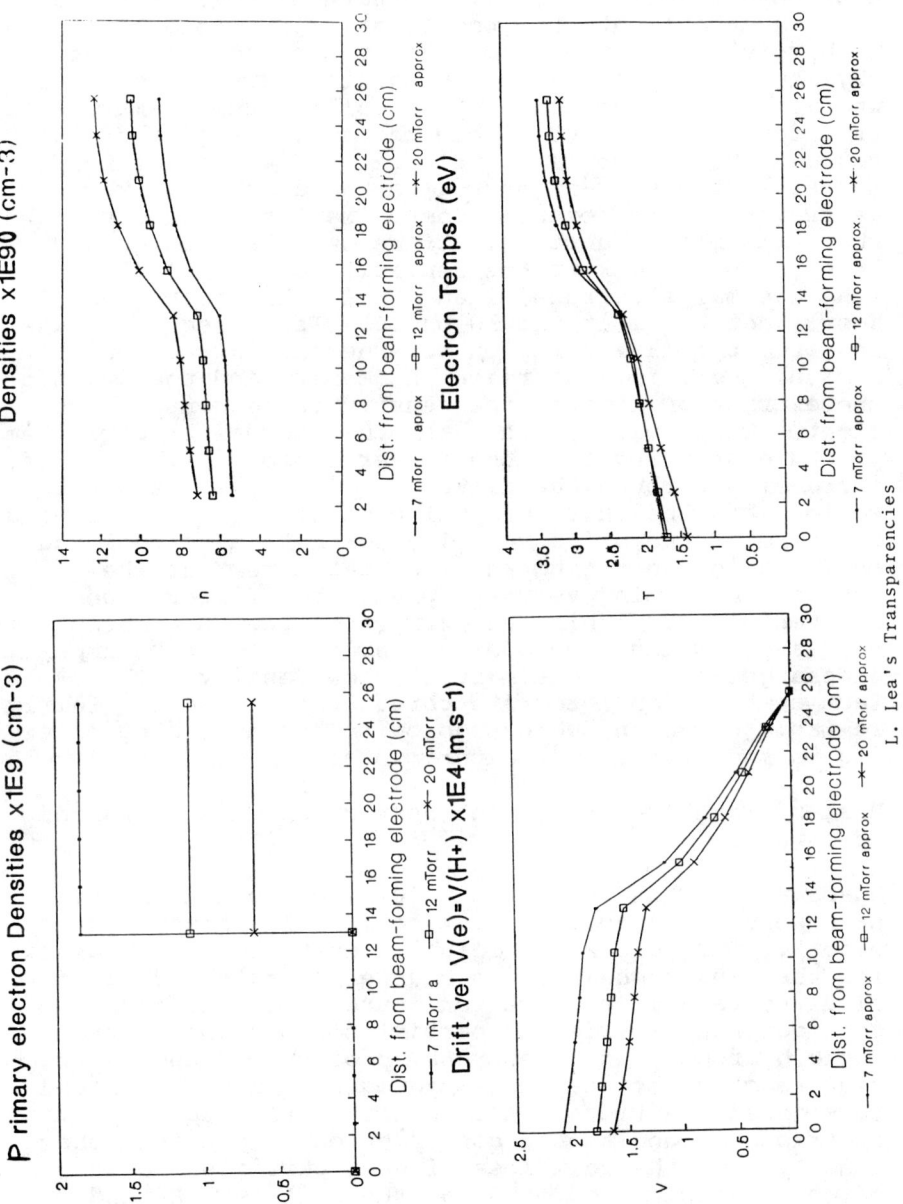

L. Lea's Transparencies

dependent form of the moment equations and they found that that was easy to solve, but of course that takes a lot more computing time. I will just show you some initial results very quickly. We have not got anything in as much detail into the model yet as we would like to. We have yet to put most of the atomic and molecular physics in, but this is just showing very crudely what we have at the moment. We are calling the primary electron density constant within the driver region of the source, the filter is about here, and we've got no primaries in the extraction region and at this very moment we're working to describe the primaries correctly rather than this block function. If we use that though and then go on to look at the densities of the electrons (the thermal electrons) I am afraid this factor of 10 should not be 9, it should be 10. We do see that passing from the driver region towards the extraction region, there is a decrease in density and then within the extraction region, the density is reasonably constant. To go along with that, the thermal electron temperature starting from the driver region of the source, decreasing through the filter and into the extraction region, and the third graph I have here is just showing the form of the drift velocity for both electrons and positive ions starting at zero at the rear of the source, increasing very steeply up the filter, then flattening off within the extraction region. These numbers are tempting to model a source of about 20 amps discharge current, i.e., fairly low density. The fact that we can integrate throughout the source fairly reliably now means we can go on to the next step of trying to model an actual experimental results.

Hiskes: Thank you Les. The next speaker is Professor Ogasawara.

Ogasawara: This is on the volume source and we still have an unsatisfactory point so I will explain our effort. Our code was originally developed by Fukumasa for the single chamber ion source, we extended it to simulate tandem negative ion source. In our model we consider nine species of particles including neutral and ionic hydrogen, and electrons, production, and destruction reactions of negative hydrogen ions and we rated considered reactions. Altogether 30 reactions: Judging interacting conditions, conservation of proton number density, and the wall loss of each particle and the effect of excitation of H_2 at the wall from H_2^+ and H_3^+ are considered. A stationary equation is solved. Next, I will explain modeling. Modeling of the magnetic field is as follows: Primary electrons are completely filtered so no primary electrons are present in the second chamber. Neutrals and hydrogen ions go almost freely through the magnetic filter, since the Larmor radius of a hydrogen ion is larger or comparable to the radius of the magnetic filter. For the hydrogen ion, we take into account the effect of magnetic field and include the effect on electron temperature by the magnetic filter by using experimental results. Next, I'll go to results of the above code. We made several simulations. The best result is a relation between the number density of the negative hydrogen ion and the arc current. The tendency of the current is very much similar to the JAERI experimental relation between negative hydrogen ion current density and the arc current. Next an unsatisfactory point; our code is zero dimensional, since we assume density and temperature to be uniformly in space. Recent source has the very same thickness in the second chamber so one-dimensional treatment must be applied. This is our weak point. The other point is that the electron energy distribution was not taken into account at first, recently we have included the energy distribution but there was a very small effect on the negative ion density, an unsatisfactory result. A qualitative agreement with the experimental results is obtained in almost all the simulations we made, however, there are unsatisfactory results for pressure dependence of ion ratio and the pressure dependence of the electron density, and our current dependence on the optimum pressure. Our recent efforts have been concentrated on these points regarding pressure. We must consider the temperature decrease. Consideration of this temperature decrease actually made the results better. In order to obtain the optimum pressure, we must take into account the increase in stripping loss of negative ions after extraction from the source. To reflect this effect in our model, we tried to increase the reaction rate of the stripping. By increasing the stripping loss rate and by considering temperature decrease, we are now having prospects for clearing the unsatisfactory pressure dependence. Actually, we want to obtain the optimum pressure. Another conclusion is the use of accurate temperature variation and accurate reaction rate is one of the most important factors in source modeling. Regarding the magnitude of a reaction rate, Dr. Skinner told me that we need a common database for reaction rates. I completely agree with him, especially in our code, we must take into account the energy equation and move to the one dimensional code. Thank You.

Hiskes: Thank you Professor Ogasawara. This then brings us to the question portion. We are now going to open up the panel to questions on electron energy distribution functions, first of all.

Holmes: A very simple question, "Do we actually need to know the electron distribution functions, can we not represent it by just two temperatures, because from the plasma modeling point of view you can't easily use an EEDF, you can only use temperatures, the transport equations don't permit us to use the other form".

Hiskes: I'll direct that question to Dr. Bretagne.

Bretagne. I think there are various methods to determine the electron distribution function. In fact, I can agree with you when you say that the electron distribution function in typical sources can be considered as the sum of two Maxwellians. But, you do not know the value of these two temperatures. You have to measure them at least. You can do this, for instance, using a Langmuir probe, but perhaps it's more simple to calculate the electron distribution function. I think that in some cases the dynamic action of EEDF is too high to offer a good determination of the tail of the election distribution.

Hiskes: Does that answer your question?

Holmes: Perhaps to follow that up by saying, we can calculate the temperature using certain plasma processes, if we happen to know the fast electron part of the EDF or temperature because as Leslie (Lea) mentioned, there is an energy transfer rate, equipartition rates between electrons: fast electrons and the ordinary electrons, and between the ordinary electrons and the ions, so we only need to know one temperature.

Hiskes: I'll just make a further comment and that is that for some rates the σv is almost independent of energy; for example, the σv rate is almost independent of energy from 30 volts to 120 volts. On the other hand, the dissociation cross section, the electron dissociation cross section on H_2 is quite energy dependant being rather steep in the 15 to 30 eV range. So that the question of the accuracy of electron energy distribution function depends to some extent on the process that we're concerned with.

Graham: I have a comment about the general work on the EEDF. Most of the time the calculations are actually being done for a very small region of the source, in fact, normally it would be less than half the source, that's the driver region. After you leave the driver region, of course you have got the filter region where the EEDF is going to be a lot different since it's been changed as you go through the filter region. Then, you go into the extractor and I'm wondering if there's any plan to look at the spatial dependance of the EEDF especially as you go through the filter region where we know from some kind of model trajectory calculations of the fast electron that things will get very interesting.

Hiskes: Not only a plan but Dr. Lea has already given us the electron energy distributions through the filter. Do you want to respond to that?

Lea: Well as far as the model that I described, we just treat the thermal electron component in the rear of the source and then passing through the filter into the extraction region. There we can see an energy change and a temperature change, we haven't done any energy distribution function measurements with Langmuir probes in our source other than straight probe measurements which give temperatures and they seem to show that you can fit a Maxwellian at most places within the source to the thermal component.

Hemsworth: Most of the modeling work is aimed at calculating the EEDF for volume sources with magnetic confinement. One can have different types of sources. In fact, you can have a different source for your driver region, for example, where the electron energy distribution can be different from that which you obtained in the volume source. Or, alternatively, one can introduce different groups of electrons in the extraction region as has been done by the people at LBL in the past, so experimentally it is actually possible to modify electron energy distribution. Is it not a reasonable suggestion to ask the modelers, what is the optimum electron energy distribution that would give the best negative ion yield?

Hiskes: That question has been posed in the past and at least with respect to the standard model, the fast electron excitation and thermal electron dissociative attachment, I think that it's safe to say that problem has been solved, that the optimum conditions have been specified. There are a couple of papers in the Journal of Applied Physics by Karo and myself, in which we have identified the optimum electron densities in the two chambers. However, the whole subject is becoming enriched with barium and molecular ions, so quite possibly it's now time to review that.

Hiskes: One more question on electron energy distribution.

Fink: Are we short on low energy electrons or are we short of high energy electrons in most of these sources?

Hiskes: When the discharge current goes up to a thousand amperes, I think we probably have saturated, with respect to the fast electron energy distribution. The thermal electron density and energy is a little less under control as one tends to get what comes out through the filter, so I would say that that's probably the area where we may be short.

Fink: There are some peculiar electron energy distribution around like you said that might tend more to our needs, so that you could alter the filter to try to get a higher percentage.

Hiskes: I think what you want is electron energy distribution at some high energy, maybe 50 volts, no spread in energy dropping abruptly to 1 volt across the filter with no energy separation, I think that's what you want.

Fink: Nothing like that around.

Hiskes: Okay, let's move on to another topic; how about atomic and molecular processes? Would someone like to raise a question about the status of atomic and molecular processes?

Young: If I were Santa Claus and could bring out a rate constant and put it in your stockings at Christmas time which one would you want, that you do not know now.

Hiskes: Who do you want to direct that question to?

Young: To anybody.

Hiskes: Mario Capitelli. If you could have any rate for your modeling which would you choose first?

Capitelli: Well you know that our model is a very complicated model, and to do a sensitivity analysis on the model, for example, is more difficult than to do the modeling. So it is very difficult to select. I was told for a long time that atomic hydrogen was a very important part in forming the distribution, in destroying the distribution. Well, in your case for example, I am sure that you can not explain your distribution by the arrival. I see that Marthe (Bacal) doesn't agree, but I think one possibility, that you can explain your distribution involving the fast atomic hydrogen. Anyway I think that the most important process is the small eV. The Big eV , (I have) one big question about the Hiskes rate, I think we can call it the Hiskes rate. The big eV is about the possibility that when you pump the singlet state you can never have some collisional mixing that destroy the pumping, so no cascading but also spreading to the other levels.

Hiskes: Did you try to calculate the collisional linking between the singlet states by electron (because in some cases you have a lot of electrons) against the radiative losses. We take into account the destruction of specific vibrational states by ionization and dissociation.

Capitelli: Not in the model, in the rates. For example, when you calculate rates you probably need to know the rates, for singlet state excitation, then the ability for Frank-Condon's factor to cascade. Well, now when you excited the level you can have some collisional linking between the electronic state (the singlet state), and I wonder if this kind of linking can decrease your rates.

Hiskes: Well, that's a question of a time scale. Once you form a particular excited electronic state it decays in about 10^{-8} seconds which is very short compared to the next collision process, so the next collision process takes probably 10^{-6} seconds.

Capitelli: Yes, I know that there is a time scale, the problem is to calculate in a good way the time scale. In some cases, because they are very near an electronic state with a singlet state, the cross section can be very high, so your time scale can overlap with the radiative time scale, I'm not sure. This is one of the problems that I'm asking myself these days.

Hiskes: Well I have not looked at that problem in recent years. When I did look at it, it looked to me like the time scales were quite distinct.

Belchenko: I have a question. Why is it necessary to excite the vibrational molecules only with light particles, like electrons? Is it possible to excite the vibrational molecules by maybe fast neutrals, protons or neutral atoms whose energy is approximately 50 eV or 100 eV?

Hiskes: The cross sections are roughly comparable for the same relative velocity, you need several keV, probably 5 - 10 keV protons to get the same excitation rate.

Capitelli: I think that this is not a problem especially when you have a translational distribution for atomic hydrogen. If it is fast enough and also if your translational distribution is not a Maxwellian one, it is one problem so you can excite because until this moment we talked of atomic hydrogen as a quencher of vibrational energy. I think that you can be right in that atomic hydrogen and can also excite a vibrational energy because of the gap. I think that when you balance the two rates: the activation and the pumping by atomic hydrogen, I think you can have some surprise in this sense, so you can pump vibrational energy also through atomic hydrogen. With protons I don't know, protons have translational distribution function which is very similar to atomic hydrogen.

Hiskes: Does somebody else on the panel want to comment on that question?

Holmes: How well do you know all the rate coefficients? Some are calculations I presume because a lot of the modeling must depend on whether we know these cross sections.

Hiskes: You mean all the rate coefficients, there is a large number of rate coefficients.

Holmes: The more important ones.

Hiskes: I think that when we get above the first few excited states, whether they be electronic or vibrational that almost all of our information is theoretical, at least guided by theoretical work.

Holmes: So there about 10%, 20%, and 50%.

Hiskes: Ranging from 0 to a factor of 2, I would guess. It is certainly true that in this business we've had to rely heavily on theoretical rate processes. Any other questions on atomic and molecular processes?

Graham: Will there be a problem in modeling deuterium or is it?

Hiskes: I do not see a problem modeling it. But I see a problem that you have more vibrational states about 50% - 60% more so you have a lot more detail to concern yourself with, but in principle, there's no problem. In principle, the first approximation there's not going to be a whole lot of difference. But I just think it involves a lot more detail. Maybe someone else would like to comment on that.

Capitelli: When you shift the vibrational distribution function to higher levels you must know the rate coefficient involving higher vibrational levels of deuterium. The problem now is to understand if there exists the same scaling laws in deuterium as compared with hydrogen. Some things have been presented by Wadehra two days ago. We have studied deuterium and our first results showed that there is no big difference in the cross section. You cannot put the data on hydrogen and shift only by the threshold. With scaling laws, the problem is more complicated. I think that for the next conference the code can give us the first result of the modeling deuterium. I am hoping the next conference will be done in three years.

Skinner: It is not a question but a suggestion. At least half a dozen groups are doing modeling work, and I really do believe that one of the best things that we can do would be to get together a set of cross sections and a computer program that we can share and to specify which cross sections we're using, because there's a lot of choice that goes into the cross sections and a lot of uncertainty. At the very least, each of us should be able to check the others work by using the same cross sections. We may not have to make the same choices in our work, but at least we should be able to work from something we have in common that we can use the other cross sections if we want to. I'm not volunteering to do it, but it is something that really needs to be done. It was started in Los Alamos for the MHD code but wasn't kept going. It's something we need to do to be able to make our research meaningful and be able to do good comparisons with our work.

Hiskes: A question. Of course, part of the reason for holding meetings of this sort is to flush out those kinds of information.

Allison: Ken Smith and Alvin Glasser have published such a compendium on cross sections.

Hiskes: Yeah, we have lots of compendiums of cross sections, I think that Daryl (Skinner) is posing the question specific to those most important processes in the codes relative to negative ions, if I understand.

Capitelli: The problem is that all of the sets of cross sections have been published on hydrogen by different people involve only the characterization of the vibrational ground state of hydrogen, so this is the big problem. This kind of a modeling requires the knowledge of the dependence of a cross section on the vibrational quantum number. Without this knowledge we cannot be sure of anything. This must be clear. And everytime we do modeling without knowing the dependence of the cross sections on the vibrational quantum number, it is only a semi-imperical calculation to reproduce some experimental data.

Cacciatore: Turning back to the accuracy of the rate constants which are used in the modeling, the rate constants of $H_2 - H_2$ that we have are accurate within 20% - 30%. Anyway, the accuracy of the rate constants is really a big problem. I don't believe this is a computational problem, this is a dynamical problem. In order to have a very accurate cross section, the first point is to have an accurate interaction potential between the two colliders. If the interaction potential is not known, it is easy to produce incorrect rate constants and a little bit better in a way. For interaction in the wall region, things may not be as good as for the interaction in the gas phase. In addition to the not well known interaction potential between the molecules in the gas phase and the atoms in the lattice surface, there is also another problem to be solved. It is important to introduce in the dynamics the coupling between the internal motion of the molecules and the internal degree of the surface; this is a very important point to consider. We know that the surface is very important: If the surface is not a flat surface, for instance, and it tears (fissures) or things like that, the reactivity of the surface is completely changed. Also the impurity is very important. Of course it is also important to see whether the surface is a clean surface or not. So to simulate from a dynamical point of view, all this physical situation, it is rather difficult. We can not just consider the molecule and or atoms, colliding with a flat, perfectly clean, surface or perhaps with some atoms absorbed on it. I think this is very difficult.

Hiskes: Thank you for those comments. I would like to invite questions on the vibrational distributions either experimental or theoretical. Does anyone want to comment on the diagnostics or on the specific modeling calculations?

Hopman: It's clear that the methods (of determining) vibrational distributions are dissimilar. I don't think that this really has to do with a measurement technique, by itself. But there must be something different in the sources that are used and the only thing that I have discovered so far in the case of the LBL source, you direct with a laser beam also an atomic beam of mercury into the source. This mercury is captured in the plasma and will stay there for a very long time. I first made a very rapid estimate that the influx is of the order of 10^7 atoms per second, so it takes a long time to build up a mercury concentration in the source. On the other hand, it also suggests that such mercury contamination can be avoided very easily. I don't know what is the rep rate of your laser. If there is a high rep rate it becomes more difficult. You can always use some wheel with a hole in it and let the laser beam pass through the hole. The mercury atoms are very slow so that the laser beam can reach the plasma before the mercury and you reduce the contamination. If there is any contamination, you reduce it by an enormous amount. My suggestion to the LBL group is to look carefully into this possibility further. These things are not equal, we have to look at our experiments and to try to find out what really is different there.

Hiskes: Thank you, I'll interpret those as comments not really appropriate. Although we have a potential response there in the back from the LBL group.

Young: We never looked directly for mercury in volume discharge. We never really thought about atomic contamination. It is not a straight line path from our VUV system into the source. The vapor pressure is relatively low at room temperature, it has to reflect off several surfaces, it would need to be a diffused scattering in order to get from the mercury oven to the source. If it were really a problem and if people really thought that there was mercury contamination, it turns out that for the wavelength regions you use in order to probe those vibrational states of mercury you can use windows. It turns out that for mercury, the wavelength in the VUV is above the lithium fluoride cutoff limit, so there is always that option if people were really worried and really felt that it was polluting our measurement. But at this point, I really have not done the calculations, we have not looked specifically for mercury and I am not sure what kind of a density of mercury one would need to study in order to predict what kind of perturbation it would do for what it's worth.

Skinner: If you want to make the best negative ion source that you can, then you want to start with the best source of vibrational excited molecules that you can make, and it seems to me that you have to pick one of the methods of exciting molecules and optimize for that mechanism. And there are at least four possible mechanisms I know about, and maybe more. One is the

Hall process of making vibrationally excited molecules through recombination on surfaces. The calculations done by Hiskes of making vibrational excited molecules through collisions of H_2^- and H_3^+ on the wall. There is the big eV process (high energy electrons). But little eV electron process, that is the thermal electrons could also give a very respectable distribution. It seems to me that each one of those processes requires a very different kind of device. The first two are on surfaces so you want lots of surface - the two that are driven by electrons maybe up to a very different kind of EEDF that minimizes things you don't want, like dissociation. So my question is: Which of these four and others that you can think of (if that's the most important of the dominant processes) can give us the best source of vibrational excited molecules i.e., to give us the highest fraction of vibrational excited molecules in the v equals 6, 7, 8, and 9.

Bretagne: I want to say some words about the respective advantages of two methods: VUV laser absorption and multiphoton absorption. I insist on the fact that it will be very interesting to use two methods on the same source for measuring the atom density and the vibrational distribution. I disagree with the comment and answer which was given this morning by Dr. Young about the difficulties in the MPI methods to determine densities. In fact, my group in Orsay succeeded in determining the densities of H atoms using MPI. We proved that this method is very sensitive.

Young: Sensitivity is the problem. As long as you have no background I would agree that MPI is much more sensitive than absorption, I think it is more difficult to quantitate the signal that you get because, inherently, it is a multiphoton process. And because it's a multiphoton process, you know that the zeroth order expectation goes as a square of the laser flux or the power density. People have found that is not always the case and not only is it not only the case, it can vary from laser system to laser system. You will read that there are comparisons within one laboratory that has access to many different kinds of lasers and they will find that you can have various power scaling depending on the laser that you use. All that aside, I do not want to denigrate the MPI technique, I think it is a marvelous technique and like I said, much more sensitive than laser absorption can be. However, laser absorption does have a very special advantage, at least in our case that it does not make any difference that there are lots of other photons around, or that there are lots of other ions around, or that there are lots of other electrons around, we are not limited by the background and that is one problem in using the MPI in doing ion source diagnostics; you can't look directly inside the source, which is where you really want to look. You need to (in the case of the FOM group) have the gas effusing outside the source and probing outside the plasma chamber, which may be a valid technique. I am not going to comment on whether it is valid or invalid. All I can say is that we look directly inside the source chamber, and that may be the biggest difference between the two experiments, that we look inside and that the other experimental group looks outside.

Stern: It's really a comment on Tony Young's comment on FOM. After all the laser technique integrates across a cross section so while you are looking at the inside you really are averaging over the entire size, and of course, if you have two different distributions at two different points in space that might account for your bimodal distribution. So both techniques are really not totally pinpointing the space resolve techniques.

Hiskes: Thank you. Now I'd like to entertain a few questions on any subject relevant to negative in generation.

Jacquot: Concerning Dr. Elizarov, it seems to me that the very important point concerning the voltage of the discharge to be less than 10 volt, we put cesium to decrease the voltage in the discharge while we know that the cesium is an enemy for the Tokamak. I was just given a number for a Tokamak like NET or ITER with 1000 cubic meters. The estimation of the total flux equivalent in milliampere of cesium into ITER to increase the Z by 0.1 is of the order of 30 mA. That means that we have to compare this 30 mA of cesium flux to 100 ampere of neutral atoms that means it is a big charge. I have two questions concerning the role of the cesium in your source. First one, can you do the same things with alkaline as sodium or lithium? I know that the ionization potential of these alkalines is high. My second question, can you test the production of H^- in a Q machine where the ionization of the cesium is produced by thermal processes on the cathode without discharge, and try to produce H^-?

Elizarov: No, cesium is not perfect. In our estimation comparing cesium with lithium and sodium in my opinion, best results will be with the cesium.

Hershcovitch: Before we get too far with this topic I would like to put something on record. My question is to Elizarov. Belchenko told me when he first came here that you measured the energy distribution of H^- ions at one point, and from your measurement, it looks like a lot of them are surface produced, is that correct?

Elizarov: Yes, we analyzed the energy of all particles. We introduced an additional surface, into the discharge with negligible increase in H^- current.

Hemsworth: I'd just like to comment or reply to Claude's (Jacquot) question on the use of cesium for the sources for ITER. The Russian proposal for the ITER injector with cesium included an estimate of the total flux of the cesium from their sources into the plasma. The ITER team also estimated the maximum allowable flux and the Russian proposal was well below the value that they would accept, so it was an acceptable flux.

Hiskes: Either Belchenko or Hopman. Are you responding to something that was said a moment ago?

Belchenko: I want to say that, of course, if you have the energy spread in the source a little wider than the voltages of discharge and if you have noisy discharge it's possible to explain that these negative ions may be produced in the volume. When you introduce a surface to the source, which has no influence on the negative ion yield, it is not possible to explain that this is not a surface process. Because we can remember the old experiment produced here at Brookhaven; they introduce an independent electrode in the Penning source, and when they applied the voltage, they had no increase and even a decrease in the negative ion yield. It is necessary to introduce a surface in the right way and to look for a very thin layer of the plasma and to support the big flux of primaries on the surfaces. And another comment to Dr. Jacquot's question about the production on other surfaces like lithium or sodium. We checked it in our surface plasma sources with a big variety of compounds, we used sodium, and we used lithium, and so on, and they, of course, produced less of a yield than cesium maybe two times, maybe three times, and so if you need to have low Z you can use metals with low Z in the surface plasma source, but you have a smaller yield by a factor of two or three.

Jacquot: But a factor of two is not dramatic. If you can put 100 milliamperes per square centimeter with cesium and if we produce 50 milliampere per square centimeters with sodium I will take that source immediately.

Belchenko: If you want to have maybe three times less density it is better to use lithium.

Hiskes: Actually an opened question.

Hopman: I actually have a question to the panel. What do they see as an explanation for the improved performance of the JAERI source as soon as cesium is introduced into it?

Hiskes: Professor Elizarov has had the experience with cesium.

Elizarov: I think that in the Japanese source the main processes are surface not volume.

Capitelli: I had some experience, a long time ago in measuring the vibrational distribution of nitrogen, not hydrogen. In a discharge of nitrogen, you put cesium, Na, and so on because of a very strong exchange, strong rate of the reaction between the alkaline with the vibrationally excited molecules. I think that when you put the cesium in a volume source, the vibrationally excited molecules decrease. It is my feeling that the production of negative ions comes from the interaction between the molecules and the surfaces because vibrationally excited molecules decrease in the source when you add alkaline metals.

Bretagne: Another thing in the discussion about the different aspects which got me involved in source improvement. The first one was about the electron energy distribution function. Certainly the introduction of cesium in the discharges changes drastically the electron energy distribution function. Without having done any calculations, it is very difficult to answer this question. We can at least try to answer this question, if we have cross sections. Even if the set of cross sections is not complete, we can try to study the effect. In fact, my opinion is that probably there is also an important effect.

Prelec: My comment should come after Uri's (Belchenko) comment of the Brookhaven Penning source. That was a very crude experiment done in I guess, '75 or '76, the first time that we tried a converter electrode in a diode type source. Due to the crudeness of the design we couldn't go very high with the voltage of that converter; it would spark over. I guess that was the explanation why we could not see any effect, the voltage may have been 10, 20 volts, so this is not an argument why that surface did not work.

Fink: What it seems to me, as I understood the experiment at JAERI, was that they maintained the discharge voltage. They have put in the cesium, they maintained the discharge current, so they must have the cooled the cathode, the filament must have been cooler to reduce the emission that would come out thermally because the activity of the cesium would make electron emission easier. And if that is the case then we did not shift the electron energy distribution. Not true?

Hiskes: No.

Hemsworth: No, cesium does not live on hot cathodes bombarded by ions.

Bretagne: No cesium on hot cathodes.

Fink: It would have to be much cooler to run those things practically off.

Mori: Okay, one comment on that cesium problem, actually at KEK we have developed the volume source also for H^- product and we already put cesium into a volume source and we have found a very interesting effect: By putting the cesium instantaneously, the plasma conditions changed. Because the arc power supply is current controlled, the voltage changed so very dramatically by putting the cesium. We could not see any increase of the H^- current at that time, but 15 minutes later, the H^- current increased by a factor of 4 without perturbing initial condition. So I strongly feel that most of the H^- beam might be produced at the surface of the plasma electrode or somewhere else, but anyhow this is a surface effect. A comment on the beam shape: Before we introduce the cesium into the plasma chamber, the beam is a little bit noisy, about 10% - 15% noisy. After we put the cesium and wait around 10 minutes to 15 minutes we couldn't see any noise, the beam is very stable, and its shape is very similar to a converter source. Okay, that is why we feel that most of the H^- beam might be produced at the surface.

Hiskes: Thank you. Would somebody from JAERI like to comment on those comments?

Okumura: We do have not enough data to conclude that the production is mainly from the surface, but, from our present experience, we feel also that it is mainly surface as we showed (in my presentation) the hysteresis of H^- ions when we increase the temperature. We also measure ion and electron temperatures and the ion species ratio. We did not observe the hysteresis on the electron temperature and ion species. The only explanation for such a hysteresis is if we increase the temperature of the plasma grid, the cesium evaporates from the surface, then we decrease the temperature, then the cesium attaches the surface, then the surface becomes very clear. I think that that is the reason why we observe the hysteresis. From those experiences, we feel very strongly that the reason for H⁻ current enhancement is a surface effect. A different problem, I would like to know that if the electron temperature decreases from 1.2 eV to .9 eV, can the H⁻ current increase by a factor of 2? Is it possible to obtain such a gain by the decrease of electron temperature?

Hiskes: I don't think dissociation cross section rates vary that rapidly in that narrow range, so it is maybe a change of electron density or something. Joe Wadehra is not here but I think I can speak on that point.

Whealton: We heard about the results that Elizarov presented on his source. He has enumerated again the philosophy of having a very low arc current under 10 eV. This is in variance with a lot of ion sources, other volume type ion sources. The question I have is, do we understand the processes? I do not personally, but do we understand the processes of negative ion formation in Elizarov's source or are there some mysteries at this time?

Hiskes: It was listed on the program, here on the subject, that was still uncertain, so why do not redirect the question to Professor Elizarov.

Elizarov: No. We do not have the experimental distribution of H_2 excited molecules. We have no dynamic electrons in our discharge, we have no experimental data in many regimes we operate. I want to note one process, a chemical process; the creation of C_sH. It is known that excited cesium has cross section for creation of C_sH of approximately 10^{-14} cm². This process maybe intense in volume, but we did not study this process only.

Jacquot: A dream of one of my colleagues in Cadarache, Dr. Fumelli, is to realize a negative ion source as you realize an electron gun, that means without plasma. My feeling is that Elizarov's source goes in this direction a little bit.

Hershcovitch: There are a lot of resemblances between Elizarov's source and our source. We also inject the plasma from a hollow cathode although in a different direction: Elizarov's has a circular geometry while we are injecting it from the side. We were injecting the cesium in a similar fashion. But I can tell you this about hollow cathode discharges: they produce very dense plasmas, as close to pure or 100% ionized plasma as one can get, unlike other source plasmas.

Bacal: It is a question to Dr. Elizarov about the philosophy of these low voltage discharge in his source.

Editor's comment: Dr. Elizarov was having difficulty expressing himself in English. Dr. Bacal was translating for him.

Hiskes: Could we have a translation?

Elizarov: My calculation is published.

Editor's comment: Dr. Elizarov is referring to a paper by Baksht, Elizarov, Ivanov and Yur'ev, Sav. J. Plasma Phys. $\underline{14}$, 56 (1988).

Bacal: Yes, but could you tell us what is the important physics, why is it better?

Elizarov: No positive ions: H^+, H_2^+, H_3^+ and atomic H. Vibrationally excited H_2, we accept many processes where we need H^- generation.

Hiskes: Can you resolve that ambiguity in Russian and translate it.

Bacal: Well he says the only positive ion is Cs^+ and there are no H_2^+, H_3^+, H^+ positive ions. Now I did not understand, why he said it is an advantage.

Hiskes: Somehow I have the impression there are no positive ions in here at all.

Elizarov: Cs^+. Plasma in our source is not traditional, the main positive charge is Cs^+ the main negative charge is H^-, our plasma is Cs^+ and H^-.

Jacquot: In Q machine you have Cs^+ with an ionization of 100%, no Cs^0, put a little bit of hydrogen and see if you can see H^-.

ACCELERATION AND NEUTRALIZATION

THE MATCHING OF SPS TO AN ELECTROSTATIC NEGATIVE ION ACCELERATOR

G.I. Dimov
Institute of Nuclear Physics, 630090, Novosibirsk, USSR

There are no proven approaches for a steady state electrostatic acceleration of multiampere ion beams with energies up to 1 MeV. At a high current density, the distance to accelerate is limited by ion space charge, and the electric field turns out to be very high. As a result, it is not certain that the acceleration can be realized. At small current densities, the cross section of the beam becomes very large, which can lead to prohibitive aberrations of the beam. Beam transport with a small aberration is easier to achieve if the cross section is changed from round to a rectangular one.

The distribution of potential along a space charge limited H⁻ ion beam is

$$U_{kV} = 69.7 \; j_A^{2/3} \; Z^{4/3} \tag{1}$$

The corresponding electric field is

$$E_{kV/cm} = -32.16 \sqrt{j_A} \sqrt{U_{kV}}. \tag{2}$$

Here j_A is ion current density in $A \cdot cm^{-2}$, and Z is the distance along the accelerator including the source extraction gap, measured in cm. At a current density of $j = 30 \; mA \cdot cm^{-2}$ the total voltage is $U = 1.5$ MV, and the acceleration distance is $Z_a = 58$ cm; the electric field at the end of the accelerator increases to $E_a = 35$ kV $\cdot cm^{-1}$.

In principle, it is possible to substantially increase the acceleration distance by changing the normal diode potential distribution (1) at the end. If the electric field is reduced from its normal value $E_1 = 26$ kV $\cdot cm^{-1}$ to 0 by means of electrodes at the potential $U_1 = 500$ keV ($Z_1 = 25.3$ cm), for $Z > Z_1$ the corresponding distribution of potential will be defined by the equation[1]:

$$Z^* = Z_1 \left[1 + \sqrt{\left\{\frac{U}{U_1}\right\}^{3/2} + 3 \frac{U}{U_1} - 4}\; \right]. \tag{3}$$

The corresponding electric field is

$$E^* = -\frac{4}{3} \frac{U_1^{3/4}}{Z_1} \sqrt{\sqrt{U} - \sqrt{U_1}}. \qquad (4)$$

In this case, and for $j = 30$ mA $\cdot$ cm^{-2}, the acceleration distance increases to $Z_a^* = 106$ cm and the electric field strength at the end of acceleration decreases to $E_a^* = 22.5$ keV $\cdot$ cm^{-1}.

It is not, however, possible to reduce the field at point Z_1 from 26 keV $\cdot$ cm^{-1} to 0 without a transition element. In the part where $E \approx 0$ along the length l_1, due to transverse repulsion the ion beam will become defocused with a corresponding focal distance

$$f_1 = 7.7 \cdot 10^{-3} \frac{U_{1kV}^{3/2}}{j_A \, l_1}. \qquad (5)$$

If $l_1 \simeq 10$ cm, then $f_1 \simeq 290$ cm and corresponding focal length of the beam at the accelerator exit is $f_a = f_1 \sqrt{U_a/U_1} \simeq 4$ m. We can, in principle, compensate this defocusing effect with a single electrostatic lens surrounding the beam in the region $Z = Z_1$.

In order to preserve the rectilinearity of the stream of ions in a limited cross section, the geometry and potentials of surrounding electrodes must ensure a potential distribution along the beam surface given by equations (1,3) and must compensate the transverse electric field due to the beam space charge. The electric field in a ribbon shaped beam is

$$E_{kV/cm} = -258.6 \frac{j_A}{\sqrt{U_{kV}}} x. \qquad (6)$$

where x is the distance from the center of the ribbon and perpendicular to it.

The problem of the geometry of acceleration electrodes becomes more complicated if it is required that accelerated ions do not hit the electrodes. Ions intercepted by the electrodes cause their erosion and can change potential distribution between electrodes in case a low-power divider is used. The latter effect may be avoided by using a multi-stage rectifier. We can avoid the erosion only be providing some clearance between the beam and the electrodes.

Multi-aperture sources without magnetic field may be attached directly to the high voltage end of the accelerator tube. In this case, for ion-optical matching, it is enough to have an average source current density corresponding to (1), uniform over the emission surface. However, in this case, accelerator performance may be unreliable due to high voltage break-downs to the multi-aperture ion-optical system of the source.

The multi-aperture negative ion sources of the SPS type operate in a transverse magnetic field of ≃ 0.5 kG. If the SPS is attached directly to the accelerator tube, magnetic field will penetrate into the tube and the curvature of the magnetic field lines will favor an accumulation of electrons on the tube axis. This can lead to the development of a Penning discharge. Moreover, it is necessary to have a space for removal of cesium atoms diffusing from the SPS. If the ion source is moved away from the accelerator tube, high voltage break-downs are eliminated and the hydrogen flow from the source into the accelerator region will be reduced due to a differential pumping. However, in this case the problem appears of the beam transport from the source to the accelerator tube and the problem of ion-optical matching becomes significantly more complicated.

Ribbon shaped beams may be obtained from SPS with the slit either perpendicular or parallel to the magnetic field.

In the first case a wide magnet with a small gap is used. The unfavorable fringe magnetic field decreases fast. The beam aberrations may be substantially reduced if the magnet gap is somewhat increased. In the second case, the magnetic field is narrow and the poles distant so that it becomes necessary to ensure the field uniformity by using ceramic magnets and transverse ferromagnetic louvers placed at the boundaries. Because of a large gap between magnetic poles, in this case the situation with the penetration of the magnetic field and with beam aberrations at the magnet exit is less favorable. However, there are some advantages: A uniform current density of H⁻ ions over the emission surface of the source can be achieved more easily, and a beam transport in the bending magnet with space charge neutralization by positive ions becomes possible.

Table 1 summarizes main SPS parameters achieved in Novosibirsk in pulsed and quasi-steady regimes[2,3]; it also shows future parameters of a steady state multi-ampere source.

The transparency of the anode which serves as the emission surface for H⁻ ions, is only 5 - 6 % due to the geometrical focusing of ions in the discharge. Because of that, the gas efficiency reaches 10 - 20%. The measured cesium flux from the source is about 10^{-4} g/hour at an H⁻ yield of 1 A. Table 1 shows that the H⁻ current density in SPS surpasses the capabilities of the electrostatic acceleration, if a special system of the ion-optical transformation of a beam between the source and the acceleration tube is not used.

Table 1

Parameters	Experiment		Extrapolation
Pulse Duration	0.8 ms	0.5 s	Steady State
H⁻ Current, A	11	1	10
Average Emission Current Density, A cm^{-2}	200	75	50 - 100
Discharge Voltage, V	150	100	100 - 120
Power per Unit of the Surface in Discharge, kW cm^{-2}	1.6	0.6	0.4 - 0.8

We shall consider one of the possible schemes to match an SPS with a 10 A H⁻ accelerator at a relatively high current density of 30 mA/cm^2. The H⁻ emission current density is taken as j_o = 40 mA/cm^2 due to some divergence of a beam during the transport. The corresponding area of the emission surface is then equal to 250 cm^2. The beam cross section is chosen as 25 x 10 cm^2, where the smaller dimension is perpendicular to the magnetic field. In this case, the beam from the source can be transported through a bending magnet, which acts as an extension of the source magnet.

Figure 1a shows how the bending magnet connects the SPS to the acceleration tube. In the radial direction, 180° bend of the beam in a uniform magnetic field transports without distortion the emission image of the beam. The longitudinal cross section of the transport system and the cross section of the magnet are shown on Figs. 1b, 1c. Magnetic field in the bending magnet is slightly stronger than the field in the source and this ensures a favorable curvature of the field lines in the ion optical system of the source to dump accompanying electrons along magnetic field lines by the extraction electric field. Beam focusing in the other direction can be achieved by a small curvature of the ion optical system of the source (for a slit system slits are directed along the smaller dimension). The bending magnet ensures an efficient capture of cesium atoms on surfaces before reaching the acceleration tube. Ferromagnetic louvers used to limit the magnetic field, have a high transparency and make possible a pumping of hydrogen and cesium vapors from the bending magnet. Positive ions, serving to neutralize the space charge of the H⁻ beam, cannot penetrate into the acceleration tube due to a repulsing electric field. Ferromagnetic inserts, located in front of the acceleration tube, change the curvature of magnetic field lines in this region so that the electrons can be dumped along the field lines by the electric field of the acceleration tube.

Ferromagnetic louvers, in the shape of magnetic equipotentials, can be added on both sides of the region of interest if it is necessary to conserve a two-dimensionality of a magnetic field at the exit of the bending magnet and in the region of ferromagnetic inserts.

The optimal magnetic field is about 400 G in the discharge chamber of a steady state SPS. The field in the bending magnet has been chosen to be B = 415 G. The average bending radius R equals 70 cm for an H⁻ ion energy of eU_o = 40 keV. The gap between poles has been chosen to be h = 31 cm in order to reduce beam aberrations at the exit of the magnet. This is 25% higher than the beam size along the magnetic field.

We shall estimate aberrations for a two-dimensional field geometry and without taking into account the effect of ferromagnetic inserts. In this case, the magnetic field is described by nondimensional equations:

$$Z = \frac{1}{2\pi} + \Phi + \frac{1}{2\pi} e^{2\pi\Phi} \cos 2\pi V; \quad (7)$$

$$y = V + \frac{1}{2\pi} e^{2\pi\Phi} \sin 2\pi V;$$

where Z is the distance from the end of magnet poles in a direction of the decreasing magnetic field, normalized to the gap width h, y is the distance from the middle plane of the magnet in units of h; Φ, and V are the magnetic flux and the potential, resp., in units of Bh. In comparison with the ions moving in the middle plane, the ions at the edge of the beam will have a bending angle smaller by $\Delta\phi$ and will have a radial shift ΔR outward.

$$\Delta\phi = \Delta\Phi \frac{h}{r}; \quad \Delta R = \Delta x \frac{h^2}{R}; \quad (8)$$

where

$$\Delta\Phi = \Phi(y,Z) - \Phi(0,Z);$$

$$(9)$$

$$\Delta x = \int_{-\infty}^{Z} \Phi(y,Z) \, dZ - \int_{-\infty}^{Z} \Phi(o,Z) \, dZ.$$

For y = 0.4 and Z = 1.66, we obtain from (7,9): $\Delta\Phi$ = 0.017; Δx = 0.091. From (8) we obtain: $\Delta\phi$ = 7.5 10^{-3} rad, ΔR = 1.25 cm. The angular deflection $\Delta\phi$ at the exit of the accelerator is reduced by the factor

$$\sqrt{U_a/U_o}$$

to a small value of 1.2 10^{-3} rad. The shift ΔR is more dangerous, However, a limitation of the source emission surface can compensate this value without a serious decrease of the beam current.

A two-dimensionality of the fringing field of the magnet does not have to be necessarily preserved because the fall of the magnetic field is faster without this requirement. However, estimates of aberrations are more complex in the latter case.

We shall consider now the H$^-$ beam space charge compensation process in the bending magnet. Fast negative ions ionize hydrogen molecules. Positive ions, drifting across the magnetic field, leave the beam. Shielding electrodes (7 on Fig. 1c) can practically prevent ions from leaving along the magnetic field lines. Electrons are rapidly dumped along the magnetic field. At the same time, ions are destroyed by hydrogen molecules, positive ions, and electrons. Cross sections of the fundamental elementary processes for H$^-$ ions with 40 keV energy are given in Table 2.

Table 2

Process	Cross Section, cm^2
1. $H^- + H_2 \rightarrow \begin{cases} H_2^+ (\simeq 0.1 \text{ eV}) + \ldots \\ 2H^+ (\simeq 4 \text{ eV}) + \ldots \end{cases}$	$\sigma_i \simeq \begin{cases} 1.2 \cdot 10^{-16} \\ 1.5 \cdot 10^{-17} \end{cases}$
2. $H^- + H_2 \rightarrow \begin{cases} H^0 + \ldots \\ H^+ + \ldots \end{cases}$	$\sigma_\ell \simeq \begin{cases} 6.6 \cdot 10^{-16} \\ 4.2 \cdot 10^{-17} \end{cases}$
3. $H^- + H_2^+ \rightarrow H^0 + H_2$ $H^- + H^+ \rightarrow H^0 + H^0$	$\sigma_{ex} \simeq \begin{cases} 10^{-15} \\ 10^{-15} \end{cases}$
4. $H^- + e \rightarrow H^0 + 2e$	$\sigma_e \simeq 5 \cdot 10^{-15}$

The transverse life time of ions with the Bohm diffusion coefficient is $\tau_\perp \simeq 3 \cdot 10^{-3}$ s for H_2^+ and $\tau_\perp \simeq 0.8 \cdot 10^{-4}$ s for H^+. The hydrogen density, necessary for neutralization, is

$$n_{H_2} > \frac{1}{\sigma_i v_b \tau} ; \qquad (10)$$

where v_b is the velocity of H$^-$ ions, and τ is the life time of positive ions.

From the relationship between the cross sections for production and life times, it follows that H_2^+ ions are dominant compensating ions. From (10) we obtain for these ions $n_{H_2} > 10^{10}$ cm^{-3}. The reduction of the H$^-$ current is

$$\frac{\Delta I}{I} \simeq - (n_{H_2} \sigma_\ell + n_{H_2^+} \sigma_{ex} + n_e \sigma_e) L; \qquad (11)$$

where L is the length of the compensated beam. We assume $n_{H_2} \simeq 3 \cdot 10^{10}$ cm^{-3} ($\simeq 10^{-6}$ Torr), $n_{H_2^+} \simeq 2 n_{H^-}$, $n_e \simeq n_{H^-}$. From (11) we obtain $\Delta I/I \simeq -0.7\%$, for $j_0 = 40$ mA/cm^2 ($n_{H^-} \simeq 10^9$ cm^{-3}) and $L \simeq 2.5$ m.

The electric field in front of the acceleration tube removes the space charge neutralization of the H$^-$ beam. Space charge of the beam, electric field of the tube and the shape of the boundary between compensated and not compensated regions of the beam will determine the motion of ions up to the boundary of the uniform accelerating field. In principle, a normal boundary surface may exist at the entrance into the acceleration tube, consistent with the electric field, and equivalent to the emission surface of the source. The stability and the possibility of controlling this surface requires a special consideration.

The Zr - powder or other appropriate material may be used for pumping of cesium vapors from the bending magnet. Zirconium has a high adsorption energy for cesium atoms: 3.22 eV for zero coverage and 2.55 eV for a monolayer coverage. It is technically possible to prepare powder with a diameter of grains $d \simeq 0.1\,\mu$. The quantity of cesium in grams, which the powder may absorb, is

$$G \simeq \frac{\pi \sigma}{d} \frac{A}{N_A} V; \qquad (12)$$

where G is the surface density of a cesium monolayer, A is the atomic mass of the cesium, N_A is the Avogadro number, V is the powder volume. The capacity of the powder is about 17 g/l for $d \simeq 10^{-5}$ cm, and $\sigma = 2.5 \cdot 10^{14}$ cm^2. The cesium flux is $\simeq 10^{-3}$ g/hour from SPS at the H$^-$ current of 10A. The 5ℓ volume of the powder is sufficient for a maximum absorption of this flux over 1 year.

For the gas efficiency of the source of about 15%, all hydrogen may be pumped around the bending magnet if the average hydrogen pressure in the magnet is $\simeq 10^{-5}$ Torr. In this case, the H$^-$ beam loss due to collisions with hydrogen molecules is $\Delta I/I \simeq 5\%$.

References

1. V.I. Gaponov, "Elektronika", Fizmatgiz, Moskva, 1960.

2. Yu.I. Belchenko, G.I. Dimov, VANT, seriya Termojaderniy sintez, 1984, vip. 1 (14), p. 42 - 47.

3. Yu.I. Belchenko, A.S. Kupriyanov, "High current surface-plasma negative ion sources with geometrical focusing" - Int. Conf. on Ion Sources, Berkeley, 1989.

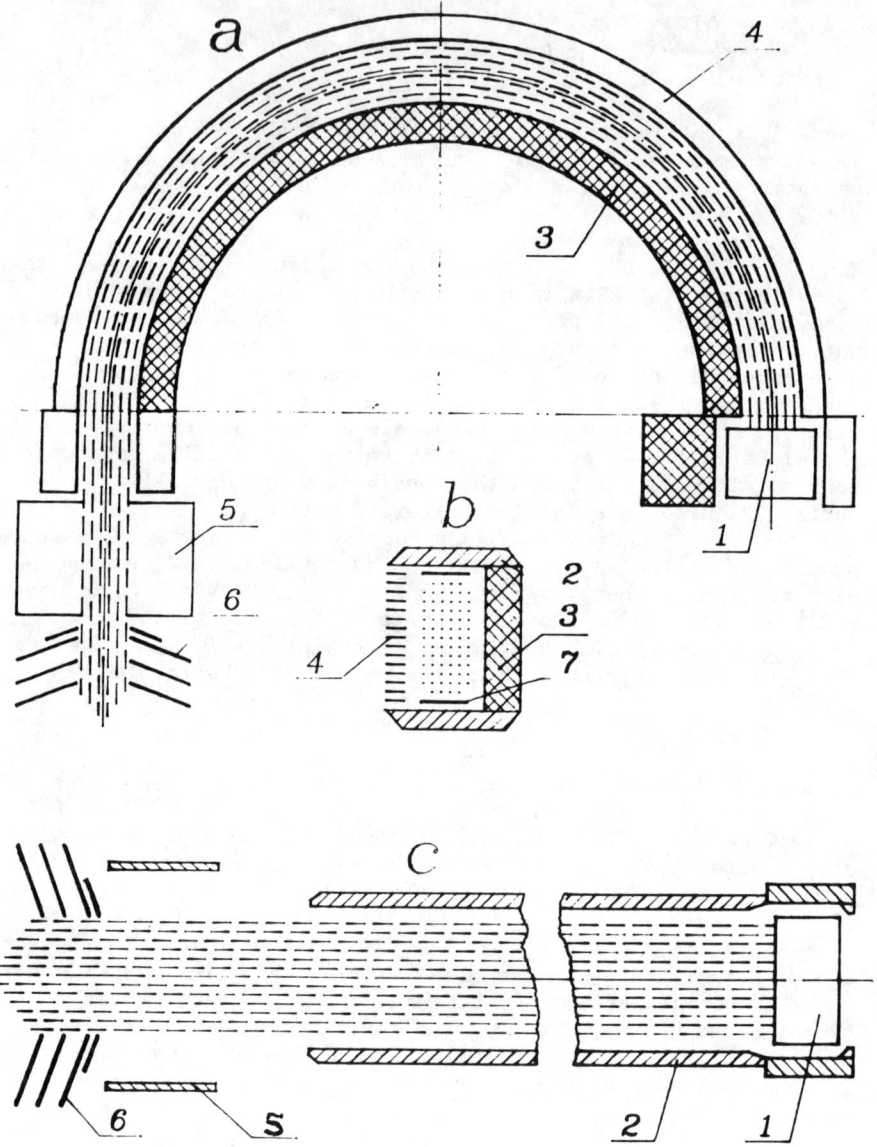

Fig. 1 a - The connection of the SPS to the accelerator tube.
b - The cross section of the bending magnet.
c - The longitudinal cross section of the beam transport from SPS to the accelerator tube.
1 - SPS; 2 - magnetic poles; 3 - ceramic magnets; 4 - ferromagnetic louvers; 5 - ferromagnetic inserts; 6 - the entrance electrode of the accelerator tube; 7 - shielding electrodes.

DISCUSSION

Jacquot: Do you assume that the space charge of the beam in the magnet is always compensated.

Dimov: Yes, it is. There is experience, with ion sources we developed for an accelerator many, many years ago. We studied this compensating process. This is a problem, but not very difficult. It needs gas density more than some limit. But not very large density because of beam losses. Then there is the problem of stability of this plasma system. There is experience with good results.

Pamela: What is the time schedule for this experiment?

Dimov: Our plan depends on Perestroika in our country. We may have a 3A ion source for one second in 1991, and maybe 10A ion source in 1993-1994, and in principle there is the possibility to accelerate quasi-steady state for 1-10 sec. in 5-6 years.

Hemsworth: The initial acceleration to 40 KeV, is that a multi-slit system?

Dimov: Multiaperture

Hemsworth: The final system goes to what energy.

Dimov: This example, which I presented, was designed for 1.5 megavolts, but practically may reach only 1.3 megavolts.

Hershcovitch: Have you looked at the transverse field focusing lens Oscar Anderson is using for your end region?

Dimov: No, I think at the end, accelerator must have the focusing lens, maybe magnetic lens or maybe electrostatic lens.

FUSION APPLICATIONS OF RF ACCELERATORS

R. THOMAE, H. DEITINGHOFF, H. HOPMAN*, H. KLEIN,
AND A. SCHEMPP

Institut für Angewandte Physik der Goethe-Universität,
Robert-Mayer-Str. 2, D-6000 Frankfurt/Main, FRG
*NET Team, c/o Max-Planck-Institut für Plasmaphysik,
Boltzmannstr. 2, D-8046 Garching bei München, FRG

ABSTRACT

High current RFQ designs have been carried out in the frequency range of 27 to 216 MHz. Particle simulations show that at 27 MHz a D^- ion beam of 0.8 A can be accelerated from 0.2 to 1.5 MeV with an efficiency (RF to negative ion beam power) of 73 %. Fourteen such RFQ's can be grouped into a module delivering 9.2 MW of neutral beam power, suitable for non-inductive current drive in next generation tokamaks. Overall system size is comparable to one based on DC acceleration. At 40 MHz, the system size would be considerably smaller but at the cost of a smaller efficiency of 62 %.

INTRODUCTION

Neutral beams have a demonstrated physics data base for non-inductive current drive (CD) and for heating of tokamak plasmas. In particular, they are proposed to drive the current in the core of next generation magnetic fusion machines like NET [1] or ITER [2]. The optimum beam energy is estimated to be around 1.3 MeV. It represents a compromise between CD efficiency and beam shine through, which causes heating of the torus wall opposite the beam injection port and which increases with increasing beam energy. The common ITER design [3] calls for a total injected beam power of 75 MW. Envisaged are beams of D^0 particles. They are obtained by neutralization of high energy D^- ions because their neutralization efficiency is much higher than that of positive deuterium ions. Therefore, a basic neutral beam injector consists of a negative ion source, an accelerator, and a neutralizer.

The objective of this paper is to study the possibility to produce the beam by means of RF acceleration techniques. The advantage of RF above DC acceleration is that high beam energies are obtained without the presence of high DC voltages. Consequently, voltage hold-off and SF_6-handling problems are absent or reduced. The disadvantage is the intrinsic lower efficiency in converting generator power into beam power.

The technique considered here is the Radio Frequency Quadrupole (RFQ) [4,5]. The RFQ makes use of the concept of spatial homogeneous focusing [6]. In this device four modulated metal rods or vanes are placed in a resonator cavity. The RF power coupled to the resonator is converted into an electric field between the electrodes. These serve to focus and accelerate the ions.

RFQ's have been developed as injectors for high energy accelerators. RFQ's are most suited to accelerate ions from 0.1 to 1 MeV/nucleon. Thus they cover the energy interval between neutrals in present-day heating beams of fusion devices and fusion α-particles. Currents range from 10 mA (demonstrated) to ≈ 1 A (anticipated) per single beam. Therefore, two major fusion applications are the generation of beams needed for heating tokamak plasmas and driving the toroidal current [7] and of diagnostic beams. An example of a diagnostic beam would be one of 3 MeV ^{3}He^0 particles with an equivalent current > 10 mA, for measurement of the α-particle distribution [8]. With its small size the RFQ is the ideal accelerator, because it is easily located near the tokamak.

We present designs of RFQ accelerators to generate the beams for non-inductive current drive. The frequency range considered is 27 to 216 MHz. The designs are made following a new procedure of varying the characteristic acceleration parameters along the structure [9]. Comparison to the usual design in which bunch length and transverse phase advance per cell are kept constant shows distinct advantages of the new procedure. The accelerator becomes shorter and with that more efficient at the same current limits and similar beam dynamics. The designs are checked by a modified version of the PARMTEQ code.

For the calculation of the electrical properties as resonance frequency, quality factor and resonance parallel resistance of the 4-Rod-RFQ's, a lumped circuit model is used which has shown good agreement with existing resonators.

RFQ DESIGNS

Design procedure

An RFQ accelerator normally is composed of a radial matching section, a shaper which linearly changes the longitudinal phase advance per cell, σ_{o1}, an adiabatic buncher in which the bunch length and the transverse focusing forces are kept constant, and the final acceleration section in which the energy gain per cell of particles is kept constant. For a given injection energy of the ions, electrode voltage U, and resonance frequency f_0, this is achieved by a certain choice of the cell parameters as aperture radius a, rod modulation m, and synchronous phase φ_s. In the conventional procedure to design the gentle buncher the internal forces, i.e. the beam space charge field, and the external forces, i.e. the quadrupole electric field, are roughly kept constant along the accelerator. This leads to the so-called bottleneck at the end of the gentle buncher, which restricts the transverse current limit. In contrast to this design procedure, throughout this paper another method is applied [9]. Here the focusing forces are parameters which are adjusted in order to keep the transverse and longitudinal current limits equal and constant. The adjustment is carried out on basis of analytical expressions for the limiting currents [10 - 12] and is repeated for each accelerating cell. Comparison of both designs carried out by PARMTEQ simulations has confirmed the advantages of the new method. For the same resonance frequency, electrode voltage, and minimum aperture radius it results in an up to 30 % shorter accelerating structure with similar beam emittance and current limits. The result is a distinctly more efficient accelerator.

Designs have been made for the acceleration of D-beams from 200 keV to 1.5 MeV in frequency range from 27 to 216 MHz. The characteristic cell parameters as a function of cell number are shown in Fig. 1, for the 27 MHz case. The figure shows that a constant limiting current can be achieved by an approximately linear variation of the modulation factor and the synchronous phase. The latter parameter always starts at -90° to ensure a large acceptance and transmission. The aperture radius decreases from a starting value, which is inversely proportional to the resonance frequency, to some 40 - 30 % smaller value in the middle of the structure from which it increases again to obtain the necessary focusing strength. The other graphs in the figure show the transverse and longitudinal phase advance per cell σ_{o1} and σ_{oT}, respectively, and the particle energy as a function of the cell number. The electrode voltage has been chosen so that at a minimum electrode distance its value is of the order of 1.8 times the Kilpatric voltage [13]. Comparison of the different designs shows that the total length of the accelerators increases with decreasing frequency and at a fixed frequency with decreasing electrode voltage. Note that the 54 MHz-design includes a 10-cell shaper which counts for 80 cm. Only for this design the influence of a shaper on transmission was investigated. It was found that the small increase in transmission was accompanied with a decrease in efficiency. Therefore, the shaper was omitted in all other designs.

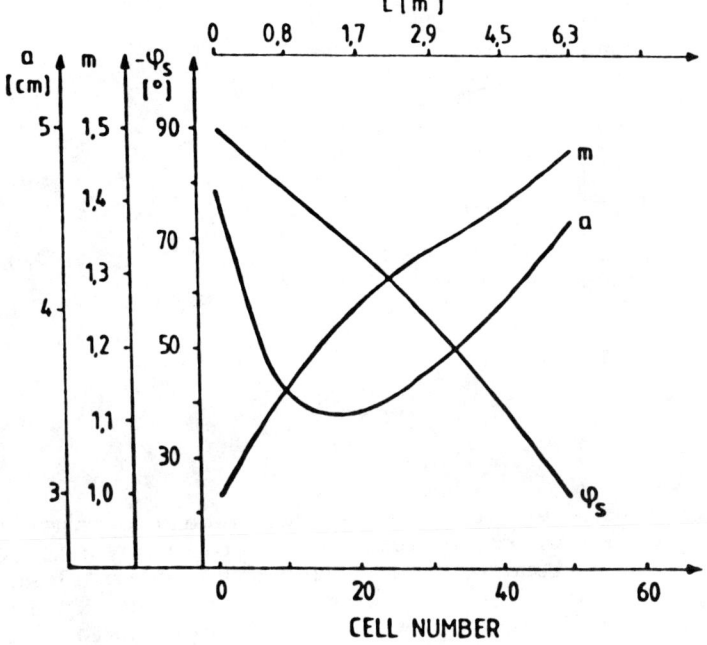

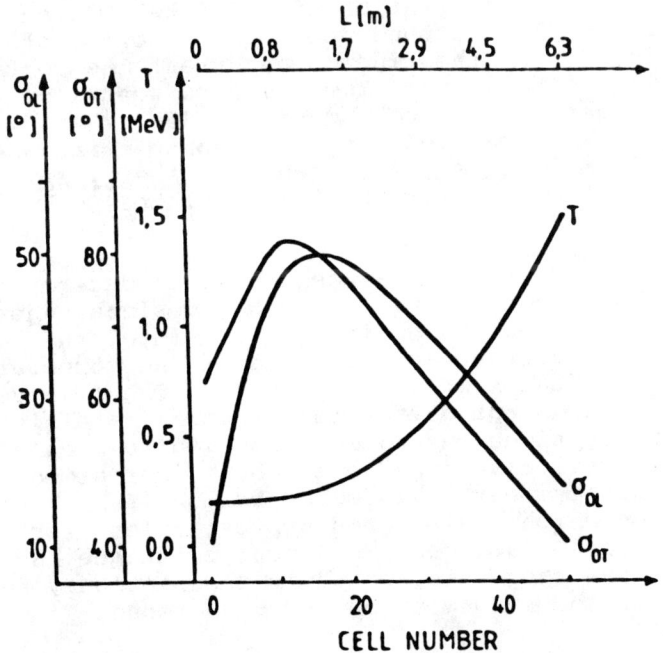

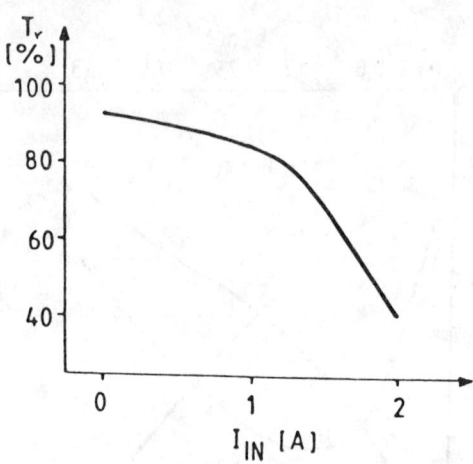

Figure 1. Presented are the following cell parameters: aperture radius a, rod modulationfactor m, synchronous phase φ_s, longitudinal and transverse phase advances per cell σ_{ol} and σ_{ot}, and particle energy T as a function of cell number for a 27 MHz-design. The electrode voltage U amounts to 300 kV. The total accelerating structure length including two $\beta\lambda/2$ matching -in and -out cells is L_t = 6.7 m. The design is made for the acceleration of a 200 keV D- ion beam. Furthermore, the dependence is shown of the transmission T_r on the injected current as obtained form PARMTEQ simulations. The zero current acceptance has been determined to α = 0.45 cm rad. The calculations are carried out for input emittances in the range from 0.05 to 0.4 cm rad. The energy spread at the exit of the accelerator is calculated to be +/- 5 %.

The PARMTEQ code [14] has been used to determine the transmission of beams through the RFQ designs. When the cell parameters had been introduced into the code, first the zero current acceptance α of the structure is determined in the x-x' and y-y' plane. The mean values of both, which do not differ more than 5 % are given in the figure captions and α is observed to scale with f_o^{-2}. All calculations have been performed for cylindrically symmetric injected beams with a Waterbag distribution which is a good approximation for beams extracted from D- sources. The input emittance has been varied in the range indicated in the captions, but no influence on the transmission has been found. The value of transmission T_r as a function of the injected current is also shown in Fig. 1. For all designs a transmission of 90 to 100 % is achieved for small injected currents. The maximum current at a transmission of 80 % for the different designs is shown in Fig. 2. Its value is found to scale with f_o^{-2}, which is due to the fact that for lower frequencies both the phase advance per cell, i.e. the focusing forces, and the accelerator acceptance are larger. In the 27 MHz design the maximum accelerated current amounts to 2.5 A, whereas in the 216 MHz-design it amounts to only 40 mA.

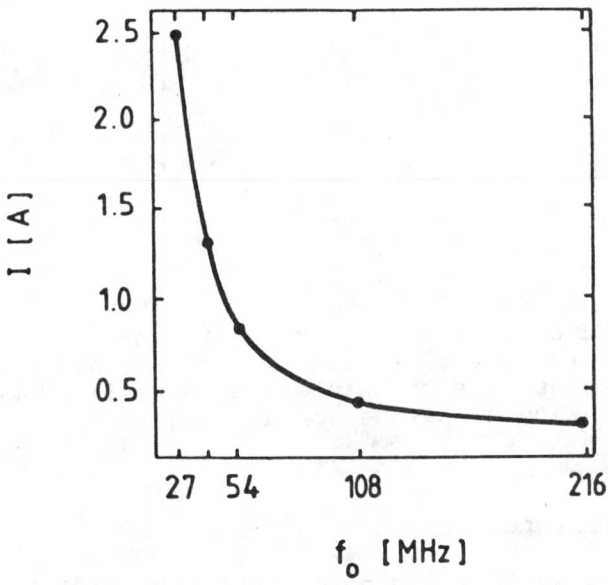

Figure 2. PARMTEQ simulation results for the accelerated current at 80 % transmission for designs at different resonance frequencies.

To determine the current capability as a function of electrode voltage, simulations have been done at a resonance frequency of 27 MHz. In Fig. 3 the accelerated current at a transmission of 80 % is shown for electrode voltages in range from 300 to 800 kV. A linear dependence is found.

All calculations are carried out with two $\beta\lambda/2$ matching-in and -out cells. The two-cell matching-in section serves to adapt the time independent emittance of the input beam to the time dependent accelerator acceptance. At the accelerator exit it is important that the beam has a minimum divergence. This is achieved by the matching-out section which transforms the time dependent beam emittance into that of a nearly parallel beam. The investigations have shown that the emittance growth at 80 % transmission is 10 to 50 %.

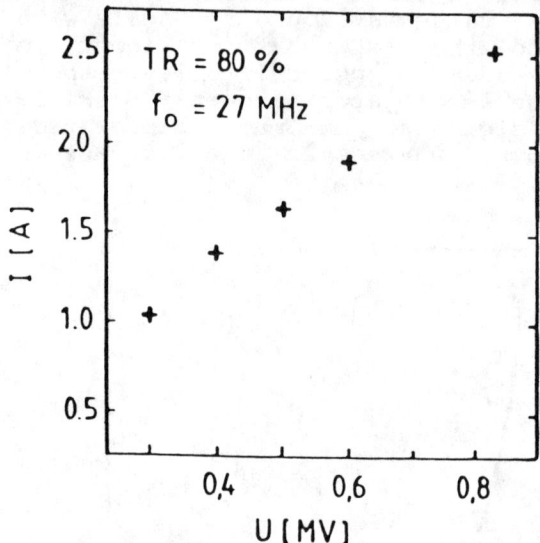

<u>Figure 3.</u> PARMTEQ simulation results for the accelerated current at 80 % transmission for different electrode voltages. The resonance frequency is 27 MHz for all cases. Each electrode voltage requires a separate design.

Electrical properties

From the cell parameters as aperture radius and cell length the capacitive loading of the accelerating structure can be calculated, providing a value is given to the electrode radius r. Since an efficient accelerator requires a small capacitive loading this value should be small. On the other hand it is well known that a proper quadrupole field only is obtained when the electrode and aperture radius are similar. Furthermore, the electrode radius determines the effectiveness of cooling the structure. For these reasons the electrode radius has been chosen between 10 and 30 % smaller than the mean aperture radius. When the capacitance of the electrodes is determined, the efficiency of the accelerator can be optimized by the choice of the geometrical dimensions of the supporting stems [15].

The accelerator efficiency η is defined as the ratio of beam power to the total RF power:

$$\eta = P_b/(P_b + P_{rf} + P_{bl}), \qquad (1)$$

where, the exit beam power P_b is taken for 80 % transmission, and P_{rf} gives the RF losses in the RFQ structure. For the calculation of P_{bl}, the power associated with beam transmission losses, half of the exit energy is used, because the simulations have shown that most of the particles get lost in the first part of the accelerator. In Table I, a number of relevant parameters is listed for all cases where a design was carried out.

<u>Table I</u>. Parameters needed for the calculation of the acceleration efficiency for several RFQ designs. I and I_{lo} are the accelerated and lost currents, respectively. J denotes the current density at the exit of the accelerator and is determined by the ratio of accelerated current and overall cross-section of the accelerator.

resonance frequency f_o,	27	27	27	40	54	108	216	MHz
electrode voltage U,	300	500	800	400	310	190	80	kV
loss power P_{rf},	240	500	1100	650	500	250	100	kW
accelerated current I,	0.8	1.3	2.0	1.1	0.6	0.3	0.05	A
current loss I_{lo},	0.2	0.3	0.5	0.3	0.2	0.06	0.01	A
power to beam P_b,	1.0	1.7	2.6	1.4	0.8	0.4	0.07	MW
beam loss power P_{bl},	130	195	325	195	130	40	10	kW
acceleration efficiency η,	73	71	65	62	56	58	39	%
total electrode length L_t,	6.7	5.6	4.9	4.5	5.0	2.2	2.1	m
stem width times height,	50x 55	50x 55	45x 55	30x 35	30x 30	20x 17	15x 8	cm²
current density J,	2.9	4.7	8.0	11.0	6.7	7.5	4.1	A/m²

From the table it is clear that the acceleration efficiency depends on the resonance frequency, as η slowly increases with decreasing frequency. At a fixed resonance frequency of 27 MHz there is a small increase in η for decreasing electrode voltage.

Discussion

The designs for 27 MHz have the advantage of having the highest efficiencies. Further, the current capability per channel is large (up to 2.5 A) and only a small number of accelerators is needed to obtain the required neutral beam power. On the other hand, the single acceleration sections have large dimensions of 0.65 x 0.5 x 5.0 m³ and are heavy ($\approx$ 2000 kg) which possibly makes handling and aligning of the accelerators more difficult. This problem is reduced in a design based on a higher frequency, but then the current per accelerator is small, which deteriorates the reliability of the injector system due to the large number of components. Further, for the highest frequencies investigated (108, 216 MHz) the electrode dimensions become unacceptably small, because mechanical rigidity can not be guaranteed.

Before beams enter the tokamak plasma, they have to pass through the port in the vessel wall. The power density in the port is determined by the beam divergence and by the mean current density at the exit of the accelerator. The latter is defined by the ratio of accelerated current and the accelerator overall cross-section and is given in the last line of Table I. This current density has a maximum of 11 A/m² at 40 MHz.

Thus we have found two selection criteria: efficiency and overall current density. We proceed by giving a more detailed comparison of the 27 and 40 MHz design in Table II. The last column in the Table gives the accelerator efficiency for the case that the structure is cooled to liquid nitrogen temperature, not including the power for producing the liquid nitrogen. A further increase of the acceleration efficiency could be obtained by using superconducting resonator cavities. Nevertheless, this increase is small due to the high beam loading. Furthermore, the low specific heat capacity of the cooling agent requires a very high flow.

Tab. II. Detailed comparison of the 40 and 27 MHz-design. The symbols η and η^* present the values for the efficiencies of accelerators at room temperature and at liquid nitrogen temperature

f_0 [MHz]	weight [kg]	D [cm]	P [kW]	therm.load [kW/m]	E_m [MV/m]	J [A/m²]	η [%]	η^* [%]
40	600	4	650	145	17.4	11.0	62	80
27	2000	8	240	36	7.7	2.9	73	83

The data in Table II show that the 27 MHz-design has the following advantages: a higher overall efficiency, a smaller electric field strength, and a smaller specific thermal load. A major disadvantage is the small mean current density. However, in the next section, in which the adaptation of the accelerator to the torus port is discussed, it will be shown that despite the small current density sufficient neutral beam power can be transmitted through a port into the plasma. Therefore, the 27 MHz-design is elected for the discussion of the source and port matching.

A funneling concept to increase the efficiency has been considered. In this concept two beams, accelerated in two low frequency RFQ's, are injected into one RFQ operating at twice the frequency. Due to the higher frequency the beam is then accelerated more efficiently and without transmission loss, providing the beams are bunched to the necessary phase width demanded by the acceptance of the higher frequency RFQ. Investigations show that for an energy gain of only 1.3 MeV, this concept is not suitable. Furthermore, to this date no experimental experience is available.

SOURCE, MATCHING AND OVERALL DIMENSIONS

The beam current required for a single 27 MHz RFQ channel is of the order of 1.2 A. This value is 20 % larger than that given in Table I, because of the properties of a D^- source, which produces in addition to the proper beam, a so-called halo beam which possibly might be lost in the RFQ [16]. In Fig. 4, a 16 aperture geometry is shown, with which the required beam can be produced. For a current density of 12 mA/cm² at the exit of the 200 kV extraction system the extracted current per channel is 74 mA at an aperture radius of 1.4 cm. The required current density at the source is 18 mA/cm², when 66 % stripping survival is assumed [16]. The figure illustrates that a cross-section of 40 by 40 cm² affords sufficient space for the 16 aperture holes, magnets, cooling channels, and mounting flange. This cross section is smaller than that of the RFQ support stem given in Table I.

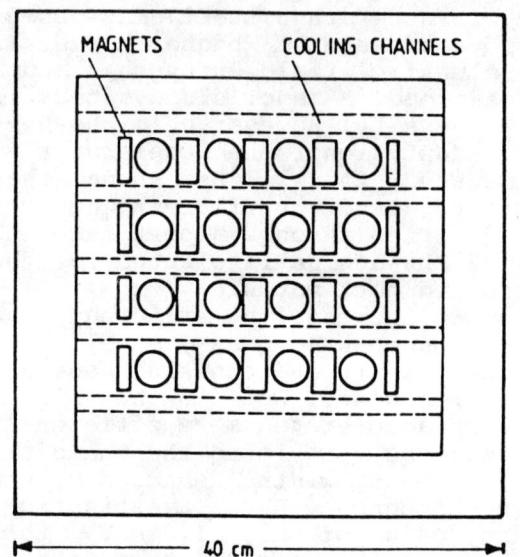

Figure 4. Possible source extraction arrangement for a 1.2 A, 200 kV D-ion beam.

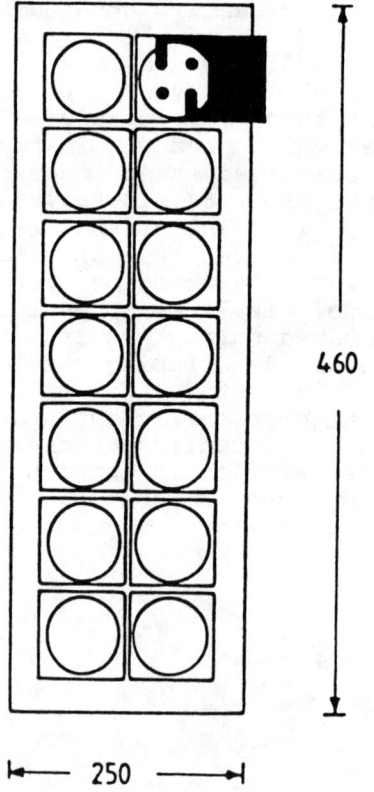

Figure 5. Possible arrangement to form a module of 14 sources and accelerators. Two rows of 7 sources are followed by a drift space and the RFQ's (only one is shown). The module cross section has the approximate dimensions of 2.4x4.6 m². It has been designed to deliver a D- power of 17 MW.

The source must produce a beam with a converging envelope angle of roughly 6° to fit into the RFQ acceptance. With the so-called beam steering, the different beamlets are directed to the accelerator entrance by means of a radial displacement of the apertures in the different extraction electrodes [17]. Calculations show that the required 6° angle is obtained for aperture displacements of 1 - 3 mm in a 200 kV tetrode-extraction system. For a 20 cm beam diameter at the source and a 6° convergence angle, the distance between source and accelerator must be roughly 1 m, which allows for sufficient pumping of the source gas.

Fig. 5 shows how sources and accelerators are combined into a module [9]. 14 sources and RFQ's are arranged in two rows. This allows for enough space at the sides of the accelerators for feeding the structures with cooling water and RF power. The height and the width of the module are 4.6 and 2.4 m, respectively. The beam power per module amounts to 17 MW of D⁻ ions.

The neutralization yield of a D⁻ ion beam in a gas cell is of the order of 55 %. Therefore, one module is capable of injecting 9.5 MW of D° particles into the plasma. In analogy to the ITER neutral beam injector design [3], we propose to place a vertical array of three modules on one port. Simple ray tracing shows that the beams produced by the three modules fit through the torus port when a distance of 30 m is taken between accelerator and port. A beam divergence of 5 mrad is taken for this estimate. In Table III, the overall dimensions of injector modules are compared for the 27 and 40 MHz case.

<u>Table III.</u> The dimensions of the 27 and 40 MHz modules.

f_o [MHz]	U [kV]	no. of modules	dim.module height x width x length [m³]	D° power [MW]
27	300	3	4.6x2.4x8.5	9.2
40	400	2	3.0x1.0x6.0	12.7

CONCLUSION

In this paper the potential of RF acceleration for current drive and neutral beam heating in TOKAMAK's has been investigated. The determination of the current capability - by means of PARMTEQ simulations - and the electrical properties for designs in the frequency range between 27 and 216 MHz has led to two advantageous designs with a resonance frequency of 27 and 40 MHz. These designs combine high efficiency, good beam dynamics, as well as good electrical properties.

A single acceleration line consists for both designs of a multi aperture source, the steering extraction system, a drift space, and the RFQ. 14 of those lines are combined into a module, in which the RFQ's are arranged so that water cooling and RF feedings can be mounted to the side.

To obtain the required D^o power of 25 MW per port, it is proposed to place three 27 MHz modules or two 40 MHz modules on one port.

The advantages of 27 MHz design are the higher overall efficiency and the more relaxed accelerator parameters as electric field strength and specific thermal load. The acceleration efficiency is calculated to be 73 % at room temperature and 83 % when the accelerators are cooled to liquid nitrogen temperature. These values do not include AC to RF conversion, nor source, pumping, of cooling power.

The 40 MHz design has the advantage that due to the higher mean current density only two modules per port have to be installed and the overall dimensions of the injector amounts to 6 m in height, 1 m in width, and 6 m in length.

The distance for both designs between accelerator exit and port entrance gives enough space for the neutralization of the beams.

It is worthwhile to mention that all designs can be improved somewhat by a more detailed design, which includes several more iteration steps for the determination of the accelerator cell parameters.

Supplementary investigations have shown that the acceleration efficiency can be increased to more than 80 % when higher exit energies of 2 -3 MeV are required.

For the verification of the overall suitability and for the realization of the injector system two experiments should to be carried out. Firstly, a beam line consisting of an ion source with a 200 kV extraction system and a single channel RFQ has to be developed. This experiment allows to investigate both the production and injection of the D$^-$ beam as well as the beam dynamics during acceleration, e.g. emittance growth and energy spread. Furthermore, the influence of beam loading and the reliability at CW operation can be studied. The beam of this accelerating line can be used for plasma diagnostics [3]. Secondly, a module with the multiple RFQ system has to be developed. In this experiment the RF properties and the coupling of the accelerators will be investigated and the mechanical construction will be tested.

REFERENCES

[1] R. Toschi, M. Chazalon, F. Engelmann, J. Nihoul, J. Räder, and E. Salpietro, Fusion Technology 14 (1988) 19.
[2] "ITER: Concept Definition", presented by K. Tomabechi at the 12th Int. Conf. on Plasma Phys. and Contr. Fusion Res., Nice, France, 12-19 October 1988.
[3] "Neutral Beam System Iterim Report", ITER CD&H Group, Report ITER-IL-4-9-14 (1989), unpublished.
[4] V.V. Vladimirski, I.M. Kapchinsky, and V.A. Teplyakov, Bulletin of Inventions 10 (1970) 75, Patent USSR 265312.
[5] H. Klein, IEEE Trans. Nucl. Sci. NS-30 (1983) 3313.
[6] I.M. Kapchinsky and V.A. Teplyakov, Prib. Tekh. Eksp. 4 (1970) 19.
[7] W.R. Becraft, J.H. Whealton, T.P. Wangler, A. Schempp, G.E. McMichael, M.A. Akerman, G.C. Barber, W.K. Dagenhart, H.H. Haselton, R.J. Raridon, K.E. Rothe, P.M. Ryan, B.D. Murphy, and W.L. Stirling, Conf. Proc. Lin. Acc., Williamsburg (1988).
[8] A.B. Izvozchikov, A.I. Kislyakov, M.P. Petrov, and A.V. Khudoleev, Report ITER-IL-PH-7-9-S-07 (1989), unpublished.
[9] A. Schempp, Conf. Proc. of EPAC, S. Tazzari, Editor, World Scientific, Rome (1988) 464.
[10] M. Reiser, Part. Acc. 8 (1978) 167.
[11] T.P. Wangler, LASL-Report LA-8388, Los Alamos (1980).
[12] P. Junior, Part. Acc. 13 (1985) 231.
[13] W.D. Kilpatrick, Rev. Sci. Instr. 28 (1957) 824.
[14] K.R. Crandall, R.H. Stokes, T.P. Wangler, Conf. Proc. LASL-Report LD-UR-79-2499, Montauk N.Y. (1979).
[15] A. Schempp, IAP-Report 84-8, Univ. Frankfurt (1984).
[16] "A Conceptual Design of a 1 MeV, 50 MW Beam Line for NET", J. Coupland, I. Gray, A. Holmes, M. Inman, L. Lea, K. Martel, and R. Parker, Culham Lab. Report on account of NET contract 315/88-7/FU UK NET (1989)
[17] A.J.T. Holmes and E. Thompson, Rev. Sci. Instr. 52, (1981) 2.

DISCUSSION

Semashko: What is your current density? Do you show 1 mA per centimeter square. Is this average current density?

Hopman: The current density means the total current accelerated in this channel divided by this area.

Semashko: Approximately 1 mA per centimeter per square we showed 10 amperes per meter square.

Hopman: This is the highest value at 40 megahertz. But we have carried out design for this value of 3 mA per square centimeter because you have the highest efficiency of 70%. And if you use liquid nitrogen cooling, this goes up to 80%.

Whealton: We have also conducted a study of RFQs for ITER application, and we come to basically the same conclusion. But we were more conservative by a factor of about 2 in current. For a particular frequency we choose 40 megahertz because of the additional feature that the divergence of the beam coming out decreases with increase in frequency. So it seemed to favor the 40 megahertz. Two very important features of the RFQ design is the pumping right after the ion source and before the accelerator. Present day volume ion sources with .2% gas efficiency, which are state of the art for high current density can be used in these accelerators, and are in architecture compatible with either application. The second point is that this particular technology is extrapolatable to higher energy so that it is not a dead end technology at the 1.3 megavolt level. If a reactor comes, in the future, which requires 3 megavolts then that technology is applicable to such application.

Hopman: About the divergence of the beam that is coming out. You are correct, of course, that improves if you go to high frequencies, but we have, on purpose, taken the lowest frequency design. We have looked to see if we can get a power through the duct and at the lowest frequency without problems. For this particular application, the emittance argument doesn't seem to play a role.

Jacquot: Does your efficiency of 62% at 40 MHz include the efficiency of the power supply.

Hopman: No, it is what I said: The rf power applied to the RFQ goes into denominator in the power in the accelerated beam and the numerator.

Okumura: Many beamlets should be focused at your RFQ entrance, and what is the acceptance of your RFQ?

Hopman: The acceptance angle is 6 degrees. So if you have a source where the apertures fill in an area of 30 x 30 centimeter, that's a 4 x 4 array in our example. At a meter distance, you can easily inject beamlets into the RFQ entrance.

Dimov: What increase in emittance do you have in one channel with a current of about one ampere?

Hopman: The particle simulation says that the emittance increases between 30 and 60 percent, compared to the entrance emittance.

INJECTOR FOR RFQ USING ELECTROSTATICALLY FOCUSED TRANSPORT AND MATCHING

O.A. Anderson, L. Soroka, J.W. Kwan, and R.P. Wells*
Accelerator and Fusion Research Division
Lawrence Berkeley Laboratory
University of California
1 Cyclotron Road, Berkeley, CA 94720

ABSTRACT

We discuss the principles and performance of a new type of high-current H⁻ injector for RFQs. The distinguishing feature of our injector is that we replace the conventional gas-neutralized transport and matching units by electrostatic focusing units. Our system prevents plasma formation along the beam instead of utilizing it. Some advantages of this approach are discussed.

1. INTRODUCTION

This paper describes the design and preliminary operation of a dc H⁻ injector for a radio frequency quadrupole (RFQ). The system tested consists of a volume production ion source [1], a 100-keV electrostatic preaccelerator [2], and a new electrostatic transport and matching system. The injector has been tested with up to 45 mA of H⁻, but simulations show that it is capable of handling beams with several times higher current. To allow for dc operation, all components are water cooled.

As discussed previously [2], the H⁻ volume source operates at a pressure of more than 10 mTorr. To minimize stripping of the H⁻ beam, we use an open accelerator and transport structure, with high conductance to pumps. Electrons from the source are removed by an electron trap mounted on the extractor grid [2].

We use electrostatic quadrupole (ESQ) focusing for our low energy beam transport (LEBT) and an axisymmetric lens for matching into the RFQ. The ESQ LEBT, described in Section 2, provides sufficient length for effective pumping while prevent-

---

*This work was supported by USASDC Contract No. MIPR W31RPD-63-A087 and by the Director, Office of Energy Research, Office of Fusion Energy, Development and Technology Division, of the U.S. Department of Energy under Contract No. DE-AC03-76SF00098.

ing the accumulation of charge and the formation of plasma in the transport channel. The final match is obtained with an axisymmetric electrostatic lens rather than with ESQ focusing, as discussed in Section 3. Experimental results are given in Section 4.

2. ELECTROSTATIC QUADRUPOLE LEBT

Our new approach for transport of intense H⁻ beams uses ESQ focusing instead of the usual combination of magnets and gas neutralization [3,4]. Some of the features of our plasma-free LEBT are: (1) The open structure gives good gas pumping capability. (2) Gas pressure can be arbitrarily low in the LEBT; gas independence improves reproducibility. (3) Gas independence permits either dc operation or pulsed beams with arbitrarily short pulse lengths. (4) Electrostatic tuning provides excellent flexibility. (5) Emittance growth from plasma noise is eliminated. (6) Also eliminated is emittance growth from sheath transitions into and out of gas neutralized regions. (7) The de-neutralization transition near the match point in the RFQ, a tricky issue in conventional designs, is avoided.

Our LEBT uses ESQ technology [5] which was developed and tested for the DOE magnetic fusion energy program [6] and adapted for the present application.

2.1 Field Strength Compared with Magnetic (Gas Neutralized) Case

Electrostatic quadrupole focusing

The quadrupole pole face electric field E_Q required to transport a specified beam current I with normalized emittance ϵ_N at beam energy qV is given by [7]

$$\frac{L^2}{a_Q^2} E_Q^2 = C_1 \frac{I}{A_o^2} V^{1/2} + C_2 \frac{\epsilon_N^2}{A_o^4} V \qquad (1)$$

for a matched beam; L is the quad cell length, a_Q is the quad aperture radius, and A_o is the mean beam radius. The constants C_1 and C_2 depend on the particle charge and mass and the electrode occupancy factor; C_2 also includes a correction for beam ripple which is usually negligible at higher energies where the ϵ_N term becomes significant [7].

The E_Q^2 external force term on the left of Eq. (1) balances the space charge and emittance pressure terms on the right. In a typical LEBT, $E_Q \approx 10$ kV/cm. For a bright beam transported at low energy, the emittance term is usually negligible, and

$$E_Q \propto V^{1/4}. \tag{1'}$$

We routinely use Eq. (1) or (1') when designing ESQ accelerators and LEBTs.

Magnetic, gas neutralized, focusing

If the electrostatic quadrupoles are replaced by magnetic quadrupoles, then

$$E_Q^2 \longrightarrow C_3 V B_Q^2.$$

For a 100 keV D⁻ beam, 10 kV/cm is approximately equivalent to 3 kG. However, in the absence of external electrostatic fields, a neutralizing plasma develops. We define the neutralization coefficient K_n, where typically $K_n \approx 0.99$, and get

$$C_3 \frac{L^2}{a_Q^2} B_Q^2 = C_1 (1 - K_n) \frac{I}{A_o^2} V^{-1/2} + C_2 \frac{\epsilon_N^2}{A_o^4}. \tag{2}$$

In Eq. (2), note that the factor $(1 - K_n)$ is very sensitive to the degree of neutralization; for example, it doubles if K_n changes from 0.99 to 0.98. We also note that K_n is a poorly known function of position and time in a magnetic LEBT, so that Eq. (2) is not useful for design purposes unless the emittance term is large enough to dominate. Thus, the design of magnetic LEBTs must be largely empirical.

2.2 Mechanical Configuration of Experimental LEBT

The electrostatic LEBT which we tested is shown in Fig. 1. The hardware is the same as used for the 200 kV Constant Current Variable Voltage (CCVV) prototype accelerator tested and described previously [5,8,9]. Two ESQ modules containing a total of five sets of quadrupoles are shown in the figure. In the accelerator prototype application, the first module is for matching and the second is for acceleration. The initial quadrupole in the matching module has been shortened to facilitate the transition from an initially round beam to a beam matched for transport or acceleration. The second module is the first of an eventual series of identical acceleration modules [9], and therefore does not have a shortened exit quadrupole.

Even though the hardware shown in Fig. 1 was not designed for LEBT applications, it proved to be perfectly adequate as a demonstration LEBT capable of handling parameters of interest in all known RFQ injector applications. Naturally, a dedicated electrostatic LEBT would be designed differently, with fewer elements and shorter overall length.

The CCVV accelerator that we adapted for our LEBT experiment was developed for applications that require constant beam current over a wide energy range. Flexibility in energy results from the use of transverse field ESQ focusing in the main accelerator rather than the use of longitudinal field focusing as in conventional Pierce columns. Therefore it is easy to reduce the output beam energy from the nominal 200 keV CCVV level to the 100 keV level used in the LEBT mode. In fact, much lower energies can be reached without loss of beam current, because the preaccelerator voltage, which determines the beam current, is always kept at the same value [5].

Operation in the LEBT mode requires not only reduction in the output energy as just mentioned, but also adjustment of the beam shape. In the CCVV accelerator mode, the exit beam is elliptical in cross section, whereas the RFQ LEBT needs to produce a round beam. This adjustment can also be made, as shown in the next section.

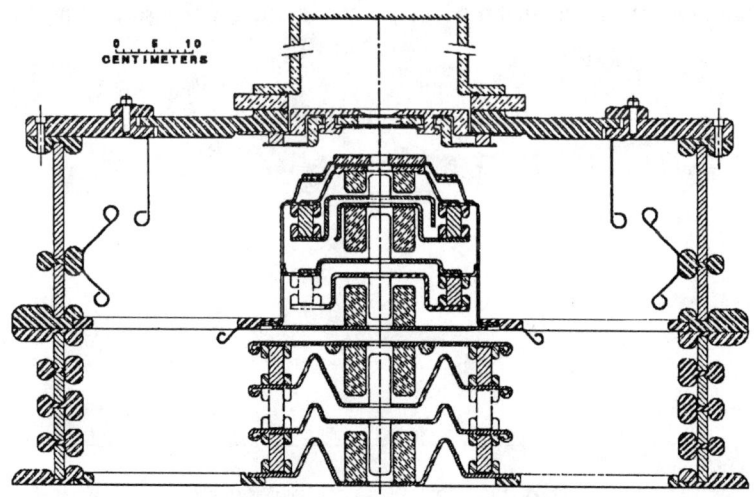

Fig. 1. ESQ system used for testing 100 keV electrostatic LEBT principle. Running in its accelerator mode, this system is the 200 keV prototype for a MeV dc accelerator with fusion energy applications.

2.3 Beam Simulations for Experimental LEBT

Fig. 2 shows an envelope simulation for the hardware of Fig. 1 running in the LEBT mode. Corresponding to a particular experimental run, the preaccelerator was assumed to inject 45 mA of H⁻ in a round beam of 0.8 cm radius into the LEBT. The normalized emittance is 0.160 π mrad-cm. The ESQ voltages were adjusted to produce a round beam at the exit with a diverging envelope (about 24 mrad) as required for the aperture lens module described in the next section. The voltage variation across the beam is small because no attempt is made to use the ESQ voltages for matching into the RFQ. That function is reserved for the aperture-lens module discussed in Section 3.

We have also recently started running self consistent 3-D particle simulations for the ESQ structure of Fig. 1, using a version of the ARGUS code [10] especially developed for our purposes by SAIC [11]. Fig. 3 shows preliminary results from a test run by SAIC for the acceleration mode. In the near future we will be able to do 3-D LEBT mode simulations at LBL which will allow exact predictions for emittance growth.

3. AXISYMMETRIC APERTURE-LENS MATCHING MODULE

A difficult problem in designing RFQ injectors is in the beam matching: a small beam diameter and large convergence angle are required at the match point just inside the RFQ entrance. We found that producing a round beam of this sort, using only ESQ

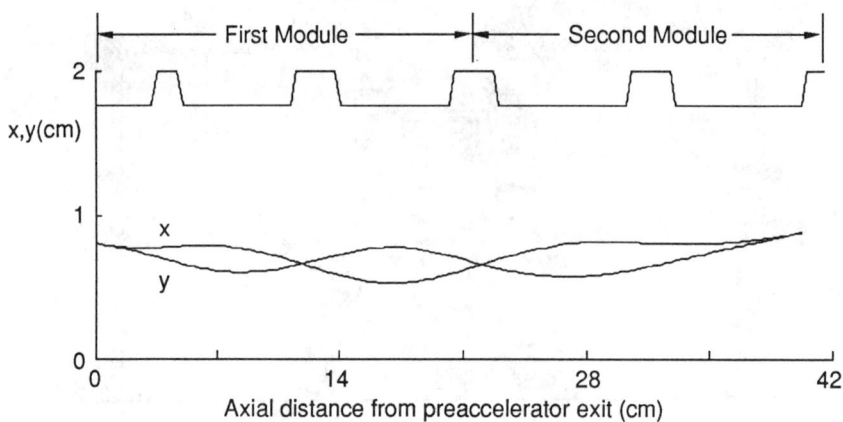

Fig. 2. Beam envelope simulation for the system of Fig. 1, running in LEBT mode. Parameters are given in the text.

units, requires relatively large focusing voltages which could produce excessive aberrations. Since low emittance growth is one of our goals, we chose the alternative of axisymmetric focusing between the LEBT and RFQ. Although axisymmetry *per se* does not guarantee improvement, a big advantage is that the design can be optimized with exact 2-D round-beam particle codes, which run much faster than 3-D particle codes. We used the self-consistent 2-D WOLF code [12] to model our matching module.

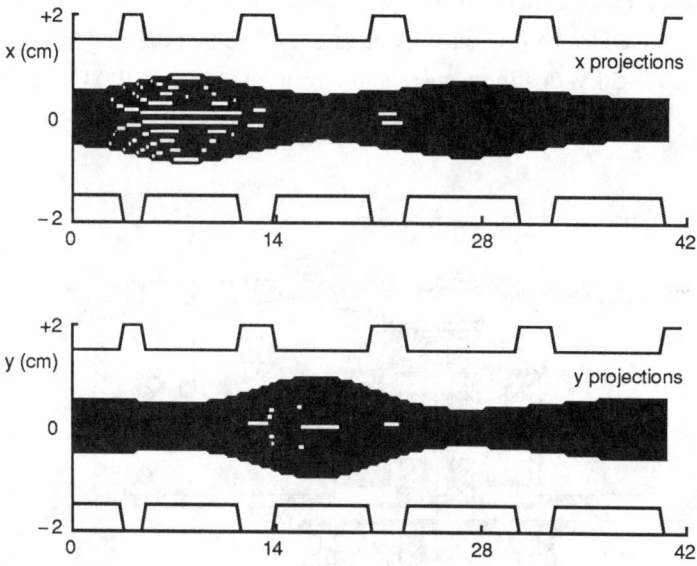

Fig. 3. Beam trajectories for the device of Fig. 1, running in accelerator mode with 200 mA of H⁻. The trajectories are shown projected in the x and y directions.

3.1 Construction of Ring Lens

Fig. 4 shows a third module added to the two ESQ modules of Fig. 1; this module is our axisymmetric lens. One sees that it is *not* a conventional einzel lens; by definition, einzel lenses have field-free drift regions, which we wish to avoid, since plasma can accumulate in such regions. On the other hand, we do not use a classical thin-plate aperture lens because we wish to avoid electric field concentrations. Our compromise design looks like a ring (near the beam), and we simply call it a ring lens.

The figure shows one of three insulators which support the ring. These insulators also serve as vacuum feed-throughs, providing both electrical current and cooling water for the ring lens electrode. This electrode is a two piece assembly with the outer ring serving as a support for an inner ring which can be shimmed and translated to accurately align the electrode bore with the beam axis. Not shown is a movable, grounded, exit electrode containing a small aperture simulating the entrance of the RFQ. The exit electrode is positioned by a remotely controlled actuator that replaces the small aperture by a large one during tune up of the ESQ portion of the LEBT.

Depending on requirements, a voltage in the range of -95 to -98 kV (with respect to ground) is applied to the ring during operation. Note that the ring is actually energized by a small 2–5 kV floating power supply connected to the -100 kV H$^-$ ion source potential.

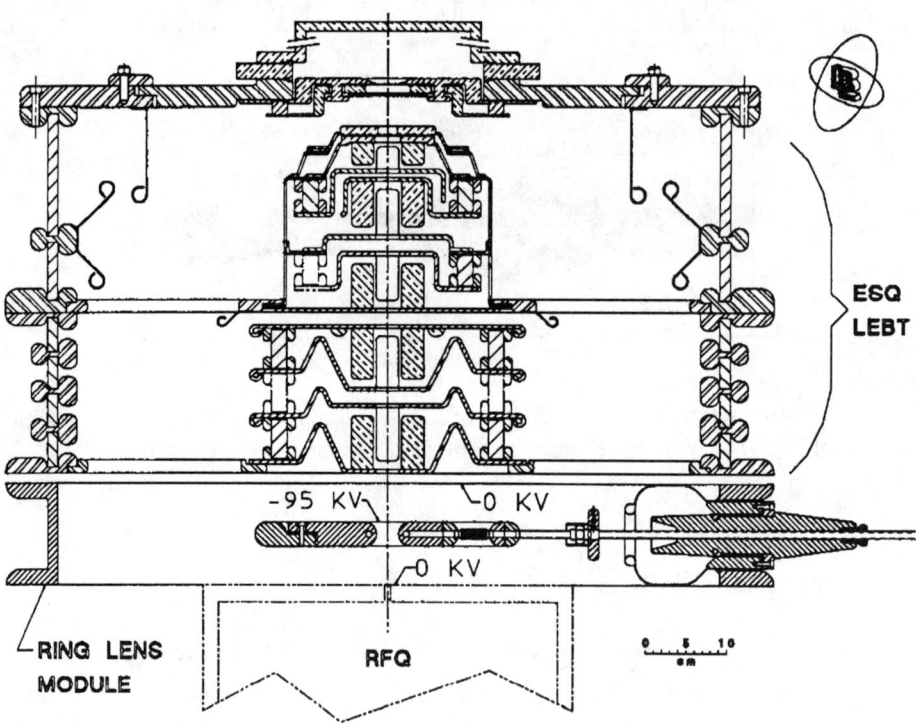

Fig. 4. LEBT shown with ring lens added. As described in the text, the ring is supported by three feed through insulators, one of which is seen here. The hypothetical RFQ entrance region shown schematically with dotted lines is replaced experimentally by the exit electrode described in the text.

3.2 Self Consistent Particle Simulation of Ring Lens

Fig. 5 shows a sample particle simulation for the ring lens system of Fig. 4, obtained with the self-consistent WOLF particle code [12]. Parameters are given in the figure; the current is larger than for Fig. 2, so this case is more stringent. The equipotentials show that the beam is decelerated from 100 keV to about 15 keV and reaccelerated to 100 keV at the RFQ input. Aberrations are small, and the emittance growth is only 2.9%.

It would be interesting to compare the aberrations for the ring lens and the ESQ focusing options for RFQ entrance matching. This will be possible in the near future using the ARGUS code [10, 11] for the ESQ option.

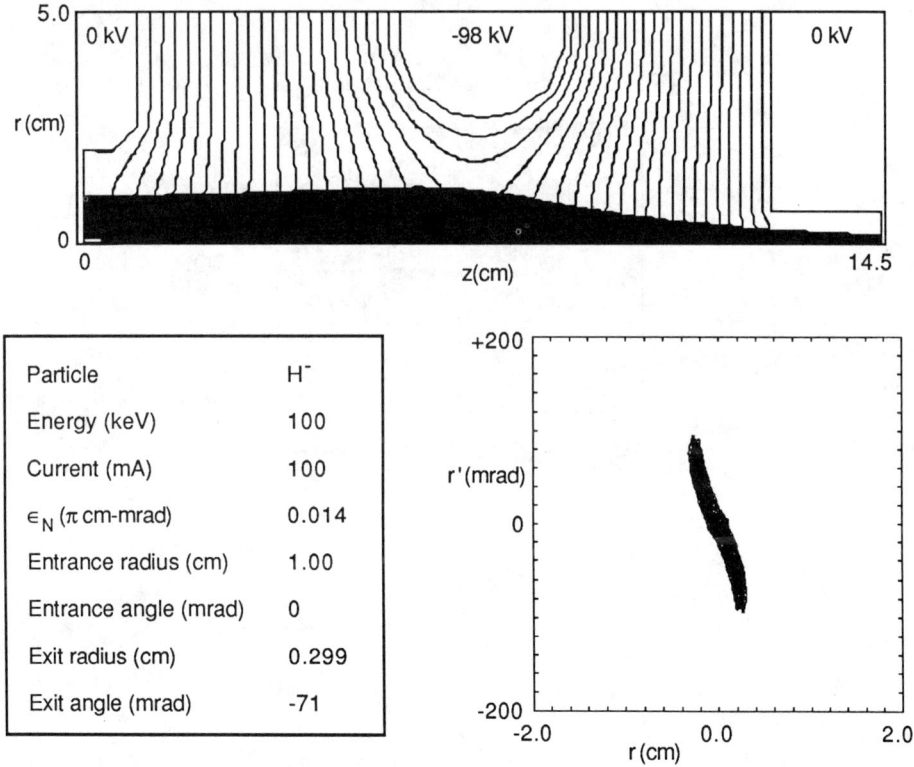

Fig. 5. Sample particle simulation of the axisymmetric lens for RFQ matching, showing the beamlet trajectories and the phase plot at 12.6 cm, the location of the simulated RFQ entrance. The beam radius is 3 mm at this point and decreases toward the RFQ match point.

4. EXPERIMENTAL RESULTS

Preliminary test results are shown in Fig. 6. These were obtained using our large aperture H⁻ volume source [13]. For the run shown, the extraction grid diameter was reduced to 1.4 cm in order to improve the current density. In this particular run, the source produced a rather asymmetric beam with a lump on one edge, as seen in Fig. 6a.

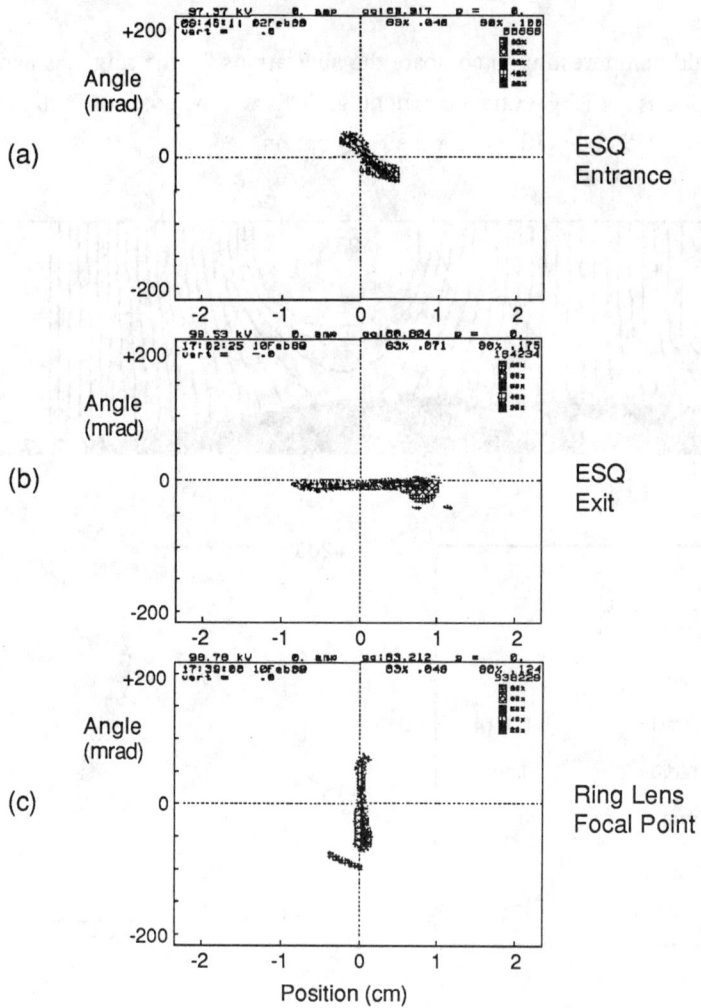

Fig. 6. Emittance scans projected back to: (a) the ESQ entrance; (b) the ESQ exit; and (c) the ring lens focal point. The larger phase space area seen in (b) is probably an instrumental effect. In (a) and (c) the normalized emittance is about 0.05 π mrad-cm.

The lump grows as it is transported through the LEBT (Fig. 6b) and appears to be somewhat disconnected at the ring lens focus (Fig. 6c). The important point, however, is that the bulk of the phase plots are essentially straight and free of aberration after passing through the LEBT and after passing through the ring lens.

In this run, the LEBT operated at 100 kV. The transported H⁻ current was 30 mA, primarily limited by the plasma generator. The beam loss, if any, was too small to measure. The beam exiting the LEBT at 10 mm radius was focused by the ring lens to a radius of 1 mm; the convergence angle was 70 mrad at the focal point.

By adjusting the ring voltage and the electrode spacings on each side of the ring, the matching module can handle a wide range of beam currents and can produce a wide range of convergence angles, up to 100 mrad.

ACKNOWLEDGMENTS

We would like to thank all the members of the MFE and HIFAR groups of the Accelerator and Research Division at LBL who contributed to this work. We are especially grateful to C.H. Kim (LBL), who gave us his variable-energy ESQ envelope code and helped us to use it.

REFERENCES

[1] K.N. Leung, K.W. Ehlers, and R.V. Pyle, "Optimization of H⁻ Production in a Magnetically Filtered Multicusp Source," Rev. Sci. Instrum. $\underline{56}$, 364 (1985).

[2] O.A. Anderson, C.F. Chan, W.S. Cooper, W.B. Kunkel, J.W. Kwan, A.F. Lietzke, C.A. Matuk, P. Purgalis, and L. Soroka, "Design of a 200 mA DC H⁻ Injector for an RFQ," Proc. of the 1987 Particle Accelerator Conference, Washington, DC, March 16-19, 1987; IEEE Cat. No. 87CH2387-9, p. 289.

[3] P. Allison and J.D. Sherman, "Operating Experience with a 100-keV, 100-mA H⁻ Injector," Proc. 3rd Int'l Symposium on the Production and Neutralization of Negative Ions and Beams, Brookhaven National Laboratory, 1983; K. Prelec, Ed.; AIP Conf. Proc. 111, 511 (1983).

[4] J.G. Alessi, et al., "AGS Preinjector Improvement," Proc. of the 1987 Particle Accelerator Conference, Washington, DC, March 16-19, 1987; IEEE Cat. No. 87CH2387-9, p. 276.

[5] O.A. Anderson, L. Soroka, C.H. Kim, R.P. Wells, C.A. Matuk, P. Purgalis, W.S. Cooper, and W.B. Kunkel, Proc. of the First European Particle Accelerator Conference, Rome, June 7-11, 1988; World Scientific Pub. Co, Singapore, p. 470 (1989).

[6] O.A. Anderson, et al., "A High Energy Neutral Beam System for Reactors," 15th Symposium On Fusion Technology, Utrecht, Netherlands, Sept. 19-23 1988; Elsevier Science Publishers B.V., p. 573 (1989).

[7] O.A. Anderson, "Integration Method for Alternating-gradient Problems with Space Charge," Lawrence Berkeley Laboratory report LBL-26123, to be submitted to Particle Accelerators.

[8] O.A. Anderson, L. Soroka, C.H. Kim, R.P. Wells, C.A. Matuk, P. Purgalis, J.W. Kwan, M.C. Vella, W.S. Cooper, and W.B. Kunkel, "Applications of the Constant-Current Variable-Voltage DC Accelerator," Nucl. Instrum. and Meth. B40/41, 877 (1989).

[9] O.A. Anderson, W.S. Cooper, W.B. Kunkel, J.W. Kwan, R.P. Wells, C.A. Matuk, P. Purgalis, L. Soroka, M.C. Vella, G.J. De Vries, and L.L. Reginato, "The CCVV High-Current Megavolt Range DC Accelerator," 1989 Particle Accelerator Conf., Chicago, March 20-23, 1989; to be published by IEEE.

[10] A. Mankofsky, "Three-Dimensional Electromagnetic Particle Codes and Applications to Accelerators," Linear Accelerator and Beam Optics Codes, C.R. Eminheizer, ed., A.I.P. Conf. Proc. No. 177, 1988, p. 137.

[11] A. Mondelli, C. Chang, A. Drobot, K. Ko, A. Mankofsky, and J. Petillo, "Application of the Argus Code to Accelerator Design Calculations," 1989 Particle Accel. Conf., Chicago, March 20-23, 1989; to be published by IEEE.

[12] W.S. Cooper, K. Halbach, and S.B. Magyary, "Computer-Aided Extractor Design," Proc. 2nd Symp. on Ion Sources and Formation of Ion Beams, Berkeley, CA; Lawrence Berkeley Laboratory report LBL-3399, p. II-1-6 (1974).

[13] J.W. Kwan, G.D. Ackerman, O.A. Anderson, C.F. Chan, W.S. Cooper, G.J. De Vries, K.N. Leung, A.F. Lietzke and W.F. Steele, "Operation of a DC Large Aperture Volume-Production Source," International Conf. on Ion Sources, Berkeley, CA, July 10-14, 1989; to be published by Rev. Sci. Instrum.

DISCUSSION

Raparia: The beam from the ion source is converging what ion source do you use?

Anderson: The beam from the preaccelerator is converging. First, the beam starts out with very low velocity, almost 0, then we accelerate. We apply 100 kV dc voltage to accelerate the beam. So that is essentially a Pierce gun and the beam is compressed a little bit by the accelerator and tends to be converging at that point.

Raparia: Do you have one einzel lens? Is it sufficient to measure the beam because you need degrees of freedom for matching in RFQ.

Anderson: This was the x direction, we also measure in the y direction. But I didn't bring that viewgraph. We tried to make them identical but because of the magnetic field and the source, the x and y are not actually identical even at the source.

Allison: But would you need 2 degrees of freedom?

Anderson: Yes, we have 5 sets of quadropules, and we have more degrees of freedom then we really need. If we were designing this system from scratch just to make a low energy beam transport, we wouldn't need that many quadropules. We have them available because we have already built this accelerator.

Mori: What was the measured output of the normalized emittance at the entrance of the RFQ?

Anderson: Are you asking about the current?

Mori: How many mA do you measure?

Anderson: There were 30 milliamperes going in, we did not measure any loss.

Mori: I'm just asking what is the normalized emittance.

Anderson: The emittance is determined physically by our type of source and it is fairly high, in the order of 0.05 π cm mrad.

Mori: .05, is it RMS?

Anderson: It is RMS.

Mori: So it is 4 or 5 times more.

Anderson: It is not very low, but the reason for that is that our source is not a very bright source.

Mori: In the future volume source, is the emittance lower, isn't it?

Anderson: Well that is what we would like to do, but we haven't succeeded yet.

Mori: So that means that there is some emittance (growth) process between the ion source and/or through the accelerator.

Anderson: If you notice the area here and here it is confirmed by the computer printout. It is essentially the same. We don't see any emittance growth throughout the system.

Mori: Throughout the system, that means 100 KeV region.
Anderson: This is all 100 KeV, we don't measure down to 1 eV where it starts at the source.

Roberts: Is that increase in area something as simple as you have reached a resolution (limit) in the momentum angular measurement. If the angular resolution is small, and the beams are parallel, if you are measuring a thing, the area looks bigger than it is. Does this show up in the other coordinate too?

Anderson: We haven't figured out if it is in the hardware or the software, but very likely it could have something to do with the angular resolution.

Hopman: In your application injecting it into an RFQ it is quite natural that you have the source at high voltage for those RFQs at ground potential. In fusion applications, one could think of having the source on ground and the neutralizer on high voltage, because what comes out is a neutral beam. So my question is, do you foresee any complications if you operate your system with the source grounded?

Anderson: I don't think we would have anymore complications than anyone else would. We don't like that idea particularly we haven't thought about it very much. We tend to favor upgrading the source at high voltage and the output at ground, but I think you could do it either way.

Whealton: That emittance corresponds to an ion temperature in excess of 5 eV doesn't it.

Anderson: Actually it is 4 eV.

Whealton: All the measurements of ion temperatures in the source are below 1 eV so it appears that the emittance growth occurs perhaps in the extraction region. I have a question and that is in the typical low energy beam transport systems. One of the objectives is to have the pressure in the RFQ about 10^{-5}, so these things are made rather long with pumping all along the edges. In this particular case it looked like it was relatively short and the pumping was not on the side but in fact the gas went through the system in this configuration. With the ion source gas efficiency on the order of less than 1%, do you expect the gas pressure in the RFQ to be higher?

Anderson: The pressure was a very good argument for operating the source at ground because that would allow you to easily pump at the source end. What we are all hoping for is that someone will develop much better sources with better gas efficiency and that would be a wonderful thing. I think you have heard talks about adding cesium and things like that, all those things are giving us some hope that there will be improvements in sources.

Whealton: You would propose having the RFQ at the 100 kilovolts.

Anderson: No, not for RFQ, your right. I was thinking for the other application.

TEST OF A COMPACT 750 keV H⁻ PREINJECTOR*

C. R. Meitzler, P. Datte, F. R. Huson, R. Kazimi, C. Kronke, S. Machida,
W. MacKay, S. Ohnuma, D. Raparia, D. Sun, P. Tompkins, J. Ziegler
The Texas Accelerator Center, 4802 Research Forest Drive,
The Woodlands, TX 77381

ABSTRACT

A 750 keV RFQ based accelerator is being developed at the Texas Accelerator Center. A modified magnetron ion source will produce 10–100 mA of 30 keV H⁻ beam. A 30 keV transport line that transports the beam from the ion source to the entrance of the RFQ without becoming neutralized has been designed and is under construction. The RFQ is a 86 cm long, four rod structure that operates at 470 MHz. Results of tests on the cold model are reported.

INTRODUCTION

A prototype of a proposed Superconducting Super Collider (SSC) preinjector is being developed by the Texas Accelerator Center (TAC). The energy of this prototype is 750 keV. The energy will be increased to 2.5 MeV which is required for SSC, after tests on this proof of principle system is completed.

A preliminary design of the 750 keV test accelerator is presented in Figure 1. Table 1 shows the design goals of this prototype. One feature of this accelerator is the short distance between the ion source and the RFQ. A second feature of the system is that its total length, from the ion source to the high energy end of the RFQ, is less than that of the RFQ at BNL[1]. In the following sections each of the components is described and the results of initial measurements, when available, are presented.

THE H⁻ ION SOURCE

The H⁻ ion source is a surface-plasma source of the magnetron type modeled after the source on the AGS preinjector[1,2] at Brookhaven. The cathode uses a spherical dimple[2] to provide geometric focusing of the H⁻ onto a 2.8 mm diameter aperture. The beam will be extracted at 30 keV. The H⁻ current will be in the range of 10-100 mA, although initially it will be limited to 10 mA due to limitations in the RFQ power supply. For the present application, the source

---

* This work was performed under the auspices of the U.S. Dept. of Energy.

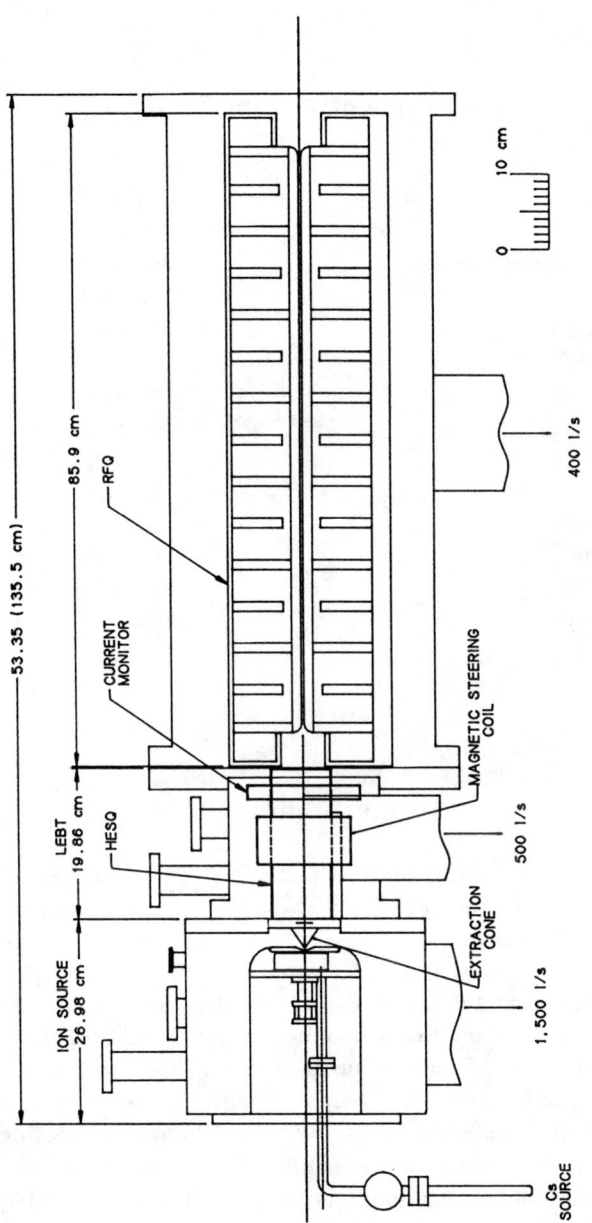

Figure 1. The 750 keV test accelerator.

must be able to operate at 10 Hz with pulse lengths that are variable between 50–100 µs. Longer 750–1250 µs pulses are available for use with an Allison[3] type emittance scanner which is under construction. A detailed description of the source is available elsewhere[4]

Table 1 Preinjector specifications

Ion Source	
Type	Magnetron
Ion	H⁻
Extraction voltage	30. kV
Current	10. mA
Pulse length	50-100 µ sec
LEBT	
Type	HESQ
Electrode voltage	7. kV
Length	22. cm
Pitch (λ)	10. cm
RFQ	
Type	Four Rod Cavity
Frequency	470 MHz
Energy	750. keV
Length	86. cm

Low Energy Beam Transport

Existing low energy beam transport (LEBT) systems use magnetic focusing provided by solenoids or quadrupoles. These LEBT systems use charge neutralization in the background gas to minimize the required focusing strength. The neutralization time must be short compared to the pulse length; otherwise, a fraction of the beam at the front of the pulse may be lost due to inadequate focusing strength. Before the beam becomes fully neutralized, the phase space ellipse rotates making it difficult to match to the RFQ acceptance. Additional problems arise due to beam-plasma instability and the fact that beam becomes charged again when it enters into the electric field of accelerating structure which sweeps away the charge neutralizing ions formed in the collisions. Improper matching in such transitions and the conversion of field energy into transverse kinetic energy associated with changes in the beam profiles may cause significant emittance growth and particle loss.

For a high energy physics accelerator such as the SSC, one needs a pulse approximately 25 - 50 μs long. Since the neutralization time is on the order of 100 μs when hydrogen is the residual gas, we cannot use the beam during the first 100 μs due to the rapid rotation of the beam ellipse. This means that one would need the pulse length from ion source to be about 125 -150 μs. The neutralization time could be reduced by introducing Xenon into the LEBT; however, full neutralization would still require that longer source pulses are used. A way to avoid the problem of neutralization would be to use a continuous electrostatic focusing channel that transports the beam from the ion source to the RFQ.

The proposed LEBT scheme is to use a helical electrostatic quadrupole channel[5] (HESQ) to transport the 30 keV beam from the ion source extractor to the entrance of the RFQ. The HESQ is a continuously twisted quadrupole. It provides spatially continuous focusing in both x and y directions. The HESQ is 22 cm long with an aperture of 2.6 cm. Figure 2 shows the beam envelope for a 10 mA beam passing through the HESQ. The input emittance for the calculation was obtained from Alessi et al.[6] at BNL which is 0.18 πmm-mrad (norm.,rms). Calculations indicate that the growth in the emittance in HESQ is 0.02 π mm-mrad.

Surrounding the HESQ are a set of box-type magnetic dipoles capable of moving the position of the beam at the RFQ entrance approximately 1 mm. A toroidal current pick-up at the HESQ exit provides current monitoring capabilities for source and transport performance.

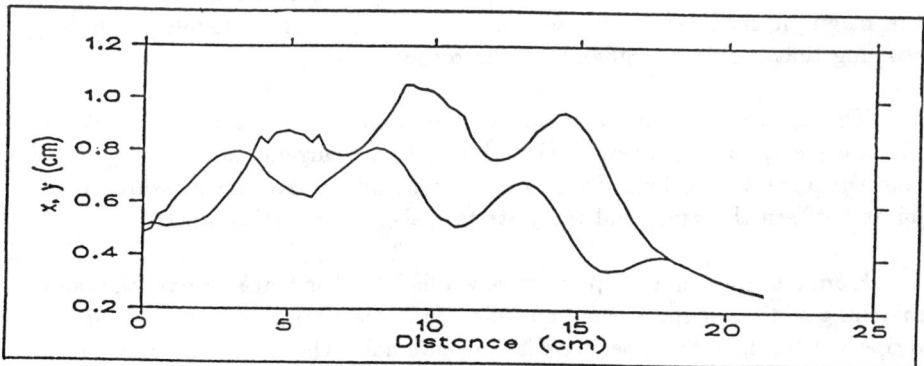

Figure 2. The beam envelope for a 10 mA beam passing through the HESQ.

750 keV RFQ

A radio-frequency quadrupole (RFQ) is the favored method for low energy acceleration, focusing and bunching of both light and heavy ion beams. Among different RFQ structures, the Frankfurt four-rod design[7] has the advantage of simpler construction and better field quality. We have introduced a variation to this geometry in which the "U" shaped supports are replaced with cylindrical cavities[8] and vane-shaped electrodes are used instead of rods. With this design we have been able to achieve higher frequency operation for a reasonable cavity size. The basic unit module consists of a cylinder with two end caps which are connected to the electrode : each end cap is connected to two opposing electrode. The structure is made up of a series of such modules next to each other. This structure retains many virtues of the regular four-rod type structure For example, the unwanted dipole mode is eliminated and since the fields are confined within the cavities, the structure can be put together and tuned before it is inserted into the vacuum vessel.

In order to simplify construction, we have changed the end plates from a circular to a square shape[9], as shown in Figure 3. In this way the structure will consist of a series of square plates supporting the four vanes and four long rectangular plates making the four sides of the structure. The corners of the cavities have been left open to have better vacuum quality. The reasons we can do this are: First, the diagonal planes going through the opposing corners are the symmetry planes of the structure; therefore, there should be no currents crossing these planes, or in other words, the B field is perpendicular to these planes. Second, the fields are weak at the corners, so leaving the corners open should not appreciably change the resonance frequency or the Q. Figure 4 shows the magnetic field for a cross-section at the middle of a module ($z = l/2$), showing that B field is negligible at the corners.

The square plates are made in two pieces and the vanes are held between the two pieces. By symmetry, there should be no current wanting to go from one half-plate to another. Therefore, we can concentrate the pressure on the joints between the vanes and the plate to make a good RF contact.

From a beam dynamics point of view this RFQ has four sections: the radial matching section, shaper, gentle bunching and acceleration sections. The design recipe for the first three sections is conventional. The design criteria for the acceleration section has been changed to make it short in length without losing its beam qualities. In the acceleration section, the modulation and the phase angle are conventionally kept constant. In the present design[10], the modulation is increased and the transverse current limit is held constant instead. This results in the same transmission efficiency and emittances, but with a higher accelerating gradient. The RFQ specifications are listed in Table 2 and the main parameters as function of length are shown in figure 5.

Table 2 RFQ specifications

Frequency	470 MHz
Injection energy	0.03 MeV
Output energy	0.75 MeV
Current	10.0 mA
Length	86. cm
Bore radius	0.25 cm
Max. Modulation	2.5
Acceptance	1.25 π mm-mrad
Intervane voltage	66.7 kV
Transmission	95 %
No. of modules	14
Width of the modules (W)	20. cm
Total power (peak)	102. kW
Beam power (peak)	7.0 kW
Structure power (peak)	95. kW

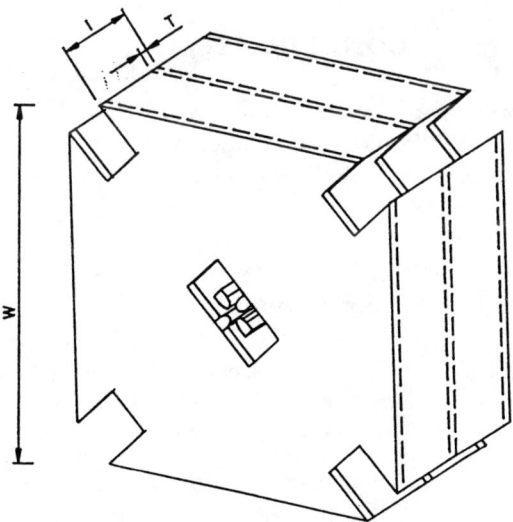

Figure 3. The 470 MHz RFQ structure.

STATUS

The design studies of the system have been largely completed, and it is entering a construction phase. Individual components are being manufactured and tested, with the full accelerator being assembled in the coming winter.

The ion source has been constructed and initial tests of the system have been performed. The magnetron has been operated; however, no beam has been extracted at this time. A measurement of the source emittance will be performed in the next few months after an electric sweep scanner type emittance head has been constructed and tested. The HESQ components have been manufactured and assembled. The steering coils, current monitor and vacuum chamber are being manufactured.

We have assembled a cold model of the RFQ. It consists of four regular modules and has unmodulated vanes. The measured resonance frequency for the lowest mode is 471.5 MHz which is less than one-half percent from the the design frequency of 470 MHz. The nearest mode was at 740 MHz, which is quite far away from the fundamental mode. This reconfirms the fact that higher mode mixing should not be a problem in this structure. We have tested a scheme for fine tuning the structure in which we insert a conductive blade in one of the corners: inserting a copper blade in one corner increased the frequency by 1 MHz. The measured unloaded Q is approximately 4400 which should increase when the RF joints between the RFQ components are improved.

CONCLUSION

The initial tests of the TAC 750 keV RFQ linac show that it will be possible to build a compact H^- RFQ for use as a preinjector to the SSC linac injectors. Bench tests of the RFQ cold model indicate that it should be able to accelerate 10 mA to 750 keV.

Acknowledgements

The authors would like to thank Jim Alessi and Chuck Schmidt for their invaluable suggestions and discussions.

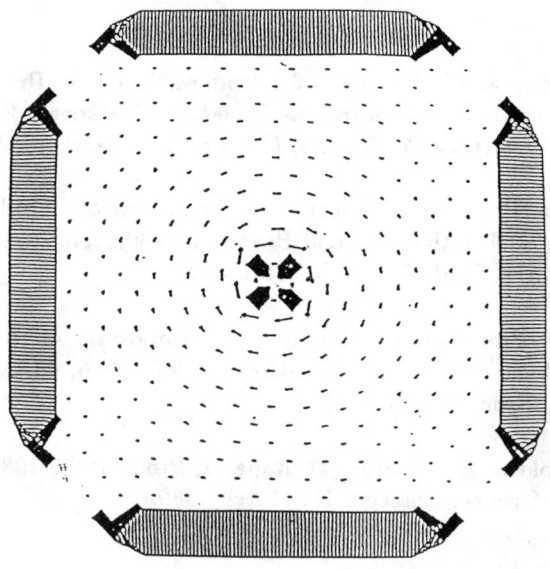

Figure 4. Plot of the magnetic field in the middle of a module.

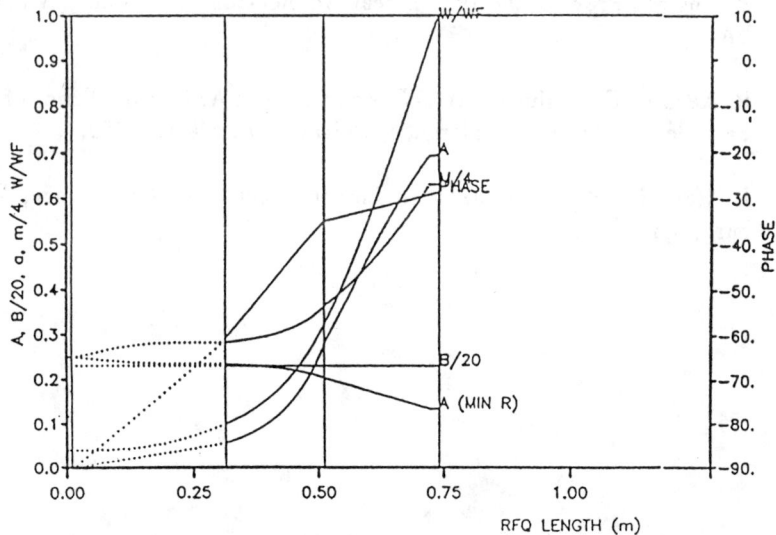

Figure 5. The main parameters of the RFQ vs. RFQ length.

REFERENCES

1. J. G. Alessi, J. M. Brennan, J. Brodowski, H. N. Brown, A. Kponou, V. LoDestro, P. Montemurro, K. Prelec, R. Witkover, Proc. of the 1989 Particle Accelerator Conference, Chicago, IL, March 1989.

2. J.G. Alessi, Proc. of the Fourth Int. Symp. on the Production and Neutralization of Negative Ions and Beams, Brookhaven, NY, 1986, ed. James G. Alessi, AIP Conf. Proc. No. **158**, 419.

3. P. Allison, Proc. of the Fourth Int. Symposium on the Production of Negative Ions and Beams, Brookhaven, NY, 1986, ed. James G. Alessi, AIP Conf. Proc. No. **158**, 465.

4. P.A. Tompkins, F.R. Huson, D. Raparia, Proc. of the 1989 Particle Accelerator Conference, Chicago, IL, March 1989.

5. D. Raparia, these proceedings.

6. J.G. Alessi and A. Kponou, private communication.

7. A. Schempp, H. Klein, H. Deitinghoff, P. Junior, M. Ferch, A. Gerhard, K. Langbein and N. Zoubek, Proc. of the 1984 Linear Accelerator Conference, 1984, 100-102.

8. R. Kazimi, Proc. of the 1988 Linear Accelerator Conference, Williamsburg, VA

9. R. Kazimi, F. R. Huson, W. W. MacKay, and A. Nassiri, Proc. of the 1989 Particle Accelerator Conference, Chicago, IL, March 1989.

10. D. Raparia, Proc. of the 1988 Linear Accelerator Conference, Williamsburg, VA

HESQ, A LOW ENERGY BEAM TRANSPORT*

Deepak Raparia

Department of Physics, University of Houston, Houston, Texas 77204-5504

and

Texas Accelerator Center, The Woodlands, Texas 77381

ABSTRACT

In this paper Helical Electrostatic Quadrupole (HESQ) channel is suggested for low energy beam transport of H^- beams from ion source to the RFQ. Being an electrostatic focusing lens, the HESQ avoids neutralization time. The HESQ lenses provide stronger first-order focusing in contrast to weak second-order focusing of einzel lenses and is also stronger in focusing than the alternating gradient focusing. In this paper, we will present analytical formalism for such channel and results of PIC simulation with space charge.

INTRODUCTION

Invention of the the RFQ has solved the long existing problems at low β-end in the drift tube linac by virtue of spatially continuous external focusing forces and transfered such problem to low energy beam transport (LEBT). The LEBT is the section of linac which provides the matching between ion source and the radio frequency quadrupole (RFQ). The LEBT consists of lenses that focus the beam into the RFQ. The beam from an ion source is relatively large in radius and divergence and must be matched to the RFQ.

Existing LEBT systems for H^- beams use magnetic focusing solenoids or permanent magnetic quadrupoles. These LEBT systems utilize charge neutralization in the background gas to minimize the required focusing strength. However, the physics of such charge neutralized beam is not fully understood. Futhermore, the neutralization time must be short compared to the pulse length; otherwise, a fraction of the beam at the front of the pulse maybe lost due to inadequate focusing. During this neutralization time the space charge force changes. Because of this the beam phase space ellipse rotates, making it difficult to match the beam to the RFQ acceptance. Additional problems arise due to beam-plasma instability and the fact that beam becomes charged again, when it enters into the electric field of accelerating structure which sweeps away the charge neutralizing ions formed in the collisions. Improper matching in such transitions and the conversion of field energy into transverse

---

* This work supported by U. S. Department of Energy under grant No. DE-FG05-87ER40374.

kinetic energy associated with changes in the beam profiles may caused significant emittance growth and particle loss.

For the Superconducting Super Collider (SSC), one need 25 - 50 μ sec pulse length while the neutralization time is of the order of 100 μ sec in case of hydrogen as residual gas. During the first 100 μ sec we cannot use the beam. That means one would need the pulse length from ion source 125 - 150 μ sec and this reduces the lifetime of ion source by factor of two or more. The neutralization time can be reduce by introducing Xenon into the LEBT; however, full neutralization will always considerably increase the pulse length required from source. The problem of neutralization can be avoided by using focusing in the form of electric forces which will be discussed in the following section.

HELICAL ELECTROSTATIC QUADRUPOLE LENSES (HESQ)

The buildup of plasma and charge neutralization cannot occur in electrostatic lenses, as in case of magnetic lenses, since the electron-ion pairs produced in collision between beam particles and background gas are swept out of the beam region by the electric field between the electrodes. Furthermore, magnetic focusing is particularly ineffective at low beta because of the velocity term in the force equation. Electric focusing, on the other hand, has no such velocity term in the force equation and should be a prime candidate for the focusing role at low velocities. The main problems in the electrostatic focusing is breakdown and aberrations. The problem of the breakdown can be overcome by providing lower but spatially continuous focusing forces, as in the helical electrostatic quadrupole (HESQ), instead of spatially discreet but higher focusing forces as in a FODO structure. The spatially continuous focusing forces also help keep the beam size smaller all the time, thus minimizing aberrations.

An alternative scheme for the higher current is to use the HESQ in the LEBT. The HESQ is nothing but a continuously twisted electrostatic quadrupole (figure 1). The HESQ lenses provide stronger first-order focusing in contrast to weak second-order focusing of Einzel lenses, an axial symmetric beam which is necessary for the RFQ matching, and stronger focusing than the alternating gradient type. The idea of helical quadrupole is quite old[1-5]. All of theses works were about 'magnetic helical quadrupole' for different purpose.

The HESQ system must satisfy following conditions. First, voltage and spacings 'd' between neighboring electrodes must be below breakdown limit[6] given by

$$V_B[kV] = 79.9 \cdot d^{2/3}[cm]. \qquad (1)$$

Second, the maximum voltage 'seen' by a particle at the edge of the beam must be less than about 20 % of the beam voltage i. e.

$$V_0/V_b \leq 20\% \qquad (2)$$

This is necessary to reduce the chromatic aberration within acceptable limit. Third, the maximum ratio of beam radius and the quad radius must be less than 0.75 i.e.

$$R_{beam}/R_{quad} \leq 0.75 \tag{3}$$

To reduce higher order effects one has to satisfy following condition

$$R_{beam} \leq \lambda/10. \tag{4}$$

Where λ is pitch of the helix. If the last condition is not satisfied then particles 'see' a weak longitutinal field which increases with radius and changes its direction along the length of the HESQ and this alternating longitudinal field provides additional focusing. This can be thought as electrostatic symmetric lenses are superimposed on quadrupole field but net the effect will be undesirable because of non-linearity.

EQUATION OF MOTION AND STABILITY DIAGRAM

In this section, we have used a special model[5] for continuously twisted quadrupole which also include the linear space charge effects i.e. only K-V distribution, for other distribution one has to take the approach of PIC code. The potential function for continuously twisted quadrupole channel can be written as

$$\Phi(r, \phi, z) = const \cdot I_2(\nu r) \cos 2(\phi - \nu z) \tag{5}$$

where ν is transverse rotation per unit length along the z-axis. In the lowest order approximation, the equations of motion are

$$x'' = -k^2 \{x \cos(2\nu z) + y \sin(2\nu z)\} + x\Delta_{sc}$$

$$y'' = -k^2 \{x \sin(2\nu z) - y \cos(2\nu z)\} + y\Delta_{sc} \tag{6}$$

Here k is related to the quadrupole strength and Δ_{sc} represent space charge defocusing force for the K-V distribution given by

$$a\Delta_a = b\Delta_b = \frac{120eI}{mc^2 \beta_r^3 (a+b)} \tag{7}$$

where a and b are the simi-axis of ellipitical beam with τ charge per unit length. Electric fields inside the beam are

$$E_x = \frac{4\tau}{4\pi\epsilon_0} \frac{x}{(a+b)}; \qquad E_y = \frac{4\tau}{4\pi\epsilon_0} \frac{y}{(a+b)} \tag{8}$$

and the effective k^2 and Δ_{sc} are

$$k^2 = k_{ESQ}^2 + \left(\frac{\Delta_b - \Delta_a}{2}\right); \qquad \Delta_{sc} = \frac{\Delta_a + \Delta_b}{2} \tag{9}$$

The particle coordinate in the rotating frame of reference is

$$u = x\cos(\nu z) + y\sin(\nu z)$$
$$v = -x\sin(\nu z) + y\cos(\nu z). \quad (10)$$

Differentiating (10) w.r.t. z one gets;

$$u' = x'\cos(\nu z) + y'\sin(\nu z) + \nu v$$
$$v' = -x'\sin(\nu z) + y'\cos(\nu z) - \nu u. \quad (11)$$

differentiating (11) w.r.t. z gives;

$$u'' = x''\cos(\nu z) + y''\sin(\nu z) + 2\nu v' + \nu^2 u$$
$$v'' = -x''\sin(\nu z) + y''\cos(\nu z) - 2\nu u' + \nu^2 v. \quad (12)$$

Equation (6) and (12) together produce

$$u'' = 2\nu v' + \nu^2 u - k^2 u + u\Delta_{sc}$$
$$v'' = -2\nu u' + \nu^2 v + k^2 v + v\Delta_{sc} \quad (13)$$

The two coupled Mathieu equation (6) have simplified to coupled linear differential equation with constant coefficients (13). The solution of (13) are

$$s_1 \equiv p^2 = \nu^2 - \Delta_{sc} + \vartheta$$
$$s_2 \equiv q^2 = \nu^2 - \Delta_{sc} - \vartheta \quad (14)$$

where

$$\vartheta = (k^4 - 4\nu^2 \Delta_{sc})^{1/2} \quad (15)$$

These solutions p^2 and q^2 should be real and positive for the stability requirement. That is

$$2\nu\sqrt{\Delta_{sc}} \leq k^2 \leq \nu^2 + \Delta_{sc} \quad (16)$$

This is represented by the shaded region in figure 2,

The general solutions of (13) can be written as

$$u = A\cos P + C_1 B\sin Q \qquad v = -C_2 A\sin P + B\cos Q \quad (17)$$

where P=pz + const and Q=qz + const and

$$C_1 = \frac{2\nu^2 + k^2 - \vartheta}{2\nu q}, \qquad C_2 = \frac{2\nu^2 - k^2 + \vartheta}{2\nu p} \quad (18)$$

Differentiating (17) one gets,

$$u' = -pA \sin P + C_1 qB \cos Q, \qquad v' = -C_2 pA \cos P - qB \sin Q \qquad (19)$$

Eliminatign P and Q leads to the two invariants

$$\frac{(u' - C_1 qv)^2}{D_1^2} + \frac{(C_1 v' + qu)^2}{D_2^2} = A^2, \qquad \frac{(C_2 u' - pv)^2}{D_1^2} + \frac{(v' + C_2 pu)^2}{D_2^2} = B^2 \qquad (20)$$

where $D_1 = p - C_1 C_2 q, D_2 = C_1 C_2 p - q$. The maximum radial excursion R is the larger of

$$C_1 A + B \qquad or \qquad A + C_2 B \qquad (21)$$

There are two normal modes each with two distinct frequencies in the rotating frame (u,v). The procedure of transforming back to lab frame introduces another frequency that of coordinate rotation. This is the reverse of the more usual situation in which a system has pure frequency in a fixed coordinates system.

PHASE SPACE PROJECTION, TRANSVERSE EMITTANCES

Unlike the FODO channel, the HESQ transport system is coupled in x and y plane. In this transport system the four dimensional mean square emittance[7]

$$\epsilon_{4d}^2 = <x^2><p_x^2> - <xp_x>^2 + <y^2><p_y^2> - <yp_y>^2$$

$$+ 2<xy><p_x p_y> - 2<xp_y><yp_x> \qquad (22)$$

is constant while its projections in x and y plane vary depending upon the coupling terms (last two terms in the above equation). Brown and Servranckx[8] shown that (1) if a beam is uncoupled at the beginning of a system, and the initial x and y emittances are equal, then at all the point downstream, the projected emittances in x and y plans will be equal to each other, independent of the magnitude of the x-y coupling at that point. Their magnitude may, however, be equal to or greater than the values of the intial, uncoupled emittances. (2) At any position downstream, where x-y coupling is present the sum of the square of projected x and y emittances will be equal to or grater than the square of the four dimensional emittance (eq 22).

EXAMPLE: HESQ at TAC

At Texas Accelerator Center, we are building a HESQ for matching circular symmetric Megnetron ion source to a 470 MHz four rod RFQ[10]. The main parameters are given in Table I. In the design quadrupole rotation is done in steps of 18 deg [figure 3]. The field calculation, which is done using RELAX3D[9], shows that the longitudinal electric field components are not more than 3-5 % anywhere in the HESQ. A PIC simulation of the beam through the HESQ is shown in the figure 4, notice that beam size is always less than equal to 1 cm, i.e., less that 70% of the aperture. Figure 5 shows the input and output phase space plot of this HESQ. Figure 6 shows the projected emittance in horizantal and verticle planes and four dimensional emittance as beam traverses through the HESQ which agrees with the theoritical results given by Brown and Servranckx[8].

Table I HESQ specifications

length	19. [cm]
voltage	7. [kV]
break down voltage	80. [kV]
pitch of helix	10. [cm]
electrode spacing	1. [cm]
bore radius	1.3 [cm]
input beam parameters :	
beam current	10. [mA]
transverse emittance (n,rms)	0.18 [π .mm.mrad]
output beam parameters:	
beam Current	10. [mA]
transverse emittance (n,rms)	0.2 [π .mm.mrad]

CONCLUSION

Use of the HESQ in H$^-$ LEBTs avoids neutralization time. It is stroger focusing system than einzel lens and alternating gradient quadrupoles. It also provides an axial symmetric beam which is necessary for the RFQ matching. Being spatially continuous focusing system, the HESQ keeps the beam size smaller all the time, thus minimizing aberrations. The HESQ is presently being developed and may have superior performance and reliability than other LEBTs.

Special thanks to S. Ohnuma and F. R. Huson for their encouragement.

REFERENCES

[1] L. C. Teng, Helical Quadrupole Magnetic Focusing Systems, Argonne National Lab Report, ANLAD-55, 1959.
[2] S. Ohnuma, TRIUMF Report, TRI-69-10, 1969.
[3] G. Salardi, E. Zanazzi and F. Uccelli, Nuclear Instruments and Method in Physics Research 59, 1968, pp 152-156
[4] R. M. Pearce, Nuclear Instruments and Method in Physics Research 83, 1970, pp 101-108
[5] R. L. Gluckstern, Proceeding of the 1979 Linear Accelerator Conference, BNL 51134, pp 245-248
[6] L. J. Laslett, Selected Works of L. Jackson Laslett, LBL-PUB-616, Vol III, p 6-49, September 1987; M. Reiser, et al, Microwave and particle beam source and propagation, SPIE Vol. 873, p 172 (1988)
[7] A. J. Dragt, R. L. Gluckstern, F. Nari, and G. Rangarajan, to be published in the Proc. of the US-CERN school on Accelerator Physics, Carpi Italy, 1988
[8] K. L. Brown and Roger V. Servranckx, SLAC report, SLAC-PUB-4679, August 1989
[9] H. Houtman and C. J. Kost, TRIUMF report, TRI-PP-83-95, September 1983.
[10] C. R. Meitzler, et al, This Conference.

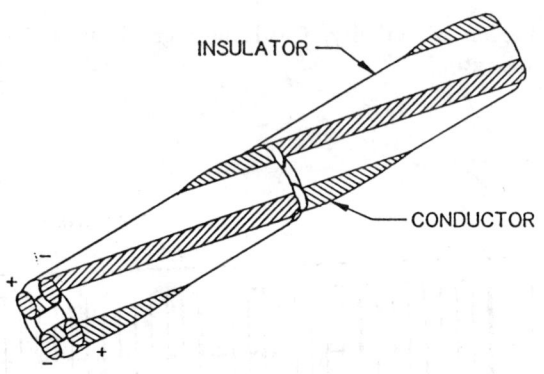

Figure 1: Conceptual design of the HESQ.

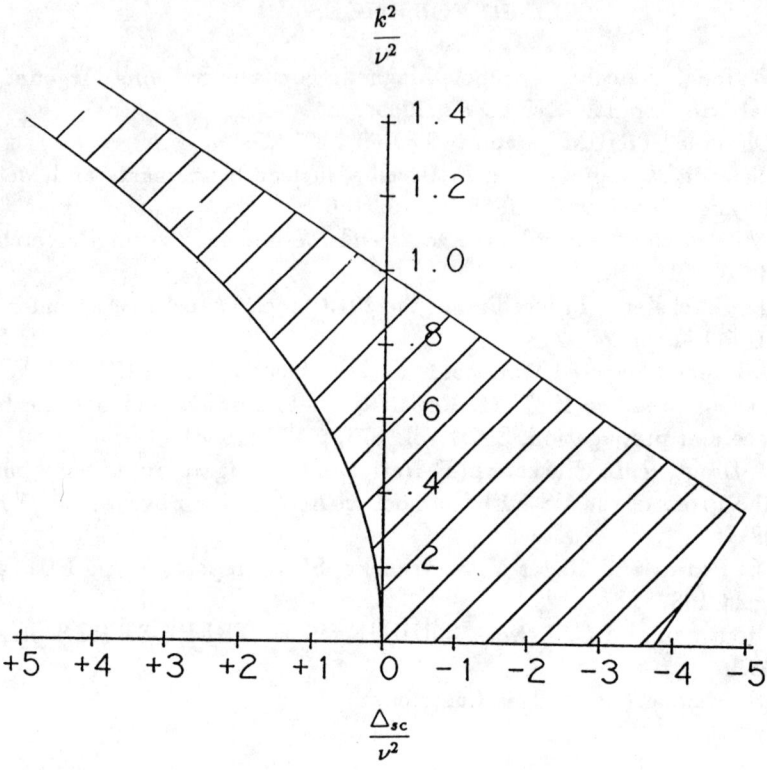

Figure 2: Region of stability, Quadrupole gradient vs Space Charge Defocusing strength.

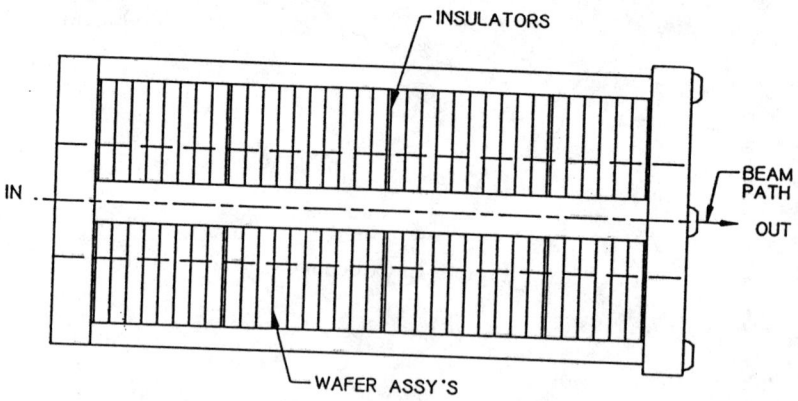

Figure 3: HESQ at TAC. Quadrupole rotation is done in steps of 18 deg.

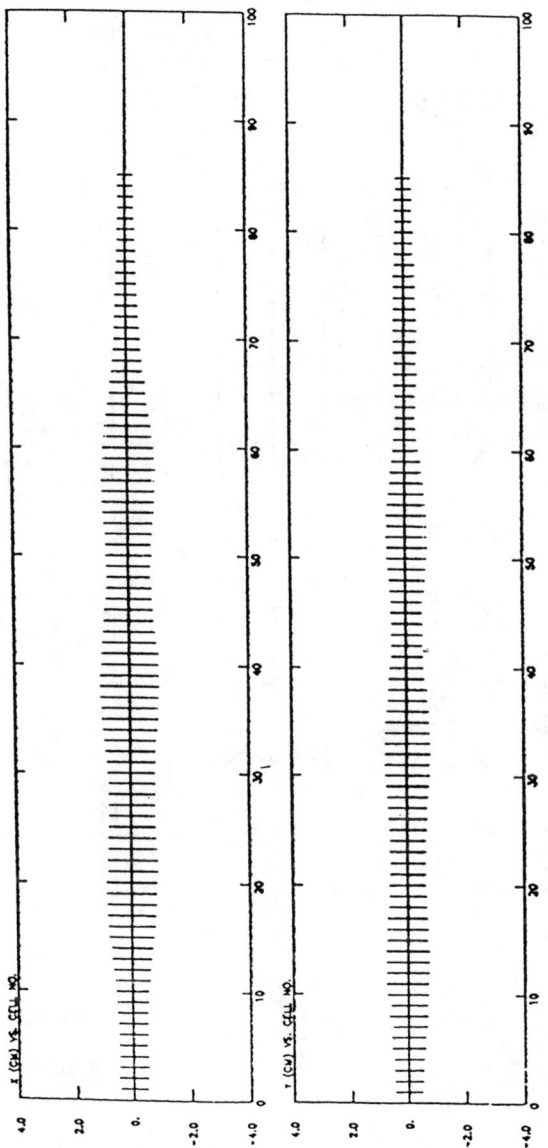

Figure 4: x and y profiles through the HESQ.

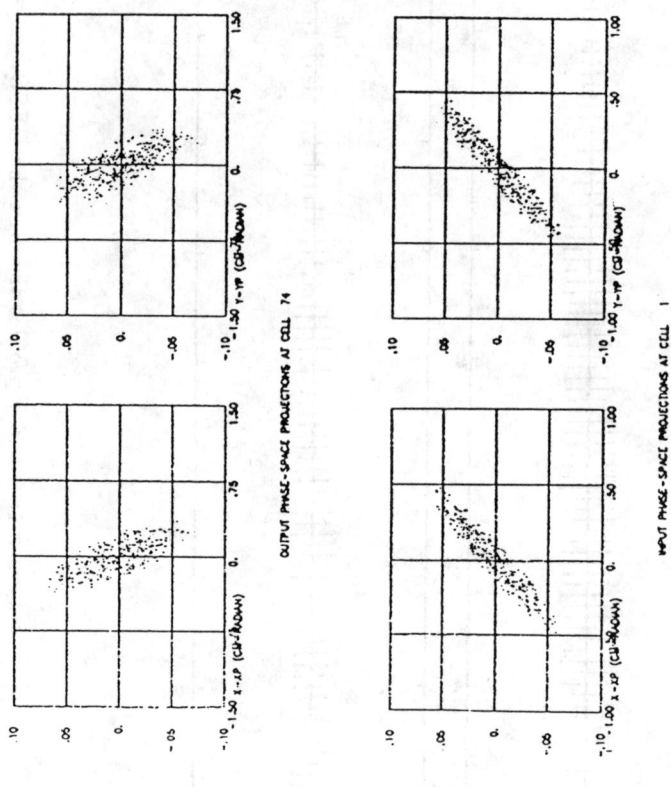

Figure 5: Input and output phase-space projection in x and y planes.

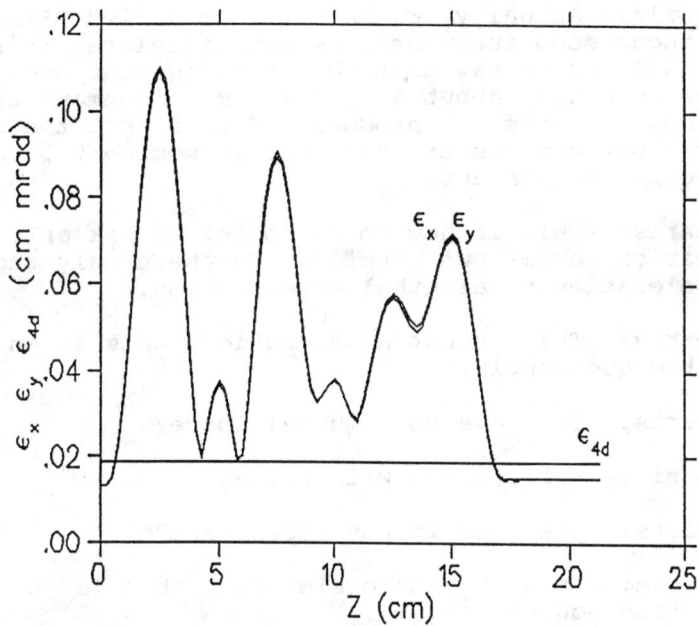

Figure 6: The projected emittences (normalized, rms) in x and y planes and four dimensional emittance as beam traverses through the HESQ.

HESQ, A Low Energy Beam Transport

DISCUSSION

Mori: I have one question about the misalignment problem of this kind of a disc shape. Do you have estimates of the admittance growth due to the misalignment of this disc.

Raparia: Actually, we have a code called RELAX3D. Due to those step functions, we have simulated this thing in RELAX3D and we saw that this misalignment does not produce more than about 3 or 4 mil misalignment of electrode. It does not produce a dipole into the beam, and for a one centimeter area beam it wouldn't be more than a couple of percents.

Roberts: This is new to me so let me ask an elementary question for my own benefit. Is there only radial acceleration no azmuthal acceleration.

Raparia: This is not acceleration, this is an electrostatic quadrupole.

Roberts: You have no azmuthal force?

Raparia: Well, there will be.

Roberts: How does it not accelerate?

Raparia: It does not accelerate. This is just a match from ion source to . . .

Roberts: I understand but inside, I'm worried about angular momemtum, you don't have any when you go in and does it do work on the beam azmuthally!

Raparia: Yes, it does, when you come out the particle is rotating.

Roberts: I understand that, but is that angular momentum now detrimental in the RFQ or in your accelerator?

Raparia: Angular momentum, well then you reduce this thing, you would keep last part of the voltage. My simulation doesn't see any angular momentum. It is very low.

Roberts: Now you rotate in the solenoid. But it doesn't rotate when it goes in, and the magnetic field doesn't do any work, so it will not rotate when it comes out. But this is different.

H⁻ SOURCE AND LOW ENERGY TRANSPORT FOR THE BNL RFQ PREINJECTOR*

J.G. Alessi, J.M. Brennan, and A. Kponou
Brookhaven National Laboratory, Upton, NY 11973

SUMMARY

An RFQ has replaced the 750 keV Cockcroft-Walton as the H⁻ preinjector for the AGS. A magnetron surface-plasma source with a circular aperture is used to produce 65 - 100 mA of H⁻ at 35 keV with a discharge current of less than 20 A. The circular symmetry of the beam is maintained in the 2 m transport to the RFQ via the use of magnetic solenoids for focusing. Currents of up to 60 mA have been obtained out of the RFQ.

I. INTRODUCTION

Since January 1, 1989, the H⁻ preinjector for the Brookhaven Alternating Gradient Synchrotron has been a 750 keV RFQ accelerator. The H⁻ source is a magnetron surface plasma source. There is a 2 meter transport line from the source to the RFQ, containing two focusing solenoids and a fast electrostatic beam chopper. The purpose of the chopper is to reduce beam losses during capture in the AGS by throwing away at 35 keV those particles which would be outside the rf bucket of the AGS. There is a 6 meter transport line from the RFQ to the linac, in order to leave a second H⁻ preinjector line and a polarized H⁻ line undisturbed.

To date, this new preinjector has approximately 5000 hours of operation, with only minimal downtime, and a very stable current of 25 mA out of the linac (200 MeV). Details of the RFQ (a 200 MHz, 4-vane type) and 750 keV transport line can be found in Ref. 1. In this paper, some features of the ion source and 35 keV beam transport will be discussed. This section is shown schematically in Fig. 1.

II. H⁻ SOURCE GEOMETRY

The magnetron surface plasma H⁻ source was chosen for the RFQ preinjector since we have 6 years of operational experience with this type source[2]. We have found it capable of operating up to 9 months continuously at a 5 Hz rep rate (0.25% duty factor). A similar source is used very successfully at FNAL.[3] Since the RFQ required a symmetric beam at the input, several years ago we began to study its operation with a circular aperture rather than the usual slit aperture. Initial studies included a 90°, n = 1/2 bending magnet after extraction.[4] In subsequent work this magnet was eliminated, and after extraction the beam was injected directly into the transport line, as shown in Fig. 1.

*Work Performed Under the Auspices of the U.S. Depart. of Energy.

The BNL RFQ Preinjector

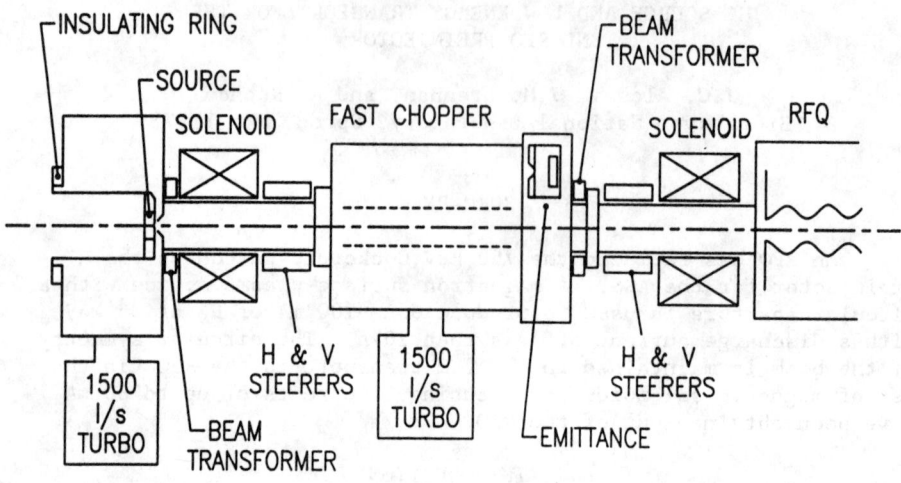

Fig. 1: Schematic of the source and 35 keV transport line.

The main parts of the source are the same as previously used.[2,3] The source operates at a 5 Hz repetition rate, and the gas, discharge, and extraction voltage are all pulsed. The discharge pulse is typically 700 μs, and the 35 kV extractor is pulsed for 500 μs, producing a 500 μs beam pulse. The source is mounted in a reentrant fashion in a vacuum box pumped by a 1500 l/s turbomolecular pump. The only opening between the source chamber and the transport line is the 2 mm diameter extractor aperture. During operation, the pressure in the source vacuum box is typically 4.5×10^{-6} T. The source transverse magnetic field is provided by SmCo magnets clamped above and below the source body. This provides a field of $\approx$ 1 kG in the discharge region. The source cathode has a spherical dimple (r = 6.25 mm) to geometrically focus the surface produced H$^-$ ions from a large cathode area into the 2.8 mm diameter anode aperture. The extractor electrode has a 2 mm aperture and a 4 mm gap between the anode and extractor. The final geometry used is shown in Fig. 2. In tests, we obtained the best performance (smallest emittance, best match to the RFQ acceptance), with a 3.5 mm gap, but the decision was made to use the 4 mm gap for normal operation in order to reduce the frequency of extractor arcs. With the 4 mm gap, the calculations showed some beam loss on the extractor electrode.

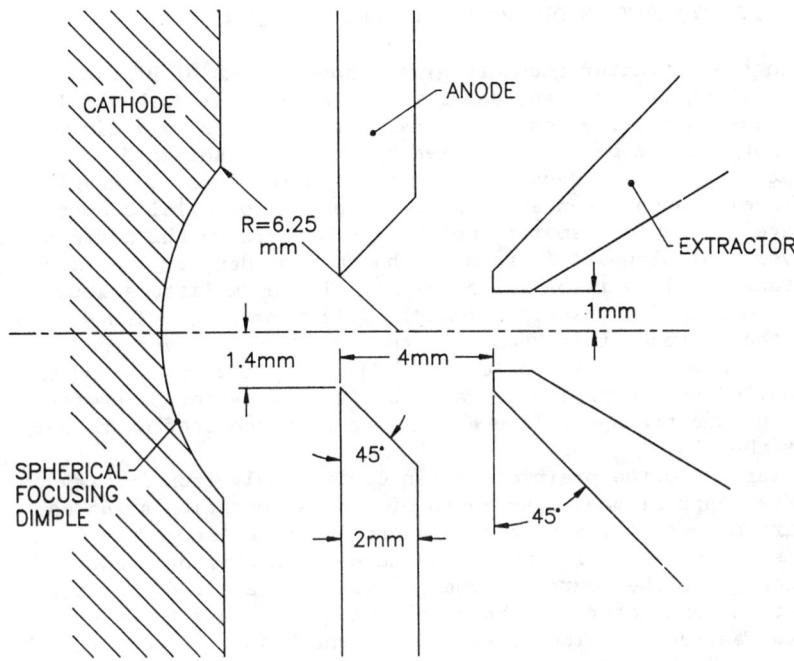

Fig. 2: Source and extractor geometry.

The source magnetic field causes the extracted electrons to be deflected and hit the extractor electrode on one side of the aperture. To reduce erosion, the tip of the extractor is made of tungsten (the bulk of the extractor electrode is made from aluminum). In spite of this, there is still significant erosion of the electrode (a narrow etched line begins to appear within days).

III. THE 35 keV BEAM TRANSPORT

The 35 keV beamline, to transport and match the beam to the RFQ, is shown in Fig. 1. Focusing in the line is provided by two 10 cm diameter, 25 cm long, pulsed magnetic solenoids. The solenoids typically operate at a current of approximately 400 A, where they produce a field of ≈ 3300 G over an effective length of 20 cm. The center of the first solenoid is 23 cm from the extractor, the two solenoids are separated by 150 cm, and the entrance of the RFQ (start of vanes) is 30 cm from the center of the second solenoid. Between the two solenoids there is a fast chopper (see Section V), and horizontal and vertical emittance heads of the slit and collector type.

IV. OPERATION OF THE SOURCE AND TRANSPORT LINE

With the extractor geometry given above, 65 mA of H⁻ was extracted, which is a current density of 1 A/cm^2. At 35 keV this beam is space charge limited at the extractor, and is very divergent, so not all the beam is captured by the first solenoid. Approximately 55 mA is transported to the middle of the 35 keV line. The emittance taken at this point shows a parallel beam of approximately 7 cm in diameter, which is symmetric in the horizontal and vertical planes. Therefore, the current density over most of the transport line is only 1.5 mA/cm^2. The normalized emittance for 90% beam fraction was approximately 0.12 π cm-mrad. The second solenoid then focuses this beam into the RFQ with a convergence angle of 130 mrad. The output of the RFQ was typically 45 - 50 mA. With a smaller extractor gap we were able to get better transmission through the transport line and RFQ, and currents of up to 60 mA out of the RFQ.

During FY'90 the preinjector ran continuously from January to June. After approximately one month of stable operation at 65 mA, the H⁻ current out of the source began to increase, and after 6 months was up by 62% to 105 mA. This increase in current occurred with no change in the source discharge, but was rather due to the erosion of the extractor electrode, allowing more beam to pass through the extractor. The divergence of the beam was increasing, however, so the current in the middle of the 35 keV line only increased by 36%, and the current out of the RFQ only increased by 22%. The emittance in the 35 keV line increased to approximately $\epsilon_n(90\%) = 0.14$ π cm-mrad. Therefore, in spite of the degradation in optics caused by the extractor erosion, we could continue to run and actually gain current as time went on.

The source typically operates at a discharge current of only 15 - 20 A at 150 V. The output is space charge limited, and the discharge current can drop to almost 10 A before the H⁻ current begins to drop (i.e. becomes emission limited). We operate at a discharge current higher than necessary in order to decrease the sensitivity to slow changes in the discharge, or to variations in the current during the pulse. It also helps keep the source temperature up at this low duty factor. The high efficiency of the source, with an extraction of 100 mA of H⁻ with only a 20 A, 150 V arc, is part of the reason that it is capable of operating stably for many months. The extracted electron current is generally less than the H⁻ current. There are approximately 0.5% missed pulses due to extractor arcs. The temperature of the cesium reservoir is typically $\leq$ 110° C.

V. FAST BEAM CHOPPER AND SPACE CHARGE EFFECTS

The purpose of the fast beam chopper in the 35 keV line is to dump at low energy those particles which would be lost during capture in the AGS. With this chopper 2.5 MHz beam bunches can be produced, with the width and phase of each bunch being individually programmable to match the moving rf buckets in the AGS during multiturn injection. Details of the use of this chopper have been given elsewhere[5], and in this section we will comment on its effect on the 35 keV beam.

The chopper is a pulsed electrostatic deflector, where voltages of ± 760 V are applied to plates above and below the beam to deflect the beam outside the opening of a 1.4 cm diameter aperture at the entrance to the RFQ. The chopper has a total length of 38 cm, and a separation between the upper and lower plates of 8 cm. At 35 keV, the flight time of the beam through the chopper is 150 ns. The chopper is therefore made up of 15 pairs of plates over the 38 cm length, which are connected as a slow-wave structure by coaxial cables, so that the voltage pulse travels plate-to-plate at the beam velocity. The chopper plates are powered by a pair of commercial high voltage pulse generators[6] having rise and fall times of less than 10 ns. They are driven by a digital delay generator.

Rise and fall times on the beam pulses of 10 ns are obtained. The minimum pulse width is approximately 80 ns, limited by the high voltage pulsers. Under normal operation (unchopped beam), the H⁻ beam is space charge neutralized in the transport line by ionization of residual gas. With the typical pressure in the transport line of 10^{-6} Torr, the neutralization time is approximately 50 μs. When the chopper voltage is applied for microseconds or longer, the neutralizing ions are swept out of the beam and the beam becomes very divergent due to space charge blowup. The result is that a much higher voltage than what one might at first expect is required to completely reject the beam before the RFQ. From studies of the beam current out of the RFQ as a function of dc voltage on the chopper, and comparisons with calculations of the beam optics in the 35 keV line under various space charge conditions, it appears that neutralization is lost only in the region between the chopper plates. This is supported by the fact that when biased grids were installed to create potential barriers at the entrance and exit of the chopper it had no effect on the performance.

A typical operation of the chopper for testing was a string of voltage pulses with 200 ns on time and 200 ns off time, for the 500 μs beam pulse. Beam is transported through the RFQ when the voltage is off, so there are no aberrations on the beam from imperfections in the electrostatic fields in the chopper region. In spite of this, emittances measured in the 35 keV line showed a distortion when the chopper was used. (The sample time of the emittance is greater than the 2.5 MHz chopping frequency, so one can observe both the deflected and undeflected beam emittances, separated in the y' plane in the emittance plot). The explanation for the distortion seems to be the following. The 200 ns voltage on-time is not long enough to sweep away the neutralizing ions, but merely displaces them relative to the H⁻ beam. When the voltage goes off, the restoring force between the H⁻ beam and these displaced positive ions gives a kick to the H⁻ beam in a direction opposite to the direction that the chopper deflects the beam. In addition, due to the displacement of the neutralizing ions, a portion of the H⁻ beam is unneutralized, and therefore becomes divergent. Both these effects are observed on the emittance. This distortion of the emittance decreases/increases as the voltage-on time decreases/increases at the 2.5 MHz frequency. When operating with the chopper, the 35 keV line can be retuned to partially compensate for this distortion, but the full intensity out of the RFQ cannot be recovered. The net effect can be a loss of 25 - 50% of the beam current out of the RFQ. Because of this loss, we plan to install a fast chopper in the 750 keV line to replace the 35 keV chopper.

ACKNOWLEDGEMENTS

We would like to thank the entire Linac staff for their work on this project. Special thanks go to Tom Russo for his operation of the source, and to John Brodowski for the mechanical design of the transport line and chopper.

REFERENCES

1. J.G. Alessi, J.M. Brennan, J. Brodowski, H.N. Brown, A. Kponou, V. LoDestro, P. Montemurro, K. Prelec, R. Witkover, R. Gough, and J. Staples, Proc. 1989 Particle Accelerator Conf., Chicago, March 20-23, 1989.

2. R.L. Witkover, Proc. Third Int. Symp. on the Production and Neutralization of Negative Ions and Beams, Brookhaven, 1983, AIP Conf. Proc. 111 (1983) 398.

3. C.W. Schmidt and C.D. Curtis, Proc. Fourth Int. Symp. on the Production and Neutralization of Negative Ions and Beams, Brookhaven, 1986, AIP Conf. Proc. 158 (1987) 425.

4. J.G. Alessi, Proc. Fourth Int. Symp. on the Production and Neutralization of Negative Ions and Beams, Brookhaven, 1986, AIP Conf. Proc. 158 (1987) 419.

5. J.M. Brennan, L. Ahrens, J. Alessi, J. Brodowski, J. Kats, W. van Asselt, Proc. 1989 Particle Accelerator Conf., Chicago, March 20 - 23, 1989.

6. Directed Energy, 718 Bonita, Ft. Collins, Colorado 80526.

PRACTICAL CONSIDERATIONS FOR A PLASMA NEUTRALIZER

K. G. Moses, J. R. Trow,* and J. C. Dooling
JAYCOR, Plasma Technology Division, Torrance, CA 90503

ABSTRACT

A beam of energetic H^- or D^- ions is efficiently converted to the uncharged state upon impact on a dense plasma. Neutralization efficiencies of the emerging beam can exceed 80%. This paper reports on a design approach to a plasma-stripping cell, illucidating several practical aspects. The plasma is formed by a low-frequency (1.4 MHz), high-power (20 - 60 kW) rf discharge in a multipole ring-cusp magnetic field. A mild steel rectangular box is used as a low-reluctance return path for the magnetic flux, a heat-transfer medium, and mechanical support. The working gases used in these studies are H_2, He, Ar, and Xe. The central density found in the resulting plasma ranged from about 10^{13} to > 10^{14} cm^{-3} with the heavier species exhibiting the higher densities. Test results performed over a pressure range of 2 - 10 mTorr are presented. The rf power required to achieve a given plasma density with this design should scale with the cell length for a given gas.

Measurements of magnetic-flux-density components in the plasma cell volume occupied by the beam path yielded results that are in good agreement with previous calculations. A two-axis Hall probe is used to measure the magnetic field in two planes, separated by a distance of 1 cm and extending 7 cm beyond the beam entrance, over one quadrant of the beam path. Using these measurements, we produce surface plots of the components and modulus of the magnetic flux density, as well as the projection of flux contours onto their respective planes. The magnitude of any magnetic-field component is < 15 G and also the modulus of B is < 15 G in the path of the beam through the cell. The magnetic field at 0.5 cm from the internal copper wall measures > 1.5 kG, while only 1.4 G is found at the same distance outside the box.

INTRODUCTION

The assertion that beams of negative ions can be neutralized more efficiently by impacting a plasma, rather than a cold gas target, is confirmed scientifically by the work of K. H. Berkner et al.[1] What remains to be done is the realization of a practical means of generating plasmas efficiently with appropriate integrated line densities (target thickness). In this work and our earlier studies,[2-4] large volumes of plasma are generated using low-frequency pulsed inductive rf discharges within a ring cusp multipole-magnetic field geometry. These

---

*Present address Applied Materials, Santa Clara, CA 95052.

plasmas exhibit sufficient line-integrated electron densities and degrees of ionization to neutralize beams of energetic negative ions whose energies exceed 500 keV. The method of plasma generation and the cell configuration used in these studies are directly applicable to higher energy neutral beam injector systems (NBIS). Innate scalability and modularity of the system design facilitates linear stacking to achieve a desired target thickness. Further, the plasma formation process is accomplished with an electrical economy consistent with increased overall electrical efficiency of the NBIS compared to that possible using a cold gas target.

The remainder of this paper is divided into four sections: first is a brief description of the present test plasma neutralizer cell; second, extension of the data base of rf discharges in hydrogen and argon[2] to include plasma-density measurements in xenon and helium discharges; third, the magnitude of residual cusp magnetic fields within the cell and the origin of a reference frame is at the center. We report the measurements of the components and magnitude of the magnetic field over one quadrant of the beam path in the neutralizer cell. The geometric symmetry of the permanent-magnet layout in the cell should insure the magnetic field measured in one quadrant is closely representative of the fields in the other quadrants. The plasma neutralizer studies presented here, although not complete, give clear direction to the remaining developmental work required for a sufficient engineering data base for designing a plasma neutralizer. The remaining issues are discussed briefly in the final section of this report.

NEUTRALIZER CELL CONSTRUCTION

A cutaway drawing of the test neutralizer cell is displayed in Fig. 1 together with the important cell dimensions. The internal volume of the cell is 21 liters, and it is bounded by walls of sheet copper 0.0767 cm thick. The copper sheets are fixed to aluminum spacers mounted on the inner walls of a mild-steel rectangular box. The aluminum spacers stand off the copper walls from contact with neodymium permanent magnets (0.75 X 0.75 X 5.0 cm) which are also mounted on the walls of the steel box. These magnets are arranged in rings of alternating magnetic poles producing multidipole fields. The dipole fields enhance the radial confinement of energetic electrons, the primary ionizing agent. Additionally, the steel box provides a flux return path for the magnetic field, a sink for the heat deposited by the plasma in the copper walls, and structural support for the entire assembly.

A three-turn inductive-loop antenna is located along the axis at the center of the cell (not shown in Fig. 1). The area of the antenna, transverse to the beam direction, is about 280 cm^2 (9 X 31 cm) and is greater than the beam apertures so that the antenna is protected from direct beam bombardment. The length of the antenna is about 13 cm. The rf system consists of a constant amplitude, tunable rf oscillator (Tektronix 191), which drives an intermediate amplifier that delivers 23 kW to the 80 kW final power amplifier. The duration of the rf pulse is 0.5 to 1.0 ms in these studies. The initiation of the discharge is aided by electrons emitted from a biased, hot-tungsten filament that is turned on prior to an rf pulse. The test gas is introduced into the cell through two ports located along the axis of the bottom wall. Fill pressure is adjusted by a voltage pulse applied to two parallel piezoelectric valves located in the gas line. The gas pressure attainable in the cell is limited by the conductance of a single valve so it was found necessary to parallel two valves to reach a pressure of 10 mTorr when xenon is used as the working gas.

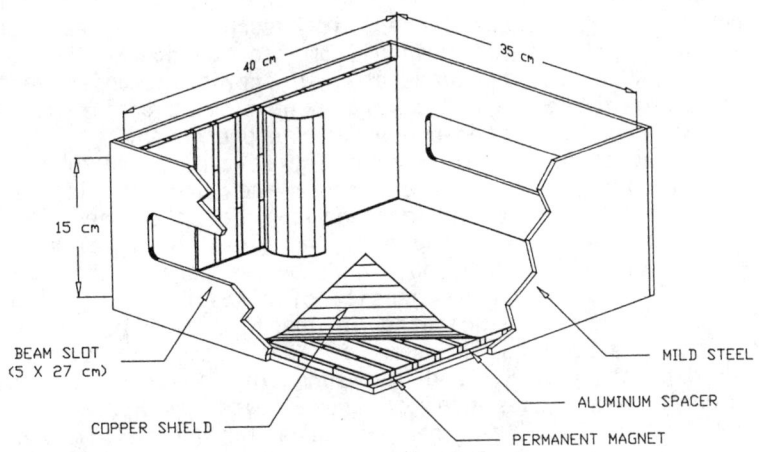

Fig. 1. Neutralizer cell construction.

The entrance and exit slots, each having an open area of 135 cm^2, are designed to accommodate a negative ion slab beam from the transverse field focusing (TFF) accelerator that was in development at LBNL at the beginning of these studies. The walls of the steel box containing the beam entry and exit slots are devoid of magnets. Their exclusion and the large beam apertures are an overly conservative test of the achievable density by the technique of inductive rf plasma production. Under these conditions the confinement time of the plasma is determined roughly by the transit time of the plasma traversing the length of the cell at the ion acoustic speed. It should be clearly understood that magnets located on the wall of the exit slot are not to be discounted for: (1) They cannot deflect the neutral components of the beam; (2) They can deflect the residual charged components, which is desirable; and (3) They should enhance confinement of the primary electrons, reducing the power requirements for achieving a given plasma density. Further, magnets placed on the wall of the beam entrance can provide additional improvement of confinement with a concomitant reduction of required rf power if beamline designers can cope with a small displacement imparted to the beam by the magnetic forces.

PLASMA DENSITY DEPENDENCE ON RF POWER

Two cylindrical Langmuir probes (0.51 mm diam X 5.1 mm) provide measurements of local electron density and temperature in the cell. The position of the probes parallel to the beam direction is variable, and measurements are taken at increments of 5.0 cm. One probe is located on the central axis of the cell, and the other is located 8.3 cm to one side. A triangular voltage waveform, having a pulse duration of 20 µs, is applied to the probes; and the probe current is recorded during each pulse. The plasma density is determined from the ion saturation current using a Laframboise fit parameter. In previous studies we found that measurements using each probe gave essentially the same results, so only the center probe data is presented.

Measurements of the plasma density at the center of the cell versus rf power applied to the antenna are shown in Figs. 2 - 4. The degree of ionization in percent can be determined from the scale on the right-hand ordinate of the plots. The values of the degree of ionization are obtained by comparing the measured plasma density to the number density of a gas at a temperature of 300 °K at the pressure indicated in the figures. In each of the figures the pressure is the same for each gas species, and data is recorded at 10, 5, and 2 mTorr in Figs. 2, 3, and 4, respectively. The gases used in these tests are H_2, He, Ar, and Xe; however, no data has been measured for He at 2 mTorr. At that low pressure, we were unable to produce a discharge in helium. The data displayed in all three figures show that higher plasma densities are achieved in discharges with heavier gas species. This observation is consistent with the assumption that the principle loss mechanism is plasma streaming out of the ends at the ion-sound speed. The exception to this trend is the relative position of the hydrogen data with respect to that of helium on the plots. A direct comparison of the ionization potential of He, 24.6 eV, compared to 13.6 eV for H appears to be a reasonable explanation for this result; but this simple reasoning neglects the influence of electron temperature and dissociation of molecular hydrogen. We reported earlier that electron temperatures are higher in lower pressure discharges, with a concomitant increase in the degree of ionization; but since we have not made a determination of the species populations, i.e., the atom, ion, and molecular ion concentrations, we cannot validate this speculation.

The driving frequency of the rf system in the experiments we report here is 1.4 MHz. We see in Fig. 2 that a central plasma density of 1.3×10^{14} cm^{-3} with an associated 41% degree of

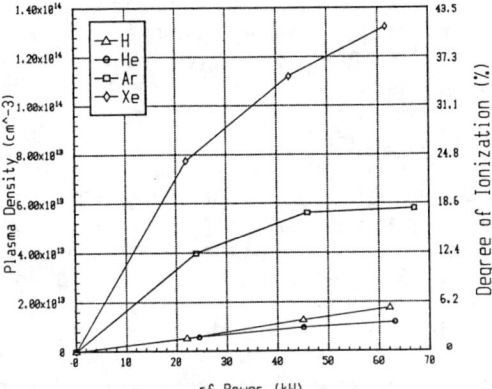
Fig. 2. Central plasma density at 10 mTorr vs. rf power.

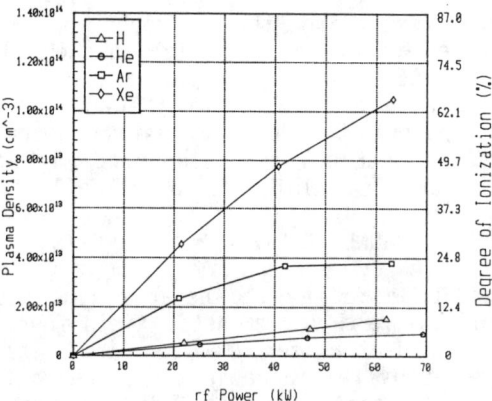
Fig. 3. Central plasma density at 5 mTorr vs. rf power.

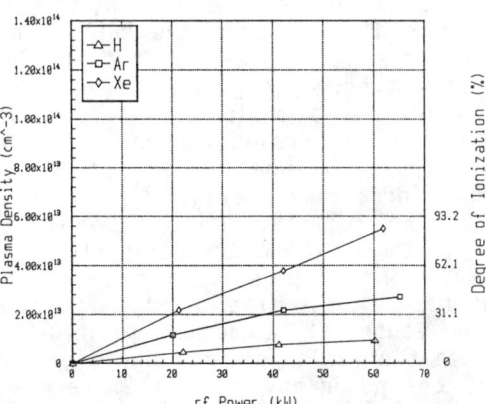
Fig. 4. Central plasma density at 2 mTorr vs. rf power.

ionization is obtained with discharges in Xe at pressures of 10 mTorr with about 61 kW applied to the antenna. Correspondingly, discharges in hydrogen and helium produce plasma densities of roughly 10^{13} cm^{-3} with about 5% ionization. The values of plasma densities and percent ionization in argon discharges lie somewhere midway between xenon and the latter two. The data measured in rf discharges at 5 and 2 mTorr, shown in Figs. 3 and 4, respectively, display similar characteristics, albeit lower values of plasma density and higher degrees of ionization. For example, we find in Fig. 3 that 5 mTorr discharges in xenon produce plasma densities slightly greater than 10^{14} cm^{-3} and 65% ionization; and in Fig. 4 we see that at 2 mTorr Xe, plasma densities are only 5.5 X 10^{13} cm^{-3}, but the degree of ionization is 85%. On the other hand, plasma densities remain close to 10^{13} cm^{-3} in hydrogen discharges at 5 and 2 mTorr with 9% and 14.9% ionization, respectively. The plots for xenon and argon discharges exhibit strong saturation of plasma density with rf power. However, hydrogen and helium discharges do not indicate saturation up to the power levels used in these studies. Hence, extending the hydrogen operation to higher levels of rf power (> 60 kW) may yield higher plasma densities. For example, the extrapolation of plasma density by a factor of two is plausible with a doubling of the applied rf power. Even at this higher power, the NBIS may still operate at a higher level of overall electrical efficiency and with a reduced level of gas loading than is possible with a gas cell.

MAGNETIC FIELD MEASUREMENTS

A key question asked by the negative ion beam community is the magnitude of the residual magnetic field found along the beam path through the neutralizer cell. It is the increase in divergence imparted by the magnetic forces which they consider to be problematic. As stated previously, the present neutralizer cell is designed to accommodate a TFF accelerator slab beam. The permanent magnets, which confine the primary electrons, are placed at a minimum distance from the beam path. However, the magnitude of magnetic multipole fields falls off rapidly with distance; and, as a consequence, they should not strongly effect the trajectories of the beam ions. To test this premise, we undertook to measure the residual magnetic fields in the beam path through the cell. A two-axis Hall probe, model YOB4-1808, with a model 640 Incremental Gaussmeter, F. W. Bell, Inc., is used to measure the magnetic fields. The magnitude of the magnetic field 0.5 cm above the exterior steel wall of the cell ranged from about 1.5 to 3 Gauss, while the maximum field in the interior of the box, at the same distance from the copper wall, is about 1,500 Gauss. The probe mounted on a Unislide, model 4469, mapped the field in two planes, separated 1 cm apart vertically, over one quadrant of the beam path within the cell. An arbitrary coordinate system is indicated in Fig. 5 to establish reference directions. The cross-section of an incident beam lies in the X,Z plane, while the direction of the beam velocity is parallel to the Y axis. The measurement planes extend 7 cm beyond the beam slot in the Y direction and from -0.2 to 11.8 cm in the X direction across the interior of the cell, as can be seen in the cutaway drawing shown in Fig. 5. Measurements in the X and Y direction are made in 1 cm intervals in both the $Z = 0$ and $Z = 1$ cm planes, as indicated in Fig. 5.

The values of the magnetic field, X, Y, and Z components in the $Z = 0$ plane, are shown in Figs. 6 - 8; and Figs. 9 - 11 show these same measurements taken in the $Z = 1$ cm plane. In the figures, the maximum value of all magnetic field components fall off slowly with distance from the $Y = 0$ axis within the cell. Once past the beam slot, the field components rapidly go to zero with increasing Y. The magnitude of the X - components of the magnetic field are very small throughout the central region of both planes (Figs. 6 and 9). However, at the edge of the beam slot, the X - components are larger; but they still remain less than 10 Gauss.

The Y - components display a sinusoidal variation in the Y direction, and they are the strongest magnetic components found in the cell. This phenomenon is evident in Figs. 7 and 9. The periodicity corresponds to the distance between successive rings of the permanent magnets with the zero crossing occurring at the Y position of the magnet rings. The amplitude of the Y - component increases slightly as we go from the $Z = 0$ plane to the $Z = 1$ plane as the latter plane is closer to the rings of permanent magnets. The maximum value of the Y - component remains less than 15 Gauss out to the edge of the beam slot.

Now, consider Figs. 8 and 10 which contain plots of the Z - components of the magnetic field in the two planes. The component of the magnetic field in the Z direction shows the same qualitative characteristics as the Y - component, except that the amplitude of the Z - component is smaller in the $Z = 0$ plane than it is in the $Z = 1$ plane. Also, the position of the maxima of the Z - component is shifted toward the Y position of the permanent magnet rings compared to the maxima of the Y - component. The Z - components from opposing permanent dipole magnets are equal in magnitude, but antiparallel and should cancel each other under the tacit assumption that the magnets are located with perfect symmetry with respect to one another in the cell. Clearly this is not the case and there is a small misalignment. However, the Z - components are small, less than 3 Gauss in the $Z = 0$ plane. Previous two-dimensional calculations[2] of the magnetic field geometry in the plane containing the beam axis and the short dimension of the box, and infinite in extent in the X,Z direction, agree very well with the measurements reported here. The results of these calculations are reproduced in Fig. 12 for comparison with the measurements.

In Figs. 13 and 14 we present the surfaces of the modulus of the magnetic field B (mod B surfaces) in the $Z = 0$ and $Z = 1$ cm

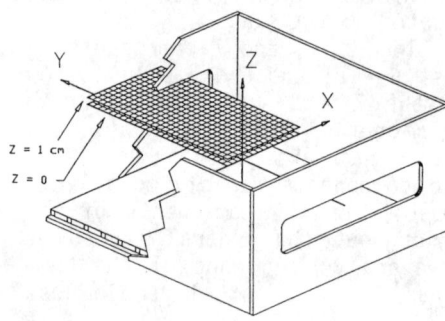

Fig. 5. Location of magnetic field measurements.

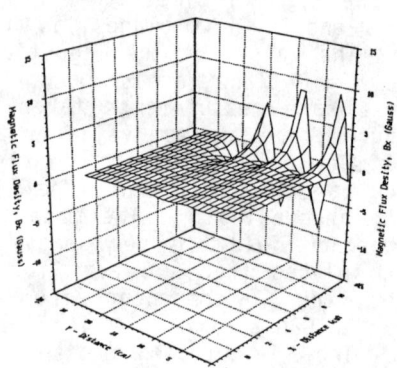

Fig. 6. X - component of magnetic flux density in Z = 0 plane.

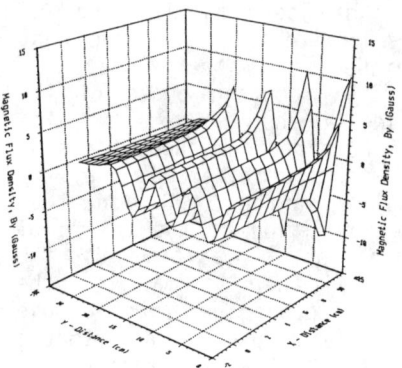

Fig. 7. Y - component of magnetic flux density in Z = 0 plane.

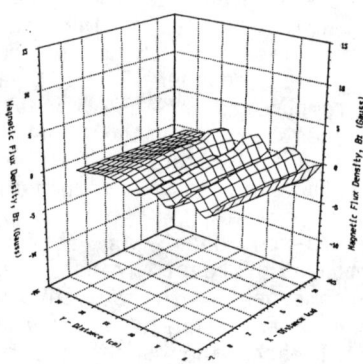

Fig. 8. Z - component of magnetic flux density in Z = 0 plane.

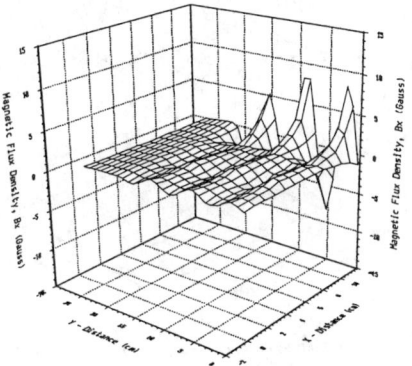

Fig. 9. X - component of magnetic flux density in plane Z = 1 cm.

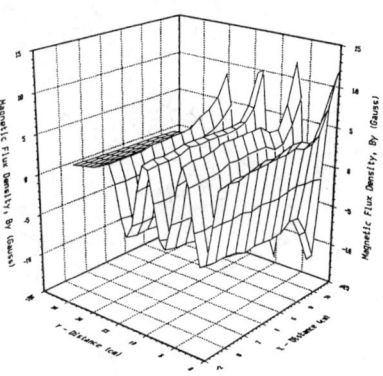

Fig. 10. Y - component of magnetic flux density in plane Z = 1 cm.

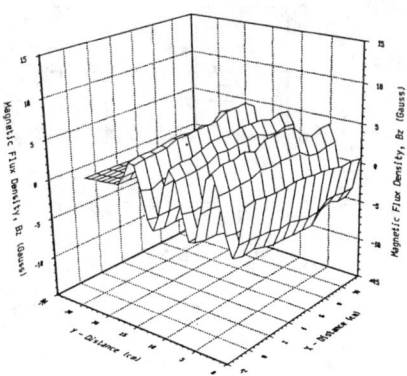

Fig. 11. Z - component of magnetic flux density in plane Z = 1 cm.

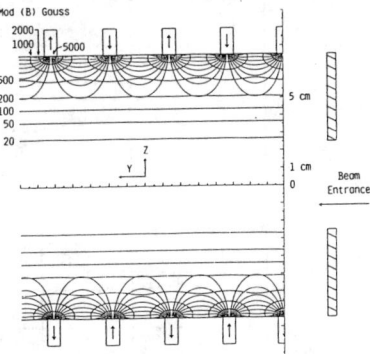

Fig. 12. 2-D magnetic field with the same narrow dimension (15 cm) but infinite in the other. The steel box is allowed for by making the magnets twice as high.

Practical Considerations for a Plasma Neutralizer

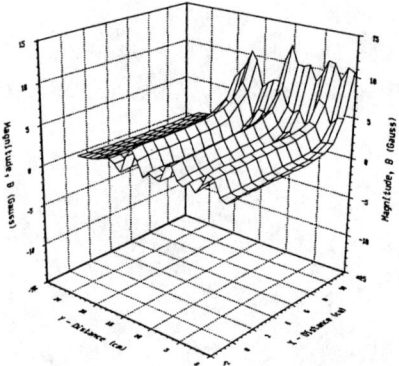

Fig. 13. Magnitude of magnetic flux density, |B|, in Z = 0 plane.

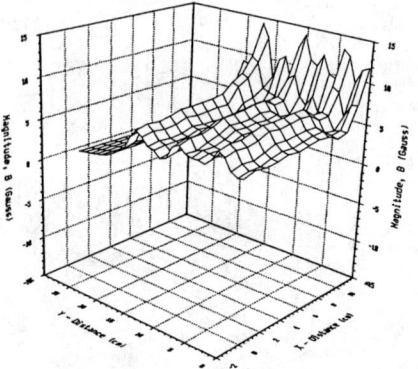

Fig. 14. Magnitude of magnetic flux density, |B|, in plane Z = 1 cm.

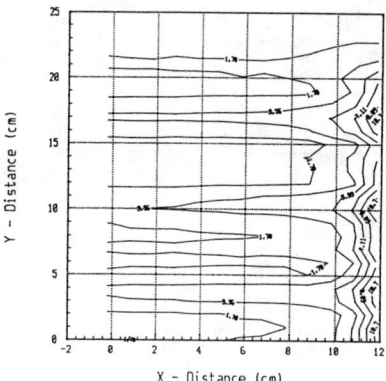

Fig. 15. Mod B contours in Z = 0 plane (Gauss).

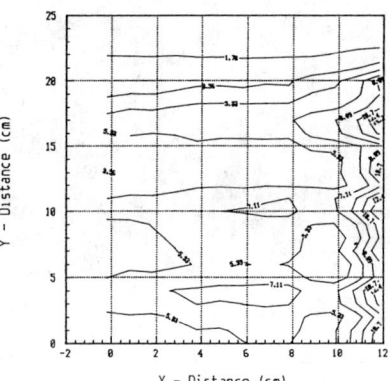

Fig. 16. Mod B contours in Z = 1 cm plane (Gauss).

planes, respectively, calculated from the field measurements in the cell. The periodicity of the field is doubled since mod B is positive definite. The value of mod B in the inner portions of the cell are generally less than 6 Gauss and at the edge of the beam slot are less than 15 Gauss. The projection of the mod B surfaces onto the X,Y plane are shown in the contour plots found in Figs. 15 and 16. The values of mod B for any contour are shown on the plots. The rapid fall off of the field from the edge of the beam slot is clearly seen on the right-hand side of the contour plots. It should also be noted that the fields in the beam path through the cell could be weakened, if so desired, by increasing the X,Z cross section of the cell without loss of primary electron confinement. The increase would probably require an increase in the applied rf power to produce an equivalent power density in the enlarged cell.

CONCLUSIONS

From the data presented above, it appears that Xe is the best choice for the working gas in a plasma neutralizer for a given rf power level. The highest plasma density and highest degree of ionization are produced in Xe discharges. Still, it is not evident that Xe is the appropriate gas species to use in a plasma neutralizer because the cell gas could contaminate the tokamak plasma. While the flow of Xe atoms from the neutralizer cell can be controlled by cryogenic panels and a carefully designed differential pumping system, the gas particles streaming directly down the beamline from the neutralizer into the tokamak could provide a continual influx of high Z impurities to the tokamak. Fortuitously, the streaming Xenon atoms will ionize rapidly at the edge of the confined plasma. By virtue of the strong magnetic fields of the tokamak, the rate of diffusion of the cold, massive Xe ions into the interior of the tokamak is very slow compared to H ions. Most of these heavy ions can be removed from the plasma edge by a divertor at a sufficiently high rate to hold the high Z contamination to an acceptable level. Alternatively, the temperature of the plasma in the scrape-off region will be reduced by strong bremsstrahlung and line radiation produced by Xe ions in this region. Cooling the scrape-off plasma should prove beneficial by reducing erosion of divertor plates caused by the impact of high-temperature plasma. Clearly, this argument requires confirmation by a thorough systems analysis and evaluation of beamline and divertor designs to establish credibility.

ACKNOWLEDGMENT

This work was supported by the U.S. Department of Energy under Grant No. DE-FG03-87ER52144.

REFERENCES

1. K. H. Berkner et al., "Plasma Neutralizers for H⁻ or D⁻ Beams," Proc. of the Second Intern'l. Symp. on the Production and Neutralization of Negative Hydrogen Ions and Beams, BNL 51304, Brookhaven National Lab. N.Y., October 1980
2. K. G. Moses and J. R. Trow, "Plasma Neutralizers for High Energy Neutral Beam Injectors," Proc. III European Workshop on Production and Application of Light Negative Ions, Amersfoort, The Netherlands, February 1988.
3. J. R. Trow and K. G. Moses, "A Dense, Low Temperature Plasma Target, Driven by an Immersed Inductive Antenna," Proc. of the Seventh APS Topical Conf. on the Appl. of RF Power to Plasmas, Kissimmee, Florida, May 1987.
4. J. R. Trow and K. G. Moses, "Characteristics of an RF Plasma Neutralizer," Proc. of the Fourth International Symp. on the Production and Neutralization of Negative Ions and Beams, Brookhaven National Laboratory, October 1986.

PLASMA NEUTRALIZERS

P.M. Vallinga and D.C. Schram
Eindhoven University of Technology, P.O. Box 513,
NL-5600 MB Eindhoven, The Netherlands

H.J. Hopman
NET team, c/o Max-Planck-Institut für Plasmaphysik, Boltzmannstr. 2,
D-8046 Garching bei München, FRG, and
FOM Institute for Atomic and Molecular Physics.
Amsterdam, The Netherlands

ABSTRACT

Presented are results on the modelling of a cascaded arc. Under suitable conditions, the plasma effusing out of the arc has a high degree of ionization. It is proposed to use this arc as the plasma source of a plasma neutralizer in neutral beam injectors for fusion research. The neutralization of a D^- beam results in the generation of an electron beam. The power in this e-beam is too small to sustain the plasma in the neutralizer for the 1.3 MeV fusion beams.

INTRODUCTION

The neutral beams that at present are being discussed for heating of the plasma and partial drive of the toroidal current in next step fusion devices like NET and ITER, have particle energies between 1 and 1.3 MeV [1,2]. These beams are obtained by the neutralization of negative ion beams, both H^- and D^-. The powers envisaged range from $\approx$ 50 (NET) to $\geq$ 75 MW (ITER). Therefore, it is important to generate these beams with the highest possible efficiency. Areas where efficiency gains are likely to be possible are the negative ion source and the neutralizer. Negative ion sources do get attention in the neutral beam community [see this conference], but this is not so with neutralizers. In the available designs of neutral beam injectors, the working hypothesis is to use a gas neutralizer. However, the use of a plasma neutralizer [3] would give a considerable increase of overall beam line efficiency, possibly from $\approx$ 42% to $\approx$ 58%.

Neutralizers for negative ion beams have two particularities that distinguish them from those for positive ion beams. Firstly, one is dealing with three charge fractions, which, in the case of a deuterium beam, are D^-, D^0, and D^+. This property necessitates a modification of energy recovery techniques when they are carried over from positive to negative ion beams [4].

Secondly, during the stripping of the negative ion, an electron is released that is travelling at approximately the same speed as the parent ion. At a D⁻ energy of 1.3 MeV the neutralization results in the generation of 350 eV electrons. In the case of positive ion beams, there is no stripping and the ionization of neutralizer atoms or molecules results in electrons of low energy only.

Fast electron generation occurs in gas as well as in plasma neutralizers. The question then arises if these electrons can be used to simplify the neutralizer. In the case of a neutralizer to which only gas is admitted, they may ionize the gas and form a plasma. To create a sufficient number of ion pairs the fast electrons must be confined, for instance with a multi-pole magnetic field. Then, the situation is analogous to a negative ion source in which the hot-cathode discharge has been replaced by a negative ion beam. The comparison allows us to predict some aspects of this neutralizer. A preliminary estimate indicates that the beam current is too small to ionize the gas to the necessary high degree of ionization [3], $\alpha \geq 50\%$.

It follows that the plasma in the neutralizer needs to be generated by external means. We propose the use of cascaded arcs [5, 6] to inject plasma into a box lined with a multi cusp magnetic field. Presented will be the most relevant result of a modelling of this arc. It is found that the calculated degree of ionization in the arc is nearly 100%. The arc power required for a full scale neutralizer is estimated at 30 kW for a 8.3 MW neutral beam module.

A cascaded arc is a wall stabilized arc, consisting of a cathode, a stack of electrically insulated cascade plates, and an anode. The arc channel is formed in the central bore of the cascade plates, which are 5 mm thick, water cooled, copper plates, that are separated by 1 mm gaps maintained by PVC spacing rings. Gas is admitted to the arc channel at the cathode side and plasma flows out of the channel through a hole (a nozzle) in the anode. The arc plasma is characterized by high electron densities, high degrees of ionization, moderate temperatures, and a low power consumption.

PLASMA NEUTRALIZER REQUIREMENTS

Culham has proposed the use of a close coupled neutralizer in their design of a neutral beam injector for NET [7]. As a result of the short distance between source and neutralizer, and because of the small divergence of a high energy negative ion beam, the individual beamlets do not merge. Therefore, neutralizers can have nearly closed front and back side, contrary to the open structures used on present fusion machines. A neutralizer can be envisaged as a box on all sides covered by permanent magnets, with $\approx$ 30 mm wide slots in the front and back wall to allow the beamlets to pass through. Slots are needed because of the demand to sweep the beamlets in the vertical plane (beam profile control). The number of slots is equal to the number of aperture columns in the source. Using a Culham type source in the case of a 8.3 MW, 1.3 MeV module (ITER NB injector concept [8]) , the neutralizer cross section would be approximately 1.5 by 1.5 m² and it would have about 10 slots [9].

In the following we give a rough estimate of the ionization rate needed to maintain the plasma in a neutralizer box of length l, and width and height d. The total ion flow to the walls (Bohm sheath criterion) is equal to,

$$\varphi \approx 0.6\, n_i\, c_s\, A_c, \qquad (1)$$

where, n_i is the plasma density, $c_s = \sqrt{k(T_e+T_i)/m}$ is the acoustic speed with m the ion mass, and the total loss area A_c is given by the product of the length and the width δ of the cusp lines between the magnet rows. We assume linear line cusps perpendicular to the velocity of the negative ion beam at a pitch D. This way the magnetic field is mainly parallel to the beam and the perturbation on the beam is minimal. In the front and back wall, the magnet configuration is determined by the slots. Finally, we have $\delta = 4\sqrt{\rho_e \rho_i}$ [10], which is four times the hybrid Larmor radius. From these considerations we obtain,

$$A_c \approx \delta\, d\,(4l + d)\,/\,D. \qquad (2)$$

A further given fact is the optimum plasma target density Π, given by [3],

$$\Pi \equiv n_i\, l \approx 2 \times 10^{19}\, m^{-2}. \qquad (3)$$

With the aid of Eqs. (1, 2, 3) we can write,

$$\varphi = 2.4\, \Pi\, \sqrt{\frac{kT_e}{m}}\, \frac{d\,(4l + d)}{D\, l}\, \sqrt{\rho_e \rho_i} \propto \frac{1}{B}\, \sqrt[4]{\frac{T_e^3 T_i}{m}} \qquad (4)$$

The scaling law demonstrates that it is beneficial to increase the B field, an approach being studied by Culham [9], or to reduce T_e. The ion mass has less influence, but switching from deuterium to argon ions, one gains a factor 2.1. As a numerical example, we take $T_e = 5$ eV, T_i (D⁺) $= 0.4$ eV, (temperatures measured in buckets [11], in which the plasma is created by energetic electrons), $B = 0.15$ T, $d = 1.5$ m, $l = 1$ m, and $D = 5$ cm. We obtain $\varphi = 2 \times 10^{22}$ s⁻¹; for an Ar⁺ plasma the same quantity is 9×10^{21} s⁻¹.

In the above derivation, wall losses were assumed dominant. With other types of plasma creation, the electron temperature may be much smaller than the 5 eV quoted in the example, leading to smaller diffusion losses. However, volume losses like three-body and radiative recombination of D^+ ions become more important, as well as loss channels involving (vibrationally excited) molecule reactions. This suggests that there is an optimum temperature for which the flux φ is minimal. This temperature may be quite different for atomic and molecular gases, because of the more complex chemistry of the latter.

PLASMA CREATION

As mentioned in the introduction, the neutralization of a negative ion beam is accompanied by the formation of a 350 eV electron beam. In the case of a module delivering a 8.3 MW, 1.3 MeV (D^0) beam, the electron current amounts to $\approx$ 6.5 A, and represents a power of $\approx$ 2.24 kW. Because the electrons are released inside the neutralizer, we assume that they are confined by the multi cusp magnetic field and loose their energy by ionization of the gas. Taking an expenditure of 64 eV per ion pair created [12], the electron beam could ionize 2.2×10^{20} atoms per second. It is clear that in the case of beams for fusion the electron beam power is too small to establish the required plasma target. The situation becomes more favourable with beams of higher energy and higher current density.

We conclude that the plasma must be created by external power. We can choose between two approaches. (1): One can inject a plasma [5, 13]. Then the neutralizer must have an open structure to allow efficient pumping of the ions that neutralize on the walls. The gas volume to be pumped is found from φ. In the case of deuterium, it is $\approx$ 280 Torr l / s. (2): One can inject power and gas separately into the neutralizer and create the plasma in situ [9, 12]. In this case, the neutralizer must be closed for gas transport to reduce the pumping requirements. Taking again 64 eV for the creation of a deuterium ion pair, the required power is found to be 200 kW.

To learn about consequences of injecting plasma from an external source we studied cascaded arcs, because of their high power efficiency.

ARC MODEL

To describe the evolution of the arc plasma as function of the axial position along the channel between cathode and anode a self consistent one-dimensional model has been set up. A two-dimensional model, which takes into account the radial profiles, has been formulated [D. Milojevic, D.C. Schram, and P.M. Vallinga, to be published]. In this paper, we present the 1-D model results obtained by numerical integration of the conservation laws for mass, momentum and energy. Calculated as function of the coordinate x are the densities and the temperatures of heavy particles and of electrons, the pressure, and the directed flow velocity. A further result is the arc voltage.

It is assumed that the heavy particle components D_2, D^0, and D^+, with densities n_2, n_1, and n_i, are closely coupled and have the same temperature T_h and drift velocity u. The electron component has a density n_e and temperature T_e. Then, the degree of ionization is defined by $\alpha \equiv n_e/(n_2 + n_1 + n_i)$, the degree of dissociation by $\beta \equiv n_1/(n_1 + n_2)$, and the reduced mass velocity by $M \equiv u/c$, with $c = \sqrt{5kT_h/3m}$.

The energy input is to the electrons and is due to Ohmic dissipation, Q_{Ohm}, of the arc current I_a in the arc channel of diameter D. The electrons loose energy by dissociation of the molecules, Q_d, by ionization of the heavy particles, Q_i, by elastic energy transfer in electron heavy particle collisions, Q_{eh}, by work performed on the plasma and leading to the plasma expansion, Q_u, and through radiative processes like the escape of line radiation emitted by excited heavy particles, the escape of continuum radiation due to free-free transitions and recombination to excited levels.

Changes in the densities of species are brought about by direct or indirect electron impact ionization and dissociation, three particle recombination, and radiative recombination. Momentum transfer is by means of elastic collisions between electrons and heavy particles. In addition, also friction between the plasma and the wall is taken into account.

When performing a calculation, the input parameters are the arc current I_a, the pressure at the channel entrance p_o, the channel diam. D, the channel length l, and the gas flow ϕ. Of these, one parameter is a dependent one because of the boundary condition that a sonic condition, $M = 1$, is reached at the end of the channel in the anode.

SOME RESULTS

The model described above has been applied to an arc burning on argon gas. Calculated values of T_e, n_e, and pressure p, agreed within 5% with values measured at some ten different points along the arc channel, providing the channel was given a diameter of 38 mm, instead of the experimental value of 40 mm [6]. This difference is related to the existance of a 0.1 mm thick wall layer. These results give confidence in the applicability of the model to a deuterium arc, for which no experimental data are available.

The simulations in general show an increase of the degree of ionization α with arc current and with channel length, and a decrease with gas flow. Fig. 1 illustrates the latter point with results for three cases in deuterium gas:

	case A	case C	case D
gas flow (scc/s)	100	150	200
arc length (mm)	118	55	31

with further, D = 3.8 mm, I_a = 95 A, and p_0 = 0.35 bar. As mentioned before, not all parameters are free. In this series of calculations, the arc length was adjusted to obtain in the anode the sonic condition M = 1. Starting values at x = 0 are α, β ≈ 0.01 and T ≈ 1000 K. The figure shows that α increases approximately linearly with x and reaches a maximum value > 0.8 for the smallest flow. Also the degree of dissociation β is shown. In case A, nearly full dissociation is obtained in the first 40% of the arc length.

In case A, the gas is heated to T_h > 10^4 K in the first 10% of the channel, in which thermal equilibrium is reached. From there on T_e and T_h increase nearly linearly from 13,000 K to 16,000 K at the anode side. The highest temperature is found for the smallest flow. The electron density is plotted in Fig. 2. The highest density is reached with the smallest flow. The initial increase in n_e is due to ionization of the gas; the decrease in the second half of the channel is related to the increase in plasma flow velocity. The combined result is a monotonic decrease of the pressure with position x. The major energy terms associated with these processes are plotted in Fig. 3 for case A. All terms are normalized to the Ohmic energy input. It is seen that the electron energy used for dissociation, Q_d, disappears at the position where β tends to one; ionization losses Q_i remain high up to the point where the electron density reaches its maximum. Beyond this position Q_i decreases, even though α still increases. Saturation of α only occurs near the end of the channel. Further is indicated the elastic energy transfer from electrons to heavy particles, Q_{eh}. This quantity becomes the dominating one at the end of the channel, where the plasma speed u increases rapidly and the sonic condition M = 1 is reached. Q_e is the sum of the three terms just discussed. In the energy balance, the radiation losses are unimportant. Together they contribute less than a few % and are not presented. Figure 4 presents the energy contributions related to the gradient terms. The energy needed for plasma acceleration is $Q_u = \frac{5}{2} kT_e n_e (\nabla u)$, and the energy due to variations in electron density and temperature are $Q_n = \frac{3}{2} kT_e u (\nabla n_e)$ and $Q_t = \frac{3}{2} n_e u (\nabla kT_e)$, respectively. Again, the terms are normalized with respect to the Ohmic input. At the end of the channel, the dominant term is the plasma acceleration. It reaches a value of 0.6 at the anode and makes up for the deficit in Fig. 3. Due to the expansion the temperature decreases at the anode (Q_t < 0).

The calculated cumulative plasma resistance, between the cathode and the anode amounts to 1.8 Ohm. With the chosen arc current of 95 A, we arrive at an arc voltage of 170 V across the column of case A. So, this arc dissipates 16 kW.

DISCUSSION

With the aim to fill a given volume with plasma, the relevant question is the maximum flow of ion pairs provided by the arc. It is approximately attained for the conditions of case A, and amounts to $\varphi_a = 3.8 \times 10^{21}$ s^{-1}. However, reducing the gas flow results in a higher degree of ionization and in relaxed pumping requirements. At $\phi = 50$ scc/s, calculations give $\alpha \approx 100\%$ and $\varphi_a = 2.4 \times 10^{21}$ s^{-1}. Taking this latter case, we find $\varphi/\varphi_a \approx 8$. So, some 8 cascaded arcs would be needed for a full sized plasma neutralizer. This number is an upper bound, because we compared two situations with different temperatures and, therefore, different ion life times. We estimated φ using $T_e = 5$ eV, whereas in the plasma jet squirting out of the cascaded arc $T_e \leq 1$ eV. The scaling law in Eq. (4) suggests that the number of arcs might be as low as 3, implying a gas flow of 150 scc/s or **115 Torr l / s**.

The advantage of a cascaded arc is the small power consumption. From the calculated plasma flow and arc power, one obtains 26 eV per ion pair, a factor 2.5 more favourable than the low pressure discharge result [12]. Using this number, the required power is estimated at **30 kW**, for three arcs of 50 scc/s gas flow.

The above estimates of gas flow and power indicate that plasma neutralizers form a realistic option besides gas neutralizers. Moreover, the scaling law suggests that many improvements are possible. Because of the important benefits possible with plasma neutralizers, such as a much smaller area occupied by the neutral beam system in next step fusion devices experimental efforts in this area deserve a strong support.

ACKNOWLEDGEMENTS. This work is part of the research program of the association agreement EURATOM-FOM, with financial support from NWO and EURATOM.

REFERENCES

[1] R. Toschi, M. Chazalon, F. Engelmann, J. Nihoul, J. Raeder, and E. Salpietro, Fusion Technology **14** (1988) 19.

[2] ITER Concept Definition, ITER Document Series no. 3, IAEA, Vienna (1989).

[3] K.H. Berkner, R.V. Pyle, S.E. Savas, and K.R. Stalder, Proc. 2-nd Int. Symp. on Production and Neutralization of Negative Ions and Beams, Brookhaven National Laboratory Report 51304, Upton NY USA (1980), p. 291.

[4] H.J. Hopman, Nuclear Fusion **29** (1988) 685.

[5] G.M.W. Kroesen, and D.C. Schram, in Prod. and Appl. of Light Negative Ions, Proc. III-rd Eur. Workshop, Euroase Congress Center, Amersfoort, The Netherlands, 17 - 19 February 1988, p. 209.

[6] G.M.W. Kroesen, J.C.M. de Haas, and D.C. Schram, submitted to Plasma Chem. Plasma Proc.

[7] J. Coupland, I. Gray, A. Holmes, M. Inman, L. Lea, K. Martel, and R. Parker, Culham Laboratory Report 3 15/88 - 7/FU UK NET (1989).

[8] ITER Conceptual Design Interim Report, to be published in ITER Document Series, IAEA, Vienna; ITER Report TN-HD-9-1 (October 1989).

[9] A.J.T. Holmes, private communication.

[10] K.N. Leung, N. Hershkowitz, and K.R. Mackenzie, Phys. Fluids **19** (1976) 1045.

[11] M. Bacal, M. Capitelli, C. Gorse, D.A. Skinner, and J. Bretagne, in Prod. and Appl. of Light Negative Ions, Proc. III-rd Eur. Workshop, Euroase Congress Center, Amersfoort, The Netherlands, 17 - 19 February 1988, p. 112.

[12] K.G. Moses, and J.R. Trow, in Prod. and Appl. of Light Negative Ions, Proc. III-rd Eur. Workshop, Euroase Congress Center, Amersfoort, The Netherlands, 17 - 19 February 1988, p. 203; ibid. Production and Neutralization of Negative Ions and Beams, 4th Int. Symp., Brookhaven National Laboratory, AIP Conf. Proc. **158**, NY (1987), p. 651.

[13] A. Hershcovitch, V. Kovarik, and K. Prelec, in Prod. and Appl. of Light Negative Ions, Proc. III-rd Eur. Workshop, Euroase Congress Center, Amersfoort, The Netherlands, 17 - 19 February 1988, p. 217.

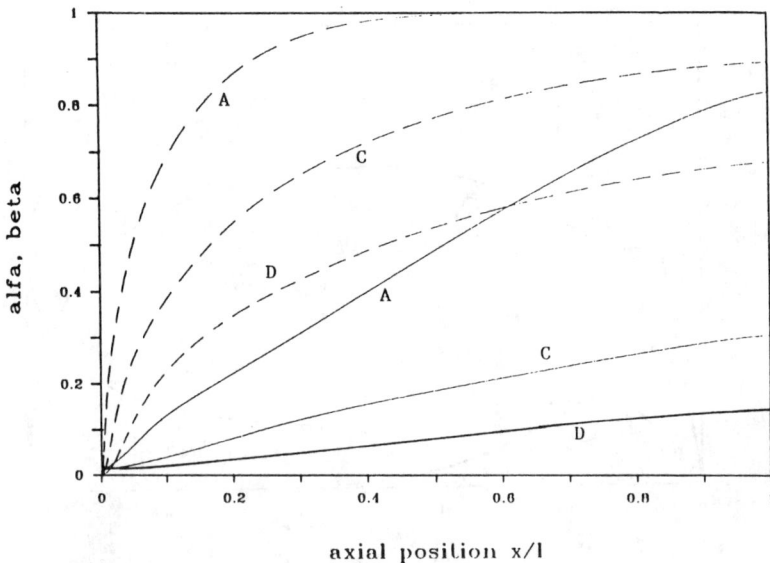

Figure 1. The calculated increase in the degree of ionization (solid lines) and dissociation (dashed lines) with axial position in a cascaded arc channel, normalized to the total arc length l. The parameters are: the arc current, I_a = 95 A; the inlet pressure, p_0 = 0.35 bar deuterium gas; gas flow and arc length are for case A) 100 scc/s and 118 mm, for case C) 150 scc/s and 55 mm, and for case D) 200 scc/s and 31 mm.

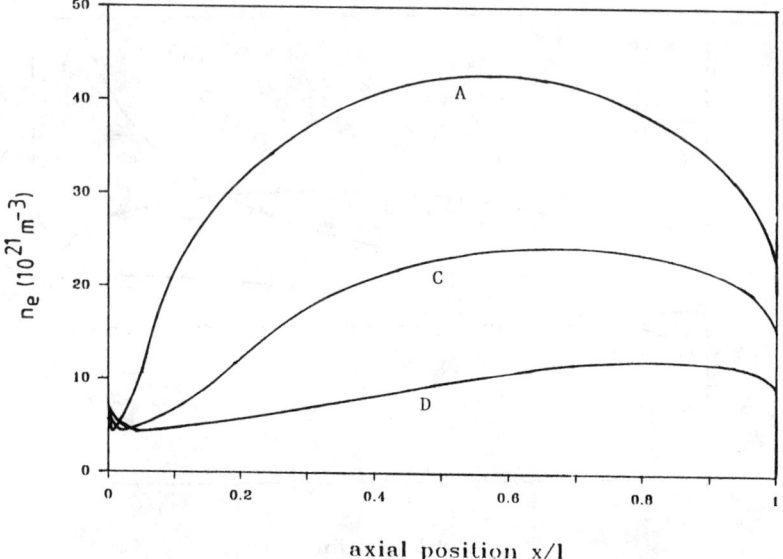

Figure 2. Axial density profiles, calculated for a cascaded arc in deuterium gas. Parameters are the same as in Fig 1.

738 Plasma Neutralizers

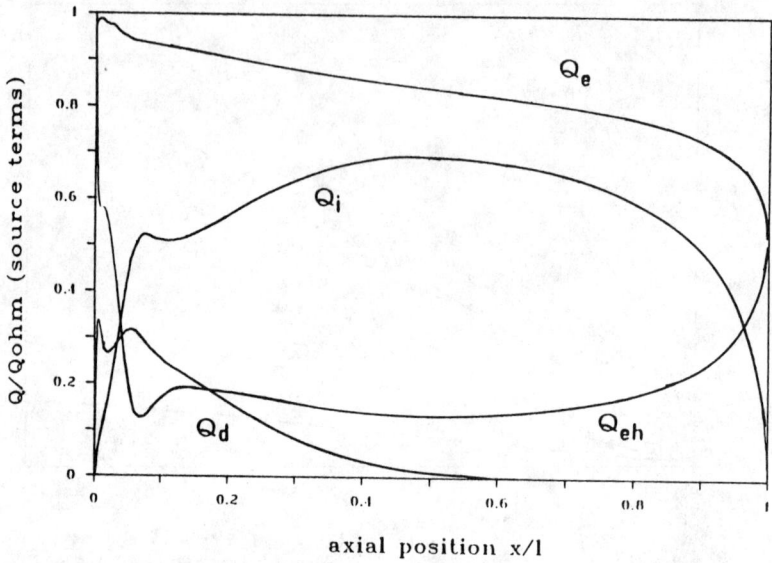

Figure 3. Various electron energy dissipation terms, normalized to the Ohmic energy input, as function of the position in the cascaded arc channel. Parameters are those of case A in Fig 1. Q_e is the sum of the three terms presented: Q_d, the dissociation loss, Q_i, the ionization loss, and Q_{eh}, the energy transferred to ions in elastic collisions.

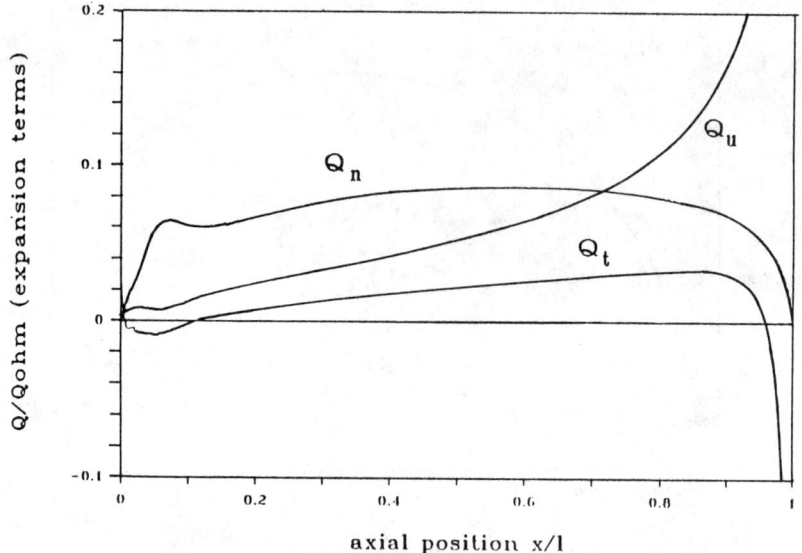

Figure 4. Terms in the electron energy equation, proportional to plasma gradients, like the density gradient Q_n, the temperature gradient Q_t, and the gradient in the directed velocity, Q_u.

DISCUSSION

Hershcovitch: I'm not asking this question to promote our past program. Why don't you consider hollow cathode discharges for which Daan Schram had an excellent program there. Hollow cathodes are much more efficient in terms of power and gas.

Hopman: In gas efficiencies matters, if the calculations give you a high degree of ionization better than 95%, it is equal to the hollow cathode. The advantage is, even though there is no hard number on it, that the lifetime of the elements in the cascaded arc is much longer than in the case of the hollow cathode. There is no need to have a tip at 3000 K like in the case of the hollow cathode. This is the main advantage that I see.

Herhcovitch: There have been experiments and studies that showed that hollow cathodes operating with LaB_6 cathodes have a very long lifetime providing there are no contaminating elements. I would suspect a neutralizer would be a clean environment where the only elements would be the plasma.

Hopman: This is something that has to be discussed.

Moses: I certainly agree with your conclusion, I think we have to get to plasma neutralizers. The question of electron temperature, what was the electron temperature in your source?

Hopman: I forgot to bring the viewgraph, I think it's 12,000 degrees, for the plasma flame that would be 1 eV?! It must be much colder than 1 eV.

Moses: We have plasma temperatures on the order of 6 to 8 eV so that is relatively cold. The reason why we have been advocating heavier gases in our work is since the ionization potential goes down, so you save that way. In a 21 liter volume, we have been able to produce densities on the order over 10^{14}, and in hydrogen in about 10^{13}. We have already used in that system the high field magnets, the neodynium magnets already so I don't think we will be able to gain much more than any other magnets at least that are available now. The question about the degree of ionization, if you look at Berkner's paper which is really the only thing we have to go by, because of the fact that the cross sections are not known for many of the cross sections we need for neutralization, you find that only 30% gets you almost to where you have to be as far as neutralization. Fifty percent is overkill but would be nice, 100% is certainly not needed. We have achieved at about 2 milliTorr keeping the pressure down on the order of about 40% ionization, so we've already exceeded that at least in xenon. The heavier gas has some other advantages for fusion applications but these are things that will probably be discussed and argued and debated for a long time to come. I can only say that Andrew's (Holmes) suggestion about putting back the magnets on the face is a very good one. We did not do that purposely because we wanted to test it out for the Berkeley beam source, and we were given specific instructions not to put any magnetic fields at the entrance and exit and we also allowed for about 125 square centimeter openings at both ends besides no magnets. So, that was a worst case situation. I agree that putting magnets on those ends would probably increase the lifetime of the plasma and reduce the power requirements greatly.

Hopman: Sure. I want to comment on one point. If you just fed power into your neutralizer and produced a plasma, and you want to reduce the pumping to have a high gas efficiency like in Andrew Holmes' proposal, then your neutralizer with a length of about one meter will have on both sides the pipes or slots with lengths of over 2 meters. In those slots, you do not have plasma but you have gas. You have to look at the overall gas target density over the full five meter length, although the plasma target is only 1 meter long. This factor makes you require a higher degree of ionization than if you would have no reduction of your gas. As far as I am aware, no one has made a study of optimum lengths of your pumping looking at those two target densities. But you have to be careful.

COMMENTS ON VELOCITY SPACE RELAXATION IN HIGH CHARGE-STATE PLASMA NEUTRALIZERS*

Ady Hershcovitch
Brookhaven National Laboratory, Upton, NY 11973

ABSTRACT

Scaling of velocity space relaxation rates, which affect the final temperature of a neutral beam, with ion charge state in plasma neutralizers is examined. Overall, this scaling indicates a substantial enhancement in emittance growth of the neutral beam with increase in the charge state of ions in plasma neutralizers.

General Considerations

Plasma neutralizers for high energy negative ions have been shown experimentally[1,2,3] to have several advantages over gas neutralizers, the most important of which is the higher optimum neutral fraction. For high energy H^-/D^-, neutralization efficiency of about 85% was determined both theoretically[4] and experimentally[2] for plasma neutralizers consisting of electrons and hydrogen ions. This value is substantially higher than the 50% neutralization efficiency of gas.

High-charge-state plasma neutralizers have been proposed[5] to neutralize very high energy (100's of MeV) H^- beams. At these energies optimum target thicknesses require very long neutralizers, since the obtainable densities of confined plasmas are limited. The use of high Z (Z designates charge state) neutralizers can, in principle, yield the same neutral fraction at a reduced line density. For fusion applications, H^-/D^- beams of about an MeV or less are needed. At this energy level, the optimum target thickness (nl) of a plasma neutralizers is in the range of a few 10^{15} ions/cm^2. Since plasma neutral targets with ion densities n of up to 10^{14} cm^{-3}, have been demonstrated[6] (including full length operational systems[2,3] used in experiments), the optimum length l of about a meter was not considered to be a problem. Higher power levels required in high Z discharges is obviously a very unattractive feature of high charge-state plasma neutralizers for fusion applications. Thus, the only possible application of high Z plasma neutralizers worth considering is for SDI, where a very large nl is needed and power constraints are not severe.

In plasma neutralizers, H^-/D^- beams undergo velocity space relaxation before stripping occurs. To properly compare neutralizers with ions of various charge states, examination of velocity space relaxation rates coupled with neutralizers length could provide the best indication of problems or benefits form neutralizing

*Work performed under the auspices of the U.S. Dept. of Energy.

with ions of various charge states. Melchert, Debus, Liehr, Olson, and Salzborn[7] measured cross sections for single and double electron removal from H^- in $H^- - Ar^{+q}$ ($q \leq 8$) for energies of 3 - 100 KeV (these results were extended[8] to 250 KeV). Based on these cross sections, it was concluded[7,8] that optimum neutral fractions of 82% - 89% are possible in high Z neutralizers and that the reduction in target thickness needed to achieve optimum neutralization scales as $Z^{1.3}$.

Velocity Space Relaxation

Scattering of H^-/D^- ions by electrons and neutralizer ions until the point of neutralization increases beam temperature i.e, it leads to emittance growth. The average point of neutralization is a certain fraction of the optimal target thickness. To properly evaluate the effect of high charge state plasma neutralizers on emittance growth, scaling of pertinent relaxation rates with Z are determined form relaxation rates derived by Trubnikov[9] and cross sections[7,8] measured by Salzborn's group.

Rates (designated by ν) of interest to velocity space relaxation of an H^- beam are associated with processes arising from the interaction of fast test particles streaming through a background of field particles having a Maxwellian distribution with a temperature much smaller than the beam energy. These processes are

slowing down (drag) $\quad \dfrac{d}{dt} V_{H^-} = - \nu_s^{H^-/e,i} V_{H^-}$

parallel diffusion $\quad \dfrac{d}{dt} [V_{H^-} - <V_{H^-}>]_{\parallel}^{2} = \nu_{\parallel}^{H^-/e,i} V_{H^-}^2$

where $<>$ denotes ensemble average of test particles (in this case, H^- ions), and the field particles are e (electrons), and i (ions of various charge states), i.e., particles that constitute the plasma neutralizer.

Since the H^- beam velocity is orders of magnitude larger than any thermal velocity ($V_{H^-} \gg V_{th_{H^-}}, V_{th_{e}}, V_{th_{i}}$), drag and parallel diffusion are insignificant in this case. However, perpendicular diffusion governed by the expression

$$\dfrac{d}{dt} [V_{H^-} - <V_{H^-}>]_{\perp}^{2} = \nu_{\perp}^{H^-/e,i} V_{H^-}^2$$

is very important since the transverse diffusion rate $\nu_{\perp}^{H^-/e,i}$ is a major contributor to increases in the perpendicular component of the beam temperature, i.e., it directly enhances emittance growth.

From Trubnikov[9], for $V_{H^-} \gg V_e, V_i$.

$$\nu_\perp^{H^-/i} = \frac{1.8 \times 10^{-7}}{E^{1.5}} n_i Z^2 \Lambda_i \qquad (1)$$

and

$$\nu_\perp^{H^-/e} = \frac{1.8 \times 10^{-7}}{E^{1.5}} n_e \Lambda_i \qquad (2)$$

where E is the H⁻ in energy in eV, and Λ is the Coulomb Logarithm. All units other than for energies and for temperatures are cgs.

At a given H⁻ energy, the target thickness nl for optimum neutralization is a calculable fixed number. Therefore, the simplest way to compare various neutralizer plasmas is to examine a fixed length neutralizer with ions of various charge states, i.e., the H⁻ transit time through any neutralizer is the same. From the experimental results by Salzborn's group[7,8], the ion density reduces as $Z^{1.3}$, and $n_e = Zn_i$, therefore, $\nu_\perp^{H^-/i}$ **in equation 1 increases as** $Z^{0.7}$ while in equation 2, $\nu_\perp^{H^-/e}$ reduces as $Z^{0.3}$. In this process for high Z neutralizers, ion scattering is the dominant effect. If, for example, a neutralizer with Z = 10 ions is considered, the enhancement in the rate at which emittance is growing is by a factor of 5 larger over that of a Z = 1 neutralizer.

Examining another extreme in which n_i is kept fixed, the target length l is reduced by $Z^{1.3}$. Therefore, the H⁻ transit time t through the neutralizer is reduced by $Z^{1.3}$. Multiplying both sides of equation 1 by $t = 1/v_{H^-}$ yields $t\nu_\perp^{H^-/i} \alpha Z^{0.7}$. This expression indicates that the enhancement in emittance growth is still proportional to $Z^{0.7}$. Therefore, it is obvious that any combinations of length and density reductions will still yield the same results. The effect of the Coulomb logarithm is not very significant and it requires, for its evaluation, information regarding the ion and electron temperatures T, as well as that of the ion mass. In all expressions of the Coulomb logarithm which contain T and Z, temperatures and charge states are on opposite sides of division lines. Consequently, increases in Z are most likely offset by increases in T, since higher plasma temperatures are needed for higher charge state neutralizers. And, Λ is proportional to the logarithm of what is likely to be an insignificant ratio.

Overall, there is a significant increase in $T_\perp$ of an H⁻ beam neutralized in a high Z neutralizer compared to a Z = 1 plasma. The dominant effect is ion scattering which scales as $Z^{0.7}$.

Finally for completeness, this exercise when repeated for drag and parallel diffusion reveals that in high Z plasma neutralizers these two processes are reduced by $Z^{0.3}$, since they are dominated by electrons. Nevertheless, as it was shown earlier in this section, drag and parallel diffusion are not important for any case of interest to neutral beams, and do not have any meaningful effect on beam quality or performance.

CONCLUSION

Based on formalism by Trubnikov and cross sections measured by Salzborn's group, there is a substantial increase in the emittance growth of an H$^-$ beam neutralized in a high Z plasma neutralizer compared to a plasma neutralizer containing singly charged ions. The effect is dominated by ion scattering whose rate scales as $Z^{0.7}$.

REFERENCES

1. G.I. Dimov, A.A. Ivanov, and R.G. Roslyakov, Sov. Phys. Techn. Phys. 22, 1091 (1976).

2. A.A. Ivanov and G.V. Roslyakov, Sov. Phys. Techn. Phys. 25, 1346 (1980).

3. A.I. Hershcovitch, B.M. Johnson, V.J. Kovarik, M. Meron, K.W. Jones, and K. Prelec, Rev. Sci. Instrum. 55, 1744 (1984).

4. K.H. Berkner, R.V. Pyle, S.E. Savas, and R.K. Stalder, Proceedings of the Second International Symposium on the Production and Neutralization of Negative Hydrogen Ions and Beams, ed. Th. Sluyters, Brookhaven, 1980, BNL-Report 51304, pp. 291 - 297.

5. A.S. Schlachter, K.N. Leung, J.W. Stearns, and R.E. Olson in production and Neutralization of Negative Ions and Beams, ed. J.G. Alessi, AIP Conf. Proc. No. 158 (1987), pp. 631 - 642.

6. K. Moses, et al, in this proceedings.

7. F. Melchert, W. Debus, M. Liehr, R.E. Olson, and E. Salzborn, Europhys. Lett. 9, pp. 433 - 439 (1989).

8. E. Salzborn in this proceedings.

9. B.A. Trubnikov, "Particle Interactions in a Fully Ionized Plasma", Reviews of Plasma Physics, Vol. 1 (Consultants Bureau, New York 1965), P. 205.

DISCUSSION

Moses: I assume that by High Z you mean charge state not mass?

Hershcovitch: That is correct.

SPACE CHARGE NEUTRALIZATION*

U. von Wimmersperg
Brookhaven National Laboratory, Upton, NY 11973

ABSTRACT

We examine the feasibility of creating an electric space-charge field of order 10^9 V/m using radial and axial focussing on a pulsed beam of 200 MeV H^- ions. When a bunch of H^- ions is focussed into a disc-shaped concentration, the electric field is uniform and the electrons are shed abruptly, leaving a neutral beam of H^o. Subsequent stripping of the H^o discs by a carbon foil produces highly ordered oscillating plasma bunches which can act as an electrode-less high gradient accelerator structure.

It is known from measurements of H^- ion dissociation due to the Lorenz force induced by transverse magnetic fields that the loss of the electron occurs exponentially with a half life related to the equivalent electric field. At 10^9 V/m the decay of the H^- ion has a half life of 5.58×10^{-13} s during which time a 200 MeV H^- beam travels a distance of 95 μm. Such an electric field can be produced by the collective space charge of tightly bunched ions. When such a bunch of H^- is focussed into a disc-shaped concentration, the electric field is uniform and the electrons are shed abruptly throughout the bunch, leaving a neutral beam of H^o.

Consider a beam produced by a 200 MHz linac, which delivers bunches spaced at time intervals of $t = 5 \times 10^{-9}$ s. If we require an electric field E at the space-charge focus, then the diameter of the charge concentration has to be of order $d = (it/\pi\epsilon_o E)^{1/2}$. If $E = 10^9$ V/m and the beam current $i = 50$ mA, then the required diameter is $d = 94.8$ μm. This would seem to be technically realizable. To this end two separate focussing strategies must be pursued simultaneously: radial focussing using a lens and axial focussing via a buncher which decelerates the head and accelerates the tail of individual ion bunches. The required disc-shaped ion concentration is achieved by making the radial focus coincide in space with the point at which the tail of the bunch overtakes the head.

Special considerations must be met at the focal point in order to produce a parallel H^o beam. The focus must generate laminar flow rather than a cross-over of trajectories. This condition can be satisfied by space-charge forces. An expression for the buncher-induced velocity difference $c\Delta\beta$ over the bunch length, which satisfies these conditions, takes a simple form in the case of an idealized bunch profile:

$$\Delta\beta/\beta = \ell/z - e^2 N \ell n(r/r_f)/(\pi\epsilon_o z \gamma mc^2 \beta^2 \tan^2 \alpha).$$

---

*Work performed under the auspices of the USA Strategic Defense Command.

Here N represents the number of H^- ions distributed with uniform density in a cylindrical bunch which has a length ℓ and radius r when it is at a distance z from the focal point. The radial focussing is represented by the angle α at radius r and it is assumed that the radial inward velocities within the bunch are such that $\dot{r} \propto r$. The radius at the laminar focus is denoted by r_f and m is the ion mass. This laminar flow condition at the focus implies zero axial momentum spread, as the energy spread introduced by the buncher is converted into electrostatic potential energy. The outgoing H^o beam then emerges parallel and with uniform momentum.

An interesting application of such a tightly bunched H^o beam is the production of highly organized oscillating plasma bunches which can act as an electrode-less high gradient accelerator structure. When a disc-shaped H^o bunch passes through a stripping foil which is just thick enough to stop the electrons coming off the hydrogen atoms, a high electric field is generated at the exit surface of the foil due to the charge concentration within the emerging proton bunch. Electrons starting from rest at the foil surface are then field-emitted and follow the proton bunch. When these electrons catch up with the protons, they overshoot the positive charge center and perform axial dipole oscillations. On a time-averaged basis the neutrality of the resulting oscillating plasma bunches is thereby restored. The frequency f of these plasma oscillations is given by $f = c(1/\beta - \beta)/8k$ with $k = mc^2/eE$, where βc is the H^o beam velocity, m and e the electron mass and charge, and E the effective electric field strength at the emerging proton bunch. For $\beta = 0.5$ and $E = 10^9$ V/m the frequency f = 110 GHz corresponding to an electromagnetic wavelength of $\lambda = 2.27$ mm. The plasma bunches travelling at $\beta = 0.5$ have a phase wavelength of $\lambda/2$. Resonant microwave coupling of the plasma bunches can be obtained by spacing them distances of integral multiples of λ apart. Acceleration of relativistic ($\beta \sim 1$) charged particle bunches can be achieved by injecting these coaxially into the oscillating plasma train with appropriate phasing.

DISCUSSION

Hopman: You indicated predictably that Lorentz stripping does not work. I don't see how the space charge stripping would work because of symmetry the electric field is radially outward. You can only take it as a small fraction of the beam in the outer annulus where due to space charge field, the electron is stripped. But on the inside, it just remains negative and is not stripped.

von Wimmesperg: What you say is perfectly correct but I did mention that if one uses the time like focusing the axial overlapping of the rear end and the front end of the pulse, then at the focus you go through a disc shape. It turns out that the electric field on a disc like is zero except at the edge. So you will have razed some beam right at the edge but otherwise you have a uniform field which reaches the critical value abruptly.

Roberts: You have taken advantage for space-like, strong focusing systems in order to get a small volume. The diameter of the fast focusing lens is kind of comparable to the focal length and that gives you small volume space like. Now to get a small volume in time like you have to have a long focal length and the slow optical system, you simultaneously used both. If you try to do the time like in your space like system, the spread in momentum to obtain that is so large over the short focal distances that the space like system will blow up on you and its going to be much worse not better. You can't simultaneously take advantage to two exclusive things. If you try this experiment, what you're going to learn is how perverse matter can really be.

von Wimmersperg: I think it is true that my model of the geometrical focusing system does not address the problem of pulling a focus in the same point for different momenta. You actually have to want the focusing strength of the element in order to do that. It's more sophisicated than I show in that lens.

Roberts: In more trouble, not less.

von Wimmersperg: It's not easy to do.

Stern: I'm wondering about another effect. In order to get that kind of a field of the order of 10^9 volts per meter, you have to start applying a rather strong field just as the beam begins to converge. That means that to some extent you're beginning to strip electrons off throughout the volume, starting with a low density of electrons in the beginning and the maximum at the focus as you have shown. But it does mean that as you go along the material doesn't remain H^- there are two species there, H^- and electrons all along, and it seems to me that what you going to get is an axial elongation of the focus so that the volume of the focalized region is going to be axially much larger due to the separation between the electrons and the H^-. Did you calculate that?

von Wimmersperg: I agree with you on that, that is what would happen if you get things wrong. What one sees from the relation which gives you the half life of the H^- ion as a function of electric field is that it is highly exponential. And so, the stripping really only occurs in the last moments as you reach the intense focus. Before that the life time of the H^- ion is very long and you will loose very few electrons until you reach that final configuration.

APPLICATIONS AND PROGRAMS

Tritium Inventory Considerations for Beam-Fuelled Tokamaks

L.R. Grisham
Princeton University Plasma Physics Laboratory
P.O. Box 451, Princeton N.J. 08543

Abstract

One of the possible applications for which negative ion based neutral beams might be used is for producing peaked density profiles in tokamaks through beam fuelling. This paper examines the implications of such beam usage upon the tritium throughput, and ways in which the required tritium inventory might be kept to a minimum.

Introduction

In recent years, design studies incorporating neutral beam injectors onto devices such as ITER have envisioned very high energy deuterium beams (1 - 1.3 MeV for ITER, for instance), where the primary purpose is to drive current in the plasma core, while providing heating in the process. Such high energy applications are in some ways quite well suited to realities of negative ion production; namely that, compared to positive deuterium ions, negative ones are harder to produce, more fragile, and are presently produced with much less gas efficiency. Consequently, applications which result in high beam powers and beam power densities through the mechanism of high particle energy are attractive to beam developers because they tend to shift part of the burden from negative ion production, which is difficult, to acceleration, which may prove to be a more tractable problem.

Nonetheless, there is another role which neutral beams might be called upon to play which would require that negative ion derived neutral beams be used at the lower end of their applicable energy range. This role would be to fuel the center of a tokamak plasma to produce the peaked density profiles which have been found in TFTR to produce enhanced energy confinement accompanied by high fusion reactivities. Such a role for neutral beams has recently been proposed [1] as a possible operating mode for the Compact Ignition Tokamak (CIT), or for a subignition Tokamak Production Reactor (TPR) [2] to produce isotopes. Since central fuelling is an important part of the beam mission in these applications, the energy would be relatively low, and the required negative ion current becomes correspondingly large, shifting the balance of developmental difficulty back from the accelerator to the source.

Although the baseline auxiliary heating plan for CIT calls for ICRH, a conceptual design was recently carried out by Grisham, Cooper, Purgalis, and Brown [3] to investigate the implications of installing negative-ion-derived neutral beams to produce beam fuelled peaked profiles. The energy requirement was relatively low for negative ions - 300 keV, but the total power requirement might be as high as 60 MW [1]. This design study is described in reference 3. The study assumed that one used circularly focused (for gas efficiency) barium surface converter sources with a gas efficiency of 4%, electrostatic accelerators, a transverse ion temperature of 10 eV after the accelerators, room temperature gas neutralizers, and magnetic deflection of residual ions. With these assumptions, even for the very constrained port geometry of CIT (a duct through the coils with a height of 101.6 cm, width of 37.5 cm, and a length of 160 cm) it appeared to be possible to inject 10 MW per port with the accelerator emitting plane located 20 meters from the tokamak vacuum vessel.

The Tokamak Production Reactor would also require beams at relatively low energies, and thus might use much the same technologies as would be appropriate for CIT. We extended the CIT study to a 150 MW beam system operating at 400 keV on TPR, resulting in a system with about 12.8 MW per beamline if gas neutralizers were used.

Tritium Throughput

A crucial consideration in determining the feasibility for these two devices, or for any beam fuelled reactor, is what the implications for the required tritium inventory would be. Since, in these applications, the beams actually provide a sizeable fraction of the particles in the hot core where most of the fusion reactions must be produced, it is important that approximately half of the beam particles be tritium. If this were not done -- that is, if the beams were entirely deuterium-then even if the target plasma started out with pure tritium, the hot core would rapidly come to be dominated by deuterons, which would result in a reduced fusion reaction rate.

The difficulty which arises in using neutral beams derived from negative ions to fuel the plasma core is that they are relatively inefficient in terms of the number of beam particles that actually enter the plasma compared to the number of particles used to produce the beam. This is a much more serious problem in the case of tritium than with deuterium. In the CIT case, with the assumptions previously cited, 30 MW of tritium beams would require a throughput of 1700 curies of tritium per second in order to deliver 34 Ci/sec of beam to the plasma. This illustration assumes that the gas neutralizers, which are isolated from the sources and accelerators by a pumping region, are fed only deuterium, even on the beamlines producing tritium beams. For 5 second beam pulses (with 0.5 sec preflow of tritium into the sources), this results in a consumption of about 9400 Ci/shot in CIT. This would be a substantial fraction of the 50,000 Ci allowed site inventory for tritium on CIT. Moreover, not all of this 50 kCi would be available in a day.

Reducing Tritium Inventory

Almost all of the 9400 Ci of tritium used in a shot (including a significant fraction of the 170 Ci that entered the plasma) would end up in the beamline cryopanels. This could be recovered by regenerating the panels, but it would now be mixed with deuterium. Under the scenario just described, where half the beamlines produce pure deuterium beams and half pure tritium, while the neutralizers are all fed deuterium to reduce the total tritium usage per shot, this effluent gas would not be readily reusable. It would first have to be liquified to separate the tritium from the deuterium. This would be expensive, time consuming, and energy-intensive. The, advantages of the scenario using isotopically pure beams and deuterium neutralizers are that it minimizes the tritium consumption per shot, allows the neutralizer line density to be tuned to the optimum for each isotope's velocity, and would permit the deuterium and tritium to be injected at different energies so that they would have the same deposition profiles. For these reasons one might choose this operating mode if one desired only a few tritium shots per day, or if the allowed site inventory were somewhat higher. This approach would certainly not be acceptable, however, for an operating scenario requiring many shots, or rapid repetition, or continuous beams. In the TPR case cited, which requires continuous beam operation, this approach would lead to a tritium throughput of 28.5 kg/day, which would lead to a huge inventory requirement if cryopanel regenerations were infrequent.

Clearly, there are other operating scenarios which, although using more tritium throughput per second, would result in lower inventory requirements for continuous or repetitive operation.

In order to minimize the required tritium inventory for continuous or rapid repeat beam systems, two goals appear to be paramount: the cryopanels must be regenerated as frequently as possible (which also means as rapidly as possible), and the gas effluent stream coming off the cryopanels must require as little processing as possible before being reintroduced to the beam system.

The first of these goals could be accomplished, albeit with some engineering complexity, by changing the design of the cryopanels which are used as the primary pumping systems for beamlines. In present day systems a regeneration cycle is slow, because the gas which comes off the panels is allowed to pass through the liquid nitrogen cooled chevrons into the large beam box, where it must be slowly pumped away at relatively low pressures. During this regeneration period, the helium panels are not cooled, which allows the temperature of the helium panels to rise toward the temperature of the liquid nitrogen chevrons due to conduction through the effluent gas and radiation from the $77^\circ K$ chevrons.

Reducing the cryopanel regeration time requires that two things be alone: the effluent gas must be confined to a smaller volume so that it can be pumped away more rapidly at higher pressure, and the temperature excursion of the helium cryocondensation panels must be limited, so that they can be quickly refilled with liquid helium at the end of the regeneration cycle.

The first of these requirements might be met by making the $77^\circ K$ chevrons movable, so that they could be used to close off the small volume between these chevrons and the helium cryocondensation panels while the effluent gas was pumped away. Sedgeley and Baxter have, in fact, demonstrated such a movable chevron system [4].

The second requirement would be met by maintaining a flow of cold helium gas through the helium cryocondensation panel during the brief regeneration cycle, with the temperature chosen just warm enough to drive the deuterium and tritium off the panels. Restricting the temperature excursion of the panels would not only reduce the subsequent cool down time; it would also reduce the thermal stresses associated with repeated regeneration cycles.

In order to have the beamlines capable of continuous operation, it would also be necessary to make them sector - regenerable. Sedgely and Baxter [4] have demonstrated the operation of a two-sided cryopump, on which the $77^\circ K$ chevrons would be moved so that one side was regenerating while the other was pumping. This technique would need to be extended to a system consisting of a planar array of pumping sectors, with some fraction of the sectors being regenerated at any given time while the remainder continued to pump.

With sector-regenerable cryopumps, the minimum circulating inventory (MCI) of tritium is then reduced from the daily throughput value by a factor R which is approximately equal to the number of regeneration cycles per sector per day. If this number R were twelve, for instance, which is almost certainly obtainable, then the MCI for the TPR case mentioned earlier would come down to about 2.4 kg. If R could be made as large as 60, then the MCI for the same TPR case would come down to 0.48 kg. An R of 60 could be obtained if each sector had a cycle consisting of 20 minutes of pumping, followed by 4 minutes of regeneration and recovery. This would give a duty factor of 80%, meaning that the overall size of the pumping array would need to be increased by 25% to obtain the same total pumping speed as would be achieved with a conventional cryopump systems in which the while array is either pumping or being regenerated at any given time. Achieving R as large as 60 in a reliable fashion is undoubtedly difficult, but it is not unreasonable if the temperature excursion of the helium cryopanels during regeneration is controlled by cold helium gas circulating within the panels.

Making R large to reduce the MCI is not in general sufficient by itself to minimize the required tritium inventory for a beam fuelling system. In particular, it will have a limited impact if there is a large holdup of tritium in a reprocessing system which liquifies the gas effluent stream to separate the deuterium and tritium. Thus, one wants to design the beam operating scenario so as to minimize the gas reprocessing.

One possibility would be to still run the sources in each beamline on either pure tritium or pure deuterium, but to run the neutralizers on an easily condensable noble gas such as argon, which could be removed from the gas effluent stream by cooled panels at temperatures when the vapor pressure of deuterium and tritium would be high. The reason for using a noble gas would be to avoid forming molecules with the hydrogen isotopes. How practical this approach would prove to be would depend upon how low the backstreaming of mixed isotopes from the tokamak could be made, and how much deuterium and tritium would get adsorbed on the noble gas frost which would form on the noble gas recovery panels in the effluent stream. With sufficient pumping in the long (about 9-10 m for CIT) ducts associated with these beams, the contamination from the tokamak could probably be controlled. In any event, the mixed deuterium and tritium in the tokamaks would eventually require isotopic separation if it were to be reused in pure isotope beams, but this would be only a small fraction (about 2% in the examples considered for CIT and TPR) of the total tritium throughput.

If deuterium or tritium were adsorbed to any significant extent on the noble gas frost, they could be relatively easily removed from the noble gas when the noble gas condensation plates were warmed up, and they could be recovered as relatively pure isotopes since, in the scenario using noble gas neutralizers, the sources on each beamline would be running on only one isotope.

Such a scenario using a noble gas neutralizer would probably be suitable for a tokamak such as CIT with a low duty factor. In the case of a continuously operating tokamak such as TPR, even the relatively small fraction of tritium that became mixed with deuterium in the tokamak would amount to a large quantity requiring isotopic separation. For a continuously operating system, it would appear that the most viable scenario would be to run all the ion sources and, if one used gas neutralizers, all of them as well, on a mixture of deuterium and tritium chosen to produce the desired mixture in the tokamak core plasma. Running with a mixture of deuterium and tritium has two disadvantages. One is that both isotopes have to be accelerated to the same energy, whereas for them to have the same fuelling profiles, one would actually like to have the same velocities instead of the same energies. The other disadvantage is that since both isotopes will be passing at the same energy, but different velocities, through the neutralizer, the line density of a gas neutralizer cannot be simultaneously optimized for both velocities. Fortunately, the drop in efficiency which this entails is only a few percent, depending upon the operating energy.

If one feeds the same mixture of deuterium and tritium to all the gas neutralizers in a beam fuelling system, then the total tritium throughput is, of course, increased (by very roughly a factor of 3 for the particular CIT and TPR examples considered). This increase is offset, however, by the very significant advantage that, in the mixed deuterium-tritium mode very little reprocessing of the gas effluent stream would be required before feeding the mixture back to the sources and neutralizers. Running mixed deuterium - tritium beams would appear to be the most practical choice for a high duty factor beam-fuelled tokamak. For a low duty factor beam fuelled tokamak, either mixed isotope beams or the pure beam approach with noble gas neutralizers might prove attractive, with the choice between them depending upon the specific characteristics of the tokamak's operating cycle and facilities. Especially for the high duty factor tokamak, the system would also need to contain some capability for removing helium ash, either with cryosorption (on frost, zeolite, or charcoal) or turbomolecular pumps.

Of course, there are also improvements that one might make to the neutralizers that would improve the gas efficiency of the total system. These gas neutralizers should operate in the free molecular flow regime, so cooling them to liquid nitrogen temperature should reduce the gas efflux from them by roughly a factor of two. This should be feasible, provided that scrapers remove the beam wings to neutralizers (which would be separated from the accelerators). Going further, plasma neutralizers could raise the beam neutralization efficiency, and might be energetically favorable if magnetic end blocking is used, as recently suggested by Holmes [5]. Substantial work has already been done toward developing plasma neutralizers [6]. Going still further, if they proved practical, photodetachment neutralizers would improve the overall gas efficiency of a beam system. While any of these improvements would reduce the tritium throughput of a beam fuelled tokamak, they would not qualitatively alter the fact that beam fuelling involves large tritium throughputs. In addition, the system gas efficiency is highly sensitive to the source gas efficiency, which might end up being either somewhat better or worse than has been assumed. Nonetheless, it appears that, as has been discussed, the required tritium inventories can be kept to manageable levels with appropriate operating and pumping scenarios.

References

1. R. Goldston, private communication (1988).
2. R. Moir, D. Jassby, and R. Budney, private communication (1989)
3. L.R. Grisham, W. Cooper, P. Purgalis, and T. Brown, to be published in Proceedings of the Thirteenth Symposium on Fusion Engineering, Knoxville (1989).
4. D.W. Sedgeley, T.H. Batzer, W.R. Call, Fusion Technology $\underline{8}$, 1229 (1985)
5. A Holmes, private communication.
6. Ken Moses, private communication.

NEGATIVE ION-BASED NEUTRAL INJECTION ON DIII-D*

L. D. Stewart, D. K. Bhadra, A. P. Colleraine, and J. Kim
General Atomics, San Diego, Ca. 92138

ABSTRACT

High energy negative ion-based neutral beam injection is a strong candidate for heating and non-inductive current drive in tokamaks. Many of the questions related to the physics and engineering of this technique remain unanswered. In this paper, we consider the possibility of negative ion-based neutral beam injection on DIII-D. We establish the desired parameter space by examining physics trades. This is combined with potential design constraints and a survey of component technology options to establish an injector concept. Injector performance is estimated assuming particular component technologies, and concept flexibility with respect to incorporating alternate technologies is described.

INTRODUCTION

Neutral beam injection (NBI) is a principal candidate for plasma heating and for noninductive current drive in tokamaks. However, there are many important NBI physics and engineering issues which still need resolution in order to lay the groundwork for the planning, design and execution of NBI in future applications such as ITER.

The DIII-D tokamak at GA has been suggested as a test-bed on which high energy NBI physics issues might be explored. We have considered the possibility of negative ion-based NBI on DIII-D. To establish the desired parameter space, we have examined the impact of beam energy on potential DIII-D NBI heating and current drive physics experiments. We have looked at potential design constraints and have surveyed component technology options to establish an injector concept. We have estimated injector performance parameters by assuming particular component technologies. Finally, because the actual injector configuration will depend on future development program direction, we have explored concept flexibility with respect to incorporating alternate technologies.

PHYSICS TRADES

The existing experimental data base for neutral beam injection into tokamaks extends to beam energies of ~100 keV: an order-of-magnitude lower than needed for ITER. A study of physics processes, while applying ~0.5 MeV injection into DIII-D, would increase confidence in theoretical predictions and in the choices for ITER design parameters.

---

*Work sponsored in part by the Department of Energy under Contract No. DE-AC03-89ER51114.

Table I lists four physics processes relating to neutral beam injection into a tokamak plasma, and indicates programmatic importance and key DIII-D experiments for each. The two bottom-line physics processes are heating and current drive, and two primary physics processes which directly affect heating and current drive efficiency are multistep ionization and shear-Alfven instability. The programmatic importance in all cases derives from the choice of neutral beams as the primary current drive engine for ITER.

Table I Physics processes

Process	Importance	Key Experiment
Heating	Determines ITER mix of beam and rf heating	Inject high E_b beams, ECH and ICH into same target
Current drive	Determines ITER power needs	Amps/watt versus E_b
Multistep ionization	Determines ITER shine-through and orbit losses	Shine-through and penetration versus E_b
Shear-Alfven instability	Establishes ITER I-drive efficiency	Search for onset in DIII-D operating space

Data on basic heating efficiency and current drive efficiency would allow some confidence in the ITER choices of beam energy, beam power, and supplemental rf type and power. A key set of experiments on plasma heating would be the injection of neutral beams, ECH, and ICH into essentially identical plasmas. Recognizing that the method of heating is a primary factor in determining plasma parameters, the approach would be to generate ITER-relevant plasmas in DIII-D using 10 to 20 MW of 80 keV neutral beams, then to apply a megawatt of ~0.5 MeV beams, ECH, or ICH.

Physics processes which influence heating and current drive effectiveness are multistep ionization and possibly the shear-Alfven instability. Many predictions of neutral beam effectiveness neglect multistep ionization and thereby underestimate beam attenuation. With a representation of the true cross section as $\sigma_{gs}(1+\delta)$, where σ_{gs} is the ground state cross section, the predicted correction factor δ is shown in Fig. 1.[1] The dotted vertical lines on Fig. 1 and on the following figures show beam energies of 80 keV, representing the existing DIII-D injectors, and 400 keV, representing a future negative ion-based injector.

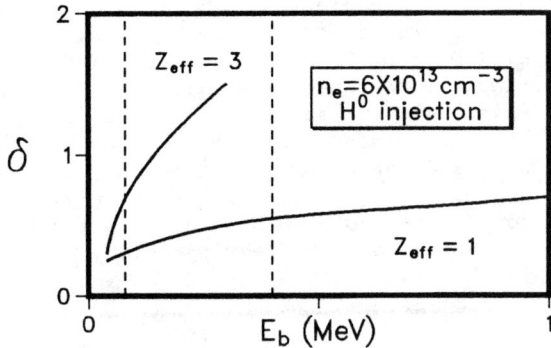

Fig. 1. Excitation-enhancement of the beam stopping cross section.

The predicted δ is ~1 for plasma conditions and beam energies of interest on DIII-D, so the real ionization cross section is predicted to be about twice that for the ground state. Because of the strong influence of δ on penetration and power deposition, particularly for ITER, there is a high premium on experimental verification. The observables on DIII-D will be Balmer alpha emission distribution and shine-through power. Figure 2 shows predicted DIII-D shine-through and orbit losses for quasi-tangential injection and for relevant plasma conditions.[2]

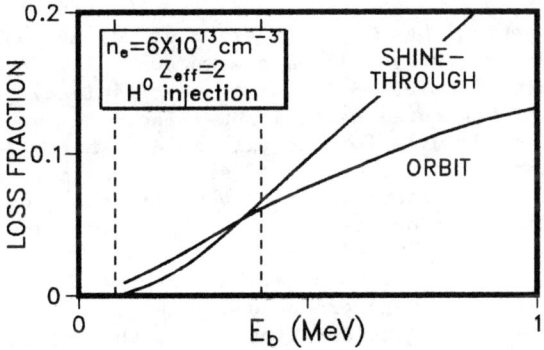

Fig. 2. NBI shine-through and orbit losses.

The theory of the shear-Alfvén[3] mode predicts that it should appear in ITER with high energy NBI.[4] What is needed now is data for benchmarking the theory and for establishing the effect of the shear-Alfvén mode on the confined plasma. Neutral beams are predicted to drive the shear-Alfvén mode when the ion thermal velocity (v_i), the electron thermal velocity (v_e), the beam ion velocity (v_b)

and the Alfven speed (v_a) are ordered as $v_i < v_a < v_b < v_e$. The conditions on v_i and v_e are met in tokamak plasmas.

However, in DIII-D with 80 keV NBI, the condition $v_a < v_b$ is attained only marginally. Because v_a scales as $B/\sqrt{n_e}$, lowering v_a entails low field-high density operation, a disruption-prone and somewhat atypical regime. Injection with ~400 keV beams, on the other hand, allows DIII-D operation in the normal, broader, and more stable operating regime. This is illustrated in Fig. 3, a density-field plot which compares the DIII-D Murakami scaling with the regimes of $v_b > 2 v_a$ for 80 and 400 keV injection.

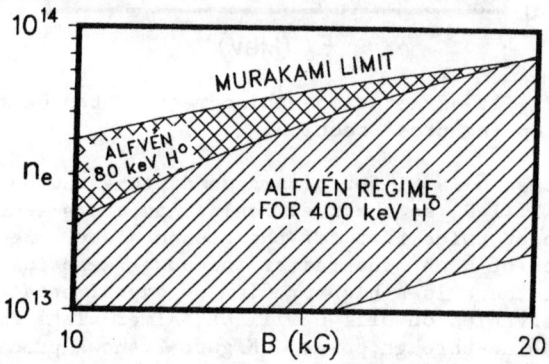

Fig. 3. DIII-D Murakami scaling and regimes of $v_b > 2 v_a$ for 80 andd 400 keV H° injection.

Calculated neutral beam current drive efficiency as a function of energy for representative DIII-D plasma conditions is shown in Fig. 4.[2] The curve labeled η_G is for gross efficiency, the driven current divided by the total neutral power incident on the plasma, and the curve labeled η_N is for net efficiency, where the neutral power has been corrected for orbit and shine-through losses. Because of their uncertainty, shear-Alfvèn effects have not been incorporated. The efficiencies show a broad maximum at about 400 keV for either H° or D° injection.

Because the initial DIII-D negative ion-based NBI system would be limited in current to that produced by one (or possibly two) sources, a more relevant consideration may be the amount of driven current for a fixed amount of beam current. This is shown in Fig. 5 to level off at about 1 MeV, with a knee at ~500 to 700 keV. So although somewhat more beam-driven current could be obtained with 1 MeV injection, there is diminishing return above about 600 keV.

INJECTOR DESIGN CONSTRAINTS

Table II lists principal constraints, their origin and their impact on the design of a negative ion-based neutral beam system for

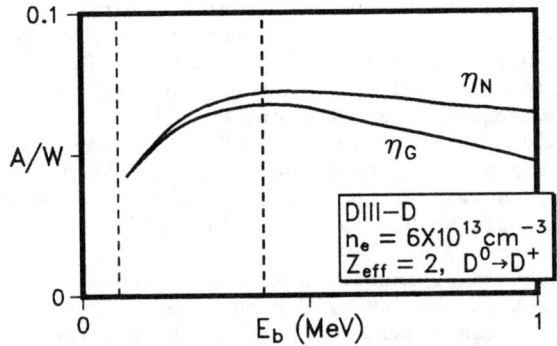

Fig. 4. NBI current drive efficiency.

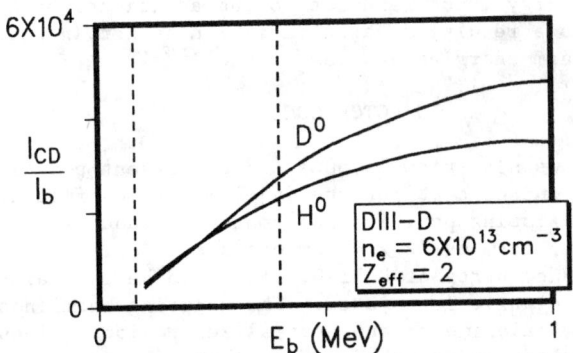

Fig. 5. NBI current drive effectiveness: driven current ÷ beam current.

Table II Constraints on injector design

Constraint	Origin	Impact
Minimize overall perturbation	DIII-D operations; total cost	Convert one or two beams initially
Try to use present cryopump	Cost	Basic beamline geometry fixed; source and stripper gas loads ~ present
Length	Building walls	Little impact on design
Shine-through	Heat dissipation in DIII-D wall armor	E_b can be at least 1 MeV
Neutrons	Deuterium beam onto dumps, calorimeter	E_b can be at least 1 MeV

DIII-D. As an overall constraint, the size of this proposed undertaking is bounded by realism on budgets and by the need to avoid interference with DIII-D experimental operations. The scale for initial modification of beamline, power and cooling systems must not be large. As a consequence, at most one beamline, and more likely one beam, is considered for conversion. For the same reasons, it is desirable to use the existing beamline cryopump. This means the beamline geometry, as well as the gas loading due to the ion source and neutralizer, should be basically similar to the existing positive ion-based injectors.

Three other potential constraints were evaluated. Building walls either are not in the way or can be moved. If care is taken to avoid very low plasma density operation, shine-through can be tolerated up to at least 1 MeV. Most neutron generation will occur in the tokamak plasma; most plasma neutron generation will be due to the present energy inputs and not to the additional high energy neutral beam. As a result, neutron radiation is not increased significantly for beam energies at least up to 1 MeV.

INJECTOR CONCEPT

Figure 6 shows side-view layouts of the present positive ion injectors and of our concept for the injector as modified for negative ions. The existing positive ion sources produce an approximately rectangular beam of 12 x 48 cm cross section. This geometry matches the entrance ports of DIII-D, and would be maintained in converting to a negative ion system. The existing beamlines have cryopanels in the calorimeter and neutralizer regions. These cryopanels are to be left undisturbed if possible. It does not appear reasonable to adapt the existing 180 degree bending magnet and charged beam dumps. At a minimum, extensive redesign would be required to accommodate the higher beam energy as well as the significant amounts of beam power in both the negative and positive components of the neutralized beam.

In the proposed negative ion system, sizes for the various beamline components have been estimated assuming specific component technologies which accommodate either a 400 keV H$^-$ beam or a 400 keV D$^-$ beam. We have chosen an extraction energy of 400 keV as a compromise which maximizes physics performance while minimizing technology extrapolation and cost. A 400 keV high voltage system at the current we anticipate (5 A) is well within present design technology for stackable, modular units.

Starting at the tokamak end of the proposed beamline, the calorimeter region with its cryopanel is unchanged. Upstream from the calorimeter is an ion separator/dump unit. The unit shown has been sized such that either magnetic or electrostatic technology could be accommodated. Upstream from the separator is a stripper with gaps at each end for pumping. The 2 meter length shown for the stripper is that estimated for H$_2$ gas. The 0.8 meters allotted for the accelerator is sufficient for Pierce or ESQ[5] technologies. The

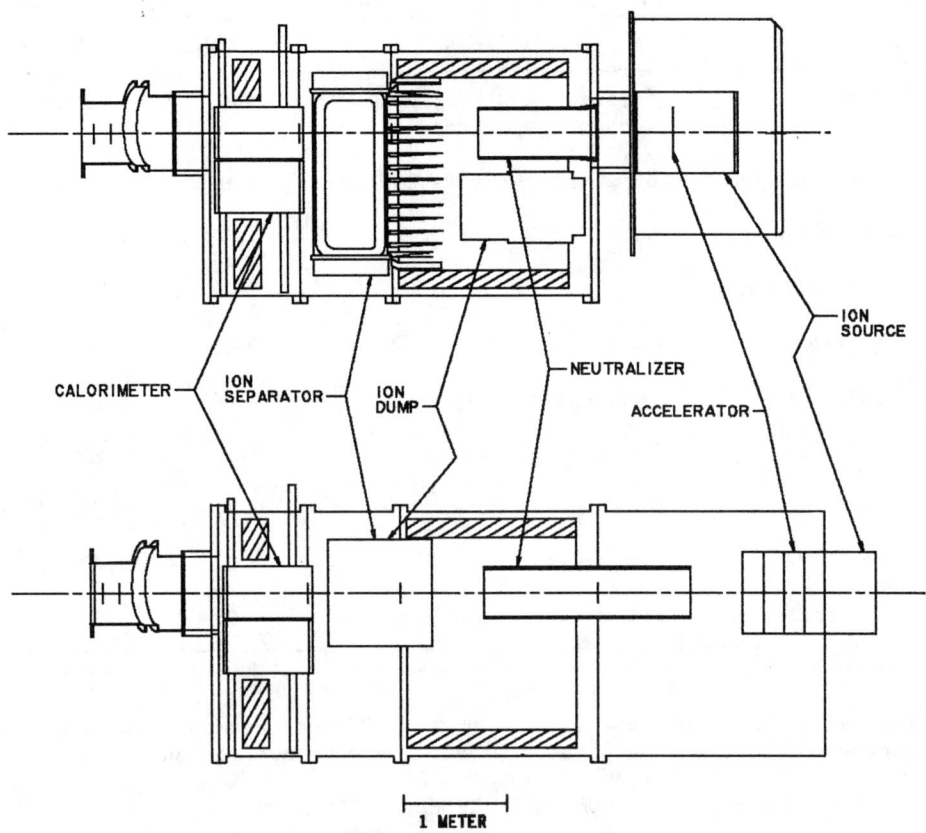

Fig. 6. Side views of the existing and proposed DIII-D beamlines.

total additional length needed for the negative ion system is provided by adding a 2 meter vacuum chamber segment. The interface of this segment with existing chamber will be as open as possible to maximize conductance between the accelerator and the main cryopanel.

Table III presents performance parameters estimated for the injector concept when assuming 10% gas efficiency for the ion source, accelerator pumping constrained by a Pierce-like geometry, and an H_2 gas stripper. The estimated neutral power includes transmission losses and the power/energy degradation due to stripping in the accelerator. The impact of this stripping is indicated by the full energy and the above-two-thirds energy (>270 keV) beam fractions in the table. The neutral, positive, and negative fractions

Table III Estimate performance parameters (one beam)

Parameter	Basis	H° Value	D° Value
Extraction energy	Physics benefits, cost, technology extrapolation	400 keV	400 keV
Ion current	H⁻ at 20 mA/cm² x 250 cm²	5.0 A	3.5 A
Neutral power	Combined loss estimates	1.1 MW	0.72 MW
Beam fractions:			
Neutral	Cross sections	0.60	0.60
Full energy	Estimate premature stripping	0.86	0.80
>270 keV	Estimate premature stripping	0.92	0.89
Positive	Cross sections	0.20	0.20
Negative	Cross sections	0.20	0.20
Transmission	Estimate	0.95	0.95

are essentially the same for the H° and D° cases because the relevant cross sections have the same scaling with nucleon energy.

COMPONENT TECHNOLOGIES

The beamline concept was formulated after surveying candidate technology options[5-9] for the principal injector components. Although no detailed component design was attempted, component sizes were estimated and initial evaluations on possible system effects of incorporating various technologies were made. Table IV indicates the component technologies which would be allowed within the constraints of Table II and the envelope of Fig. 6.

The first option entered in each column is the technology assumed in deriving the performance estimate of Table III. Substitution of a volume source, with its significantly lower gas efficiency, would cause increased premature stripping within the accelerator, resulting in additional accelerator loading, more power in the less-than-full energy portion of the neutral beam, and less injected neutral power. A capillary stripper-separator-dump[6] would decrease the extractor beamlet packing density such that a larger source would be needed or less power would be generated. The use of

Table IV Allowable injector component options

Source	Accelerator	Stripper	Separator/Dump
Surface*	Pierce*	Standard gas*	Magnetic*
Volume	ESQ	Cooled gas	Electrostatic
		Capillary gas	Capillary tube
		Plasma	

*Basis for performance estimate.

a plasma stripper would yield a larger neutral fraction and therefore more injected neutral power.

Of all the component technologies surveyed, only an RFQ accelerator is questionable as a candidate for application on the proposed negative ion system. There is rf power on-site at about the right frequency, but the ~4 meters required for existing high-current RFQ concepts[7,8] would violate length constraints. Although these RFQ concepts are for considerably higher energies (several MeV), the gentle-bunching function in an RFQ prevents a straightforward decrease in length with decreasing energy. We speculate that a shorter RFQ could be designed which compromises bunching and therefore has reduced acceleration efficiency. This possibility has not been explored.

SUMMARY

A concept for a negative ion-based NBI for DIII-D has been described and has the following attributes:

- It would extend NBI physics data base on heating and current drive for more relevance to ITER.

- It would have an acceptable impact on DIII-D operations.

- With the possible exception of an RFQ, there would be no restrictions on the incorporation of the many candidate negative ion component technologies.

The authors gratefully acknowledge W. S. Cooper of LBL for discussions and ideas.

REFERENCES

1. C. D. Boley, R. K. Janev and D. E. Post, Phys. Rev. Lett. $\underline{52}$, 534 (1984).
2. D. K. Bhadra and Jinchoon Kim, Proc. of the 13th Symp. on Fusion Engineering, Knoxville (1989).
3. H. L. Berk, W. Horton, Jr., M. N. Rosenbluth and, P. H. Rutherford, Nuclear Fusion $\underline{15}$, 819 (1975).
4. R. B. Campbell, paper 2C27, 1988 Sherwood Theory Conference, Gatlinburg, 1988.
5. O. A. Anderson, Proc. of the Tenth Int. Conf. on the Application of Accelerators in Research and Industry, Denton, Texas (1988).
6. A. J. T. Holmes, Proc. of the International Conference on Ion Sources, LBL (1989).
7. R. H. Stokes, T. P. Wangler, and K. R. Crandall, Proc. of the 1987 IEEE Particle Accelerator Conf., Chicago (1987).
8. W. R. Becraft, J. H. Whealton, T. P. Wangler, A. Schempp, G. E. McMichael, et al., Proc. of the 1988 Linear Accelerator Conf., Williamsburg (1988).
9. W. S. Cooper, "A Neutral Beam Development Plan for CIT," unpublished (1988).

ASSESSMENT OF POSSIBLE IMPLICATIONS OF A NEUTRAL BEAM CONFIGURATION FOR CIT

L.R. Grisham
Princeton University Plasma Physics Laboratory
P.O. Box 451
Princeton, New Jersey 08543

W.S. Cooper and P. Purgalis
Lawrence Berkeley Laboratory
Berkeley, CA

T. Brown
Grumman Corporation

Abstract

The present plans for the Compact Ignition Tokamak (CIT) call for the auxiliary heating to be done by RF techniques. However, Goldston has proposed that an alternative scenario might involve running TFTR type "supershot" plasmas with peaked density profiles. This would require central fuelling by deuterium and tritium beams. We have performed a conceptual study of a system to inject negative ion based 300 keV beams of D and T using the available port geometry. We report on the resulting conceptual design, the tritium usage during operation, and the implications for the test cell configuration. Work supported by U.S. DOE Contract Numbers DE-AC02-76CHO3073 and DE-ACO3-76SF00098.

Introduction

This paper is being published in Proceedings of the Thirteenth Symposium on Fusion Engineering, Knoxville (1989).

DISCUSSION

Moses: Larry, why hasn't a plasma neutralizer been considered, I'm sure you have considered it but what are your reasons for not using it.

Grisham: Okay, first of all this was aimed at actually not a very friendly audience. This CIT plan was entirely to use rf, and in order to present as small a cross section for getting a shot at it, we tried to view one of the most relatively pessimistic assumptions one could use: this is assuming a not very exciting neutralization efficiency. I mean nobody is going to argue that we are going to get lower neutralization efficiency than this. And, in the case of CIT, it was something that had to be ready, (well, I guess it is not quite clear now exactly when it has to be ready) but at least in principle at the time, it had to be ready relatively shortly and so certainly the least controversial thing to assume was that you use gas. Everybody believes you can put gas into tubes, and we probably know what the thoughput is. If we are able to demonstrate that you could do it with plasma neutralizers, then all this makes it better and a stronger case. The real battle between us and the rf advocates is not actually over whether any of these numbers are 20% too good or 30% too bad. It's just whether they want to have beams on there at all. Our argument was that rf hadn't been shown to work in H modes, and their argument was that beams hadn't been shown to work in perpendicular injection, and unfortunately, we each have since gotten firmer ground to stand under on that. There is no policy more implicit in choosing gas other than the fact that we didn't want it to get involved in people arguing over, for instance, whether the neutralizer had been shown to work or what was the particular power efficiency we choose for the neutralizer.

Moses: There are still reasons for using it besides just the neutralization efficiency. You could increase your gas efficiency because your target becomes much smaller and it just seems to make a lot more sense engineering wise because your beam lines become much smaller and your walls don't have to be moved. I think you would have a much stronger argument for it. Plus the fact you could turn it around and say to the rf people, we will generate the neutralizer plasma by rf.

Grisham: From the point of view of moving a wall, it turns out again, the reason for going out to that distance is to get a large enough ballistic arc, to get the power through the duct, and in addition, you then need enough pumping around there for the gas influx from the sources. In fact, that's the reason for not putting these arrays closer together. Changing any of these parameters much would not greatly affect where the box could go. Like I said, you can't move it a whole lot from the back because of the divergence. But the thing that would help is certainly if you got a higher neutralizer efficiency, like a plasma neutralizer, it definitely would make it more attractive. And in particular it would make it more attractive because if your were neutralizing now 80% tritium instead of 57%, it would certainly cut the tritium throughput. Again though, if you look at these numbers, even if you cut the tritium throughput by that factor, it doesn't actually effect the operating scenario. We're still using a huge amount of tritium to fuel the plasma and instead of having a net fueling efficiency of 2% in that case, maybe we would have a net fueling efficiency of maybe 2.75% which would be a move in the right direction. But it still would mean that we would have a choice between either doing a very few shots in a day or else going to a more exotic type of handling system. I mean all these things are important but they don't affect the fundamental way that you would end up running.

Whealton: Using this conservative approach what did you assume for the current densities and the convertor and the compressions together?

Grisham: I assume that I believe 5% conversion back at the barium surface convertor which is a number Berkeley considered reasonable, based on the FOM results. Again, if that turned out to be 4% or 6% it wouldn't change it a lot.

Whealton: What's the current density and compression?

Grisham: I've forgotten exactly what the factor was, but the compression was such that if you didn't do any clipping when it went into the accelerator channel, you would actually end up with an ion temperature of about 14 or 15 eV. This was assuming that you clipped the wings some or reexpanded it slightly at the end of the channel in order to end up with 10 eV by the time you got out of the accelerator. By conservative, I didn't really mean dead simple, I just meant that we weren't assuming for instance, things like photodetachment neutralizers or enormous improvements in the gas efficiency.

Hemsworth: I have several questions. The first one is if you use 300 kV for your D^- injector and surely you ought to have one in 450 kV for your T^- injector.

Grisham: To tell the truth, what we really should have had was 300 for the T^- and somewhat lower for the D^-. This was a compromise between Rob (Goldstone) and I on the energy, where we agreed for purposes for looking at this model that we would sort of adopt 300. The problem with negative ion based systems, of course, is when you go towards vanishingly low energies. It gets really hard to do, I mean at one point there was a thought that maybe only 100 or 200 kilovolts D^- which, if you are trying to get a set amount of power in, it is very complicated. But you are quite right, you would probably actually prefer slightly different energies but we didn't include that in this. You're probably also going to object to the fact that in this experiment that we did looking at D and H in a beam on your test bed and there are all kinds of complications attendant on that. The assumption here was that with negative ion based system you all you no longer have all these cross products, that you do with positive, plus if you are designing it with that in mind you can design the dumps to handle it. The two different momentum components.

Hemsworth: One more comment and then another question please. You started off by saying you could either have surface or volume sources, and I think that the experience with deuterium as opposed to hydrogen would say that if you really wanted tritium source then you don't have a volume source. That would be my comment. The other question is, "Could you be a little more specific on the regeneration of your cryopanels", because that, to my knowledge, is a very fierce problem to regenerate cryopanels at the rate of 12 or 20 times per day.

Grisham: Yes, I think that would be a tremendous problem. Although it's interesting you say that, when we talked about this at Livermore, at a very tiny workshop on Tokamak production reactor there and I showed that viewgraph that implied that it was tentative about whether one could do 24 regenerations a day, it was afterwards criticized by Ralph Moir for being tentative about it. There is obviously enormous engineering difficulties in doing something like this because now it means that you've got a large number of moving parts in vacuum which, of course, the first thing we all want to do is vacuum weld everything. The sort of thing which Livermore demonstrated back a few years ago, was a system where you had a two sided pump with chevrons on each side and the chevrons were designed so that by moving half of them against the other half you could close one side or the other side. And, they sort of regenerated that in about 4 minutes. Now they had engineering difficulties in that these chevrons it was hard to get them perfectly leak tight so they didn't dump some gas through. But it did look like that was feasible to do. What hasn't been demonstrated is whether you could build a system that could stand that much thermal cycling: 24 times a day, day in, day out. If you manage to keep the temperature excursions small, which you really want to do if your going to do this fast, the thermal cycling should not be too great but I agreed it would be difficult to do. Whether you really need to do it that often of course depends on the constraints of what you are trying to do. The Tokamak production reactor is kind of the most extreme case one could think of for a D^- system, I mean 75 megawatts of this is tritium beams at 400 kilovolts. For something more modest, where you did not have such an enormous current in tritium, you could imagine not recycling it as often. It's pretty clear though for any system using tritium beams that your going to have to come up with some sort of dynamic recycling system because let's say if you look at any of these numbers and you argue this way or that and you adjust them even by factors of 10 they are still big numbers. But anyway this scheme has been demonstrated by Livermore in a less ambitious way. When I was asked to look at what the implications would be for this Tokamak production reactor, I spent like an hour looking at it and saying that's easy, you can't do it. I also knew how long we take regenerating our cryopanels, which is a weekend sort of project. And then, I talked to Livermore some and discovered that involved a lot of engineering in there. It certainly doesn't have to be a slower process as we do. I agree for something like Tokamak production reactor, developing the pumps so that they work reliably, is probably of the same magnitude of effort as getting the rest of the beam line developed, and probably more of an extrapolation from where we are now.

NEGATIVE ION BEAM PROGRAMS AT JAERI

Y. Ohara

Japan Atomic Energy Research Institute,
Naka-machi, Naka-gun, Ibaraki-ken, 311-01 JAPAN

Abstract

Three negative ion beam programs for nuclear fusion research are proposed at the Japan Atomic Energy Research Institute(JAERI): a 500keV, 10MW deuterium neutral beam injection(NBI) for heating and current drive in the JT-60U tokamak, a 0.5/1MeV, 50MW deuterium beam injection for the Fusion Experimental Reactor(FER), and a 2MeV, 20mA lithium beam injection for ion temperature measurement in the JT-60U. The 500keV negative-ion-based NBI system for the JT-60U will be developed in the first step, followed by the development of the reactor relevant 0.5/1MeV NBI system. This two step development mitigates the risk in developing a MeV class NBI system.

Introduction

A high energy neutral beam injection(NBI) system based on negative ion beams is considered to be a primary heating and current drive system for next fusion machines, and to hold the key to realize steady state burning plasmas. One of the most important components in this system is a high power negative ion source which can reliably deliver a 0.5 - 1MeV negative deuterium beam. We started to develop high power negative ion sources in 1984, soon after finishing R&D for the JT-60 NBI system based on the positive ion beams. Since then, the negative ion beam power has been increased step by step using volume production type[1] ion sources. In our recent experiments, a 75keV, 3.4A negative hydrogen ion beam has been produced using the Multi-Ampere Volume Source.[2] In addition, a 50keV, 7.8A negative hydrogen ion beam has also been produced successfully for a duration of 100ms by introducing cesium vapor into the Volume Source.[3] Owing to these recent progress of the negative ion source, utilization of the negative-ion-based NBI system has become realistic. The development of the positive and negative ion sources at JAERI is shown in Fig.1.

In the present paper, negative ion beam programs at JAERI are described together with our R&D strategy.

Negative-Ion-Based NBI Systems Planned at JAERI

We have three programs on the development of the negative-ion-based NBI systems for nuclear fusion research. The first one is to develop a 500keV deuterium beam injection system for the JT-60U scheduled in 1990. The second one is a 0.5/1MeV NBI system for the

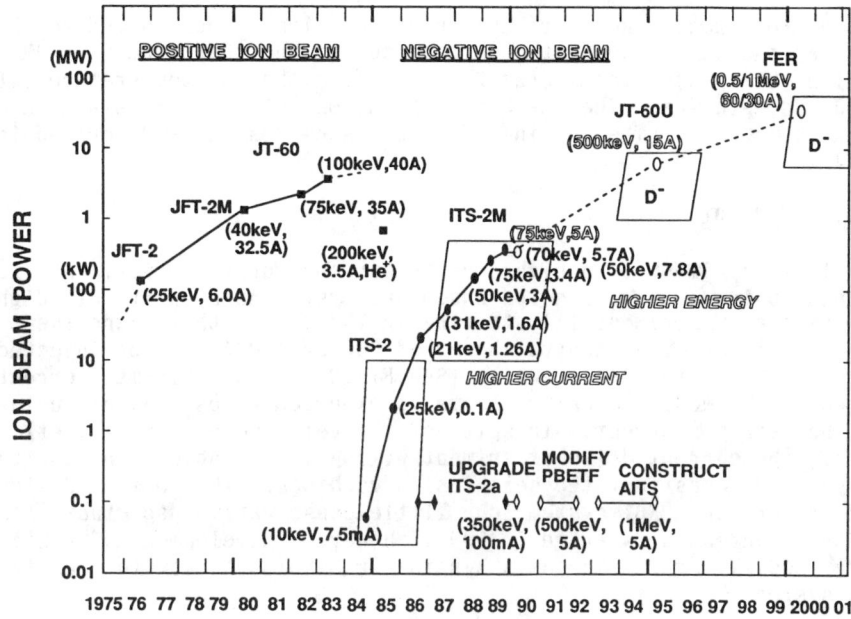

Fig.1 Development of high power positive ion sources and negative ion sources at JAERI.

TABLE I. Development plan of the negative-ion-based NBI systems.

	JT-60U N-NBI	FER - NBI	Li beam system for JT-60U
TO BE OPERATIONAL	1993	2000	1994
BEAM SPECIES	D⁻/ H⁻	D⁻/ H⁻	Li⁻
BEAM ENERGY	500 keV	0.5 / 1 MeV	2 MeV
INJECTION POWER	10 MW	50 MW	20 mA
PULSE DURATION	5 s	1000 s	0.2 s x 5
BEAMLINES	2	2	1
BEAM CURRENT / SOURCE (Full Energy)	11 A	60 / 30 A	50 mA
	1. CURRENT DRIVE (2MA) 2. PENETRATION 3. ALFVEN INSTABILITY 4. PLASMA ROTATION 5. ALPHA PARTICLE HEATING (1MW)	1. HEATING 2. CURRENT DRIVE 3. CURRENT PROFILE CONTROL Q = 5	1. ION TEMPERATURE 2. CURRENT PROFILE

Fusion Experimental Reactor(FER) in Japan which is considered to be built at the beginning of the next century. The third one is a 2MeV lithium beam injection system for measuring the ion temperature of the JT-60U plasmas. The basic specifications of each NBI system are shown in Table I. The outline of each system design is described in Ref.4.

(1)JT-60U N-NBI

In the JT-60U, a high density current drive experiment is planned using a negative-ion-based NBI system as a part of High Performance Experiment III starting in 1993.[5] In this experiment, hydrogen or deuterium neutral beams of 500keV, 10MW will be injected tangentially with two beamlines(See Fig.2). The plasma current driven in the medium density plasma is expected to be 2MA excluding the bootstrap current. In spite of the relatively low injection power, the current drive experiment will give valuable data base to design and construct the next fusion machines. The basic design concept of this NBI system is a little conservative so that its construction can be started in 1991. However, development of a high energy 500keV accelerator as shown in Fig.3 hold the key to realize this system.

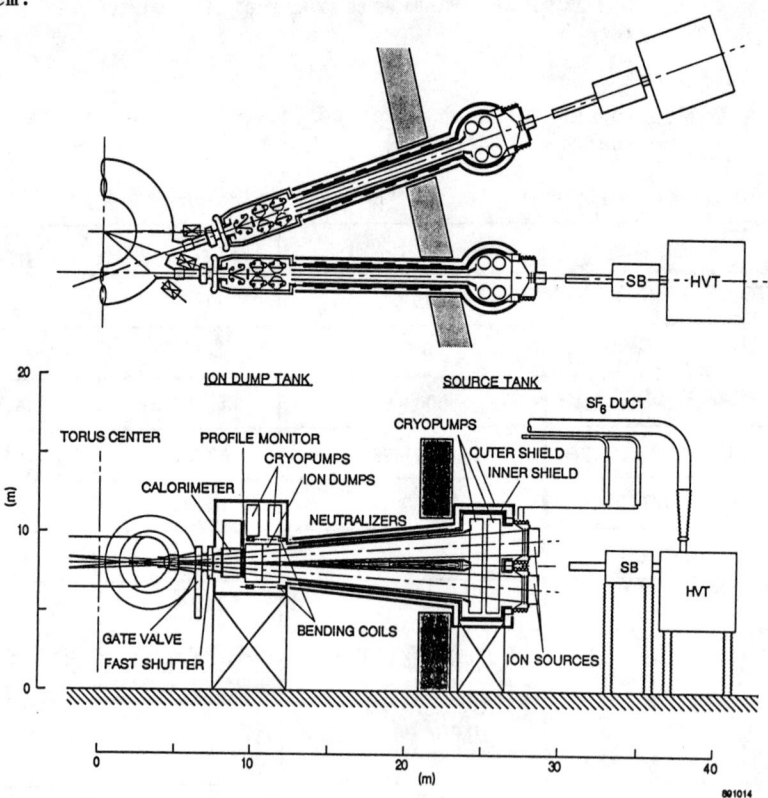

Fig.2 Layout of the JT-60U N-NBI beamline.

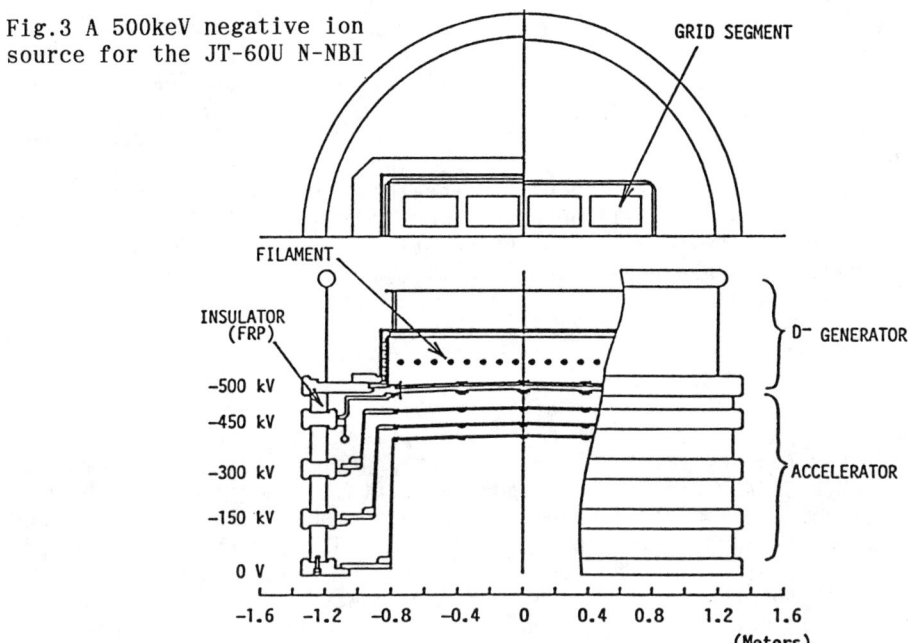

Fig.3 A 500keV negative ion source for the JT-60U N-NBI

Performances of the ion source are determined on the basis of the achieved values in our recent experiments on the volume production type ion source. The Multi-Ampere Source at JAERI has achieved the H^- current density of about 25mA/cm^2 at the plasma boundary, where the source pressure is 2.1Pa.[2] However, the design value of the H^- current density at the plasma boundary is determined to be a moderate value of 12mA/cm^2 at a lower source pressure of 1.2Pa, so as to reduce the electron current and the stripping loss of the ions in the accelerator. This also improve the voltage holding in the acceleration gaps. On the other hand, the D^- current density is expected to be 80% of the H^- current density.[6] Hence, the design value of the D^- current density is determined to be 9mA/cm^2 at a source pressure of 1.2Pa. This value requires the ion extraction area of 52cm x 150cm to produce a 500keV, 11A deuterium ion beam. Utilization of cesium vapor is not considered since the degradation of the voltage holding in the accelerator due to the cesium contamination is not made clear yet. The electron beam accompanied by the negative ions can be suppressed effectively by biasing the plasma grid with respect to the anode.[6] Therefore, the long pulse operation is possible with conventional cooling channels in the extraction grid, though the ion current density decreases by about 20%. Each grid in the extractor/accelerator is composed of eight segments, which are oriented to the center of the injection port. Each segment has about 180 apertures whose diameter is 14mm. The ion beams extracted at the voltage of 50kV are accelerated up to 500keV in the three-stage accelerator. The beam energy can be changed from 200keV to 500keV by switching on or off the applied voltage to each acceleration stage. In order to restrain the electrons produced in the acceleration gaps from leaking into the downstream stage, permanent magnets are mounted on each grid in such a manner that the transverse magnetic field can be produced across each aperture. This improve the voltage holding and reduce the heat loading in the acceleration grids. Since the ion backstream from the neutralizer plasma is small, the ion suppressor grid is eliminated. The insulator of the accelerating column is composed of three voidless FRP cylinders whose outer diameter is 2.5meters. A full scale cylinder whose stand-off voltage is 200kV has been ordered. We are preparing to manufacture a full scale prototype 500keV ion source in 1991.

(2)FER-NBI

A new design concept of the negative-ion-based NBI system has been proposed early in 1986,[7,8] and applied to the design study of the FER in FY1986.[9] The NBI system injects deuterium beams of 500keV, 50MW through two tangential ports, on each of which one beamline is mounted. Though the required beam energy for plasma heating is 500keV, the energy can be raised up to 1MeV for current drive experiments. The design has been updated reflecting the R&D results on the negative ion sources. However, the basic design concept is not changed compared to the first proposal. In order to make the NBI system more attractive, we are considering three options. In the first option, the source operating pressure is decreased down to 0.5Pa from 1.2Pa of the standard design value, while the current density is fixed to be 25mA/cm^2 at the plasma emitter. This naturally reduce the stripping loss in the accelerator and also reduce the gas flow rate. The higher current density imposes higher electric field on the acceleration gaps in order to obtain good beam optics. We aim at lower operating pressure rather than higher current density. The cesium-introduced Multi-Ampere Source has a possibility to satisfy this first option, though a bakable accelerator must be developed and the activated cesium must be managed. The second option is to apply a plasma neutralizer instead of the conventional gas neutralizer. Needless to say, the plasma neutralizer improves the system efficiency remarkably if highly-charged ions can be produced efficiently in the plasma cell. The third option is to utilize the soft landing beam dump[10] whose concept is similar to the energy recovery system. The soft landing system will reduce the heat flux on the beam dump surface and surface damage such as blistering. We continue to investigate the plasma neutralizer and the soft landing beam dump as well as the low pressure negative ion source.

(3)Lithium Beam Diagnostic System for JT-60U

The total heating power in the JT-60U amounts to 60MW including the neutral beam injection power by the above-mentioned negative-ion-based NBI system. Such a high heating power increases the high energy ions like alpha particles produced by the D-^{3}He reaction. Since the behaviour of these ions influences the global energy confinement time, the diagnosis of these high energy ions is important in understanding the behaviour of these ions.[11] This leads us to develop an active beam diagnostic system using high energy lithium ion beams. In order to make clear the R&D target, we are now designing a system which can inject 2MeV, 20mA lithium neutral beams into the JT-60U.

In order to realize this system, a 100mA lithium negative ion beam must be extracted from the extraction area of about 100cm^2. We have just started testing a lithium ion source whose structure is similar to the volume production type negative hydrogen ion source.

R&D Plan of Negative Ion Beams

We have three plans about the negative-ion-based neutral beam injection system as mentioned above; the JT-60U N-NBI, the FER-NBI and the lithium beam diagnostic system for the JT-60U. Since the completion of the JT-60U N-NBI is scheduled to be in 1993, we are making efforts to develop it, particularly the 500keV negative ion accelerator. The total ion beam current obtained experimentally at JAERI is almost one third of the required value for the JT-60U. Additionally, a 200keV, 3.5A helium beam diagnostic system has already been developed for the JT-60.[12] Hence, it will no longer be difficult to construct a 200keV x MW class negative-ion-based NBI system(See Fig.4). In order to increase the beam energy, we started to construct a 350kV x 100mA test stand(ITS-2a), where the beam optics and voltage holding characteristics in a high energy accelerator will be investigated. On the basis of the experiences in both the ITS-2M and ITS-2a test stands, we plan to construct a 500keV, 5A test stand at the facility of the JT-60 prototype injector unit(PBETF), where the full scale prototype ion source for the JT-60U will be tested. In order to develop the ion source for the FER, we plan to construct the Advanced Injector Test Stand(AITS) in 1994, where 500keV/10A or 1MeV/5A negative ion beams can be accelerated. On the other hand, we continue to study the plasma neutralizer and ion beam soft landing system in the ITS-2M test stand. The R&D schedule and future plans are shown in Fig.5. Development of the JT-60U N-NBI to obtain good plasma current drive in the JT-60U is an appropriate and necessary step towards the next fusion machines.

Conclusions

1. We will develop the high energy NBI system in two steps. In the first step, a 500keV NBI system for the JT-60U will be developed in 1993. On the basis of the experiences obtained in the first step, a reactor relevant 1MeV NBI system will be developed for the FER.
2. We have started to develop a lithium negative ion source which can be utilized for the active beam diagnostic system for the JT-60U.

3. In order to develop a high energy accelerator which can deliver a convergent negative hydrogen beam stably, a 350keV, 0.1A test stand(ITS-2a) is now under construction.
4. In order to confirm the performances of the 500keV ion source for the JT-60U, a 500keV, 5A test stand(PBETF) will be constructed. An 1MeV, 5A test stand(AITS) is planned to develop an 1MeV accelerator for the FER.

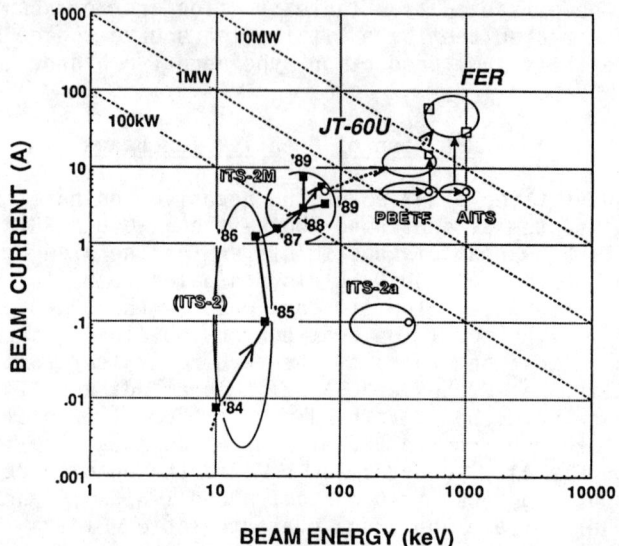

Fig.4 Increase of ion beam current and energy to the JT-60U and the FER

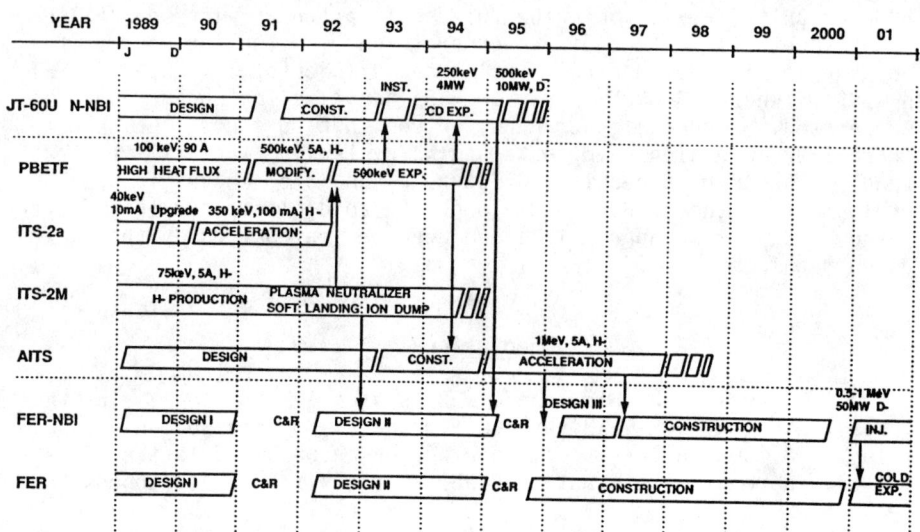

Fig.5 Long range plan of the negative ion source development at JAERI

Acknowledgement

The author would like to thank Dr. M. Seki, Dr. M. Kuriyama, Dr. S. Tanaka, Dr. Y. Okumura and other members of the NBI group for their valuable discussions and comments. He is also grateful to Dr. T. Kunieda, Dr. S. Matsuda, Dr. S. Shimamoto and Dr. M. Tanaka for their continuous encouragement.

References

[1] M.Bacal and G.W.Hamilton, "H$^-$ and D$^-$ production in plasmas," Phys. Rev. Lett. 42, 1538 (1979).
[2] M.Hanada, et al., "A 14cm x 36cm volume negative ion source producing multi-ampere H$^-$ ion beams" presented at the Int. Conf. on Ion Sources, Berkeley, June 10-14(1989)C50.
[3] Y.Okumura, et al.: "Cesium mixing in the multi-ampere volume H$^-$ ion source" presented at the 5th Int. Symp. on the Production and Neutralization of Negative Ions and Beams, Brookhaven, Oct.30-Nov.3. 1989.
[4] Y.Ohara, et al.: "Recent activities of negative ion beams at JAERI" presented at the 13th Symp. on Fusion Engineering, Knoxville, Oct.2-6, 1989, 08-O-05
[5] JT-60 Team presented by H.Kishimoto, "Recent progress in JT-60 experiments" in Proceedings of the 12th Int. Conf. on Plasma Physics and Controlled Nuclear Fusion Research, Nice, Oct 12-19 1988, IAEA-CN-50/A-1-4.
[6] T.Inoue, et al., "Comparison of H$^-$ and D$^-$ production in a magnetically filtered multicusp source" presented at the Int. Conf. on Ion Sources, Berkeley, June 10-14(1989)C49.
[7] H.Horiike, et al., "Conceptual design of negative-ion-based 500keV 20MW neutral beam injector," Japan Atomic Energy Research Institute Report JAERI-M86-064 (1986)[in Japanese]/UCRL-Trans-12153.
[8] Y.Ohara, et al., "Design of a 500keV 20MW negative-ion-based neutral beam injection system for fusion experimental reactor," in Proceedings of the 12th Symp. on Fusion Engineering, Monterey, Oct.12-16, 1987, pp.298-301.
[9] S.Yamamoto, et al., "Tokamak reactor operation scenario based on plasma heating and current drive by negative-ion-based neutral beam injector" in Proceedings of the 11th Int. Conf. on Plasma Physics and Controlled Nuclear Fusion Research, Kyoto, Nov.13-20, 1986, IAEA-CN-47/H-I-3.
[10] M.Araki, et al.: "Design study of a beam energy recovery system for a negative-ion-based neutral beam injector" To be submitted for publication in Fusion Technology.
[11] A.S.Schlachter and W.S.Cooper, "Proposed neutral-beam diagnostics for fast confined alpha particles in a burning plasma" in Proceedings of the 4th Int. Symp. on the Production and Neutralization of Negative Ions and Beams, Brookhaven, Nov. 1986, pp.727-738.
[12] T.Itoh, et al., "Development of 200keV 3.5A helium beam injector" in Proceedings of the 4th Int. Symp. on Heating in Toroidal Plasmas, Rome, March 21-28, 1984, pp.1081-1086.

DISCUSSION

Hopman: It was a pleasant surprise for me to see the source on the ground and the neutralizer at high voltage, what are the arguments for the FER or the JT60 upgrade injector to have the source at high voltage, and the neutralizes at ground.

Ohara: We utilize the existing plasma neutralizer, I think this is very difficult to utilize a system with the source at ground since it is very hard to suppress power surges.

Jacquot: For the ITS test stand that you concentrated on for the moment, what type of power supply did you use, what kind of power supply: 350 kilowatt 0.1 Ω.

Ohara: Commercial dc power supply.

Pamela: You didn't tell us about using cesium or not in the sources for JT60 upgrade.

Ohara: The present design does not use cesium. But there is another possibility, with the use of cesium the number of ion sources can be decreased. About two sources maybe possibly inject in the 10 MW. But we have no experience with the voltage holding characteristic when we have some cesium in the accelerator.

Belchenko: What kind of pumping are you going to use in the dc neutral beam line?

Ohara: We plan to use three cryopanels. Two cryopanels in operation and other one panels is regenerated so we can prepare part of the cryo system steady state.

Semashko: What do you think about the efficiency of a soft landing beam dump.

Ohara; I think that it is a very important technology to make the system compact. We inject high power through a limited area. We have to increase the heat flux of the ion beam in this case. It is quite difficult to handle the ion beams so in this case, the soft landing system would be really important.

Pamela: I noticed that in your FER design there is an important change compared to the previous design since now you have a shorter neutralizer in the beam dump about 30 meters away from the tokamak, and in the soft landing system you need a magnetic field. In the previous design, you are using the fringe field from the Tokamak, what about this new design?

Ohara: We are now designing the soft landing system. Magnetic field would be produced by a pulse when we utilize the soft landing system in the present design. In the former design, the beam dump needs to be located in the reactor room and its very difficult to maintain the ion beam dump. So we removed the beam dump outside the reactor room.

Bacal: I wanted to ask about the titanium pumps which we saw in the design of the lithium injector, what is the philosophy of titanium pumps.

Ohara: Hydrogen gas would be utilized for neutralization of lithium negative ions. Titanium pumps can evacuate the hydrogen at high pumping speed. But, argon cannot be pumped by titanium pumps and nitrogen gas cannot be utilized because voltage holding in the accelerated may be degraded.

Concluding Remarks

Ron Hemsworth

I find it an honor to be asked to actually summarize this Conference. As it is the last talk of the Conference, I will try to keep it on a reasonable time scale, I'm sure you are all anxious to get away for the weekend and then perhaps to get back to work to apply the ideas that must have been inspired by some of the talks at the Conference. By a quick count of the number of participants, there are over a hundred active participants in this Conference, and they come from something like ten countries or more and major contingents coming from Japan, Europe, and USSR, as well as the home team from the USA. This I think shows the overall importance that negative ions have in our world at the moment with a total of forty four oral presentations, more than twenty five posters, and the long panel discussion of last night. In my summary I will not try to go through all those papers and try to summarize them: I just made a selection of what I consider the more important aspects. I apologize to those people whose work I'm sure is good and interesting but I don't have time to mention. I would like to start actually by making a complaint. The main reason why negative ions have such an importance is because the next generation of fusion machines want to use high energy neutral beams; and high energy implies using negative ions. But in fusion devices, we intend to use deuterium and tritium as the partners, and that implies deuterium beams and not hydrogen beams so it comes to me as a surprise to see people keep talking about hydrogen all the time. I would hope that next time we come back to this Conference, in three years time, that people have learned that D is much more important than H and address that problem. I'll also make this comment several times throughout my talk. It is interesting to observe that both surface sources and volume sources still seem to be in contention as to providers of our basic negative ion. We started the Conference with surface sources, had two rather nice papers by Aart Kleyn and Ron van Os, and they both presented this viewgraph which shows the model Ron van Os has applied in his thesis to explain H^- formation of surfaces where this shows the potential energy for the hydrogen atom. When the atom is a long way from the metal surface, the electron in an H^- would sit at this level which is the electron affinity. When the system is close to a metallic surface the image forces pushed this down, and if we have a low work function surface such as the metal shown here, the electron has the possibility to sit either on the H atom or to remain in the conduction band. And if you manage to hit this H^- while the electron is resonant around the atom, then you can knock it off and have your H^-; this model was quite successfully applied to the data that Kleyn showed and Ron van Os showed. What I could not get out of this, is there an isotope effect? Does deuterium look different from hydrogen? Aart Kleyn talked exclusively about hydrogen and the only D^- that I could find in the two talks was that Ron van Os said that they had achieved 20 milliamps per square centimeter with D^- using a barium surface. I couldn't find if there was any comparison within the same device to see if there was an isotope effect. I did talk to Yuri Belchenko who gave the talk about his negative ion beams including 1.1 amps of H^-. He has told me that when he operates deuterium for long pulses, he actually only gets half that value; so contrary to your comment Henk (Hopman), there seems to be a strong isotope effect at least in some sources. There is obviously a lot of work to do there and again I suggest that had you done the work with the right isotope in the first place, perhaps you wouldn't need to worry too much. Now, there have been quite a few papers on modeling of volume sources, and here we have a much more complicated situation than that of surface sources. I've stolen this viewgraph from Darryl Skinner and it shows the tremendous, complex system of equations one has to look at in order to model these volume sources, it shows only some of the equations. My favorite comment again, is that D doesn't appear on here. We all know, I think, that the main mechanism which is proposed for the production of H^- in volume sources starts with the creation of vibrationally excited hydrogen and ends with dissociated attachment to form the H^-. Now in these models, I was interested in the comments from Dr. Capitelli and his paper. He put up this viewgraph again, hydrogen dominated, which lists the reactions which are important in creating vibrational excited hydrogen. Most of the data he presented in his talk was calculation of cross sections for the dissociation of hydrogen. He used the Gryzinski method to normalize experimental measurements for dissociation cross section and ended up with a value which was two times that which had been previously used. In doing the same analysis for vibrational excitation, he ended up with only half the value that had been previously calculated by John Hiskes who had fitted the data to the Abnecio calculations. What I didn't find, which is what an experimentalist is always looking for, is the sensitivity of the model to this factor of two change in these two cross sections. The question was actually asked during a panel session, and we did not get a satisfactory answer there either. I would hope that when we come back next time, we get some answers on this sensitivity analysis. However, all is not doom and gloom. We have had some nice examples of improvements in our understanding of the modeling, and I found this particular new piece from John (Hiskes) on the creation of vibrationally excited hydrogen by the neutralization of H_2^+ and H_3^+ on the surface. In John's calculation over a range of a tenth of an eV to a hundredth of an eV in energy, you actually end up with 80% of the particles coming off being vibrationally excited. This seems to be clearly a source of vibrationally excited hydrogen which is not included in the models and hopefully will need some improvement, if once we do get it in. Another part which I found rather nice was the paper by Dr. Wadehra. He presented this viewgraph showing the excitation cross section from $v = 0$ to $v = 4$ where, by having a much better computer system, was able to calculate many more points. He has managed to show this structure in the cross sections which is in qualitative agreement with the experiment, and as the models rely heavily on theoretical calculations of the cross sections, this sort of agreement is rather important. The question was asked at the Panel Session last night about the isotope effects. There was at least one paper again by Dr. Wahedra where the isotope effect was mentioned and this viewgraph shows his calculations for the cross section as a function of vibrational state for reactions between electron and a hydrogen isotope H, D, or T. What he has done is to show that the cross section scaling is variant as the mass of the isotope to the $-v/2$ power, v is the vibrational quantum number, i.e., it has quite a strong isotope effect that needs to be put into some of your modeling. The theoreticians seem to be doing quite a good job in some areas at least in improving our knowledge of cross sections for modeling of volume sources. However, I feel that there is scope for some improvement in experimental data which also can be put into these models. Richard Hall had done a study of available data in available data in the literature for recombination coefficients as a function of the Debye temperature. What I wanted to pick out was the variations found experimentally. For example, the recombination of nickel has almost a factor of 3 variation in experimental data. It's clearly a whole area for experimentalists to work in to give good data to the modelers. There have been at this Conference, a very large number of papers with good diagnostics of the systems which will enable us to have a better understanding. I'm going to select one particular set. There are many other examples here that I could have chosen, those very nice laser photo-detachment measurement systems that the group under Marthe Bacal has been using for measurement of H^- temperature, and emission diagnostics for H atom densities in the source. The most impressive data was that of the VUV absorption spectroscopy which yielded curves like this for H atom density in the LBL source and this sets a target for the theoreticians and modelers to try to reproduce data like this showing the atomic densities as a function of arc current, and to produce a whole series of different parameters scans. The data which seem to cause the most comment and concern was the one that did not show any enhancement of upper vibrational levels, everyone knows what this is, I think. I would like to move away from the production to talk about the beam systems. One thing that I noticed at this Conference is that there was not very much on beams. The only new beam data, which is the single most outstanding result of the Conference, was the one from the JAERI group where they introduced cesium into their large volume source and have this very impressive 7.8 amps of extracted H^-. It's unfortunately H^-, isn't it? No one is really interested in H^-, we want D^- but it is still impressive. Although we managed to extract and accelerate these beams, the paper by John Whealton exposed some of our ignorance on this, this is taken from John's talk. Here he's plotted the normalized extracted current against the normalized voltage, and the data shown here is the experimental data from Culham and these are John's calculations. In this region we see that there are significant discrepancies. Obviously there is still quite a lack of understanding, particularly about negative ions. We get good fits to positive ions and I know I won't make any comment on that because I think it's quite a controversial subject. Again, what we want to produce ultimately are neutral beams and I found calculations from the data taken by Dr. Salzborn, and again picked up by Ady (Hershcovitch) to be very important. Dr. Salzborn mentioned his neutralization cross sections for H^- on highly charged argon. Here is the calculation from the last Brookhaven Conference for neutralization, one can achieve for various charge states with the assumption that the cross section varied as q squared. However, the measured values are varing as q to the 1.3. This set of curves, with these sort of peaks on them, are encouraging in that it still shows we have the high efficiency but it also shows that it's going to be much more difficult to make such a plasma neutralizer. Maybe that's over emphasizing it but it isn't easy to start with, and the last thing we really wanted was to make it more difficult. Now for the final part, I would like to look at things which have been proposed for the future. There are so many small systems, and people have plans for different models and

Future Programmes

EC

$$\left.\begin{array}{ll} 4A & D^- & 200\,kV \\ 2A & D^o \end{array}\right\} Culham$$

$$\left.\begin{array}{l} 1\,MV\;\; 4A\;\; D^-\;\; Test \\ Stand\;\; Feasability\;\; Study \end{array}\right\} Cadarache$$

→ End 1991   Then?

USSR

$$\left.\begin{array}{l} 3A \left\{\begin{array}{c} D^- \\ {}_{||}\,? \\ H^- \end{array}\right\} \to 1991 \\ 10A \hspace{3em} \to 1994 \end{array} \quad 1.5\,MV \right\} N'vsobirsk$$

USA

$$\left.\begin{array}{lll} 300\,kV & 60\,MW & CTC \\ 400\,kV & 150\,MW & TPR \end{array}\right\} \begin{array}{c} 50/50 \\ D^o/T^o \end{array}$$

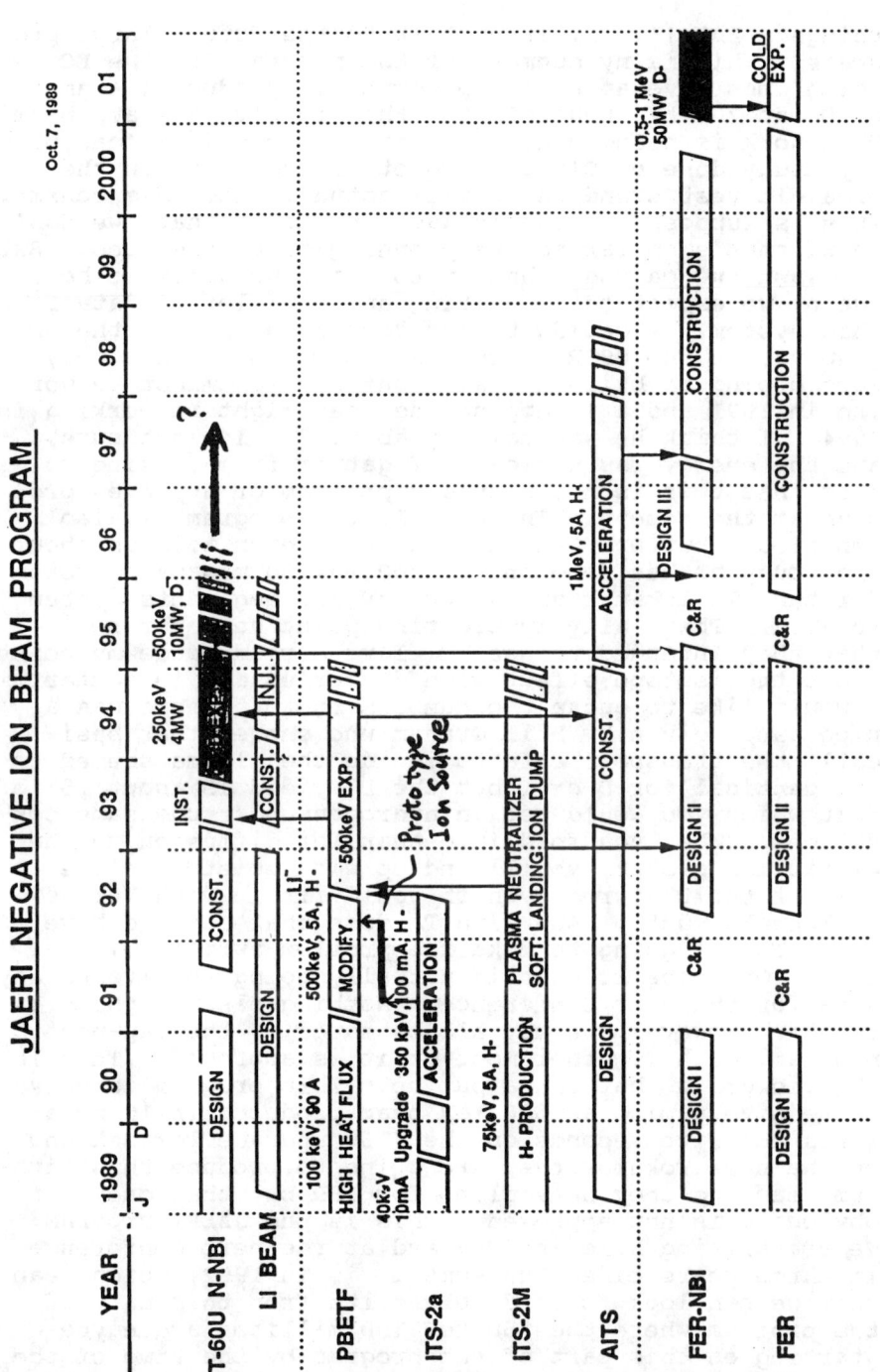

things. But I'll limit my talk to the future large programs. This is my summary of the program for the EC which does have an active program for producing 4 amps of D^- at 200 kV, neutralizing that to give a 2 amp beam. That work is being done at Culham. There is a feasibility study done by Claude Jacquot of Cadarache on the 1 megavolt test stand which will actually use this scheme. This is supposed to end in 1991 and after that, we don't know, that's as far as the planning in Europe goes. But at least one can say that by the time we assemble here again, we expect to be getting an awful lot of data from this system. I couldn't find very much on what the program was in the USSR. Professor Dimov said that they were hoping to build a 3 amp source which might be working in 1991 and a 10 amp source that might be working in 1994. I think he was talking about H^-, I'm not sure, and the energy was unclear. I gather from talking to them that this is not a funded program or approved program at the moment. In the USA, the program is also ambitious, but Dr. Grisham from Princeton told us about the study he has done on the 300 keV 60 megawatt system for the CTC Tokamak and a 400 keV 150 megawatts system for TPR. The really interesting point here for me is that they intended to use T. I've already made my point about the isotope effect with deuterium and in deuterium I would like to quote the numbers there. There was a nice paper given by Bill Graham who showed that basically the transport effects inside the plasma seemed to be identical for D or T but the D^- yield is about 75% of that which you would get in hydrogen under the same conditions. When you fold into that the diffusion to the extraction system, you'll end up with something like half of the D^- current which would fit in with Yuri's (Belchenko) data. And with T, does that mean we have 1/3? That's going to make the problem awfully difficult. Only the EC part is actually going to give us any data for the next Conference. Nothing else on there is funded. However, we may all be saved by the Japanese program which I gather most of it is approved. This is the viewgraph that Ohara put up. This program is very impressive aiming at 500 keV beam produced in 10 megawatts for five seconds on the JT60 upgrade Tokamak and on the same Tokamak they are going to produce this lithium beam and they have plans for FER but that one obviously is not approved. This is the JAERI program: We are sitting here in 1989 and at the next Conference in three years time, I assume it is in 1992, which means that we can look forward to results from this part of the program where the 350 keV 100 milliamp are maybe starting on this part of the program by the time of the next meeting. So at the next meeting on big beam systems, we should have some data from the Japanese program and the European program. I would like to summarize that the standard of papers presented at this meeting and the standard of presentation has been extremely good, with the exception that I keep complaining about that you work with the wrong isotope. I hope that we will proceed to the same sort of level at the next Conference. For the objectives, I hope that goes without saying, that we want better sources by the next Conference and better models of them. We expect to have 4 amp of D^- working at Culham, and the 100 milliamp 350 keV working at Japan and be on our way to the 5 amp system. As a final comment, I would like to congratulate BNL for organizing this very nice, superb Conference in pleasant surroundings. Thank you.

APPENDICES

APPENDIX I: LIST OF PARTICIPANTS

Dr. James G. Alessi
AGS Department
Bldg. 930
Brookhaven National Laboratory
Upton, NY 11973

Dr. Paul Allison
Los Alamos National Laboratory
AT-2
MS H818
P.O. Box 1663
Los Alamos, NM 87545

Dr. Gerald D. Alton
Physics Division
P.O. Box 2008
Oak Ridge National Laboratory
Oak Ridge, TN 37831-6368

Dr. Ignacio Alvarez
Instituto de Fisica,
UNAM
Laboratorio de Cuernavaca
Apdo. Postal 139-B
62191
Cuernavaca, Mor.
MEXICO

Dr. Oscar A. Anderson
Lawrence Berkeley Laboratory
MS 4-230
1 Cyclotron Road
Berkeley, CA 94720

Dr. Akira Ando
National Institute for Fusion Science
Nagoya, 464-01
JAPAN

Dr. Marthe Bacal
Ecole Polytechnique
Laboratoire de Physique
  des Millieu Ionises
91128 Palaiseau Cedex
FRANCE

Dr. Yuri I. Belchenko
Institute of Nuclear Physics
Sibirian Branch of Academy
  of Sc. USSR
Prospekt Labrentieba 11
630090 Novosibirsk-90
USSR

Dr. John Benjamin
Physics Department
Bldg. 901A
Brookhaven National Laboratory
Upton, NY 11973

Dr. Jack Bretagne
Laboratoire de Physique des Gaz
  et des Plasmas
Universite de Paris-Sud
91405 Orsay Cedex
FRANCE

Dr. Anne-Marie Bruneteau
Ecole Polytechnique
Laboratories de Physique
  des Milieux Ionises
91128 Palaiseau Cedex
FRANCE

Dr. Jack Bruneteau
Ecole Polytechnique
Laboratories de Physique
  des Milieux Ionises
91128 Palaiseau Cedex
FRANCE

Dr. Mario Cacciatore
Department of Chemistry
CNR University of BARI
Via Amendolia 173
70176
Bari
ITALY

Dr. Mario Capitelli
Department of Chemistry
University of Bari
Via Amendola 173
70126 Bari
ITALY

Dr. Gunther Clausnitzer
University of Giessen
Strahlenzentrum
WEST GERMANY

Dr. Emile Conard
IBA
Chemin Du Cyclotron 2
B-1348
Louvain Le Neuve
BELGIUM

Dr. Stephen Cox
Culham Laboratory
AEA Technology
Abingdon, Oxford
ENGLAND

Dr. Basil DeVito
AGS Department
Brookhaven National Laboratory
Building 911A
Upton, NY 11973

Dr. Pascal Devynck
Ecole Polytechnique
Laboratoire de Physique
  des Millieu Ionises
91128 Palaiseau Cedex
FRANCE

Dr. Gennadii I. Dimov
Institute of Nuclear Physics
Sibirian Branch of Acad.
  of Sc.
Prospekt Lavrentieba 11
63090 Novosibirsk-90
USSR

Dr. Dieter Eckhartt
Max Planck Inst. fur Plasma Physik
Boltzmannstrase 2
D-8046 Garching bei Munchen
WEST GERMANY

Dr. Leonid Elizarov
Kurchatov Institute of Atomic Energy
Ulitza Kirchatova
123 182 Moscow
USSR

Dr. Joel H. Fink
Negion, Inc.
4023 East Avenue
Hayward, CA 94542

Dr. Amnon Fisher
Department of Physics
Univ. of California at Irvine
Irvine, CA 92717

Dr. William G. Graham
Physics Department
Queens University
Belfast BT 71 NN
NORTHERN IRELAND

Dr. Larry R. Grisham
Plasma Physics Laboratory
Princeton University
P.O. Box 451
Princeton, NJ 08540-2202

Dr. Richard Hall
Laboratoire de Dynamique
  Moleculuire et Atomique
Univ. Pierre et Marie Curie
4 Place Jussieu T12-E5
75252 Paris Cedex 05
FRANCE

Dr. Ron S. Hemsworth
JET Joint Undertaking
Abingdon OX142PF
GREAT BRITAIN

Dr. Ady Hershcovitch
AGS Department
Building 911B
Brookhaven National Laboratory
Upton, NY 11973

Dr. John R. Hiskes
Lawrence Livermore Laboratory
P.O. Box 5511
MS L-630
Livermore, CA 94550

Dr. Andrew J. Holmes
Culham Laboratory
Applied Physics & Tech. Div.
Abingdon, Oxfordshire
OX14 3DB
ENGLAND

Dr. Henk J. Hopman
The NET Team, IPP
Max-Planck-Institut fur Plasma
Boltzmannstr. 2
D8046 Garching bei Munchen
WEST GERMANY

Dr. Claude Jacquot
DRFC/STIF C.E.N. de Cadarache
B.P. 1
13115 St. Paul Lez Durance
FRANCE

Dr. Brant Johnson
BNL-DAS
Bldg. 815
Brookhaven National Laboratory
Upton, NY 11973

Dr. Keith Jones
BNL-DAS
Bldg. 815
Brookhaven National Laboratory
Upton, NY 11973

Dr. Yves Jongen
Ion Beam Applications
Chemin Du Cyclotron 2
B-1348
Louvain La Neuve
BELGIUM

Dr. Arnold M. Karo
Dept. of Chem. & Materials Sci.
L-325
Lawrence Livermore Laboratory
P.O. Box 808
Livermore, CA 94550

Dr. Robert King
Culham Laboratory
Abingdon
Oxon, OX14 3DB
ENGLAND

Dr. Aart W. Kleyn
FOM-Instituut voor Atoom-
 en Molecuulfysica
Kruislaan 407
1098 SJ Amsterdam
Watergraafsmeer
THE NETHERLANDS

Ms. Maryanne Kmit
AGS Department
Bldg. 911B
Brookhaven National Laboratory
Upton, NY 11973

Dr. Ahovi Kponou
AGS Department
Bldg. 911B
Brookhaven National Laboratory
Upton, NY 11973

Dr. Tsutomu Kuroda
National Institute for Fusion
 Science
Furo-cho, Chikusa-ku
Nagoya
Nagoya 464
JAPAN

Dr. Ronald Lankshear
Bldg. 902C
Brookhaven National Laboratory
Upton, NY 11973

Dr. Stephen Laycock
Ion Beam Applications
Chemin Du Cyclotron, 2
B-1348 Louvain-la-Neuve
BELGIUM

Dr. S.W. Lee
Dept. of Physics
Engineering Physics
Stevens Institute of Tech.
Castle Point Station
Hoboken, NJ 07030

Dr. Leslie M. Lea
Culham Laboratory
Abingdon Oxfordshire
Ox14 3DB
ENGLAND

Dr. Dimitrios Lianos
7515 Clubfield Drive SW
Huntsville, AL 35807

Dr. Derek Lowenstein
AGS Department
Brookhaven National Laboratory
Bldg. 911B
Upton, NY 11973

Dr. Ronald Martin
Argonne National Laboratory
9700 South Cass Avenue
Bldg. 207
Argonne, IL 60439

Dr. Roy McAdams
Culham Laboratory
Abindgon, OXON
OX14 3DB
GREAT BRITAIN

Dr. Steven Melnychuk
Stevens Institute of Technology
Physics Department
Castle Point Station
Hoboken, NJ 07030

Dr. Charlie Meitzler
TAC/SHSU
2319 Timberlock Drive
The Woodlands, TX 77380

Dr. Mati Meron
DAS Department
Brookhaven National Laboratory
Bldg. 815
Upton, NY 11973

Dr. H. Harvey Michels
United Technologies Research
Applied Physics Dept.
MS 92
East Hartford, CN 06108

Dr. Yoshiharu Mori
KEK
National Laboratory for HEP
Oho 1-1
Tsukuba-Shi
Ibaraki-Ken 305
JAPAN

Dr. Robert Morris
U.S. Air Force
Geophysics Laboratory
Hanscom AFB, MA 01731

Dr. Kenneth Moses
JAYCOR
Plasma Tech. Div.
3547 Voyager Street
Suite 104
Torrance, CA 90503-1667

Dr. Alwyn A. Mullan
University of Ulster
Physics Department
Coleraine, BT5215A
NORTHERN IRELAND

Dr. Masatada Ogasawara
Faculty of Science and Technology
KEIO University
Yokohama
JAPAN 223

Dr. Yoshihiro Ohara
Plasma Heating Laboratory 1
Japan Atomic Energy Res. Instit.
Naka-Machi, Naka-Gun
Ibaraki-Ken
311-01
JAPAN

Dr. Yoshikazu Okumura
Japan Atomic Res. Instit.
Naka-Machi, Naka-Gun
Ibaraki-Ken
311-01
JAPAN

Dr. John Orthel
GHGA
336 Paseo Pacifica
Encinitas, CA 92024

Dr. Jerome Pamela
DRFC
C.E.N de Cedarache
13108 St. Paul-Lez-Durance
FRANCE

Dr. John Phillips
U.S. Army Strategic Defense
P.O. Box 1500
CSSD-H-VE
Huntsville, AL 35807-3801

Dr. Krsto Prelec
AGS Department
Brookhaven National Laboratory
Bldg. 911B
Upton, NY 11973

Dr. Antonio Raino
Dipartimento Di Fisica
University de Bari
Bari
Italy

Dr. Deepak Raparia
Texas Accelerator Center
4802 Research Forest Dr.
Bldg. #2
The Woodlands, TX 77381

Dr. Thomas G. Roberts
Physical Dynamics Corporation
2815 Bently Street, SE
Huntsville, AL 35801

Dr. Erhard Salzborn
Institut fur Kernphysik
Universitat Giessen
Leihgesterner WEG 217
D-6300 Giessen
WEST GERMANY

Dr. Charles Schmidt
MS-307
Fermilab
Batavia, IL 60510

Dr. Paul W. Schmor
TRIUMF
4004 Wesbrook Mall
University of British Columbia
Vancouver, BC   V6T 2A3
CANADA

Dr. Milos Seidl
Dept. of Physics
Engineering Physics
Stevens Institute of Tech.
Castle Point Station
Hoboken, NJ   07030

Dr. Nikolai Semashko
I.V. Kurchatov Institute
   of Atomic Energy
Ploshad Kurchatova 46
Moscow 123182
USSR

Dr. Joseph Sherman
MS H-818, AT-2
Los Alamos National Laboratory
P.O. Box 1663
Los Alamos, NM 87545

Dr. Darryl A. Skinner
Ecole Polytechnique
Laboratory PMI
91128 Paliseau Cedex
FRANCE

Dr. Theo Sluyters
AGS Department
Bldg. 911B
Brookhaven National Laboratory
Upton, NY 11973

Dr. H. Vernon Smith
AT-2, MS-H818
Los Alamos National Laboratory
P.O. Box 1663
Los Alamos, NM 87545

Dr. C. Lewis Snead
DNE Department
Brookhaven National Laboratory
Bldg. 902C
Upton, NY 11973

Dr. Eng-Kie Souw
Brookhaven National Laboratory
Bldg. 701
Upton, NY 11973

Dr. Raul Stern
Physics Department
University of Colorado
Boulder, CO 80309

Dr. Anthony Taylor
Argonne National Laboratory
9700 South Cass Avenue
Bldg. 207
Argonne, IL 60439

Dr. Matthew F. Thornton
Culham Laboratory
Abingdon
Oxon OX14 EDB
ENGLAND

Dr. Alan E. Todd
Grumman Space Systems
Princeton Corporate Center
4 Independence Way
Princeton, NJ 08540-6620

Dr. Vincenzo Valentino
I.N.F.N. Dipartimento Di Fisica
Universita Di Bari
Bari
ITALY

Dr. Ron van Os
Lawrence Berkeley Laboratory
1 Cyclotron Road
MS 4/230
Berkeley, CA 04720

Dr. Peter Vanier
Brookhaven National Laboratory
Bldg. 902C
Upton, NY 11973

Dr. U. von Wimmersperg
Brookhaven National Laboratory
Bldg. 902C
Upton, NY 11973

Dr. Jogindra Wadehra
Department for Physics/Astronomy
Wayne State University
Detroit, MI 48202

Dr. John H. Whealton
Bldg. 9201-2
Oak Ridge National Laboratory
P.O. Box X
Oak Ridge, TN 37831-8071

Dr. Richard L. Witkover
AGS Department
Bldg. 911B
Brookhaven National Laboratory
Upton, NY 11973

Dr. Rob L. York
Los Alamos National Laboratory
MP-DO, Mail Stop H823
Los Alamos, NM 87545

Dr. Anthony T. Young
Lawrence Berkeley Laboratory
MS 4/230
Berkeley, CA 94720

Dr. F.B. Yousif
Department of Physics
The University of
   of Western Ontario
London, Ontario
CANADA N6A 3K7

Dr. Dick Yuan
4004 Wesbrook Mall
TRIUMF VBC
BC, V6T 2A3
CANADA

Dr. Anatoli Zelenski
Institute for Nuclear Research
Moscow, USSR

Dr. Martin Zucker
Nuclear Energy Dept.
Bldg. 902C
Brookhaven National Laboratory
Upton, NY 11973

APPENDIX II: LIST OF SESSIONS

FIFTH INTERNATIONAL SYMPOSIUM ON THE PRODUCTION
AND NEUTRALIZATION OF NEGATIVE IONS AND BEAMS
BROOKHAVEN NATIONAL LABORATORY
BROOKHAVEN CENTER

OCTOBER 30 - NOVEMBER 3, 1989

MONDAY, OCTOBER 30, 1989

9:00 a.m.   Opening Remarks
D. Lowenstein, Chairman, AGS Department
A. Hershcovitch, Symposium Chairman

Session I
Surface Production H^-/D^- Sources and their Fundamental Processes
K. Prelec Presiding

9:20 a.m.   Surface Processes in the Production of Negative Hydrogen Ions
A.W. Kleyn (FOM - Institute) (40)

10:00 a.m. H^- Formation from Non-Cesiated Converter-Type Negative Ion Sources
C.F.A. van Os, K.N. Leung, A.T. Lietzke, J.W. Stearns, and W.B. Kunkel (LBL) (25)

10:25 a.m.   BREAK

10:55 a.m.   Surface Production of Negative Hydrogen Ions by Reflection of Hydrogen Atoms From Low Work Function Surfaces
M. Seidl, S.T. Melnychuk, W.E. Carr, A.E. Souzis, and J. Isenberg (Stevens Institute of Technology) (20)

11:15 a.m.   Long Pulsed SPS with Geometric Focusing
Yu. Belchenko (Institute of Nuclear Physics, Novosibirsk)(30)

11:45 p.m.   Adaptation of SPS with Electrostatic Accelerator
G.I. Dimov (Institute of Nuclear Physics, Novosibirsk) (30)

12:15 p.m.   LUNCH

MONDAY, OCTOBER 30, 1989

Session II Fundamental Processes
M. Bacal Presiding

2:00 p.m.	Ion - Ion Collisions Involving H^- Ions E. Salzborn (University of Giessen) (30)
2:30 p.m.	Recombination of Atomic Hydrogen on Metal Surfaces R.I. Hall (Universite P et M Curie), I. Cadez (Institute of Physics, Belgrade), M. Landau, F. Pichou, C. Schermann (Universite P et M Curie) (30)
3:00 p.m.	Vibrational Relaxation of Highly Exited H_2 Molecules in Gas Phase and Gas Surface Interaction M. Cacciatore, M. Capitelli, and G.D. Billing (University of Bari) (15)
3:15 p.m.	BREAK
3:45 p.m.	Coupled Solution of Boltzmann Equation and Non Equilibrium Vibrational Kinetics in H_2 Volume Sources M. Capitelli, C. Grose, M. Cacciatore, R. Alberto, and P. Cives (University of Bari) (30)
4:15 p.m.	Generation of H^-, $H_2(v")$, and H Atom by H_2^+ and H_3^+ Ions Incident Upon Barium Surfaces J.R. Hiskes, and A.M. Karo (LLL) (30)
4:45 p.m.	Electron Energy Distributions and Vibrational Population Distributions J.R. Hiskes (LLL) (20)
5:05 p.m.	Energetics of Negative Ion Formation by Dissociative Attachment to Light Molecules H.H. Michels (United Technological Research Center), and J.M. Wadehra (Wayne University) (20)
5:25 p.m.	Isotope Effect in Vibrational Excitation of H_2 by Low Energy Electron Impact D.E. Atems, and J.M. Wadehra (Wayne University) (20)
6:00 p.m.	Wine and Cheese Reception (Courtesy of AUI) (South Dining Room, Brookhaven Center)

TUESDAY, OCTOBER 31, 1989

Session III
Negative Ion Sources
A. Holmes Presiding

9:00 a.m.	Cesium Mixing in the Multi-Ampere Volume H⁻ Ion Source Y. Okumura, M. Hanada, T. Inoue, H. Kojima, Y. Matsuda, Y. Ohara, M. Seki, and R. Watanabe (JAERI) (30)
9:30 a.m.	Negative Ion Production in an Ion Source Operating in H_2 and D_2 W.G. Graham (Queen's University, Belfast) and A.A. Mullan (University of Ulster, Coleraine) (20)
9:50 a.m.	Optimization of the Sheet Plasma Negative Ion Source A. Ando, T. Kuroda, Y. Oka, O. Kaneko, Y. Takeiri, T. Kawamoto, and A. Karita (Nagoya University) (15)
10:05 a.m.	BREAK
10:25 a.m.	H⁻ Volume Source K. Prelec (Brookhaven National Laboratory) (20)
10:45 a.m.	H⁻ Optically Pumped Source A. Zelenski (INR, Moscow) (30)
11:15 a.m.	Intense Negative Heavy Ion Sources Y. Mori (KEK) (40)
11:55 a.m.	Effects of Electron Suppression Fields on D⁻ Production M.F. Thornton, A.J.T. Holmes, and L.M. Lea (Culham) (20)
12:15 p.m.	An Axial Geometry High Intensity Heavy Negative Ion Source G.D. Alton (ORNL) (20)
12:35 p.m.	LUNCH

TUESDAY, OCTOBER 31, 1989

Session IV Posters
J. Alessi Presiding

2:00 –
5:30 p.m.

1. Direct Comparison of Theoretically and Experimentally Determined EEDF's
J. Bretagne (University of Paris Sud, Orsay), W.G. Graham (Queens University, Belfast) and M.B. Hopkins (Dublin City University)

2. Measurement of Atomic Temperature and Density of Emission Spectroscopy
A-M Bruneteau, R. Leroy, P. Berlemont, G. Hollos, M. Bacal, (Ecole Polytechnique), and J. Bretagne (University of Paris Sud, Orsay)

3. Enhancement of Negative Ion Extraction and Electron Suppression by a Magnetic Field
J. Bruneteau, R. Leroy, M. Bacal (Ecole Polytechnique) and J.R. Whealton (ORNL)

4. Mass Spectrometry in a Multicusp Tandem Ion Source
A.A. Mullan (University of Ulster) and W.G. Graham (Queen's University, Belfast)

5. Plasma Parameters Measurements in a Low Frequency rf H_2 Discharge
C.A. Anderson (University of Ulster, Coleraine), and W.G. Graham (Queen's University, Belfast)

6. Radiofrequency Trap for Tests on Production and Excitation of Ions
G. Brautti, A. Boggia, A. Raino, V. Stagno, V. Variale, and V. Valentino (University of Bari)

7. Steady-State Production of H^- Ions
P.M. Golovinsky, V.P. Goretsky, A.N. Mosijuk, I.S. Soloshenko, A.F. Tarasenko, and A.I. Tschedrin (Institute of Physics of Ukr. SSR Academy of Science, Kiev, USSR)
Presented by N.N. Semashko

8. Physics Test of an Electron Suppressor with Variable Electric and Magnetic Fields
R. King, R. McAdams, A.F. Newman, and A.J.T. Holmes (Culham)

9. Operation of a Large Negative Ion Source in Deuterium
L.M. Lea, A.J.T. Holmes, and M.F. Thornton (Culham)

Posters (Continued)

10. Transport Processes Through Magnetic Filter of Negative Ion Sources
    M. Ogasawara, T. Yamakawa, F. Sato, and Y. Okumura (KEIO University, Yokohama)

11. H^- Formation from H_3^+ Collisions in Noble Gases
    I. Alvarez, H. Martinez, J. deUrquijo, and C. Cisneros (Laboratory de Cuernavaca, University of Mexico)

12. Gas Phase Alkali-Hydrogen Interactions in Negative Ion Sources
    H.H. Michels (United Technologies Research Center) and J.M. Wadehra (Wayne University)

13. Dissociative Recombination of D_3^+
    F.B. Yousif and J.B.A. Mitchell (University of Western Ontario)

14. Vibrational Population of H_2 Produced by a Discharge
    C. Schermann (Universite P et M Curie), I. Cadez (Institution of Physics, Belgrade), R.I. Hall, M. Landau, and F. Pichou (Universite P et M Curie)

15. Translational Energy Dependences of Rate Constants for Collisional Detachment of NO^- as a Function of Temperature
    R.A. Morris[1], A.A. Viggiano, and J.F. Paulson (Geophysics Laboratory, Hanscom AFB)
    [1]On Contract to GL from Systems Integration Eng. Inc., Lex., MA)

16. Magnetic Field of a Toroidal H^- Volume Source
    C.R. Meitzler (Texas Accelerator Center)

17. Practical Consideration for a Plasma Neutralizer
    K.G. Moses, J.R. Trow, and J.C. Dooling (JAYCOR)

18. Electron - Stripping From 2 t 7 MeV H^- and $H°$ Beams on Noble Gas Targets
    B.M. Johnson, M. Meron, and K.W. Jones (Brookhaven National Laboratory)

19. Computerized Analysis of H_2 Plasma in a Small H^- Cusp Source
    D.H. Yuan, K. Jayamanna, and P.W. Schmor (TRIUMF)

20. A Small Intense DC H^- Cusp Source at TRIUMF
    D.H. Yuan, K. Jayamanna, and P.W. Schmor (TRIUMF)

21. Simulation of Charge Particle Dynamics in Gas Media of Fusion Reactor Injector
    V.P. Sidorov and Si Yu. Udovichenko (Vekva Institute, Sukhumi, USSR)
    Presented by Yu. I. Belchenko

Posters (Continued)

22. Continuously Operated Negative Ion Surface Plasma Source
    A.A. Bashkeev, and V.G. Dudnikov (Institue of Nuclear Physics, Novosibirsk)
    Presented by Yu. I. Belchenko

23. Pulsed Negative Ion Diode
    H. Lindenbaum, R. Prohaska, and A. Fisher (U.C. Irvine)

24. Recent Progress in the BNL Intense Polarized H^- Source Program
    A. Kponou, J.G. Alessi, B. DeVito, and A. Hershcovitch (BNL)

25. H^- Source and Low Energy Transport for the BNL RFQ Preinjector
    J.G. Alessi, J.M. Brennan, and A. Kponou (BNL)

26. Emittance Measurements on a Volume H^- Source
    J.G. Alessi (BNL)

27. Performance of the BNL Negative Heavy Ion Source
    J.A. Benjamin (BNL)

28. Participation of the Division of Physics of Beams in the APS Spring Meeting
    M. Month (BNL)

WEDNESDAY, NOVEMBER 1, 1989

Session V(a)
Applications and Systems
T. Kuroda Presiding

9:00 a.m.   Conceptual Design Study of Negative Ion Beams For CIT
            L. Grisham (PPPL); W. Copper, and P. Purgalis (LBL) (30)

9:30 a.m.   Negative Ion Beam Programs at JAERI
            Y. Ohara (JAERI) (30)

10:00 a.m. European Community D^- Injector Development Program
            A.J.T. Holmes (Culham) (30)

10:30 a.m. BREAK

Session V(b)
Sources and Development Programs
T. Kuroda Presiding

11:00 a.m. Stationary H^- Plasma Volume with Cesium Hollow Cathode
            S.P. Antipov, L.I. Elizarov, M.I. Martynov, and
            V.M. Chesnokov (Kurchatov Institute) (20)

11:20 a.m. Selection of Conditions for Production of Maximum H^- Beam
            Current Density From a Multicusp Source
            N.N. Semashko (Kurchatov Institute) (20)

12:00 p.m. LUNCH

3:30 p.m.   **Leave for Cruise and Banquet**
            (Buses depart from Brookhaven Center)

THURSDAY, NOVEMBER 2, 1989

Session VI
Diagnostics
W. Graham Presiding

9:00 a.m.   Spectroscopy of Hydrogen-Cesium Discharge Plasma
            V. Antsiferov, V. Beskorovainyi, Yu. Belchenko,
            G. Dereviakin, A. Maksimov, P. Sova, and L. Skripal
            (Institute of Nuclear Physics, Novosibirsk) (30)

9:30 a.m.   Laser Diagnostics of H$^-$ Ion Sources
            A.T. Young and G.C. Stutzin (LBL) (30)

10:00 a.m. BREAK

10:30 a.m. H° Temperature and Density Measurements in a Penning Surface
            - Plasma H$^-$ Ion Source II
            H.V. Smith Jr., P. Allison, F.J. Pitcher, R.R. Stevens, Jr.,
            and G. Worth (LANL); G.C. Stutzin, A.T. Young,
            A.S. Schlachter, K.N. Leung, and W.B. Kunkel (LBL) (20)

10:50 a.m. Dynamics of Negative Hydrogen Ions in a Volume Source
            R.A. Stern (University of Colorado) and M. Bacal (Ecole
            Polytechnique) (30)

11:20 a.m. Measurement of H$^-$ Thermal Energy by Two Laser Pulse Photo
            Detachment
            M. Bacal, P. Berlemont, J. Bruneteau, R. Leroy, P. Devynck,
            (Ecole Polytechnique); and R.A. Stern (University of
            Colorado) (30)

11:50   LUNCH

THURSDAY, NOVEMBER 2, 1989

Session VII
Extraction Losses In and Modeling of Volume Sources
P. Allison Presiding

2:00 p.m. Extraction Induced RMS Emittance Growth for Volume Negative Ion Sources
J.H. Whealton (ORNL) (40)

2:40 p.m. Modeling of Volume Hydrogen Negative Ion Sources
D.A. Skinner, P. Berlemont, and M. Bacal (Ecole Polytechnique) (40)

3:10 p.m. BREAK

3:30 p.m. Modeling of JAERI H⁻ Tandem Volume Sources
J.H. Fink (Negion, Inc.) (30)

4:00 p.m. Generalized Multibody Computer Simulations of Plasma Wall Desorption and Energy Transfer
A.M. Karo, J.R. Hiskes, and T.M. DeBoni (LLL) (30)

5:00 p.m. Panel Session: Extraction Loss and Modeling of Volume Sources

Panel Members: J.R. Hiskes, Moderator; J. Bretagne, M. Capitelli, L. Elizarov, and L.M. Lea, and M. Ogasawara

FRIDAY, NOVEMBER 3, 1989

Session VIII
Acceleration, Neutralization, and Detection
Yu. Belchenko Presiding

8:30 a.m.	Fusion Applications of rf Accelerators R. Thomae, H. Klein (Goethe University, Frankfurt), and H.J. Hopman (NET Team and FOM) (30)
9:00 a.m.	RFQ Injector With Electrostatically Focused Transport and Matching O.A. Anderson, L. Soroka, J.W. Kwan, and R.P. Wells (LBL) (30)
9:30 a.m.	Conceptual Design of a 1 MeV Test Stand C. Jacquot (Nuclear Research Center, Cadarache) (20)
9:50 a.m.	Test of a Compact 750 KeV H$^-$ Preinjector C.R. Meitzler, P. Datte, F.R. Huson, R. Kazimi, C. Kronke, S. Machida, W. MacKay, S. Oknuma, D. Raparia, D. Sun, P. Tomkins, and J. Zeigler (TAC) (10)
10:00 a.m.	BREAK
10:20 a.m.	HESQ, A Low Energy Transport System D. Raparia (Texas Accelerator Center) (20)
10:50 a.m.	Plasma Neutralizers P. Vallinga, D. Schram (Eindhoven U.T.) and H.J. Hopman (NET Team and FOM) (30)
11:20 a.m.	Comments on Velocity Space Relaxation in High Z Neutralizers A. Hershcovitch (Brookhaven National Laboratory) (10)
11:30 a.m.	Space Charge Neutralization U. von Wimmersperg (Brookhaven National Laboratory) (15)
11:45 a.m.	Neutral Beam Detectors U. von Wimmersperg (Brookhaven National Laboratory) (15)
12:00 p.m.	Concluding Remarks R.S. Hemsworth (JET)

Author Index

A

Afanasyev, P. N., 614
Alessi, J. G., 385, 526, 711
Allison, P., 462
Alton, G. D., 412
Alvarez, I., 135
Anderson, C. A., 278
Anderson, O. A., 676
Ando, A., 223
Antipov, S. P., 184
Antsiferov, V. V., 427
Astapkovich, A. M., 614
Atems, D. E., 121

B

Bacal, M., 266, 474, 489, 504, 557
Bashkeev, A. A., 329
Belchenko, Yu. I., 198, 427
Berlemont, P., 489, 504, 557
Beskorovaynyy, V. V., 427
Bhadra, D. K., 761
Billing, G. D., 62
Boggia, A., 285
Brautti, G., 285
Brennan, J. M., 711
Bretagne, J., 129, 504, 626
Brown, T., 771
Bruneteau, A. M., 504
Bruneteau, J., 266, 489

C

Cacciatore, M., 62, 74
Cadez, I., 49, 159
Capitelli, M., 62, 74, 626
Carr, W. E., 30
Celiberto, R., 74
Chesnokov, V. M., 184
Cisneros, C., 135
Cives, P., 74
Colleraine, A. P., 761
Cooper, W. S., 771

D

Datte, P., 690
DeBoni, T. M., 585
Debus, W., 40
Deitinghoff, H., 660
Derevyankin, G. E., 427
de Urquijo, J., 135
DeVito, B., 385
Devynck, P., 489
Dimov, G. I., 651
Dooling, J. C., 717
Dudnikov, V. G., 329

E

Elizarov, L. I., 184, 626

F

Fink, J. H., 572
Fisher, A., 354
Fukumoto, S., 392

G

Golovinsky, P. M., 340
Goretsky, V. P., 340
Graham, W. G., 129, 214, 278, 516
Grisham, L. R., 753, 771
Gorse, C., 74

H

Hall, R. I., 49, 159
Hanada, M., 169
Hardy, J. R., 585
Hemsworth, R., 786
Hershcovitch, A., 385, 741
Hiskes, J. R., 88, 95, 585, 626
Hollos, G., 504
Holmes, A. J. T., 233, 244
Hopkins, M. B., 129
Hopman, H. J., 660, 729
Huson, F. R., 690

I

Ikegami, K., 392
Inoue, T., 169

J

Jayamanna, K., 323, 603

K

Kaneko, O., 223
Karita, A., 223
Karo, A. M., 88, 585
Kawamato, T., 223
Kazimi, R., 690
Kim, J., 761
King, R. F., 255
Klein, H., 660
Kleyn, A. W., 3
Kojima, H., 169
Kokhanovskii, S. A., 373
Konieczny, C., 489
Kponou, A., 385, 711
Kronke, C., 690
Krylov, A. I., 290
Kunkel, W. B., 17, 450, 462
Kupriyanov, A. S., 198
Kuroda, T., 223
Kuznetsov, V. V., 290
Kwan, J. W., 676

L

Landau, M., 49, 159
Lea, L. M., 233, 244, 626
Lee, S. W., 30
Leroy, R., 266, 489, 504
Liehr, M., 40
Lietzke, A. F., 17
Lindenbaum, H., 354
Leung, K. N., 17, 450, 462

M

Machida, S., 690
MacKay, W., 690

Maximov, A. M., 427
Martinez, H., 135
Martynov, M. I., 184
Matsuda, Y., 169
McAdams, R., 255
McDonald, M., 323
Meitzler, C. R., 298, 385, 690
Melchert, F., 40
Melnychuk, S. T., 30
Meszaros, P. S., 539
Michels, H. H., 114, 142
Mitchell, J. B. A., 152
Morales, A., 135
Mori, Y., 392
Moses, K. G., 717
Mosijuk, A. N., 340
Mullan, A. A., 214, 516

N

Naylor, G. O. R., 244
Newman, A. F., 255

O

Ogasawara, M., 596, 626
Ohara, Y., 169, 776
Oka, Y., 223
Ohnuma, S., 690
Okumura, Y., 169, 596

P

Penkin, D. V., 290
Pichou, F., 49, 159
Pitcher, E. J., 462
Polushkin, V. G., 373
Prelec, K., 304
Prohaska, R., 354
Purgalis, P., 771

R

Rainò, A., 285
Raparia, D., 690, 699
Raridon, R. J., 539

Rostoker, N., 354
Rothe, K. E., 539

S

Salzborn, E., 40
Sato, F., 596
Schempp, A., 660
Schermann, C., 49, 159
Schlachter, A. S., 462
Schram, D. C., 729
Schmor, P. W., 323, 603
Seidl, M., 30
Seki, M., 169
Semashko, N. N., 290
Sheperd, G., 354
Sidorov, V. P., 614
Skripal', L. P., 427
Skinner, D. A., 557
Smith, Jr., H. V., 462
Soloshenko, I. A., 340
Soroka, L., 676
Sova, P. G., 427
Stagno, V., 285
Stearns, J. W., 17
Stern, R. A., 474, 489
Stevens, Jr., R. R., 462
Stewart, L. D., 761
Stutzin, G. C., 450, 462
Sun, D., 690
Svistunov, Yu. A., 614

T

Takagi, A., 392
Takeiri, Y., 223
Tarasenko, A. F., 340
Thomae, R., 660
Thornton, M. F., 233, 244
Tompkins, P., 690

Trow, J. R., 717
Tschedrin, A. I., 340

U

Udovichenko, S. Yu., 614
Ueno, A., 392

V

Vallinga, P. M., 729
Valentino, V., 285
Van der Donk, D., 152
van Os, C. F. A., 17
Variale, V., 285
Vishnevskii, K. N., 373
von Wimmersperg, U., 534, 746

W

Wadehra, J. M., 114, 121, 142
Watanabe, K., 169
Wells, R. P., 676
Whealton, J. H., 266, 539
Worth, G. T., 462

Y

Yamakawa, T., 596
Young, A. T., 450, 462
Yousif, F. B., 152
Yuan, D. H., 323, 603

Z

Ziegler, J., 690
Zelenskii, A. N., 373
Zuev, Yu. V., 614

AIP Conference Proceedings

		L.C. Number	ISBN
No. 101	Positron-Electron Pairs in Astrophysics (Goddard Space Flight Center, 1983)	83-71926	0-88318-200-9
No. 102	Intense Medium Energy Sources of Strangeness (UC-Sant Cruz, 1983)	83-72261	0-88318-201-7
No. 103	Quantum Fluids and Solids – 1983 (Sanibel Island, Florida)	83-72440	0-88318-202-5
No. 104	Physics, Technology and the Nuclear Arms Race (APS Baltimore –1983)	83-72533	0-88318-203-3
No. 105	Physics of High Energy Particle Accelerators (SLAC Summer School, 1982)	83-72986	0-88318-304-8
No. 106	Predictability of Fluid Motions (La Jolla Institute, 1983)	83-73641	0-88318-305-6
No. 107	Physics and Chemistry of Porous Media (Schlumberger-Doll Research, 1983)	83-73640	0-88318-306-4
No. 108	The Time Projection Chamber (TRIUMF, Vancouver, 1983)	83-83445	0-88318-307-2
No. 109	Random Walks and Their Applications in the Physical and Biological Sciences (NBS/La Jolla Institute, 1982)	84-70208	0-88318-308-0
No. 110	Hadron Substructure in Nuclear Physics (Indiana University, 1983)	84-70165	0-88318-309-9
No. 111	Production and Neutralization of Negative Ions and Beams (3rd Int'l Symposium, Brookhaven, 1983)	84-70379	0-88318-310-2
No. 112	Particles and Fields – 1983 (APS/DPF, Blacksburg, VA)	84-70378	0-88318-311-0
No. 113	Experimental Meson Spectroscopy – 1983 (Seventh International Conference, Brookhaven)	84-70910	0-88318-312-9
No. 114	Low Energy Tests of Conservation Laws in Particle Physics (Blacksburg, VA, 1983)	84-71157	0-88318-313-7
No. 115	High Energy Transients in Astrophysics (Santa Cruz, CA, 1983)	84-71205	0-88318-314-5
No. 116	Problems in Unification and Supergravity (La Jolla Institute, 1983)	84-71246	0-88318-315-3
No. 117	Polarized Proton Ion Sources (TRIUMF, Vancouver, 1983)	84-71235	0-88318-316-1
No. 118	Free Electron Generation of Extreme Ultraviolet Coherent Radiation (Brookhaven/OSA, 1983)	84-71539	0-88318-317-X
No. 119	Laser Techniques in the Extreme Ultraviolet (OSA, Boulder, Colorado, 1984)	84-72128	0-88318-318-8

No. 120	Optical Effects in Amorphous Semiconductors (Snowbird, Utah, 1984)	84-72419	0-88318-319-6
No. 121	High Energy e^+e^- Interactions (Vanderbilt, 1984)	84-72632	0-88318-320-X
No. 122	The Physics of VLSI (Xerox, Palo Alto, 1984)	84-72729	0-88318-321-8
No. 123	Intersections Between Particle and Nuclear Physics (Steamboat Springs, 1984)	84-72790	0-88318-322-6
No. 124	Neutron-Nucleus Collisions – A Probe of Nuclear Structure (Burr Oak State Park - 1984)	84-73216	0-88318-323-4
No. 125	Capture Gamma-Ray Spectroscopy and Related Topics – 1984 (Internat. Symposium, Knoxville)	84-73303	0-88318-324-2
No. 126	Solar Neutrinos and Neutrino Astronomy (Homestake, 1984)	84-63143	0-88318-325-0
No. 127	Physics of High Energy Particle Accelerators (BNL/SUNY Summer School, 1983)	85-70057	0-88318-326-9
No. 128	Nuclear Physics with Stored, Cooled Beams (McCormick's Creek State Park, Indiana, 1984)	85-71167	0-88318-327-7
No. 129	Radiofrequency Plasma Heating (Sixth Topical Conference, Callaway Gardens, GA, 1985)	85-48027	0-88318-328-5
No. 130	Laser Acceleration of Particles (Malibu, California, 1985)	85-48028	0-88318-329-3
No. 131	Workshop on Polarized ^{3}He Beams and Targets (Princeton, New Jersey, 1984)	85-48026	0-88318-330-7
No. 132	Hadron Spectroscopy–1985 (International Conference, Univ. of Maryland)	85-72537	0-88318-331-5
No. 133	Hadronic Probes and Nuclear Interactions (Arizona State University, 1985)	85-72638	0-88318-332-3
No. 134	The State of High Energy Physics (BNL/SUNY Summer School, 1983)	85-73170	0-88318-333-1
No. 135	Energy Sources: Conservation and Renewables (APS, Washington, DC, 1985)	85-73019	0-88318-334-X
No. 136	Atomic Theory Workshop on Relativistic and QED Effects in Heavy Atoms	85-73790	0-88318-335-8
No. 137	Polymer-Flow Interaction (La Jolla Institute, 1985)	85-73915	0-88318-336-6
No. 138	Frontiers in Electronic Materials and Processing (Houston, TX, 1985)	86-70108	0-88318-337-4
No. 139	High-Current, High-Brightness, and High-Duty Factor Ion Injectors (La Jolla Institute, 1985)	86-70245	0-88318-338-2
No. 140	Boron-Rich Solids (Albuquerque, NM, 1985)	86-70246	0-88318-339-0
No. 141	Gamma-Ray Bursts (Stanford, CA, 1984)	86-70761	0-88318-340-4

No. 142	Nuclear Structure at High Spin, Excitation, and Momentum Transfer (Indiana University, 1985)	86-70837	0-88318-341-2
No. 143	Mexican School of Particles and Fields (Oaxtepec, México, 1984)	86-81187	0-88318-342-0
No. 144	Magnetospheric Phenomena in Astrophysics (Los Alamos, 1984)	86-71149	0-88318-343-9
No. 145	Polarized Beams at SSC & Polarized Antiprotons (Ann Arbor, MI & Bodega Bay, CA, 1985)	86-71343	0-88318-344-7
No. 146	Advances in Laser Science–I (Dallas, TX, 1985)	86-71536	0-88318-345-5
No. 147	Short Wavelength Coherent Radiation: Generation and Applications (Monterey, CA, 1986)	86-71674	0-88318-346-3
No. 148	Space Colonization: Technology and The Liberal Arts (Geneva, NY, 1985)	86-71675	0-88318-347-1
No. 149	Physics and Chemistry of Protective Coatings (Universal City, CA, 1985)	86-72019	0-88318-348-X
No. 150	Intersections Between Particle and Nuclear Physics (Lake Louise, Canada, 1986)	86-72018	0-88318-349-8
No. 151	Neural Networks for Computing (Snowbird, UT, 1986)	86-72481	0-88318-351-X
No. 152	Heavy Ion Inertial Fusion (Washington, DC, 1986)	86-73185	0-88318-352-8
No. 153	Physics of Particle Accelerators (SLAC Summer School, 1985) (Fermilab Summer School, 1984)	87-70103	0-88318-353-6
No. 154	Physics and Chemistry of Porous Media—II (Ridge Field, CT, 1986)	83-73640	0-88318-354-4
No. 155	The Galactic Center: Proceedings of the Symposium Honoring C. H. Townes (Berkeley, CA, 1986)	86-73186	0-88318-355-2
No. 156	Advanced Accelerator Concepts (Madison, WI, 1986)	87-70635	0-88318-358-0
No. 157	Stability of Amorphous Silicon Alloy Materials and Devices (Palo Alto, CA, 1987)	87-70990	0-88318-359-9
No. 158	Production and Neutralization of Negative Ions and Beams (Brookhaven, NY, 1986)	87-71695	0-88318-358-7
No. 159	Applications of Radio-Frequency Power to Plasma: Seventh Topical Conference (Kissimmee, FL, 1987)	87-71812	0-88318-359-5
No. 160	Advances in Laser Science–II (Seattle, WA, 1986)	87-71962	0-88318-360-9

No. 161	Electron Scattering in Nuclear and Particle Science: In Commemoration of the 35th Anniversary of the Lyman-Hanson-Scott Experiment (Urbana, IL, 1986)	87-72403	0-88318-361-7
No. 162	Few-Body Systems and Multiparticle Dynamics (Crystal City, VA, 1987)	87-72594	0-88318-362-5
No. 163	Pion–Nucleus Physics: Future Directions and New Facilities at LAMPF (Los Alamos, NM, 1987)	87-72961	0-88318-363-3
No. 164	Nuclei Far from Stability: Fifth International Conference (Rosseau Lake, ON, 1987)	87-73214	0-88318-364-1
No. 165	Thin Film Processing and Characterization of High-Temperature Superconductors	87-73420	0-88318-365-X
No. 166	Photovoltaic Safety (Denver, CO, 1988)	88-42854	0-88318-366-8
No. 167	Deposition and Growth: Limits for Microelectronics (Anaheim, CA, 1987)	88-71432	0-88318-367-6
No. 168	Atomic Processes in Plasmas (Santa Fe, NM, 1987)	88-71273	0-88318-368-4
No. 169	Modern Physics in America: A Michelson-Morley Centennial Symposium (Cleveland, OH, 1987)	88-71348	0-88318-369-2
No. 170	Nuclear Spectroscopy of Astrophysical Sources (Washington, D.C., 1987)	88-71625	0-88318-370-6
No. 171	Vacuum Design of Advanced and Compact Synchrotron Light Sources (Upton, NY, 1988)	88-71824	0-88318-371-4
No. 172	Advances in Laser Science–III: Proceedings of the International Laser Science Conference (Atlantic City, NJ, 1987)	88-71879	0-88318-372-2
No. 173	Cooperative Networks in Physics Education (Oaxtepec, Mexico 1987)	88-72091	0-88318-373-0
No. 174	Radio Wave Scattering in the Interstellar Medium (San Diego, CA 1988)	88-72092	0-88318-374-9
No. 175	Non-neutral Plasma Physics (Washington, DC 1988)	88-72275	0-88318-375-7
No. 176	Intersections Between Particle Land Nuclear Physics (Third International Conference) (Rockport, ME 1988)	88-62535	0-88318-376-5
No. 177	Linear Accelerator and Beam Optics Codes (La Jolla, CA 1988)	88-46074	0-88318-377-3

No. 178	Nuclear Arms Technologies in the 1990s (Washington, DC 1988)	88-83262	0-88318-378-1
No. 179	The Michelson Era in American Science: 1870–1930 (Cleveland, OH 1987)	88-83369	0-88318-379-X
No. 180	Frontiers in Science: International Symposium (Urbana, IL, 1987)	88-83526	0-88318-380-3
No. 181	Muon-Catalyzed Fusion (Sanibel Island, FL, 1988)	88-83636	0-88318-381-1
No. 176	Intersections Between Particle and Nuclear Physics (Third International Conference) (Rockport, ME 1988)	88-62535	0-88318-376-5
No. 177	Linear Accelerator and Beam Optics Codes (La Jolla, CA 1988)	88-46074	0-88318-377-3
No. 178	Nuclear Arms Technologies in the 1990s (Washington, DC 1988)	88-83262	0-88318-378-1
No. 179	The Michelson Era in American Science: 1870–190 (Cleveland, OH 1987)	88-83369	0-88318-379-X
No. 180	Frontiers in Science: International Symposium (Urbana, IL 1987)	88-83526	0-88318-380-3
No. 181	Muon-Catalyzed Fusion (Sanibel Island, FL 1988)	88-83636	0-88318-381-1
No. 182	High T_c Superconducting Thin Films, Devices, and Application (Atlanta, GA 1988)	88-03947	0-88318-382-X
No. 183	Cosmic Abundances of Matter (Minneapolis, MN 1988)	89-80147	0-88318-383-8
No. 184	Physics of Particle Accelerators (Ithaca, NY 1988)	89-83575	0-88318-384-6
No. 185	Glueballs, Hybrids, and Exotic Hadrons (Upton, NY 1988)	89-83513	0-88318-385-4
No. 186	High-Energy Radiation Background in Space (Sanibel Island, FL 1987)	89-83833	0-88318-386-2
No. 187	High-Energy Spin Physics (Minneapolis, MN 1988)	89-83948	0-88318-387-0
No. 188	International Symposium on Electron Beam Ion Sources and their Applications (Upton, NY 1988)	89-84343	0-88318-388-9
No. 189	Relativistic, Quantum Electrodynamic, and Weak Interaction Effects in Atoms (Santa Barbara, CA 1988)	89-84431	0-88318-389-7
No. 190	Radio-frequency Power in Plasmas (Irvine, CA 1989)	89-45805	0-88318-397-8
No. 191	Advances in Laser Science–IV (Atlanta, GA 1988)	89-85595	0-88318-391-9

No. 192	Vacuum Mechatronics (First International Workshop) (Santa Barbara, CA 1989)	89-45905	0-88318-394-3
No. 193	Advanced Accelerator Concepts (Lake Arrowhead, CA 1989)	89-45914	0-88318-393-5
No. 194	Quantum Fluids and Solids—1989 (Gainesville, FL, 1989)	89-81079	0-88318-395-1
No. 195	Dense Z-Pinches (Laguna Beach, CA, 1989)	89-46212	0-88318-396-X
No. 196	Heavy Quark Physics (Ithaca, NY, 1989)	89-81583	0-88318-644-6
No. 197	Drops and Bubbles (Monterey, CA, 1988)	89-46360	0-88318-392-7
No. 198	Astrophysics in Antarctica (Newark, DE, 1989)	89-46421	0-88318-398-6
No. 199	Surface Conditioning of Vacuum Systems (Los Angeles, CA, 1989)	89-82542	0-88318-756-6
No. 200	High T_c Superconducting Thin Films: Processing, Characterization, and Applications (Boston, MA, 1989)	90-80006	0-88318-759-0
No. 201	QED Stucture Functions (Ann Arbor, MI, 1989)	90-80229	0-88318-671-3
No. 202	NASA Workshop on Physics From a Lunar Base (Stanford, CA, 1989)	90-55073	0-88318-646-2
No. 203	Particle Astrophysics: The NASA Cosmic Ray Program for the 1990s and Beyond (Greenbelt, MD, 1989)	90-55077	0-88318-763-9
No. 204	Aspects of Electron-Molecule Scattering and Photoionization (New Haven, CT, 1989)	90-55175	0-88318-764-7
No. 205	The Physics of Electronic and Atomic Collisions (XVI International Conference) (New York, NY, 1989)	90-53183	0-88318-390-0
No. 206	Atomic Processes in Plasmas (Gaithersburg, MD, 1989)	90-55265	0-88318-769-8
No. 207	Astrophysics from the Moon (Annapolis, MD, 1990)	90-55582	0-88318-770-1
No. 208	Current Topics in Shock Waves (Bethlehem, PA, 1989)	90-55617	0-88318-776-0
No. 209	Computing for High Luminosity and High Intensity Facilities (Santa Fe, NM, 1990)	90-55634	0-88318-786-8